ADVANCES IN CHEMICAL PHYSICS

VOLUME LXXVII

Advances in CHEMICAL PHYSICS

EDITED BY

I. PRIGOGINE

University of Brussels
Brussels, Belgium
and
University of Texas
Austin, Texas

AND

STUART A. RICE

Department of Chemistry
and
The James Franck Institute
The University of Chicago
Chicago, Illinois

VOLUME LXXVII

A WILEY-INTERSCIENCE PUBLICATION

John Wiley & Sons, Inc.

NEW YORK / CHICHESTER / BRISBANE / TORONTO / SINGAPORE

An Interscience® Publication

Library of Congress Cataloging Number: 58-9955

ISBN 0–471–51609–0

Printed in the United States of America

10987654321

CONTRIBUTORS TO VOLUME LXXVII

HÉCTOR D. ABRUÑA, Department of Chemistry, Baker Laboratory, Cornell University, Ithaca, New York

DAVID L. ANDREWS, School of Chemical Sciences, University of East Anglia, Norwich, England

D. M. ANDERSON, Physical Chemistry, Chemical Center, Lund, Sweden

V. V. BARELKO, Institute of Chemistry, USSR Academy of Sciences, Moscow Region, USSR

CHARLES W. BAUSCHLICHER, Jr., NASA Ames Research Center, Moffett Field, California

H. T. DAVIS, Chemical Engineering and Materials Science Department University of Minnesota Minneapolis, Minnesota

ANDREW E. DEPRISTO, Department of Chemistry, Iowa State University, Ames, Iowa

DAVID M. HANSON, Department of Chemistry, State University of New York, Stony Brook, New York

KEVIN P. HOPKINS, School of Chemical Sciences, University of East Anglia, Norwich, England

A. N. IVANOVA, Institute of Chemistry, USSR Academy of Sciences, Moscow Region, USSR

ABDELKADER KARA, Department of Chemistry, Iowa State University, Ames, Iowa

CHARLES M. KNOBLER, Department of Chemistry and Biochemistry, University of California, Los Angeles, California

STEPHEN R. LANGHOFF, NASA Ames Research Center, Moffett Field, California

P. A. MONSON, Physical Chemistry Laboratory, South Parks Road, Oxford, England, U.K. and Department of Chemical Engineering, University of Massachusetts, Amherst, Massachusetts

G. P. MORRISS, Research School of Chemistry, Australian National University, Canberra, Australia

J. C. C. NITSCHE, Department of Mathematics, University of Minnesota, Minneapolis, Minnesota

L. E. SCRIVEN, Chemical Engineering and Materials Science Department, University of Minnesota, Minneapolis, Minnesota

PETER R. TAYLOR, ELORET Institute, Sunnyvale, California

YU. E. VOLODIN, Institute of Chemistry, USSR Academy of Sciences, Moscow Region, USSR

V. N. ZVYAGIN, Institute of Chemistry, USSR Academy of Sciences, Moscow Region, USSR

INTRODUCTION

Few of us can any longer keep up with the flood of scientific literature, even in specialized subfields. Any attempt to do more and be broadly educated with respect to a large domain of science has the appearance of tilting at windmills. Yet the synthesis of ideas drawn from different subjects into new, powerful, general concepts is as valuable as ever, and the desire to remain educated persists in all scientists. This series, *Advances in Chemical Physics*, is devoted to helping the reader obtain general information about a wide variety of topics in chemical physics, which field we interpret very broadly. Our intent is to have experts present comprehensive analyses of subjects of interest and to encourage the expression of individual points of view. We hope that this approach to the presentation of an overview of a subject will both stimulate new research and serve as a personalized learning text for beginners in a field.

I. Prigogine
Stuart A. Rice

CONTENTS

CHEMISTRY INDUCED BY CORE ELECTRON EXCITATION

DAVID M. HANSON

Department of Chemistry
State University of New York
Stony Brook, New York

CONTENTS

I. INTRODUCTION

Radiation chemistry generally has been concerned with secondary and tertiary reactions caused by energetic electrons that result from ionization by high-energy radiation or bombardment by high-energy particles. Recently there has been a surge of interest in the primary steps in radiation chemistry. The term *primary steps* refers to the excitation and relaxation processes of a single molecule exposed to high-energy radiation or particles. This review focuses on steps that begin with the excitation or ionization of a core electron in a molecule. This chemistry is relevant to understanding the effects of radiation on biological systems and on materials used in space, in nuclear reactors, or with the techniques of x-ray and electron lithography and microscopy. X-ray lithography and associated processes of etching and deposition appear to be providing a path into the world of ultra-microelectronics. Core electron spectroscopy and photochemistry also are enhancing our knowledge of surface and interface structure and reactivity.

Many of the unique properties of synchrotron radiation are contributing to this research. These properties include a continuous intensity distribution to very high photon energies, high intensity and brightness, small divergence, polarization, and pulsed time structure. The techniques of electron spectroscopy, mass spectroscopy, and fluorescence spectroscopy are being used with monochromatic radiation for selective excitation to elucidate the processes of interest. Figure 1 shows the spectral distribution for the VUV Ring at the National Synchrotron Light Source. Such a facility clearly opened the region of photon energies up to about 1000 eV for investigation. This energy range matches the binding energies (Ley and Cardona 1979), listed in Table 1, of core electrons associated with the common elements forming molecules of chemical and biological interest.

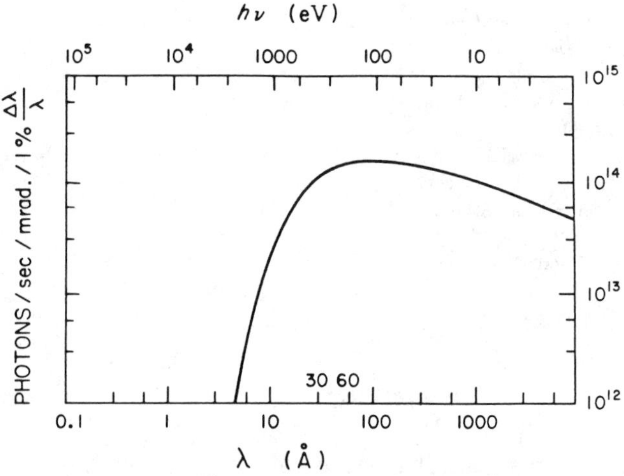

Figure 1. Spectral distribution of the VUV storage ring at the National Synchrotron Light Source, presently operating at 750 MeV with peak beam currents of 900 mA.

TABLE I
Core Electron Binding Energies

Atom	Orbital	Energy (eV)
C	1s	284
N	1s	410
O	1s	543
F	1s	697
Si	2s	150
	2p	99
P	2s	189
	2p	136
S	2s	231
	2p	163
Cl	2s	270
	2p	201

Existing grating monochromators, designed for the soft x-ray region with resolutions of $E/\Delta E = 100$ to 1000 or more, can provide a high flux ($> 10^9$/sec mm^2) of monochromatic photons, which is more than adequate to selectively excite core electrons in molecules in several different ways. It is possible to excite the 1s, 2s, or 2p electrons of one element or another in a molecule by using monochromatic radiation because the differences in binding energies are tens to hundreds of electron volts. Even more interesting is the possibility shown in Fig. 2 of selectively exciting different atoms of the same element that are in different chemical environments of the molecule. Chemical shifts of the binding energy of, say, carbon 1s electrons, are known from ESCA studies (Siegbahn et al. 1969) to be as large as 10 eV. Finally, different final states can be reached in the excitation process with different chemical consequences. These states include configurations with the core electron excited to a valence orbital, a Rydberg orbital, a shape resonance, or the ionization continuum. These various

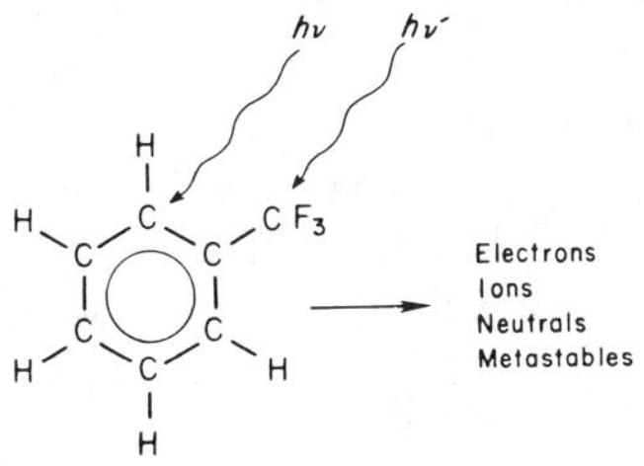

Figure 2. Illustration of atomic site selective excitation by soft x-ray photons.

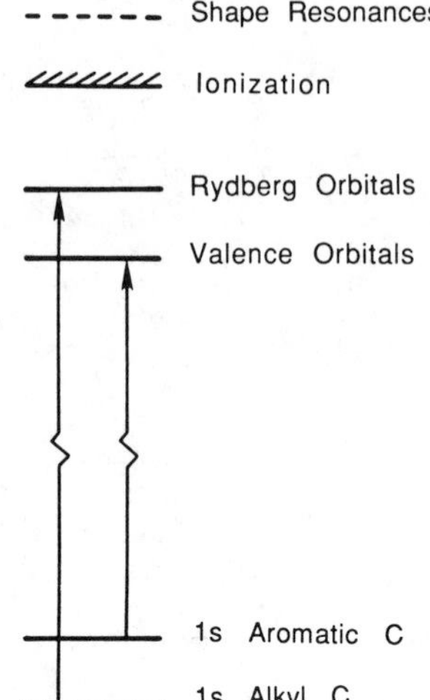

Figure 3. Energy level diagram illustrating chemical shifts of the core levels and different final state configurations that can be reached by core electron excitation.

selective excitation processes are illustrated in Fig. 3. The objective of the research is to determine what happens to molecules following such excitation by using the techniques of electron, mass, and fluorescence spectroscopy to monitor, characterize, and correlate the electrons, ions, neutrals, and metastables that are produced as a consequence.

The following picture is emerging from this research. Core electron excited states have finite lifetimes and unique characters. The electronic and chemical relaxations are coupled and depend upon the atomic and electronic structure of the molecule, the atomic site of the core hole, and the configuration of the core hole excited state. Experiments leading to an understanding of these dependences are just beginning to be done.

II. CORE HOLE EXCITED STATES

A. Spectroscopic Investigations

The term *core hole excited state* refers to the situation where a core electron in an atom or molecule has been excited or ionized leaving a vacancy in a core, i.e. nonvalence, orbital. These states have been investigated by x-ray absorption or electron energy loss measurements. Comprehensive reviews of the latter are available (Brion et al. 1982; King and Read 1985; Hitchcock 1989). An x-ray absorption spectrum can be separated into two regions. The EXAFS region (extended x-ray absorption fine structure) extends from about 25 eV to several hundred eV above a core ionization threshold or edge. In this region the dominant excitation process is ionization of the core electron.

The scattering of the outgoing electron wave from surrounding atoms, which generally can be described by a single scattering theory, produces an interference pattern in the absorption spectrum that provides structural information about the atomic site (Teo and Joy 1981; Lee et al. 1981; Prins and Koningsberger 1987). The NEXAFS region (near edge x-ray absorption fine structure), which also is called XANES (x-ray absorption near edge structure), extends approximately 25 eV on either side of the ionization edge. In this region the single-scattering model does not apply. Spectrai features

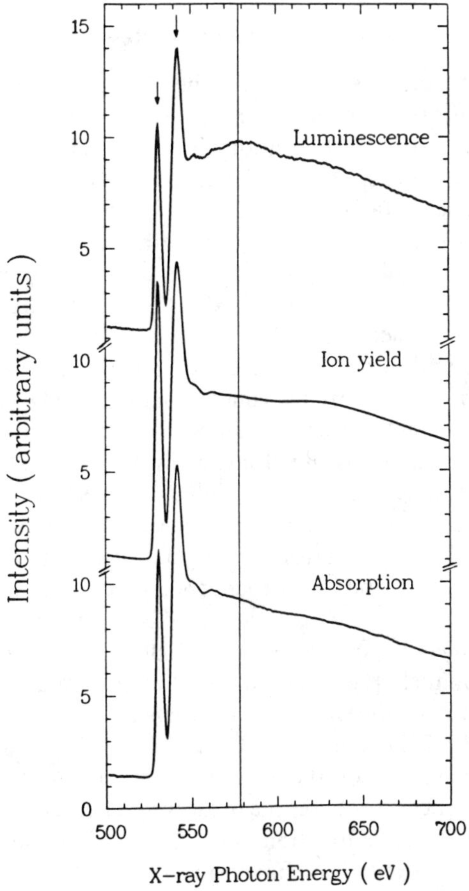

Figure 4. X-ray spectra of molecular oxygen at a pressure of 0.1 torr near the oxygen K-edge. From top to bottom: the total optical luminescence yield, the total ion current, and the absorption coefficient.

are due to resonances that are described by multiple-scattering and molecular orbital theories.

The soft x-ray absorption spectrum of oxygen is shown in Fig. 4 together with the total optical luminescence yield and the total ion yield. Differences between these spectra demonstrate that the relaxation is excitation energy dependent, i.e., the relaxation depends upon the nature of the core hole excited state. The two sharp peaks at 531 and 542 eV below the ionization edge at 543 eV are assigned to core electron transitions to pi and sigma antibonding molecular orbitals. The increase in the luminescence around 580 eV reflects the possibility for shake-up transitions, and the feature around 620 eV has been assigned to an EXAFS oscillation (Yang et al. 1988).

B. Description of the Resonances

Below the ionization edge, features in the absorption spectra called *discrete resonances* are due predominantly to transitions of the core electron to bound states that result from configurations in which the core electron has been promoted to a valence or Rydberg orbital or an appropriate linear combination. These states are degenerate with the ionization continua of other configurations and therefore can autoionize in an Auger-like process where one electron falls into the core hole and another electron is emitted.

Above the ionization edge, the NEXAFS features are due to transitions to quasi-bound molecular orbitals (also called shape or continuum resonances). In addition, multiple electron excitation processes, in which valence electrons also are excited (shake-up) or ionized (shake-off) can occur. Shape resonances result from the shape of the molecular potential, which temporarily traps the outgoing electron. Two theoretical models have been used to describe the energy position, intensity, spectral lineshape, and other properties of the shape resonances. These other properties include the population of vibrational states, the angular distribution of photoelectrons and Auger electrons, and the dependence of the angular distribution on vibrational state. One model uses a multiple-scattering formalism while the other uses a molecular orbital approach. Both approaches are directed at the same goal, namely the construction of an accurate wavefunction with a large amplitude near the molecule at an energy above the ionization edge (Dehmer et al. 1982; Dehmer 1984; Lynch et al. 1984; Langhoff 1984; Natoli 1983). The dependence of the wavefunction on energy and internuclear separation is critical because the observed characteristics of the shape resonance are a sensitive function of energy and internuclear separation.

The diversity of excitation channels in the NEXAFS region offers many opportunities for research. The near-edge structure, which depends upon geometry, coordination number, and bond order and ionicity, is sensitive to the local bonding environment of the atom. The molecular orbital picture of the resonances provides insight regarding the potential that NEXAFS has for characterizing bonds around particular atoms. Generally both bonding and antibonding molecular orbitals are associated with a bond. Observation of a resonance assigned to a particular antibonding orbital, e.g. by polarization of the absorption or characteristics of the relaxation, identifies the presence of a particular bond. While the bonding structure of simple molecules is sufficiently well-understood that this identification is not likely to be important, the situation for atoms in complicated or disordered environments or molecular species on surfaces or at interfaces is not so well-defined. NEXAFS reveals the presence of different types of bonds (Stohr et al. 1987; Stohr and Outka 1987a) and can serve to identify chemical reactions occurring on surfaces. A quantitative correlation between the energy of the resonances and bond lengths has been noted (Stohr et al. 1983; Stohr et al. 1984; Hitchcock et al. 1984; Sette et al. 1984; Piancastelli et al. 1987a; Hitchcock and Stohr 1987; Piancastelli et al. 1987b). The coupling between the resonance and the electromagnetic field is anisotropic so the orientation of bonds can be derived from the angular dependence of the absorption with polarized x-rays (Stohr and Outka 1987b).

Since the resonances have distinct characteristics, the relaxation of the core hole excited state and the dissipation of the excitation energy by electron emission, dissociation, luminescence, and the excitation of rotational and vibrational motion will vary from one resonance to another. These relaxation processes are referred to as the photoexcitation or photoionization dynamics, and characterizing the dynamics can aid in identifying and assigning the resonances. For example, it sometimes is difficult to assign bound state resonances to valence or Rydberg orbitals and features above ionization edges to shape resonances or shake-up structure; the characteristics of the electronic relaxation and the subsequent chemistry may serve as an identifying fingerprint. The electronic decay associated with above-threshold features in the 1s absorption spectra of benzene and ethylene recently has been used to identify and assign these features to doubly excited states and shape resonances (Piancastelli et al. 1989).

III. ELECTRONIC RELAXATION

The decay of a core hole excited state is dominated by autoionization or Auger processes for the light elements of the periodic table. The contribution from x-ray fluorescence is small, amounting to less than 1% for carbon and reaching 10% for argon. For the first series of the periodic table, the lifetime of a core hole excited state is about 10 femtoseconds, corresponding to an Auger transition rate of around 10^{14}/sec. In special cases discussed in Section IV.A, a molecule in a core hole excited state may dissociate prior to

TABLE II
Core Hole Relaxation Processes

ID	Initial	Final	Charge
PA	c(1)v(*n*)u(1)	c(2)v(*n* – 1)u(0) + e^-	+1
SA	c(1)v(*n*)u(1)	c(2)v(*n* – 2)u(1) + e^-	+1
PDA	c(1)v(*n*)u(1)	c(2)v(*n* – 2)u(0) + $2e^-$	+2
SDA	c(1)v(*n*)u(1)	c(2)v(*n* – 3)u(1) + $2e^-$	+2
CA	c(2)v(*n* – 2)u(1)	c(2)v(*n* – 2)u(0) + e^-	+2
AD	c(1)v(*n*)	c(2)v(*n* – 2) + e^-	+2
ASU	c(1)v(*n*)	c(2)v(*n* – 3)u(1) + e^-	+2
DAD	c(1)v(*n*)	c(2)v(*n* – 3) + $2e^-$	+3
ICD	c(1)v(*n*)s(1)	c(2)v(*n* – 1)s(0) + e^-	+1

PA = participant autoionization; SA = spectator autoionization; PDA = participant double autoionization, SDA = spectator double autoionization; CA = cascade autoionization; AD = Auger decay; ASU = Augur decay with shake-up; DAD = double Auger decay, ICD = interchannel decay.

the electronic relaxation, which then occurs in a fragment particle. The lifetime of the core hole state is known from theoretical calculations (Callan 1961; McGuire 1969; Higashi et al. 1982) and from measurements of linewidths in high resolution spectra. A lifetime of 10 fs is sufficient for vibrational motion to be defined, and vibrational structure is observed in core electron spectra (Gelius et al. 1974; King et al. 1977). The time evolution of the core hole excited state and its effect on vibrational populations and vibronic structure in spectra is a topic of current research (Flores-Riveros et al. 1985; Correia et al. 1985; Carroll and Thomas 1987; Carroll et al. 1987; Poliakoff et al. 1988; Murphy et al. 1988; Carroll and Thomas 1988).

Table II, which is adapted from a previous description (Nenner 1987; Nenner et al. 1988), presents some of the electronic relaxation processes associated with the decay of a core hole. In these equations, c represents a core orbital, v an occupied valence orbital, u an unoccupied valence or Rydberg orbital, and s represents a shape resonance orbital. The term "orbital" is used simply to mean a one-electron wavefunction. An electron in a continuum orbital free from the influence of the molecular potential is represented as e^-.

The case of resonance excitation to a bound state is considered first:

$$c(2)v(n)u(0) — h\nu \rightarrow c(1)v(n)u(1)$$

Several decay channels are possible. The most important are listed in Table II and are labeled PA (participant autoionization), SA (spectator autoionization), PDA (participant double autoionization), SDA (spectator double autoionization), and CA (cascade autoionization).

In participant autoionization (PA), the core electron excited to the unoccupied orbital is removed in an Auger-like process, and the molecule is left with a missing valence electron in what is called a 1-hole state. In spectator autoionization (SA), which also has been called resonance Auger decay, the electron in the highest energy orbital is not involved. SA leaves the molecule in a 2-hole 1-electron state because two electrons are missing from lower energy orbitals, and one electron is in an excited orbital. Both these autoionization processes leave the molecule with a $+1$ charge.

The final states in autoionization of a bound state resonance also can be reached by direct photoemission. Since in PA the molecule is left with a single hole in the valence shell, lines in the electron energy spectrum are expected at the energy of the bound state resonance minus the binding energies of the various valence electrons as measured in a standard photoelectron spectrum. In SA the molecule is left with two valence holes and an electron in a higher orbital. These states are reached by shake-up in direct photoemission. The intensities in photoemission and autoionization spectra will be different because the transition matrix elements differ, but matching the transition energies can serve to aid assignments (Eberhardt et al. 1984; Eberhardt et al. 1986c; Freund and Liegener, 1987).

Relaxation involving emission of more than one electron also occurs, e.g., participant double autoionization (PDA) and spectator double autoionization (SDA). These channels leave the molecule with a $+2$ charge and in 2-hole and 3-hole 1-electron states, respectively. Since relaxation of the core hole may leave the molecule with a deep valence hole or with an electron in a high energy orbital, additional autoionization, called delayed, subsequent or cascade autoionization (CA) can occur. An example is given in Table II.

Shape resonances rapidly decay into the continuum

$$c(2)v(n) \text{—} h\nu \rightarrow c(1)v(n)s(1) \rightarrow c(1)v(n)s(0) + e^-$$

This diagram represents ionization of a core electron via a shape resonance orbital. In the first step the core electron is excited to an orbital above the ionization threshold that has a large amplitude in the vicinity of the molecule. The second step represents separation of the electron from the molecule. The subsequent relaxation then is identical to the situation where the core electron is ionized directly. These channels, depicted in Table II, are Auger decay (AD), Auger decay with shake-up (ASU), and double Auger decay (DAD). Auger decay also can be followed by cascade autoionization.

For AD the molecule is left in a 2-hole state with a $+2$ charge, but the $+1$ ion also can be produced by coupling of the shape resonance to the valence electron ionization continuum. The core hole shape resonance then relaxes by interchannel decay (ICD). AD and ICD can be distinguished by the ionic products and the electron energies. AD produces the +2 ion and a low-energy photoelectron and a high-energy Auger electron. ICD produces a +1 ion and a high-energy photoelectron (Poliakoff et al. 1989a).

The time between the escape of the electron from the shape resonance, or from the molecule if there is no shape resonance, and Auger decay of the core hole may be sufficiently short to allow the outgoing photoelectron to participate in the Auger process. This participation can be pictured as an interaction between the Auger electron and the photoelectron and is studied under the rubric "post-collision interaction" (Schmidt 1987).

Higher-order processes contribute in both the excitation of the core electron and relaxation of the core hole. These multi-electron processes are significant because of the strong perturbation caused by the creation or the annihilation of a core hole. In these processes, additional electrons are excited (shake-up) or ionized (shake-off). Two electron processes, double autoionization and double Auger decay, were mentioned above. The final states reached in core hole decay may be excited states and also may autoionize. It is clear that excitation of a core electron and the relaxation of the core hole provide many paths leading to multiple-electron excited states. These states have a unique chemistry relative to the single-electron excited states produced by arc lamp, laser, or vacuum ultraviolet (VUV) excitation.

The study of core hole autoionization has been termed deexcitation electron spectroscopy (DES) (Eberhardt et al. 1986b, 1986c). This spectroscopy provides information about the dynamics of excitation and decay and about charge transfer and screening for coordinated molecules (Eberhardt 1986c). Single-electron emission in autoionization, Auger decay, or cascade autoionization can be detected readily by electron spectroscopy since well-defined peaks in an electron energy spectrum are produced by these processes. Double-electron emission events are more difficult to detect because the kinetic energy is shared by the two electrons to produce a broad continuous distribution. Observation of low-energy electrons (threshold electrons) or electron-electron or electron-ion coincidence techniques are needed to pick out the multiple-electron processes (Nenner et al. 1988).

Studies have demonstrated that the electronic relaxation depends upon the location of the core hole site and the configuration of the core hole excited state. The important feature is that the core hole is localized on a specific atom and this localization is projected onto the valence electrons in the decay process. Molecular Auger spectra thus present a view of molecular electronic structure from the perspective of particular atoms in a molecule. The spectra therefore can serve to identify particular molecules and functional groups, to distinguish between localized and delocalized bonding, and to measure orbital atomic populations for various atoms in a molecule (Rye and Houston 1984). This localization and the projection onto the valence electrons also give rise to the possibility that atom-specific chemistry can result from core electron excitation.

IV. MOLECULAR CHEMISTRY

Relaxation of a core hole excited state in a molecule often results in the removal of more than one electron from bonding valence molecular orbitals, and as a result fragmentation of the molecule occurs with high probability. Generally, gas-phase molecules fragment much more efficiently with core electron excitation than with valence excitation. Studies of this fragmentation are being directed at answering several questions. Is the dissociation selective, and in what sense? What are the mechanisms of bond rupture—electronic or vibrational, dissociation or predissociation? What is the energy distribution in the fragments? What is the angular distribution of the fragments? What is the time sequence of ionization and dissociation from the core hole excited state? Are stable doubly and multiply charged molecular ions formed that can be identified and characterized? Are negatively charged ions formed? What information about electronic processes such as post-collision interaction, interchannel coupling, multiple electron ionization, and cascade autoionization can be gleaned from the fragments?

Initial studies were conducted with x-rays from conventional sources in the 1960s and with electron impact excitation in the 1970s. This research is discussed in Sections IV.B and IV.I, respectively. Core electrons from essentially all the elements in a molecule are ionized with x-rays from conventional sources. It is not possible to excite core electrons from specific elements to specific final states. Measurements of energy loss in electron impact excitation serve to identify the core hole state that was produced, but an electron–ion coincidence measurement is necessary to relate the fragmentation to a specific excitation event. A prohibitive price in terms of signal level is required for such an experiment (Hitchcock 1989). Interest was renewed in the early 1980s when it was recognized, from studies of electron- and photon-stimulated desorption of ions from molecular solids (Hanson et al. 1982; Stockbauer et al. 1982a, 1982b; Kelber and Knotek 1982, 1983, 1984; Rosenberg et al. 1981, 1983, 1985), that the use of synchrotron radiation offered unparalleled opportunities for progress in this field. The first experiments using synchrotron radiation to study the fragmentation associated with the excitation of core electrons in gas-phase molecules were reported in 1983 (Eberhardt et al. 1983a). It is not feasible to review all the work in this area that has been done since. Table III, however, provides a list, by molecule, of relevant papers.

Research to date has revealed four general types of mechanisms for fragmentation following core electron excitation. These mechanisms are

TABLE III
Studies of Molecular Fragmentation

Molecule	References
Acetic acid	Eberhardt and Sham 1984
Acetone	Eberhardt et al. 1983a; Nelson et al. 1987
Benzene	Eberhardt and Sham 1984
Benzene, chloro-	Eberhardt and Sham 1984
Carbon dioxide	Carlson and Krause 1972; Van Brunt et al. 1972; Hitchcock et al. 1979
Carbon disulfide	Hayes 1987
Carbon monoxide	Van der Wiel et al. 1970; Van der Wiel and El-Sherbini 1972; Carlson and Krause 1972; Van Brunt et al. 1972; Kay et al. 1977; Ramaker 1983b; Eberhardt et al. 1983a, 1984, 1986a, 1986b,1987; Hitchcock et al. 1988
Carbon tetrafluoride	Carlson and Krause 1972; Lapiano-Smith et al. 1989
Chromium hexacarbonyl	Eberhardt et al. 1984
Ethane, 1,1,2-trichloro-	Eberhardt and Sham 1984
Ethane, 1,1,1-trifluoro-	Muller-Dethlefs et al. 1984
Ethyl iodide	Carlson and White 1968
Ethylene	Beckmann et al. 1985
Ethylenes, fluoro-substituted	Beckmann et al. 1985
Hydrogen bromide	Morin and Nenner 1986
Hydrogen iodide	Carlson and White 1966; Morin and Nenner 1987
Iron pentacarbonyl	Eberhardt et al. 1984
Lead, tetramethyl-	Carlson and White 1968; Nagaoka et al. 1987, 1989
Methane	Saito and Suzuki 1986a
Methyl bromide	Morin and Nenner 1987; Nenner et al. 1988
Methyl iodide	Dujardin et al. 1986; Morin and Nenner 1987
Nitric oxide	Carlson and Krause 1972; Van Brunt et al. 1972
Nitrogen	Van der Wiel et al. 1970; Van der Wiel and El-Sherbini 1972; Carlson and Krause 1972; Van Brunt et al. 1972; Eberhardt et al. 1983b, 1986b, 1987; Saito and Suzuki 1986b, 1987, 1988a, 1988b; Yagishita et al. 1989
Nitrous oxide	Hirsch et al. 1973; Hitchcock et al. 1979; Murakami et al. 1986; Murphy and Eberhardt 1988
Oxygen	Carlson and Krause 1972; Van Brunt et al. 1972
Silane	de Souza et al. 1986
Silicon tetrafluoride	Lapiano-Smith et al. 1989; Lablanquie et al. 1989
Silane, tetramethyl-	Morin et al. 1986b
Sulfur dioxide	Dujardin et al. 1989
Sulfur hexafluoride	Hitchcock et al. 1978
Tin, tetramethyl-	Ueda et al. 1989
Water	Ramaker 1983a

classified and discussed below in terms of ultra-fast dissociation, Coulomb explosion, molecular orbital depopulation, and valence bond depopulation. The important details of specific mechanisms associated with this general classification have yet to be elucidated.

A. Ultra-fast Dissociation

Although Auger decay of a core hole occurs on a 10-femtosecond time scale, it is possible in opportune cases for dissociation to occur from the core hole excited state. The force driving this dissociation can be understood in terms of the core-equivalent model. In this model, an atom with a core hole and a nuclear charge Z has the properties of the $Z + 1$ atom or ion because of the loss of shielding by the missing core electron. The core equivalent model predicts a term value for a resonance equal to that of the $Z + 1$ atom, and the valence properties of the $Z + 1$ atom determine whether the core hole state will be bound or dissociative. The term value is the position of the resonance relative to the ionization edge. Consequently, for example, in HBr, excitation of a Br 3d electron to an antibonding orbital transforms Br into Ar and the HBr bond becomes unstable. The repulsive potential surface imparts a high velocity to the H atom, which then escapes, and the electronic relaxation of the core hole occurs in the bromine atom (Morin and Nenner 1986). Evidence for this mechanism consists of structure in electron energy spectra that cannot be explained by ionic states of the parent molecule but can be explained in terms of the ionic states of a fragment.

Ultra-fast dissociation of the core hole excited state also was considered for silane (de Souza et al. 1986), methyl bromide, hydrogen iodide, and methyl iodide (Morin and Nenner 1987; Nenner et al. 1988). When the hydrogen atom is replaced by a methyl group, the time scale for dissociation is lengthened, and autoionization competes more effectively with the dissociation so direct dissociation of the core hole state becomes less important. Such ultra-fast dissociation therefore needs to be considered for light atoms or groups, primarily hydrogen atoms, that are bonded to atomic centers whose valence cannot tolerate an increase in the atomic number. This valence condition can be called "saturation". As a consequence of a saturated valence, increasing the atomic number by one in the equivalent core analogy requires that the number of bonds be reduced by one. For example, by this definition, the valence of nitrogen in NH_3 is saturated, and the valence of nitrogen in NO is not saturated. Ultra-fast dissociation therefore would be expected for NH_3 but not for NO because OH_3 is unstable and O_2 is stable.

B. Coulomb Explosion

Research on the fragmentation of molecules following core electron excitation was initiated by T. A. Carlson and R. M. White in 1966 (Carlson and White

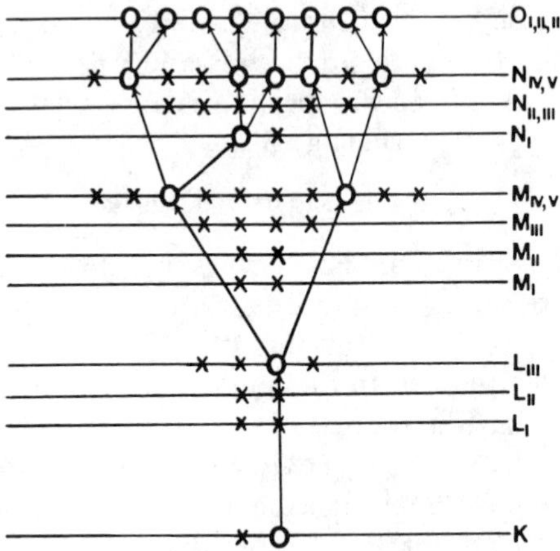

Figure 5. An illustration of the relaxation of a K-shell core hole in xenon by x-ray fluorescence (L[111] → K) followed by an Auger cascade to produce Xe^{8+}.

1966). They irradiated hydrogen and deuterium iodides and methyl iodide with 8–9 keV x-rays from a tungsten source filtered through 10 mil of Be. The spectral distribution of the source convoluted with the ionization cross-sections for different shells caused most of the inner-shell vacancies to be produced in the L shell of iodine. The resulting cascade of Auger processes, for example as shown in Fig. 5, produced a highly charged molecular ion that flew apart in what was termed a *Coulomb explosion.* Techniques of mass spectroscopy were used to determine the kinetic energies and relative abundances of the fragment ions. The ion kinetic energy distributions for HI and DI provided information about the lifetime of the Auger processes. For the case of methyl iodide, the abundance of molecular ions was extremely low, less than 1%. The Auger cascade resulting from a vacancy in the L shell of iodine serves to ionize all of the constituent atoms and completely fragments the molecule. The most probable reaction as deduced from the most abundant ions is

$$CH_3I \rightarrow C^{2+} + I^{5+} + 3H^+ + 10e^-$$

The kinetic energies that result from the Coulomb repulsion of these ions starting from the initial positions of the atoms in the molecule were calculated and found to be in qualitative agreement with the measured kinetic energy distributions, thus supporting the idea that most of the ionization occurs while the fragments are close to each other and the recoil energies result from the Coulomb repulsion of these ions.

TABLE IV
Extent of Ionization and Maximum Ionic Charge

Molecule	I^{+n}	e^{-}_{total}
HI	7	8
CH_3I	5	10
C_2H_5I	2	11

Similar studies on ethyl iodide and tetramethyl lead, where vacancies were created in the L shell of iodine and the M and L shells of lead, gave results that were consistent with this concept of a Coulomb explosion (Carlson and White 1968). More than 97% of the fragments were atomic ions, and the fragment kinetic energies were close to those calculated by assuming an instantaneous Coulomb repulsion of the ions. Table IV shows how the extent of ionization increases and the localization of charge on the core hole atom decreases as the size of the molecule increases. In HI the most abundant iodine ion has a charge of +7, whereas in ethyl iodide it is +2. In HI a total of 8 electrons are ionized while in ethyl iodide it is 11. Values for methyl iodide are between these extremes. The valence holes created in the Auger decay evidently delocalize over the molecule, and additional autoionization processes are possible in larger molecules.

These studies later were extended to molecules containing only elements from the first series of the periodic table (Carlson and Krause 1972). For these molecules, with only two electronic shells, an Auger cascade cannot occur. Each K shell vacancy produced by x-ray ionization of a 1s electron produces a single Auger event with the total loss of two valence electrons. Additional electrons can be lost in shake-off ionization and double Auger decay, which were estimated to contribute about 20% to the the observed ion yields.

For the diatomic molecules that were studied—nitrogen, oxygen, nitric oxide, and carbon monoxide—the concept of a Coulomb explosion appears to be relevant. The yield of atomic ions is high, 93% to 97%, and the ion kinetic energies of around 7 eV for +1 ions and about twice this value for +2 ions are consistent with the Coulomb repulsion model. For the polyatomic molecules the situation is different. The yield of atomic ions drops to 85% for carbon dioxide and to 74% for carbon tetrafluoride. For excitation of a core to bound state resonance in nitrous oxide, involving the terminal nitrogen atom, the yield of atomic ions is only 63% (Murakami et al. 1986). These molecules do not simply explode following excitation of a core electron.

To emphasize this point, the Coulomb explosion mechanism is defined in the following terms. The Coulomb explosion mechanism for fragmentation of a molecule following core electron excitation is characterized by an extremely high yield ($>90\%$) of atomic ions that are produced with high kinetic energies. In this mechanism, essentially all of the electronic core hole relaxation channels place the molecule on potential energy surfaces that are repulsive for all bonds in the molecule because the Coulomb forces between positive charges dominate. This mechanism is expected to account for the fragmentation of molecules following the excitation of deep core electrons where Auger cascades produce highly charged species. In a large molecule limit, only some fraction of the molecule will be blown apart by a Coulomb explosion. It has not been determined yet when the large molecule limit is reached, and the extent of fragmentation for various bonding structures in the large molecule limit is not known.

In cases where the yield of molecular ions is higher than 10% and where the fragmentation pattern depends upon the atomic site of the core hole, the dissociation processes clearly depend upon the electronic structure of the molecule and the details of the electronic relaxation, i.e. not all pathways produce essentially the same result. The mechanism then may involve vibrational dissociation or electronic or vibrational predissociation as well as direct dissociation. Even in these cases, some of the electronic relaxation channels may rupture all the bonds in a molecule and high-kinetic-energy fragments can be produced. Such channels sometimes are labeled a "Coulomb explosion," but this terminology should not be confused with the more specific use of the term that is proposed above.

C. Molecular Orbital Depopulation

The yield of fragment ions produced by decay of the nitrogen pi resonances, $N(1s^{-1}, 3\pi^1)$ in nitrous oxide was first investigated using electron impact and electron–ion coincidence techniques (Hitchcock et al. 1979). It was found that excitation of the central nitrogen 1s electron produced about 20% more atomic ions than excitation of the terminal nitrogen 1s electron, and a core hole on the terminal nitrogen produced a high yield of N^+ and NO^+ and a low yield of $N_2{}^+$. One interpretation of these observations was that the core hole relaxation processes in this molecule were localized to some extent and bonds around the site of the core hole were broken preferentially.

This study was repeated using synchrotron radiation for excitation and extended to include the oxygen pi resonance (Murakami et al. 1986). Based on this concept of localized relaxation, excitation of an oxygen 1s electron was expected to produce a yield of atomic ions similar to that for excitation of the terminal nitrogen, and the yield of O^+ and $N_2{}^+$ should be high and the yield of NO^+ should be low. Contrary to this expectation, 75% of the ions produced with oxygen excitation were atomic ions, identical to the result obtained for excitation of the 1s electron of the central nitrogen, and the yield of NO^+ was slightly larger than the yield of N_2^+. These observations clearly are not consistent with the idea of localized bond rupture around the site of the core hole.

The fragmentation of nitrous oxide following site-selective excitation of 1s electrons of all three atoms to the lowest unoccupied molecular orbital clearly shows a dependence on the atomic site of excitation as shown in Fig. 6 and Table V. The fragmentation patterns are not consistent with the idea of a Coulomb explosion as defined in the preceding section, or with the concept of localized bond rupture around the atomic site of excitation. The

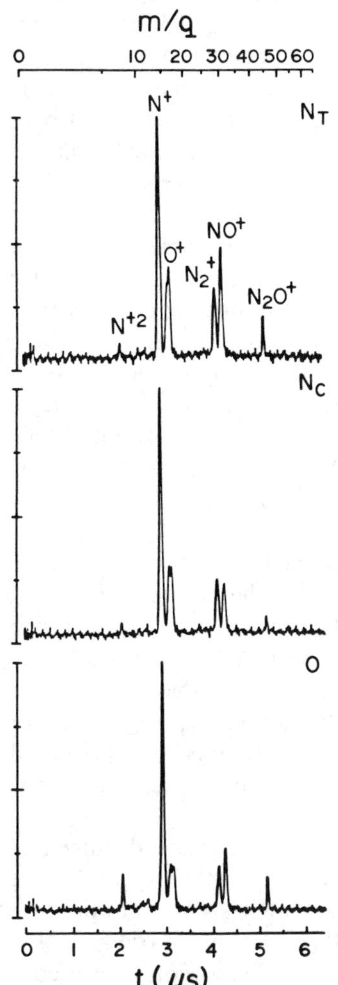

Figure 6. Time-of-flight mass spectra for N_2O obtained with excitation at 401 eV (top), 405 eV (middle), and 536 eV (bottom), corresponding to the $2p\pi^* \longleftarrow 1s$ transitions involving the 1 s electrons of the terminal and central nitrogen atoms and the oxygen atom, respectively.

TABLE V
Relative Ion Yields as a Percentage of Total Yield for Different Core Hole Sites

Site	N^{2+}	N^+	O^+	N_2^+	NO^+	N_2O^+
N_T	1	39	23	14	20	4
N_C	1	50	24	12	11	2
O	4	50	20	10	12	3

Coulomb explosion mechanism is identified by a high yield of atomic ions (greater than 90% and an insensitivity to the site of the core hole). For nitrous oxide, the yield of atomic ions varies from 62% to 75%, depending on the site of the core hole. The concept of localized relaxation and bond rupture does not fit the data because this concept predicts a similarity for the extent of fragmentation obtained with the core hole on either the terminal nitrogen or the oxygen. In contrast to this prediction, it is observed that the fragmentation patterns for core holes on the central nitrogen and oxygen are similar.

The differences in the mass spectra for different core hole sites can be understood qualitatively in terms of the atomic populations and the overlap populations of the valence molecular orbitals. The atomic populations of an orbital give a quantitative measure of the electron density on a particular atom and indicate which orbitals are involved in the autoionization or Auger decay process (Jennison 1978). The overlap populations provide a quantitative measure of the bonding properties of these orbitals. Bonds will tend to be broken if the orbitals involved in the autoionization are bonding orbitals; bonds will tend to remain if the orbitals are anti-bonding or nonbonding orbitals.

This population analysis successfully explained the observations and produced a key prediction (Murakami et al. 1986). The 2π orbital is N—N bonding and N—O anti-bonding. This orbital has a high atomic population on the terminal nitrogen and a low atomic population on the central nitrogen. Consequently, this orbital should be involved in the autoionization to a considerable extent when the core hole is on the terminal nitrogen but not when it is on the central nitrogen, and as a result, the ratio NO^+/N_2^+ should be larger in the N_T mass spectrum than in the N_C mass spectrum, as was observed.

Recent Auger electron–ion coincidence measurements have documented that the lowest energy state produced by Auger decay of the oxygen 1s core hole, which is a $2\pi^{-2}$ state, produces rupture of the N—N bond to produce N^+ and NO^+ ions and does not rupture the N—O bond to produce N_2^+ and O^+ ions (Murphy and Eberhardt 1988). This observation is in accord with the predictions of the overlap population analysis described in the preceding paragraph.

The principle characteristics of autoionization spectra associated with the three different core hole sites in nitrous oxide also are consistent with the population analysis (Larkins et al. 1988). Electrons in orbitals with large atomic populations on a particular atom are removed preferentially in the autoionization of the core hole excited state for that atomic site. These spectra are dominated by spectator autoionization, which produces 2-hole, 1-electron states. Valence holes in the 2π, 6σ, and 7σ orbitals are produced when the core hole is on the terminal nitrogen, and the largest atomic population for these orbitals is on the terminal nitrogen. Autoionization of the central nitrogen core hole primarily produces 1π valence holes, and this orbital makes the largest contribution to the atomic population of the central nitrogen. The 1π and 6σ orbitals have large atomic populations on oxygen, and these orbitals are involved in oxygen core hole autoionization. The theoretical analysis of the Auger (Larkins 1987) and autoionization spectra (Larkins et al. 1988) is based on one-electron molecular orbitals and does not require localized valence hole states, which seem to be needed for other systems.

D. Valence Bond Depopulation

The term *valence bond depopulation* is used to identify a mechanism of fragmentation in which a localized picture of bonding may be more appropriate for describing the electronic and chemical relaxation accompanying decay of a core hole than a molecular orbital picture. In this context, valence bond depopulation refers to an Auger final state in which the two valence holes produed by Auger decay are highly correlated and localized on the atomic site of the core hole. This phenomenon is called Coulomb localization and is discussed in Section V.A. Such localization has been invoked to explain the Auger spectra of some metals (Antonides et al. 1977; Sawatzky 1977; Sawatzky and Lenselink 1980; Cini 1976, 1977, 1978, 1979), semiconductors (Bassett et al. 1982; Jenninson 1982), covalently bonded insulators (Ramaker 1980) and gas phase molecules (Thomas and Weightman 1981; Weightman et al. 1983; Rye et al. 1980; Kelber et al. 1981; Rye and Houston 1983, 1984), and electron- and photon-stimulated desorption of ions from solid surfaces (Jennison et al. 1982, Feibelman 1981, Ramaker et al. 1981, 1982, Madden et al. 1982). Direct evidence for Coulomb localization has been found in studies of core hole relaxation in carbon and silicon tetrafluorides (Lapiano-Smith et al. 1989).

The Auger spectra of CF_4 and SiF_4 have been analyzed in terms of the concept of Coulomb localization. Contributions from localized two-hole states were used to explain the presence of six rather than three main peaks in the Si(LVV) spectrum, and the similarity of the F(KVV) spectrum to that of Ne. The contribution from the localized states was larger for SiF_4 relative to CF_4. This difference was attributed to the larger size, greater F–F distance, and higher bond ionicity of SiF_4, all of which decrease the interactions that lead to valence hole delocalization, but the theoretical description of the Auger final states is not yet complete (Rye and Houston 1983, 1984; Aksela et al. 1986; Ferrett et al. 1988; de Souza et al. 1989).

In the study of fragmentation following core electron excitation, the observation of F^{2+} ions following excitation of a fluorine 1s electron in silicon tetrafluoride provides direct evidence for the formation of a localized two-hole state in Auger decay and for its persistence on the time scale of dissociation. In contrast no F^{2+} was observed for carbon tetrafluoride. The absence of this doubly charged fragment from CF_4 is consistent with the stronger intramolecular interactions in this molecule as described in the previous paragraph (Lapiano-Smith et al. 1989).

The first evidence that fragmentation of a molecule in the gas phase could be localized to some extent around the site of the core hole was reported for acetone (Eberhardt et al. 1983a). Partial ion yield spectra were obtained, and the pi resonance was a prominent feature in only the C^+ and O^+ spectra. The pi resonance is associated with excitation of the 1s electrons of the carbon atom in the carbonyl group. The 1s to pi excitation therefore appears to cause the molecule to break around the CO group, producing C^+, O^+, and neutral CH_3 groups.

Such atomic site selectivity was not found for excitation of oxygen 1s electrons in acetone, however (Nelson et al. 1987). Mass spectra obtained with excitation of an oxygen 1s electron to the pi resonance, a shape resonance, or to the ionization continuum revealed that neither carbon or oxygen atomic ions are significant products of oxygen 1s excitation. Rather, the production of methyl ion fragments is enhanced at the oxygen threshold, and the enhancement appears to follow the oxygen 1s partial absorption cross-section. At this time, it is not clear why these observations for carbon excitation and for oxygen excitation differ.

E. Energy Constraints

Under the appropriate conditions, lineshapes in time-of-flight mass spectra can be determined by ion velocity distributions. An analysis of these lineshapes therefore can provide information about the kinetic energy released in a fragmentation process. For core electron excitation induced fragmentation, this approach was first applied to the fragmentation of CO (Eberhardt et al. 1984). Such an analysis also has been conducted for the case of nitrogen to provide information about the kinetic energy distributions of N^+ as a function of photon energy (Saito and Suzuki 1986b, 1988a) and the anisotropy of fragmentation caused by the alignment of molecules in the excitation step (Saito and Suzuki 1988b).

Knowledge of the kinetic energy released in the fragmentation together with the minimum or threshold energy needed to form particular products makes it possible to examine the energetics of the fragmentation (Eberhardt et al. 1984), and at this point, the most important constraint in determining which fragmentation channels are associated with particular core hole decay routes appears to be the conservation of energy. The kinetic energy, T, of the fragments is given by the following equation:

$$T = E(\text{exc}) - E(\text{elec}) - E(\text{form}) - E(\text{int}),$$

where E(exc) is the initial excitation energy, E(elec) is the energy carried away by the electronic ionization associated with decay of the core hole, E(form) is the minimum energy needed to form the separated fragments, and E(int) is the energy stored in internal degrees of freedom of the fragments. For the case of diatomic molecules, where fragmentation produces only atoms or atomic ions, E(int) only involves excited electronic states, which in particular cases may not be important since they may be widely separated energetically, and near threshold not enough energy is available to produce electronic excitation.

The threshold energy for a particular fragmentation channel satisfies the condition $T = 0$, and a particular fragmentation channel can be associated with a particular electronic relaxation route only if E(elec) for that route is less than a critical value, E_C, which is obtained when $E(\text{int}) = 0$, that is,

$$E_C(\text{elec}) < E(\text{exc}) - E(\text{form})$$

If E(elec) is larger than E_C(elec), the electronic relaxation has carried away too much energy to allow the fragmentation to occur. For core electron ionization, E(exc) corresponds to the binding energy of the core electron since any excess energy, above the binding energy, is carried away by the photoelectron. For core electron excitation to a bound-state resonance that decays by autoionization, E(exc) is the transition energy to this resonance. Observed thresholds for the production of particular ions may be several eV's higher than the threshold obtained from the above equation with $T = 0$ because the fragments may be produced close to each other high on a repulsive potential and hence will appear with considerable kinetic energies. These considerations have been illustrated for the case of carbon monoxide (Eberhardt et al. 1984; Nelson 1987) and nitrogen (Saito and Suzuki 1988a) and further tested by electron–ion coincidence experiments described in the following section.

F. Coincidence Experiments

Since the initial core electron excitation rapidly relaxes by autoionization or Auger decay, the concomitant fragmentation or other chemistry usually results from the final states reached by this electronic relaxation. Understanding this chemistry then involves relating this chemistry to and understanding the autoionization and Auger decay. The correlation between fragmentation and electronic decay was first shown explicitly for nitrogen (Eberhardt et al. (1983b) and actual coincidence measurements for Auger decay and fragmentation were reported somewhat later (Eberhardt et al. 1986a, b, 1987).

These measurements qualitatively confirm the expectations stemming from the conservation of energy constraints described in the previous section. For the case of nitrogen, low-kinetic-energy N^+ ions are observed in coincidence with high-energy Auger electrons. Lower-energy Auger electrons produce coincidences with higher-kinetic-energy ions and with N^{2+} ions. The association of particular Auger electron energies with the formation of particular fragments with a characteristic kinetic energy was analyzed quantitatively in terms of the spacing and qualitative shape of potential energy curves and the excited-state configuration of the products as a function of the electronic configuration of the molecule before fragmentation (Eberhardt et al. 1986b, 1987). The results for carbon monoxide are qualitatively similar and have been discussed in detail (Eberhardt et al. 1986a, 1987).

The fragmentation of nitrogen following core electron excitation was investigated by using the photoion-photoion coincidence technique. The data were analyzed in terms of fragment kinetic energies and the electronic states of the molecular ions and fragments. Both consistencies and discrepancies with previous work are described (Saito and Suzuki 1987).

Extensive ion yield spectra, mass spectra, and ion–ion coincidence data have been acquired for carbon monoxide at both the carbon and oxygen K ionization edges. Dissociative multiple ionization efficiencies, ion branching ratios, and kinetic energy distributions were derived. The results were related to electron energy spectra and potential energy curves for states of CO^{2+} (Hitchcock et al. 1988).

The sensitivity of fragmentation to the nature of resonances was shown for tetramethylsilane by using the photoion–photoion coincidence technique (Morin et al. 1986). In this technique, fragmentation channels are identified by the coincidences of pairs of ions in the time-of-flight mass spectrometer. For tetramethylsilane, two resonances were observed in the ion pair yield curves. The lower energy resonance, below the ionization edge, had been shown from electron spectra to decay into a one-hole state, but the production of ion pairs at the resonance reveals that this resonance also decays by double or multiple ionization.

The key to understanding the fragmentation induced by core electron excitation and ionization lies in Auger electron–ion coincidence experiments. The design of early experiments of this type was not quite optimum: undulator sources were used so the excitation energy could not be tuned to resonances, electron energies were measured in the presence of an ion extraction field thus degrading the electron energy resolution, and ion-ion coincidences were not obtained. Electron–multiple ion coincidence measurements are needed to separate and identify different fragmentation channels associated with the same autoionization or Auger decay channel as defined by the resolution in the electron energy measurement.

G. Angular Distributions

Since synchrotron radiation is polarized, an anisotropic orientational distribution of molecules can be produced by core electron excitation. This distribution is determined by the orientation of the transition dipole moment with respect to the electric vector of the radiation. Information about this alignment and subsequent time evolution is contained in the angular distributions of fragmentation products resulting from the excitation. Details have been worked out and described in several references for the case of valence electron excitation and ionization (Zare 1972; Yang and Bersohn 1974). If the fragmentation is fast relative to the rotational period and if the axial recoil approximation is valid (Zare 1988), then the angular distribution serves to identify the symmetry of the core hole excited state and provides an important characterization of how the electronic wavefunction varies through a shape resonance. An analysis of the angular distribution of fragments resulting from core hole decay for nitrogen recently has been reported (Saito and Suzuki 1988b; Yagishita et al. 1989).

H. Fluorescence Detection

Fluorescence in the visible and ultraviolet regions of the spectrum provides a convenient means for detecting decay produts, both neutral and ionic, produced in excited electronic states. Fluorescence spectra with resolved rotational and vibrational structure provide information about the energy spacing between electronic states, about the structure and bonding properties of these states, and about the populations of rotational and vibrational levels, which can characterize the populating mechanisms associated with decay of the core hole excited state. The production of the doubly charged molecular cation by decay of the core hole is of particular interest because little is known about the properties of these ions and because the fluorescence decay

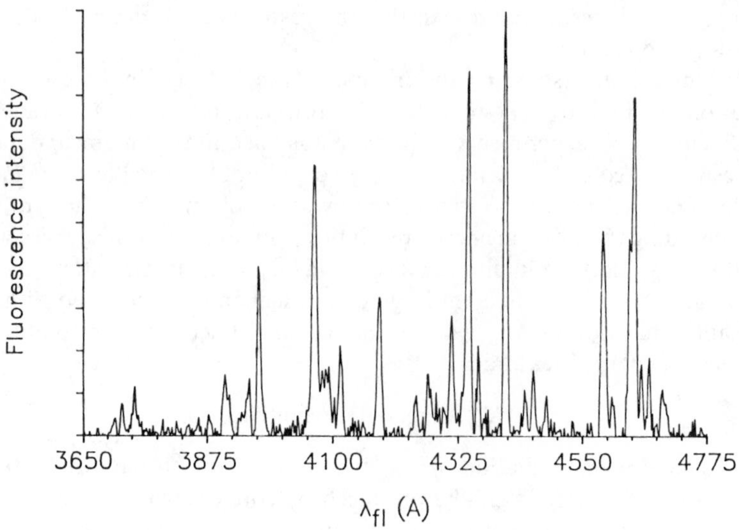

Figure 7. Fluoresence lines assigned to O_2^{2+}.

of the Auger final state essentially allows Auger spectroscopy to be done with the high resolution of an optical spectrometer.

The first observation of dispersed luminescence following core electron excitation by monochromatic synchrotron radiation was reported for the case of oxygen (Tohji et al. 1986; Yang et al. 1988). Two series of bands were observed and assigned to O_2^+ and O_2^{2+} luminescence. The O_2^+ bands are the result of excitation by secondary processes. The O_2^{2+} bands recently (Poliakoff et al. 1989b) have been resolved into individual vibronic transitions as shown in Fig. 7. Double zeta Gaussian type molecular orbital, configuration interaction calculations (Yang et al. 1989) are entirely consistent with the assignment of the O_2^{2+} bands to the $B^3\Pi g \rightarrow A^3\Sigma_u^+$ transition, but the number of lines in the high-resolution spectrum indicates that additional electronic transitions are present as well.

Vibrationally resolved fluorescence of N_2^+ has provided information regarding the decay of core electron resonances in nitrogen (Poliakoff et al. 1988, 1989a). Vibrational branching ratios probe bond-length-dependent properties of core electron resonances, interference effects resulting from lifetime broadening of the vibrational levels of the core hole excited state, and interchannel coupling of ionization continua. With the resolution of rotational levels, the partitioning of angular momentum also can be examined.

I. Electron Impact Excitation and Fragmentation

During the 1970s, electron–ion coincidence techniques were used to obtain information about molecular fragmentation following core electron excitation by high-energy electron impact. Because the excitation step in electron impact involves inelastic scattering, it is necessary to detect and measure the energy of the scattered electron in order to determine the excitation energy and identify the molecular state produced. Fragmentation products in a mass spectrometer then must be detected in coincidence with the scattered electrons in order to relate the fragmentation to a specific excited state. The data acquisition rate in such a coincidence experiment is not very favorable. Nevertheless, it was possible by these means to obtain, for the first time, information about the fragmentation associated with excitation of core electron resonances. Studies of nitrogen and carbon monoxide (Van der Wiel et al. 1970; Van der Wiel and El-Sherbini 1972; Kay et al. 1977), sulfur hexafluoride (Hitchcock et al. 1978), and carbon dioxide and nitrogen dioxide (Hitchcock et al. 1979) were reported. The general conclusions from these early studies were that fragmentation associated with excitation of a bound-state resonance differs from that associated with ionization to the continuum, while excitation of a shape resonance does not. These characteristics are a consequence of the nature of the bound-state resonance, which can decay to a $+1$ ion, and the lifetime of a shape resonance, which generally is shorter than the lifetime of the core hole excited state. The shape resonance decays to an ionic state that is very similar, if not identical, to that reached by direct photoionization of the core electron. The subsequent Auger decay produces final states in the $+2$ ion with their characteristic fragmentation patterns.

Autoionization spectra resulting from specific resonances can be obtained by electron-electron coincidence measurements (Haak et al. 1984; Ungier and Thomas 1983, 1984, 1985). To associate a fragmentation pattern with a particular core hole excited state and a particular autoionization or Auger decay channel, a double-coincidence experiment must be done using electron impact excitation. The energy of the scattered electron must be determined, the energy of the emitted electron must be determined, and the ions produced in coincidence with these two events must be determined. The difficulties inherent in these kinds of experiments have been aptly summarized by Hitchcock (1989), "If you can do it by photons, don't waste your time with electron-coincidence techniques."

The electron impact technique has provided valuable information regarding core electron excitation spectra. A laboratory-based apparatus can provide high-quality spectra with a resolution comparable if not superior to that obtained from monochromators used with synchrotron sources. The contributions that this techniques has made and promises to make have been discussed recently (Hitchcock 1989). The bottom line is that relaxation processes are best studied with a synchrotron source, but a synchrotron source is not needed to obtain spectra of core hole excited states.

J. Negative Ion Formation

The research described in the above discussion focused on the positively charged ions that are produced by core electron excitation and ionization. This focus is reasonable because a core hole excited state quickly relaxes by autoionization or Auger decay to produce positively charged species. The probability for producing high-energy-charge transfer states that might lead to negative ions should be small, and such states may quickly relax to a more equitable charge distribution. It recently has been reported that the ratio of negative to positive ions for the case of sulfur 2p ionization in SO_2 is indeed small, 10^{-5}. Since the negative ions are produced in the continuum, it was concluded that there must be some highly excited states of the ${SO_2}^+$ ion that dissociate or predissociate into SO^{2+} and O^-. Confirmation of this possibility was obtained from direct double ionization thresholds. Although the yield of negative ions is small, they may serve as an important new probe of resonances associated with electronegative atoms in a molecule (Dujardin et al. 1989). Electron-stimulated desorption of negative ion from surfaces also has been observed (Johnson et al. 1988).

K. Transition Stress Theory

The forces acting on the atomic nuclei of a molecule when electrons are excited or ionized, and the electronic configuration changes from that for the ground state, can be analyzed in terms of the transition stress. The objective of the analysis is to have a relatively simple method for predicting which bonds in a polyatomic molecule break as a result of the multiple electron changes in the electronic configuration that result from core electron excitation or ionization and subsequent autoionization or Auger decay. The concept of transition stress is being developed and tested first for diatomic molecules (Zhang and Hanson 1986, 1987a, b, 1989).

The orbital stress for an electron is defined as the average magnitude of the forces exerted by that electron on the two nuclei in a diatomic molecule. When the electronic configuration changes and an electron is promoted from one orbital, i, to another, f, a transition stress on the molecule is produced. The transition stress is the difference between the orbital stress of orbital f and that of orbital i. The transition stresses are additive for multiple electron excitation. Both the nuclear separation and the electronic wavefunctions change to remove this stress. These changes are called the nuclear relaxation and electronic relaxation, respectively. Transition stresses for many electronic states of several diatomic molecules and ions were calculated. It was found that the ratio of the nuclear relaxation to the electronic relaxation is essentially constant for any molecule, independent of the electronic configuration or state. In some cases, this ratio is the same for ionic states of the molecule as well. The bond lengths and vibrational force constants of excited states are related to the transition stress, and the rotational constants are linear functions of the transition stress. The theoretical relationships and calculated values are in good agreement with experimental values and with the results of *ab initio* calculations.

The transition stress theory is based on concepts of one-electron molecular orbital theory. The transition stresses are calculated for the orbitals and virtual orbitals of the molecule in its lowest energy configuration. With this information, the bond lengths, rotational constants, and vibrational frequencies of excited states of the molecule and its ions can be predicted successfully. This success indicates that the theory can account, with reasonable accuracy, for the changes in the forces exerted on the nuclei when the electronic configuration changes. When the repulsive forces are sufficiently large, bond rupture occurs. The criterion for bond rupture requires that the nuclear relaxation causes the bond length to exceed the sum of Pauling's nonbonded radii of the atoms (Zhang and Hanson 1989).

V. SOLID STATE AND SURFACE CHEMISTRY

A. Desorption Induced by Electronic Transitions

Desorption induced by electronic transitions (DIET) is a ubiquitous phenomenon that has been widely studied in recent years. Materials in space, in atomic reactors, and in synchrotron beam lines are subject to a constant flux of radiation and particles that erodes the surfaces and affects the stability of the material or the quality of the ambient atmosphere or vacuum. Photon- and electron-stimulated desorption also have become two of the many tools that are used for surface analysis. These techniques have been used to probe both the atomic and electronic structure of surfaces. Excitation of core electrons in the substrate, the absorbate, and in atomic or molecular films is a significant process in DIET. Since this topic has been covered in several recent monographs (Tolk et al. 1983; Brenig and Menzel 1985; Stulen and Knotek 1988) and review articles (Madey et al. 1984; Knotek 1982, 1984; Menzel 1986; Avouris et al. 1987; Avouris and Walkup 1989), it will not be covered in detail here. The relationships between excitation, relaxation, and fragmentation processes occurring in the gas phase and on surfaces and in solids have been discussed in detail for the cases of water (Ramaker 1983a) and carbon monoxide (Ramaker 1983b). Two important concepts relevant to chemistry induced by core electron excitation have emerged from this research. One is multiple-electron excitation and the other is electronic localization.

As has been described, the creation and relaxation of a core hole excited state involving shake-up, shake-off, and single and multiple electron autoionization and Auger decay, generally leaves the system with two or more holes in bonding orbitals, which essentially means that bond has disappeared and often the system is left on a repulsive potential curve. The importance of such multiple-electron excited states has been discussed for bond rupture and desorption for covalently bonded species (Ramaker 1983a, b, c; Madey et al. 1981; Jaeger et al. 1982; Treichler et al. 1985).

Electronic localization is important in the valence bond depopulation mechanism discussed in Section IV.D. Electronic localization can result from geometrical distortions or from the contraction of orbitals due to the loss of shielding by the removal of a core or valence electron. For two-hole states that result from autoionization or Auger decay of the core hole state, a more significant localization mechanism, called Coulomb localization, is present. The concept of Coulomb localization developed from Auger spectra of solids (Cini 1976, 1977, 1978, 1979; Sawatzky 1977; Antonides et al. 1977; Sawatzky and Lenselink 1980). The essential idea is that Auger decay of a core hole produces two valence holes associated with a single atom because the core hole was localized on that atom. The separation of the two holes by one-hole hopping is retarded because a large amount of potential energy, U, the effective hole–hole interaction, must be dissipated or converted into kinetic energy. The rate for two-hole hopping depends upon the ratio of V, the hopping matrix element, to U, the hole–hole repulsion. In many systems $U \gg V$ and the two holes are effectively localized at the site of the core hole for a time sufficiently long to allow desorption or other chemistry to occur (Ramaker et al. 1982; Ramaker 1983c).

The origin of Coulomb localization for solids can be easily understood in physical terms. If U, the hole–hole repulsion, is larger than the electronic bandwidth, the separation is forbidden because the states lie in the bandgap, i.e. holes with the high velocities needed to absorb the potential energy decrease cannot be produced (Ramaker et al. 1982). For molecules, the situation is not so simple (Dunlap et al 1981), but quantum mechanical calculations have supported the relevance of the concept (Thomas and Weightman 1981; Weightman et al. 1983), and such localization has been used to explain molecular Auger spectra (Rye et al. 1980; Kelber et al. 1981; Rye and Houston 1983, 1984). The physics of the situation for molecules is that the Auger final state with two holes on one atom has a very much higher energy than the state where the holes are separated. The relaxation of one state to the other may be controlled by the electronic coupling, in which case it will depend upon the energy separation, U, and the interaction matrix element, V. The calculations show that $U \gg V$ so the relaxation will be slow. The relevance of Coulomb localization is important because it may lead to localized chemistry, which may be useful in technology, for example, in the modification of surface chemistry for materials in microelectronics technology. This application is discussed below.

B. Microelectronics Applications

The chemical modification of surfaces and films is essential in the manufacture of integrated circuits. Patterns and material must be transferred to substrate surfaces and removed from them. As the dimensions of advanced electronic devices shrink to the submicron and molecular scale, properties of the processing techniques become increasingly critical. Characteristics such as low temperatures, low levels of induced damage and contamination, and molecular scale selectivity and controllability are essential. The use of radiation-induced chemistry is desirable if it can be conducted at lower temperatures and with smaller levels of damage and contamination than particle impact and thermal methods.

Radiation from synchrotron sources has several attractive features in this regard. The short wavelengths provide high spatial resolution in resist and lithography applications and are not absorbed significantly by some surface contaminants. High photon fluxes assure that the necessary photochemistry can occur in a reasonable amount of time. The high photon energies allow excitation of valence electrons of small molecules and also core electrons. Direct double ionization of valence electrons also can occur. As discussed above, the excitation of core electrons produces new chemistry different from that caused by the excitation of valence electrons directly, and has the potential of selectivity and control on an atomic scale.

The utility of synchrotron sources in x-ray lithography in providing high spatial resolution in the manufacture of submicron electron devices is widely recognized. Companies in several nations, most notably West Germany (Cosy MicroTec), England (Oxford Instruments), and Japan (Sumitomo Heavy Industries, among others) are in the process of constructing compact or laboratory-scale synchrotron sources for commercial lithography applications. It is clear that we are on the threshold of a major advance in the application of commercial synchrotron radiation sources. Initially, the application is directed at the lithography market, but the availability of laboratory synchrotron sources will undoubtedly lead to other applications as well (Williams 1988).

A series of experiments relevant to polymer materials used in lithography has demonstrated that photon-energy-selected chemistry can be induced by monochromatic synchrotron radiation in films of polystyrene (Hanson et al.

1985). It was found that creation of a core hole via the pi resonance of the benzene ring in polystyrene does not result in bond rupture and ion desorption, whereas excitation at slightly higher energies does.

The first study of chemical vapor deposition induced by synchrotron radiation was reported by Kyuragi and Urisu (1987a). Silicon nitride films were formed from silane and nitrogen gases at room temperature and below. The merits of this process are the use of low temperatures and the production of a high-purity, low-hydrogen-content film with a sharp edge profile. The sharp edge profile is indicative of a surface reaction. In current processing technology, nitridation is accomplished at temperatures around 1000°C with ammonia or hydrazine gases. The high temperatures are necessary to thermally desorb hydrogen and regenerate surface dangling bonds that dissociate the ammonia (Bozso and Avouris 1986). Such high temperatures also promote degrading effects like diffusion, which become increasingly harmful for submicron structures. The formation of silicon nitride by synchrotron radiation to photodecompose ammonia adsorbed on silicon also has been reported (Cerrina et al. 1987).

The utility of high-energy excitation to facilitate chemistry relevant to microelectronics also has been demonstrated using electron-beam irradiation. It was found that if electron-stimulated desorption was used to remove the hydrogen, then the silicon nitride could be formed at 90 K (Bozso and Avouris 1986). The feasibility of electron-induced chemical vapor deposition was extended to the production of hydrogenated amorphous silicon, silicon dioxide, silicon oxynitride, and silicon nitride films on Si(100) surfaces at 100 K by using disilane, oxygen, nitric oxide, and ammonia, respectively. The electron-beam energy was in the range of 300 to 1000 eV (Bozso and Avouris 1988a, 1988b).

The first study of synchrotron radiation induced etching was reported by Kyuragi and Urisu (1987b). A carbon film was irradiated in the presence of oxygen gas. The etch rate depended linearly on the incident beam power and increased as the square root of the oxygen pressure, indicative of a surface reaction. Urisu and Kyuragi (1987) reported studies of synchrotron radiation etching of silicon and silicon dioxide with SF_6 and SF_6/O_2. The results were completely different from those obtained by plasma etching with these reagents. In plasma etching, the etch rate of Si by SF_6 is an order of magnitude or more faster that the rate for SiO_2, and the rate increases when oxygen is added to the atmosphere. With synchrotron radiation, both Si and SiO_2 etch at about the same rate, but the rate for Si drops to zero with the addition of a small amount of oxygen, while the rate for SiO_2 decreases only slightly. This quenching effect is an asset because usually the etching process is used to remove an SiO_2 film and minimal silicon loss is desirable. The depth profiles reveal that the etching of SiO_2 is mainly a surface reaction, the edges are sharp, while the etching of Si occurs over a broad area, which means that gas-phase excitation is dominant. The SiO_2 etching rate also decreases with increasing temperature, indicating that this reaction is radiation-induced and not thermal. These characteristics indicate that the reaction mechanisms with synchrotron radiation are quite different from those for plasma etching. The dependence of the reactions on photon energy has not been analyzed in detail, but the preliminary indication is that the etching rate is enhanced by photons in the range of a few tens to a few hundred angstroms (Urisu et al. 1989). These wavelengths serve to excite core electrons of silicon and deep valence electrons, and doubly ionize valence electrons. Multiple-electron excited states are produced by the core electron excitation and also by double ionization of valence electrons.

The above experiments on etching and CVD were conducted using white light from the synchrotron. Consequently, both valence and core electrons were excited and ionized and chemical selectivity and control were not sought, but the feasibility of atomic site selective surface modification has been demonstrated in other cases (Yarmoff et al. 1988). A silicon single-crystal surface was fluorinated by exposure to XeF_2. Photoemission spectra clearly reveal the presence of SiF and SiF_3 species on the surface by the difference of 2 eV in the binding energies of the 2p electrons for these compounds. It therefore is possible to selectively excite silicon atoms in a particular oxidation state by tuning monochromatic synchrotron radiation to the 2p core electron binding energy of that state. The F^+ ions that desorb when this is done are the ones bonded to the selected silicon atom. This chemical specificity was demonstrated by the difference in the angular distributions of the desorbing ions. F^+ ions from the monofluoride desorb normal to the surface, while those from the trifluoride species desorb in an off-normal direction. The relationship between bond angles and desorption directions is well-known (Madey 1986). It thus is possible to activate certain silicon sites, those forming SiF, by using monochromatic synchrotron radiation to remove the fluorine atom. The first example of bonding site selectivity was reported for the photon-stimulated desorption of O^+ from $NaWO_3$ (Benbow et al. 1982). Thresholds for O^+ desorption from $NaWO_3$ were observed to correspond to electron binding energies of W but not to those of Na.

VI. CONCLUDING COMMENTS

There are many prospects for research involving chemistry induced by core electron excitation. Opportunities also exist in each of the steps leading to the chemistry. The areas include core electron excitation, autoionization and Auger decay, and fragmentation and other chemical reactions.

The types of states that give rise to resonance structure in soft x-ray absorption spectra and electron energy loss spectra are known, but specific assignments generally are based on analogy and appeal to approximate theoretical calculations. Measurements of symmetry-dependent properties such as angular distributions of electrons and fragments need to be made in order to give the assignments a stronger experimental foundation. Characterization of the spectral structure in terms of the relaxation processes (autoionization, Auger decay, and fragmentation) that are associated with it also may provide definitive experimental evidence for assignments. By these means, discrete resonances can be identified as involving valence or Rydberg orbitals, or mixtures of these, and continuum resonances can be associated with valence orbitals or shake-up states. Contributions from shake-off processes also can be identified by observing products characteristic of multiple ionization.

While much effort has been expended on understanding the Auger spectra of molecules, only a few autoionization spectra of core hole excited states have been obtained and analyzed. A detailed understanding of the electronic relaxation must be developed because this is essential to understanding the subsequent chemistry that occurs.

While fragmentation is the dominant chemical reaction induced by core electron excitation of molecules in the gas phase, other reactions such as rearrangements can be expected, and this possibility needs to be investigated. The extent of a Coulomb explosion in a large molecule is not known, and the role of Coulomb localization in the chemistry of isolated molecules needs to be examined further. Electron–multiple ion coincidence experiments are essential in the study of the chemistry because it is necessary to relate specific electronic decay channels to particular fragmentation patterns as identified by the several ions that are produced.

Primary chemical reactions induced on surfaces and in thin films by the presence of a core hole have been investigated only for the case of desorption. Experiments relevant to microelectronics technology have demonstrated that surface chemistry induced by synchrotron radiation can be unique, and the details of the processes involved need to be elucidated.

The detection of fluorescence provides a high-resolution probe of the electronic decay channels and serves to detect neutral and ionic products in excited states. This technique has been implemented in only a few cases. Techniques for detecting ground-state neutral products of core hole decay also need to be applied.

It is clear that the chemical reactions induced by core electron excitation are atom-specific. They generally depend upon the atomic site of the core hole and the electronic configuration of the core hole excited state. In this sense, the chemistry is selective. The chemistry can be selected by tuning the photon energy to produce particular core hole excited states. In another sense, it is not selective because there usually are many pathways for the decay of each core hole state. The utility of this chemistry then will involve finding conditions where one pathway may dominate or where all pathways lead to the desired result.

Acknowledgment

Support for this research from the National Science Foundation under Grant CHE-8703340 is gratefully acknowledged.

References

Aksela, S., K. H. Tan, H. Aksela, and G. M. Bancroft, 1986, *Phys. Rev.* **A33**, 258.

Antonides, E., E. C. Janse, and G. A. Sawatzky, 1977, *Phys. Rev.* **15**, 1669.

Avouris, Ph. and R. E. Walkup, 1989, *Ann. Rev. Phys. Chem.* **40**, 173.

Avouris, Ph., F. Bozso, and R. E. Walkup, 1987, *Nuc. Instrum. Methods* **B27**, 136.

Bassett, P. J., T. E. Gallon, M. Prutton, and J. A. D. Matthew, 1982, *Surf. Sci.* **33**, 213.

Beckmann, H. O., W. Braun, H. W. Jochims, E. Ruhl, and H. Baumgartel, 1985, *Chem. Phys. Lett.* **121**, 499.

Benbow, R. L., M. R. Thuler, and Z. Hurych, 1982, *Phys. Rev. Lett.* **49**, 1264.

Bozso, F. and Ph. Avouris, 1986, *Phys. Rev. Lett.* **57**, 1185.

Bozso, F. and Ph. Avouris, 1988a, *Appl. Phys. Lett.* **53**, 1095.

Bozso, F. and Ph. Avouris, 1988b, *Phys. Rev.* **B38**, 3943.

Brenig, W. and D. Menzel, Eds., 1985, *Desorption Induced by Electronic Transitions II* (Springer-Verlag, Berlin).

Brion, C. E., S. Daviel, R. Sodhi, and A. P. Hitchcock, 1982, in *X-Ray and Atomic Inner Shell Physics*, B. Crasemann, Ed. (American Institute of Physics, New York).

Callan, E. J., 1961, *Phys. Rev.* **124**, 793.

Carlson, T. A. and M. O. Krause, 1972, *J. Chem. Phys.* **56**, 3206.

Carlson, T. A. and R. M. White, 1966, *J. Chem. Phys.* **44**, 4510.

Carlson, T. A. and R. M. White, 1968, *J. Chem. Phys.* **48**, 5191.

Carroll, T. X. and T. D. Thomas, 1987, *J. Chem. Phys.* **86**, 5221.

Carroll, T. X. and T. D. Thomas, 1988, *J. Chem. Phys.* **89**, 5983.

Carroll, T. X., S. E. Anderson, L. Ungier, and T. D. Thomas, 1987, *Phys. Rev. Lett.* **58**, 867.

Cerrina, F., B. Lai, G. M. Wells, J. R. Wiley, D. G. Kilday, and G. Margaritondo, 1987, *Appl. Phys. Lett.* **50**, 533.

Cini, M., 1976, *Solid State Commun.* **20**, 605.

Cini, M., 1977, *Solid State Commun.* **24**, 681.

Cini, M., 1978, *Phys. Rev.* **B17**, 2788.

Cini, M., 1979, *Surf. Sci.* **87**, 483.

Correia, N., A. Flores-Riveros, H. Agren, K. Helenelund, L. Asplund, and U. Gelius, 1985, *J. Chem. Phys.* **83**, 2035.

Dehmer, J. L., 1984, in *Resonances*, D. G. Truhlar, Ed. (American Chemical Society, Washington, D.C.).

Dehmer, J. L., D. Dill, and A. C. Parr, 1982, in *Photophysics and Photochemistry in the Vacuum Ultraviolet*, S. McGlynn et al., Eds. (Reidel, Dordrecht).

de Souza, G. G. B., P. Morin, and I. Nenner, 1989, *J. Chem. Phys.* **90**, 7071.

de Souza, G. G. B., P. Morin, and I. Nenner, 1986, *Phys. Rev.* **A34**, 4770.

Dujardin, G., L. Hellner, B. J. Olsson, M. J. Besnard-Ramage, and A. Dadouch, 1989, *Phys. Rev. Lett.* **62**, 745.

Dujardin, G., L. Hellner, D. Winkoun, and M. J. Besnard, 1986, *Chem. Phys.* **105**, 291.

Dunlap, B. I., F. L. Hutson, and D. E. Ramaker, 1981, *J. Vac. Sci. Technol.* **18**, 556.

Eberhardt, W., 1987, *Physica Scripta* **T17**, 28.

Eberhardt, W. and T. K. Sham, 1984, *SPIE Proc.* **447**, 143.

Eberhardt, W., T. K. Sham, R. Carr, S. Krummacher, M. Strongin, S. L. Weng, and D. Wesner, 1983a, *Phys. Rev. Lett.* **50**, 1038.

Eberhardt, W., J. Stohr, J. Feldhaus, E. W. Plummer, and F. Sette, 1983b, *Phys. Rev. Lett.* **51**, 2370.

Eberhardt, W., C. T. Chen, W. K. Ford, and E. W. Plummer, 1984, in *Desorption Induced by Electronic Transitions II*, W. Brenig and D. Menzel (Springer-Verlag, Berlin).

Eberhardt, W., E. W. Plummer, I. W. Lyo, R. Reininger, R. Carr, W. K. Ford, and D. Sondericker, 1986a, *Aust. J. Phys.* **39**, 633.

Eberhardt, W., E. W. Plummer, C. T. Chen, R. Carr, and W. K. Ford, 1986b, *Nucl. Instr. and Meth.* **A246**, 825.

Eberhardt, W., E. W. Plummer, C. T. Chen, and W. K. Ford, 1986c, *Aust. J. Phys.* **39**, 853.

Eberhardt, W., E. W. Plummer, I. W. Lyo, R. Carr, and W. K. Ford, 1987, *Phys. Rev. Lett.* **58**, 207.

Feibelman, P. J., 1981, *Surf. Sci.* **102**, L51.

Ferrett, T. A., M. N. Piancastelli, D. W. Lindle, P. A. Heimann, and D. A. Shirley, 1988, *Phys. Rev.* **A38**, 701.

Flores-Riveros, A., N. Correia, H. Agren, L. Pettersson, M. Backstrom, and J. Nordgren, 1985, *J. Chem. Phys.* **83**, 2053.

Freund, H. J. and C. M. Liegener, 1987, *Chem. Phys. Lett.* **134**, 70.

Gelius, U., S. Svensson, H. Siegbahn, E. Basiliev, A. Faxalv, and K. Siegbahn, 1974, *Chem. Phys. Lett.* **28**, 1.

Haak, H. W., G. A. Sawatzky, L. Ungier, J. K. Gimzewski, and T. D. Thomas, 1984, *Rev. Sci. Instrum.* **55**, 696.

Hanson, D. M., R. Stochbauer, and T. E. Madey, 1982, *J. Chem. Phys.* **77**, 1569.

Hanson, D. M., S. L. Anderson, M. C. Nelson, G. P. Williams, and N. Lucas, 1985, *J. Phys. Chem.* **89**, 2235.

Hayes, R. G., 1987, *J. Chem. Phys.* **86**, 1683.

Higashi, M., E. Hiroike, and T. Nakajima, 1982, *Chem. Phys.* **68**, 377.

Hirsch, R. G., R. J. Van Brunt, and W. D. Whitehead, 1973, *J. Chem. Phys.* **59**, 3863.

Hitchcock, A. P., 1989, *Ultramicroscopy*, in press.

Hitchcock, A. P. and J. Stohr, 1987, *J. Chem. Phys.* **87**, 3253.

Hitchcock, A. P., C. E. Brion, and M. J. Van der Wiel, 1978, *J. Phys.* **B11**, 3245.

Hitchcock, A. P., C. E. Brion, and M. J. Van der Wiel, 1979, *Chem. Phys. Lett.* **66**, 213.

Hitchcock, A. P., S. Beaulieu, T. Steel, J, Stohr, and F. Sette, 1984, *J. Chem. Phys.* **80**, 3927.

Hitchcock, A. P., P. Lablanquie, P. Morin, E. Lizon, A. Lugrin, M. Sioon, P. Thiry, and I. Nenner, 1988, *Phys. Rev.* **A37**, 2448.

Jaeger, R., R. Treichler, and J. Stohr, 1982, *Surface Sci.* **117**, 533.

Jennison, D. R., 1978, *Phys. Rev. Lett.* **40**, 807.

Jennison, D. R., 1982, *J. Vac. Sci. Technol.* **20**, 548.

Jennison, D. R., J. A. Kelber, and R. R. Rye, 1982, *Phys. Rev.* **B25**, 1384.

Johnson, A. L., S. A. Joyce, and T. E. Madey, 1988, *Phys. Rev. Lett.* **61**, 2578.

Kay, R. B., Ph. E. Van der Leeuw, and M. J. Van der Wiel, 1977, *J. Phys.* **B10**, 2521.

Kelber, J. A. and M. L. Knotek, 1982, *Surf. Sci.* **121**, L499.

Kelber, J. A. and M. L. Knotek, 1983, *J. Vac. Sci. Technol.* **A1**, 1149.

Kelber, J. A. and M. L. Knotek, 1984, *Phys. Rev.* **B30**, 400.

Kelber, J. A., D. R. Jennison, and R. R. Rye, 1981, *J. Chem. Phys.* **75**, 652.

King, G. C. and F. H. Read, 1985, in *Atomic Inner Shell Physics*, B. Crasemann, Ed. (Plenum, New York).

King, G. C., F. H. Read, and M. Tronc, 1977, *Chem. Phys. Lett.* **52**, 50.

Knotek, M. L., 1982, in *X-Ray and Atomic Inner Shell Physics*, B. Crasemann, Ed. (American Institute of Physics, New York).

Knotek, M. L., 1984, *Rep. Prog. Phys.* **47**, 1499.

Kyuragi, H. and T. Urisu, 1987a, *J. Appl. Phys.* **61**, 2035.

Kyuragi, H. and T. Urisu, 1987b, *Appl. Phys. Lett.* **50**, 1254.

Lablanquie, P., A. C. A. Souza, G. G. B. de Souza, P. Morin, and I. Nenner, 1989, *J. Chem. Phys.* **90**, 7078.

Langhoff, P. W., 1984, in *Resonances*, D. G. Truhlar, Ed. (American Chemical Society, Washington, D.C.).

Lapiano-Smith, D. A., C. I. Ma, K. T. Wu, and D. M. Hanson, 1989, *J. Chem. Phys.* **90**, 2162.

Larkins, F. P., 1987, *J. Chem. Phys.* **86**, 3239.

Larkins, F. P., W. Eberhardt, I. W. Lyo, R. Murphy, and E. W. Plummer, 1988, *J. Chem. Phys.* **88**, 2948.

Lee, P. A., P. H. Citrin, P. Eisenberger, and B. M. Kincaid, 1981, *Rev. Mod. Phys.* **53**, 769.

Ley, L. and M. Cardona, Eds., 1979, *Photoemission in Solids II* (Springer-Verlag, Berlin).

Lynch, D. L., V. McKoy, and R. R. Lucchese, 1984, in *Resonances*, D. G. Truhlar, Ed. (American Chemical Society, Washington, D.C.).

Madden, H. H., D. R. Jennison, M. M. Traum, G. Margaritondo, and N. G. Stoffel, 1982, *Phys. Rev.* **B26**, 896.

Madey, T. E., 1986, *Science* **234**, 316.

Madey, T. E., R. Stockbauer, S. A. Flodstrom, J. F. van der Veen, F. J. Himpsel, and D. E. Eastman, 1981, *Phys. Rev.* **B23**, 6847.

Madey, T. E., D. E. Ramaker, and R. Stockbauer, 1984, *Ann. Rev. Phys. Chem.* **35**, 215.

McGuire, E. J., 1969, *Phys. Rev.* **185**, 1.

Menzel, D., 1986, *Nucl. Instrum. Methods* **B13**, 507.

Morin, P. and I. Nenner, 1986, *Phys. Rev. Lett.* **56**, 1913.

Morin, P. and I. Nenner, 1987, *Physica Scripta* **T17**, 171.

Morin, P., G. G. B. de Souza, and I. Nenner, 1986, *Phys. Rev. Lett.* **56**, 131.

Muller-Dethlefs, K., M. Sander, L. A. Chewter, and E. W. Schlag, 1984, *J. Phys. Chem.* **88**, 6098.

Murakami, J., M. C. Nelson, S. L. Anderson, and D. M. Hanson, 1986, *J. Chem. Phys.* **85**, 5755.

Murphy, R. and W. Eberhardt, 1988, *J. Chem. Phys.* **89**, 4054.

Murphy, R., I. W. Lyo, and W. Eberhardt, 1988, *J. Chem. Phys.* **88**, 6078.

Nagaoka, S., S. Suzuki, and I. Koyano, 1987, *Phys. Rev. Lett.* **58**, 1524.

Nagaoka, S., I. Koyano, K. Ueda, E. Shigemasa, Y. Sato, A. Yagishita, T. Nagata, and T. Hayaishi, 1989, *Chem. Phys. Lett.* **154**, 363.

Natoli, C. R., 1983, in *EXAFS and Near Edge Structure*, A. Bianconi, L. Incoccia, and S. Stipcich, Eds. (Springer-Verlag, Berlin).

Nelson, M. C., 1987, *Fragmentation of Small Molecules Induced by Soft X-Rays*, Ph.D. Dissertation, State University of New York at Stony Brook.

Nelson, M. C., J. Murakami, S. L. Anderson, and D. M. Hanson, 1987, *J. Chem. Phys.* **86**, 4442.

Nenner, I., 1987, in *Giant Resonances in Atoms, Molecules, and Solids*, J. P. Connerade, J. M. Esteva, and R. C. Karnatak, Eds. (Plenum, New York).

Nenner, I., P. Morin, M. Simon, P. Lablanquie, and G. G. B. de Souza, 1988, in *Desorption Induced by Electronic Transitions III*, R. H. Stulen and M. Knotek.

Piancastelli, M. N., D. W. Lindle, T. A. Ferrett, and D. A. Shirley, 1987a, *J. Chem. Phys.* **86**, 2765.

Piancastelli, M. N., D. W. Lindle, T. A. Ferrett, and D. A. Shirley, 1987b, *J. Chem. Phys.* **87**, 3255.

Piancastelli, M. N., T. A. Ferrett, D. W. Lindle, L. J. Medhurst, P. A. Heimann, S. H. Liu, and D. A. Shirley, 1989, *J. Chem. Phys.* **90**, 3004.

Poliakoff, E. D., L. A. Kelly, L. M. Duffy, B. Space, P. Roy, S. H. Southworth, and M. G. White, 1988, *J. Chem. Phys.* **89**, 4048.

Poliakoff, E. D., L. A. Kelly, L. M. Duffy, B. Space, P. Roy, S. H. Southworth, and M. G. White, 1989a, *Chem. Phys.* **129**, 65.

Poliakoff, E. D., L. A. Kelly, D. Lapiano-Smith, C. I. Ma, K. Wu, and D. M. Hanson, 1989b, to be published.

Prins, R. and D. Koningsberger, Eds., 1987, *Principles, Applications, and Techniques of EXAFS, SEXAFS, and XANES* (Wiley, New York).

Ramaker, D. E., 1980, *Phys. Rev.* **B21**, 4608.

Ramaker, D. E., 1983a, *Chem. Phys.* **80**, 183.

Ramaker, D. E., 1983b, *J. Chem. Phys.* **78**, 2998.

Ramaker, D. E., 1983c, in *DIET I*, N. H. Tolk, M. M. Traum, J. C. Tully, and T. E. Madey, Eds. (Springer-Verlag, Berlin).

Ramaker, D. E., C. T. White, and J. S. Murday, 1981, *J. Vac. Sci. Technol.* **18**, 748.

Ramaker, D. E., C. T. White, and J. S. Murday, 1982, *Phys. Lett.* **89A**, 211.

Rosenberg, R. A., V. Rehn, V. O. Jones, A. K. Green, C. C. Parks, G. Loubriel, and R. H. Stulen, 1981, *Chem. Phys. Lett.* **80**, 488.

Rosenberg, R. A., P. R. LaRoe, V. Rehn, J. Stohr, R. Jaeger, and C. C. Parks, 1983, *Phys. Rev.* **B28**, 3026.

Rosenberg, R. A., P. J. Love, P. R. LaRoe, V. Rehn, and C. C. Parks, 1985, *Phys. Rev.* **B31**, 2634.

Rye, R. R. and J. E. Houston, 1983, *J. Chem. Phys.* **78**, 4321.

Rye, R. R. and J. E. Houston, 1984, *Acc. Chem. Res.* **17**, 41.

Rye, R. R., D. R. Jennison, and J. E. Houston, 1980, *J. Chem. Phys.* **73**, 4867.

Saito, N. and I. H. Suzuki, 1986a, *Chem. Phys.* **108**, 327.

Saito, N. and I. H. Suzuki, 1986b, *Chem. Phys. Lett.* **129**, 419.

Saito, N. and I. H. Suzuki, 1987, *J. Phys. B: At. Mol. Phys.* **20**, L785.

Saito, N. and I. H. Suzuki, 1988a, *Int. J. Mass Spectrom. Ion Phys.* **82**, 61.

Saito, N. and I. H. Suzuki, 1988b, *Phys. Rev. Lett.* **61**, 2740.

Sawatzky, G. A., 1977, *Phys. Rev. Lett.* **39**, 504.

Sawatzky, G. A. and A. Lenselink, 1980, *Phys. Rev.* **B21**, 1790.

Schmidt, V., 1987, *J. Phys.* Colloq. C9/1, 401.

Sette, F., J. Stohr, and A. P. Hitchcock, 1984, *J. Chem. Phys.* **81**, 4906.

Siegbahn, K., et al., 1969, *ESCA Applied to Free Molecules* (North–Holland, Amsterdam).

Stockbauer, R., E. Bertel, and T. E. Madey, 1982a, *J. Chem. Phys.* **76**, 5639.

Stockbauer, R., D. M. Hanson, S. A. Flodstrom, and T. E. Madey, 1982b, *Phys. Rev.* **B26**, 1885.

Stohr, J. and D. A. Outka, 1987a, *J. Vac. Sci. Technol.* **5**, 919.

Stohr, J. and D. A. Outka, 1987b, *Phys. Rev.* **B36**, 7891.

Stohr, J., J. L. Gland, W. Eberhardt, D. Outka, R. J. Madix, F. Sette, R. J. Koestner, and U. Doebler, 1983, *Phys. Rev. Lett.* **51**, 2414.

Stohr, J., F. Sette, and A. L. Johnson, 1984, *Phys. Rev. Lett.* **53**, 1684.

Stohr, J., D. A. Outka, K. Baberschke, D. Arvanitis, and J. A. Horsley, 1987, *Phys. Rev.* **B36**, 2976.

Stulen, R. H. and M. L. Knotek, Eds., 1988, *Desorption Induced by Electronic Transitions III* (Springer-Verlag, Berlin).

Teo, B. K. and D. C. Joy, Eds., 1981, *EXAFS Spectroscopy, Techniques, and Applications* (Plenum, New York).

Thomas, T. D. and P. Weightman, 1981, *Chem. Phys. Lett.* **81**, 325.

Tohji, K., D. M. Hanson, and B. X. Yang, 1986, *J. Chem. Phys.* **85**, 7492.

Tolk, N. H., M. M. Traum, J. C. Tully, and T. E. Madey, Eds., 1983, *Desorption Induced by Electronic Transitions I* (Springer-Verlag, Berlin).

Treichler, R., W. Riedl, W. Wurth, P. Feulner, and D. Menzel, 1985, *Phys. Rev. Lett.* **54**, 462.

Ueda, K., E. Shigemasa, Y. Sato, S. Nagaoka, I. Koyano, A. Yagishita, T. Nagata, and T. Hayaishi, 1989, *Chem. Phys. Lett.* **154**, 357.

Ungier, L. and T. D. Thomas, 1983, *Chem. Phys. Lett.* **96**, 247.

Ungier, L. and T. D. Thomas, 1984, *Phys. Rev. Lett.* **53**, 435.

Ungier, L. and T. D. Thomas, 1985, *J. Chem. Phys.* **82**, 3146.

Urisu, T. and H. Kyuragi, 1987, *J. Vac. Sci. Technol.* **B5**, 1436.

Urisu, T., H. Kyuragi, Y. Utsumi, J. Takahashi, and M. Kitamura, 1989, *Rev. Sci. Instrum.* **60**, 2024, 2157.

Van Brunt, R. J., R. W. Powell, R. G. Hirsch, and W. D. Whitehead, 1972, *J. Chem. Phys.* **57**, 3120.

Van der Wiel, M. J., Th. M. El-Sherbini, and C. E. Brion, 1970, *Chem. Phys. Lett.* **7**, 161.

Van der Wiel, M. J. and Th. M. El-Sherbini, 1972, *Physica* **59**, 453.

Weightman, P., T. D. Thomas, and D. R. Jennison, 1983, *J. Chem. Phys.* **78**, 1652.

Williams, G. P., 1988, *Synchrotron Radiation News* 1, No. 2, 21.

Yagishita, A., N. Maezawa, M. Ukai, and E. Shigemasa, 1989, *Phys. Rev. Lett.* **62**, 36.

Yang, S. C. and R. Bersohn, 1974, *J. Chem. Phys.* **61**, 4400.

Yang, B. X., D. M. Hanson, and K. Tohji, 1988, *J. Chem. Phys.* **89**, 1215.

Yang, H., J. L. Whitten, and D. M. Hanson, 1989, to be published.

Yarmoff, J. A., A. Taleb-Ibrahimi, F. R. McFeely, and Ph. Avouris, 1988, *Phys. Rev. Lett.* **60**, 960.

Zare, R. N., 1972, *Mol. Photochem.* **4**, 1.

Zare, R. N., 1988, *Augular Momentum* (Wiley, New York).

Zhang, Y. and D. M. Hanson, 1986, *Chem. Phys. Lett.* **127**, 33.

Zhang, Y. and D. M. Hanson, 1987a, *J. Chem. Phys.* **86**, 347.

Zhang, Y. and D. M. Hanson, 1987b, *J. Chem. Phys.* **86**, 666.

Zhang, Y. and D. M. Hanson, 1989, *Chem. Phys.* **138**, 71.

SYNERGISTIC EFFECTS IN TWO-PHOTON ABSORPTION: THE QUANTUM ELECTRODYNAMICS OF BIMOLECULAR MEAN-FREQUENCY ABSORPTION

DAVID L. ANDREWS AND KEVIN P. HOPKINS

School of Chemical Sciences
University of East Anglia
Norwich, England

CONTENTS

I. INTRODUCTION

There is a wide range of phenomena in chemical physics resulting from synergistic interactions of atoms and molecules (Schuster et al. 1980). The term *synergistic* denotes a mechanistic cooperativity, reflected in the fact that complex systems often behave in a very different way from a simple sum of their individual parts. One of the most well-known synergistic effects at the atomic or molecular level is the concerted photoemission process which forms the basis for laser action (Haken 1985). At the other end of the molecular spectrum, synergistic interactions in biomolecules are known to be responsible for many of the control and regulation mechanisms in living organisms (Haken 1977).

In connection with photoabsorption, it has long been known that a collisional interaction between two atoms or molecules irradiated with light of a suitable frequency can result in their simultaneous excitation. The first observations of this effect were made in infrared studies on compressed gases (Ketelaar 1959). Indeed, the effect has been proposed as an explanation for some of the spectral features of planetary atmospheres, where high pressures of gaseous mixtures naturally occur (Danielson 1974). Experimental studies have mostly focused on interaction-induced optical transitions in gases (Yakovlenko 1973; Geltman 1976, 1987; Gallagher and Holstein 1977; Green et al. 1979; Brechignac et al. 1980; Débarre and Cahuzac 1986) or in crystals containing rare earth ions (Varsanyi and Dieke 1961; Dexter 1962; Tulub and Patzer 1968; Nakazawa and Shionoya 1970; Last et al. 1987). Although such processes involve pairs of atoms in collision or close proximity, each absorption involves only a single photon.

In condensed matter, cooperative effects of a different kind have long been established in connection with the photoproduction of relatively long-lived triplet states, either in individual molecules in the case of liquids (Parker and Hatchard 1962; Nosworthy and Keene 1964), or in triplet excitons in crystalline solids (Avakian and Merrifield 1968; Groff et al. 1970). Because of their long lifetimes, the concentrations of such species can build up to a point where triplet–triplet annihilation ensues, either by molecular diffusion or by exciton migration (Craig and Walmsley 1964), resulting in the generation of highly excited states which may subsequently decay through short-wavelength luminescence (delayed fluorescence) (Siebrand 1965; Smith 1968; Arnold et al. 1970; Sasaki and Hayakawa 1978). These processes often have a quadratic dependence on the intensity of irradiation, and thus exhibit features characteristic of two-photon absorption. However, since the separate photon absorption processes and also the subsequent triplet–triplet annihilation are uncorrelated and individually satisfy the requirements of energy conservation, such phenomena cannot be regarded as truly synergistic in any normal sense of the word.

Where singlet states are concerned, lifetimes are generally too short for concentrations to build up to a point where collisional annihilation is significant, if conventional irradiation sources are employed. With laser light of sufficient intensity, however, such effects are indeed observable and are thought to play a part in the dynamics of energy transfer in photosynthetic systems, for example (Van Grondelle 1985). With laser excitation it is also possible to observe nonlinear optical effects in which two or more photons are absorbed by each atomic or molecular pair. The theoretical prediction of such a process was first made by Rios Leite and De Araujo (1980) in a paper concerned with cooperative absorption by atom pairs in solids. However, the first experimental observation made shortly afterwards by White (1981) came from laser excitation studies of gaseous mixtures of barium and thallium. Atoms of both species were found to be simultaneously promoted to excited states by a concerted process involving the pairwise absorption of laser photons, a process which thus acquires the character of *mean-frequency absorption.*

Following theoretical developments (Andrews and Harlow 1983, 1984a), *bimolecular* mean-frequency absorption was next discovered by researchers working on the photodynamics of charge transfer reactions. For example, Ku et al. (1983) performed studies on gaseous mixtures of xenon and chlorine passed through a laser fluorescence cell. The proposed reaction mechanism

$$\mathrm{Xe} + \mathrm{Cl_2} + 2\hbar\omega \rightarrow \{\mathrm{Xe}\text{—}\mathrm{Cl_2}^{**} \rightarrow \mathrm{Xe}^{+}\text{—}\mathrm{Cl_2}^{-}\} \rightarrow \mathrm{XeCl} + \mathrm{Cl} \qquad (1.1)$$

was attributed to excitation from the van der Waals ground state potential to the ion-pair potential via configurational interaction of the Xe and Cl_2. This discovery clearly delineates the underlying mechanism for the process as one in which the synergistic excitation of the two absorbing species results from molecular proximity rather than any collisional effect. The reaction mechanism (1.1) has been further corroborated by Apkarian and co-workers, who have also extended the study to charge transfer reactions in solid and liquid xenon (Fajardo and Apkarian 1986, 1987, 1988; Wiedeman et al. 1987, 1988). It has additionally been shown to be the predominant reaction route in the case of $Xe:Cl_2$ van der Waals complexes generated in seeded molecular beams (Boivineau et al. 1986a, b).

Recently, a new type of synergistic photoabsorption process involving two-frequency excitation has been the subject of renewed theoretical interest (Andrews and Hopkins 1987, 1988a, 1988b). Here the two chemical centers which undergo concerted excitation may or may not be chemically similar, and can represent either distinct chromophores within a single molecule, loosely bound systems such as van der Waals molecules or solute particles within a coordination shell of solvent molecules, or else completely separate molecules. Where the two centers are chemically identical, the term *bicimer* (Locke and Lim 1987) appropriately describes the result of the excitation. Two-beam two-photon absorption in lanthanide (III) compounds has been the subject of a recent study by Sztucki and Strek (1988); no experimental studies of synergistic two-beam effects have yet been reported at the time of writing.

While the effects of interest are most readily studied by specifically designed two-beam laser experiments discussed in detail below, it also transpires that the quantum uncertainty mechanism involved may play a significant role in other photoabsorption processes where optical nonlinearity is not immediately apparent. This is particularly the case in connection with studies based on white or broadband light, and the effects may be manifest in the appearance of anomalous features in the corresponding absorption spectra, especially those obtained using ultrashort laser pulses (Andrews 1988). Here, intensity-dependent lineshapes or extinction coefficients and the appearance of ostensibly extraneous spectral lines may all be attributable to the effects of synergistic photoabsorption. It thus appears that synergism in two-photon absorption may have a more general significance than has hitherto been recognized.

II. QUANTUM UNCERTAINTY AND PROBABILITY CONSIDERATIONS

Bimolecular mean-frequency absorption is a nonlinear process which results entirely from the effects of molecular proximity, although wavefunction

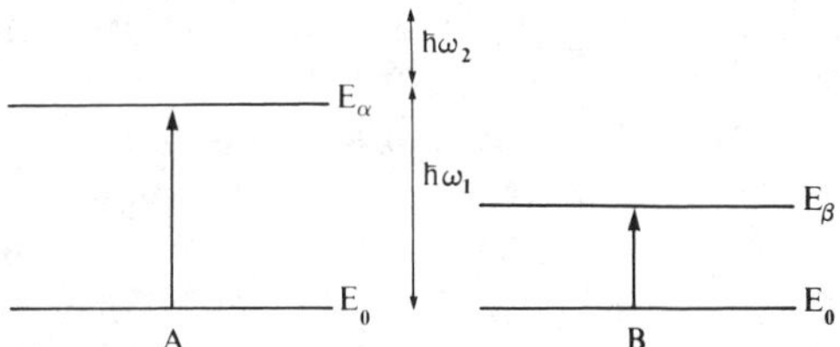

Figure 1. Energy-level diagram for the most general case of synergistic two-photon absorption. The absorption of two photons with different frequencies ω_1 and ω_2 effects transitions to states $|\alpha\rangle$ and $|\beta\rangle$ in two chemically different molecules A and B, subject only to overall energy conservation.

overlap is in no way directly implicated. The two chemical centers involved in the process may be either distinct chromophores in a single molecule, or completely separate molecules. The two participating centers undergo a concerted excitation through the absorption of two laser photons, and in the most general case the process can be represented by the equation

$$\mathrm{A} + \mathrm{B} + \hbar\omega_1 + \hbar\omega_2 \rightarrow \mathrm{A}^* + \mathrm{B}^* \tag{2.1}$$

which is restricted only by the energy conservation requirement

$$E_{\alpha 0} + E_{\beta 0} = \hbar\omega_1 + \hbar\omega_2 \tag{2.2}$$

as illustrated in Fig. 1. It is assumed that both molecules A and B are initially in their ground states, and that they are promoted during the absorption process to excited vibronic states designated by the asterisks in Eq. (2.1), and α and β in Eq. (2.2). As will be shown below, this general two-molecule two-photon interaction encompasses four particular types of interaction which are of special interest. However, the basic theory can first be developed with reference to the totally general case.

A. Energy–Time Uncertainty and Synergistic Absorption

To understand the physical mechanism underlying these processes, it is helpful to first consider the broad significance of the energy–time Uncertainty Principle for photoabsorption processes. It is well-known that for a molecular excited state with an average lifetime δt and an average energy displacement δE from the ground state, there exists the relation (Finkel 1987)

$$\delta t \delta E \geqslant h/4 \tag{2.3}$$

The application of this result to the interaction involved in a normal photoabsorption process in which individual atoms or molecules absorb single photons amounts to a statement that for a time interval δt, the mismatch between the energy gain of the absorber and the energy of the absorbed photon cannot exceed δE. For any normal timescale of observation, this condition ensures that only photons whose energy closely matches a transition energy of the absorber can in fact be absorbed.

The implication of Eq. (2.3) for a concerted photoabsorption process involving the coupling of two atomic or molecular excitations is less obvious, however. In this case, Eq. (2.3) determines the timescale within which the excess energy δE absorbed by one center needs to be conveyed to a center with a corresponding negative energy mismatch, in order to fulfil the requirement for long-term energy conservation at each center. There are in fact two distinct mechanisms for a concerted absorption of light based on this Uncertainty Principle, as detailed below.

1. *Cooperative Absorption Mechanism*

Consider first a molecule A in an initial state $|0\rangle$ undergoing a transition to an excited state $|\alpha\rangle$ through absorption of light with circular frequency ω_1. If ω_1 is off-resonant with respect to the transition frequency, there is a mismatch in energy by an amount

$$\delta E = \hbar\omega_1 - E_{\alpha 0} \tag{2.4}$$

Application of the uncertainty relation (2.3) shows that it is impossible to constrain conservation of energy over a timescale less than

$$\tau = \tfrac{1}{4}(\omega_1/2\pi - E_{\alpha 0}/h)^{-1} \tag{2.5}$$

The transition is therefore allowed provided the local energy mismatch exists for a time not exceeding τ.

A suitable mechanism for compensation of the energy mismatch is provided through absorption by a second molecule B of a photon with frequency ω_2 given by

$$\omega_2 = -\omega_1 + (E_{\alpha 0} + E_{\beta 0})/\hbar \tag{2.6}$$

Thus by the cooperative absorption of the photon pair, the total energy absorbed is $(E_{\alpha 0} + E_{\beta 0})$, the sum of the two molecular transition energies, and overall energy conservation is therefore achieved.

Clearly there needs to exist a mechanism for conveyance of the mismatch energy from one molecule to another (see Fig. 2), so that over longer times each molecule can individually satisfy the requirements of energy

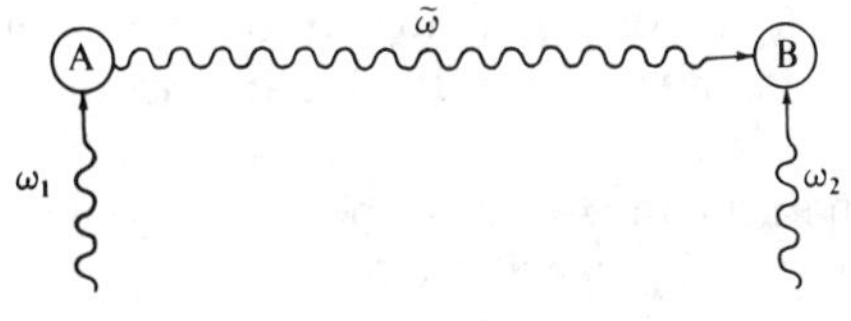

Figure 2. A cooperative mechanism for synergistic two-photon absorption. Molecule A absorbs a photon of frequency ω_1 and molecule B a photon of frequency ω_2, with the mismatch energy propagated from A to B by a virtual photon $\tilde{\omega}$.

conservation. Within the framework of quantum electrodynamics (QED), virtual photon coupling (Feynman 1961) provides the mechanism for cooperative absorption of this type. The concept of virtual photons originated in connection with nuclear physics (Dodge 1985), but the formalism has increasingly found application in chemical physics, for example in the theory of intermolecular interactions (Craig and Thirunamachandran 1982; Vigoureux 1983; Grossel et al. 1983; Power and Thirunamachandran 1983; Andrews and Sherborne 1987; Barron and Johnston 1987; Van Labeke et al. 1988) and the calculation of atomic energy shifts (Compagno et al. 1983, 1985).

By introducing the formalism of virtual photon coupling, the timescale for cooperative absorption, τ, can be interpreted in terms of a range of propagation for which the exchanged photon has virtual character. Thus the distance R between two molecules that cooperate in the absorption process must be subject to the condition

$$R \leqslant R_{\max} \geqslant \tfrac{1}{4}hc(\hbar\omega_1 - E_{\alpha 0})^{-1} \tag{2.7}$$

As an example, for a circular frequency mismatch of 2.5×10^{13} Hz, we have the constraint $R_{\max} \geqslant 3\,\mu$m. With a smaller mismatch, R can obviously become very large compared to molecular dimensions, and in the limit where the mismatch is zero, there is no longer any restriction on the molecular separation. This corresponds to the case where two entirely uncorrelated absorption processes occur.

2. *Distributive Absorption Mechanism*

A secondary mechanism for the absorption of nonresonant frequencies involves the concerted absorption of two photons by a single molecule, with virtual photon conveyance of the excess energy to the second absorber (see Fig. 3). As with conventional two-photon absorption, there is no need for

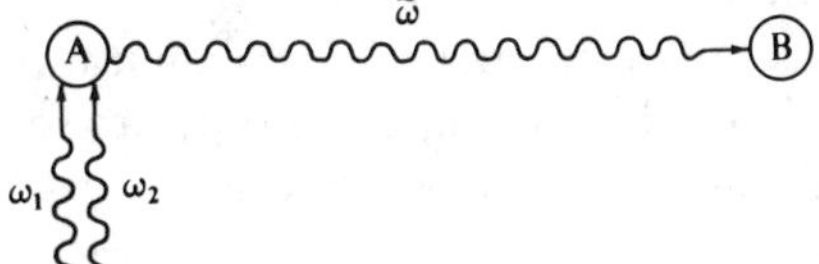

Figure 3. A distributive mechanism for synergistic two-photon absorption. Molecule A absorbs two photons of frequencies ω_1 and ω_2, and the mismatch energy propagates to molecule B by a virtual photon $\tilde{\omega}$.

the first molecule to possess an energy level corresponding to the energy of either one of the absorbed photons, and thus no identifiable intermediate state is populated, as would be the case in a two-step process. In this case, forthwith termed the distributive mechanism, the two absorbed photons may again have differing energies given by $\hbar\omega_1 = (E_{\alpha 0} + \delta E)$ and $\hbar\omega_2 = (E_{\beta 0} - \delta E)$, but here the excess energy absorbed by the first molecule now equals the transition energy for the second molecule, $E_{\beta 0}$. The corresponding limit on the range of intermolecular distance is then given by

$$R \leqslant R_{\max} \geqslant \tfrac{1}{4}hc/E_{\beta 0} \tag{2.8}$$

If the photon energy $\hbar\omega_1$ is anywhere near to the excitation energy $E_{\alpha 0}$, it is clear from Eqs. (2.7) and (2.8) that where virtual photon coupling is involved, the distributive mechanism will only be effective over a much shorter range than the cooperative mechanism. For example, if $E_{\alpha 0}/h = 5 \times 10^{14}$ Hz, we have $R_{\max} \geqslant 0.15\ \mu$m.

B. Probability Considerations

One of the first considerations when comparing the cooperative and distributive mechanisms for synergistic photoabsorption is a difference in the probability aspect of the two processes. At first sight, the requirement of the distributive mechanism for two laser photons to be absorbed in a concerted process at a single molecule appears to render the effect significantly less probable than the cooperative mechanism, which has the apparently looser requirement for two photons to be absorbed by molecules at any two different points in the sample. A simple statistical treatment of each process based on a Poisson distribution, the most appropriate form of distribution for laser light (Louisell 1973), in fact shows that for any given pair of molecules the conditions for the distributive process are met half as often as those of the cooperative process. If the mean number of photons per molecular volume is denoted by m, the probability of finding n photons is given by $P_n = (m^n/n!)e^{-m}$: hence for cooperative absorption we have $P_1^2 = m^2 e^{-2m}$, and for distributive absorption $P_2 = m^2 e^{-2m}/2$ (Andrews 1985).

However in most samples of chemical interest there is normally more than one pair of molecules to consider. With N molecules, there are clearly $\frac{1}{2}N(N-1)$ pairs which can participate in a cooperative absorption process, but $N(N-1)$ to participate in a distributive process. Hence, overall the photon statistics do not provide a basis for differentiating the significance of the two mechanisms. However, as shown below, the selection rules for the two processes differ markedly, and often result in a single mechanism being exclusively operative.

III. SELECTION RULES

As seen above, synergistic two-photon absorption can in principle take place by either or both of the mechanisms, where (i) each laser photon is absorbed by a different molecule (the cooperative mechanism), or (ii) both laser photons are absorbed by a single molecule (the distributive mechanism). In each case, the energy mismatch for the molecular transitions is transferred between the molecules by means of a virtual photon that couples with each molecule by the same electric-dipole coupling as the laser photons. The result, however, is a significant difference in the selection rules applying to the two types of processes.

In the cooperative case, the two molecular transitions are separately allowed under well-known two-photon selection rules, since each molecule absorbs one laser photon and either emits or absorbs a virtual photon. In the same way, the distributive case provides for excitation through three-and one-photon allowed transitions, and may thus lead to excitation of states that are formally two-photon forbidden. (In general, it is sufficient to stipulate that both transitions involved in the distributive mechanism are one-photon allowed since, with the rare exception of icosahedrally symmetric molecules, all transitions which are one-photon allowed are of necessity also three-photon allowed (Andrews and Wilkes 1985).)

Since, on the whole, these processes are of most interest for molecules of fairly high symmetry, it can safely be assumed that in most cases one mechanism alone is involved in the excitation to a particular pair of excited states α and β. Certainly this is rigorously true for centrosymmetric species, where, under the cooperative mechanism, both transitions must preserve parity ($g \leftrightarrow g, u \leftrightarrow u$), but under the distributive mechanism parity reversal ($u \leftrightarrow g$) results at each center. Only in the case of solutions where solute–solvent interactions can reduce excited-state symmetry is this rule weakened (Mohler and Wirth 1988). The assumption that only one mechanism can be operative for any given bimolecular mean-frequency transition gives the advantage of considerably simplifying the form of the rate equations.

One other feature is worth noting at this point, and it concerns the case where the synergistic pair has a fixed mutual orientation, even if the pair itself rotates freely. While the local symmetry of each of the absorbers A and B determines the selection rules for the transitions they undergo, the symmetry of their relative juxtaposition also plays a role in determining the polarization characteristics of their synergistic photoabsorption. This is principally manifest in the occurrence of two-photon circular dichroism where the A—B pair has definite handedness, as will be demonstrated in Section IX. Thus it transpires that not only the local symmetry, but also the global symmetry of the rotating pair is significant in determining the response to particular polarizations of light.

IV. CLASSIFICATION OF SYNERGISTIC TWO-PHOTON PROCESSES

As mentioned above, there are four specific cases of bimolecular mean frequency absorption that are of special interest. These are distinguished by the type of mechanism (cooperative or distributive) involved and whether the photons absorbed have the same or different frequencies. The latter condition is in most cases determined by whether a single laser beam or two laser beams are employed for the excitation. We consider first the single-beam cases.

A. Single-Frequency Excitation

In single-beam bimolecular photoabsorption, the two absorbed photons have the same frequency, and it is the synergistic interaction between two non-identical centers that is of interest. This interaction provides the mechanism for energy exchange such that an overall process

$$\mathrm{A} + \mathrm{B} + 2\hbar\omega \rightarrow \mathrm{A}^* + \mathrm{B}^* \tag{4.1}$$

can take place even when the individual transitions $\mathrm{A} \rightarrow \mathrm{A}^*$ and $\mathrm{B} \rightarrow \mathrm{B}^*$ are forbidden on energy grounds, as illustrated in Fig. 4. From a phenomenological viewpoint, the process evidently has the characteristics of mean-frequency photoabsorption. For this effect to be experimentally observable, ω must be chosen to lie in a region where neither A nor B displays absorption, and we thus have

$$\hbar\omega = \tfrac{1}{2}(E_{\alpha 0} + E_{\beta 0}) \tag{4.2}$$

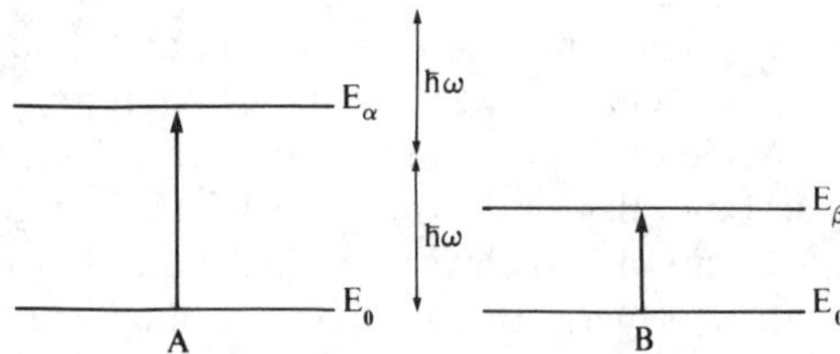

Figure 4. Energy-level diagram for single-beam synergistic absorption. The pairwise absorption of photons with frequency ω effects transitions to states $|\alpha\rangle$ and $|\beta\rangle$ in two different molecules.

$$\hbar\omega \neq E_{\alpha 0}, E_{\beta 0} \tag{4.3}$$

So, for two chemically distinct molecules or chromophores A and B, with well-characterized vibronic excited states α and β, a proximity-induced two-photon absorption process can be induced by tuning the exciting laser to a frequency equivalent to a mean of molecular excitation frequencies for the two molecules.

The two mechanisms for single-beam two-photon absorption are most clearly visualized with the aid of the calculational aids known as time-ordered diagrams. Figure 5(a) shows a typical time-ordered diagram for a single-beam cooperative two-photon absorption process. Diagrams such as this, originated by Feynman (Feynman 1949; Ward 1965; Wallace 1966), are utilized in QED along with the pertinent equations in order to derive the necessary quantum mechanical probability amplitudes. They can be regarded as a form of space-time diagram with time progressing vertically. The straight lines represent molecules (or atoms) and the wavy lines photons. It is worth emphasizing that such diagrams are purely illustrative devices, representing the ultimately indeterminable sequences of photon creation and annihilation events involved in a particular process. Thus no importance should be attached to the precise displacements on either the vertical or horizontal axis: it is the ordering of radiation-molecule interactions which each diagram represents that is of importance.

The sequence of events depicted by Fig. 5(a), for example, is as follows: center A first absorbs a real photon of wave vector **k** and polarization **e**, and thereby undergoes a virtual transition to an intermediate excited state $|r\rangle$; a virtual photon of wave vector $\boldsymbol{\varkappa}$, polarization $\boldsymbol{\varepsilon}$, and frequency $\tilde{\omega} = c|\varkappa|$

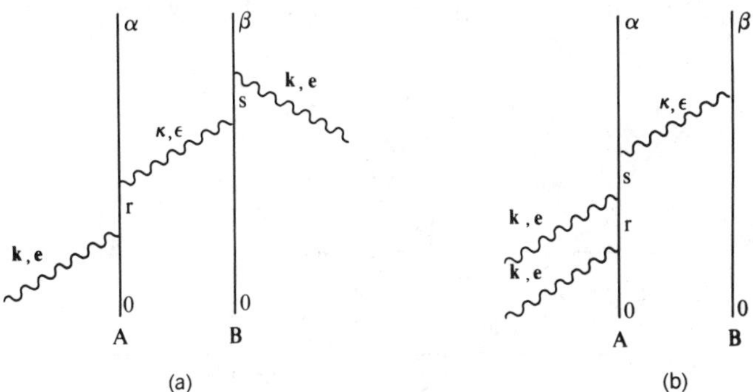

Figure 5. Typical time-ordered diagrams for single-beam two-photon absorption: (a) shows one of the diagrams associated with the cooperative mechanism, and (b) one of the diagrams for the distributive mechanism.

is then created as A adopts its final state $|\alpha\rangle$; the virtual photon propagates from center A to center B where it is annihilated, resulting in the promotion of B to an excited state $|s\rangle$; finally B absorbs a second real (laser) photon, and thereby attains its final state $|\beta\rangle$. In total, there are 24 such time-ordered diagrams associated with this process, each corresponding to a topologically different sequence of photon events.

Fig. 5(b) is an illustration of a typical time-ordered diagram for the distributive mechanism. It clearly illustrates the distinguishing feature of this mechanism, in that both laser photons are absorbed by one center; one immediate consequence of this is the above-mentioned difference in the selection rules. Once again, there are 24 such diagrams to be taken into account in the rate calculations. Although development of the theory follows along similar lines for both the cooperative and distributive cases, the dissymmetry of the distributive mode leads to a more complicated rate equation, and when the rate for van der Waals molecules is considered, a new type of rotational average is required. One result of this is, as shown below, the manifestation of two-photon circular dichroism.

B. Two-Frequency Excitation

In other cases of interest, the two centers have identical chemical composition and are excited by the absorption of two different photons, as, for example, from two different laser beams with frequencies ω_1 and ω_2. This process can be represented by the equation

$$A + A + \hbar\omega_1 + \hbar\omega_2 \rightarrow A^* + A^* \tag{4.4}$$

for which the energetics are shown in Fig. 6. Again, for the synergistic process to be observable, the frequencies ω_1 and ω_2 must be chosen in a region where single-photon absorption cannot lead to the excitation of either center.

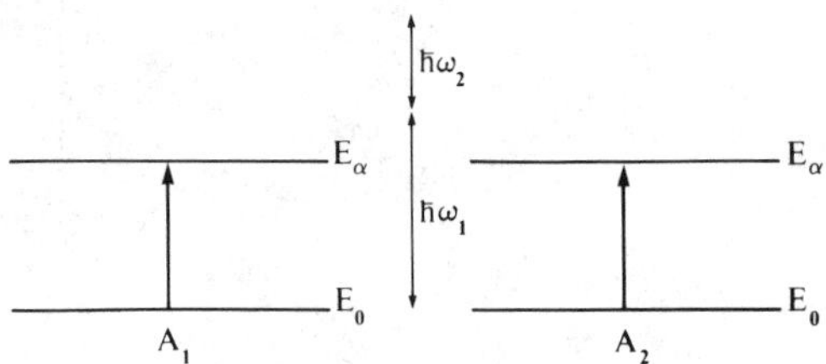

Figure 6. Energy-level diagram for two-beam synergistic absorption. The concerted absorption of one photon from each beam (frequencies ω_1 and ω_2) effects transitions to state $|\alpha\rangle$ in two chemically identical molecules.

Thus we have

$$\tfrac{1}{2}(\hbar\omega_1 + \hbar\omega_2) = E_{\alpha 0} \tag{4.5}$$

$$\hbar\omega_1, \hbar\omega_2 \neq E_{\alpha 0} \tag{4.6}$$

The former relation, Eq. (4.5), indicates the fact that this cooperative process again has the characteristics of mean-frequency absorption: here, however, it is the molecular excitation frequency which equals the mean of the two photon frequencies.

From an experimental point of view, the double-beam process allows a greater flexibility, in that some choice can be exercized in the individual frequencies of the two laser beams, subject to the satisfaction of Eq. (4.5). It is therefore normally possible to avoid tuning either laser to a frequency that might swamp the bimolecular process with conventional single-photon absorption. The utilization of resonances with intermediate energy levels (see Section X) is also facilitated by the ability to tune one of the lasers while keeping the mean value of the two laser frequencies fixed.

The absorption of two different photons also significantly increases the number of polarization variables arising in the rate equations. The ability of the experimentalist to independently vary the polarization and experimental configuration of the two laser beams allows a choice of values for these polarization parameters which significantly increases the amount of information that can be derived from the spectra. In contrast to the case of single-beam excitation, this also affords the opportunity to observe the induction of circular dichroism in a system of achiral molecules.

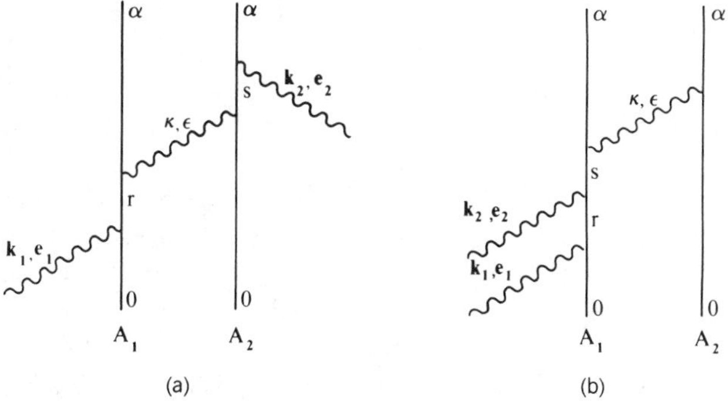

Figure 7. Typical time-ordered diagrams for two-beam two-photon absorption: (a) shows one of the diagrams associated with the cooperative mechanism, and (b) one of the diagrams for the distributive mechanism.

A typical time-ordered diagram for the cooperative two-beam process is illustrated in Fig. 7(a). Although it bears a close resemblance to Fig. 5(a), the appearance of two laser photons with differing frequencies (ω_1 and ω_2) produces a total of 48 time-ordered diagrams, as opposed to the 24 that occur in the single-beam case. Another 48 diagrams are required for representation of the corresponding distributive case, as typified by Fig. 7(b), where the virtual photon carries the entire excitation energy from one molecule to the other. The comments made in the preceding section regarding the single- and double-beam cooperative mechanisms again apply here, and once more the corresponding difference in the selection rules associated with the distributive mechanism ensues.

In passing, we note that parametric four-wave mixing processes could in principle contribute to the effects described above. Thus, in the single-beam case, the CARS process $\omega + \omega \rightarrow \omega_1 + \omega_2$ at a single center could also result in the synergistic excitation of two chemically different centers. Equally in the double-beam case, the four-wave interaction $\omega_1 + \omega_2 \rightarrow \omega + \omega$ followed by absorption of the frequency ω, could contribute to the excitation of a pair of neighboring molecules of the same species. However, both of these four-wave interactions will be relatively ineffectual unless (one of) the emission frequencies is stimulated by an additional source; moreover, the processes described here are not associated with the wave-vector matching characteristics of CARS and related phenomena.

V. QUANTUM ELECTRODYNAMICS

The theory of synergistic effects in two-photon absorption is based upon the standard methods of molecular quantum electrodynamics (QED). The fundamental development of these methods is delineated in the book by Craig and Thirunamachandran (1984), which has established the rigorous use of SI units in molecular QED. The same conventions are adopted in this review, representing a departure from two early papers (Andrews and Harlow 1983, 1984a) in which equations were cast in Gaussian units.

To develop the basic formalism, we initially consider the most general case as represented by Eq. (2.1), i.e., the process in which two molecules labelled A and B are excited to final states $|\alpha\rangle$ and $|\beta\rangle$, respectively, through the absorption of two different laser photons. It is assumed that these are derived from two separate beams simultaneously traversing the sample. The first beam is characterized by wave vector $\mathbf{k}_1$, polarization $\mathbf{e}_1$ and frequency ω_1 ($=c|\mathbf{k}_1|$), while beam 2 is similarly characterized by $\mathbf{k}_2$, $\mathbf{e}_2$ and ω_2, and the absorption process satisfies the energy conservation relation

$$\hbar\omega_1 + \hbar\omega_2 = E_{\alpha 0} + E_{\beta 0}. \tag{5.1}$$

The specific cases where: (i) A and B are identical molecules, or (ii) the two absorbed photons are identical are considered in detail subsequently.

A. Hamiltonian Representation

We begin by writing down the quantum electrodynamical Hamiltonian for the system comprising the radiation and the two molecules A and B. We adopt the Hamiltonian given by the Power–Zienau–Woolley transformation, which may be expressed as follows (Power and Zienau 1959; Woolley 1971)

$$H = H_{\mathrm{rad}} + \sum_{\zeta = A,B} H_{\mathrm{mol}}(\zeta) + \sum_{\zeta = A,B} H_{\mathrm{int}}(\zeta) \tag{5.2}$$

Here H_{rad} is the radiation field Hamiltonian given by

$$H_{\mathrm{rad}} = (\varepsilon_0/2) \int \{\varepsilon_0^{-2}\mathbf{d}^{\perp 2}(\mathbf{r}) + c^2\mathbf{b}^2(\mathbf{r})\}\, d^3\mathbf{r} \tag{5.3}$$

where $\mathbf{d}^{\perp}$ and $\mathbf{b}$ are the transverse electric field and magnetic field operators, respectively; $H_{\mathrm{mol}}(\zeta)$ is the nonrelativistic Schrödinger operator for the molecule ζ; and $H_{\mathrm{int}}(\zeta)$ is the Hamiltonian representing the molecular interaction with the radiation.

It is important to note that there is no intermolecular Coulomb potential in Eq. (5.2); in the Power–Zienau–Woolley formalism all intermolecular interactions are mediated by a coupling to the radiation field. It is for this reason that any process involving the transfer of energy from one molecule to another must be calculated on the basis of an exchange of virtual photons between the two molecules. The quantum electrodynamical method automatically incorporates the same kind of retardation effects as those which, for example, modify the R^{-6} dependence of the exchange interaction to R^{-7} at large distances (Casimir and Polder 1948). Similar variations in inverse power with distance have recently been noted in connection with electron–atom scattering (Au 1988).

The explicit form of the interaction Hamiltonian $H_{\mathrm{int}}(\zeta)$ consists of a series of multipolar terms, but for most purposes the electric-dipole (E1) approximation is sufficient. Although the results are calculated within this approximation for each molecular center ζ, detailed analysis of the coupling provides results equivalent to the inclusion of higher-order multipole terms for the pair. The same assumption underlies the well-known coupled-chromophore model of optical rotation (Kuhn 1930; Boys 1934; Kirkwood 1937). The Hamiltonian for the system may thus be written as

$$H_{\mathrm{int}}(\zeta) = -\varepsilon_0^{-1}\boldsymbol{\mu}(\zeta)\cdot\mathbf{d}^{\perp}(\mathbf{R}_\zeta) \tag{5.4}$$

where $\boldsymbol{\mu}(\zeta)$ is the electric-dipole moment operator for molecule ζ located at position $\mathbf{R}_\zeta$. The transverse electric displacement operator for the radiation field, $\mathbf{d}^\perp$, can be written as a summation over radiation modes:

$$\mathbf{d}^\perp(\mathbf{r}) = \mathrm{i}\sum_{\mathbf{k},\lambda} \{\hbar c k \varepsilon_0/2V\}^{1/2}\{\mathbf{e}^{(\lambda)}(\mathbf{k})a^{(\lambda)}(\mathbf{k})e^{\mathrm{i}\mathbf{k}\cdot\mathbf{r}} - \bar{\mathbf{e}}^{(\lambda)}(\mathbf{k})a^{+(\lambda)}(\mathbf{k})e^{-\mathrm{i}\mathbf{k}\cdot\mathbf{r}}\} \tag{5.5}$$

Here $\mathbf{e}^{(\lambda)}(\mathbf{k})$ is the unit polarization vector for the mode characterized by propagation vector $\mathbf{k}$ and polarization λ, with a frequency given by $\omega = \mathrm{c}|\mathbf{k}|$; $a^{+(\lambda)}(\mathbf{k})$ and $a^{(\lambda)}(\mathbf{k})$ are, respectively, the corresponding creation and annihilation operators, and V is the quantization volume.

B. Time-Dependent Perturbation Theory

The basis for the rate calculations is time-dependent perturbation theory. This is utilized to calculate an absorption rate Γ from the Fermi Golden Rule (Schiff 1968):

$$\Gamma = (2\pi/\hbar)|M_{fi}|^2\rho_f \tag{5.6}$$

where the transition matrix element (probability amplitude) is given by

$$\begin{aligned} M_{fi} = \langle f|H_{\text{int}}|i\rangle &+ \sum_r \frac{\langle f|H_{\text{int}}|r\rangle\langle r|H_{\text{int}}|i\rangle}{E_i - E_r} \\ &+ \sum_{s,r} \frac{\langle f|H_{\text{int}}|s\rangle\langle s|H_{\text{int}}|r\rangle\langle r|H_{\text{int}}|i\rangle}{(E_i - E_s)(E_i - E_r)} \\ &+ \sum_{t,s,r} \frac{\langle f|H_{\text{int}}|t\rangle\langle t|H_{\text{int}}|s\rangle\langle s|H_{\text{int}}|r\rangle\langle r|H_{\text{int}}|i\rangle}{(E_i - E_t)(E_i - E_s)(E_i - E_r)} + \cdots \end{aligned} \tag{5.7}$$

All states appearing in this expression are states of the system that comprises the radiation and the two molecules A and B. The symbols r, s, and t denote intermediate states, while i and f represent the initial and final states. The summations are performed over all intermediate states of the system, i.e., all states excluding the initial and final states. These are represented by

$$|i\rangle = |0; 0; n_1; n_2; 0\rangle \tag{5.8}$$

$$|f\rangle = |\alpha; \beta; (n_1 - 1); (n_2 - 1); 0\rangle \tag{5.9}$$

where the sequence in the ket denotes: |the state of A; the state of B; the number of photons in beam 1; the number of photons in beam 2; the number of virtual photons⟩.

From Eqs. (5.4) and (5.5) it follows that each appearance of H_{int} is associated with the creation or annihilation of a photon. It is thus readily apparent that the first non-zero contribution from Eq. (5.7) is the fourth-order term, corresponding to four separate photon creation and annihilation events; these comprise the two annihilations of real photons from the incident light, and the creation and annihilation of the virtual photon which couples the two molecules.

The complete set of interaction sequences incorporated in the fourth-order term for M_{fi} are accounted for by 96 time-ordered diagrams, in the general case, 48 of which are associated with the cooperative mechanism and 48 with the distributive mechanism: examples of each type are shown in Fig. 8. The matrix element contribution for the cooperative mechanism corresponding to Fig. 8(a), for example, when evaluated using Eq. (5.7) gives the following result;

$$\begin{aligned} M_{fi}^{(1)} = \sum_r \sum_s \sum_\varkappa \sum_\varepsilon &\langle \alpha; \beta; n_1 - 1; n_2 - 1; 0 | H_{\text{int}} | \alpha; s; n_1 - 1; n_2; 0 \rangle \\ &\times \langle \alpha; s; n_1 - 1; n_2; 0 | H_{\text{int}} | \alpha; 0; n_1 - 1; n_2; 1 \rangle \\ &\times \langle \alpha; 0; n_1 - 1; n_2; 1 | H_{\text{int}} | r; 0; n_1 - 1; n_2; 0 \rangle \\ &\times \langle r; 0; n_1 - 1; n_2; 0 | H_{\text{int}} | 0; 0; n_1; n_2; 0 \rangle \\ &\times \{ (E_{0\alpha} + E_{0s} + \hbar\omega_1)(E_{0\alpha} + \hbar\omega_1 - \hbar c\varkappa)(E_{0r} + \hbar\omega_1) \}^{-1}, \end{aligned} \tag{5.10}$$

It is important to re-emphasize that no single time-ordered diagram represents a physically distinguishable process: the diagrams are ultimately only calculational aids based on the approximations of perturbation theory.

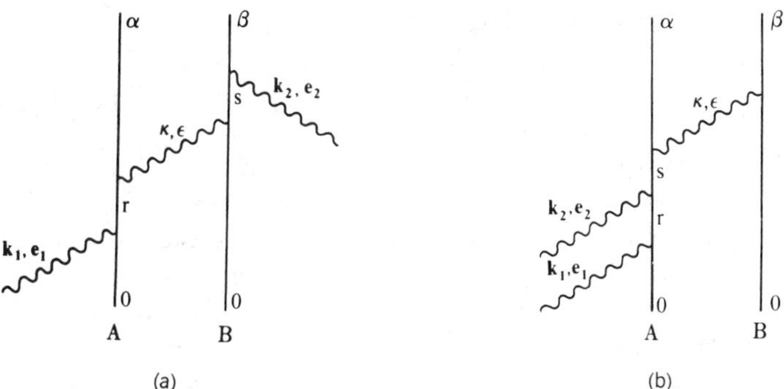

Figure 8. Typical time-ordered diagrams for the most general case of synergistic absorption with two-beam excitation of a chemically dissimilar pair of molecules: (a) relates to the cooperative mechanism and (b) the distributive mechanism.

In fact, the photon creation and annihilation events at each molecule appear simultaneous, as far as real experimental measurements with finite time resolution are concerned. However, the time–energy uncertainty relation does permit short-lived states that are not properly energy-conserving. This helps explain why it is necessary to include diagrams corresponding to time sequences in which a virtual photon is created before either real photon arrives. It nonetheless transpires that such apparently unphysical cases produce the smallest contributions to the matrix element.

Working in a similar way, it is found that there are 24 contributions arising from diagrams of the distributive type represented in Fig. 8(b), where both real photons are absorbed at center A, and a virtual photon conveys the energy mismatch to B; the final 24 distributive contributions arises from the mirror-image case where both real photons are absorbed at B and the virtual photon propagates to A. The addition of all 96 matrix element contributions then produces the complete fourth-order result for M_{fi}.

Once the summation over virtual photon wave-vectors and polarizations in Eq. (5.10) is performed, the result can be cast in terms of a retarded resonance electric dipole–electric dipole interaction tensor $V_{kl}(\omega, \mathbf{R})$ (Power and Thirunamachandran 1983; Andrews and Sherborne 1987), using the identity

$$\left(\frac{\hbar c}{2V\varepsilon_0}\right)\sum_{\varkappa,\varepsilon} \varkappa \varepsilon_k \bar{\varepsilon}_l \left[\frac{e^{i\varkappa.\mathbf{R}}}{(\hbar\omega - \hbar c\varkappa)} - \frac{e^{-i\varkappa.\mathbf{R}}}{(\hbar\omega + \hbar c\varkappa)}\right] = V_{kl}(\omega, \mathbf{R}) \tag{5.11}$$

where $\mathbf{R}$ denotes the vector displacement $\mathbf{R}(B) - \mathbf{R}(A)$, and $V_{kl}(\omega, \mathbf{R})$ is given by;

$$\begin{aligned} V_{kl}(\omega, \mathbf{R}) = (1/4\pi\varepsilon_0 R^3)[\{(\delta_{kl} - 3\hat{\mathbf{R}}_k\hat{\mathbf{R}}_l)(e^{i\omega R/c} - (i\omega R/c)e^{i\omega R/c}) \\ - (\delta_{kl} - \hat{\mathbf{R}}_k\hat{\mathbf{R}}_l)(\omega R/c)^2 e^{i\omega R/c}\}] \end{aligned} \tag{5.12}$$

which has the property $\bar{V}_{kl}(\omega, R) = V_{kl}(-\omega, R)$. The leading term of Eq. (5.12) in square brackets, with R^{-3} dependence, designates a static interaction, the following term with R^{-2} dependence an inductive interaction, and the last term a radiative interaction. The effect of incorporating higher-order multipolar contributions to the interaction tensor has recently been discussed by Thirunamachandran (1988), and Vigoureux et al. (1987) have shown how to adapt the formalism to include coupling with surface modes in the case of adsorbate molecules.

C. Tensor Formulation

The matrix element for the absorption process may now be expressed in terms of molecular tensors $S^{\alpha 0}$, $S^{\beta 0}$, $\chi^{\alpha 0}$, $\chi^{\beta 0}$, as follows, using the convention of implied summation over repeated tensor indices:

$$\begin{aligned} M_{fi} = K\{ & e_{1i}e_{2j}S_{ik}^{\alpha 0}(\omega_1)S_{jl}^{\beta 0}(\omega_2)V_{kl}([\omega_{\beta 0}-\omega_2],\mathbf{R})\exp(i[\mathbf{k}_1\cdot\mathbf{R}_A+\mathbf{k}_2\cdot\mathbf{R}_B]) \\ & + e_{1i}e_{2j}S_{ik}^{\beta 0}(\omega_1)S_{jl}^{\alpha 0}(\omega_2)V_{kl}([\omega_{\beta 0}-\omega_1],\mathbf{R})\exp(i[\mathbf{k}_2\cdot\mathbf{R}_A+\mathbf{k}_1\cdot\mathbf{R}_B]) \\ & + e_{1i}e_{2j}\chi_{ijk}^{\alpha 0}(A,\omega_1,\omega_2)\mu_l^{\beta 0}V_{kl}(\omega_{\beta 0},\mathbf{R})\exp(i[\mathbf{k}_1+\mathbf{k}_2]\cdot\mathbf{R}_A) \\ & + e_{1i}e_{2j}\chi_{ijk}^{\beta 0}(B,\omega_1,\omega_2)\mu_l^{\alpha 0}\bar{V}_{kl}(\omega_{\alpha 0},\mathbf{R})\exp(i[\mathbf{k}_1+\mathbf{k}_2]\cdot\mathbf{R}_B)\} \end{aligned} \qquad (5.13)$$

where

$$K = -(\hbar c/2V\varepsilon_0)(n_1 n_2 k_1 k_2)^{1/2} \qquad (5.14)$$

(A slightly different formula applies when the frequencies ω_1 and ω_2 become equal, as will be seen in the next section.) The parameters $\omega_{\alpha 0}$ and $\omega_{\beta 0}$ in Eq. (5.13) are defined by $\hbar\omega_{\alpha 0} = E_{\alpha 0}$ and $\hbar\omega_{\beta 0} = E_{\beta 0}$, and effectively represent the conservation energy transferred between the two centers by the virtual photon.

The first two terms in Eq. (5.13) arise from the cooperative mechanism, while the distributive mechanism gives rise to the third and fourth terms. Deriving the general rate for a proximity-induced two-photon absorption process from the square modulus of the result is an elaborate procedure producing sixteen terms, including cross-terms associated with quantum mechanical interference between the cooperative and distributive mechanisms. However, in view of the selection rules discussed earlier, it is not generally necessary to perform this calculation since each of the four specific mechanisms for two-photon absorption under consideration can, at most, have only two terms of Eq. (5.13) contributing to the matrix element.

At this stage, it is appropriate to describe the detailed structure and properties of the molecular tensors appearing in Eq. (5.13). The explicit form of the second-rank molecular response tensor $\mathbf{S}_{ij}^{f0}(\omega)$ is

$$\mathbf{S}_{ij}^{fo}(\omega) = \sum_r \left[\frac{\mu_i^{fr}\mu_j^{ro}}{(E_{fr}-\hbar\omega)} - \frac{\mu_j^{fr}\mu_i^{ro}}{(E_{ro}-\hbar\omega)}\right] \qquad (5.15)$$

It is readily shown that this is identically equal to the electronic Raman scattering tensor for the Raman transition $|f\rangle \leftarrow |0\rangle$. Note that the first term in Eq. (5.15) dominates if there exists a state $|r\rangle$ such that $E_{fr} \approx \hbar\omega$; the second term dominates if there is a state such that $E_{r0} \approx \hbar\omega$. Such cases lead to resonance enhancement and will be discussed in a more detailed manner later. The molecular tensors $\mathbf{S}_{ik}^{\beta 0}(\omega_1)$, $\mathbf{S}_{jl}^{\alpha 0}(\omega_2)$, etc., are to be understood as generalizations of the above equation.

The third-rank molecular tensor χ_{ijk}^{f0} is defined as

$$\begin{aligned}\chi_{ijk}^{f0}(\omega_1,\omega_2)=\sum_{s,r}\Bigg[&\frac{\mu_i^{ro}\mu_j^{sr}\mu_k^{fs}}{(E_{os}+\hbar\omega_1+\hbar\omega_2)(E_{or}+\hbar\omega_1)}+\frac{\mu_i^{ro}\mu_j^{fs}\mu_k^{sr}}{(E_{fs}-\hbar\omega_2)(E_{or}+\hbar\omega_1)}\\&+\frac{\mu_i^{sr}\mu_j^{fs}\mu_k^{ro}}{(E_{fr}-\hbar\omega_1-\hbar\omega_2)(E_{fs}-\hbar\omega_2)}+\frac{\mu_i^{sr}\mu_j^{ro}\mu_k^{fs}}{(E_{os}+\hbar\omega_1+\hbar\omega_2)(E_{or}+\hbar\omega_2)}\\&+\frac{\mu_i^{fs}\mu_j^{ro}\mu_k^{sr}}{(E_{or}+\hbar\omega_2)(E_{fs}-\hbar\omega_1)}+\frac{\mu_i^{fs}\mu_j^{sr}\mu_k^{ro}}{(E_{fr}-\hbar\omega_1-\hbar\omega_2)(E_{fs}-\hbar\omega_1)}\Bigg]\end{aligned} \tag{5.16}$$

This molecular response tensor is a more general form of three apparently different tensors which have featured in previous work on multiphoton processes; one is the χ_{ijk}^{f0} tensor arising in the single-frequency distributive two-photon absorption (Andrews and Harlow 1984a), and another is the $\mathbf{T}_{ijk}$ tensor which appears in the theory of three-photon absorption (Andrews and Wilkes 1985). It is also exactly identical to the two-frequency hyper-Raman transition tensor $\boldsymbol{\beta}_{ijk}^{f0}$ (Andrews 1984).

In subsequent development of the synergistic rate equations, the parametric dependence on frequency of $\mathbf{S}_{ij}^{f0}$ and χ_{ijk}^{f0} is left implicit so as to avoid unnecessary congestion, and $\omega_1=\omega_2=\omega$ for single-beam cases. It is not generally necessary to label tensors A or B to denote at which center they are to be evaluated, since the superscript α or β designates whether each tensor represents interactions at center A or B. In the double-beam case it proves necessary for calculational purposes to distinguish tensors associated with the two (chemically identical) centers by labeling them A_1 and A_2.

VI. RATE EQUATIONS FOR RIGIDLY ORIENTED SYSTEMS

The rate of a general, proximity-induced, two-photon absorption process can be calculated by combining the Fermi Golden rule, Eq. (5.6), with Eq. (5.13). For each of the four specific cases to be studied, the selection rules normally dictate that only two of the four terms in Eq. (5.13) arise for any given mechanism, thus reducing the number of terms in the rate equation to just four. A brief derivation of the rate of absorption for each case is presented below.

A. Single-Frequency Cooperative Two-Photon Absorption

The single-beam cooperative mechanism applies where a single laser beam is utilized to excite the sample, and where transitions in both species A and

B are forbidden as single-photon processes, but allowed under two-photon selection rules. This mechanism is represented by the time-ordered diagram of Fig. 5(a), and is governed by the energy conservation relations of Eqs (4.2) and (4.3).

The relevant transition matrix element can be obtained by examination of Eq. (5.13). Setting $\omega_1 = \omega_2 \equiv \omega$, and dropping forbidden contributions from the third and fourth terms, the matrix element becomes:

$$M_{fi} = K' e_i e_j S_{ik}^{\alpha 0}(\omega) S_{jl}^{\beta 0}(\omega) V_{kl}([\omega_{\beta 0} - \omega], \mathbf{R}) \exp(i\mathbf{k}\cdot[\mathbf{R}_A + \mathbf{R}_B]) \quad (6.1)$$

where K' is given by

$$K' = -(\hbar ck/2V\varepsilon_0) n^{1/2}(n-1)^{1/2} \quad (6.2)$$

This constant is essentially the limit of K as given by Eq. (5.14), in the case where the two absorbed photons become identical; however, the factor $(n_1 n_2)^{1/2}$ is replaced by $n^{1/2}(n-1)^{1/2}$ since the photon annihilation operator acts twice on the same radiation mode. As will be seen below, this difference is ultimately reflected in a dependence on the coherence properties of the laser source, which is uniquely associated with single-beam processes. It is also worth observing that although the first two terms of Eq. (5.13) become identical if the two absorbed photons are derived from the same beam, inclusion of a factor of 2 in Eq. (6.1) would amount to double-counting the time-ordered diagrams, and is therefore not appropriate.

The rate for the process may now be calculated from the Fermi rule, and is given by;

$$\begin{aligned}\Gamma_{1c} = (2\pi\rho_f/\hbar)K'^2 \{ & e_i e_j \bar{e}_m \bar{e}_n S_{ik}^{\alpha 0}(\omega) S_{jl}^{\beta 0}(\omega) \bar{S}_{mo}^{\alpha 0}(\omega) \bar{S}_{np}^{\beta 0}(\omega) \\ & \times V_{kl}([\omega_{\beta 0} - \omega], \mathbf{R}) \bar{V}_{op}([\omega_{\beta 0} - \omega], \mathbf{R}) \}\end{aligned} \quad (6.3)$$

This S.I. result corresponds to the expression originally derived in Gaussian units by Andrews and Harlow (1983). As it stands, all molecular and radiative parameters in the above expression are referred to a single Cartesian frame of reference, making the result directly applicable only to rigidly oriented molecules or solids. For application to fluid samples, further processing of the result to account for molecular orientation is necessary before this result and others below can be applied; the procedure is discussed in detail in Section VII.

To complete the calculation, certain photon statistical features of the result must be examined and recast in terms of experimentally determinable parameters. The problem lies in the constant K' which, defined through

Eq. (6.2) in terms of photon number n and quantization volume V, is not directly amenable to experimental application. Moreover, since the quantization volume is merely a theoretical artefact, it must invariably cancel out in any final rate equation. However the ratio of photon number and quantization volume is directly related to the mean irradiance I; the relationship is as follows;

$$I = n\hbar c^2 k/V \tag{6.4}$$

A secondary problem lies in the fact that the number states employed in Section V, which formally represent radiation for which there is a precise nonfluctuating value for the number of photons, provide only a poor representation of coherent laser light. More physical results can be obtained by considering the photon number to be subject to fluctuations which satisfy particular types of statistical distribution. By suitably weighting rate equations calculated on the basis of number states, various kinds of radiation can then be modeled. The result is a replacement of the factor $n(n-1)$ by $g_{12}^{(2)}\bar{n}^2$, where $\bar{n}$ is the mean photon number and $g_{12}^{(2)}$ is the degree of second-order coherence of the beam (Loudon 1983). This quantity depends on the detailed statistical properties of the light source, provided the two absorptions which it correlates occur within the coherence length of the radiation. For coherent light $g_{12}^{(2)}$ takes the value of unity; for thermal light it equals 2, and other values typify different kinds of photon distribution. Taking these factors into account, it transpires that K'^2 is more properly expressed in terms of the mean irradiance $\bar{I}$ and the degree of second-order coherence by the quite general result:

$$K'^2 = (1/4c^2\varepsilon_0^2)\bar{I}^2 g_{12}^{(2)} \tag{6.5}$$

B. Single-Frequency Distributive Two-Photon Absorption

In this mechanism, two-photon transitions are forbidden and the excitation of the participating molecules occurs through one- and three-photon allowed transitions. Both the real (laser) photons are absorbed by one molecule, excitation of its partner resulting from the virtual photon coupling. Because of the difference in selection rules from the previous case, the first two terms of Eq. (5.13) are now zero, and contributions arise only from the third and fourth terms. It must also be noted that setting the two absorbed photon frequencies to be equal in Eq. (5.16) to produce $\chi^{\alpha 0}_{(ij)k}(\omega, \omega)$ introduces index symmetry into the tensor, as indicated by the brackets embracing the first two indices. A factor of $\frac{1}{2}$ must then be introduced into the definition of this tensor in order to avoid over-counting contributions. The transition matrix element for this mechanism is then as follows:

$$M_{fi} = K' e_i e_j \{\chi^{\alpha 0}_{(ij)k} \mu^{\beta 0}_l V_{kl}(\omega_{\beta 0}, \mathbf{R}) \exp(2i\mathbf{k}\cdot\mathbf{R}_A) + \chi^{\beta 0}_{(ij)k} \mu^{\alpha 0}_l \bar{V}_{kl}(\omega_{\alpha 0}, \mathbf{R}) \exp(2i\mathbf{k}\cdot\mathbf{R}_B)\} \quad (6.6)$$

Taking the square modulus of this result to evaluate the overall rate from Eq. (5.6) again introduces a factor of K'^2, which can be evaluated from Eq. (6.5) subject to replacement of $g^{(2)}_{12}$ by $g^{(2)}_{11}$, (which is independent of coherence length), the subscripts in the latter case denoting the fact that both laser photons are absorbed by the same molecule. The final result for the rate is then given by;

$$\begin{aligned}\Gamma_{1d} = (2\pi\rho_f/\hbar)K'^2 e_i e_j \bar{e}_m \bar{e}_n \{ & \chi^{\alpha 0}_{(ij)k} \mu^{\beta 0}_l \bar{\chi}^{\alpha 0}_{(mn)o} \bar{\mu}^{\beta 0}_p V_{kl}(\omega_{\beta 0}, \mathbf{R}) \bar{V}_{op}(\omega_{\beta 0}, \mathbf{R}) \\ & + \chi^{\beta 0}_{(ij)k} \mu^{\alpha 0}_l \bar{\chi}^{\beta 0}_{(mn)o} \bar{\mu}^{\alpha 0}_p \bar{V}_{kl}(\omega_{\alpha 0}, \mathbf{R}) V_{op}(\omega_{\alpha 0}, \mathbf{R}) \\ & + \chi^{\beta 0}_{(ij)k} \mu^{\alpha 0}_l \bar{\chi}^{\alpha 0}_{(mn)o} \bar{\mu}^{\beta 0}_p \bar{V}_{kl}(\omega_{\alpha 0}, \mathbf{R}) \bar{V}_{op}(\omega_{\beta 0}, \mathbf{R}) \exp(2i\mathbf{k}\cdot\mathbf{R}) \\ & + \chi^{\alpha 0}_{(ij)k} \mu^{\beta 0}_l \bar{\chi}^{\beta 0}_{(mn)o} \bar{\mu}^{\alpha 0}_p V_{kl}(\omega_{\beta 0}, \mathbf{R}) V_{op}(\omega_{\alpha 0}, \mathbf{R}) \exp(-2i\mathbf{k}\cdot\mathbf{R}) \} \end{aligned} \quad (6.7)$$

The tensor indices in Eq. (6.7) are now referred to a molecular frame of reference arbitrarily centered on molecule A, thereby reducing the value of R_A to zero. The result is again equivalent to a result given previously in Gaussian units (Andrews and Harlow 1984a), although it is important to note that only the real part of the retarded resonance electric dipole-electric dipole interaction tensor, V_{kl}, was employed in the original derivation. Subsequent studies have shown that it is in fact necessary to employ this interaction tensor in its complete, complex form, rather than simply utilizing the real part (Power and Thirunamachandran 1983). The persistence of phase factors in the rate equation should be noted, and the physical significance of this will become evident in a later section.

C. Two-Frequency Cooperative Two-Photon Absorption

The transition matrix element for double-beam cooperative two-photon absorption can be obtained by setting $A \equiv B$ and $\alpha \equiv \beta$ in Eq. (5.13), and discarding the forbidden contributions associated with the third and fourth terms. The vectors $\mathbf{R}_A$ and $\mathbf{R}_B$ now become $\mathbf{R}_{A1}$ and $\mathbf{R}_{A2}$, leading to a matrix element given by;

$$\begin{aligned} M_{fi} = K e_{1i} e_{2j} \{ & S^{\alpha 0}_{ik}(\omega_1) S^{\alpha 0}_{jl}(\omega_2) V_{kl}([\omega_{\alpha 0} - \omega_2], \mathbf{R}) \exp(i[\mathbf{k}_1\cdot\mathbf{R}_{A1} + \mathbf{k}_2\cdot\mathbf{R}_{A2}]) \\ & + S^{\alpha 0}_{ik}(\omega_1) S^{\alpha 0}_{jl}(\omega_2) V_{kl}([\omega_{\alpha 0} - \omega_1], \mathbf{R}) \exp(i[\mathbf{k}_2\cdot\mathbf{R}_{A1} + \mathbf{k}_1\cdot\mathbf{R}_{A2}]) \} \end{aligned} \quad (6.8)$$

Incorporating this into the Fermi Golden Rule yields a rate given by

$$\Gamma_{2c} = (2\pi\rho_f/\hbar)K^2 e_{1i}e_{2j}\bar{e}_{1m}\bar{e}_{2n}S_{ik}^{\alpha0}(\omega_1)S_{jl}^{\alpha0}(\omega_2)\bar{S}_{mo}^{\alpha0}(\omega_1)\bar{S}_{np}^{\alpha0}(\omega_2)$$
$$\times \{V_{kl}([\omega_{\alpha0} - \omega_2], \mathbf{R})\bar{V}_{op}([\omega_{\alpha0} - \omega_2], \mathbf{R})$$
$$+ V_{kl}([\omega_{\beta0} - \omega_1], \mathbf{R})\bar{V}_{op}([\omega_{\beta0} - \omega_1], \mathbf{R})$$
$$+ V_{kl}([\omega_{\alpha0} - \omega_2], \mathbf{R})\bar{V}_{op}([\omega_{\beta0} - \omega_1], \mathbf{R})\exp(i\mathbf{R}\cdot(\mathbf{k}_2 - \mathbf{k}_1))$$
$$+ V_{kl}([\omega_{\beta0} - \omega_1], \mathbf{R})\bar{V}_{op}([\omega_{\alpha0} - \omega_2], \mathbf{R})\exp(-i\mathbf{R}\cdot(\mathbf{k}_2 - \mathbf{k}_1))\} \tag{6.9}$$

Using similar arguments to those outlined above, K^2 can be expressed in terms of the mean irradiances $\bar{I}_1$ and $\bar{I}_2$ of the two laser beams as follows:

$$K^2 = (1/4c^2\varepsilon_0^2)\bar{I}_1\bar{I}_2 \tag{6.10}$$

and it is evident that there is no explicit dependence on the coherence properties of the sources. This is based on the assumption that there is no coherence correlation between the two radiation modes absorbed by the sample, which will normally be the case where two laser beams are used for the excitation: this is another feature that distinguishes the process from the single-beam case.

D. Two-Frequency Distributive Two-Photon Absorption

Finally, the transition matrix element for double-beam distributive two-photon absorption can be obtained from the third and fourth terms in Eq. (5.13), leading to a matrix element

$$M_{fi} = Ke_{1i}e_{2j}\{\chi_{ijk}^{\alpha0}(A_1)\mu_l^{\alpha0}(A_2)V_{kl}(\omega_{\alpha0}, \mathbf{R})$$
$$+ \chi_{ijk}^{\alpha0}(A_2)\mu_l^{\alpha0}(A_1)\bar{V}_{kl}(\omega_{\alpha0}, \mathbf{R})\exp(i[\mathbf{k}_1 + \mathbf{k}_2]\cdot\mathbf{R})\} \tag{6.11}$$

The rate is then given by

$$\Gamma_{2d} = (2\pi\rho_f/\hbar)K^2 e_{1i}e_{2j}\bar{e}_{1m}\bar{e}_{2n}$$
$$\times \{\chi_{ijk}^{\alpha0}(A_1)\mu_l^{\alpha0}(A_2)\bar{\chi}_{mno}^{\alpha0}(A_1)\bar{\mu}_p^{\alpha0}(A_2)V_{kl}(\omega_{\alpha0}, \mathbf{R})\bar{V}_{op}(\omega_{\alpha0}, \mathbf{R})$$
$$+ \chi_{ijk}^{\alpha0}(A_2)\mu_l^{\alpha0}(A_1)\bar{\chi}_{mno}^{\alpha0}(A_2)\bar{\mu}_p^{\alpha0}(A_1)\bar{V}_{kl}(\omega_{\alpha0}, \mathbf{R})V_{op}(\omega_{\alpha0}, \mathbf{R})$$
$$+ \chi_{ijk}^{\alpha0}(A_2)\mu_l^{\alpha0}(A_1)\bar{\chi}_{mno}^{\alpha0}(A_1)\bar{\mu}_p^{\alpha0}(A_2)\bar{V}_{kl}(\omega_{\alpha0}, \mathbf{R})$$
$$\times \bar{V}_{op}(\omega_{\alpha0}, \mathbf{R})\exp(i(\mathbf{k}_1 + \mathbf{k}_2)\cdot\mathbf{R})$$
$$+ \chi_{ijk}^{\alpha0}(A_1)\mu_l^{\alpha0}(A_2)\bar{\chi}_{mno}^{\alpha0}(A_2)\bar{\mu}_p^{\alpha0}(A_1)V_{kl}(\omega_{\alpha0}, \mathbf{R})$$
$$\times V_{op}(\omega_{\alpha0}, \mathbf{R})\exp(-i(\mathbf{k}_1 + \mathbf{k}_2)\cdot\mathbf{R})\} \tag{6.12}$$

where once again the factor K^2 is as given in (6.10). The rate expressions given above represent the pre-averaged results. That is, they are directly applicable to systems where the two centers participating in the absorption process are rigidly held in fixed orientation with respect to the laser beam(s) *and* with respect to one another. These results are also a suitable starting point for the derivation of rate expressions applicable to fluid systems, where each center is free to rotate with respect to the beam(s) and also with respect to each other. This requires a series of rotational averaging procedures which are described in detail in the next section.

VII. RATE EQUATIONS FOR FLUID MEDIA

In fluid phase studies of mean-frequency absorption, it is necessary to take account of the effect of molecular tumbling on the rate of photoabsorption. Assuming that rotational structure is not resolvable, as will be the case in liquids, the necessary rotational averaging can be performed using a classical procedure based on the ergodic theorem. We first consider the case in which the relative orientation of the two centers A and B is fixed, but the A—B system is free to move in the laser beam or beams. The initial result thereby obtained is then applicable to van der Waals molecules, and to polyatomic compounds in which A and B represent independent chromophores. The results may also be extended to solutions in an obvious way, thus describing coordination shell interactions between solute and solvent molecules through the addition of contributions relating to each A—B pair. To derive results that are appropriate for any such systems, it is necessary to perform a rotational average of the rate equations given in Section VI.

This principal rotational average is accomplished by first defining two Cartesian reference frames: one is a laboratory-fixed frame, (denoted by p), in which the laser polarization and propagation vectors are fixed, and the second frame (denoted by a), in which the molecular tensors and the vector $\mathbf{R}$ are fixed, is located on the A—B system. For convenience, the a frame is defined as having its origin at center A. The specific type of rotational average required depends on the exact nature of the pre-averaged result. For terms which do not carry an exponential phase factor, straightforward tensor averaging methods are employed (Andrews and Thirunamachandran 1977). However, where a phase factor persists in the result, a phased tensor averaging procedure is required (Andrews and Harlow 1984b). This reflects the fact that the phase of the laser light will generally be different at the two absorbing centers.

A. Single-Frequency Cooperative Two-Photon Absorption

The simplest case is that of single-beam cooperative two-photon absorption, which is the only process where no phased averages arise. The pre-averaged rate, which is suitable for two centers that are rigidly held in a fixed orientation with respect to the laser beam and with respect to each other, has been given earlier as Eq. (6.3). Setting this equation in the two reference frames a and p defined above leads to

$$\begin{aligned}\Gamma_{1c} = (2\pi\rho_f/\hbar)K'^2\{&e_{p_i}e_{p_j}\bar{e}_{p_m}\bar{e}_{p_n}l_{p_ia_i}l_{p_ja_j}l_{p_ma_m}l_{p_na_n}\\ &\times S^{\alpha0}_{a_ia_k}(\omega)S^{\beta0}_{a_ja_l}(\omega)\bar{S}^{\alpha0}_{a_ma_o}(\omega)\bar{S}^{\beta0}_{a_na_p}(\omega)\\ &\times V_{a_ka_l}([\omega_{\beta0}-\omega],\mathbf{R})\bar{V}_{a_oa_p}([\omega_{\beta0}-\omega],\mathbf{R})\}\end{aligned} \tag{7.1}$$

where the first letter of each pair of indices denotes the frame in which the following index is set, and $l_{p_\alpha a_\alpha}$ is the direction cosine between the p_α and a_α directions. Now, since all the vectors and tensors are referred to Cartesian frames in which they are invariant with respect to molecular rotation, the rotational averaging may be effected by averaging over the direction cosines alone. The required result for this calculation is (Andrews and Thirunamachandran 1977)

$$\begin{aligned}&\langle l_{p_ia_i}l_{p_ja_j}l_{p_ma_m}l_{p_na_n}\rangle\\ &= (1/30)\begin{bmatrix}\delta_{p_ip_j} & \delta_{p_mp_n}\\ \delta_{p_ip_m} & \delta_{p_jp_n}\\ \delta_{p_ip_n} & \delta_{p_jp_m}\end{bmatrix}^T\begin{bmatrix}4 & -1 & -1\\ -1 & 4 & -1\\ -1 & -1 & 4\end{bmatrix}\begin{bmatrix}\delta_{a_ia_j} & \delta_{a_ma_n}\\ \delta_{a_ia_m} & \delta_{a_ja_n}\\ \delta_{a_ia_n} & \delta_{a_ja_m}\end{bmatrix}\end{aligned} \tag{7.2}$$

Application of this result to Eq. (7.1) yields

$$\begin{aligned}\langle\Gamma_{1c}\rangle = (\pi\rho_f/15\hbar)K'^2 &V_{kl}([\omega_{\beta0}-\omega],\mathbf{R})\bar{V}_{op}([\omega_{\beta0}-\omega],\mathbf{R})\\ &\times\{(4\eta-2)S^{\alpha0}_{ik}(\omega)S^{\beta0}_{il}(\omega)\bar{S}^{\alpha0}_{jo}(\omega)\bar{S}^{\beta0}_{jp}(\omega)\\ &+(3-\eta)S^{\alpha0}_{ik}(\omega)S^{\beta0}_{jl}(\omega)\bar{S}^{\alpha0}_{io}(\omega)\bar{S}^{\beta0}_{jp}(\omega)\\ &+(3-\eta)S^{\alpha0}_{ik}(\omega)S^{\beta0}_{jl}(\omega)\bar{S}^{\alpha0}_{jo}(\omega)\bar{S}^{\beta0}_{ip}(\omega)\}\end{aligned} \tag{7.3}$$

where the brackets around Γ_{1c} signify the averaged result, and where

$$\eta = (\mathbf{e}\cdot\mathbf{e})(\bar{\mathbf{e}}\cdot\bar{\mathbf{e}}) \tag{7.4}$$

The parameter η takes the limiting values $\eta = 0$ for circularly polarized light and $\eta = 1$ for plane polarized light. In Eq. (7.3) and subsequently, the first of each pair of tensor indices, i.e., the one which indicates in which frame the particular tensor is rotationally invariant, has been dropped since all the indices are now related to the same system frame.

Equation (7.3) is the simplest of the results for synergistic photoabsorption presented in this review. The corresponding results for the three remaining mechanisms are appreciably more complex, and are more conveniently written using a shorthand notation. For consistency, it is therefore appropriate to re-express Eq. (7.3) in terms of this generalized notation as follows;

$$\langle\Gamma_{1c}\rangle = (2\pi\rho_f/\hbar)K'^2 \sum_{p,q} g^{pq}_{(4;0)} A'^{(0;p)} T_c^{(0;q)}(\alpha, \beta, \alpha, \beta; \omega, \omega, \omega, \omega; \omega_{\beta 0} - \omega, \omega_{\beta 0} - \omega) \tag{7.5}$$

where the $g^{pq}_{(4;0)}$ are numerical coefficients derived from the theory of tensor averaging (Andrews and Ghoul 1982), and A', T_c are the specific polarization and molecular tensor parameters for this particular mechanism. In general these are defined by:

$$A'^{(j;p)}(\hat{\mathbf{u}}) = e_{1i} e_{2j} \bar{e}_{1k} \bar{e}_{2l} U^{(4;j;p)}_{ijkl}(\hat{\mathbf{u}}) \tag{7.6}$$

and

$$\begin{aligned} &T_c^{(j;q)}(\alpha, \beta, \alpha', \beta'; \omega_1, \omega_2, \omega'_1, \omega'_2; \omega_{\alpha o} - \omega_1, \omega_{\alpha' 0} - \omega'_1) \\ &\quad = S^{\alpha 0}_{\lambda\pi}(\omega_1) S^{\beta 0}_{\mu\rho}(\omega_2) \bar{S}^{\alpha' 0}_{\nu\sigma}(\omega'_1) \bar{S}^{\beta' 0}_{o\tau}(\omega'_2) \\ &\quad \times V_{\pi\rho}([\omega_{\alpha 0} - \omega_1], \mathbf{R}) \bar{V}_{\sigma\tau}([\omega_{\alpha' 0} - \omega'_1], \mathbf{R}) W^{(4;j;q)}_{\lambda\mu\nu o}(\hat{\mathbf{R}}) \end{aligned} \tag{7.7}$$

where $U^{(4;j;p)}_{ijkl}(\hat{\mathbf{u}})$ and $W^{(4;j;q)}_{\lambda\mu\nu o}(\hat{\mathbf{R}})$ are irreducible tensor projections (Andrews and Harlow 1984b) and the explicit results for A' and T_c are presented in Tables I and II.

The result, Eq. (7.5), is directly applicable to the case of single-beam cooperative two-photon absorption where the mutual orientation of A and B is fixed, but the A—B system may rotate with respect to the laser beam (the rotating pair case). We now consider the situation in which the two molecules involved in the interaction are free to take up any separation and mutual orientation; this may be termed the free molecules case. Two further averages are now required to derive a suitable rate expression. The first step involves specifying a molecule-fixed Cartesian reference frame b in which the molecular tensors of center B are rotationally invariant, and then performing a rotational average with respect to the a frame. This step accounts for the rotation of center B relative to center A. The second step is carried out by specifying an r frame in which the **R** vectors are rotationally invariant, and subsequently averaging over the orientation of this frame with respect to the a frame; this step accounts for the random orientation of the vector **AB** relative to center A.

It is worth noting that this full triple-averaging procedure is required to account for any fluid sample composed of randomly orientated free molecules; the overall procedure, due to Schipper (1981), is summarized in Table III. Although each of the three stages of the averaging procedure is essential, the order in which they are conducted in a particular case is not important.

TABLE I
Explicit Form of the Polarization Parameters $A'^{(j;p)}(\hat{\mathbf{u}})$

j	p	$A'^{(j;p)}$
0	1	$(\mathbf{e}\cdot\mathbf{e})(\bar{\mathbf{e}}\cdot\bar{\mathbf{e}})$
0	2	1
0	3	1
1	1	0
1	2	$(\mathbf{e}\times\bar{\mathbf{e}})\cdot\hat{\mathbf{u}}$
1	3	$(\mathbf{e}\times\bar{\mathbf{e}})\cdot\hat{\mathbf{u}}$
1	4	$(\mathbf{e}\times\bar{\mathbf{e}})\cdot\hat{\mathbf{u}}$
1	5	$(\mathbf{e}\times\bar{\mathbf{e}})\cdot\hat{\mathbf{u}}$
1	6	0
2	1	$(-\frac{1}{3})(\mathbf{e}\cdot\mathbf{e})(\bar{\mathbf{e}}\cdot\bar{\mathbf{e}})$
2	2	$-\frac{1}{3}$
2	3	$-\frac{1}{3}$
2	4	$-\frac{1}{3}$
2	5	$-\frac{1}{3}$
2	6	$(-\frac{1}{3})(\mathbf{e}\cdot\mathbf{e})(\bar{\mathbf{e}}\cdot\bar{\mathbf{e}})$
3	1	$(-\frac{1}{5})(\mathbf{e}\times\bar{\mathbf{e}})\cdot\hat{\mathbf{u}}$
3	2	$(-\frac{1}{5})(\mathbf{e}\times\bar{\mathbf{e}})\cdot\hat{\mathbf{u}}$
3	3	0
4	1	$(\frac{1}{35})[(\mathbf{e}\cdot\mathbf{e})(\bar{\mathbf{e}}\cdot\bar{\mathbf{e}})+2]$

TABLE II
Explicit Form of Cooperative Molecular Invariants[a]

j	q	$Z^{(j;q)}_{\pi\rho\sigma\tau}$
0	1	$S^{\alpha0}_{\lambda\pi}(\omega_1)S^{\beta0}_{\mu\rho}(\omega_2)\bar{S}^{\alpha'0}_{\mu\sigma}(\omega'_1)\bar{S}^{\beta'0}_{\mu\tau}(\omega'_2)$
0	2	$S^{\alpha0}_{\lambda\pi}(\omega_1)S^{\beta0}_{\mu\rho}(\omega_2)\bar{S}^{\alpha'0}_{\lambda\sigma}(\omega'_1)\bar{S}^{\beta'0}_{\mu\tau}(\omega'_2)$
0	3	$S^{\alpha0}_{\lambda\pi}(\omega_1)S^{\beta0}_{\mu\rho}(\omega_2)\bar{S}^{\alpha'0}_{\mu\sigma}(\omega'_1)\bar{S}^{\beta'0}_{\lambda\tau}(\omega'_2)$
1	1	$\varepsilon_{\lambda\mu\nu}\hat{R}_\nu S^{\alpha0}_{\lambda\pi}(\omega_1)S^{\beta0}_{\mu\rho}(\omega_2)\bar{S}^{\alpha'0}_{o\sigma}(\omega'_1)\bar{S}^{\beta'0}_{o\tau}(\omega'_2)$
1	2	$\varepsilon_{\lambda\mu\nu}\hat{R}_\nu S^{\alpha0}_{\lambda\pi}(\omega_1)S^{\beta0}_{o\rho}(\omega_2)\bar{S}^{\alpha'0}_{\mu\sigma}(\omega'_1)\bar{S}^{\beta'0}_{o\tau}(\omega'_2)$
1	3	$\varepsilon_{\lambda\mu\nu}\hat{R}_\nu S^{\alpha0}_{\lambda\pi}(\omega_1)S^{\beta0}_{o\rho}(\omega_2)\bar{S}^{\alpha'0}_{o\sigma}(\omega'_1)\bar{S}^{\beta'0}_{\mu\tau}(\omega'_2)$
1	4	$\varepsilon_{\lambda\mu\nu}\hat{R}_\nu S^{\alpha0}_{o\pi}(\omega_1)S^{\beta0}_{\lambda\rho}(\omega_2)\bar{S}^{\alpha'0}_{\mu\sigma}(\omega'_1)\bar{S}^{\beta'0}_{o\tau}(\omega'_2)$
1	5	$\varepsilon_{\lambda\mu\nu}\hat{R}_\nu S^{\alpha0}_{o\pi}(\omega_1)S^{\beta0}_{\lambda\rho}(\omega_2)\bar{S}^{\alpha'0}_{o\sigma}(\omega'_1)\bar{S}^{\beta'0}_{\mu\tau}(\omega'_2)$
1	6	$\varepsilon_{\lambda\mu\nu}\hat{R}_\nu S^{\alpha0}_{o\pi}(\omega_1)S^{\beta0}_{o\rho}(\omega_2)\bar{S}^{\alpha'0}_{\lambda\sigma}(\omega'_1)\bar{S}^{\beta'0}_{\mu\tau}(\omega'_2)$
2	1	$\hat{R}_\mu\hat{R}_\nu S^{\alpha0}_{\lambda\pi}(\omega_1)S^{\beta0}_{\lambda\rho}(\omega_2)\bar{S}^{\alpha'0}_{\mu\sigma}(\omega'_1)\bar{S}^{\beta'0}_{\nu\tau}(\omega'_2)$ $-(\frac{1}{3})S^{\alpha0}_{\lambda\pi}(\omega_1)S^{\beta0}_{\lambda\rho}(\omega_2)\bar{S}^{\alpha'0}_{\mu\sigma}(\omega'_1)\bar{S}^{\beta'0}_{\mu\tau}(\omega'_2)$
2	2	$\hat{R}_\mu\hat{R}_\nu S^{\alpha0}_{\lambda\pi}(\omega_1)S^{\beta0}_{\mu\rho}(\omega_2)\bar{S}^{\alpha'0}_{\lambda\sigma}(\omega'_1)\bar{S}^{\beta'0}_{\nu\tau}(\omega'_2)$ $-(\frac{1}{3})S^{\alpha0}_{\lambda\pi}(\omega_1)S^{\beta0}_{\mu\rho}(\omega_2)\bar{S}^{\alpha'0}_{\lambda\sigma}(\omega'_1)\bar{S}^{\beta'0}_{\nu\tau}(\omega'_2)$
2	3	$\hat{R}_\mu\hat{R}_\nu S^{\alpha0}_{\lambda\pi}(\omega_1)S^{\beta0}_{\mu\rho}(\omega_2)\bar{S}^{\alpha'0}_{\nu\sigma}(\omega'_1)\bar{S}^{\beta'0}_{\lambda\tau}(\omega'_2)$ $-(\frac{1}{3})S^{\alpha0}_{\lambda\pi}(\omega_1)S^{\beta0}_{\mu\rho}(\omega_2)\bar{S}^{\alpha'0}_{\mu\sigma}(\omega'_1)\bar{S}^{\beta'0}_{\lambda\tau}(\omega'_2)$

TABLE II (*Continued*)

j	q	$Z^{(j;q)}_{\pi\rho\sigma\tau}$
2	4	$\hat{R}_\lambda \hat{R}_\nu S^{\alpha 0}_{\lambda\pi}(\omega_1) S^{\beta 0}_{\mu\rho}(\omega_2) \bar{S}^{\alpha' 0}_{\mu\sigma}(\omega'_1) \bar{S}^{\beta' 0}_{\nu\tau}(\omega'_2)$
		$-(\frac{1}{3}) S^{\alpha 0}_{\lambda\pi}(\omega_1) S^{\beta 0}_{\mu\rho}(\omega_2) \bar{S}^{\alpha' 0}_{\mu\sigma}(\omega'_1) \bar{S}^{\beta' 0}_{\lambda\tau}(\omega'_2)$
2	5	$\hat{R}_\lambda \hat{R}_\nu S^{\alpha 0}_{\lambda\pi}(\omega_1) S^{\beta 0}_{\mu\rho}(\omega_2) \bar{S}^{\alpha' 0}_{\nu\sigma}(\omega'_1) \bar{S}^{\beta' 0}_{\mu\tau}(\omega'_2)$
		$-(\frac{1}{3}) S^{\alpha 0}_{\lambda\pi}(\omega_1) S^{\beta 0}_{\mu\rho}(\omega_2) \bar{S}^{\alpha' 0}_{\lambda\sigma}(\omega'_1) \bar{S}^{\beta' 0}_{\mu\tau}(\omega'_2)$
2	6	$\hat{R}_\lambda \hat{R}_\mu S^{\alpha 0}_{\lambda\pi}(\omega_1) S^{\beta 0}_{\mu\rho}(\omega_2) \bar{S}^{\alpha' 0}_{\nu\sigma}(\omega'_1) \bar{S}^{\beta' 0}_{\nu\tau}(\omega'_2)$
		$-(\frac{1}{3}) S^{\alpha 0}_{\lambda\pi}(\omega_1) S^{\beta 0}_{\lambda\rho}(\omega_2) \bar{S}^{\alpha' 0}_{\mu\sigma}(\omega'_1) \bar{S}^{\beta' 0}_{\mu\tau}(\omega'_2)$
3	1	$(\frac{1}{5})\varepsilon_{\lambda\mu\nu}\{5\hat{R}_\xi \hat{R}_o \hat{R}_\nu S^{\alpha 0}_{\lambda\pi}(\omega_1) S^{\beta 0}_{\xi\rho}(\omega_2) \bar{S}^{\alpha' 0}_{\mu\sigma}(\omega'_1) \bar{S}^{\beta' 0}_{o\tau}(\omega'_2)$
		$- \hat{R}_o S^{\alpha 0}_{\lambda\pi}(\omega_1) S^{\beta 0}_{\nu\rho}(\omega_2) \bar{S}^{\alpha' 0}_{\mu\sigma}(\omega'_1) \bar{S}^{\beta' 0}_{o\tau}(\omega'_2)$
		$- \hat{R}_o S^{\alpha 0}_{\lambda\pi}(\omega_1) S^{\beta 0}_{o\rho}(\omega_2) \bar{S}^{\alpha' 0}_{\mu\sigma}(\omega'_1) \bar{S}^{\beta' 0}_{\nu\tau}(\omega'_2)$
		$- \hat{R}_\nu S^{\alpha 0}_{\lambda\pi}(\omega_1) S^{\beta 0}_{o\rho}(\omega_2) \bar{S}^{\alpha' 0}_{\mu\sigma}(\omega'_1) \bar{S}^{\beta' 0}_{o\tau}(\omega'_2)\}$
3	2	$(\frac{1}{5})\varepsilon_{\lambda\mu\nu}\{5\hat{R}_\xi \hat{R}_o \hat{R}_\nu S^{\alpha 0}_{\xi\pi}(\omega_1) S^{\beta 0}_{\lambda\rho}(\omega_2) \bar{S}^{\alpha' 0}_{\mu\sigma}(\omega'_1) \bar{S}^{\beta' 0}_{o\tau}(\omega'_2)$
		$- \hat{R}_o S^{\alpha 0}_{\nu\pi}(\omega_1) S^{\beta 0}_{\lambda\rho}(\omega_2) \bar{S}^{\alpha' 0}_{\mu\sigma}(\omega'_1) \bar{S}^{\beta' 0}_{o\tau}(\omega'_2)$
		$- \hat{R}_o S^{\alpha 0}_{o\pi}(\omega_1) S^{\beta 0}_{\lambda\rho}(\omega_2) \bar{S}^{\alpha' 0}_{\mu\sigma}(\omega'_1) \bar{S}^{\beta' 0}_{\nu\tau}(\omega'_2)$
		$- \hat{R}_\nu S^{\alpha 0}_{o\pi}(\omega_1) S^{\beta 0}_{\lambda\rho}(\omega_2) \bar{S}^{\alpha' 0}_{\mu\sigma}(\omega'_1) \bar{S}^{\beta' 0}_{o\tau}(\omega'_2)\}$
3	3	$(\frac{1}{5})\varepsilon_{\lambda\mu\nu}\{5\hat{R}_\xi \hat{R}_o \hat{R}_\nu S^{\alpha 0}_{\xi\pi}(\omega_1) S^{\beta 0}_{o\rho}(\omega_2) \bar{S}^{\alpha' 0}_{\lambda\sigma}(\omega'_1) \bar{S}^{\beta' 0}_{\mu\tau}(\omega'_2)$
		$- \hat{R}_o S^{\alpha 0}_{\nu\pi}(\omega_1) S^{\beta 0}_{o\rho}(\omega_2) \bar{S}^{\alpha' 0}_{\lambda\sigma}(\omega'_1) \bar{S}^{\beta' 0}_{\mu\tau}(\omega'_2)$
		$- \hat{R}_o S^{\alpha 0}_{o\pi}(\omega_1) S^{\beta 0}_{\nu\rho}(\omega_2) \bar{S}^{\alpha' 0}_{\lambda\sigma}(\omega'_1) \bar{S}^{\beta' 0}_{\mu\tau}(\omega'_2)$
		$- \hat{R}_\nu S^{\alpha 0}_{o\pi}(\omega_1) S^{\beta 0}_{o\rho}(\omega_2) \bar{S}^{\alpha' 0}_{\lambda\sigma}(\omega'_1) \bar{S}^{\beta' 0}_{\mu\tau}(\omega'_2)\}$
4	1	$\hat{R}_\lambda \hat{R}_\mu \hat{R}_\nu \hat{R}_o S^{\alpha 0}_{\lambda\pi}(\omega_1) S^{\beta 0}_{\mu\rho}(\omega_2) \bar{S}^{\alpha' 0}_{\nu\sigma}(\omega'_1) \bar{S}^{\beta' 0}_{o\tau}(\omega'_2)$
		$-(\frac{1}{7})\{\hat{R}_\mu \hat{R}_\nu S^{\alpha 0}_{\lambda\pi}(\omega_1) S^{\beta 0}_{\lambda\rho}(\omega_2) \bar{S}^{\alpha' 0}_{\mu\sigma}(\omega'_1) \bar{S}^{\beta' 0}_{\nu\tau}(\omega'_2)$
		$+ \hat{R}_\mu \hat{R}_\nu S^{\alpha 0}_{\lambda\pi}(\omega_1) S^{\beta 0}_{\mu\rho}(\omega_2) \bar{S}^{\alpha' 0}_{\lambda\sigma}(\omega'_1) \bar{S}^{\beta' 0}_{\nu\tau}(\omega'_2)$
		$+ \hat{R}_\mu \hat{R}_\nu S^{\alpha 0}_{\lambda\pi}(\omega_1) S^{\beta 0}_{\mu\rho}(\omega_2) \bar{S}^{\alpha' 0}_{\nu\sigma}(\omega'_1) \bar{S}^{\beta' 0}_{\lambda\tau}(\omega'_2)$
		$+ \hat{R}_\lambda \hat{R}_\nu S^{\alpha 0}_{\lambda\pi}(\omega_1) S^{\beta 0}_{\mu\rho}(\omega_2) \bar{S}^{\alpha' 0}_{\mu\sigma}(\omega'_1) \bar{S}^{\beta' 0}_{\nu\tau}(\omega'_2)$
		$+ \hat{R}_\lambda \hat{R}_\nu S^{\alpha 0}_{\lambda\pi}(\omega_1) S^{\beta 0}_{\mu\rho}(\omega_2) \bar{S}^{\alpha' 0}_{\nu\sigma}(\omega'_1) \bar{S}^{\beta' 0}_{\mu\tau}(\omega'_2)$
		$+ \hat{R}_\lambda \hat{R}_\nu S^{\alpha 0}_{\lambda\pi}(\omega_1) S^{\beta 0}_{\nu\rho}(\omega_2) \bar{S}^{\alpha' 0}_{\mu\sigma}(\omega'_1) \bar{S}^{\beta' 0}_{\mu\tau}(\omega'_2)\}$
		$+(\frac{1}{35})\{S^{\alpha 0}_{\lambda\pi}(\omega_1) S^{\beta 0}_{\lambda\rho}(\omega_2) \bar{S}^{\alpha' 0}_{\mu\sigma}(\omega'_1) \bar{S}^{\beta' 0}_{\mu\tau}(\omega'_2)$
		$+ S^{\alpha 0}_{\lambda\pi}(\omega_1) S^{\beta 0}_{\mu\rho}(\omega_2) \bar{S}^{\alpha' 0}_{\lambda\sigma}(\omega'_1) \bar{S}^{\beta' 0}_{\mu\tau}(\omega'_2)$
		$+ S^{\alpha 0}_{\lambda\pi}(\omega_1) S^{\beta 0}_{\mu\rho}(\omega_2) \bar{S}^{\alpha' 0}_{\mu\sigma}(\omega'_1) \bar{S}^{\beta' 0}_{\lambda\tau}(\omega'_2)\}$

[a] Defined by

$$T_c^{(j;q)}(\alpha, \beta, \alpha', \beta'; \omega_1, \omega_2, \omega'_1, \omega'_2; \omega_{\alpha 0} - \omega_1, \omega_{\alpha' 0} - \omega'_1)$$
$$= S^{\alpha 0}_{\lambda\pi}(\omega_1) S^{\beta 0}_{\mu\rho}(\omega_2) \bar{S}^{\alpha' 0}_{\nu\sigma}(\omega'_1) \bar{S}^{\beta' 0}_{o\tau}(\omega'_2)$$
$$\times V_{\pi\rho}([\omega_{\alpha 0} - \omega_1], \mathbf{R}) \bar{V}_{\sigma\tau}([\omega_{\alpha' 0} - \omega'_1], \mathbf{R}) W^{(4;j;q)}_{\lambda\mu\nu o}(\hat{\mathbf{R}})$$

and here given by

$$T_c^{(j;q)} = Z^{(j;q)}_{\pi\rho\sigma\tau} V_{\pi\rho}([\omega_{\alpha 0} - \omega_1], \mathbf{R}) \bar{V}_{\sigma\tau}([\omega_{\alpha' 0} - \omega'_1], \mathbf{R})$$

TABLE III
Rotational Averaging Scheme[a]

Vector and tensor quantities	e	$(S^{\alpha 0}, X^{\alpha 0}, \mu^{\alpha 0})$	$(S^{\beta 0}, X^{\beta 0}, \mu^{\beta 0})$	V	Result
Pre-averaging	p	p	p	p	
	$\downarrow$	$\updownarrow$	$\updownarrow$	$\updownarrow$	
Rotational averaging $a \leftrightarrow p$	p	a	a	a	Rotating pair
	$\downarrow$	$\downarrow$	$\updownarrow$	$\downarrow$	
Rotational averaging $b \leftrightarrow a$	p	a	b	a	
	$\downarrow$	$\downarrow$	$\downarrow$	$\updownarrow$	
Rotational averaging $r \leftrightarrow a$	p	a	b	r	Free molecules

[a]The Cartesian reference frames p, a, b, and r are defined as follows: p denotes a laboratory-fixed frame in which the laser polarization vectors are fixed; a signifies a frame of reference that has been chosen to have its origin at center A; similarly b has its origin at center B; r represents a frame in which the $\mathbf{R}$ vectors are rotationally invariant.

Ultimately, equivalent rate expressions result regardless of the order in which the averages are calculated. For instance, it is possible to conduct an alternative series of rotational averages in which first A and B are averaged with respect to $\mathbf{R}$, and then $\mathbf{R}$ is averaged with respect to the laboratory frame. In performing the lengthy calculations associated with this work, it is useful to have the option to use whichever route proves to be shorter, and least mathematically complex, for the particular case under study. Applying such procedures to Eq. (7.5) gives the rate of single-beam cooperative two-photon absorption for a completely fluid sample with no molecular orientational correlation. The result is an equation that can be written in exactly the same form as Eq. (7.5), but where each $T_c^{(j;q)}$ is replaced by its fully averaged counterpart $\langle\langle T_c(j;q) \rangle\rangle$. It is worth noting that this procedure has no effect on the polarization dependence, which is entirely determined by the parameters A'.

While a similar method of calculation is required for each of the processes under consideration, important differences arise in all other cases because an exponential phase factor persists in the pre-averaged results. This leads to the necessity of employing phased rotational averaging. Such averaging procedures are described elsewhere, and for a detailed discussion of the subject the reader is referred to Andrews and Harlow (1984b). Results are given below for both rotating pairs and free molecules undergoing bimolecular mean-frequency absorption by each of the three remaining mechanisms.

B. Single-Frequency Distributive Two-Photon Absorption

To obtain a result applicable to a rotating pair requires the performance of a rotational average on Eq. (6.7). Whereas the first two terms of this equation require a straightforward fourth rank average, terms three and four require a phased fourth rank average. The result for the pair rate may be expressed as follows;

$$\langle \Gamma_{1d} \rangle = \left(\frac{2\pi\rho_f}{\hbar} \right) K'^2 \left\{ \sum_{p,q} g^{pq}_{(4;0)} A'^{(0;p)}(\hat{\mathbf{s}}) T^{(0;q)}_d(\alpha, \beta, \alpha, \beta; A, B, A, B; \omega_{\beta 0}, \omega_{\beta 0}) \right.$$

$$+ \sum_{j=0} \sum_{p,q} \frac{(2j)!}{2^j (j!)^2} i^j j_j(|\mathbf{s}| R) g^{pq}_{(4,j)} A'^{(j;p)}(\hat{\mathbf{s}})$$

$$\left. \times T^{(j;q)}_d(\alpha, \beta, \beta, \alpha; A, B, B, A; \omega_{\beta 0}, -\omega_{\alpha 0}) \right\}$$

$$+ \{\alpha, A, -\omega_{\alpha 0} \leftrightarrow \beta, B, \omega_{\beta 0}\} \tag{7.8}$$

where $\mathbf{s}$ is the wave-vector sum for the two absorbed photons ($= 2\mathbf{k}$), and

$$T^{(j;q)}_d(\alpha, \beta, \alpha', \beta'; A_1, A_2, A_3, A_4; \omega_{\alpha 0}, \omega'_{\alpha 0})$$

$$= \chi^{\alpha 0}_{\lambda\mu\pi}(A_1) \mu^{\beta 0}_{\rho}(A_2) \bar{\chi}^{\alpha' 0}_{\nu o \sigma}(A_3) \bar{\mu}^{\beta' 0}_{\tau}(A_4) V_{\pi\rho}(\omega_{\alpha 0}, \mathbf{R}) \bar{V}_{\sigma\tau}(\omega'_{\alpha 0}, \mathbf{R}) W^{(4;j;q)}_{\lambda\mu\nu o}(\hat{\mathbf{R}}) \tag{7.9}$$

where the j_n denote spherical Bessel functions of order n, and the general form of the molecular parameters T_d is given in Table IV. As is evident on inspection of the polarization parameters $A'^{(j;p)}(\hat{\mathbf{s}})$ in Table 1, remembering that $\hat{\mathbf{s}} = \hat{\mathbf{k}}$, the involvement of terms with $j \neq 0$ in the rate equation leads to a dependence on the two variables $\eta = (\mathbf{e} \cdot \mathbf{e})(\bar{\mathbf{e}} \cdot \bar{\mathbf{e}})$, and $\zeta = (\mathbf{e} \times \bar{\mathbf{e}}) \cdot \hat{\mathbf{k}}$, the relationship between which is discussed in Section IX. Once again, the corresponding rate equation for the case of a perfect fluid consisting of free molecules is obtained by carrying out two further averages on each molecular term in the above expression.

C. Two-Frequency Cooperative Two-Photon Absorption

For two-frequency excitation, the averaged result for a rotating pair is given by

$$\langle \Gamma_{2c} \rangle = \left(\frac{2\pi\rho_f}{\hbar} \right) K^2 \left\{ \sum_{p,q} g^{pq}_{(4;0)} A^{(0;p)}(\hat{\mathbf{u}}) \right.$$

$$\times T^{(0;q)}_c(\alpha, \alpha, \alpha, \alpha; \omega_1, \omega_2, \omega_1, \omega_2; \omega_{\alpha 0} - \omega_1, \omega_{\alpha 0} - \omega_1)$$

$$+ \sum_{j=0} \sum_{p,q} \frac{(2j)!}{2^j (j!)^2} i^j j_j(|\mathbf{u}| R) g^{pq}_{(4,j)} A^{(j;p)}(\hat{\mathbf{u}})$$

$$\left. \times T^{(j;q)}_c(\alpha, \alpha, \alpha, \alpha; \omega_1, \omega_2, \omega_1, \omega_2; \omega_{\alpha 0} - \omega_1, \omega_{\alpha 0} - \omega_2) \right\}$$

$$+ \{\omega_{\alpha 0} - \omega_1 \leftrightarrow \omega_{\alpha 0} - \omega_2\} \tag{7.10}$$

TABLE IV
Explicit Form of the Distributive Molecular Invariants[a]

j	q	$Z^{(p;q)}_{\pi\rho\sigma\tau}(A_1, A_2, A_3, A_4)$
0	1	$\chi^{\alpha 0}_{\lambda\lambda\pi}(A_1)\mu^{\beta 0}_{\rho}(A_2)\bar{\chi}^{\alpha' 0}_{\mu\mu\sigma}(A_3)\bar{\mu}^{\beta' 0}_{\tau}(A_4)$
0	2	$\chi^{\alpha 0}_{\lambda\mu\pi}(A_1)\mu^{\beta 0}_{\rho}(A_2)\bar{\chi}^{\alpha' 0}_{\lambda\mu\sigma}(A_3)\bar{\mu}^{\beta' 0}_{\tau}(A_4)$
0	3	$\chi^{\alpha 0}_{\lambda\mu\pi}(A_1)\mu^{\beta 0}_{\rho}(A_2)\bar{\chi}^{\alpha' 0}_{\mu\lambda\sigma}(A_3)\bar{\mu}^{\beta' 0}_{\tau}(A_4)$
1	1	$\varepsilon_{\lambda\mu\nu}\hat{R}_{\nu}\chi^{\alpha 0}_{\lambda\mu\pi}(A_1)\mu^{\beta 0}_{\rho}(A_2)\bar{\chi}^{\alpha' 0}_{oo\sigma}(A_3)\bar{\mu}^{\beta' 0}_{\tau}(A_4)$
1	2	$\varepsilon_{\lambda o\nu}\hat{R}_{\nu}\chi^{\alpha 0}_{\lambda\mu\pi}(A_1)\mu^{\beta 0}_{\rho}(A_2)\bar{\chi}^{\alpha' 0}_{o\mu\sigma}(A_3)\bar{\mu}^{\beta' 0}_{\tau}(A_4)$
1	3	$\varepsilon_{\lambda o\nu}\hat{R}_{\nu}\chi^{\alpha 0}_{\lambda\mu\pi}(A_1)\mu^{\beta 0}_{\rho}(A_2)\bar{\chi}^{\alpha' 0}_{\mu o\sigma}(A_3)\bar{\mu}^{\beta' 0}_{\tau}(A_4)$
1	4	$\varepsilon_{\mu o\nu}\hat{R}_{\nu}\chi^{\alpha 0}_{\lambda\mu\pi}(A_1)\mu^{\beta 0}_{\rho}(A_2)\bar{\chi}^{\alpha' 0}_{o\lambda\sigma}(A_3)\bar{\mu}^{\beta' 0}_{\tau}(A_4)$
1	5	$\varepsilon_{\mu o\nu}\hat{R}_{\nu}\chi^{\alpha 0}_{\lambda\mu\pi}(A_1)\mu^{\beta 0}_{\rho}(A_2)\bar{\chi}^{\alpha' 0}_{\lambda o\sigma}(A_3)\bar{\mu}^{\beta' 0}_{\tau}(A_4)$
1	6	$\varepsilon_{\mu o\nu}\hat{R}_{\nu}\chi^{\alpha 0}_{\lambda\lambda\pi}(A_1)\mu^{\beta 0}_{\rho}(A_2)\bar{\chi}^{\alpha' 0}_{\mu o\sigma}(A_3)\bar{\mu}^{\beta' 0}_{\tau}(A_4)$
2	1	$\hat{R}_{\mu}\hat{R}_{\nu}\chi^{\alpha 0}_{\lambda\lambda\pi}(A_1)\mu^{\beta 0}_{\rho}(A_2)\bar{\chi}^{\alpha' 0}_{\mu\nu\sigma}(A_3)\bar{\mu}^{\beta' 0}_{\tau}(A_4)$
		$-(\frac{1}{3})\chi^{\alpha 0}_{\lambda\lambda\pi}(A_1)\mu^{\beta 0}_{\rho}(A_2)\bar{\chi}^{\alpha' 0}_{\mu\mu\sigma}(A_3)\bar{\mu}^{\beta' 0}_{\tau}(A_4)$
2	2	$\hat{R}_{\mu}\hat{R}_{\nu}\chi^{\alpha 0}_{\lambda\mu\pi}(A_1)\mu^{\beta 0}_{\rho}(A_2)\bar{\chi}^{\alpha' 0}_{\lambda\nu\sigma}(A_3)\bar{\mu}^{\beta' 0}_{\tau}(A_4)$
		$-(\frac{1}{3})\chi^{\alpha 0}_{\lambda\mu\pi}(A_1)\mu^{\beta 0}_{\rho}(A_2)\bar{\chi}^{\alpha' 0}_{\lambda\mu\sigma}(A_3)\bar{\mu}^{\beta' 0}_{\tau}(A_4)$
2	3	$\hat{R}_{\mu}\hat{R}_{\nu}\chi^{\alpha 0}_{\lambda\mu\pi}(A_1)\mu^{\beta 0}_{\rho}(A_2)\bar{\chi}^{\alpha' 0}_{\nu\lambda\sigma}(A_3)\bar{\mu}^{\beta' 0}_{\tau}(A_4)$
		$-(\frac{1}{3})\chi^{\alpha 0}_{\lambda\mu\pi}(A_1)\mu^{\beta 0}_{\rho}(A_2)\bar{\chi}^{\alpha' 0}_{\mu\lambda\sigma}(A_3)\bar{\mu}^{\beta' 0}_{\tau}(A_4)$
2	4	$\hat{R}_{\lambda}\hat{R}_{\nu}\chi^{\alpha 0}_{\lambda\mu\pi}(A_1)\mu^{\beta 0}_{\rho}(A_2)\bar{\chi}^{\alpha' 0}_{\mu\nu\sigma}(A_3)\bar{\mu}^{\beta' 0}_{\tau}(A_4)$
		$-(\frac{1}{3})\chi^{\alpha 0}_{\lambda\mu\pi}(A_1)\mu^{\beta 0}_{\rho}(A_2)\bar{\chi}^{\alpha' 0}_{\mu\lambda\sigma}(A_3)\bar{\mu}^{\beta' 0}_{\tau}(A_4)$
2	5	$\hat{R}_{\lambda}\hat{R}_{\nu}\chi^{\alpha 0}_{\lambda\mu\pi}(A_1)\mu^{\beta 0}_{\rho}(A_2)\bar{\chi}^{\alpha' 0}_{\nu\mu\sigma}(A_3)\bar{\mu}^{\beta' 0}_{\tau}(A_4)$
		$-(\frac{1}{3})\chi^{\alpha 0}_{\lambda\mu\pi}(A_1)\mu^{\beta 0}_{\rho}(A_2)\bar{\chi}^{\alpha' 0}_{\lambda\mu\sigma}(A_3)\bar{\mu}^{\beta' 0}_{\tau}(A_4)$
2	6	$\hat{R}_{\lambda}\hat{R}_{\mu}\chi^{\alpha 0}_{\lambda\mu\pi}(A_1)\mu^{\beta 0}_{\rho}(A_2)\bar{\chi}^{\alpha' 0}_{\nu\nu\sigma}(A_3)\bar{\mu}^{\beta' 0}_{\tau}(A_4)$
		$-(\frac{1}{3})\chi^{\alpha 0}_{\lambda\lambda\pi}(A_1)\mu^{\beta 0}_{\rho}(A_2)\bar{\chi}^{\alpha' 0}_{\mu\mu\sigma}(A_3)\bar{\mu}^{\beta' 0}_{\tau}(A_4)$
3	1	$(\frac{1}{5})\varepsilon_{\lambda\mu\nu}\{5\hat{R}_{\nu}\hat{R}_{\xi}\hat{R}_{o}\chi^{\alpha 0}_{\lambda\xi\pi}(A_1)\mu^{\beta 0}_{\rho}(A_2)\bar{\chi}^{\alpha' 0}_{\mu o\sigma}(A_3)\bar{\mu}^{\beta' 0}_{\tau}(A_4)$
		$-\hat{R}_{o}\chi^{\alpha 0}_{\lambda\nu\pi}(A_1)\mu^{\beta 0}_{\rho}(A_2)\bar{\chi}^{\alpha' 0}_{\mu o\sigma}(A_3)\bar{\mu}^{\beta' 0}_{\tau}(A_4)$
		$-\hat{R}_{o}\chi^{\alpha 0}_{\lambda o\pi}(A_1)\mu^{\beta 0}_{\rho}(A_2)\bar{\chi}^{\alpha' 0}_{\mu\nu\sigma}(A_3)\bar{\mu}^{\beta' 0}_{\tau}(A_4)$
		$-\hat{R}_{\nu}\chi^{\alpha 0}_{\lambda o\pi}(A_1)\mu^{\beta 0}_{\rho}(A_2)\bar{\chi}^{\alpha' 0}_{\mu o\sigma}(A_3)\bar{\mu}^{\beta' 0}_{\tau}(A_4)\}$

TABLE IV (*Continued*)

j	q	$Z^{(p;q)}_{\pi\rho\sigma\tau}(A_1, A_2, A_3, A_4)$
3	2	$(\frac{1}{5})\varepsilon_{\lambda\mu\nu}\{5\hat{R}_\nu\hat{R}_\xi\hat{R}_o\chi^{\alpha 0}_{\xi\lambda\pi}(A_1)\mu^{\beta 0}_\rho(A_2)\bar{\chi}^{\alpha' 0}_{\mu o\sigma}(A_3)\bar{\mu}^{\beta' 0}_\tau(A_4)$
		$-\hat{R}_o\chi^{\alpha 0}_{\nu\lambda\pi}(A_1)\mu^{\beta 0}_\rho(A_2)\bar{\chi}^{\alpha' 0}_{\mu o\sigma}(A_3)\bar{\mu}^{\beta' 0}_\tau(A_4)$
		$-\hat{R}_o\chi^{\alpha 0}_{o\lambda\pi}(A_1)\mu^{\beta 0}_\rho(A_2)\bar{\chi}^{\alpha' 0}_{\mu\nu\sigma}(A_3)\bar{\mu}^{\beta' 0}_\tau(A_4)$
		$-\hat{R}_\nu\chi^{\alpha 0}_{o\lambda\pi}(A_1)\mu^{\beta 0}_\rho(A_2)\bar{\chi}^{\alpha' 0}_{\mu o\sigma}(A_3)\bar{\mu}^{\beta' 0}_\tau(A_4)\}$
3	3	$(\frac{1}{5})\varepsilon_{\lambda\mu\nu}\{5\hat{R}_\nu\hat{R}_\xi\hat{R}_o\chi^{\alpha 0}_{\xi o\pi}(A_1)\mu^{\beta 0}_\rho(A_2)\bar{\chi}^{\alpha' 0}_{\lambda\mu\sigma}(A_3)\bar{\mu}^{\beta' 0}_\tau(A_4)$
		$-\hat{R}_o\chi^{\alpha 0}_{\nu o\pi}(A_1)\mu^{\beta 0}_\rho(A_2)\bar{\chi}^{\alpha' 0}_{\lambda\mu\sigma}(A_3)\bar{\mu}^{\beta' 0}_\tau(A_4)$
		$-\hat{R}_o\chi^{\alpha 0}_{o\nu\pi}(A_1)\mu^{\beta 0}_\rho(A_2)\bar{\chi}^{\alpha' 0}_{\lambda\mu\sigma}(A_3)\bar{\mu}^{\beta' 0}_\tau(A_4)$
		$-\hat{R}_\nu\chi^{\alpha 0}_{oo\pi}(A_1)\mu^{\beta 0}_\rho(A_2)\bar{\chi}^{\alpha' 0}_{\lambda\mu\sigma}(A_3)\bar{\mu}^{\beta' 0}_\tau(A_4)\}$
4	1	$\hat{R}_\lambda\hat{R}_\mu\hat{R}_\nu\hat{R}_o\chi^{\alpha 0}_{\lambda\mu\pi}(A_1)\mu^{\beta 0}_\rho(A_2)\bar{\chi}^{\alpha' 0}_{\nu o\sigma}(A_3)\bar{\mu}^{\beta' 0}_\tau(A_4)$
		$-(\frac{1}{7})\{\chi^{\alpha 0}_{\lambda\lambda\pi}(A_1)\mu^{\beta 0}_\rho(A_2)\bar{\chi}^{\alpha' 0}_{\nu o\sigma}(A_3)\bar{\mu}^{\beta' 0}_\tau(A_4)\hat{R}_\nu\hat{R}_o$
		$+\chi^{\alpha 0}_{\lambda\nu\pi}(A_1)\mu^{\beta 0}_\rho(A_2)\bar{\chi}^{\alpha' 0}_{\lambda o\sigma}(A_3)\bar{\mu}^{\beta' 0}_\tau(A_4)\hat{R}_\nu\hat{R}_o$
		$+\chi^{\alpha 0}_{\lambda\nu\pi}(A_1)\mu^{\beta 0}_\rho(A_2)\bar{\chi}^{\alpha' 0}_{o\lambda\sigma}(A_3)\bar{\mu}^{\beta' 0}_\tau(A_4)\hat{R}_\nu\hat{R}_o$
		$+\chi^{\alpha 0}_{\nu\lambda\pi}(A_1)\mu^{\beta 0}_\rho(A_2)\bar{\chi}^{\alpha' 0}_{\lambda o\sigma}(A_3)\bar{\mu}^{\beta' 0}_\tau(A_4)\hat{R}_\nu\hat{R}_o$
		$+\chi^{\alpha 0}_{\nu\mu\pi}(A_1)\mu^{\beta 0}_\rho(A_2)\bar{\chi}^{\alpha' 0}_{o\mu\sigma}(A_3)\bar{\mu}^{\beta' 0}_\tau(A_4)\hat{R}_\nu\hat{R}_o$
		$+\chi^{\alpha 0}_{\nu o\pi}(A_1)\mu^{\beta 0}_\rho(A_2)\bar{\chi}^{\alpha' 0}_{\lambda\lambda\sigma}(A_3)\bar{\mu}^{\beta' 0}_\tau(A_4)\hat{R}_\nu\hat{R}_o\}$
		$+(\frac{1}{35})\{\chi^{\alpha 0}_{\lambda\lambda\pi}(A_1)\mu^{\beta 0}_\rho(A_2)\bar{\chi}^{\alpha' 0}_{\mu\mu\sigma}(A_3)\bar{\mu}^{\beta' 0}_\tau(A_4)$
		$+\chi^{\alpha 0}_{\lambda\mu\pi}(A_1)\mu^{\beta 0}_\rho(A_2)\bar{\chi}^{\alpha' 0}_{\lambda\mu\sigma}(A_3)\bar{\mu}^{\beta' 0}_\tau(A_4)$
		$+\chi^{\alpha 0}_{\lambda\mu\pi}(A_1)\mu^{\beta 0}_\rho(A_2)\bar{\chi}^{\alpha' 0}_{\mu\lambda\sigma}(A_3)\bar{\mu}^{\beta' 0}_\tau(A_4)\}$

[a]Defined by

$$T_d^{(j;q)}(\alpha, \beta, \alpha', \beta'; A_1, A_2, A_3, A_4; \omega_{\alpha 0}, \omega_{\alpha 0})$$
$$= \chi^{\alpha 0}_{\lambda\mu\pi}(A_1)\mu^{\beta 0}_\rho(A_2)\bar{\chi}^{\alpha' 0}_{\nu o\sigma}(A_3)\bar{\mu}^{\beta' 0}_\tau(A_4)V_{\pi\rho}(\omega_{\alpha 0}, \mathbf{R})\bar{V}_{\sigma\tau}(\omega_{\alpha 0}, \mathbf{R})W^{(4;j;q)}_{\lambda\mu\nu o}(\hat{\mathbf{R}}).$$

and here given by

$$T_d^{(j;q)} = Z^{(p;q)}_{\pi\rho\sigma\tau}V_{\pi\rho}(\omega_{\alpha 0}, \mathbf{R})\bar{V}_{\sigma\tau}(\omega_{\alpha 0}, \mathbf{R})$$

where the wave-vector difference $\mathbf{u} = (\mathbf{k}_2 - \mathbf{k}_1)$ and

$$A^{(j;p)}(\hat{\mathbf{u}}) = e_{1_i} e_{2_j} \bar{e}_{1_k} \bar{e}_{2_l} U^{(4;j;p)}_{ijkl}(\hat{\mathbf{u}}) \tag{7.11}$$

the explicit values for which are given in Table V. As before, subsequent rotational averaging of the molecular parameters T_c produces the result applicable to a system of randomly oriented molecules.

D. Two-Frequency Distributive Two-Photon Absorption

The results for the distributive case of two-photon absorption can be expressed in a similar manner. The polarization parameters $A^{(j;p)}$ are identical in form to those tabulated in Table V, and the wave-vector sum **s** is now

TABLE V
Explicit Form of the Polarization Parameters $A^{(j;p)}(\hat{\mathbf{u}})$

j	p	$A^{(j;p)}(\hat{\mathbf{u}})$
0	1	$(\mathbf{e}_1 \cdot \mathbf{e}_2)(\bar{\mathbf{e}}_1 \cdot \bar{\mathbf{e}}_2)$
0	2	1
0	3	$(\mathbf{e}_1 \cdot \bar{\mathbf{e}}_2)(\mathbf{e}_2 \cdot \bar{\mathbf{e}}_1)$
1	1	$(\bar{\mathbf{e}}_1 \cdot \bar{\mathbf{e}}_2)(\mathbf{e}_1 \times \mathbf{e}_2) \cdot \hat{\mathbf{u}}$
1	2	$(\mathbf{e}_1 \times \bar{\mathbf{e}}_1) \cdot \hat{\mathbf{u}}$
1	3	$(\mathbf{e}_2 \cdot \bar{\mathbf{e}}_1)(\mathbf{e}_1 \times \bar{\mathbf{e}}_2) \cdot \hat{\mathbf{u}}$
1	4	$(\mathbf{e}_1 \cdot \bar{\mathbf{e}}_2)(\mathbf{e}_2 \times \bar{\mathbf{e}}_1) \cdot \hat{\mathbf{u}}$
1	5	$(\mathbf{e}_2 \times \bar{\mathbf{e}}_2) \cdot \hat{\mathbf{u}}$
1	6	$(\mathbf{e}_1 \cdot \mathbf{e}_2)(\bar{\mathbf{e}}_1 \times \bar{\mathbf{e}}_2) \cdot \hat{\mathbf{u}}$
2	1	$(\mathbf{e}_1 \cdot \mathbf{e}_2)(\hat{\mathbf{u}} \cdot \bar{\mathbf{e}}_1)(\hat{\mathbf{u}} \cdot \bar{\mathbf{e}}_2) - (\frac{1}{3})(\mathbf{e}_1 \cdot \mathbf{e}_2)(\bar{\mathbf{e}}_1 \cdot \bar{\mathbf{e}}_2)$
2	2	$(\hat{\mathbf{u}} \cdot \mathbf{e}_2)(\hat{\mathbf{u}} \cdot \bar{\mathbf{e}}_2) - \frac{1}{3}$
2	3	$(\mathbf{e}_1 \cdot \bar{\mathbf{e}}_2)(\hat{\mathbf{u}} \cdot \mathbf{e}_2)(\hat{\mathbf{u}} \cdot \bar{\mathbf{e}}_1) - (\frac{1}{3})(\mathbf{e}_1 \cdot \bar{\mathbf{e}}_2)(\mathbf{e}_2 \cdot \bar{\mathbf{e}}_1)$
2	4	$(\mathbf{e}_2 \cdot \bar{\mathbf{e}}_1)(\hat{\mathbf{u}} \cdot \mathbf{e}_1)(\hat{\mathbf{u}} \cdot \bar{\mathbf{e}}_2) - (\frac{1}{3})(\mathbf{e}_1 \cdot \bar{\mathbf{e}}_2)(\mathbf{e}_2 \cdot \bar{\mathbf{e}}_1)$
2	5	$(\hat{\mathbf{u}} \cdot \mathbf{e}_1)(\hat{\mathbf{u}} \cdot \bar{\mathbf{e}}_1) - \frac{1}{3}$
2	6	$(\bar{\mathbf{e}}_1 \cdot \bar{\mathbf{e}}_2)(\hat{\mathbf{u}} \cdot \mathbf{e}_1)(\hat{\mathbf{u}} \cdot \mathbf{e}_2) - (\frac{1}{3})(\mathbf{e}_1 \cdot \mathbf{e}_2)(\bar{\mathbf{e}}_1 \cdot \bar{\mathbf{e}}_2)$
3	1	$(\mathbf{e}_1 \times \bar{\mathbf{e}}_1) \cdot \hat{\mathbf{u}}(\hat{\mathbf{u}} \cdot \mathbf{e}_2)(\hat{\mathbf{u}} \cdot \bar{\mathbf{e}}_2) - (\frac{1}{5})[(\mathbf{e}_1 \times \bar{\mathbf{e}}_1) \cdot \mathbf{e}_2(\hat{\mathbf{u}} \cdot \bar{\mathbf{e}}_2) + (\mathbf{e}_1 \times \bar{\mathbf{e}}_1) \cdot \bar{\mathbf{e}}_2(\hat{\mathbf{u}} \cdot \mathbf{e}_2)$ $+ (\mathbf{e}_1 \times \bar{\mathbf{e}}_1) \cdot \hat{\mathbf{u}}]$
3	2	$(\mathbf{e}_2 \times \bar{\mathbf{e}}_1) \cdot \hat{\mathbf{u}}(\hat{\mathbf{u}} \cdot \mathbf{e}_1)(\hat{\mathbf{u}} \cdot \bar{\mathbf{e}}_2) - (\frac{1}{5})[(\mathbf{e}_2 \times \bar{\mathbf{e}}_1) \cdot \mathbf{e}_1(\hat{\mathbf{u}} \cdot \bar{\mathbf{e}}_2) + (\mathbf{e}_2 \times \bar{\mathbf{e}}_1) \cdot \bar{\mathbf{e}}_2(\hat{\mathbf{u}} \cdot \mathbf{e}_1)$ $+ (\mathbf{e}_2 \times \bar{\mathbf{e}}_1) \cdot \hat{\mathbf{u}}(\mathbf{e}_1 \cdot \bar{\mathbf{e}}_2)]$
3	3	$(\bar{\mathbf{e}}_1 \times \bar{\mathbf{e}}_2) \cdot \hat{\mathbf{u}}(\hat{\mathbf{u}} \cdot \mathbf{e}_1)(\hat{\mathbf{u}} \cdot \mathbf{e}_2) - (\frac{1}{5})[(\bar{\mathbf{e}}_1 \times \bar{\mathbf{e}}_2) \cdot \mathbf{e}_1(\hat{\mathbf{u}} \cdot \mathbf{e}_2) + (\bar{\mathbf{e}}_1 \times \bar{\mathbf{e}}_2) \cdot \mathbf{e}_2(\hat{\mathbf{u}} \cdot \mathbf{e}_1)$ $+ (\bar{\mathbf{e}}_1 \times \bar{\mathbf{e}}_2) \cdot \hat{\mathbf{u}}(\mathbf{e}_1 \cdot \mathbf{e}_2)]$
4	1	$(\hat{\mathbf{u}} \cdot \mathbf{e}_1)(\hat{\mathbf{u}} \cdot \mathbf{e}_2)(\hat{\mathbf{u}} \cdot \bar{\mathbf{e}}_1)(\hat{\mathbf{u}} \cdot \bar{\mathbf{e}}_2) - (\frac{1}{7})[(\mathbf{e}_1 \cdot \mathbf{e}_2)(\hat{\mathbf{u}} \cdot \bar{\mathbf{e}}_1)(\hat{\mathbf{u}} \cdot \bar{\mathbf{e}}_2) + (\hat{\mathbf{u}} \cdot \mathbf{e}_2)(\hat{\mathbf{u}} \cdot \bar{\mathbf{e}}_2)$ $+ (\mathbf{e}_1 \cdot \bar{\mathbf{e}}_2)(\hat{\mathbf{u}} \cdot \bar{\mathbf{e}}_1)(\hat{\mathbf{u}} \cdot \mathbf{e}_2) + (\mathbf{e}_2 \cdot \bar{\mathbf{e}}_1)(\hat{\mathbf{u}} \cdot \mathbf{e}_1)(\hat{\mathbf{u}} \cdot \bar{\mathbf{e}}_2) + (\hat{\mathbf{u}} \cdot \mathbf{e}_1)(\hat{\mathbf{u}} \cdot \bar{\mathbf{e}}_1)$ $+ (\bar{\mathbf{e}}_1 \cdot \bar{\mathbf{e}}_2)(\hat{\mathbf{u}} \cdot \mathbf{e}_1)(\hat{\mathbf{u}} \cdot \mathbf{e}_2)] + (\frac{1}{35})[(\mathbf{e}_1 \cdot \mathbf{e}_2)(\bar{\mathbf{e}}_1 \cdot \bar{\mathbf{e}}_2) + 1 + (\mathbf{e}_1 \cdot \bar{\mathbf{e}}_2)(\mathbf{e}_2 \cdot \bar{\mathbf{e}}_1)]$

given by $\mathbf{s} = (\mathbf{k}_1 + \mathbf{k}_2)$; the resultant rate equation may be expressed as

$$\langle \Gamma_{2d} \rangle = \left(\frac{2\pi\rho_f}{\hbar} \right) K^2 \left\{ \sum_{p,q} g^{pq}_{(4;0)} A^{(0;p)}(\hat{\mathbf{s}}) \right.$$

$$\times\, T_d^{(0;q)}(\alpha, \alpha, \alpha, \alpha; A_1, A_2, A_1 A_2; \omega_{\alpha 0}, \omega_{\alpha 0})$$

$$+ \sum_{j=0} \sum_{p,q} \frac{(2j)!}{2^j (j!)^2} i^j j_j(|\hat{\mathbf{s}}| R) g^{pq}_{(4,j)} A^{(j;p)}(\hat{\mathbf{s}})$$

$$\left. \times\, T_d^{(j;q)}(\alpha, \alpha, \alpha, \alpha; A_1, A_2, A_2, A_1; \omega_{\alpha 0}, -\omega_{\alpha 0}) \right\}$$

$$+ \{A_1, V \leftrightarrow A_2, \bar{V}\}. \tag{7.12}$$

Once again, the above result may be adapted to obtain the fully averaged rate by replacing each $T_d(j;q)$ by its fully averaged counterpart $\langle\langle T_d^{(j;q)} \rangle\rangle$.

VIII. RANGE-DEPENDENCE

The results of Sections VI and VII are applicable over an unrestricted range of separations between the two participating centers, and it should be recalled that the quantum electrodynamical formalism employed for their derivation automatically incorporates relativistic retardation effects. It is, therefore, instructive to examine the detailed dependence of the rate equations on intermolecular separation. The study of the limiting short- and long-range behavior of the rates is facilitated through the identification of two regions known respectively as the near-zone and the far-zone, in each of which well-characterized asymptotic behavior can be identified. The limiting ranges of these zones is determined, for each process, by the character of the complex retarded resonance electric dipole–electric dipole interaction tensor $V_{kl}(\omega, \mathbf{R})$, and also the nature of any exponential terms appearing in the pre-averaged rate equation.

The general form of the interaction tensor $V_{kl}(\omega, \mathbf{R})$, applicable for the entire range of pair separations, is given by Eq. (5.12). The appropriate near-zone form of the interaction can be identified with the static (zero-frequency) result, $V_{kl}(0, \mathbf{R})$, obtained in the limit where $\omega R/c \ll 1$, and is given by:

$$\lim_{\omega R/c \ll 1} V_{kl}(\omega, \mathbf{R}) = V_{kl}(0, \mathbf{R}) = \frac{1}{4\pi\varepsilon_0 R^3}(\delta_{kl} - 3\hat{R}_k \hat{R}_l) \tag{8.1}$$

which is traceless. The far-zone form of the interaction tensor is taken where

$\omega R/c \gg 1$, and thus long-range behavior is determined by the limit

$$\lim_{\omega R/c \gg 1} V_{kl}(\omega, \mathbf{R}) = \frac{\omega^2}{4\pi c^2 \varepsilon_0 R}(\hat{R}_k \hat{R}_l - \delta_{kl}) e^{i\omega R/c} \tag{8.2}$$

which is purely transverse with respect to the intermolecular vector **R**. Since the frequency parametrization of the interaction tensor is different for each of the four cases of synergistic two-photon absorption under consideration, the conditions under which the limiting near- or far-zone behavior ensues also differ markedly.

The exponential phase factors which appear in the pre-averaged rate equations (6.7), (6.9), and (6.12), i.e., all cases except that of single-beam cooperative absorption, Eq. (6.3), introduce further considerations. Here, what might be termed near-zone behavior is obtained when these exponentials can be approximated by unity. In the corresponding rotationally averaged results of Section VII, the corresponding level of approximation results in retention of just the leading terms involving the j_0 spherical Bessel functions, since each $j_j(\alpha)$ falls off as α^{-j} in the near-zone. Consequently only the $j = 0$ terms contribute appreciably to the sums over j in the second terms of Eqs. (7.8), (7.10), and (7.12), thereby significantly reducing the complexity of the results. Overall, the near-zone and far-zone limits to the rate equations are therefore subject to conditions imposed both by the nature and form of interaction tensor and also of any phase factors which may be present for each case. A complete breakdown of the boundary conditions for the four processes described in this review is given explicitly in Table VI.

Having defined the short- and long-range limits, it is now possible to examine the general behavior of the rate equations for synergistic photoabsorption within these regions. In the near-zone, since the limiting (static) form of the complex retarded resonance electric dipole–electric dipole

TABLE VI
Near- and Far-Zone Boundary Conditions Resulting from the Form of the Phase Factors and Parametrization of the Interaction Tensor

Two-photon absorption process	Near-zone	Far-zone
Single-beam cooperative	$\lvert\omega_{\alpha 0} - \omega\rvert R/c \ll 1$	$\lvert\omega_{\alpha 0} - \omega\rvert R/c \gg 1$
Single-beam distributive	$2kR \ll 1$	$(\omega_{\alpha 0}R/c, \omega_{\beta 0}R/c) \gg 1$
	$\Rightarrow (\omega_{\alpha 0}R/c, \omega_{\beta 0}R/c) \ll 1$	$\Rightarrow 2kR \gg 1$
Double-beam cooperative	$\lvert\omega_{\alpha 0} - \omega_1\rvert R/c \ll 1$	$\lvert\omega_{\alpha 0} - \omega_1\rvert R/c \gg 1$
	$\lvert\mathbf{k}_1 - \mathbf{k}_2\rvert R \ll 1$	$\lvert\mathbf{k}_1 - \mathbf{k}_2\rvert R \gg 1$
Double-beam distributive	$\lvert\mathbf{k}_1 + \mathbf{k}_2\rvert R \ll 1$	$\omega_{\alpha 0}R/c \gg 1$
	$\omega_{\alpha 0}R/c \ll 1$	$\lvert\mathbf{k}_1 + \mathbf{k}_2\rvert R \gg 1$

interaction (Eq. 8.1) has an R^{-3} dependence, the corresponding near-zone rate expressions with quadratic dependence on V vary as R^{-6}. This kind of distance dependence is well known in the Förster theory of dipolar resonance energy transfer (Förster 1949). Thus, the rate of bimolecular mean-frequency absorption involving pairs of free molecules falls off very rapidly within the near-zone.

For the far-zone case where molecules are separated by comparatively large distances, the rate equations for the three cases with radiative phase factors become complicated since all the spherical Bessel functions make comparable contributions to the result. Here an inverse-square variation with molecular separation results from the quadratic dependence on the long-range limit of V, expressed by Eq. (8.2). It is therefore worth noting that although the rate of synergistic absorption continues to fall off as the separation between interacting molecules increases, it does not fall off anywhere near as quickly as might be expected from an examination of the near-zone behavior. While inverse-square rate dependences of this type are rare and therefore often suspect in atomic and molecular physics, the far-zone behavior reflects an exact asymptotic equivalence to the classical results for radiative energy transfer; a detailed proof has been given by Andrews and Hopkins (1988b). This comparison establishes the correctness of the long-range distance dependence of the virtual photon coupling.

It is interesting to note that it is not essential for both centers involved in a distributive absorption process to be simultaneously located at the focus of the two laser beams. Naturally, one of the two absorbing centers must be irradiated by both beams, but the second center absorbs only a virtual photon and need not, therefore, be within the volume of sample irradiated by either beam. This has an unusual consequence for the case where the two centers are discrete molecules. The number of potential partners, A_2, which may be involved in the distributive excitation of any particular molecule A_1 greatly exceeds the number available for cooperative excitation, and, assuming a uniform sample density, increases with the square of the intermolecular distance. Since the long-range form of the rate equation has an inverse-square dependence on the separation, the total of all the contributions from partner molecules within a shell of given thickness centered on A_1 is, in the long-range limit, independent of the shell radius. This surprising result is a molecular analogue of the astrophysical problem known as Olber's Paradox, which poses the question of why the sky is not uniformly bright with starlight. This paradox arises in a similar way since, although starlight intensity drops off with the square of distance, the number of stars in a homogeneous universe also increases quadratically with distance from any given reference point. In both cases, the resolution of the paradox is connected with a consideration of the neglected effects of light scattering.

One of the main differences between the cooperative and distributive mechanisms lies in the range over which the limiting near-zone behavior occurs. In the cooperative cases, the near-zone form of the interaction tensor applies where $R\Delta\omega/c \ll 1$, $\Delta\omega$ representing the energy mismatch propagated by the virtual photon (see Table VI). The supplementary condition $|\mathbf{k}_1 - \mathbf{k}_2|R \ll 1$ in the double-beam case is less restrictive, except in an

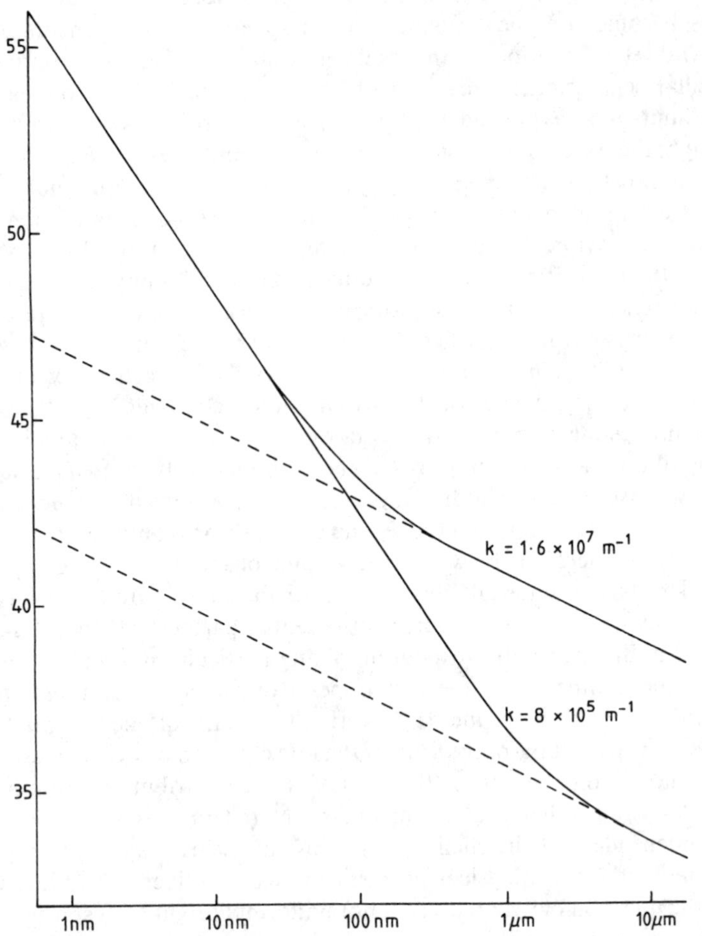

Figure 9. Logarithmic plots on an arbitrary vertical scale of the function $V_{ij}(ck, \mathbf{R})\bar{V}_{ij}(ck, \mathbf{R})$ (solid lines) which describes the dependence on intermolecular distance R of synergistic photoabsorption in free molecules. The values of k for the upper and lower curves are typical of the distributive and cooperative mechanisms, respectively. The broken lines show the asymptotic k^4R^{-2} behavior, whose relative displacement is 5.20 ($= \log(20)^4$).

experimentally unusual configuration where the two photons are absorbed from essentially counterpropagating beams. However, in the distributive cases, the most severe constraints for near-zone behavior are imposed by the condition $|\mathbf{s}|R \ll 1$, where $\mathbf{s}$ is the wave-vector sum for the two absorbed photons. This automatically guarantees satisfaction of the condition $\omega_{\alpha 0}R/c \ll 1$, again except in counterpropagating beam configurations.

In certain cases, significant differences in the extent of the near-zone result from this distinction, especially when the excited-state energies are large but similar. Figure 9 illustrates this point with a log-log plot of the general function $V_{ij}(ck, \mathbf{R})\bar{V}_{ij}(ck, \mathbf{R})$ which occurs in all triply averaged rate equations. The upper curve is plotted for a value of $k = 1.6 \times 10^7\,\text{m}^{-1}$, corresponding to distributive conveyance of an electronic energy $E_{\beta 0}$ with a wavelength of about 400 nm. The lower curve with $k = 8 \times 10^5\,\text{m}^{-1}$ corresponds to the cooperative mechanism where only an electronic energy difference (nominally $E_{\beta 0}/20$) is conveyed; here the difference equates to a vibrational energy with a wavenumber of around $1250\,\text{cm}^{-1}$. At short distances the two graphs are indistinguishable and display the near-zone R^{-6} dependence. However the extent of the near-zone for the former case is much shorter, with the limiting far-zone R^{-2} behavior already established at $R = 1\,\mu\text{m}$; for the latter case, far-zone behavior obtains at $R = 10\,\mu\text{m}$. The result of this difference is that the long-range rates (which vary with k^4) differ by a factor of $(20)^4 = 160{,}000$ in favor of the distributive mechanism (Andrews 1989).

An even more striking illustration concerns the classic case of synergistic $2_0^1 4_0^3$ transitions in a mixture of formaldehyde and deuterioformaldehyde, where the absorption wavenumbers are 30,340.15 and $30{,}147.62\,\text{cm}^{-1}$, respectively (Moule and Walsh 1975). Here the near-zone for single-beam two-photon absorption extends to values of R up to $\sim 10\,\mu\text{m}$: however for the distributive mechanism, near-zone behavior extends only to ~ 100 nm, and the far-zone R^{-2} behavior is already dominant at $10\,\mu\text{m}$. Here the distributive mechanism is favored by a factor of $\sim 6 \times 10^8$.

IX. POLARIZATION DEPENDENCE

In the experimental studies that have identified synergistic two-photon processes, little attention has yet been paid to effects of laser polarization. However, judicious control of polarization should enable more detailed information to be derived from the recorded spectra, particularly in the case of two-beam excitation where more polarization parameters are variable. This section provides a detailed theory of the polarization dependence of synergistic two-photon absorption in fluid media, including an analysis of some unusual dichroic effects.

A. Single-Beam Polarization Parameters

In terms of polarization analysis, the single-beam cases yield little of interest, since only one independently variable polarization parameter, ζ, arises. This is defined by the relation

$$\zeta = (\mathbf{e} \times \bar{\mathbf{e}}) \cdot \hat{\mathbf{k}} \tag{9.1}$$

Although rate equations have been cast in terms of both ζ and a second parameter, η, defined by Eq. (7.4),

$$\eta = (\mathbf{e} \cdot \mathbf{e})(\bar{\mathbf{e}} \cdot \bar{\mathbf{e}}) \tag{9.2}$$

it can in fact be shown that the two are related through the equation

$$\eta = 1 + \zeta^2 \tag{9.3}$$

Both parameters distinguish the degree of helicity of the incident radiation. For instance, η assumes the value of unity for plane polarized light and zero for circularly polarized light, and may therefore be regarded as a direct measure of the degree of ellipticity of the laser beam. However, only ζ, which is zero for plane polarized light, differentiates the sense of handedness in circular or elliptical polarizations, and for the circularly polarized cases we have

$$\zeta^{(L)} = -i \quad \text{and} \quad \zeta^{(R)} = i \tag{9.4}$$

On casting the rate equations entirely in terms of ζ, it becomes evident that the result for the single-beam cooperative case contains only terms in ζ^2 and numerical terms, while additional terms linear in ζ occur in the single-beam distributive case. Hence, the odd-j terms in Eq. (7.8) only contribute to the result when circularly or elliptically polarized incident radiation is employed, and their sign is then dependent on the handedness of that incident radiation. A direct consequence of this is the exhibition of two-photon circular dichroism in the distributive absorption process for pairs of molecules with fixed mutual orientations: no such effect can occur under the cooperative mechanism.

B. Two-Photon Circular Dichroism

Two-photon circular dichroism (Meath and Power 1987) was first predicted over a decade ago (Tinoco 1975; Power 1975; Andrews 1976), and is closely related to two-photon resonance effects in optical rotatory dispersion, the first observations of which have recently been made by Gedanken and Tamir (1988). The extent of circular dichroism in distributive absorption can be expressed through the ratio of the difference in rates for left- and right-circular polarization and the mean rate, i.e.,

$$\Delta^{(L/R)} = \frac{\Gamma^{(L)} - \Gamma^{(R)}}{\frac{1}{2}(\Gamma^{(L)} + \Gamma^{(R)})} \tag{9.5}$$

which, using Eqs. (7.8) and (9.4), gives

$$\Delta^{(L/R)} = \frac{cj_1(2kR) + ej_3(2kR)}{a + bj_0(2kR) + dj_2(2kR) + fj_4(2kR)} \tag{9.6}$$

where the explicit expressions for the coefficients of the spherical Bessel functions are as follows:

$$\begin{aligned} a = {} & 56(3\chi^{\alpha 0}_{(ij)k}\bar{\chi}^{\alpha 0}_{(ij)o} - \chi^{\alpha 0}_{(ii)k}\bar{\chi}^{\alpha 0}_{(jj)o})\mu^{\beta 0}_l \bar{\mu}^{\beta 0}_p V_{kl}(\omega_{\alpha 0}, \mathbf{R})\bar{V}_{op}(\omega_{\alpha 0}, \mathbf{R}) \\ & + (\alpha, \omega_{\alpha 0} \leftrightarrow \beta, \omega_{\beta 0}), \end{aligned} \tag{9.7}$$

$$\begin{aligned} b = {} & 56(3\chi^{\beta 0}_{(ij)k}\bar{\chi}^{\alpha 0}_{(ij)o} - \chi^{\beta 0}_{(ii)k}\bar{\chi}^{\alpha 0}_{(jj)o})\mu^{\alpha 0}_l \bar{\mu}^{\beta 0}_p V_{kl}(\omega_{\beta 0}, \mathbf{R})\bar{V}_{op}(\omega_{\alpha 0}, \mathbf{R}) \\ & + (\alpha, \omega_{\alpha 0} \leftrightarrow \beta, \omega_{\beta 0}), \end{aligned} \tag{9.8}$$

$$\begin{aligned} c = {} & 672\varepsilon_{imt}\hat{R}_t\chi^{\beta 0}_{(ij)k}\bar{\chi}^{\alpha 0}_{(jm)o}\mu^{\alpha 0}_l \bar{\mu}^{\beta 0}_p V_{kl}(\omega_{\beta 0}, \mathbf{R})\bar{V}_{op}(\omega_{\alpha 0}, \mathbf{R}) \\ & + (\alpha, \omega_{\alpha 0} \leftrightarrow \beta, \omega_{\beta 0}), \end{aligned} \tag{9.9}$$

$$\begin{aligned} d = {} & 20(8\chi^{\beta 0}_{(ii)k}\bar{\chi}^{\alpha 0}_{(jj)o} - 12\chi^{\beta 0}_{(ij)k}\bar{\chi}^{\alpha 0}_{(ij)o} - 12\chi^{\beta 0}_{(ij)k}\bar{\chi}^{\alpha 0}_{(jj)o}\hat{R}_i\hat{R}_j - 12\chi^{\beta 0}_{(ii)k}\bar{\chi}^{\alpha 0}_{(mn)o}\hat{R}_m\hat{R}_n \\ & + 36\chi^{\beta 0}_{(ij)k}\bar{\chi}^{\alpha 0}_{(jm)o}\hat{R}_i\hat{R}_m)\mu^{\alpha 0}_l \bar{\mu}^{\beta 0}_p V_{kl}(\omega_{\beta 0}, \mathbf{R})\bar{V}_{op}(\omega_{\alpha 0}, \mathbf{R}) \\ & + (\alpha, \omega_{\alpha 0} \leftrightarrow \beta, \omega_{\beta 0}), \end{aligned} \tag{9.10}$$

$$\begin{aligned} e = {} & 168\varepsilon_{imt}\hat{R}_t(5\chi^{\beta 0}_{(ij)k}\bar{\chi}^{\alpha 0}_{(mn)o}\hat{R}_j\hat{R}_n - \chi^{\beta 0}_{(ij)k}\bar{\chi}^{\alpha 0}_{(jm)o})\mu^{\alpha 0}_l \bar{\mu}^{\beta 0}_p V_{kl}(\omega_{\beta 0}, \mathbf{R})\bar{V}_{op}(\omega_{\alpha 0}, \mathbf{R}) \\ & + (\alpha, \omega_{\alpha 0} \leftrightarrow \beta, \omega_{\beta 0}), \end{aligned} \tag{9.11}$$

$$\begin{aligned} f = {} & 6(35\chi^{\beta 0}_{(ij)k}\bar{\chi}^{\alpha 0}_{(mn)o}\hat{R}_i\hat{R}_j\hat{R}_m\hat{R}_n - 5\chi^{\beta 0}_{(ii)k}\bar{\chi}^{\alpha 0}_{(mn)o}\hat{R}_m\hat{R}_n - 5\chi^{\beta 0}_{(ij)k}\bar{\chi}^{\alpha 0}_{(mm)o}\hat{R}_i\hat{R}_j \\ & - 20\chi^{\beta 0}_{(ij)k}\bar{\chi}^{\alpha 0}_{(in)o}\hat{R}_j\hat{R}_n + \chi^{\beta 0}_{(ii)k}\bar{\chi}^{\alpha 0}_{(jj)o} \\ & + 2\chi^{\beta 0}_{(ij)k}\bar{\chi}^{\alpha 0}_{(ij)o})\mu^{\alpha 0}_l \bar{\mu}^{\beta 0}_p V_{kl}(\omega_{\beta 0}, \mathbf{R})\bar{V}_{op}(\omega_{\alpha 0}, \mathbf{R}) \\ & + (\alpha, \omega_{\alpha 0} \leftrightarrow \beta, \omega_{\beta 0}). \end{aligned} \tag{9.12}$$

The result given by Eq. (9.6) is applicable to a rotating pair with arbitrary separation of the centers A and B. However, in such a case where the intermolecular distance R is fixed, it is generally appropriate to take the near-zone limit, based on the arguments discussed in the last section. By taking the dominant terms in the numerator and denominator of Eq. (9.6), we thus obtain the simpler result that

$$\Delta^{(L/R)} \simeq (x/y)kR, \tag{9.13}$$

where

$$\begin{aligned} x = 448\varepsilon_{imt}\hat{R}_t[\{&\chi^{\beta 0}_{(ij)k}\bar{\chi}^{\alpha 0}_{(mj)o}\mu^{\alpha 0}_k\bar{\mu}^{\beta 0}_o - 3\chi^{\beta 0}_{(ij)k}\bar{\chi}^{\alpha 0}_{(mj)o}\mu^{\alpha 0}_k\bar{\mu}^{\beta 0}_p\hat{R}_o\hat{R}_p \\ &- 3\chi^{\beta 0}_{(ij)k}\bar{\chi}^{\alpha 0}_{(mj)o}\mu^{\alpha 0}_l\bar{\mu}^{\beta 0}_o\hat{R}_k\hat{R}_l + 9\chi^{\beta 0}_{(ij)k}\bar{\chi}^{\alpha 0}_{(mj)o}\mu^{\alpha 0}_l\bar{\mu}^{\beta 0}_p\hat{R}_k\hat{R}_l\hat{R}_o\hat{R}_p\} \\ &+ \{\alpha \leftrightarrow \beta\}], \end{aligned} \tag{9.14}$$

$$\begin{aligned} y = 56[\{ &- \chi^{\alpha 0}_{(ii)k}\bar{\chi}^{\alpha 0}_{(mm)o}\mu^{\beta 0}_k\bar{\mu}^{\beta 0}_o + 3\chi^{\alpha 0}_{(ii)k}\bar{\chi}^{\alpha 0}_{(mm)o}\mu^{\beta 0}_k\bar{\mu}^{\beta 0}_p\hat{R}_o\hat{R}_p \\ &+ 3\chi^{\alpha 0}_{(ii)k}\bar{\chi}^{\alpha 0}_{(mm)o}\mu^{\beta 0}_l\bar{\mu}^{\beta 0}_o\hat{R}_k\hat{R}_l - 9\chi^{\alpha 0}_{(ii)k}\bar{\chi}^{\alpha 0}_{(mm)o}\mu^{\beta 0}_l\bar{\mu}^{\beta 0}_p\hat{R}_k\hat{R}_l\hat{R}_o\hat{R}_p \\ &+ 3\chi^{\alpha 0}_{(ij)k}\bar{\chi}^{\alpha 0}_{(ij)o}\mu^{\beta 0}_k\bar{\mu}^{\beta 0}_o - 9\chi^{\alpha 0}_{(ij)k}\bar{\chi}^{\alpha 0}_{(ij)o}\mu^{\beta 0}_k\bar{\mu}^{\beta 0}_p\hat{R}_o\hat{R}_p \\ &- 9\chi^{\alpha 0}_{(ij)k}\bar{\chi}^{\alpha 0}_{(ij)o}\mu^{\beta 0}_l\bar{\mu}^{\beta 0}_o\hat{R}_k\hat{R}_l + 27\chi^{\alpha 0}_{(ij)k}\bar{\chi}^{\alpha 0}_{(ij)o}\mu^{\beta 0}_l\bar{\mu}^{\beta 0}_p\hat{R}_k\hat{R}_l\hat{R}_o\hat{R}_p \\ &- \chi^{\beta 0}_{(ii)k}\bar{\chi}^{\alpha 0}_{(mm)o}\mu^{\alpha 0}_k\bar{\mu}^{\beta 0}_o + 3\chi^{\beta 0}_{(ii)k}\bar{\chi}^{\alpha 0}_{(mm)o}\mu^{\alpha 0}_k\bar{\mu}^{\beta 0}_p\hat{R}_o\hat{R}_p \\ &+ 3\chi^{\beta 0}_{(ii)k}\bar{\chi}^{\alpha 0}_{(mm)o}\mu^{\alpha 0}_l\bar{\mu}^{\beta 0}_o\hat{R}_k\hat{R}_l - 9\chi^{\beta 0}_{(ii)k}\bar{\chi}^{\alpha 0}_{(mm)o}\mu^{\alpha 0}_l\bar{\mu}^{\beta 0}_p\hat{R}_k\hat{R}_l\hat{R}_o\hat{R}_p \\ &+ 3\chi^{\beta 0}_{(ij)k}\bar{\chi}^{\alpha 0}_{(ij)o}\mu^{\alpha 0}_k\bar{\mu}^{\beta 0}_o - 9\chi^{\beta 0}_{(ij)k}\bar{\chi}^{\alpha 0}_{(ij)o}\mu^{\alpha 0}_k\bar{\mu}^{\beta 0}_p\hat{R}_o\hat{R}_p \\ &- 9\chi^{\beta 0}_{(ij)k}\bar{\chi}^{\alpha 0}_{(ij)o}\mu^{\alpha 0}_l\bar{\mu}^{\beta 0}_o\hat{R}_k\hat{R}_l + 27\chi^{\beta 0}_{(ij)k}\bar{\chi}^{\alpha 0}_{(ij)o}\mu^{\alpha 0}_l\bar{\mu}^{\beta 0}_p\hat{R}_k\hat{R}_l\hat{R}_o\hat{R}_p\} \\ &+ \{\alpha \leftrightarrow \beta\}] \end{aligned} \tag{9.15}$$

Systems involving two centers with fixed mutual orientation have been shown to exhibit other circular differential effects with a similar linear dependence on the separation. Examples include circular differential Rayleigh and Raman scattering (Barron and Buckingham 1974; Andrews and Thirunamachandran 1978), optical rotation (Barron 1975) and two-photon circular dichroism in which only one chromophore is excited (Andrews 1976). As might be expected, on performing subsequent rotational averages for the case where A and B are randomly oriented, the odd-j terms in the rate equations vanish and consequently no circular dichroism is displayed.

C. Double-Beam Polarization Studies

In terms of polarization analysis, double-beam cooperative and double-beam distributive two-photon absorption display identical behavior since essentially the same set of polarization parameters, $A^{(j;p)}$, arises in each. The only difference is that the role of the wave-vector difference $\mathbf{u}$ in the cooperative case is played by the wave-vector sum $\mathbf{s}$ in the distributive case. For any particular laser beam and polarization geometry, the values of the nineteen $A^{(j;p)}$ which arise can be directly calculated and inserted into the relevant rate equations to give the corresponding rates. Consideration is given here to a number of polarization combinations in which the two laser beams are co-propagating. This represents the experimentally most useful configuration since it maximizes the volume of sample traversed by both beams.

The values of $A^{(j;p)}$ for seven experimentally useful combinations of plane and circular polarizations are listed in Table VII. Where both laser beams are plane polarized, the values of the polarization parameters are given for the cases where the polarization planes of the two beams are mutually parallel (column 1) or perpendicular (column 2). The first point to notice is that in

TABLE VII
Values of the Polarization Parameters $A^{(j;p)}(\hat{\mathbf{u}})$ for Seven Polarization Combinations with Co-propagating Beams

j	p	$e_1^P \parallel e_2^P$	$e_1^P \perp e_2^P$	$e_1^P; e_2^L$	$e_1^P; e_2^R$	$e_1^L; e_2^L$	$e_1^R; e_2^R$	$e_1^L; e_2^R$
0	1	1	0	$\frac{1}{2}$	$\frac{1}{2}$	0	0	1
0	2	1	1	1	1	1	1	1
0	3	1	0	$\frac{1}{2}$	$\frac{1}{2}$	1	1	0
1	1	0	0	$i/2$	$-i/2$	0	0	$-i$
1	2	0	0	0	0	$-i$	i	$-i$
1	3	0	0	$-i/2$	$i/2$	$-i$	i	0
1	4	0	0	$-i/2$	$i/2$	$-i$	i	0
1	5	0	0	$-i$	i	$-i$	i	i
1	6	0	0	$-i/2$	$i/2$	0	0	i
2	1	$-\frac{1}{3}$	0	$-\frac{1}{6}$	$-\frac{1}{6}$	0	0	$-\frac{1}{3}$
2	2	$-\frac{1}{3}$	$-\frac{1}{3}$	$-\frac{1}{3}$	$-\frac{1}{3}$	$-\frac{1}{3}$	$-\frac{1}{3}$	$-\frac{1}{3}$
2	3	$-\frac{1}{3}$	0	$-\frac{1}{6}$	$-\frac{1}{6}$	$-\frac{1}{3}$	$-\frac{1}{3}$	0
2	4	$-\frac{1}{3}$	0	$-\frac{1}{6}$	$-\frac{1}{6}$	$-\frac{1}{3}$	$-\frac{1}{3}$	0
2	5	$-\frac{1}{3}$	$-\frac{1}{3}$	$-\frac{1}{3}$	$-\frac{1}{3}$	$-\frac{1}{3}$	$-\frac{1}{3}$	$-\frac{1}{3}$
2	6	$-\frac{1}{3}$	0	$-\frac{1}{6}$	$-\frac{1}{6}$	0	0	$-\frac{1}{3}$
3	1	0	0	0	0	$i/5$	$-i/5$	$-i/5$
3	2	0	0	$-i/2$	$i/2$	$i/5$	$-i/5$	0
3	3	0	0	$-i/2$	$i/2$	0	0	$i/5$
4	1	$\frac{3}{35}$	$\frac{1}{35}$	$\frac{2}{35}$	$\frac{2}{35}$	$\frac{2}{35}$	$\frac{2}{35}$	$\frac{2}{35}$

both cases the odd-j parameters vanish, so that the leading j_0 terms represent a particularly good approximation in the near-zone. Secondly, the difference in the values of $A^{(j;p)}$ for the two cases indicates a linear dichroism which is a known characteristic of two-photon absorption (Monson and McClain 1970).

1. *Natural Two-Photon Circular Dichroism*

In each of the remaining polarization conditions, there are generally contributions from all values of j, and the rate expressions are accordingly somewhat more complex. The most interesting feature of these results is once again the appearence of circular dichroism. Columns 3 and 4 of Table VII give the values for the polarization parameters where beam 1 is plane polarized, and beam 2 is circularly polarized with either left- or right-handed helicity. While the even-j values of $A^{(j;p)}$ are the same for either handedness, the odd-j values change sign when the helicity is reversed. Hence, chiral discrimination is manifest; obviously the same remarks apply to the case where beam 1 is circularly polarized and beam 2 plane polarized. Equally comparing columns 5 and 6 illustrates a circular dichroism associated with two circularly polarized beams of the same handedness; once again reversing the helicity of the entire radiation field changes the sign of the odd-j polarization parameters. These manifestations of chirality are only observed for the case where the two centers A_1 and A_2 have a fixed mutual orientation; it is also true that the corresponding molecular properties $T^{(1;q)}$ and $T^{(3;q)}$ disappear unless the two centers are dissymmetrically juxtaposed. The result is therefore a circular dichroism associated with the well-known coupled-oscillator model of a chiral system (Barron 1982). Not surprisingly, the chirality disappears when A_1 and A_2 are free to rotate independently, since as noted above the odd-j terms then vanish.

Since the circular dichroism is generally associated with coupled groups in the near-zone range of distances, the explicit results for the Kuhn dissymmetry factors can be obtained from the leading j_1/j_0 terms in the rate expressions, and are as follows:

$$g_1 \equiv \frac{\Gamma(p,L) - \Gamma(p,R)}{\frac{1}{2}[\Gamma(p,L) + \Gamma(p,R)]} \tag{9.16}$$

$$= \frac{(\omega_1 - \omega_2)R(T^{(1;1)} - T^{(1;3)} - T^{(1;4)} - 2T^{(1;5)} - T^{(1;6)})}{c(T^{(0;1)} + 6T^{(0;2)} + T^{(0;3)})} \tag{9.17}$$

$$g_2 \equiv \frac{\Gamma(L,L) - \Gamma(R,R)}{\frac{1}{2}[\Gamma(L,L) + \Gamma(R,R)]} \tag{9.18}$$

$$= \frac{(\omega_1 - \omega_2)R(T^{(1;2)} + T^{(1;3)} + T^{(1;4)} + T^{(1;5)})}{c(2T^{(0;1)} - 3T^{(0;2)} - 3T^{(0;3)})} \tag{9.19}$$

Here $\Gamma(p, L)$ refers to the rate of absorption with beam 1 plane polarized and beam 2 left-handedly polarized, and so forth. The molecular parameters, $T^{(j;q)}$, appearing in Eqs. (9.17) and (9.19) are different for the two cases (cooperative and distributive), and their values for the particular process under study can be read off from Tables II and IV with the limiting short-range form of the interaction potential given by Eq. (8.1). Again we note the linear dependence on the group separation in Eqs. (9.17) and (9.19).

2. *Laser-Induced Circular Dichroism*

Finally, consideration is given to the results appearing in column 7 of Table VII, which apply to the situation in which the two laser beams have circular polarizations of opposite handedness. The results here differ from those in either columns 5 or 6 in each value of j. This represents the fact that changing the helicity of one beam produces a dichroism associated with a discrimination of the handedness of the A_1—A_2 pair dressed by the chirality of the other circularly polarized beam. This again is a known feature of two-photon absorption (Thirunamachandran 1979), and one which persists even when the pair is not held in a fixed mutual orientation. In this case the dissymmetry factor has both numerator and denominator given by the leading j_0 terms, and the explicit result is as follows:

$$g_1 \equiv \frac{\Gamma(L, L) - \Gamma(L, R)}{\frac{1}{2}[\Gamma(L, L) + \Gamma(L, R)]} \tag{9.20}$$

$$= \frac{5(T^{(0;3)} - T^{(0;1)})}{(T^{(0;1)} + 6T^{(0;2)} + T^{(0;3)})} \tag{9.21}$$

X. RESONANCE ENHANCEMENT

Since the rate of bimolecular mean-frequency absorption will usually be somewhat less than the rate of two-photon absorption at any single center, the various possibilities of utilizing resonance enhancement to increase the rate of the synergistic process merit serious consideration. Resonance enhancement results from the fact if there is a suitable spacing of molecular energy levels not directly involved in the excitation scheme, a substantial increase in the absorption rate can nonetheless ensue, due to the effects of dispersion in the molecular susceptibility tensors. In order to determine the conditions under which resonance enhancement occurs it is necessary to return to the definitions of the molecular response tensors $\mathbf{S}^{f0}$ and χ^{f0} given by Eqs. (5.15) and (5.16) respectively. Enhancement occurs where one or more of the denominators in these equations approaches zero, although

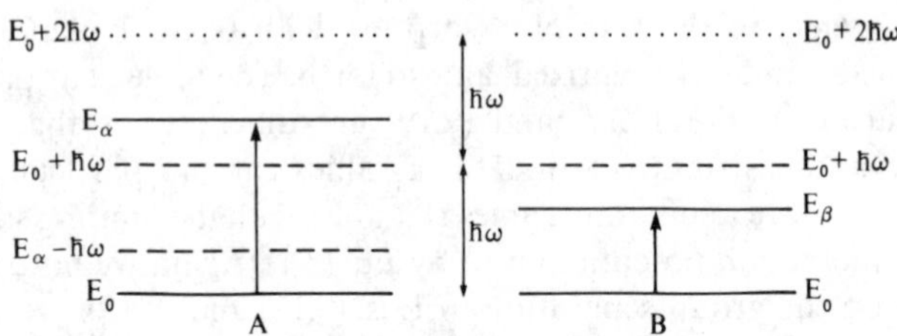

Figure 10. Resonance energy levels for single-beam synergistic absorption. The levels indicated by dashed lines represent resonances that can be exploited in absorption based on either the cooperative or the distributive mechanism; the dotted lines represent a resonance condition that applies to only the distributive case.

phenomenological damping factors, omitted here for clarity, in fact prevent infinite response at resonance.

For the single-beam cases, where the laser photon frequency is at the mean of two molecular transition frequencies, resonance enhancement occurs if either species involved in the process has an energy level matching one of those indicated in Fig. 10 by the broken lines. In the double-beam cases, the requirement for the mean of the two laser frequencies to match the molecular transition frequency provides additional freedom for one of the lasers to be tuned to one of the resonance levels shown in Fig. 11. Below we examine each case in more detail.

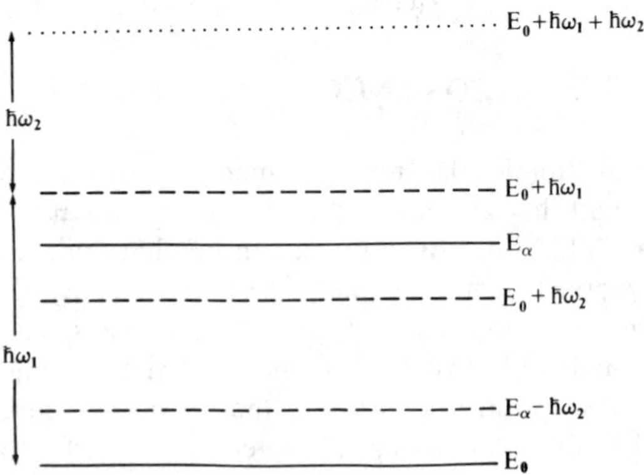

Figure 11. Resonance energy levels for two-beam synergistic absorption by pairs of chemically identical molecules. The levels indicated by dashed lines represent resonances that can be exploited in absorption based on either the cooperative or the distributive mechanism; the dotted line represents a unique resonance condition that applies to only the distributive case.

A. Single-Frequency Excitation

Considering first the cooperative mechanism, it is clear that there are two possible resonance mechanisms to consider for the tensor $S^{\alpha 0}$ as given by Eq. (5.15), corresponding to the cases where either the first or second term dominates the expression. Thus the first term dominates if there exists a state $|r\rangle$ such that $E_{\alpha r} \approx \hbar\omega$, i.e., where $E_r \approx (E_\alpha - \hbar\omega)$. Similarly, the second term dominates if there is a state such that $E_{r0} \approx \hbar\omega$, i.e., where $E_r \approx (E_0 + \hbar\omega)$. In principle, similar remarks apply to $S^{\beta 0}$, except that if $E_{\alpha 0} > E_{\beta 0}$ as in Fig. 10, then the condition $E_{\beta s} \approx \hbar\omega$ cannot be satisfied for any $|s\rangle$ if the

TABLE VIII
Resonance Conditions for Synergistic Photoabsorption[a]

Tensor	Resonant intermediate energy levels	Additional conditions	Other processes favored by position of intermediate energy level
	Single-beam cooperative mechanism		
$S^{\alpha 0}(\omega)$	$E_r \approx E_\alpha - \hbar\omega$	(if $E_\alpha > E_\beta$)	
	$E_r \approx E_0 + \hbar\omega$		Single-photon absorption
$S^{\beta 0}(\omega)$	$E_r \approx E_\beta - \hbar\omega$	(if $E_\beta > E_\alpha$)	
	$E_r \approx E_0 + \hbar\omega$		Single-photon absorption
	Single-beam distributive mechanism		
$\chi^{\alpha 0}(A, \omega, \omega)$	$E_s \approx E_\alpha - \hbar\omega$	(if $E_\alpha > E_\beta$)	
	$E_r \approx E_\alpha - 2\hbar\omega$		
	$E_r \approx E_0 + \hbar\omega$		Single-photon absorption
	$E_s \approx E_0 + 2\hbar\omega$		Two-photon absorption
	Double-beam cooperative mechanism		
$S^{\alpha 0}(\omega_1)$	$E_r \approx E_\alpha - \hbar\omega_1$	(if $\omega_1 < \omega_2$)	
	$E_r \approx E_0 + \hbar\omega_1$		Single-photon absorption
$S^{\alpha 0}(\omega_2)$	$E_r \approx E_\alpha - \hbar\omega_2$	(if $\omega_2 < \omega_1$)	
	$E_r \approx E_0 + \hbar\omega_2$		Single-photon absorption
	Double-beam distributive mechanism		
$\chi^{\alpha 0}(A, \omega_1, \omega_2)$	$E_s \approx E_\alpha - \hbar\omega_1$	(if $\omega_1 < \omega_2$)	
	$E_s \approx E_\alpha - \hbar\omega_2$	(if $\omega_2 < \omega_1$)	
	$E_r \approx E_\alpha - \hbar\omega_1 - \hbar\omega_2$		
	$E_r \approx E_0 + \hbar\omega_1$		Single-photon absorption
	$E_r \approx E_0 + \hbar\omega_2$		Single-photon absorption
	$E_s \approx E_0 + \hbar\omega_1 + \hbar\omega_2$		Two-photon absorption

[a]Positions of intermediate energy levels that produce resonances in the molecular tensors, conditions imposed by energy constraints, and the nature of competing absorption processes

initial state of B is the ground state. Hence, for the single-beam cooperative case, there are just three possibilities, as indicated by the broken lines below the E_α level in Fig. 10. Clearly if $E_{\beta 0} > E_{\alpha 0}$ there are two resonance conditions that may apply for center B, and one for A. These possibilities are summarized in Table VIII.

The two conditions where A has a state $|r\rangle$ or B has a state $|s\rangle$ with energy $\sim(E_0 + \hbar\omega)$ lead to resonance enhancement provided the corresponding transition moments μ^{r0}, μ^{s0} are non-zero. However, in such circumstances real transitions to these states are allowed by single-photon absorption, and it is likely that these effects will swamp observation of the cooperative process under consideration. The third resonance condition, where $E_{\alpha r} \approx \hbar\omega$, is of much more interest; this corresponds to having an energy level of center A at approximately $(E_\alpha - \hbar\omega)$. Here, provided the resonant level lies sufficiently far above the ground state not to be appreciably thermally populated, then there is no question of single-photon absorption from the laser beam at frequency ω, but there is nonetheless a genuine resonance amplification of the tensor $S^{\alpha 0}$. Observation of the cooperative absorption process is therefore facilitated if the energy levels of A happen to lie in positions which favor this possibility.

For the distributive mechanism, the possibilities for resonance behavior can be ascertained by reference to the form of the molecular response tensor $X^{\alpha 0}(\omega, \omega)$ as given by Eq. (5.16). As noted earlier, the same tensor is involved in hyper-Raman scattering, where a figure of 10^6 has been given as the order of enhancement under typical resonance conditions (Long and Stanton 1970). Clearly such a large increase will be a significant consideration in experimental studies. There are now, in principle, four different denominator factors to consider, two of which coincide with those of the cooperative mechanism and lead to the same resonance conditions. Although one additional type of resonance is predicted for states $|r\rangle$ of energy $(E_\alpha - 2\hbar\omega)$ or $(E_\beta - 2\hbar\omega)$, it is clear from the energy level diagram that neither condition can arise if the concerted absorption process takes place from the ground state.

The only other new type of resonance behavior specifically associated with the distributive case is one that applies to intermediate states $|s\rangle$ with energy $E_s \approx (E_0 + 2\hbar\omega)$. However, this leads to competition from simple two-photon absorption $|s\rangle \leftarrow |0\rangle$, so that again the synergistic effect would most likely be swamped, particularly if $2\hbar\omega$ exceeds the ionization energy of A or B. Thus it transpires that for single-beam synergistic absorption mediated by either mechanism, the most useful resonance condition corresponds to the case where there is a state of energy $\sim(E_\alpha - \hbar\omega)$, where there is no possibility of competition from either single-photon or two-photon absorption, and significant rate increases can be expected.

B. Two-Frequency Excitation

For two-frequency synergistic absorption, the various resonance possibilities are also summarized in Table VIII. For cooperative mean-frequency absorption, there are again three distinct possibilities for resonance enhancement of the process, as shown by the lowest three broken lines in Fig. 11. If the centers involved in the process possess energy levels close to $(E_0 + \hbar\omega_1)$ or $(E_0 + \hbar\omega_2)$, then although the molecular tensors $S^{\alpha 0}(\omega_1)$ or $S^{\alpha 0}(\omega_2)$ are resonantly enhanced, direct competition from single-photon absorption will mask the synergistic process. However, in the third resonance condition, where a molecular energy level exists close to $(E_\alpha - \hbar\omega_2)$ (assuming that ω_2 is the lower of the laser frequencies), then the first term of $S^{\alpha 0}(\omega_2)$, as given by the ω_2 analogue of Eq. (5.15), becomes resonantly enhanced and thus leads to an increased cooperative absorption rate. In this case, again provided the resonant level is not significantly populated, there is no possibility of competition from one-photon absorption.

In the two-frequency distributive case, the molecular tensor $\chi^{\alpha 0}(\omega_1, \omega_2)$ has resonance conditions similar to those for $\chi^{\alpha 0}(\omega, \omega)$. As in the single-beam case, two of the proposed resonance conditions would be likely to allow the process to be masked by single-photon absorption; a third leads to the possibility of conventional two-photon absorption, and a fourth cannot be satisfied if the centers involved are initially in their ground states. The remaining condition $(E_s \approx (E_\alpha - \hbar\omega_1)$ if $\omega_1 < \omega_2$, or $E_s \approx (E_\alpha - \hbar\omega_2)$ if $\omega_2 < \omega_1)$ remains the only truly useful resonance. Naturally, since the energetics of the excitation process are constrained only by a condition on the sum of the photon frequencies, there is a wide scope for choosing laser frequencies specifically with the aim of exploiting this type of resonance possibility.

XI. COMPARISON WITH CONVENTIONAL TWO-PHOTON ABSORPTION

The rate of conventional (single-center) two-photon absorption depends on the square of the focussed laser intensity, and as long ago as 1968 Gontier and Trahin showed that in the absence of accidental resonances an intensity factor of (I/I_o) is introduced for each additional photon involved in a multiphoton atomic excitation process. The constant I_o is a characteristic irradiance whose value depends on the sample, and corresponds to the situation where perturbation theory breaks down and all multiphoton processes become equally feasible. A similar treatment of molecules leads to an intensity factor per photon of $\gamma = (I/I_M)$, where I_M is an irradiance that would lead to ionization or dissociation, and would therefore have a typical value in the region of $10^{18\pm4}\,\mathrm{Wm}^{-2}$ (Eberly et al. 1987). This figure certainly exceeds the level of irradiance applied in most laser spectroscopy experiments, where values of γ seldom exceed 10^{-4}. Nonetheless two-photon spectra are readily obtained even with appreciably lower intensities. With this in mind, it is instructive to compare the likely rates of conventional and nearest-neighbor synergistic two-photon absorption.

Only by experiment or detailed ab initio calculations can quantitative values be obtained for the various tensor parameters involved in the rate equations. Unfortunately, there are no studies to date which have reported the necessary numerical values, although the synergistic effects are now experimentally well-documented. Estimation of the more general significance of many of the results presented in previous sections must therefore proceed from a different basis. As shown in early work on cooperative photoabsorption (Andrews and Harlow 1983), neighboring molecules can in fact be expected to display a synergistic absorption rate approaching the rate of two-photon absorption by individual molecules, a result which is more readily calculated. This can be argued as follows. A comparison of the short-range limit of the rate equation for cooperative absorption and the corresponding rate equation for normal two-photon absorption shows that the former contains an additional factor of the order of $\rho = S^{\alpha 0}/R^3$. Far from accidental resonances, the molecular tensor should be similar in magnitude to the polarizability, since it is constructed in the same way from products of electric dipole transition moments divided by energy mismatch factors. Molecular polarizabilities, at least for small molecules, have well-documented values, and are mostly similar in magnitude to the cube of molecular diameter. Hence when R represents a nearest-neighbor distance, the factor ρ approaches the value of unity, and the cooperative absorption rate is comparable with that of conventional two-photon absorption. Similar arguments apply in the case of distributive absorption.

Obviously, any possibilities of resonance, as discussed in Section X, can further enhance the synergistic photoabsorption rates. While most of the appropriate resonance conditions are held in common with single-center two-photon absorption, the case of resonance at a level one photon in energy below the final excited state of either molecule participating in a synergistic process is a unique feature. This not only opens up important new possibilities for rate enhancement, but it does so without the associated complications of competing absorption processes. Thus it appears that synergistic effects should be generally observable at the levels of laser irradiance typically employed for studies of conventional two-photon absorption, and it should not be necessary to utilize exceptionally intense laser sources where higher-order optical nonlinearities might become a problem.

XII. SYNERGISTIC EFFECTS IN THE ABSORPTION OF WHITE LIGHT

A. Spectrophotometry with Broadband Sources

Having established the detailed theory underlying synergistic effects in two-photon laser spectroscopy, we now consider the broader implications of these effects in connection with conventional absorption processes. Most modern spectrophotometry and color science is based on the principle that the optical response of a substance to a given wavelength of light is independent of any other wavelengths that may be present. Thus it is normally assumed that an absorption spectrum obtained using broadband light and a multichannel spectrometer would be identical to the spectrum obtained using a tunable monochromatic light source. Indeed, this is the principle underlying Fourier transform spectroscopy.

Using results established in previous sections, it can however be shown that this assertion represents only an approximation to the truth. Absorption from a white light or other broadband source in fact allows photon pairs of differing frequencies to be concertedly absorbed by molecules in close proximity (Andrews 1988). In this section, it is shown that this may result in a change to the appearance of absorption spectra, and yield an absorption law that departs from normal Beer–Lambert behavior. One particular case in which such effects may be expected to arise is where supercontinuum laser radiation is employed for spectroscopic purposes.

The generation of a white-light supercontinuum by passing mode-locked pulses of laser light through certain media was first reported in 1970 (Alfano and Shapiro 1970). The phenomenon results from a process of self-phase modulation associated with intensity-dependent refraction, although a number of other mechanisms can contribute to the effect; a useful summary is provided in a recent review by Alfano (1986). Continuum generation has been shown to occur in a wide variety of materials, and is readily producible in water. The pulses of light so generated are often referred to as constituting an ultrafast supercontinuum laser source (USLS) (Manassah et al. 1984), or picosecond continuum for short, since pulse durations are typically on the picosecond or femtosecond (Fork et al. 1983) timescale. The term *superbroadening* is also used to describe the continuum formation (Reintjes 1984).

The laser supercontinuum source has found numerous applications in the physical, chemical, and biological sciences. Many studies have concerned elementary photochemical and photobiological reactions such as those involved in the primary processes of photosynthesis and vision (von der

Linde 1977; Eisenthal 1977; Peters and Leontis 1982; Fleming 1986). These studies are mostly based on use of the continuum to probe the absorption characteristics of transient species, in order to obtain information on their decay kinetics. The processing of data from such experiments is generally based on the implicit assumption that the absorption of white light is subject to the normal Beer–Lambert law, in that the absorption at any particular frequency is assumed to be linearly proportional to the intensity of the probe light at that frequency.

B. Rate Equations for Continuum Excitation

To place into proper perspective the role of synergistic effects in the absorption of white light, it is worth first setting down the basic equations for the normal absorption process observed with monochromatic light. Consider an ensemble of molecules in an initial state $|i\rangle$, certain of which are promoted to an excited state $|f\rangle$ through absorption of light with circular frequency ω_0, i.e., we have $E_{fi} = \hbar\omega_0$. Assuming that the transition is electric dipole-allowed, the rate of (single-photon) absorption is given by;

$$\Gamma_1 = (\hbar^2 c\varepsilon_0)^{-1}\pi K_1 I(\omega_0) \qquad (12.1)$$

where $I(\omega_0)$ is defined as the irradiance per unit circular frequency (ω) interval at frequency ω_0, and K_1 is given by

$$K_1 = |\boldsymbol{\mu}^{fi}\cdot\mathbf{e}|^2 \qquad (12.2)$$

$\boldsymbol{\mu}^{fi}$ being the transition dipole moment for the $|f\rangle \leftarrow |i\rangle$ transition, and $\mathbf{e}$ the unit polarization vector of the incident light.

The result of Eq. (12.1) is more often expressed in terms of the Einstein B-coefficient; casting the result in terms of radiant energy density per unit frequency (ω) interval introduces an additional factor of $c/2\pi$, and a further factor of $\frac{1}{3}$ results from rotational averaging. However, to facilitate subsequent comparison with the rate equation for cooperative absorption, the above result is given in the form it takes prior to rotational averaging. The single most important feature to note at this stage is the linear dependence of the absorption rate on the irradiance, a dependence manifest in the characteristic exponential decay of intensity with time and hence also with distance travelled through the sample (the Beer–Lambert Law).

For cooperative absorption, two photons with frequencies $\omega = (\omega_0 + \Omega)$ and $\omega' = (\omega_0 - \Omega)$, the sum of whose energies equals the sum of the $|f\rangle \leftarrow |i\rangle$ transition energies for two different molecules, are absorbed in the concerted process illustrated by Fig. 2. The rate of absorption by the pair is essentially that given by Eq. (6.9), recast in terms of an irradiance with a large bandwidth:

$$\Gamma_{2c} = (2\hbar^2 c^2 \varepsilon_0^2)^{-1}\pi \int_0^\infty K_2(\omega_0, \Omega) I(\omega_0 + \Omega) I(\omega_0 - \Omega) d\Omega \tag{12.3}$$

where

$$K_2(\omega_0, \Omega) = |e_i e_j S_{ik}^{fi}(\omega_0 + \Omega) S_{jl}^{fi}(\omega_0 - \Omega)\{V_{kl}(\Omega, \mathbf{R}) + \bar{V}_{kl}(\Omega, \mathbf{R}) \exp(i\Delta\mathbf{k}\cdot\mathbf{R})\}|^2 \tag{12.4}$$

The result is simplified by the physically reasonable assumption that the two absorbed photons have the same polarization.

To the extent that the parameter $K_2(\omega_0, \Omega)$ is approximately frequency-independent over any range of frequencies well away from resonance (in other words, whenever dispersion effects are small), the rate contribution Γ_{2c} has a direct dependence on the frequency-domain autocorrelation function of the incident light. Since USLS radiation is pulsed, it is useful to express the result in terms of the time-dependence of the irradiance $I(t)$ through the Fourier transform

$$I(\omega) = (2\pi)^{-1/2} \int_{-\infty}^{\infty} I(t) \exp(i\omega t) dt \tag{12.5}$$

Simple manipulation of the integral in Eq. (12.3) then reveals its equivalence to the time-domain integral

$$K_2(\omega_0) \int_{-\infty}^{\infty} I^2(t) \exp(2i\omega_0 t) dt \tag{12.6}$$

In the distributive mechanism two photons with frequencies $(\omega_0 + \Omega)$ and $(\omega_0 - \Omega)$ undergo concerted absorption at the same center, and the energy mismatch E_{fi} is conveyed to another molecule by virtual photon coupling, as in Fig. 3. In this case, using exactly similar methods, the following rate equation is obtained:

$$\Gamma_{2d} = (2\hbar^2 c^2 \varepsilon_0^2)^{-1}\pi \int_0^\infty K_{2'}(\omega_0, \Omega) I(\omega_0 + \Omega) I(\omega_0 - \Omega) d\Omega \tag{12.7}$$

where

$$K_{2'}(\omega_0, \Omega) = |e_i e_j \chi_{(ij)k}^{fi}(\omega_0 + \Omega, \omega_0 - \Omega) \mu_l^{fi} V_{kl}(\omega_0, \mathbf{R})\{1 + \exp(i\mathbf{u}\cdot\mathbf{R})\}|^2 \tag{12.8}$$

and once again a link can be established with the autocorrelation function of the USLS light.

The Beer–Lambert exponential decay law for conventional (single-photon) absorption results from the elementary relation

$$-dI(\omega, z)/dz \propto I(\omega, z) \tag{12.9}$$

where z represents the distance the light has travelled through the absorbing sample. Since this is directly proportional to the propagation time within the sample, Eq. (12.9) is a result which follows directly from Eq. (12.1). When intense continuum light such as that provided by USLS radiation is absorbed, cooperative and distributive processes produce a correction term that necessitates the replacement of Eq. (12.9) by a result of the form

$$-dI(\omega, z)/dz \propto [I(\omega, z) + \chi \int K(\omega, \Omega) I(\omega + \Omega, z) I(\omega - \Omega, z) d\Omega] \tag{12.10}$$

where $K = K_2 + K_{2'}$. Clearly, in this case exponential decay is no longer to be expected.

One of the most significant implications of the result is that an absorption spectrum measured with intense white light may be significantly different from the spectrum that would be observed using tunable monochromatic radiation. In particular, there should be a decrease in the apparent width of many lines in any absorption spectrum measured with broadband radiation. This is because, for any sample transition of frequency ω_0, photons of appreciably off-resonant frequency $(\omega_0 \pm \Omega)$ can be cooperatively absorbed and result in the excitation of two separate molecules, provided selection rules permit. In fact the Lorentzian linewidth of the concerted absorption process is readily shown to be approximately $0.64 \times$ the ordinary absorption linewidth, if the probe radiation is assumed to be of nearly constant intensity in the frequency region of interest. Nonetheless, the observed linewidth would not be reduced to quite this extent, because of the additional and invariably stronger response associated with normal single-photon absorption.

C. Implications for Spectroscopy with USLS Radiation

It is difficult in general terms to assess the magnitude of the correction represented by the frequency integral, although under optimum conditions it may indeed be comparable with the first term in the square brackets (Andrews 1988). The most significant feature of Eq. (12.10) is undoubtedly the fact that the absorption by a sample at any given frequency is directly influenced by the intensity of light at other frequencies. Equation (12.10) thus represents an infinite set of coupled integro-differential equations, whose solution depends on the detailed spectral distribution of the USLS continuum light (Manassah et al. 1985a, b, 1986; Manassah 1986) and also the spectral response of the sample, as represented by the all-embracing constant K.

Precise values for the focused irradiances produced by USLS pulses are not currently available in the literature, but the intensity levels are certainly high enough for two-photon absorption to be experimentally observable, and by extension it appears that the synergistic process may produce important contributions to conventional single-photon absorption spectra. Other factors that contribute to the significance of cooperative or distributive two-photon absorption are the involvement of non-neighboring molecules, and any enhancement of the molecular response tensors through incidental one-photon resonances, as discussed in Section X. In many situations, it is therefore likely that mean-frequency absorption will play a significant role in modifying the apparent form of absorption spectra.

As noted above, the use of USLS light for probing absorption may result in a modified linewidth in the spectrum. Other more significant changes may also be expected, however. This can be illustrated as follows. Consider the case of an electronic transition which displays vibronic structure associated with a certain molecular vibration. For simplicity, let us confine attention to the (0–0) band, assuming that the vibrational frequencies in the ground and excited electronic states are similar. Although cooperative and distributive processes allow the absorption of any pair of photons whose energy sum equals that of the (0–0) excitation energies for two different molecules, the rate of each process is resonantly enhanced if either photon energy matches that of another transition.

In particular, a pair of photons whose energies match the (0–1) and (1–0) transitions can be cooperatively absorbed and so actually result in (0–0) transitions in two separate molecules. Thus, because of the resonance enhancement associated with a (0–1)-frequency photon, one should expect increased absorption at both the (0–1) and (1–0) frequencies, even if the $v = 1$ level in the electronic ground state is not appreciably populated. Features of this kind have been noted in recent USLS-probe experiments on spectral hole burning in dye solutions (Alfano et al. 1974). Although it is unlikely that cooperative absorption has any direct bearing on these studies in view of the large mean separation of the dye molecules in solution, it might be expected to become much more significant in studies of molecular crystals, for example.

In assessing the wider significance of mean-frequency absorption for flash photolytic experiments based on USLS radiation, perhaps the most important factor to consider is the enormously wide range of possibilities for synergistic absorption leading to the simultaneous excitation of more than one excited state. For example, with a single-component sample, it should be possible to observe the process

$$2\mathrm{A} + \hbar\omega + \hbar\omega' \rightarrow \mathrm{A}^* + \mathrm{A}^\ddagger \tag{12.11}$$

the double dagger denoting some other excited state than that denoted by the asterisk. If the sample is heterogeneous or contains more than one chemical species, there exists the even more general possibility of simultaneously exciting two chemically different species:

$$A + B + \hbar\omega + \hbar\omega' \rightarrow A^* + B^* \tag{12.12}$$

The reaction represented by Eq. (12.12) is potentially very significant for flash photolytic studies of processes in complex biological systems, where A and B may even be chemically different chromophores within a single large biomolecule. The theory underlying these processes can be developed from the basic results in Section V, and the associated rates of absorption should be comparable to those of the closely related synergistic absorption processes discussed earlier in this review. However, in polyatomic molecules with complex vibronic structures, the number of pairs of transitions that can be excited through absorption of two photons with the correct energy sum may be enormously large, so that cooperative absorption may exert a very significant effect on the appearance of the absorption spectrum. The numerous possibilities for resonance enhancement of the molecular response tensors at certain frequencies should also be borne in mind.

In conclusion, it is worth reiterating that the anomalous absorption effects described here may be manifest in any experiments that employ sufficiently high-intensity broadband radiation. To this extent, anomalies may be observable in experiments not specifically involving USLS light. In particular, the continued advances in techniques of laser pulse compression have now resulted in the production of femtosecond pulses only a few optical cycles in duration (Knox et al. 1985; Brito Cruz et al. 1987; Fork et al. 1987) which necessarily have a very broad frequency spread, as the time/energy uncertainty principle shows. Thus, mean-frequency absorption may have a wider role to play in the absorption of femtosecond pulses. If this is correct, it raises further questions over the suitablity of absorption-based techniques for their characterization.

XIII. CHEMICAL ASPECTS OF SYNERGISTIC PHOTOABSORPTION

A. General Considerations

Although the most thoroughly documented studies of synergistic two-photon absorption relate to matrix isolation studies, the process should certainly be manifest in other phases of matter. For samples in either a gas or liquid state, it might be argued that such effects are essentially collisional phenomena. It is certainly true that the probability of any such synergistic process is very significantly enhanced for pairs of molecules separated by small distances, a dependence which must be associated with a high degree of pressure sensitivity. However, although these processes are predominantly effective over short distances, they are nonetheless induced by proximity rather than collision. This much is made clear by the fact that irrespective of any energy shifts or selection rule weakening which collisions might induce, the absorption processes that occur at each center are separately forbidden on energy grounds: moreover, there are significant contributions from molecules in the far-zone region. By the same token, orbital overlap need not be implicated in synergistic processes involving chromophore pairs.

The synergistic processes described in this review can potentially be manifest in a very wide range of chemical and physical effects, and we now turn to a consideration of some of the broader implications. Although there has been tacit assumption through most of our discussion that the excitation of the two centers is electronic or vibronic in nature, purely vibrational excitation is also possible if suitable wavelengths are employed, and suitable Born–Oppenheimer development of the molecular tensors enables the theory to be directly applied to such cases. In the case where both centers gain one quantum of vibrational energy, then the irradiation frequency for the synergistic single-beam process is the mean of these two vibrational frequencies. Vibrational modes that can participate in this process by the cooperative mechanism must be Raman–allowed by virtue of the two-photon selection rules; in the distributive case, the vibrations must be infrared-allowed. In both cases, the limiting near-zone behavior applies up to distances comparable to infrared wavelengths. In the case of two-beam excitation, similar remarks apply.

In considering the single-beam cases, moreover, we have concentrated on the case where A and B differ, even if only by isotopic constitution. However, the theory can also apply to the case where A and B are identical, but the excited states $|\alpha\rangle$ and $|\beta\rangle$ differ. For a single-component phase, the effect should be observable as the absorption of two photons with energy halfway between that of the two excited states; this applies equally to free molecules or loosely associated van der Waals dimers. For a two-component phase containing a mixture of A and B molecules, it should thus be possible to observe synergistic effects associated with not only A–B, but also A–A and B–B interactions, and hence several new bands should appear in the two-photon absorption spectrum.

B. Methods of Observation

We now consider some of the methods by which the synergistic effects described above may be observed. As with conventional two-photon absorption, direct monitoring of a reduction in laser intensity as a result of the photoabsorption process is unlikely to be the most practicable method for studying the effect. However, fluorescent decay from any of the excited states is much more easily measured since it should be against a zero background; note that fluorescence from only one excited species is normally sufficient to demonstrate the occurrence of a synergistic process. For the sake of argument, let us suppose that the excitation of the state $|\alpha\rangle$ is monitored. Radiationless decay within the vibrational manifold of the initially populated electronic state will generally mean that the fluorescence will be Stokes-shited in frequency away from $E_{\alpha 0}/\hbar$. Nevertheless, since the same excited state $|\alpha\rangle$ is also accessible either through single-photon absorption at frequency $E_{\alpha 0}/\hbar$ (in the distributive case), or through single-beam two-photon absorption at frequency $E_{\alpha 0}/2\hbar$ (in the cooperative case), its decay characteristics can be ascertained in a separate single-beam experiment and then used to monitor the synergistic effect.

Other methods can be devised according to the photochemistry of the sample, and as an example we consider the case of a mixture of H_2CO and D_2CO. These molecules have absorption lines associated with the $2_0^1 4_0^3$ transition at 30,340.15 and 30,147.62 cm^{-1}, respectively (Moule and Walsh 1975). Irradiation with a single beam of laser light at the mean of these two frequencies should thus lead to a synergistic two-photon process, involving both the cooperative and distributive channels, in which both species are simultaneously excited. The narrow bandwidth of any standard laser source should ensure that neither species is independently excited by a conventional single-photon absorption process. Evidence for cooperative two-photon absorption is then provided by detection of the decomposition product CO resulting from the reactions:

$$H_2CO^* - \begin{cases} \rightarrow H_2 + CO \\ \rightarrow H + H + CO \end{cases} \qquad (13.1)$$

$$D_2CO^* - \begin{cases} \rightarrow D_2 + CO \\ \rightarrow D + D + CO \end{cases} \qquad (13.2)$$

The experimental verification of this reaction scheme remains one of the main challenges presented by the theory of synergistic photoabsorption.

C. Charge Transfer Reactions

Most of the experimental work on synergistic photoabsorption has concerned the studies of charge transfer between molecular halogens and rare gas atoms. This type of two-photon induced process can in general be expressed as

$$X_2 + Rg + 2\hbar\omega \rightarrow Rg^+X_2^- \qquad (13.3)$$

The subsequent reactions

$$Rg^+X_2^- \rightarrow Rg^+X^- + X \tag{13.4}$$

$$Rg^+X^- + Rg \rightarrow Rg_2^+X^- \tag{13.5}$$

generally lead to exciplexic emission which characterizes completion of the reaction, as illustrated in Fig. 12. Reactions of this type have been studied in gas-phase experiments (Yu et al. 1983; Ku et al. 1983), and in $Rg:X_2$ van der Waals complexes generated in free jet expansions (Boivineau et al. 1986; Jouvet et al. 1987). However, the most extensive studies have been performed by Apkarian and co-workers in condensed phases, involving both liquid solutions and doped rare gas matrices (vide infra). The broad absorption by halogens in the UV-visible range offers plenty of scope for the exploitation of single-photon resonances to enhance the prospects of observing synergistic effects. The advantage is offset by the fact that these resonances are mostly associated with dissociative transitions to turning points of strongly repulsive potentials, resulting in uncorrelated absorption processes that complicate the experiments. Careful analyses have been required to prove unequivocally the

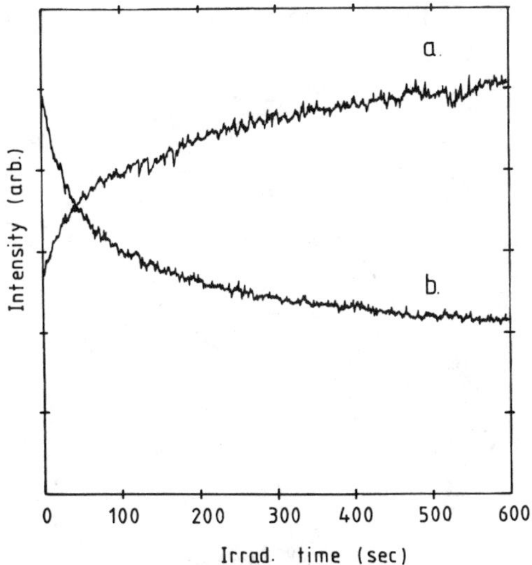

Figure 12. Permanent dissociation of Cl_2 in solid xenon irradiated at 308 nm, as a function of irradiation time (1:500 Cl_2:Xe solid at 13 K). (a) Growth of the 573 nm $Xe_2{}^+Cl^-$ $(4^2\Gamma \rightarrow 1, 2^2\Gamma)$ exciplexic emission intensity. (b) Decrease of the Cl_2 $(A' \rightarrow X)$ recombinant emission intensity at 800 nm in the same sample. Redrawn from Fajardo, Withnall et al. (1988) by permission of Harwood Academic Publishers GmbH.

existence of a synergistic mechanism for the charge transfer. Various nonsynergistic mechanisms can dominate under different conditions; the task has been to demonstrate the occurrence of charge transfer even under conditions where these alternative mechanisms cannot apply.

The first possibility to eliminate is the absorption of a second photon by the halogen in its photoexcited state, followed by charge transfer:

$$X_2^* + Rg \rightarrow Rg^+X_2^- \tag{13.6}$$

Working with F_2, Cl_2, and Br_2 in wavelength regions where such resonances are absent, it has nonetheless been shown that the charge transfer reactions persist and cannot be accounted for by two-photon absorption in the free halogens (Fajardo and Apkarian 1988a, b). A second mechanism which could contribute is photodissociation of the molecular halogen followed by photo-induced charge transfer between the atomic species:

$$X + Rg + \hbar\omega \rightarrow Rg^+X^- \tag{13.7}$$

This latter mechanism can be ruled out in a number of ways through judicious choice of experimental conditions, as most notably demonstrated by studies of the photo-induced harpoon reactions of ICl in liquid and solid xenon (Okada et al. 1989). Here it has been shown that XeI is produced even at wavelengths well beyond the threshold for atomic charge transfer between

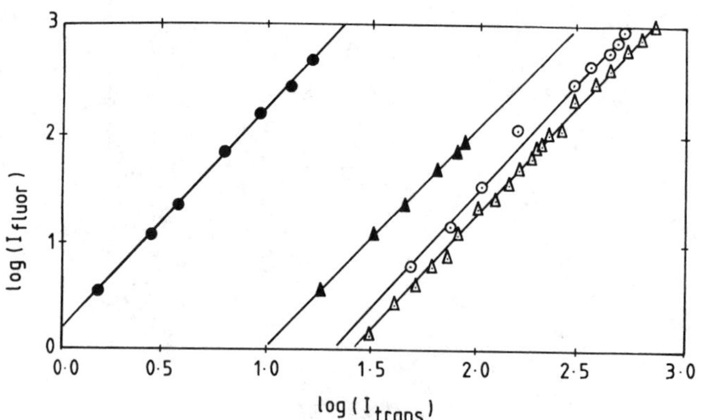

Figure 13. Logarithmic plots of exciplexic fluorescence intensities versus transmitted laser intensity in a 1.3 mM solution of Cl_2 in Xe, at various wavelengths: 340 nm (solid circles), 360 nm (solid triangles), 308 nm (open circles), 368 nm (open triangles). The corresponding slopes of the best fit lines are 2.07, 1.98, 2.11, and 2.08, respectively. Redrawn from Fajardo, Withnall et al. (1988) by permission of Harwood Academic Publishers GmbH.

the rare gas and halogen atoms. It transpires, however, that resonances associated with the final state in ICl arise in accordance with the operation of a synergistic mechanism.

The explicit characterization of synergistic two-photon absorption in liquid solutions has recently been described by Fajardo, Withnall et al. (1988). Here a clear distinction from the effects of any sequential absorption has been made on the basis of kinetic considerations. In particular, the intensity of fluorescence by the exciplex Xe^+Cl^- has been shown to depend quadratically on laser fluence over a range of wavelengths, exactly as one expects with a coherent two-photon process. Any sequential mechanism would only be expected to produce this kind of dependence if the cross-sections for photodissociation and atomic charge transfer were similar in magnitude. In fact, both cross-sections vary by at least an order of magnitude over the wavelength range studied, yet the dependence on fluence satisfies a power law with exponent 2.05 ± 0.7 throughout, as shown in Fig. 13. Synergistic two-photon absorption is thus the only mechanism that can satisfactorily account for all the experimental observations.

Acknowledgement

We would like to express our thanks to V. A. Apkarian for his detailed communication of recent experimental results on synergistic effects in photo-induced charge transfer.

References

Alfano, R. R., 1986, in *Proc. Int. Conf. Lasers' 85*, C. P. Wang, Ed. (STS Press, McLean, Virginia) pp. 110–122.

Alfano, R. R. and S. L. Shapiro, 1970, *Phys. Rev. Lett.* **24**, 585–587.

Alfano, R. R., J. I. Gersten, G. A. Zawadzkas, and N. Tzoar, 1974, *Phys. Rev.* **A10**, 698–708.

Andrews, D. L., 1976, *Chem. Phys.* **16**, 419–424.

Andrews, D. L., 1984, *Mol. Phys.* **52**, 969–972.

Andrews, D. L., 1985, *Am. J. Phys.* **53**, 1001–1002.

Andrews, D. L., 1988, *Phys. Rev.* **A38**, 5129–5139.

Andrews, D. L., 1989, *Phys. Rev.* **A40**, 3431–3433.

Andrews, D. L. and W. A. Ghoul, 1982, *Phys. Rev.* **A25**, 2647–2657.

Andrews, D. L. and M. J. Harlow, 1983, *J. Chem. Phys.* **78**, 1088–1094.

Andrews, D. L. and M. J. Harlow, 1984a, *J. Chem. Phys.* **80**, 4753–4760.

Andrews, D. L. and M. J. Harlow, 1984b, *Phys. Rev* **A29**, 2796–2806.

Andrews, D. L. and K. P. Hopkins, 1987, *J. Chem. Phys.* **86**, 2453–2459.

Andrews, D. L. and K. P. Hopkins, 1988a, *J. Mol. Struc.* **175**, 141–146.

Andrews, D. L. and K. P. Hopkins, 1988b, *J. Chem. Phys.* **89**, 4461–4468.

Andrews, D. L. and B. S. Sherborne, 1987, *J. Chem. Phys.* **86**, 4011–4017.

Andrews, D. L. and T. Thirunamachandran, 1977, *J. Chem. Phys.* **67**, 5026–5033.

Andrews, D. L. and T. Thirunamachandran, 1978a, *Proc. Roy. Soc. Lond.* **A358**, 297–310.

Andrews, D. L. and T. Thirunamachandran, 1978b, *Proc. Roy. Soc. Lond.* **A358**, 311–319.

Andrews, D. L. and P. J. Wilkes, 1985, *J. Chem. Phys.* **83**, 2009–2014.

Arnold, S., W. B. Whitten, and A. C. Damask, 1970, *J. Chem. Phys.* **53**, 2878–2884.

Au, C. K., 1988, *Phys. Rev.* **A38**, 7–12.

Avakian, R. and R. E. Merrifield, 1968, *Mol. Cryst.* **5**, 37–77.

Barron. L. D., 1975, *J. Chem. Soc. Faraday Trans.* 2, **71**, 293–300.

Barron, L. D., 1982, *Molecular Light Scattering and Optical Activity* (Cambridge University Press, Cambridge) pp. 249–263.

Barron, L. D. and A. D. Buckingham, 1974, *J. Am. Chem. Soc.* **96**, 4769–4773.

Barron, L. D. and C. J. Johnston, 1987, *Mol. Phys.* **62**, 987–1001.

Boivineau, M., J. leCalvé, M. C. Castex, and C. Jouvet, 1986a, *Chem. Phys. Lett.* **128**, 528–531.

Boivineau, M., J. leClavé, M. C. Castex, and C. Jouvet, 1986b, *J. Chem. Phys.* **84**, 4712–4713.

Boys, S. F., 1934, *Proc. Roy. Soc. Lond.* **A144**, 655–674.

Brechignac, C., Ph. Cahuzac, and P. E. Toschek, 1980, *Phys. Rev.* **A21**, 1969–1974.

Brito Cruz, C. H., R. L. Fork, and C. V. Shank, 1987, *Conf. Lasers and Electro-Optics* (Baltimore, Maryland).

Casimir, H. B. G. and D. Polder, 1948, *Phys. Rev.* **73**, 360–372.

Compagno, G., R. Passante, and F. Persico, 1983, *Phys. Lett.* **98A**, 253–255.

Compagno, G., F. Persico, and R. Passante, 1985, *Phys. Lett.* **112A**, 215–219.

Craig, D. P. and T. Thirunamachandran, 1982, *Adv. Quantum Chem.* **16**, 97–160.

Craig, D. P. and T. Thirunamachandran, 1984, *Molecular Quantum Electrodynamics* (Academic, New York).

Craig, D. P. and S. H. Walmsley, 1964, *Excitons in Molecular Crystals* (Benjamin, New York).

Danielson, R. E., 1974, *Astrophys. J.* **192**, L107–L110.

Débarre, A. and Ph. Cahuzac, 1986, *J. Phys.* **B19**, 3965–3973.

Dexter, D. L., 1962, *Phys. Rev.* **126**, 1962–1967.

Dodge, W. R., 1985, *Nucl. Instrum. Meth. Phys. Res.* **B10/11**, 423–431.

Eberly, J. H., P. Maine, D. Strickland, and G. Mourou, 1987, *Laser Focus* **23** (10), 84–90.

Eisenthal, K. B., 1977, in *Ultrashort Light Pulses—Picosecond Techniques and Applications*, S. L. Shapiro, Ed. (Springer-Verlag, Berlin) pp. 275–315.

Fajardo, M. E. and V. A. Apkarian, 1986, *J. Chem. Phys.* **85**, 5660–5681.

Fajardo, M. E. and V. A. Apkarian, 1987, *Chem. Phys. Lett.* **134**, 55–59.

Fajardo, M. E. and V. A. Apkarian, 1988a, *J. Chem. Phys.* **89**, 4102–4123.

Fajardo, M. E. and V. A. Apkarian, 1988b, *J. Chem. Phys.* **89**, 4124–4136.

Fajardo, M. E., R. Withnall, J. Feld, F. Okada, W. Lawrence, L. Wiedeman, and V. A. Apkarian, 1988, *Laser Chem.* **9**, 1–26.

Feynman, R. P., 1949, *Phys. Rev.* **76**, 769–789.

Feynman, R. P., 1961, *Quantum Electrodynamics* (Benjamin/Cummings, Reading, Massachusetts).

Finkel, R. W., 1987, *Phys. Rev.* **A35**, 1486–1489.

Fleming, G. R., 1986, *Chemical Applications of Ultrafast Spectroscopy* (Oxford University Press, Oxford) p. 80.

Fork, R. L., C. V. Shank, C. Hirlimann, R. Yen, and W. J. Tomlinson, 1983, *Opt. Lett.* **8**, 1–3.

Fork, R. L., C. H. Brito Cruz, P. C. Becker, and C. V. Shank, 1987, *Opt. Lett.* **12**, 483–485.

Förster, Th., 1949, *Z. Naturforsch.* **4A**, 321–327.

Gallagher, A. and T. Holstein, 1977, *Phys. Rev.* **A16**, 2413–2431.

Gedanken, A. and M. Tamir, 1988, *Rev. Sci. Instrum.* **58**, 950–952.

Geltman, S., 1976, *J. Phys.* **B9**, L569–L574.

Geltman, S., 1987, *Phys. Rev.* **A35**, 3775–3783.

Gontier, Y. and M. Trahin, 1968, *Phys. Rev.* **172**, 83–87.

Green, W. R., J. Lukasik, J. R. Willison, M. D. Wright, J. F. Young, and S. E. Harris, 1979, *Phys. Rev. Lett.* **42**, 970–973.

Groff, R. P., P. Avakian, and R. E. Merrifield, 1970, *Phys. Rev.* **B1**, 815–817.

Grossel, P. H., J. M. Vigoureux, and D. Van Labeke, 1983, *Phys. Rev.* **A28**, 524–531.

Haken, H., 1977, *Synergetics* (Springer-Verlag, Berlin).

Haken, H., 1985, *Light Vol. 2: Laser Light Dynamics* (North-Holland, Amsterdam) p. 9.

Jouvet, C., M. Boivineau, M. C. Duval, and B. Soep, 1987, *J. Phys. Chem.* **91**, 5416–5422.

Ketelaar, J. A. A., 1959, *Spectrochim. Acta* **14**, 237–248.

Kirkwood, J.G., 1937, *J. Chem. Phys.* **5**, 479–491.

Knox, W. H., R. L. Fork, M. C. Downer, R. H. Stoler, C. V. Shank, and J. A. Valdmanis, 1985, *Appl. Phys. Lett.* **46**, 1120–1121.

Ku, J. K., G. Inoue, and D. W. Setser, 1983, *J. Phys. Chem.* **87**, 2989–2993.

Kuhn, W., 1930, *Trans. Faraday Soc.* **26**, 293–308.

Last, I., Y. S. Kim, and T. F. George, 1987, *Chem. Phys. Lett.* **138**, 225–230.

Locke, R. J. and E. C. Lim, 1987, *Chem. Phys. Lett.* **134**, 107–109.

Long, D. A. and L. Stanton, 1970, *Proc. Roy. Soc. Lond.* **A318**, 441–457.

Loudon, R., 1983, *The Quantum Theory of Light*, 2nd. edition (Oxford University Press, Oxford), pp. 105–111.

Louisell, W. H., 1973, *Quantum Statistical Properties of Radiation* (Wiley, New York), pp. 176–180.

Manassah, J. T., 1986, *Phys. Lett.* **A117**, 5–9.

Manassah, J. T., P. P. Ho, A. Katz, and R. R. Alfano, 1984, *Photonics Spectra* **18**, (11), 53–59.

Manassah, J. T., R. R. Alfano, and Mustafa, M. A. 1985a, *Phys. Lett.* **107A**, 305–309.

Manassah, J. T., M. A. Mustafa, R. R. Alfano, and P. P. Ho, 1985b, *Phys. Lett.* **113A**, 242–247.

Manassah, J. T., M. A. Mustafa, R. R. Alfano, and P. P. Ho, 1986, *IEEE J. Quantum Electron.* **QE-22**, 197–204.

Meath, W. J. and E. A. Power, 1987, *J. Phys.* **B20**, 1945–1964.

Mohler, C. E. and M. J. Wirth, 1988, *J. Chem. Phys.* **88**, 7369–7375.

Monson, P. R. and W. M. McClain, 1970, *J. Chem. Phys.* **53**, 29–37.

Moule, D. C. and A. D. Walsh, 1975, *Chem. Rev.* **75**, 67–84.

Nakazawa, E. and S. Shionoya, 1970, *Phys. Rev. Lett.* **25**, 1710–1712.

Nosworthy, J. and J. P. Keene, 1964, *Proc. Chem. Soc.* 114.

Okada, F., L. Wiedeman, and V. A. Apkarian, 1989, *J. Phys. Chem.* **93**, 1267–1272.

Parker, C. A. and C. G. Hatchard, 1962, *Proc. Roy. Soc. Lond.* **A269**, 574–584.

Peters. K. S. and N. Leontis, 1982, in *Biological Events Probed by Ultrafast Laser Spectroscopy*, R. R. Alfano, Ed. (Academic, New York) pp. 259–269.

Power, E. A., 1975, *J. Chem. Phys.* **63**, 1348–1350.

Power, E. A. and T. Thirunamachandran, 1983, *Phys. Rev.* **A28**, 2671–2675.

Power, E. A. and S. Zienau, 1959, *Philos. Trans. Roy. Soc. London Ser.* **A251**, 427–454.

Reintjes, J. F., 1984, *Nonlinear Optical Parametric Processes in Liquids and Gases* (Academic, Orlando, Florida) pp. 355–357.

Rios Leite, J. R. and C. B. De Araujo, 1980, *Chem. Phys. Lett.* **73**, 71–74.

Sasaki, A. and S. Hayakawa, S. 1978, *Jpn. J. Appl. Phys.* **17**, 283–289.

Schiff, L. I., 1968, *Quantum Mechanics* (McGraw-Hill, New York), 3rd edition, pp. 283–285.

Schipper, P. E., 1981, *Chem. Phys.* **57**, 105–119.

Schuster, P., A. Karpfen, and A. Beyer, 1980, in *Molecular Interactions*, H. Ratajczak and W. J. Orville-Thomas, Eds. (Wiley, Chichester) pp. 117–149.

Siebrand, W., 1965, *J. Chem. Phys.* **42**, 3951–3954.

Smith, G. C., 1968, *Phys. Rev.* **166**, 839–847.

Sztucki, J. and W. Strek, 1988, *Chem. Phys.* **124**, 177–186.

Thirunamachandran, T., 1979, *Proc. Roy. Soc. London* **A365**, 327–343.

Thirunamachandran, T., 1988, *Phys. Scripta* **T21**, 123–128.

Tinoco, I., 1975, *J. Chem. Phys.* **62**, 1006–1009.

Tulub, A. V. and K. Patzer, 1968, *Phys. Stat. Sol.* **26**, 693–700.

Van Grondelle, R., 1985, *Biochim. Biophys. Acta* **811**, 147–195.

Van Labeke, D., P. H. Grossel, and J. M. Vigoureux, 1988, *J. Chem. Phys.* **88**, 3211–3215.

Varsanyi, F. and G. H. Dieke, 1961, *Phys. Rev. Lett.* **7**, 442–443.

Vigoureux, J. M., 1983, *J. Chem. Phys.* **79**, 2363–2368.

Vigoureux, J. M., P. Grossel, D. Van Labeke, and C. Girard, 1987, *Phys. Rev.* **A35**, 1493–1502.

Von der Linde, D., 1977, in *Ultrashort Light Pulses—Picosecond Techniques and Applications*, S. L. Shapiro, Ed. (Springer-Verlag, Berlin) pp. 203–273.

Wallace, R., 1966, *Mol. Phys.* **11**, 457–470.

Ward, J. F., 1965, *Rev. Mod. Phys.* **37**, 1–18.

White, J. C., 1981, *Opt. Lett.* **6**, 242–244.

Wiedeman, L., M. E. Fajardo, and V. A. Apkarian, 1987, *Chem. Phys. Lett.* **134**, 55–59.

Wiedeman, L., M. E. Fajardo, and V. A. Apkarian, 1988, *J. Phys. Chem.* **92**, 342–346.

Woolley, R. G., 1971, *Proc. Roy. Soc. Lond.* **A321**, 557–572.

Yakovlenko, S. I., 1973, *Sov. Phys. JETP* **37**, 1019–1022.

Yu. Y. C., D. W. Setser, and H. Horiguchi, 1983, *J. Phys. Chem.* **87**, 2199–2209.

ACCURATE QUANTUM CHEMICAL CALCULATIONS

CHARLES W. BAUSCHLICHER, JR. AND STEPHEN R. LANGHOFF

NASA Ames Research Center
Moffett Field, California

PETER R. TAYLOR

ELORET Institute
Sunnyvale, California

CONTENTS

I. INTRODUCTION

An important goal of quantum chemical calculations is to provide an understanding of chemical bonding and molecular electronic structure, and over the last thirty years this has been largely realized. A second goal, the prediction of energy differences to "chemical accuracy" (about 1 kcal/mole), has been much harder to attain. First, the computational resources required to achieve such accuracy are very large, and second, it is not straightforward to demonstrate that an apparently accurate result, in terms of agreement with experiment, does not result from a cancellation of errors. Therefore, in addition to performing very elaborate electronic structure calculations, calibration and the assignment of realistic uncertainties are also required.

Recent advances in electronic structure methodology,[1] coupled with the power of vector supercomputers, have made it possible to solve a number of electronic structure problems exactly using the full configuration interaction (FCI) method within a subspace of the complete Hilbert space. These exact results can be used to benchmark approximate techniques that are applicable to a wider range of chemical and physical problems. They thus provide the necessary calibration of existing techniques for electronic structure calculations, and can be used to determine the origins of any deficiencies. In fact, as we shall show, the calibrations indicate that most of the methods for generating many-electron wave functions perform very well, and that a major source of error in even the best calculations arises at a more elementary level, in the selection of the atomic expansion basis sets. In this review, we will discuss the use of FCI wave functions to benchmark simpler computational methods, and new approaches to constructing atomic basis sets that substantially reduce the basis set errors while keeping the overall computational effort manageable.

In the following section, we will review briefly the methodology of many-electron quantum chemistry. In Section III we consider in more detail methods for performing FCI calculations, and in Section IV we discuss the application of FCI methods to several three-electron problems in molecular physics. In Section V we describe a number of benchmark applications of FCI wave functions. In Section VI we discuss atomic basis sets and the development of improved methods for handling very large basis sets: these are then applied to a number of chemical and spectroscopic problems in Section VII, to transition metals in Section VIII, and to problems involving potential energy surfaces in Section IX. Although the experiences described in these sections give considerable grounds for optimism about our general ability to perform accurate calculations, there are several problems that have proved less tractable, at least with current computer resources, and we discuss these and possible solutions in Section X. Our conclusions are given in Section XI.

II. QUANTUM CHEMICAL METHODOLOGY

In the clamped-nucleus Born–Oppenheimer approximation, with neglect of relativistic effects, the molecular Hamiltonian operator in atomic units takes the form

$$\hat{H} = -\frac{1}{2}\sum_{i=1}^{n}\nabla_i^2 - \sum_{A=1}^{N}\sum_{i=1}^{n} Z_A r_{Ai}^{-1} + \sum_{i>j=1}^{n} r_{ij}^{-1} + \sum_{A>B=1}^{N} Z_A Z_B R_{AB}^{-1} \qquad (1)$$

in the absence of external fields. The terms in Eq. (1) comprise the electron kinetic energy, the nuclear–electron attraction, the electron repulsion, and the nuclear repulsion, respectively. (Relaxing the assumptions of fixed nuclei and nonrelativistic motion can be investigated perturbationally, at least for lighter elements, as discussed briefly later.) Our goal is to approximate solutions to the time-independent Schrödinger wave equation

$$\hat{H}\Psi = E\Psi \tag{2}$$

for this Hamiltonian. Although some limited progress has been made in approaching Eq. (2) by analytical methods,[2] this is a very difficult procedure that is not yet suited to the production of chemical results. Instead, most methods make (implicit or explicit) use of basis set expansion techniques: the unknown eigenfunctions of Eq. (2) are expressed in terms of a set of n-particle basis functions $\{\Phi\}$. Again, while it is possible to consider rather exotic functional forms for the Φ, involving, say, interelectronic coordinates, by far the most common approach is to construct each Φ using a product of molecular orbitals (MOs—one-electron functions) $\{\psi\}$:

$$\Phi_K = \hat{O} \prod_{i=1}^{n} \psi_i \tag{3}$$

Here a given function Φ_K involves an n–fold product of MOs, to which is applied some projection operator or operators $\hat{O}$. As electrons are fermions, the solutions to Eq. (2) will be antisymmetric to particle interchange, and it is usually convenient to incorporate this into the n-particle basis, in which case the Φ will be Slater determinants. The Hamiltonian given in Eq. (1) is also spin-independent and commutes with all operations in the molecular point group, so that projection operators for particular spin and spatial symmetries could also appear in $\hat{O}$. The Φ obtained in this way are generally referred to as configuration state functions (CSFs).

The molecular orbitals are usually obtained as linear combinations of a one-particle basis

$$\psi_i = \sum_{\mu} \chi_\mu C_{\mu i} \tag{4}$$

The one-particle basis functions $\{\chi\}$ are often referred to as atomic orbitals (AOs). The MO coefficients $\mathbf{C}$ are obtained by solving an electronic structure problem simpler than that of Eq. (2), such as the independent particle (Hartree–Fock) approximation, or using a multiconfigurational Hartree–Fock approach.[1] This has the advantage that these approximations generally provide a rather good estimate of the solutions to Eq. (2)—perhaps 99% of the total energy or more—thus suggesting that an analysis of the many-electron problem and (possibly) computational schemes for attacking it can be formulated around them. Löwdin[3] defined the "correlation energy" as the difference between the exact energy obtained from Eq. (2) and the Hartree–Fock energy, and the term "correlation problem" is widely used to refer to the problem of computing this energy difference.

The most obvious use of the n-particle basis $\{\Phi\}$ in solving Eq. (2) is the linear configuration interaction (CI) expansion

$$\Psi = \sum_K \Phi_K c_K \tag{5}$$

If the one-particle basis is complete, the use of all possible Φ—termed complete CI—in Eq. (5) will yield the exact eigenvalues and eigenfunctions of Eq. (2). Incidentally, as the use of Eq. (5) corresponds to the variational problem of making the energy stationary with respect to variations of the coefficients c_K, subject to normalization of Ψ, any guess at the c_K will yield an upper bound to the true energy. In practice, of course, a complete one-particle space is infinite, and so the complete CI problem would also be infinite in dimension. If we choose a finite, truncated one-particle space, but approximate Ψ as in Eq. (5) using all the possible n-particle basis functions, we have a full CI (FCI) wave function. The FCI wave function can be regarded as the exact solution to a Schrödinger equation projected onto the finite subspace generated by the truncated one-particle basis.

FCI wave functions have many convenient properties, for example, the results are independent of any unitary transformation on the MOs used to construct the n-particle basis, and as the only approximation made in solving Eq. (2) is the truncation of the one-particle basis, we may identify any discrepancies between calculation and experiment as arising from this truncation (assuming the Born–Oppenheimer approximation and neglect of relativity are valid). Hence it would appear that FCI calculations are an ideal approach for solving the correlation problem. The difficulty, of course, is that the factorial dependence of the length of the FCI expansion on the number of electrons correlated and the number of MOs used creates insuperable computational difficulties for most problems of chemical interest. Even by restricting the correlation treatment to a subset of the electrons, by ignoring correlation effects involving 1s electrons in first-row systems, for instance, the expansions will usually be impractically long. To obtain the dissociation energy of the N_2 molecule to within 5 kcal/mole of experiment, for example, would require an FCI expansion of some 10^{14} CSFs, correlating only the ten valence electrons. Thus we must develop schemes for truncating both the one-particle and n-particle spaces to arrive at computationally feasible wave functions.

A simple way to implement n-particle space truncation is to use the uncorrelated wave function (which as noted above is a very substantial fraction of the exact wave function) to classify terms in the n-particle space. If we consider the Hartree–Fock determinant, for example, we can construct all CSFs in the full n-particle space by successively exciting one, two,... electrons from the occupied Hartree–Fock MOs to unoccupied MOs. For cases in which a multiconfigurational zeroth-order wave function is required, the same formal classification can be applied. Since only singly and doubly excited CSFs can interact with the zeroth-order wave function via the Hamiltonian in Eq. (1), it is natural to truncate the n-particle expansion at this level, at least as a first approximation. We thus obtain single and double excitations from Hartree–Fock (denoted SDCI) or its multiconfigurational reference analog, multireference CI (MRCI).

Truncated CI methods are only one of the popular approaches to the correlation problem. Coupled-cluster (CC) schemes[4] use a different formal ansatz from Eq. (5), based on an exponential operator

$$\Psi = \exp(T)\Psi_0 \tag{6}$$

Here T comprises excitation operators that again excite one, two,... electrons from MOs occupied in the zeroth-order function Ψ_0 to unoccupied MOs. The use of the exponential operator guarantees that the energy is additively separable, that is, that the energy of two noninteracting systems is the sum of the two separate system energies. This holds for Eq. (6) irrespective of what truncation is applied to T (say, to only single and double excitations). This size-consistency property is not shared by truncated CI. On the other hand, if Eq. (6) is substituted into the variation principle, the resulting equations for the cluster amplitudes (the weights with which the given single, double,... excitation terms appear) are too complicated for practical solution, and CC methods instead solve a nonlinear, nonvariational equation system for the amplitudes. Hence the CC energy is not a variational upper bound.

Another popular approach to the correlation problem is the use of perturbation theory.[5] Ψ_0 can be taken as an unperturbed wave function associated with a particular partitioning of the Hamiltonian; perturbed energies and wave functions can then be obtained formally by repeatedly applying the perturbation operator to Ψ_0. Probably the commonest partitioning is the Møller–Plesset scheme,[5] which is used where Ψ_0 is the closed-shell or (unrestricted) open-shell Hartree–Fock determinant. Clearly, the perturbation energies have no upper bound properties but, like the CC results, they are size-consistent.

A particular advantage of the CI approach, as opposed to coupled-cluster or perturbation theoretic techniques, is that it can be readily formulated to handle the case in which Ψ_0 is multiconfigurational in character. Corresponding multireference CC or perturbation theory approaches are much less well developed. As we shall see, there are many important chemical applications in which a multireference approach is mandatory, and we shall therefore concentrate mainly on the CI method in this article.

The classification of CSFs into single, double,... excitations is straightforward and unambiguous for a closed-shell Hartree–Fock reference function. In open-shell or multireference cases, there are more possibilities for defining these excited CSFs, some of which interact with the reference space in the lowest order of perturbation theory, and some of which do not. It is very common to exclude the latter excitations. This is termed restricting the wave function to the first-order interacting space.[6]

The choice of reference space for MRCI calculations is a complex problem. First, a multiconfigurational Hartree–Fock (MCSCF) approach must be chosen. Common among these are the generalized valence-bond method (GVB)[7] and the complete active space SCF (CASSCF) method.[8] The latter actually involves a full CI calculation in a subspace of the MO space—the *active space*. As a consequence of this full CI, the number of CSFs can become large, and this can create very long CI expansions if all the CASSCF CSFs are used as reference CSFs. This problem is exacerbated when it becomes necessary to correlate valence electrons in the CI that were excluded from the CASSCF active space. It is very common to select reference CSFs, usually by their weight in the CASSCF wave function. Even more elaborate than the use of a CASSCF wave function as the reference space is the second-order CI, in which the only restriction on the CSFs is that no more than two electrons occupy orbitals empty in the CASSCF wave function. Such expansions are usually too long for practical calculations, and they seldom produce results different from a CAS reference space MRCI.

When more than one state of a system is to be investigated, it is possible to perform separate MCSCF calculations, followed by MRCI calculations, on each state. This, however, can be a very expensive process, and if transition properties between states are desired, such as transition dipole moments for spectroscopic intensities, the nonorthogonality between the MCSCF orbitals for the different states creates complications. A simple alternative is to perform an MCSCF optimization of a single average energy for all states of interest.[9] All states are thereby described using a common set of MOs. Although these MOs are obviously not optimum for any of the states, experience shows this has little effect on the final MRCI results.

In some cases, even the use of severely truncated reference CSF spaces can give rise to an MRCI expansion that is too long for practical calculations.

Siegbahn has developed an approximate scheme for handling such cases: the weights of the singly- and doubly-excited CSFs are estimated perturbationally and frozen, then scale factors multiplying the weights of groups of CSFs are optimized in a so-called contracted CI (CCI) calculation.[10] Such a calculation requires much less effort than the corresponding MRCI calculation, and the results obtained are generally good. It may be possible in some cases to perform a large number of CCI calculations, say, to characterize a potential energy surface, and to scale the results based on MRCI calibration points to improve the accuracy. Another approach to reducing the number of variational parameters in the CI wave function is the internally contracted CI approach.[11] Here single and double excitations are made not from individual reference CSFs but from the MCSCF wave function, which is treated as a single term. The coefficients of the reference CSFs are held at their MCSCF values, but this restriction can be relaxed if desired. Internal contraction normally produces results similar to the uncontracted MRCI calculation, but at less cost.

As noted briefly above, truncated CI expansions do not produce size-consistent results. Several approaches of varying sophistication have been developed to address this problem. The single-reference Davidson correction[12]

$$\Delta E = E_{\text{corr}}(1 - c_0^2) \tag{7}$$

is a simple perturbation theoretic estimate of the energy effects of higher than double excitations. Here E_{corr} is the correlation energy from single and double excitations and c_0 is the coefficient of the Hartree–Fock determinant. Equation (7) is very often extended heuristically to the multireference case[13] using the form

$$\Delta E = E_{\text{corr}}\left(1 - \sum_R c_R^2\right). \tag{8}$$

Here E_{corr} is the difference between the MRCI energy and the energy obtained with the reference CSFs, and c_R is the coefficient of reference CSF R in the MRCI wave function. We shall use the notation $(+Q)$ to denote the use of the corrections given by Eq. (7) or (8). A problem with such approximations is that as there are no useful bounds on the adjusted energy expression, some numerical "noise" can be introduced across a large set of calculations (say, for a potential energy surface), while the MRCI values themselves would be much smoother.[14] (This "noise" can also be a problem with those MRCI methods that apply selection to all CSFs used, rather than just reference CSFs.) A more sophisticated approach to the size-consistency problem is to modify the CI energy expression so that size-consistent results are obtained. The earliest such schemes are the various coupled-electron pair approximations (CEPA),[15] including the linearized coupled-cluster approach of Cizek.[16] The single-reference Davidson correction can be viewed as a first approximation to the solution of the linearized CC equations. More recently, Ahlrichs and co-workers introduced the coupled-pair functional (CPF),[17] which has many similarities with the physical reasoning behind the CEPA schemes, and the averaged CPF (ACPF) approach[18] which can be extended to the multireference case. An advantage of these methods is that it is easier to apply the powerful techniques of analytic derivative theory to cases in which a functional is made stationary.[19] The modified CPF(MCPF) approach of Chong and Langhoff[20] is a development of the single-reference CPF method that often gives better results than the original when the single-reference description is a poor approximation.

The wave function that is used to define the reference CSFs is usually the one used to optimize the MO coefficients, but the ultimate accuracy of the MOs clearly derives from the quality of the AO basis chosen at the outset. In the vast majority of current calculations, the AOs are taken as fixed linear combinations of Gaussian-type functions centered on the nuclei, but the number of functions used, the range of angular momenta, etc., all vary widely among different calculations. We shall discuss a number of these aspects in this review, as the one-particle basis set plays a fundamental role in determining the accuracy of a given calculation.

In summary, then, CI approaches to the correlation problem involve a choice of atomic basis set, the optimization of a Hartree–Fock or MCSCF wave function, the selection of CI CSFs (usually by selecting one or more reference CSFs and all single and double excitations from them), and the optimization of the CI wave function. There are two potential sources of error in this prescription—the choices of one-particle and n-particle spaces. It is these sources of error we shall review in depth in the present work.

III. FULL CONFIGURATION INTERACTION METHODS

The use of the full CI (FCI) procedure as an exact test of approximate methods has been a key element in improving the accuracy of calculations, and we therefore consider the evolution of FCI methodology in some detail.

In the earliest CI programs, the Hamiltonian matrix elements were computed individually over a list of CSFs that could be chosen more or less arbitrarily.[21] (This approach is commonly referred to as conventional CI.) A given Hamiltonian matrix element H_{KL} can be expanded as

$$H_{KL} = \sum_{p}\sum_{q} A_{pq}^{KL}(p|h|q) + \sum_{p}\sum_{q}\sum_{r}\sum_{s} B_{pqrs}^{KL}(pq|rs) \qquad (9)$$

in terms of MO one- and two-electron integrals $(p|h|q)$ and $(pq|rs)$ and coupling coefficients A and B. The latter depend on the occupation and spin coupling of the MOs $p, q, \ldots$ in the CSFs Φ_K and Φ_L. The matrix elements were stored on disk or tape, and after all the elements had been computed the desired eigenvalues and eigenvectors were extracted in a subsequent step. Clearly, this approach could be used to perform an FCI calculation. However, the factorial increase in length of the CI expansion with number of electrons and orbitals made this possible for only very small calculations, such as the special case of three-electron systems.

The development of the direct CI approach[22] eliminated the requirement for large peripheral storage for the Hamiltonian matrix and greatly reduced the time to compute the eigenvalues and eigenvectors, as much redundant work in the conventional CI approach was eliminated. In direct CI the residual vector

$$\sigma_K = \sum_L H_{KL} c_L \tag{10}$$

is computed from the expression

$$\sigma_K = \sum_L \sum_p \sum_q A^{KL}_{pq}(p|h|q)c_L + \sum_L \sum_p \sum_q \sum_r \sum_s B^{KL}_{pqrs}(pq|rs)c_L \tag{11}$$

which results from substituting Eq. (9) in Eq. (10). In this way the Hamiltonian matrix elements themselves never appear explicitly. However, in the earliest formulations of direct CI, each problem (that is, closed-shell single-reference, doublet, etc.) required the development of a new program, as the A and B values in Eq. (11) were derived by hand and coded into the program. The first direct CI program for full CI calculations with more than two electrons was a three-electron FCI program developed by Siegbahn.[23] This program was able to handle relatively large one-particle basis sets, but three-electron problems are neither common enough nor representative enough for such a code to provide many benchmarks.

An important development in FCI methodology was Handy's observation[24] that if determinants were used as the n-particle basis functions, rather than CSFs, the Hamiltonian matrix element formulas could be obtained very simply. Specifically, it was possible to create an ordering of the determinants such that for each determinant, a list of all other determinants with which it had a non-zero matrix element could easily be determined. Further, it was easy to evaluate the coupling coefficients A and B in Eq. (11) (the only non-zero values being ± 1) and therefore to compute the matrix elements. While this greatly expanded the range of FCI calculations that could be carried out, the algorithm is essentially scalar in nature and therefore rather inefficient on modern supercomputers. Nonetheless, this method supplied some important benchmarks, as discussed below.

Like any general direct CI method, the shape-driven graphical unitary group approach[25] can also be used to perform FCI calculations for some cases, but as in Handy's determinantal scheme, matrix elements involving higher than double excitations from the reference CSF are processed using scalar algorithms. Like the Handy determinantal scheme, this approach has been used to provide several benchmarks.

A major advance in the efficiency of FCI calculations was the introduction of a factorized direct CI algorithm by Siegbahn.[26] This involves formulating the FCI calculation as a series of matrix multiplications: an ideal algorithm for exploiting the power of current vector supercomputers. This algorithm is fundamental to our present ability to perform FCI benchmarks, and we discuss it in detail. We consider only the two-electron contribution to σ given by Eq. (11), which can be written as

$$\sigma_K = \sum_p \sum_q \sum_r \sum_s \sum_L (pq|rs) B^{KL}_{pqrs} c_L \tag{12}$$

Siegbahn noted that by inserting a resolution of the identity, B^{KL}_{pqrs} can be factorized explicitly into products of one-electron coupling coefficients:

$$B^{KL}_{pqrs} = \sum_J A^{KJ}_{pq} A^{JL}_{rs} \tag{13}$$

This approach had been used earlier to compute the coupling coefficients themselves;[27] but if instead the factorization of B is inserted into Eq. (12) the calculation of σ can be completely vectorized. First, the product of one set of coupling coefficients and c is collected in $\mathbf{D}$,

$$D^J_{rs} = \sum_L A^{JL}_{rs} c_L \tag{14}$$

This can be implemented as vector operations; it should be noted that there are very few non-zero A values and this sparseness can be exploited in evaluating Eq. (14). A matrix multiplication of this intermediate matrix $\mathbf{D}$ and a block of the integrals is performed:

$$E^J_{pq} = \sum_{rs} (pq|rs) D^J_{rs} \tag{15}$$

Finally, the intermediate array $\mathbf{E}$ is merged with the second set of coupling coefficients to form a contribution to σ,

$$\sigma_K = \sum_J \sum_{pq} A^{KJ}_{pq} E^J_{pq} \tag{16}$$

All steps in the calculation of σ are vectorizable, and perform well on vector supercomputers. For large basis sets the matrix multiplication step in Eq. (15) will dominate the computational effort.

This factorized direct CI scheme still has some difficulties, however. In particular, values of the coupling coefficients A are required. While the lists of A values are extremely sparse, for very large calculations there will be many non-zero values, which must either be precomputed and stored, or computed on the fly. As Siegbahn's intended application was in generating FCI wave functions for use in CASSCF optimizations, he stored the coupling coefficients on a disk file. For very large calculations, disk space becomes the limiting factor, not the CPU power of the computer being used. It is not feasible to perform useful FCI benchmark calculations using this technique in such a form. The modified strategy developed by Knowles and Handy[28] overcame this limitation: these authors showed that by using determinants instead of CSFs, all non-zero A^{KL}_{pq} values (which are ± 1) can be computed almost trivially on the fly. The CI vector is much longer when determinants are used instead of CSFs, but this is not a major handicap when large memory computers such as the CRAY-2 are employed. Thus, the factorized, determinantal FCI scheme allows FCI calculations to be performed for very large expansions and hence to be used for benchmarking approximate methods. In essence, the scheme arises from a trade-off between memory (and CPU time) and disk storage: such a trade-off is not an uncommon feature of programming for modern supercomputers.

The most time-consuming step in the Knowles and Handy determinantal FCI scheme, as noted above, is a series of matrix multiplications, so the code performs very efficiently on Cray computers. Calculations as large as 28,000,000 determinants have been performed on the CRAY-2;[29] the **c** and σ vectors and some scratch arrays must be held in memory, so about 60 million words were required. In fact, memory is no longer the resource limitation on current supercomputers, it is disk space that becomes critical. The Davidson diagonalization process[30] used to obtain the lowest eigenvalues and eigenvectors requires the **c** and σ vectors from the previous iterations. Therefore, just as storing the coupling coefficients could exhaust the disk storage long before the CPU time became prohibitive in Siegbahn's factorized scheme, so can the storage of the previous **c** and σ vectors in the determinantal algorithm. As Davidson noted in the derivation of his iterative diagonalization algorithm,[30] it is possible to fold all the previous vectors into one vector—in effect, the calculation is restarted with a new trial vector corresponding to the best guess available from a fixed number of iterations. While this can slow convergence slightly, it limits the disk storage required. In practice, with the CPU power of, say, the CRAY-2 and the high degree of vectorization in the FCI scheme, as compared to the speed and limits in disk space, it becomes mandatory to use folding of the vectors to make the largest calculations feasible.

Several recent developments should further improve the efficiency of FCI calculations. Olsen et al.[31] have derived and implemented a factorized determinantal scheme formulated somewhat differently from that of Knowles and Handy. It appears that the new scheme should be especially efficient for large basis sets and few electrons correlated, which is the situation generally encountered in benchmark calculations. Another development concerns the form of the n-particle basis. Although the use of determinants avoids any storage of coupling coefficients, it generates a longer CI vector than would the use of CSFs and, more importantly, can lead to the collapse of an excited state to a lower state of a different spin symmetry due to numerical rounding in the iterative diagonalization.[32] This can be a problem, not only in FCI benchmarks, but also in the FCI step in a CASSCF calculation, where one may wish to extract many roots in the study of spectroscopic problems. Recently, Malmqvist et al.[33] have proposed a method of using CSFs instead of determinants, but still avoiding the storage of the coupling coefficients on disk. Their approach seems ideal for the CASSCF problem where the number of active orbitals is limited, but may not be suitable for the large FCI calculations used for benchmarking.

It is clear that FCI wave function optimization is currently an active area of research, and the best method of performing FCI calculations may not yet have been achieved. Indeed, in very recent work Knowles[34] has shown how sparseness in $\mathbf{c}$ and σ can be exploited to reduce the computational effort, and Knowles and Handy[35] have been able to perform pilot calculations for a 400,000,000 determinant FCI. Harrison and Zarrabian[36] have modified the factorization used in the original Knowles and Handy scheme so as to reduce the dimension of the intermediate space that appears in the resolution of the identity; this approach has been used to perform FCI calculations with over 50,000,000 determinants. It should be feasible to perform calculations with well over 100,000,000 determinants with this approach. Thus it may well become possible to perform even larger benchmarks in the near future. As we will show below, this is important because for some problems there is a coupling of the one- and n-particle spaces, and larger FCI calculations will lead to an improved understanding of a larger range of problems.

IV. FCI APPLICATION CALCULATIONS

The simplest nontrivial FCI calculation is the three-electron case, for which the *n*-particle problem can be solved exactly for a very large one-particle space. Since relativistic effects, which are neglected, are expected to be very small for the systems considered in this work, the results should be nearly exact, and hence should compare well with experiment. In this section, we present three examples where three-electron FCI calculations have yielded accurate results of chemical interest.

The $H_2 + H \rightarrow H + H_2$ exchange reaction has been studied by Siegbahn and Liu[37] using the FCI approach in a large one-particle basis set. The potential energy surface for this reaction is estimated to be within 1 kcal/mole of the exact surface, and has been widely used to evaluate scattering methods.

FCI calculations have also been used to assess whether H_2 has a positive electron affinity (EA).[38] The impetus for this study was a qualitative calculation suggesting that H_2 might have a small EA if the "extra" electron attached into the $2\sigma_g$ orbital.[39] Before asking experimentalists to investigate this possibility, it seemed worthwhile to carry out FCI calculations in a large basis set, as this requires only a few hours of computer time, including convergence tests of the one-particle basis. FCI calculations using a large [5s 4p 3d 1f] Gaussian basis set, which yields an EA of H that is 0.006 eV smaller than the best value and a D_e of H_2 that is only 0.03 eV too small, conclusively show that H_2 does not have a positive EA. Thus, it is unlikely that a bound state of H_2^- can be observed experimentally.

Another example of a three-electron FCI application is the study of the ^{2}D Rydberg series in the Al atom.[40] The ground state of Al is $^2P(3s^2 3p^1)$. The valence occupation $3s^1 3p^2$ gives rise to a ^{2}D state, and it was suggested that one of the lowest terms in the $3s^2 nd^1(^2D)$ Rydberg series might correspond to this occupation, although this is contrary to some experimental evidence.[41,42] Theoretical studies indicated some mixing of the $3s^1 3p^2$ configuration into the lower terms of the ^{2}D Rydberg series.[43,44] Further, experimental observations by Garton[41] suggested that a state lying beyond the ionization limit of the ^{2}D series is derived principally from the $3s^1 3p^2$ occupation. Although this problem is amenable to a CASSCF/MRCI calculation, the FCI calculations eliminate any concern that there is a bias in the treatment of the *n*-particle space that could lead to an incorrect mixing of occupations. FCI calculations in a [7s 6p 11d 3f] basis set yield excitation energies for the first eight terms in the Rydberg series that are in excellent agreement with experiment. This agreement as well as direct calculation indicates that core–valence correlation has little effect on the term energies, thereby justifying a three-electron treatment. Since this FCI calculation solves the n-particle problem exactly using a large one-particle basis set, and the term energies agree well with experiment, the wave functions should reflect the true configurational mixing in these states. Although an FCI treatment is independent of the orbital basis, we take the d orbitals from the $3s^2(^1S)$ state of Al^+ that corresponds most closely to the Rydberg series to simplify the interpretation. Analysis of the FCI wave function indicates that the percentage contribution of the $3s^13p^2$ configuration in the first member of the Rydberg series is 24%. The contribution of $3s^13p^2$ rapidly decreases so that the contributions to the fourth and sixth terms are only 6.1% and < 1%, respectively. There is 17% of $3s^13p^2$ occupation not accounted for in the Rydberg series, which is concentrated in a state just above the IP limit as suggested by experiment. Thus the FCI calculations have definitively resolved the nature of the $3s^13p^2(^2D)$ state.

V. FCI BENCHMARK CALCULATIONS

The factorial growth of the FCI treatment with basis set and number of electrons clearly limits its applicability. However, FCI calculations can be performed in moderate-sized (double-zeta plus polarization (DZP) or better) basis sets to calibrate approximate methods of including electron correlation. Thus there is considerable incentive to utilize these FCI methods and the capabilities of modern supercomputers to perform large-scale benchmark calculations. In this section we consider several examples of the insight obtained from FCI benchmark calculations. In the sections dealing with applications, we consider additional FCI benchmark calculations in conjunction with specific applications.

Early FCI benchmark calculations were performed by Handy and co-workers.[45,46]. Using a double-zeta (DZ) basis set, they considered stretching the O—H bond lengths in the H_2O molecule to 1.5 and 2.0 times their equilibrium values. The FCI results showed that even the restricted Hartree–Fock (RHF)-based fourth-order many-body perturbation theory (MBPT) approach,[47,48] which includes the effects of single, double, triple, and quadruple excitations, did not accurately describe the stretching of the bond; the error increased from 0.6 kcal/mole at r_e to 10.3 kcal/mole with the bonds stretched to twice their equilibrium values. Although the MBPT method is rigorously size-consistent and contains the effects of higher than double excitations, it does not describe the bond-breaking process well, because the RHF reference becomes a poor zeroth-order description of the system as the bond is stretched. Size-consistent methods that include double excitations iteratively—infinite-order methods such as the coupled-cluster approach—do better. However, only methods that account for the multireference character in the wave function as the bonds are broken, such as the CASSCF/MRCI method, provide an accurate description at all bond lengths.[49]

While these earlier studies helped calibrate methods for the approximate treatment of the correlation problem, there were insufficient studies to draw general conclusions about the accuracy of approximate methods. Also, the one-particle basis sets were not large enough to eliminate concerns about possible couplings of the approximations in the one- and n-particle treatments. Much more extensive benchmarking has recently been possible with advances in computers and methods.[29,32,50–71] Unlike the earlier work, the basis sets are of at least DZP quality and a sufficient number of systems have been studied that trends are distinguishable. The FCI studies discussed in the remainder of this work use a version of the Knowles and Handy program. As discussed in Section III, this method is based on the work of Siegbahn and vectorizes very well.

The first application of the recent series of benchmarks was the calculation of the total valence correlaton energy of the Ne atom.[50] In this work, the FCI was compared to several single-reference-based approaches. These calculations showed that both the $+Q$ correction and the CPF approach gave reasonable estimates for the energy lowering of quadruple excitations, and that the accuracy of the different treatments varied with basis set. The calculations further showed that higher than quadruple excitations were not important in any basis set. It must be remembered that Ne is exceptionally well described by a single-reference configuration. It is clear from the results of the H_2O calculations that to obtain the best possible agreement with the FCI it will be necessary to go beyond the single-reference-based approaches.[72] However, the Ne results do suggest that for some problems, single-reference-based approaches that include an estimate of higher excitations should yield reasonably accurate results.

To evaluate single-reference and multireference CI approaches to the correlation problem further, we compare in Table I the FCI $^1A_1$–$^3B_1$ separation in CH_2 with various truncated CI results.[53] Since the $^1A_1$ and $^3B_1$ states are derived nominally from the ^3P and ^5S states of carbon, respectively, they involve different bonding mechanisms that result in a substantial correlation contribution to the separation: the SCF separation is over 14 kcal/mole too large. The error of 2.7 kcal/mole at the SDCI level is still much larger than required for chemical accuracy (≈ 1 kcal/mole). The inclusion of the contribution of unlinked higher excitations through either the Davidson correction $(+Q)$ or the CPF method reduces the error substantially. The origin of the error in the SCF/SDCI treatment is the second important configuration, arising from the double excitation $3a_1^2 \rightarrow 1b_1^2$, in the $^1A_1$ state. If the orbitals for the $^1A_1$ state are optimized in a two-configuration SCF calculation, and correlation is included by performing an MRCI

TABLE I
$^1A_1$–$^3B_1$ Separation in CH_2 (kcal/mole) Using a DZP Basis Set and Correlating the Six Valence Electrons

Method	Separation	Error
SCF[a]	26.14	14.17
SCF[a]/SDCI	14.63	2.66
SCF/SDCI+Q	12.35	0.38
CPF	12.42	0.45
TCSCF[b]/MRCI	12.20	0.23
TCSCF/MRCI+Q	12.03	0.06
CASSCF[c]/MRCI	11.97	0.00
CASSCF/MRCI+Q	11.79	−0.18
FCI	11.97	—

[a] The SCF occupations are $1a_1^2 2a_1^2 3a_1^2 1b_2^2$ and $1a_1^2 2a_1^2 1b_2^2 3a_1^1 1b_1^1$.
[b] SCF treatment for $^3B_1$ state, two-configuration MCSCF treatment for $^1A_1$ state (SCF configuration and $3a_1^2 \rightarrow 1b_1^2$ excitation).
[c] Active space comprises the C 2s 2p and H 1s orbitals .

calculation based on both these reference configurations, the error is about half that of the SDCI + Q or CPF treatments. The error is reduced to only 0.06 kcal/mole if the multireference analog of the $+Q$ correction is added. After the $3a_1^2 \rightarrow 1b_1^2$ excitation, the next most important correlation effect is that associated with the C—H bonds. If this correlation effect is accounted for in the CASSCF zeroth-order reference for a subsequent MRCI calculation, essentially perfect agreement between the MRCI and FCI is observed. That is, a well-defined CASSCF/MRCI treatment accounts for all of the differential correlation effects. It is interesting to note that adding the multireference $+Q$ correction to this MRCI energy results in overestimating the effect of higher excitations and the separation becomes smaller than the FCI result.

We next consider the spectroscopic constants[59] for the ground state of N_2. The values obtained at various levels of correlation treatment with six valence electrons correlated are summarized in Table II. The SDCI calculation yields a bond length that is in good agreement with the FCI, but the error in D_e is 0.45 eV, even when size-consistency problems are minimized by using the $^7\Sigma_u^+$ state of N_2 to represent two ground-state N(^{4}S) atoms at infinite separation. Although the addition of quadruple excitations—either variationally (SDQCI) or by the $+Q$, CPF, or MCPF approximations—further reduces the error in D_e, it remains too large for chemical accuracy. If both triple and quadruple excitations are included, the spectroscopic constants are all in good agreement with the FCI. However, this level of treatment is prohibitively expensive in a large one-particle basis set, and even

TABLE II
Spectroscopic Constants for the $^1\Sigma_g^+$ State of N_2

	DZP basis 6 electrons correlated		
Method	r_e (a_0)	ω_e (cm^{-1})	D_e (eV)
SDCI	2.102	2436	8.298
SDCI+Q	2.115	2373	8.613
SDTCI	2.107	2411	8.462
SDTQCI	2.121	2343	8.732
SDQCI	2.116	2361	8.586
CPF	2.112	2382	8.526
MCPF	2.114	2370	8.556
MRCI	2.123	2334	8.743
MRCI+Q	2.123	2333	8.766
FCI	2.123	2333	8.750

this wave function does not dissociate correctly to ground-state atoms, as this requires six-fold excitations relative to the SCF configuration at r_e. The spectroscopic constants computed from an MRCI treatment based on a CASSCF wave function are in excellent agreement with the FCI. Furthermore, this treatment agrees with the FCI for all r values. The addition of the $+Q$ correction does not affect r_e or ω_e, but it makes D_e too large compared with the FCI. A comparison of the computational difficulties encountered in describing the triple bond in N_2 with the single bond in CH_2 confirms that electron-dense systems put a much greater demand on the computational methodology.

In addition to the total energy of the system, it is desirable to carry out FCI calibration studies of properties such as dipole moments, polarizabilities, and electrostatic forces. For example, in the $O + OH \rightarrow O_2 + H$ reaction,[73] the preferred approach of the O atom is determined by the dipole–quadrupole interaction. At long distances, this favors a collinear approach to the H atom, whereas for reaction to occur the O atom must migrate to the O end of OH. An accurate description of weakly interacting systems such as van der Waals complexes requires a quantitative description of dipole–induced-dipole or induced-dipole–induced-dipole interactions. Further, the dipole moment and polarizability functions of a molecule determine its infrared and Raman spectral intensities.

As a first example of an FCI calibration of properties, we present in Table III a study of the polarizability of F^-.[56] As in the previous examples, the SDCI treatment contains significant error. The inclusion of an estimate of higher excitations improves the results; in this case, CPF is superior to the $+Q$ correction. In the multireference case, two different approaches were

TABLE III
Polarizability of F^- (a_0^3)

DZP + diffuse spd basis 8 electrons correlated, $\alpha = d^2E/dF^2$	
Method	α
SDCI	13.965
SDCI+Q	15.540
CPF	16.050
MRCI[a]	16.134
MRCI+Q	16.346
MRCI′[b]	16.034
MRCI′+Q	16.303
FCI	16.295

[a] The MRCI treatment is a CAS reference CI based on a CASSCF calculation that included the 2p electrons and the 2p and 3p orbitals in the active space.
[b] The MRCI treatment is a CAS reference CI based on a CASSCF calculation that included the 2s and 2p electrons and the 2s, 2p, 3s, and 3p orbitals in the active space.

used. In the first, the CASSCF included the 2p electrons and the 2p and 3p orbitals in the active space, and all CASSCF CSFs were used as references for the CI, in which the 2s and 2p electrons were correlated. Results obtained in this way are denoted MRCI in Table III. A more elaborate CASSCF calculation, with the 2s and 2p electrons and the 2s, 3s, 2p and 3p orbitals in the active space, was also performed; the use of all these CASSCF CSFs as references gives results denoted as MRCI′ in Table III. The MRCI and MRCI′ results do not agree as well with the FCI as do the MRCI + Q or MRCI′ + Q result. The + Q correction does not overshoot as it did for N_2 and CH_2, in part because of the larger number of electrons correlated here. As noted above, when only six electrons are correlated, the CASSCF/MRCI accounts for such a large percentage of the correlation energy that the + Q correction overestimates the remaining correlation. For more than six electrons, or for cases where the zeroth-order wave function used is less satisfactory than was the CASSCF for N_2 and CH_2, the + Q correction is a better approximation. This is especially true for such quantities as electron affinities that involve large differential correlation effects. Thus the + Q correction substantially improves the agreement with the FCI for the electron affinity of fluorine,[29] even when large CASSCF active spaces and MRCI wave functions are employed.

VI. ANO BASIS SETS

In the previous section, we showed that the CASSCF/MRCI approach yields results in excellent agreement with FCI, that is, results that are near the

n-particle limit. We may therefore expect excellent agreement with experiment when the CASSCF/MRCI approach is used in conjunction with extended one-particle basis sets. It has become clear[74–76] that, until recently, the basis set requirements for achieving the one-particle limit at the correlated level were commonly underestimated, both in the number of functions required to saturate the space for each angular momentum quantum number and in the maximum angular momentum required. For the segmented basis sets that are widely used in quantum chemistry,[77] improving the basis set normally involves replacing a smaller primitive basis set with a larger one. It is then seldom possible to guarantee that the smaller basis spans a subspace of the larger set, and it is thus difficult to establish how results obtained with different basis sets relate to convergence of the one-particle space. Ideally, the possibility of differences in primitive basis sets would be eliminated by using a single (nearly complete) primitive set, contracted in different ways such that the smaller contracted sets are subsets of the larger one. Such an approach is best implemented using a general contraction scheme, such as the one proposed by Raffenetti[78] for contracting valence orbitals at the SCF level. However, using atomic SCF orbitals to define the contraction is not necessarily suitable for handling the correlation problem, and provides no means to contract polarization functions, which require large primitive sets for accurate results. Calculations on molecular systems have shown[79] that natural orbitals (NO) provide an efficient method of truncating the orbital space in correlated treatments. Almlöf and Taylor[80] have proposed an NO procedure for contracting atomic basis sets suitable for use in correlated molecular calculations. This atomic natural orbital (ANO) approach is an efficient method for contracting large primitive valence and polarization basis sets. It has the advantage that the natural orbital occupation numbers provide a criterion for systematically expanding the basis set.

These ideas are illustrated for the N atom and N_2 in Table IV. As the contraction of the (13s 8p 6d) primitive set is expanded from [4s 3p 2d] to [5s 4p 3d] to [6s 5p 4d], the correlation energy systematically converges to that of the uncontracted results. The same is true for the (4f) and (2g) polarization sets. When these ANO sets are applied to N_2, the same systematic convergence of D_e is observed. An additional advantage of the ANO contractions is that the optimum atomic description they provide reduces the problem of basis set superposition error (SE),[81,82] in which basis functions on one atom improve the description of another atom, causing a spurious lowering of the computed molecular energy. Superposition error is a particular problem in calculations that aim at high accuracy, and it is highly desirable to minimize its effects.

In order to treat atomic states with different character equally, the ANOs can be averaged to yield a compromise set.[83] This is useful for systems with

TABLE IV
$N(^4S)/N_2(^1\Sigma_g^+)$ Extended Basis Total Energies (E_H), Correlation Energies (E_{corr})(E_H) and "Dissociation Energies" (eV)

N atom

Basis set	$E_{SCF} + 54.$	E_{corr}
(13s 8p 6d)	–0.400790	–0.111493
[6s 5p 4d]	–0.400779	–0.111321
[5s 4p 3d]	–0.400769	–0.110925
[4s 3p 2d]	–0.400725	–0.109066
(13s 8p 6d 4f)	–0.400790	–0.121385
[5s 4p 3d 2f]	–0.400769	–0.120499
[4s 3p 2d 1f]	–0.400725	–0.117584
[5s 4p 3d 2f] (2g)[a]	–0.400769	–0.122472
[5s 4p 3d 2f 1g]	–0.400769	–0.122138

N_2 molecule[b]

Basis set	$E_{SCF} + 108.$	E_{corr}	D_e(SCF)	D_e(SDCI)
(13s 8p 6d)	–0.986307	–0.338118	5.03	8.16
[6s 5p 4d]	–0.985913	–0.337304	5.02	8.14
[5s 4p 3d]	–0.984833	–0.335395	4.99	8.08
[4s 3p 2d]	–0.983483	–0.329330	4.95	7.98
(13s 8p 6d 4f)	–0.989318	–0.365735	5.11	8.45
[5s 4p 3d 2f]	–0.988031	–0.362548	5.07	8.38
[4s 3p 2d 1f]	–0.986230	–0.353283	5.03	8.24
[5s 4p 3d 2f](2g)[a]	–0.988458	–0.370808	5.09	8.51
[5s 4p 3d 2f 1g]	–0.988322	–0.369270	5.08	8.48

[a] 2 uncontracted g sets.
[b] r(N–N) = 2.1 a_0, 10 electrons correlated.

different charges, such as F and F^- or transition metals where the radial expectation values of the 3d and 4s orbitals are very different for the $3d^n4s^2$, $3d^{n+1}4s^1$, and $3d^{n+2}$ occupations. Generating the average ANOs is analogous to the state averaging used to define compromise orbitals suitable for describing molecular states of different character. A [5s 4p 3d 2f 1g] contraction based on the average of F and F^- has an SDCI level EA that agrees with the uncontracted (13s 9p 6d 4f 2g) basis set result to within 0.01 eV. This can be compared with a 0.1 eV error for the same size basis set that is contracted for F alone, but with the outermost (the most diffuse) s and p primitive functions uncontracted. The results are better if the contraction is based on F^- instead of F, but not as good as using the average ANOs. The loss in total energy as a result of averaging is also very small. ANO basis sets averaged for different states should thus supply a more uniform description in cases in which there is charge transfer or ionic/covalent mixing. The success for transition metals is equally good: the average ANOs yield separations between the lowest states that are virtually identical to the uncontracted basis set. It should be noted, however, that it may still be necessary to uncontract the most diffuse primitive functions and/or add extra diffuse functions to describe properties such as the dipole moments and polarizabilities[84,85] that are sensitive to the outer regions of the charge density. For example, in the Ni atom the polarizability is in excellent agreement with the uncontracted basis set if the outermost s and p functions are uncontracted, but half the uncontracted value if these functions are included in the ANO contraction.

Dunning[86] has recently suggested that accurate results can be obtained with smaller primitive polarization sets than those used in the ANO studies, thereby reducing the integral evaluation time. This approach is consistent with the ANO procedure in that the primitives are optimized at the CI level for the atoms. Since his basis sets contain uncontracted primitive polarization functions and segmented valence sets with the outermost functions free, the basis sets are accurate for properties as well as the energy. For example, Dunning's results for OH are in good agreement with calculations employing ANO sets. However, these conclusions may be system-dependent as indicated by the work of Ahlrichs and co-workers on the N_2[87] and F_2[74] molecules. Despite optimizing the polarization sets at the CI levels for the molecules themselves, the energies for these two homonuclear diatomics did not converge as quickly with expansion of the primitive sets as with ANOs. We conclude that it might be possible to reduce the size of the primitive sets, but this requires more study.

VII. RESULTS FOR SPECTROSCOPIC CONSTANTS

The FCI benchmark calculations discussed in Section V show that a CASSCF/MRCI treatment is capable of accurately reproducing the FCI results, at least when six electrons or fewer are correlated. Further, the ANO basis sets discussed in Section VI show that it is now possible to contract nearly complete primitive sets to manageable size with only a small loss in accuracy. Therefore, a six-electron CASSCF/MRCI treatment performed in a large ANO basis set is expected to reproduce accurately the FCI result in a complete one-particle basis set, and hence should accurately reproduce experiment. For some systems containing eight electrons, the MRCI agrees well with the FCI, while for others that have very large correlation effects, such as systems with significant negative ion character, it is necessary to include the $+Q$ correction to achieve chemical accuracy. FCI calibration for systems with more than eight electrons have not been possible, but we expect that the $+Q$ correction will improve the results in most cases. In Sections VII–IX we illustrate several calculations that have achieved unprecedented accuracy by combining FCI benchmarks and ANO basis sets.

As discussed previously, FCI calculations for CH_2 show that the CASSCF/MRCI treatment accurately accounts for the differential correlation contribution to the CH_2 $^1A_1$–$^3B_1$ separation. In Table V, this level of treatment is performed using increasingly accurate ANO basis sets.[88] It is interesting to note that although the [4s 3p 2d 1f/3s 2p 1d] basis set contains fewer contracted functions than the largest segmented basis sets previously applied to this problem,[89] it produces a superior result for the separation. The largest

TABLE V
Theoretical Study of the $^1A_1$–$^3B_1$ Separation in CH_2 and SiH_2

ANO basis set	Separation (kcal/mole)[a]
CH_2	
[3s 2p 1d/2s 1p]	11.33
[4s 3p 2d 1f/3s 2p 1d]	9.66
[5s 4p 3d 2f 1g/4s 3p 2d]	9.24
Expt+theory (T_e)	9.28 (±0.1)[b]
Core-Valence	+0.35[c]
Born-Oppenheimer	−0.1[d]
Relativistic	−0.06[a]
SiH_2	
[6s 5p 1d/3s 1p]	-19.46
[6s 5p 2d 1f/3s 2p 1d]	-20.15
[6s 5p 3d 2f 1g/4s 3p 2d]	-20.29
Relativistic	−0.30
Zero-point	−0.33[e]
Expt(T_0)	-21.0±0.7[f]

[a] Ref. 88.
[b] Ref. 90.
[c] Ref. 66.
[d] Ref. 91.
[e] Ref. 95.
[f] Ref. 96.

ANO basis set used gives a separation in good agreement with, but smaller than, the T_e value deduced from a combination of theory and experiment.[90] From the convergence of the result with expansion of the ANO basis set, it is estimated that the valence limit is about 9.05 ± 0.1 kcal/mole. The remaining discrepancy with experiment is probably mostly due to core–valence correlation effects. However, as the valence correlation treatment is nearly exact, finer effects such as Born–Oppenheimer breakdown[91] and relativity must also be considered. While FCI calculations have shown that a very high level of correlation treatment is required for an accurate estimate of the CV contribution to the separation, theoretical calculations[66] indicate that CV correlation will increase the separation by at most 0.35 kcal/mole (see later discussion). Therefore, it is now possible to achieve an accuracy of considerably better than one kcal/mole in the singlet–triplet separation in methylene.

An analogous study[88] for SiH_2 indicates that the singlet–triplet separation can also be accurately computed for this second-row molecule (see Table V). In fact, the differential valence correlation contribution to T_e converges faster with basis set expansion than for CH_2. Note also that the ${}^1A_1$ state of SiH_2 is better described by a single reference than is the ${}^1A_1$ state of CH_2. However, it now becomes necessary to include the dominant relativistic contributions, namely the mass–velocity and Darwin terms,[92] via first-order perturbation theory[93] or by using an effective core potential,[94] if chemical accuracy is to be achieved. Once relativistic effects and zero-point energy[95] have been accounted for, the theoretical T_0 value of 20.92 kcal/mole is in excellent agreement with the higher experimental value.[96]. These calculations rule out the alternative value of -18.0 kcal/mole for the ${}^1A_1$–${}^3B_1$ separation based on a lower value for the ionization potential of the ${}^1A_1$ state of SiH_2. Thus, in addition to establishing the T_e of SiH_2, these calculations also establish the IP of the ${}^1A_1$ state to be 9.15 eV, the higher of two recent experimental values (see discussion in Ref. 96). We should note that Balasubramanian and McLean[97] independently came to the same conclusions. In their calculations, a more conventional segmented one-particle basis set was used and therefore the accuracy of the SiH_2 calculations was not sufficient to definitively determine the T_e value. However, by treating both CH_2 and SiH_2 to about the same accuracy, they were able to predict an accurate T_e value for SiH_2 using the error in their CH_2 calculations. The advantage of the ANO basis sets is that they provide a convenient method of systematically approaching the basis set limit. This reduces the need to use an analogous better-known system for calibration. This is advantageous as it can be difficult to find such a system for comparison.

As the calculations on CH_2 and SiH_2 show, theory is now capable of computing energy separations between electronic states to better than a few hundred wavenumbers for both first- and second-row systems. Thus, theory has considerable utility for determining the ground state when two nearly degenerate states are present, which is the case for the Al_2, Si_2 and N_2^{2+} molecules. In the subsequent discussion, we describe how FCI benchmark calculations combined with ANO basis sets have allowed a definitive prediction of the ground state for these molecules.

Three possible ground states have been suggested for Al_2 based on previous theoretical calculations[98-101] and experimental studies.[102-105] These are the $^3\Sigma_g^-(\pi_u^2)$ and $^3\Pi_u(\sigma_g^1\pi_u^1)$ states with two one-electron bonds, and the $^1\Sigma_g^+(\sigma_g^2)$ state with one two-electron bond. All of these states correlate with the $^2P(3p^1)$ ground state of two Al atoms. The accuracy of the CASSCF/MRCI treatment for Al_2 was confirmed using FCI calculations in a small basis set.[62] In order to confirm that the one- and n-particle spaces were not strongly coupled, the FCI calibrations were performed in two different one-particle basis sets. For both basis sets, the FCI and CASSCF/MRCI T_e values agreed. CASSCF/MRCI calculations,[62] including relativistic effects, were then performed in a large ANO basis, showing that the $^3\Pi_u$ state of Al_2 lies just 174 cm^{-1} below the $^3\Sigma_g^-$ state. The $^1\Sigma_g^+$ state was found to lie much higher in energy. While it is straightforward to carry out CASSCF/MRCI calculations with six valence electrons correlated, even in very large ANO basis sets, it is far more difficult to include the Al 2s and 2p electrons at this level of correlation treatment. However, the FCI benchmark calculations also showed that the CPF state separations were qualitatively correct. Since the CPF method is size-consistent, it is expected that increasing the number of electrons correlated will not significantly affect its accuracy. It was, therefore, possible to use the CPF approach to show that the L-shell and L–M intershell correlation has little effect on the A—X separation, although it shortens the bond length slightly. Indirect support for the theoretical prediction of a $^3\Pi_u$ ground state comes from the failure to observe the $^3\Sigma_u^- \leftarrow {}^3\Sigma_g^-$ band system (well known from emission studies) in a jet-cooled beam of aluminum clusters.[106] Further support comes from the work of Cai et al.[107] who have observed two absorption bands that they have interpreted in terms of a ground state with a vibrational frequency of 284 cm^{-1}. This is in good agreement with the theoretical prediction of 277 cm^{-1} for the $X^3\Pi_u$ state[62] and in clear disagreement with the known frequency (350.0 cm^{-1})[108] for the $A^3\Sigma_g^-$ state.

C_2, Si_2, and N_2^{2+} are valence isoelectronic, but only for C_2 has there been an experimental determination of the ground state.[108] Since these molecules contain multiple bonds, in addition to calibrating the n-particle treatment with the FCI approach, we also performed extensive calculations on C_2 to compare with experiment. Despite the very large correlation energy associated with multiple bonds, the CASSCF/MRCI approach was in excellent agreement with the FCI for the spectroscopic constants. When this approach was applied to C_2 using an extensive ANO basis set, excellent agreement (to within $60\,cm^{-1}$) was obtained with the experimental separations between the $X^1\Sigma_g^+$, $a^3\Pi_u$ and $b^3\Sigma_g^-$ states.[71] Similar theoretical calculations[71] applied to the Si_2 molecule resulted in a determination of the ground state as $X^3\Sigma_g^-$, and the prediction that the $A^3\Pi_u$ state lies $440 \pm 100\,cm^{-1}$ higher. The error bars are assigned based on the C_2 calculations, the difference between the MRCI and FCI calculations, and the convergence of the energy separations with expansion of the ANO basis set. Application of the same CASSCF/MRCI treatment to N_2^{2+}[109] yields a $^1\Sigma_g^+$ ground state in analogy with C_2. In fact, the $a^3\Pi_u$–$X^1\Sigma_g^+$ separation is very similar to that found in C_2, although the $b^3\Sigma_g^-$ state is higher lying in N_2^{2+}. An analysis of both the C_2 and N_2^{2+} wave functions indicates that there is a change in the relative importance of the major CSFs between the CASSCF and MRCI wave functions. This is one reason that such high levels of correlation treatment are needed for these systems. The FCI calibration calculations were instrumental in showing that the CASSCF/MRCI calculations accurately accounted for the effects of valence correlation on the energy separations. The ability of theory to accurately treat these systems is a major advantage, as spectroscopic determination of the ground states is very difficult.

In addition to the calculation of accurate separations, the calculation of properties is important for an understanding of spectroscopy. We now consider the accurate calculation of the dipole moment function of the $X^2\Pi$ ground state of OH to determine the strength of the rotational–vibrational bands (Meinel system),[110] which are observed in the night sky, in oxygen-supported flames, in the photosphere of the sun, in OH stars, and in interstellar space. Recently it has been conjectured that the surface-originating glow observed on the Atmosphere Explorer Satellites[111] (and possibly the Space Shuttle) is at least partially due to vibrationally excited OH radicals arising from the collision of the surface with $O(^3P)$ atoms with a relative translational energy of 5 eV.

To determine an accurate electric dipole moment function (EDMF) for the $X^2\Pi$ state of OH requires a very high level of correlation treatment, since it is necessary to properly account for the O^- character in the wave function. To calibrate approximate methods, an FCI dipole moment was computed[61] at five representative r values using a [4s3p2d/2s1p] Gaussian basis set. Of the variety of approximate methods compared with the FCI, the CASSCF/MRCI treatment reproduced the FCI results best, with an error of only $0.001\,a_0$ in the position of the dipole moment maximum. The MRCI spectroscopic constants are also in excellent agreement with the FCI. At this level of correlation treatment, it did not make a substantial difference whether the dipole moment was evaluated as an expectation value or as an energy derivative. Having identified an approximate correlation treatment that accurately reproduced the FCI EDMF in a realistic one-particle basis set, this treatment was then carried out in an extended [6s 5p 4d 2f 1g/4s 3p 2d] ANO basis set. This calculation gave a permanent dipole moment that is within 0.01 D of the experimental value[112] for $v = 0$, and within 10% of the accurate experimental value for the difference in dipole moments between $v = 0$ and $v = 1$. Rotational–vibrational line strengths determined from this theoretical EDMF are expected to be more accurate than those deduced from a variety of experimental sources.

Theory has made a substantial contribution to the determination of accurate electronic transition moment functions (TMF) and radiative lifetimes for atomic and molecular systems. (See, for example, the recent review articles by Werner and Rosmus[113] and Oddershede.[114]) However, only recently has it become possible to evaluate unambiguously, using FCI benchmarks, the quality of approximate correlation methods for determining TMFs. The OH ultraviolet system (A–X) is important in many applications, such as combustion diagnostics, and is amenable to FCI calculations. Further, there is a large variation in the measured lifetimes, which range (with 1 standard deviation) from 625 ± 25 ns for experiments based on the Hanle effect[115] to 760 ± 20 ns[116] using the high frequency deflection (HFD) technique. The lifetimes measured by laser excitation fluorescence (LEF) range from 686 ± 14 ns[117] and 693 ± 10 ns[118] to 721 ± 5 ns.[119]

We have studied[63] the convergence of the A–X TM of OH with respect to convergence of both the one-particle and n-particle spaces. The FCI calibration studies indicated that the state-averaged (SA)–CASSCF/MRCI calculations reproduced the FCI value to within 0.2%, but that this required including a δ orbital in the CASSCF active space. Further, a basis set exploration showed that the [6s 5p 4d 2f 1g/4s 3p 2d] ANO basis set employed in our study is probably within 1% of the basis set limit for the transition moment. Thus the computed radiative lifetime of 673 ns for the $A^2\Sigma^+(v' = 0, N' = 1)$ state of OH is expected to be a lower bound and accurate to about 2%. This value is in excellent agreement with two of the LEF values. The theoretical calculations are sufficiently accurate to rule out the somewhat lower value determined by Hanle effect studies, and the higher values determined for the $v' = 0$, $N' = 1$ level by the HFD technique. The HFD lifetimes for higher N' values, however, are in relatively good agreement with theory—see Fig. 1 where we have plotted the lifetime for $v' = 0$ as a function of rotational level. Since the lifetimes are expected to increase monotonically with N' in the regime where the lifetimes are unaffected by predissociation, theory suggests that there may be a systematic error in the HFD lifetimes for small N', perhaps due to collisional quenching.

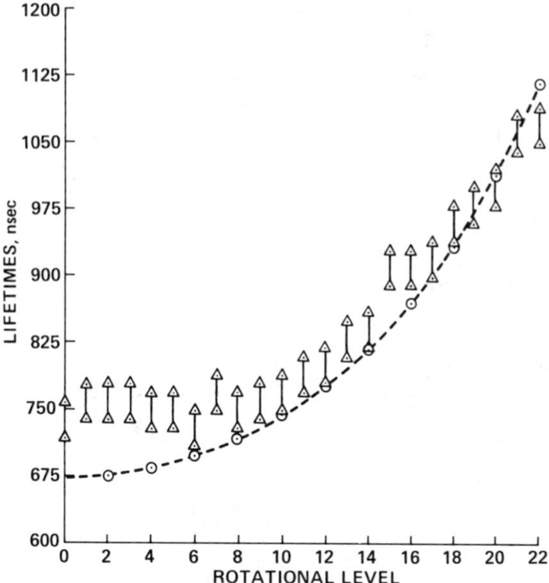

Figure 1. Comparison of the theoretical radiative lifetimes[63] for different rotational levels of the $v = 0$ level of the $A^2\Sigma^+$ state of OH (dashed line) with the high-frequency deflection measurements (F_1 component) of Brzozowski et al.[116] shown with error bars. Reprinted from C. W. Bauschlicher and S. R. Langhoff, *J. Chem. Phys.* **87**, 4665 (1987), by permission of the American Institute of Physics.

The OH applications discussed here involve transitions between valence states. The application of the SA-CASSCF/MRCI approach to valence–Rydberg transitions poses additional problems. The requirement of adding diffuse orbitals to the one-particle basis set to describe the Rydberg character is well known, so we focus on the requirements for the *n*-particle treatment.[68] An excellent system for such a calibration is the AlH molecule, which can be treated very well as a four-electron system, since L–M intershell correlation introduces only a small bond contraction.[120,121] The FCI study of AlH included the $X^1\Sigma^+$ and $A^1\Pi$ valence states and the $C^1\Sigma^+$ Rydberg state (see Fig. 2). A relatively small basis set was used, but one that was capable of describing the Rydberg character of the C state. The $A^1\Pi$–$X^1\Sigma^+$ transition moment is well described by an SA-CASSCF procedure that includes the Al 3s and 3p and H 1s orbitals in the CASSCF active space. As the $C^1\Sigma^+$ state is derived from the $^2S(3s^24s^1)$ state of Al, the active space must be expanded beyond the valence orbitals. When the 4s is added to the active space, all MRCI properties except the dipole moment of the C state are in good agreement with the FCI. Since the $C^1\Sigma^+$ state is very diffuse, small

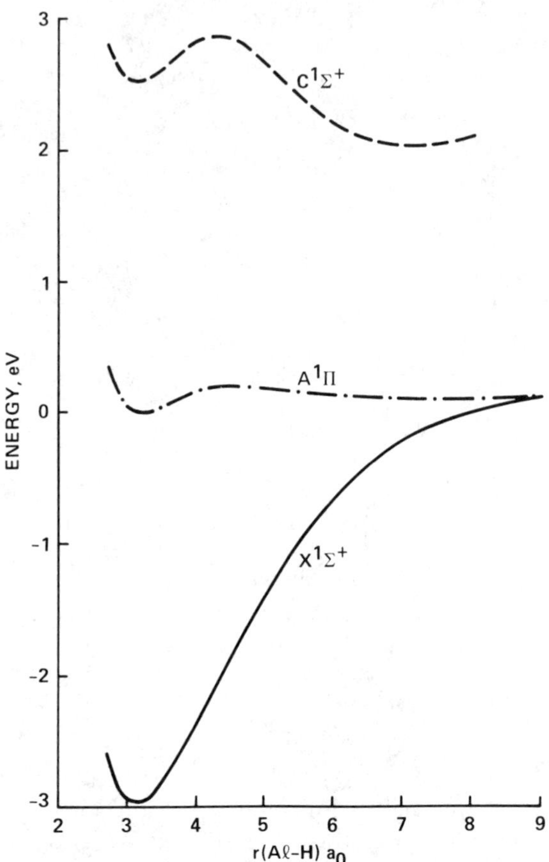

Figure 2. The FCI potential energy curves for the $X^1\Sigma^+$, $A^1\Pi$, and $C^1\Sigma^+$ states of AlH. Reprinted from C. W. Bauschlicher and S. R. Langhoff, *J. Chem. Phys.* **89**, 2116 (1988), by permission of the American Institute of Physics.

changes in the shape of the Rydberg orbital lead to very large changes in the dipole moment. The CASSCF active space must be increased by two σ, two π, and one δ orbital in order to compute an accurate dipole moment for the $C^1\Sigma^+$ state. Thus, the SA-CASSCF/MRCI approach is well suited to the study of transitions that involve Rydberg character, with only the minor complication that one or two extra orbitals must be added to the active space. However, if the Rydberg state dipole moment function is required, more substantial augmentation of the active space is needed.

We now turn to the accurate calculation of dissociation energies (D_e), which is a more challenging task than either properties or energy separations between states, because the calculations must be sufficiently flexible to describe the very different molecular and atomic correlation effects equally. Thus only methods that compute a very large fraction of the total valence correlation energy can accurately account for the differential correlation contribution to D_e. We first consider the calculation of D_e for singly-bonded systems.

The dissociation energy of NH is of astrophysical interest for the determination of its abundance in comets and the sun. It is also of interest in modeling the kinetics of rich ammonia flames. However, the experimental determinations since 1970 have ranged from 3.21 to 3.78 eV and even the recommended values are in disagreement.[122,123] Quite recently, a lower bound of 3.32 ± 0.03 for D_0 was determined using two-photon photolysis of NH_3[124]. The best estimate is obtained by combining this lower bound with an upper bound of 3.47 eV, determined from the predissociation of the $c^1\Pi$ state.[125] Thus the D_0 of NH is an order of magnitude more uncertain than that of CH or OH, which leads to unacceptably large errors in the astrophysical or kinetic modeling. Since CASSCF/MRCI calculations for OH in a large ANO basis set gave a D_e value within 0.03 eV of the accurate experimental value,[108] analogous calculations for the $X^3\Sigma^-$ state of NH are expected to yield results of comparable accuracy. Thus an accurate D_e value can be computed for NH by using the FCI approach to calibrate the *n*-particle treatment and then applying the same level of theory to CH and OH where accurate D_e values are known experimentally. The FCI calibration calculations[60] showed that the CASSCF/MRCI treatment gave equally good D_e values for CH, NH, and OH. The D_0 values for CH, NH, and OH, obtained using the CASSCF/MRCI approach in a large ANO basis set, are summarized in Table VI. A comparison of theory and experiment for CH and OH indicates that the calculations underestimate the D_0 values by about 0.03 eV. Since comparable accuracy is expected for all three systems, we can

TABLE VI
Summary of the D_0 Values for the $X^3\Sigma^-$ State of NH, in eV

	CH	NH	OH
MRCI[a]	3.433	3.344	4.360
Recommended[a]		3.37±0.03	
Experiment[b]	3.465	>3.29 <3.47	4.393

[a] Ref. 60.
[b] Refs. 108, 124 and 125.

accurately estimate the D_0 value of NH by adding 0.03 eV to the calculated value. The directly computed value cannot be too small, as the ANO basis sets have virtually no superposition error, and it is highly unlikely that the error in NH is twice that of CH or OH, so we are able to assign an uncertainty of 0.03 eV to our recommended value of 3.37 eV. This theoretical prediction is consistent with, but just slightly less than, a very recent experimental determination.[126]

The calculation of a D_e for a multiply-bonded system or those with many electrons is even more difficult than for the singly-bonded first-row hydrides such as NH or OH. We have carried out a systematic study[127] of the D_e values of N_2, O_2 and F_2 to evaluate how errors in the application of the computational methodology vary with the degree of multiple bonding. FCI calibration calculations in a realistic basis set are only possible for N_2 and O_2, and then only with the restriction that the correlation treatment is restricted to the 2p electrons. For both the $X^1\Sigma_g^+$ state of N_2 and the $X^3\Sigma_g^-$ state of O_2, the CASSCF/MRCI treatment correlating the 2p electrons accounts for essentially all of the correlation effects on the spectroscopic constants (see Table II). For N_2 this treatment in a large ANO basis set produces a D_e value that is larger than experiment[108] (see Table VII). Since this basis set has virtually no CI superposition error, this conclusively shows that 2s correlation reduces the D_e value. Subsequent MRCI calculations including 2s correlation confirm this conclusion, as the MRCI(10) (that is, ten electrons correlated) D_e value is 0.16 eV less than experiment. The decrease in D_e when the 2s electrons are correlated can be explained in terms of an important atomic correlation effect that has no analog in the molecular system, namely the 2s → 3d excitation with a recoupling of the 2p electrons. The fact that the loss of this atomic effect between infinite separation and

TABLE VII
Spectroscopic Constants for the $^1\Sigma_g^+$ State of N_2

Method	[5s 4p 3d 2f 1g] basis r_e (Å)	ω_e (cm^{-1})	D_e (eV)
MRCI(6)	1.096	2382	10.015
MRCI(6)+Q	1.096	2382	10.042
MRCI(10)	1.101	2343	9.723
MRCI(10)+Q	1.102	2336	9.745
Expt[a]	1.098	2359	9.905

[a] Ref. 108.

r_e could lead to an MRCI(6) value that was too large had been suspected,[128] but these calculations dramatically demonstrate the effect. At the MRCI(10) level (based on the CASSCF reference space from the six-electron calculation—CASSCF(6)) the error in D_e is about 4 kcal/mole, or larger than the 1 kcal/mole desired for chemical accuracy. Recent calculations[129] show that part of the error in the MRCI(10) calculation arises from inadequacies in the n-particle treatment: the MRCI reference space must include all possible distributions of the 2s and 2p electrons in the 2s and 2p orbitals (note that while the MRCI reference space is expanded, the orbitals are taken from the CASSCF(6) calculation), and part of the error is due to limitations in the one-particle space. Of the remaining one-particle errors, adding more ANOs from the previous primitive set (s through g functions) is the most important factor. Further saturation of the s through g spaces and higher angular momentum functions (h and i type functions) are of about the same importance as the errors in the n-particle space. Ideally one would like to use an FCI calculation to calibrate the errors in the n-particle space, including the N 2s electrons. However, this is not possible currently. Nevertheless, the FCI calculations and the systematic improvement of the one-particle basis sets obtained using ANO sets give new insight into the requirements for computing the D_e of N_2.

The results for the D_e value of O_2 are in better agreement with experiment than for N_2; an error of only 1.7 kcal/mole, or only about twice that found for the hydrides. For F_2, FCI calibrations are not currently possible, even with 2s correlation excluded. However, as in N_2, the insight obtained from other benchmark calculations[29] suggests that to describe the F^-F^+ character in the wave function an "extra" set of π orbitals should be added to the CASSCF active space. With this addition to the active space, the results are significantly improved: even the CASSCF results are now in good agreement with experiment, and the MRCI results in a large basis set yield a D_e with an error of only 0.5 kcal/mole. The agreement is nearly perfect with the addition of the $+Q$ correction. Thus, while the electron-dense systems are computationally more difficult, relatively accurate results can still be obtained using MRCI methods. The accuracy is improved with the inclusion of the $+Q$ estimate for the higher excitations when the number of electrons correlated becomes large. Multiple bonds place tighter demands on both the one- and n-particle treatments, but the calculations are becoming increasingly accurate as the use of large CI expansions becomes possible.

In the remainder of this section, we consider problems where the understanding of the approximations in the one- and n-particle spaces have allowed important chemical problems to be solved. The CN radical is observed in flames, comets, and stellar atmospheres. The $A^2\Pi$–$X^2\Sigma^+$ red system is particularly important as a nitrogen abundance indicator for red giant stars.[130] If the oscillator strength of the red system and the ground-state dissociation energy of CN are both accurately known, the observed emission intensity in a stellar atmosphere can be converted into elemental abundances. Unfortunately, there is a large variation in the experimental determinations of both quantities. Thus a systematic theoretical study to provide reliable values for these quantities was undertaken.[131]

As discussed for N_2, the computation of accurate bond strengths for multiply-bonded systems is a difficult undertaking, requiring both extensive one-particle basis sets and a high level of correlation treatment. However, by performing analogous calculations for the C_2, N_2, and NO molecules where the D_0 values are much better known, the calculated D_0 for CN could be extrapolated by comparing the experimental and theoretical values for these other multiply-bonded molecules. The s–p promotion and hybridization that can occur for C, but not for N or O, could lead to a different requirement in the *n*-particle treatment. The possibility that s–p hybridization introduces an error into the computed D_e value of CN was excluded by computing the D_e of CN indirectly from the D_e of CN^- using the experimental EA of C[132] and CN.[133] As the two routes agree even though the s–p near degeneracy is not present in C^-, this effect does not appear to introduce any perceptible error in the CN D_e value. In this way, we were able to arrive at a final D_0 value of 7.65 ± 0.06 eV, where the estimated error bars represent 80% confidence limits. By increasing the error bars to ± 0.10 eV, we believe that the theoretical estimate has a 99% probability of encompassing the correct value.

The calibration of the lifetime of the CN red system was done in two ways: the CN violet system, which is well-characterized experimentally, was studied, and the transitions in the isoelectronic N_2^+ molecule analogous to the red and violet system in CN were studied (the Meinel[134] and first negative systems[135]). For the CN violet, the N_2^+ Meinel, and the N_2^+ first negative systems, the theoretical radiative lifetimes agree well with experiment.[136–138] Therefore, analogous calculations are expected to yield an accurate radiative lifetime of the CN red system as well. Our computed lifetime of 11.2 μs for the $v = 0$ level of the $A^2\Pi$ state is in excellent agreement with a previous theoretical study.[139] Nevertheless, it is significantly longer than the value of 8.50 ± 0.45 μs, obtained[140] by extrapolation of time decays to zero pressure following photodissociation of C_2N_2. It is difficult to reconcile this difference, but it is conceivable that there are other decay mechanisms such as intersystem crossing between the $X^2\Sigma^+$ and $A^2\Pi$ states that shorten the experimentally observed lifetime. Other indirect experimental measurements of the intensities,[141,142] such as measured oscillator strengths, are consistent with theory or suggest an even longer lifetime. Note also that the lifetimes deduced from analysis of the solar spectrum are consistent with theory if the theoretical D_0 value of 7.65 eV is used in the analysis. Thus the theoretical lifetimes and D_0 value are consistent with the solar spectrum model of Sneden and Lambert.[130]

The analysis of cometary data requires knowing the vibrational transition band strengths in the $X^2\Sigma^+$ state of CN. Two very different values for the Einstein coefficient (A) of the fundamental 1–0 vibrational band have been reported: one based on analysis of cometary data[143] and the other from measurements in a King furnace.[144] Using the CASSCF/MRCI dipole moment function, the computed A_{10} value[145] was in excellent agreement with the value measured in the King furnace. The small uncertainty in the computed value suggests that some of the assumptions in the model used to analyze the cometary data are in error.

One of the most important chemiluminescent phenomena is the so-called Lewis–Rayleigh afterglow that occurs from the recombination of ground-state nitrogen atoms.[146] Since most of the emission occurs in the first positive bands (1+) of N_2, which cannot be directly populated in the recombination of ground-state atoms, a precursor state must be involved. Berkowitz, Chupka, and Kistiakowsky (BCK)[147] have proposed the $A'^5\Sigma_g^+$ state as the precursor, whereas Campbell and Thrush[148] invoke instead the $A^3\Sigma_u^+$ state. The main objection to the BCK theory was that the binding energy of the $A'^5\Sigma_g^+$ state (estimated[149] at the time to be $\approx 850\,cm^{-1}$) was not sufficient to maintain an appreciable steady-state population.

Recent theoretical calculations[150] on the $A'^5\Sigma_g^+$ state potential again illustrate that accurate results for the binding energy require both extensive treatments of electron correlation and a large one-particle basis set. On the basis of CASSCF/MRCI calculations using a large ANO basis set, the $A'^5\Sigma_g^+$ potential was found to have an inner well with a depth of about $3450\,cm^{-1}$ and a substantial barrier ($\approx 500\,cm^{-1}$) to dissociation. Figure 3 shows the $B^3\Pi_g$, $A^3\Sigma_u^+$, and $A'^5\Sigma_g^+$ potential curves and their vibrational levels in the region of interest. The explanation for most of the experimental observations is apparent from these potentials. For example, the barrier in the $A'^5\Sigma_g^+$ state yields a quasibound level that allows intersystem crossing from the A′ state to $v' = 13$ of the $B^3\Pi_g$ state. The observation that $v' = 13$ is populated by the same mechanism as $v' \leqslant 12$, and that emission from $v' \leqslant 13$ is similar, but markedly different from $v' > 13$, even though $v' = 13$ is above the energy of the separated atoms, is consistent with the barrier in the theoretical potential. At 300 K the maximum intensity originates from $v' = 11$ in the $B^3\Pi_g$ state. This arises since the A′ state is vibrationally relaxed before intersystem crossing to the B state, i.e., the lowest vibrational level of the A′ state most efficiently crosses to $v' = 11$ in the B state. At 77 K, an outer van der Waals well in the A′ state (not shown in the figure) allows for tunneling to the higher vibrational levels of the inner well of the A′ state. Since there are

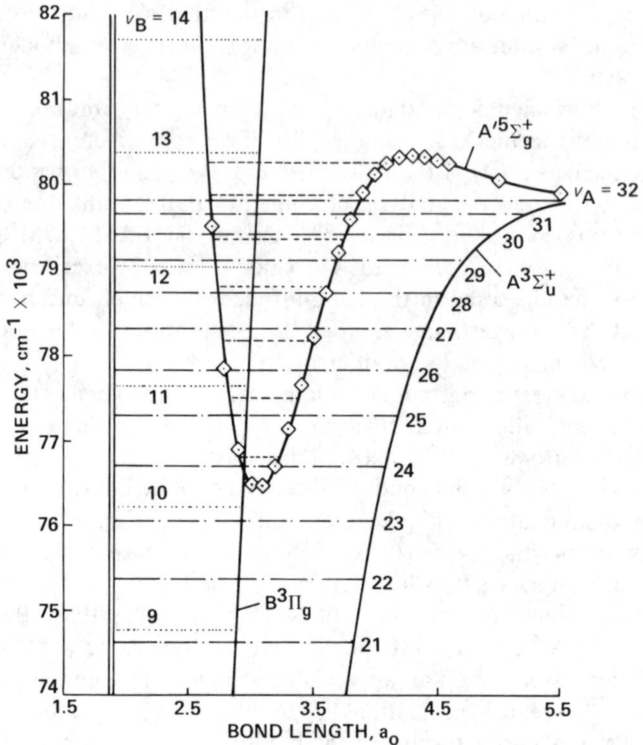

Figure 3. Potential energy curves and vibrational levels for the $A^3\Sigma_u^+$, $A'^5\Sigma_g^+$ and $B^3\Pi_g$ states of N_2. Reprinted from H. Partridge, S. R. Langhoff, C. W. Bauschlicher, D. W. Schwenke, *J. Chem. Phys.* **88**, 3174 (1988) by permission of the American Institute of Physics.

fewer collisions at this reduced temperature, intersystem crossing to the $v' = 12$ level of the $B^3\Pi_g$ state is more rapid than collisional relaxation, causing the maximum intensity in the $B^3\Pi_g$ state emission to increase from $v' = 11$ to $v' = 12$. At 4 K the barrier in the A′ state leads to a cutoff in the emission from $v' = 10 - 12$. However, the $A^3\Sigma_u^+$ state has no barrier and therefore it populates the $B^3\Pi_g$ state even at 4 K. We note that the $A^3\Sigma_u^+$ and $B^3\Pi_g$ potential curves cross at $v = 16$ in the A state and $v' = 6$ in the B state, thereby giving a maximum in emission for the $v' = 6$ levels of the B state at 4 K. Therefore, the calculations indicate that the $A^3\Sigma_u^+$ and $A'^5\Sigma_g^+$ states are both important precursors in populating the B state. The deep well in the $A'^5\Sigma_g^+$ state, revealed for the first time by these calculations, eliminates most of the objections to the original BCK theory.

Part of the emission in the Lewis–Rayleigh afterglow involves the

TABLE VIII
Band Positions and Relative Intensities of the HIR System of N_2

Band	Band positions (nm)		Relative intensities	
	Theory	Experiment[a]	Theory	Experiment[a]
0-0	806[b]	806	3.5	4.0
0-1	854	855	1.0[c]	1.0
0-2	906	907	0.6	
0-3	960	963	0.3	
0-4	1016	1023	0.1	
1-0	753	752	6.0[c]	6.0
1-1	795	795	0.2	
1-2	840	840	0.7	<0.5
1-3	885	887	1.6	
1-4	932	938	1.4	
2-0	707	706	10.0[c]	10.0
2-1	745	744	8.8	8.0
2-2	783	783	6.6	5.0
2-3	823	824	0.3	<0.5
2-4	864	868	0.7	
3-0	668	667	1.5[c]	1.5
3-1	701	700	10.4	9.0
2-3	735	735	0.6	<0.5
3-3	771	771	4.8	3.0
3-4	806	809	2.5	

[a]Ref. 152.
[b]The 0-0 bands were shifted into coincidence; this required a shift of 567 cm^{-1} in the theoretical T_e.
[c]These bands were normalized to experiment by adjusting the vibrational populations. This requires that the relative populations of v'=0-3 be 0.073, 0.295, 1.00, and 0.725, respectively.

Hermann infrared system (HIR). This band system consists of a group of unclassified multi-headed bands in the region 700–970 nm. Although the system was unassigned, it was known[151] that the bands were produced by either a triplet or quintet transition and that the system was readily generated from the energy pooling reaction between metastable $N_2(A^3\Sigma_u^+)$ molecules. As part of our theoretical study[150] of the N_2 afterglow, we found that the HIR band positions and intensities corresponded exceptionally well to our theoretical values for the $C''^5\Pi_u$–$A'^5\Sigma_g^+$ transition (see Table VIII). This, combined with the fact that the energetics are consistent with an energy pooling reaction preferentially populating the $v' = 3$ level of the $C''^5\Pi_u$ state,[152] provided rather convincing evidence for the assignment of the HIR system to this transition. This assignment has now been confirmed[153] by a rotational analysis of the $C''^5\Pi_u$–$A'^5\Sigma_g^+$ band system produced in emission in a supersonic jet discharge.

VIII. TRANSITION METAL SYSTEMS

The accuracy of calculations on transition metal systems has lagged behind that of the first and second rows of the periodic table.[154] One problem for the transition metal compounds is that an SCF wave function is often a much poorer representation of the system than for those compounds only composed of first- and second-row elements. In fact, for some transition metal systems, large CASSCF calculations are required even for a qualitatively correct description, and a quantitative description requires lengthy CI expansions. In addition to extensive *n*-particle basis requirements, experience has shown that transition metals also require considerably larger one-particle basis sets than for the first and second rows. As transition metals have such stringent one- and *n*-particle requirements, FCI benchmark calculations and ANO basis sets have given considerable insight into how to improve the accuracy of calculations on transition metal systems.

A common feature of transition metal calculations is that several atomic asymptotes are involved in the bonding: this mixing is much more prevalent for transition metal systems than non-transition metal systems and much harder to describe accurately. For example, in the transition metal hydrides it has been shown that the d^ns^2, $d^{n+1}s^1$, and d^{n+2} asymptotes all contribute to the bonding in the low-lying states. The importance of these asymptotes depends on the atomic separations. Further, the magnitude of the d–d exchange energy and the relative size of the s and d orbitals are both important in determining the bonding in the molecular system.

As a first example of calculations on transition metals, we consider the $^5D(3d^64s^2)$–$^5F(3d^74s^1)$ separation in the Fe atom. FCI benchmark calculations[58] of this separation, in a basis set that includes an f function, show that for Fe the single-reference SDCI procedure accounts for most of the differential correlation energy in the 3d and 4s shells. When this SDCI procedure is carried out in a [7s 6p 4d 4f 2g] basis set and the dominant relativistic effects accounted for, the separation is 1.25 eV: this is significantly larger than the experimental separation[155] of 0.875 eV. Adding the 3s and 3p electrons to the SDCI treatment reduces the separation to 1.06 eV. It is highly likely that most of the error in the 16-electron calculation comes from errors in the *n*-particle treatment. However, approximate methods of accounting for higher excitations cannot be reliably used for this problem; the 8-electron FCI benchmark calculations showed that the $+Q$ correction and the CPF method incorrectly estimate the differential effect of the higher excitations. This is due to the different character of 4s and 3d correlation energy contributions: for example, c_0 is reduced much more by 4s–4s correlation than an equivalent amount of 3d–3d correlation. Thus, when two states are compared, the $+Q$ correction is reliable only if the state with the larger correlation energy also has the smaller c_0 value. The $+Q$ correction may therefore fail to give reliable results for transition metal systems when different types of correlation effects are included, even though the correction works well in first-row systems when eight or so electrons are correlated. This indicates that caution must be used when applying approximate treatments that account for higher excitations to single-reference-based transition metal calculations. Finally, to compute accurate atomic separations in Fe, and presumably for the metal atoms left of it in the first transition row, it will be necessary to correlate the 3s and 3p electrons.

As noted above, the bonding in transition metal systems frequently involves a mixture of several atomic asymptotes. A study[65] of TiH has shown that a CASSCF/MRCI treatment agrees well with the FCI for the spectroscopic constants, but is much poorer for one-electron properties such as the dipole moment. This difference arises because the bonding in the $X^4\Phi$ state of TiH involves a mixture of the Ti $3d^24s^2$ and $3d^34s^1$ atomic occupations. The $3d^34s^1$ occupation forms a Ti 4s–H 1s bond that is polarized toward the H, thus yielding a large dipole moment of Ti^+H^- polarity, while in the $3d^24s^2$ occupation, 4s–4p hybridization occurs, with one hybrid orbital polarized toward the H, and the second polarized away from the H. The movement of charge toward the hydrogen is balanced by the movement away, thus the dipole moment associated with the $3d^24s^2$ asymptote is quite small. Since both asymptotes form strong bonds to the hydrogen, but have very different dipole moments, the energetics are less sensitive to the mixing than the total dipole moment. The FCI studies show that the CASSCF/MRCI method may not yield a reliable dipole moment, since the CASSCF generally includes only the nondynamical correlation and the MRCI is not able to fully account for the orbital relaxation associated with dynamical correlation. However, the benchmark calculations show that the MRCI properties converge to the FCI result if natural orbital iterations are performed, which we denote NO–MRCI. This is different from first-row systems where NO iterations were not required; for example, the CASSCF/MRCI properties for OH are in excellent agreement with the FCI without NO iterations. Therefore, in the study of transition metals at least one NO iteration should be performed to test the convergence of properties.

Since the calculations on TiH showed that the NO–MRCI properties were in good agreement with the FCI, it is expected that this level of treatment in a large one-particle basis should yield very accurate results. NiH is one of the few transition metal systems where the dipole moment (2.4 ± 0.1 D) has been measured.[156] At the SCF level, the dipole moment of NiH is much too large (about 4 D), because the wave function is biased strongly in favor of the $3d^94s^1$ asymptote of Ni.[157] This bias is not fully overcome with low levels of correlation treatment, and at the SDCI level the dipole moment (3.64 D) remains much too large. A CASSCF treatment allows the Ni $3d^84s^2$ and $3d^94s^1$ asymptotes to mix, but because the $3d^84s^2$ state lies over one eV too low at this level, the CASSCF mixes too much $3d^84s^2$ character into the wave function resulting in a dipole moment that is too small (1.74 D). Inclusion of more extensive correlation in the MRCI approach gives a better balance between $3d^84s^2$ and $3d^94s^1$ and a dipole moment (2.15 D) that is closer to experiment. However, the converged NO–MRCI dipole moment of 2.59 D is slightly larger than experiment.

The accurate treatment of weakly-bound systems is very difficult, since correlation is required to describe the dispersion, but superposition errors (SE) can be significant at high levels of correlation treatment.[82,158] To eliminate the SE, very large primitive sets, including many sets of polarization functions, are required. Until recently, large valence primitive sets were unavailable for the transition metal atoms, making it uncertain whether it was worth studying weakly-bound transition metal complexes. Partridge[159] has recently optimized large primitive basis sets for the first transition row atoms. When these (20s 12p 9d) basis sets are supplemented with a diffuse d function to describe the $3d^{n+1}4s^1$ occupation and three p functions to describe the 4p orbital, they are of the same quality as the (13s 8p) sets for first-row atoms. These large basis sets have a triple-zeta 4s function, and when flexibly contracted and supplemented with extensive polarization sets they yield very small superposition errors. Using the ANO procedure these basis sets can be contracted to a manageable size. As an example we consider the Ni—H_2O system.[85] This study used a (20s 15p 10d 6f)/[7s 6p 4d 3f] contraction for Ni, a (14s 9p 6d 4f)/[6s 5p 4d 2f] contraction for O, and an (8s 6p)/ [4s 3p] contraction for H. At the SDCI level, the total superposition error is only 0.53 kcal/mole or about 10% of the binding energy. The Ni superposition error was only about twice that of the water. As discussed in Ref. 85, previous studies of the binding energy had obtained spuriously large values as a result of superposition error. Thus with the current large primitive basis sets and improved contraction schemes, it is now possible to study weakly bound transition metal systems.

As a further example of accurate calculations for a transition metal system, consider the difficulty of determining the ground state of FeH. The laser photodetachment spectroscopic studies of FeH^- by Stevens et al.[160] have been interpreted in terms of a $^4\Delta$ ground state and a low-lying $^6\Delta$ state only 0.25 eV higher. However, the $^4\Delta$–$^4\Delta$ infrared system, which has been observed in the absorption spectra of heated mixtures of iron metal and hydrogen, has not yet been observed at low temperatures in matrix studies.[161] Although early theoretical calculations[162] placed the $^6\Delta$ state lower, qualitative arguments, based on the r_e values of the $^4\Delta$ and $^6\Delta$ states of FeH and the $^5\Delta$ ground state of FeH^- in relation to the width of the peaks in the photodetachment spectra, supported the interpretation of a $^4\Delta$ ground state by Stevens et al.[160] However, more accurate calculations are now possible using ANO basis sets and large-scale MRCI calculations.[163] For example, a CASSCF/MRCI + Q calculation, correlating the eight valence electrons in a large [8s 7p 5d 3f 2g/4s 3p 2d] ANO Gaussian basis set, predicts the $^4\Delta$ state to be 0.06 eV lower than the $^6\Delta$ state. (Note that the + Q correction should be qualitatively correct since the differential contribution from c_0^2 and the correlation energy work in the same direction). The $^4\Delta$–$^6\Delta$ separation is further increased to 0.16 eV when a correction for 3s and 3p inner-shell correlation is included based on MCPF calculations. Most of the remaining discrepancy with the experimental estimate of 0.25 eV for the separation is probably due to underestimating the differential effects of inner-shell correlation. Therefore, these theoretical predictions provide strong support for the interpretation of the photodetachment spectra of Stevens et al.[160] in which the $^6\Delta$ state is placed above the $^4\Delta$ ground state. Nevertheless, FeH represents a very difficult case for theory, namely the presence of two nearly degenerate states of different multiplicity. Very high levels of correlation treatment are required to quantitatively account for the separation as the correlation energy is substantially larger for the lower spin state. The failure to observe the $^4\Delta$–$^4\Delta$ infrared system in absorption at low temperature is probably due to the broadening of this already weak band system by the matrix.[160]

The difficulty in computing an accurate $^4\Delta$–$^6\Delta$ separation in FeH arises mostly from the large 3d–3d exchange energy resulting from the compact nature of the 3d orbital. Although the large loss in exchange energy from reversing the electron spin is balanced by the greater bonding in the $^4\Delta$ state of FeH, this latter effect requires high levels of correlation treatment to describe. Thus, low levels of correlation treatment favor the $^6\Delta$ state, resulting in the incorrect ordering of states. This situation can be contrasted with the $^1A_1$–$^3B_1$ separation in CH_2. More accurate results are possible in the latter case, because the differential exchange and bonding contributions are much smaller.

IX. THE CALCULATION OF POTENTIAL ENERGY SURFACES

In the previous sections, we have discussed the application of the most accurate theoretical methods for the calculation of spectroscopic constants. Another important area of computational chemistry is dynamics.[164] A variety of methods have been developed for the determination of scattering cross sections, rate constants, and product state distributions. These dynamical methods, either classical or quantum mechanical, require knowing the form of the potential energy surface (PES). Therefore, the accuracy of the predictions of these dynamical methods ultimately depends on the accuracy of the PES. It is clear that single-reference-based methods will not be able to describe the bond breaking and bond formation that occur during a reaction. They may, however, yield accurate surfaces for non-reactive collisions. The multireference-based approaches appear to be the only alternative for the accurate calculation of general PESs. Given the kind of insight that the FCI calibration has given into the n-particle approximations for spectroscopic problems and the high accuracy that the ANO basis sets have shown for these classes of problems, it is important to test these advances on the calculation of PESs. Unfortunately, it is not possible to directly compare a theoretical PES with experimental observations, and therefore some of our conclusions will have to wait for confirmation by application of dynamical studies to our PESs.

The first system we consider is the isotope exchange reaction for H_2 and D_2. Experimental studies of this reaction are very difficult to perform since impurities allow the formation of H atoms, which then results in a radical exchange reaction.[165] The early theoretical studies were flawed by the assumption that the reaction proceeded by a four-center symmetry-forbidden process.[166] While the later theoretical studies[167] focused on the symmetry-allowed termolecular process

$$2H_2 + D_2 \rightarrow 2HD + H_2 \tag{17}$$

they suffered from limitations in both the one- and n-particle basis sets and contained an uncertainty of perhaps 10 kcal/mole in the barrier height. However, recent calculations[70] have reduced this uncertainty to less than 1 kcal/mole. In that study, all calculations were performed on six-electron systems in order to minimize size-consistency errors in the comparisons, so the H_2 results to be quoted were obtained from calculations on three separated H_2 molecules. An FCI treatment in a DZP basis set shows that the CASSCF/MRCI D_e of H_2 is only 0.1 kcal/mole smaller than the FCI, while applying the $+Q$ correction to the six-electron calculation results in a D_e value for each H_2 that is 0.1 kcal/mole too large. There is essentially no difference in the saddle-point geometry between the FCI, MRCI, or MRCI + Q calculations, and the barrier heights are also in good mutual agreement: 70.9(MRCI), 70.4(MRCI + Q), and 70.4(FCI). The MRCI calculations were then repeated in an ANO basis set that yields a D_e value for H_2 at the MRCI + Q level (for a system of three separated H_2 molecules) that is only 0.3 kcal/mole smaller than experiment. The barrier height at the MRCI and MRCI + Q levels is 66.5 and 67.4 kcal/mole, respectively. Based on the FCI calibration in the DZP basis set, we expect that the true barrier height lies between these values. As the error in the barrier height should be smaller than the error for the process of dissociating the three H_2 molecules to six H atoms, the barrier should be accurate to better than 1 kcal/mole, which is comparable to or better than experiment.

The $F + H_2 \rightarrow FH + H$ reaction has received considerable attention from theory[168–174] as it is an excellent system for the calibration of methods: there is experimental data for the reaction rate as a function of temperature, information on product vibrational energy distribution, and the reaction thresholds are known for both H_2 and D_2.[175,176] In Table IX, several different treatments are compared to the FCI barrier height and exothermicity;[55,168,169] in all of these treatments only seven electrons (F 2p and H 1s) are correlated. The smallest MRCI treatment has the F $2p\sigma$ and H 1s orbitals active in the CASSCF and MRCI (denoted MRCI(300), since there are three active orbitals of a_1 symmetry). This calculation yields a barrier height that is 0.68 kcal/mole higher than the FCI value. The inclusion of the $+Q$ correction improves the barrier height, but it is now slightly too small. The sign of the error in the exothermicity also changes with the addition of the $+Q$ correction. Such problems with the MRCI(300) treatment are not unexpected since the HF wave function is known to contain significant H^+F^- character. For accurate results it is therefore necessary to improve the description of the electron affinity (EA) of F by expanding the active space to include $2p_\pi \rightarrow 2p'_\pi$ correlation. With such a (322) active space, a very large number of CSFs would arise in a CASSCF reference MRCI wave function, so it becomes necessary to select reference CSFs according to their CASSCF coefficients as described in Section II. With this expanded active space, the MRCI(322) + Q results are in excellent agreement with the FCI, provided that the threshold for including CASSCF CSFs as references is no larger than 0.025. Further expansion of the active space improves the results, but the $+Q$ correction is now too large. The ACPF method was also used to estimate the effect of higher excitations. The ACPF results are better than MRCI(322)(0.025), but not as good as MRCI(322)(0.025) + Q. The differences are relatively small, however. It was also found in an ACPF treatment in a large basis set that a CSF with a coefficient of less than 0.025 in the CASSCF wave function appeared with a much larger coefficient in the ACPF wave function near the saddle point. This configuration corresponds to an ionic $F^-H_2^+$ contribution, and its inclusion (making 13 reference occupations) has little effect in the small basis set that does not adequately describe F^-. However, it has an important effect in the larger basis set, where F^- is described much better. In addition, we note that singles, doubles, triples, and

TABLE IX

FCI Calibration of the Classical Barrier Height of $F + H_2 \rightarrow HF + H$[a]

A. At the FCI saddle point

	barrier	exothermicity
FCI	4.50	28.84
MRCI(300)	5.18	28.57
MRCI(300)+Q	4.43	29.12
MRCI(322)(0.05)	5.00	29.12
MRCI(322)(0.05)+Q	4.32	29.21
MRCI(322)(0.025)[b]	4.73	29.17
MRCI(322)(0.025)+Q[b]	4.51	28.80
ACPF(322)(0.025)(12-Ref)[b]	4.56	28.89
ACPF(322)(0.025)(13-Ref)[c]	4.57	28.89
MRCI(322)(0.01)	4.71	29.19
MRCI(322)(0.01)+Q	4.54	28.84
MRCI(522)(0.025)	4.55	29.41
MRCI(522)(0.025)+Q	4.32	29.31

B. At the optimized saddle-point geometry[d]

	r(F-H)	r(H-H)	ΔE_b	ΔE_{rx}
FCI	2.761	1.467	4.50	28.84
MRCI(300)	2.740	1.476	5.16	28.57
CPF	2.801	1.467	4.40	26.47
MRCI(300)+Q	2.795	1.467	4.42	29.12
MRCI(322)(0.025)[b]	2.761	1.474	4.70	29.17
MRCI(322)(0.025)+Q[b]	2.755	1.475	4.49	28.80
ACPF(322)(0.025)[c]	2.760	1.475	4.56	28.89
SDTQCI[e]	2.763	1.465	4.45	

[a] Energies in kcal/mole and bond lengths in a_0. All calculations are done using the [4s 3p 1d/2s 1p] basis set and correlating seven electrons. The barrier is referenced to F. . .$H_2(50a_0)$, and the exothermicity is computed using HF. . .$H(50a_0)$.

[b] The 12 reference configurations chosen from the CASSCF wave function.

[c] Includes the additional configuration found to be important in the ACPF calculation — see the text.

[d] Geometry optimized using a biquadratic fit to a grid of nine points.

[e] Scuseria and Schaefer, Ref. 173.

quadruples away from an SCF reference (SDTQCI) has a comparable error to the ACPF, but of opposite sign.

As we noted in the beginning of this section, single-reference-based techniques are not expected to be quantitative, but they can still be very useful in preliminary calculations. In this regard, we note that although the CPF method is quite accurate for the barrier height and saddle-point geometry, it is significantly poorer than the MRCI for the exothermicity.

In order to compute an accurate barrier height, the basis set is expanded from the [4s 3p 1d/2s 1p] set used for the FCI calibration to a [5s 5p 3d 2f 1g/4s 3p 2d] ANO set. In this large basis set, the spectroscopic constants for H_2 are in almost perfect agreement with experiment. The MRCI(222) + Q treatment of HF, which is analogous to the MRCI(322) treatment of $F + H_2$, yields an excellent r_e, but a D_e value that is 1.22 kcal/mole (0.05 eV) too small. The CI superposition error for F is 0.15 kcal/mole; this is even smaller than that obtained using the large Slater-type basis set from Ref. 170. An analysis of the wave function shows that the saddle point resembles $F + H_2$ more than HF + H. Therefore, the barrier height should be accurate to better than 1 kcal/mole.

The theoretical results for the classical saddle-point and barrier height are summarized in Table X. Based on the FCI calibration calculations, the MRCI(322)(2p) + Q calculations with 7 electrons correlated (i.e., excluding F 2s correlation) are expected to reproduce the result of an FCI calculation in a nearly complete one-particle basis set. Since F 2s correlation decreases the barrier, this MRCI(322)(2p) + Q barrier, when corrected for the CI superposition error (SE), represents an absolute upper bound of 2.52 kcal/mole for the barrier height. The ACPF(322)(2p) barrier height is slightly higher than the MRCI(322)(2p) + Q barrier height as expected from the FCI calibration calculations. This illustrates that the one and n-particle basis sets are not strongly coupled, as both methods of estimating the higher excitations agree in two different basis sets.

The inclusion of F 2s correlation decreases the barrier height, and increases the magnitude of the + Q correction. Unlike the seven-electron treatment, the ACPF(322) and MRCI(322) + Q results do not agree as well when nine electrons are correlated. Unfortunately, it is not possible to calibrate this level of treatment using the FCI approach in a realistic one-particle basis set. However, the ACPF wave function shows that there is an $F^-H_2^+$ contribution to the wave function that is not accounted for by simply adding the extra π orbitals to the active space. As noted previously, it is necessary to add an additional CSF to the reference list to bring the MRCI and ACPF barrier heights into agreement. The additional CSF significantly changes the ACPF barrier, but does not affect the MRCI barrier and increases the MRCI + Q barrier only slightly. The agreement of the two methods gives strong support for a conservative estimate of 2.00 kcal/mole for the barrier height, after correction for SE. A more realistic value for the barrier height should account for basis set incompleteness and the underestimation of the effects of higher excitations by the + Q correction: we assume that the + Q estimate is too small, since the difference between the size-consistent ACPF(322)(2p) and ACPF(322) treatments is larger than the difference between the MRCI(322)(2p) + Q and MRCI(322) + Q treatments. Our best

TABLE X
Theoretical Studies of the Classical Saddle-Point Geometry (a_0) and Energetics (kcal/mole) for the F + H_2 Reaction

Basis[a]	Level of treatment	r^b_{HF}	r^b_{HH}	barrier[c]	exothermicity[c]
12-Ref[d]					
A	MRCI(322)(2p)[e]	2.899	1.455	2.99	33.96
A	MRCI(322)(2p)+Q	2.910	1.456	2.42	33.42
A	MRCI(322)(2p)	(2.95)	(1.45)	2.96	
A	MRCI(322)(2p)+Q	(2.95)	(1.45)	2.40	
A	ACPF(322)(2p)	(2.95)	(1.45)	2.45	
A	MRCI(322)	2.914	1.451	2.63	31.61
A	MRCI(322)+Q	2.950	1.450	1.66	30.47
A	ACPF(322)	(2.95)	(1.45)	1.17	
A	CCI(322)	...[f]	...[f]	2.79	31.8
A	CCI(322)+Q	...[f]	...[f]	2.02	30.7
A+H(f)[g]	CCI(322)	...[f]	...[f]	2.73	
A+H(f)	CCI(322) +Q	...[f]	...[f]	1.95	
A−F(g)	CCI(322)	2.879	1.447	2.89	
A−F(g)	CCI(322)+Q	2.909	1.445	2.14	
13-Ref[h]					
A	MRCI(322)(2p)	(2.95)	(1.45)	2.96	33.96
A	MRCI(322)(2p)+Q	(2.95)	(1.45)	2.51	33.42
A	ACPF(322)(2p)	2.914	1.453	2.61	33.54
A	MRCI(322)	(2.95)	(1.45)	2.59	31.61
A	MRCI(322)+Q	(2.95)	(1.45)	1.81	30.47
A	ACPF(322)	2.967	1.447	1.85	30.58
Expt.[i]					31.73

[a] "A" denotes the [5s 5p 3d 2f 1g/4s 3p 2d] basis described in Ref. 168.
[b] The saddle-point geometries in parentheses have not been optimized.
[c] The barrier is referenced to F...$H_2(50a_0)$, and the exothermicity is computed using HF...$H(50a_0)$.
[d] The 12 reference configurations chosen from the CASSCF wave function.
[e] (2p) indicates a seven-electron treatment (i.e. 2s correlation is excluded).
[f] The MRCI(300)+Q saddle point geometry is used, r(F-H)=2.921 a_0 and r(H-H)=1.450 a_0.
[g] Denotes that a function of this angular momentum type has been added.
[h] Includes the additional configuration found to be important in the ACPF calculation — see the text.
[i] From data in Ref. 108.

empirical estimate of 1.60 kcal/mole is obtained by omitting the SE correction, adding instead 0.1 kcal/mole for basis set incompleteness and a further correction of 0.1 kcal/mole for an underestimation of the $+Q$ reduction to the classical barrier height. Thus, based solely on estimates from *ab initio* calculation, the barrier height should be between 1.60 and 2.00 kcal/mole.

Our estimate is very similar to that made by Truhlar and co-workers[171] using the scaled external correlation (SEC) method, but different from that suggested by Schaefer[174] (i.e., a barrier height greater than 2.35 kcal/mole) or the Monte Carlo calculations[172] that yield a barrier of 4.5 ± 0.6 kcal/mole. The lower bound for the barrier suggested by Schaefer[174] appears to be supported by SDTQCI results[173] (correlating only 7 electrons), that yield a barrier of 2.88 kcal/mole. Further, no differential effect of F 2s correlation is observed at the SDTQCI level in a small basis set. However, the basis sets used in the SDTQCI calculations are far from the one-particle limit. If an estimate for errors in the one-particle basis is included, the SDTQCI results are found to be in good agreement with the MRCI(322)(2p) + Q and ACPF(322)(2p) methods, as expected based on the FCI calibrations. Therefore, the only difference between the MRCI (or ACPF) and SDTQCI is the effect of F 2s correlation on the barrier. In Table XI we give the results of FCI benchmark calculations correlating 9 electrons in a DZ + p_F (double-zeta plus diffuse p on fluorine) basis. As we discussed above, for realistic benchmarking a larger basis set is required. This is clear from the fact that at the seven-electron level in the larger basis set, the barrier height is MRCI > ACPF > MRCI + Q > FCI > SDTQCI, while in the smaller basis set it is SDTQCI > MRCI > FCI > ACPF > MRCI + Q. These calculations are only qualitative, but they do suggest that the SDTQCI calculation underestimates the differential effect of the F 2s correlation. Thus, even a single-reference-based technique that correctly includes both triple and quadruple excitations is inferior to the CASSCF/MRCI approach. We also feel that the deficiencies of the Monte Carlo method are related to the use of a single reference to fix the nodes: the *ab initio* results clearly show that the wave function near the saddle point has a much greater multireference character than the wave function for either the products or reactants.

While the MRCI(322) + Q (or ACPF(322)) calculations in the ANO basis set are more reliable than any previous results, considerable computer time would be required to compute a global surface at this level. The barrier

TABLE XI

Study of the 2s Effect on the F + H_2 Barrier Height (kcal/mole)[a] in the DZ + p_F Basis Set

Level of treatment	MRCI	MRCI+Q	ACPF	SDTQCI	FCI
7 electron	8.78	8.74	8.75	8.82	8.77
9 electron	6.68	6.50	6.53	6.82	6.64
Δ	2.10	2.24	2.22	2.00	2.13

[a] The saddle point geometries are taken from the MRCI(2p)+Q and MRCI+Q calculations in this basis set.

height was therefore investigated using the contracted CI (CCI) approach. In the same large ANO basis set, the CCI + Q barrier is 0.4 kcal/mole higher than the corresponding MRCI + Q value. Modification of the basis set was also investigated at the CCI level: f polarization functions on H were found to lower the barrier by only 0.07 kcal/mole, while eliminating the g function on F increased the barrier by 0.12 kcal/mole. These observations are consistent with the contention that the basis set is effectively complete. The CCI calculation in this basis set is sufficiently inexpensive that much larger regions of the PES can be investigated. Of course, given the difference between the MRCI and CCI barrier heights, some account would have to be taken of the errors in CCI treatment; this might involve adjusting the parameters in the fitted potential based on the MRCI(322) + Q calculation or on information deduced from experiment.

While the CCI + Q PES scaled using the MRCI(322) + Q results should be accurate, direct comparison with experiment is difficult. To facilitate comparison, we have employed canonical variational transition state theory[177] at the classical and adiabatic barrier using the CCI + Q potential for both $F + H_2$ and $F + D_2$. These calculations account for the zero-point energy and include a tunneling correction. The results of these calculations are summarized in Table XII. As expected, the zero-point and tunneling corrections are different for H_2 and D_2. At the classical saddle point, the barrier heights for H_2 and D_2 differ by 0.2 kcal/mole, whereas at the adiabatic saddle point the barriers are essentially the same. The observation[175,176] of nearly identical thresholds for H_2 and D_2 also provides strong support for using the adiabatic barrier. In order to bring the computed threshold into agreement with experiment,[175,176] we must lower the CCI + Q classical barrier by 0.7–0.8 kcal/mole. This produces a barrier height of 1.3–1.4 kcal/mole, or after accounting for the errors associated with various approximations, a barrier height of 1.0–1.5 kcal/mole. This is in good agreement with the estimate made directly from the various multireference treatments, and also with that deduced in recent calculations by Truhlar and co-workers,[171] although it disagrees with the value inferred by Schaefer[174] from most previous calculations.

The previous two applications consider dynamical problems where only the ground-state potential energy curve is of interest. However, this is not always the case—many situations involve a curve crossing. While most scattering formalisms are developed in a diabatic representation, a theoretical PES is computed in the adiabatic representation. Hence when curve crossings (or more complicated phenomena for polyatomic systems) occur, both potentials must be accurately represented in the crossing region, and nonadiabatic coupling matrix elements (NACMEs) will be required to define the unitary transformation between the diabatic and adiabatic representa-

TABLE XII
Zero-Point and Tunnelling Effects on the Barrier Height of the $F + H_2$ and $F + D_2$ Reactions

$F + H_2$ surface

	Classical Barrier		Adiabatic Barrier	
	CCI	CCI+Q	CCI	CCI+Q
r_{HF}, a_0	2.879	2.909	3.070	3.155
r_{HH}, a_0	1.447	1.445	1.425	1.421
Barrier, kcal/mole	2.888	2.143	2.639	1.860
Sym. stretch, cm^{-1}	3706	3768	4074	4178
Bend, cm^{-1}	68.5	45.9	68.5	45.9
Asym. stretch[a], cm^{-1}	$692i$	$605i$	$530i$	$371i$
Zero-point correction[b], kcal/mole	−0.602	−0.643	−0.076	−0.057
E barrier + zero point, kcal/mole	2.286	1.500	2.563	1.803
Tunnelling correction, kcal/mole	−0.54	−0.47	−0.42	−0.29
Threshold, kcal/mole	1.75	1.03	2.14	1.51

$F + D_2$ surface

	Classical Barrier		Adiabatic Barrier	
	CCI	CCI+Q	CCI	CCI+Q
r_{HF}, a_0	2.879	2.909	3.010	3.075
r_{HH}, a_0	1.447	1.445	1.430	1.427
Barrier, kcal/mole	2.888	2.143	2.761	1.997
Sym. stretch, cm^{-1}	2623	2667	2811	2876
Bend, cm^{-1}	37.7	19.1	37.7	19.1
Asym. stretch[a], cm^{-1}	$512i$	$448i$	$428i$	$334i$
Zero-point correction[b], kcal/mole	−0.488	−0.532	−0.220	−0.233
E barrier + zero point, kcal/mole	2.400	1.611	2.541	1.764
Tunnelling correction, kcal/mole	−0.40	−0.35	−0.34	−0.26
Threshold, kcal/mole	2.00	1.26	2.20	1.50

[a]From the normal mode analysis at the classical barrier, and computed from the curvature along the Eckart potential at the adiabatic barrier.
[b]For $H_2(D_2)$ we used ω_e=4401(3116) cm^{-1}, respectively, from Ref. 108.

tions. Until recently, NACMEs were computed either using finite difference methods[178] or via approximations to avoid computing matrix elements between nonorthogonal wave functions.[179,180] However, Lengsfield, Saxe, and Yarkony[181,182] have recently developed an efficient method of evaluating NACMEs based on state-averaged MCSCF wave functions and analytic derivative methods. This should provide NACMEs of similar overall accuracy to that obtained for the adiabatic potentials.

To accurately describe curve crossings requires, in addition to the NACMEs, an equivalent treatment of both states involved in the crossing. In curve crossings where the molecular orbitals for the two states are similar, such as interactions between valence states derived from different asymptotic limits, the CASSCF/MRCI approach would be expected to describe both potentials accurately irrespective of which state is used for the orbital optimization. However, when the character of the two states is very different, such as valence–Rydberg mixing[68] or interaction between states derived from ionic and covalent limits,[69] it is more difficult to achieve equivalent accuracy for the lowest adiabatic state on either side of the crossing point. This is commonly the case for charge-exchange reactions, such as $N^+ + N_2 \rightarrow N + N_2^+$, or chemi-ionization processes such as $M + X \rightarrow M^+ + X^-$, where the optimal molecular orbitals for the ionic and neutral solutions differ greatly. To gain additional insight into the computational requirements for describing the potentials in the region of curve crossings, we have studied[69] the $Li + F \rightarrow Li^+ + F^-$ chemi-ionization process using the FCI approach. In LiF, the lowest adiabatic state at short r values, namely the ionic $X^1\Sigma^+$ state, dissociates adiabatically to neutral ground-state atoms. There is an avoided crossing at the point where the energy difference between the F electron affinity (EA) and the Li ionization potential exactly balances the $1/r$ electrostatic stabilization. Since the CASSCF description of F^- is poor,[29] the CASSCF estimate for the bond distance at the crossing point is unrealistically small. When orbitals from the ground-state CASSCF wave function are used to construct an MRCI wave function, the CASSCF description of the crossing point will compromise the accuracy of the MRCI description. This problem is not easily resolved by expanding the CASSCF active space, as very large active spaces are required to obtain a good description of the atomic electron affinities. However, by performing instead an SA–CASSCF calculation, in which the orbitals are optimized for the average of the two lowest $^1\Sigma^+$ states in LiF (the ionic and neutral states), the orbital bias is eliminated and the MRCI treatment is in excellent agreement with the FCI. It is also important to note that this averaging does not significantly degrade the description of the system near r_e. Thus, state averaging appears to be an excellent method of achieving equal accuracy for two potential curves in a curve-crossing region, and should also perform well for polyatomic systems. The utility of state averaging as a means of obtaining a good compromise set of orthogonal molecular orbitals for use in an MRCI wave function has also been found to be an excellent route to computing accurate electronic transition moments,[63] as noted above. Thus, there is much in common between methods that account accurately for differential correlation effects on a PES and those that yield accurate spectroscopic constants and molecular properties.

Application of the SA–CASSCF/MRCI approach to the LiF curve crossing in a (13s 9p 6d 4f)/[4s 3p 2d 1f] F and (14s 7p 4d)/[5s 4p 2d] Li basis set yields smooth dipole moment and potential energy curves, as expected. These calculations were able to demonstrate that the previous failure of an analysis of the curve crossing was due to poor adiabatic potentials, not the use of the qualitative Rittner model.[180] The NACMEs deduced from the SA–CASSCF/MRCI calculation are in good agreement with other determinations.[183] The limitations in these calculations arise from the difficulty in computing the EA of F. As we discuss in the next section, this is still a very challenging task. As our ability to compute electron affinities improves, the accuracy of such ionic-covalent curve crossings will also improve, as the FCI benchmarks have shown that the SA-CASSCF/MRCI calculations are correctly describing this region of the potential.

The applications considered in this section illustrate that current methods are yielding more accurate potential energy surfaces, although it is often too expensive to compute entire PESs using very extensive MRCI calculations. However, with advances in surface fitting techniques it may be possible to merge accurate calibration calculations at critical points on the PES with global calculations using less computationally expensive methods. For example, we have used the CCI method in a smaller basis set to map out the $F + H_2$ potential.[168] This more approximate level of treatment agrees with the best calculations to within about 1 kcal/mole and is therefore expected to be semi-quantitatively correct. Scaling of these results can thus be expected to yield accurate global surfaces.

X. SYSTEMS WITH A STRONG COUPLING OF THE ONE- AND *n*-PARTICLE SPACES

In the previous sections, we have illustrated how the accuracy of *ab initio* calculations has been greatly enhanced by recent developments. However, there are a few systems where even these new advances have not fully answered the question as to which is the appropriate quantum chemical treatment. These problems involve correlation of core electrons and computation of electron affinities, and we first consider core correlation.

As noted above, a high level of correlation treatment is required to fully account for the effect of 3s and 3p correlation on the ^5D–^5F separation in Fe atom.[58] It is impossible to study inner-shell correlation effects for transition metal systems using FCI benchmark calculations, and furthermore errors associated with other approximations are generally as large. However, the calculations for the $^1A_1$–$^3B_1$ separation in CH_2[88] indicate that the treatment of the valence correlation is nearly exact and one of the largest remaining errors is the neglect of correlating the C 1s electrons. In addition, as correlating the core in C adds only two extra electrons, it is amenable to FCI benchmark calculations. Thus we have investigated[66] the effect of core correlation on the $^1A_1$–$^3B_1$ separation in CH_2, as well as the C ^3P–^5S

separation. Correlating the core not only places more stringent requirements on the n-particle treatment, it also increases the demands on the one-particle basis set.

The study of the error associated with the ANO contraction procedure on the ^3P–^5S separation in the C atom shows that while the ANO procedure works very well for valence correlation, the convergence of the basis set with contraction level for the core–core (CC) and core–valence (CV) correlation is very slow. This is related to the fact that the change in c_0 for a given correlation energy lowering, and hence the size of the natural orbital occupation numbers associated with the core correlation, is smaller than those associated with valence correlation energy. Thus, when both core and valence correlation are included, it becomes impossible to order the NOs based on their contribution to the correlation energy using the occupation numbers, and therefore it is difficult to use the ANO procedure effectively. This is analogous to the origin of the failure of the $+Q$ correction in describing higher excitations in the Fe atom. The contraction problem is reduced somewhat when the nearly constant CC correlation is deleted and only the CV correlation is included, but the convergence with contraction level is still much slower than when only valence correlation is included. Note that elimination of the CC correlation also reduces the requirements on the primitive basis set.

In addition to problems associated with the one-particle basis set, the FCI benchmarks show that very high levels of correlation treatment are required to correctly balance the CV (or CC) and valence correlation. This usually involves increasing the active space such that a larger fraction of the valence correlation energy is included in the zeroth-order wave function. In fact, for an MRCI calculation to account for all of the CV effect, the zeroth-order wave function has to account for essentially all of the valence correlation. This leads to a large expansion in the valence part of the problem, which when coupled with the larger basis set requirements and additional electrons being correlated, leads to a much larger computational problem than the study of valence correlation only. Failing to expand the valence treatment adequately can lead to an overestimation of the differential CV contribution even though the total CV correlation energy may be underestimated.

The effect of CV correlation on the $^1A_1$–$^3B_1$ separation in CH_2 was estimated by using a totally uncontracted C s and p basis set to avoid the contraction problem. Using the C ^3P–^5S separation as a calibration, it is likely that the largest possible valence active space that could be used in the CH_2 calculations would lead to a slight overestimation of the CV effect. However, even this calculation resulted in a CI expansion of about 1.5 million CSFs. The benchmark studies[66] of CV correlation indicate that it increases the $^1A_1$–$^3B_1$ separation in CH_2 by about 0.35 kcal/mole (see Table V).

Unfortunately, such CV calculations are too large to be routinely carried out for more complex systems. Thus there is considerable motivation for developing less expensive ways of including this effect, such as the CV operator approach being developed by Meyer and co-workers.[184]

The nuclear hyperfine interaction, which depends on the spin density at the nucleus, is very sensitive to CC and CV correlation. To determine even qualitatively correct results for the nitrogen atom hyperfine coupling constant requires correlating the 1s orbital.[185] Very high levels of correlation treatment as well as large flexibly contracted basis sets are required for quantitative results. Further there is a strong coupling of the one- and *n*-particle basis sets, which makes it difficult to perform definitive FCI calibration calculations.[67] To achieve quantitative agreement between the MRCI and FCI hyperfine constants for the N atom, it was necessary to include the 2s, 2p, 3s, 3p and 3d orbitals in the CASSCF active space. This is consistent with the calculations to determine the C ^3P–^5S separation, which showed that the zeroth-order reference must account for almost all of the valence correlation energy in order for the MRCI to accurately account for CV and CC effects. Also, the basis set must include several f functions to be within 5% of experiment. In spite of the strong coupling between the basis set and correlation treatment, the FCI calculations demonstrated that diffuse s functions contribute significantly to the spin density, because of an important class of configurations involving $2s \rightarrow 3s$ promotion with a spin recoupling. Thus, the study of the isotropic hyperfine constant of the N atom indicates that quantitative agreement with experiment requires very large one-particle basis sets (flexibly contracted in the core and augmented with diffuse and polarization functions), as well as an extensive treatment of electron correlation (i.e., one that correlates the core and accounts for essentially all of the valence correlation energy).

The last application that we discuss that has a strong coupling between the one- and *n*-particle spaces is the calculation of the electron affinities (EAs) of atomic oxygen[52] and fluorine.[29] We have noted earlier for OH and $F + H_2$ that extra orbitals have to be added to the CASSCF active space to describe the negative ion character. The determination of accurate EAs is difficult because the negative ions have more correlation energy than the neutrals and the correlation effects are different. The one-particle requirements are also greater for the negative ions in that the basis sets must be augmented with diffuse functions. For example, in O/O^- if only the 2p electrons are correlated, improving the level of treatment from an SCF/SDCI to a CASSCF/MRCI that includes the 2p and 2p′ orbitals in the active space increases the EA from 1.06 to 1.26 eV. Further expansion of the active space, to include the 3d orbital, increases the EA by only 10% of the effect of adding the 2p′. When the 2s and 2p electrons are correlated, the change in the EA is even larger when the 2p′ orbital is added to the active space, from 0.99 at the SCF/SDCI level to 1.28 eV at the CASSCF/MRCI level. In fact, at the SDCI level the inclusion of 2s correlation decreases the EA, while at the CASSCF/MRCI level 2s correlation increases it. It is highly likely that the effect of adding the 3d orbital to the active space will increase when both 2s and 2p correlation are included. Thus, there is a strong coupling between the importance of 2s correlation and the level of correlation treatment. Less extensive studies on F/F^- suggest that the $+Q$ correction helps in accounting for this 2s effect, but large reference spaces are still required. The dimension of the n-particle problem is further increased by the fact that tight thresholds must be employed in selecting the reference space based on the CASSCF wave functions to avoid significant discrepancies with the unselected CAS reference space MRCI. Thus the study of atomic EAs indicate that as molecular systems become increasingly ionic, larger theoretical treatments are required to obtain equivalent accuracy for the neutral and ion.

For the three applications discussed in this section, the FCI benchmark calculation as well as the CASSCF/MRCI calculations in the large basis sets have given insight into the reason why such extensive calculations are required. Unfortunately, this level of treatment is not routinely possible. It is possible that a CV operator approach[184] will allow the inclusion of this effect without greatly expanding the calculation. Since the $+Q$ correction appears to improve the results for the systems considered in this section, it is also possible that improved methods of estimating the higher excitations will eliminate the need to expand the zeroth-order valence treatment over that which is required for only the valence correlation.

XI. CONCLUSIONS

The cumulative experience derived from FCI benchmark calculations indicates that the more sophisticated n-particle space treatments in use, particularly the CASSCF/MRCI method, generally provide a close approximation to the solution of the correlation problem. In cases in which bonds are not broken, or excited states are not considered, in light atom molecules, single-reference-based treatments of exact or approximate coupled-cluster type also generate high-quality wave functions. With current supercomputers, it is perfectly feasible to consider CC or MRCI expansions of more than 10^6 terms, which opens up large areas of chemistry and spectroscopy to accurate investigation. Indeed, the main factor that determines the accuracy of such treatments is the completeness of the one-particle space. Although there has been justifiable pessimism in the past about the slow convergence towards basis set completeness with respect to both radial saturation and angular quantum number, recent developments in optimized basis sets, particularly the use of atomic natural orbital contractions, have shown that results of full chemical accuracy (better than 1 kcal/mole even in the dissociation energy of N_2) can be achieved with manageable basis sets and reasonable MRCI expansions.

In this review, we have shown how FCI benchmark calculations can be used not only to provide general criteria by which approximate *n*-particle space treatments can be judged, but also to provide detailed calibration as to which treatments are appropriate in specific cases. In this way it is possible to attach confidence limits to the computed results obtained when such a treatment is applied in a large ANO basis. This ability to estimate realistic "error bars" for calculated quantities is very important when the results are to be used in interpreting experiment or in other calculations, and we can expect more use to be made of it in the future. We have described several applications of this technique to problems in spectroscopy (such as the D_e values for NH and CN) and chemistry ($F + H_2$ barrier height).

For the future, there is reason for considerable optimism about the methodology and application of accurate quantum chemistry. A new generation of FCI programs should increase the size of FCI benchmark calculations by more than an order of magnitude, while the constant advances in more conventional methodologies, even in areas thought to be exhausted, will continue. Further, new methods, particularly the application of multi-reference coupled-cluster approaches, can be expected to find wider use, and some of the more exotic techniques that are currently being explored may begin to contribute to meaningful chemical applications. When combined with the wider availability and increasing power of supercomputers and minisupercomputers, the prospects for accurate quantum chemical calculations look bright indeed.

Acknowledgments

We would like to thank P. J. Knowles and N. Handy for giving us the original version of the FCI code. We would also like to acknowledge collaborations with the groups in Lund (B. O. Roos), Minneapolis (J. Almlöf), and Stockholm (P. E. M. Siegbahn). In addition, a number of our colleagues at NASA-Ames have contributed to this work. We would like to especially thank L. A. Barnes, T. J. Lee, H. Partridge, and D. W. Schwenke. The authors would also like to thank the NAS facility for access to the CRAY-2 and CRAY Y-MP, and to the Central Computing Facility for access to the CRAY X-MP/14se. Finally, we are grateful to the *Journal of Chemical Physics* for permission to reproduce Figs. 1–3.

References

1. K. P. Lawley, Ed., *Advances in Chemical Physics: Ab Initio Methods in Quantum Chemistry*, Wiley, New York, 1987, Vol. 67 and 69.
2. K. McIsaac and E. N. Maslen, *Int. J. Quantum Chem.* **31**, 361 (1987), and references therein.

3. P. O. Löwdin, *Adv. Chem. Phys.* **2**, 207 (1959).
4. R. J. Bartlett, *Ann. Rev. Phys. Chem.* **32**, 359 (1981).
5. J. A. Pople, J. S. Binkley and R. Seeger, *Int. J. Quantum Chem. Symp.* **10**, 1 (1976).
6. A. D. McLean and B. Liu, *J. Chem. Phys.* **58**, 1066 (1973).
7. F. W. Bobrowicz and W. A. Goddard, in *Methods of Electronic Structure Theory*, H. F. Schaefer, Ed., Plenum Press, New York, 1977, p. 79.
8. B. O. Roos, *Adv. Chem. Phys.* **69**, 399 (1987).
9. K. K. Docken and J. Hinze, *J. Chem. Phys.* **57**, 4928 (1972).
10. P. E. M. Siegbahn, *Int. J. Quantum Chem.* **23**, 1869 (1983).
11. H.-J. Werner and P. J. Knowles, *J. Chem. Phys.* **89**, 5803 (1988).
12. S. R. Langhoff and E. R. Davidson, *Int. J. Quantum Chem.* **8**, 61 (1974).
13. M. R. A. Blomberg and P. E. M. Siegbahn, *J. Chem. Phys.* **78**, 5682 (1983).
14. R. J. Bartlett, I. Shavitt and G. D. Purvis, *J. Chem. Phys.* **71**, 281 (1979), where the addition of the $+Q$ correction was shown to significantly increase the standard deviation for the fitted potential.
15. R. Ahlrichs, *Comput. Phys. Commun.* **17**, 31 (1979).
16. J. Cizek, *J. Chem. Phys.* **45**, 4256 (1966).
17. R. Ahlrichs, P. Scharf and C. Ehrhardt, *J. Chem. Phys.* **82**, 890 (1985).
18. R. J. Gdanitz and R. Ahlrichs, *Chem. Phys. Lett.* **143**, 413 (1988).
19. T. Helgaker and P. Jørgensen, *Adv. Quantum Chem.* **19**, 183 (1988).
20. D. P. Chong and S. R. Langhoff, *J. Chem. Phys.* **84**, 5606 (1986).
21. I. Shavitt, in *Methods of Electronic Structure Theory*, H. F. Schaefer, Ed., Plenum, New York, 1977, p. 189.
22. B. O. Roos and P. E. M. Siegbahn, in *Methods of Electronic Structure Theory*, H. F. Schaefer, Ed., Plenum, New York, 1977, p. 277.
23. P. E. M. Siegbahn, in *Quantum Chemistry: The State of the Art*, V. R. Saunders and J. Brown, Eds., Science Research Council, Didcot Oxon., 1975, p. 81.
24. N. C. Handy, *Chem. Phys. Lett.* **74**, 280 (1980).
25. P. Saxe, D. J. Fox, H. F. Schaefer and N. C. Handy, *J. Chem. Phys.* **77**, 5584 (1982).
26. P. E. M. Siegbahn, *Chem. Phys. Lett.* **109**, 417 (1984).
27. M. J. Downward and M. A. Robb, *Theor. Chim. Acta.* **46**, 129 (1977).
28. P. J. Knowles and N. C. Handy, *Chem. Phys. Lett.* **111**, 315 (1984).
29. C. W. Bauschlicher and P. R. Taylor, *J. Chem. Phys.* **85**, 2779 (1986).
30. E. R. Davidson, *J. Comput. Phys.* **17**, 87 (1975).
31. J. Olsen, B. O. Roos, P. Jørgensen and H. J. Jensen, *J. Chem. Phys.* **89**, 2185 (1988).
32. C. W. Bauschlicher and P. R. Taylor, *J. Chem. Phys.* **86**, 2844 (1987).
33. P.-Å. Malmqvist, A. P. Rendell and B. O. Roos, *J. Phys. Chem.*, in press.
34. P. J. Knowles, *Chem. Phys. Lett.* **155**, 513 (1989).
35. P. J. Knowles and N. C. Handy, *J. Chem. Phys.* **91**, 2396 (1989).
36. R. J. Harrison and S. Zarrabian, *Chem. Phys. Lett.* **158**, 393 (1989).
37. P. E. M. Siegbahn and B. Liu, *J. Chem. Phys.* **68**, 2457 (1978).
38. C. W. Bauschlicher, *J. Phys. B* **21**, L413 (1988).
39. R. D. Harcourt, *J. Phys. B* **20**, L617 (1987).

40. P. R. Taylor, C. W. Bauschlicher and S. R. Langhoff, *J. Phys. B* **21**, L333 (1988).
41. W. R. S. Garton, in *Proceedings of the Fifth Conference on Ionization Phenomena in Gases*, Amsterdam, North-Holland, 1962, p. 1884.
42. K. B. S. Eriksson, and I. B. S. Isberg, *Arkiv Fysik* **23**, 527 (1963).
43. A. W. Weiss, *Adv. At. Mol. Phys.* **9**, 1 (1973).
44. A. W. Weiss, *Phys. Rev. A* **9**, 1524 (1974).
45. P. Saxe, H. F. Schaefer and N. C. Handy, *Chem. Phys. Lett.* **79**, 202 (1981).
46. R. J. Harrison and N. C. Handy, *Chem. Phys. Lett.* **96**, 386 (1983).
47. R. J. Bartlett, H. Sekino and G. D. Purvis, *Chem. Phys. Lett.* **98**, 66 (1983).
48. W. D. Laidig and R. J. Bartlett, *Chem. Phys. Lett.* **104**, 424 (1984).
49. F. B. Brown, I. Shavitt and R. Shepard, *Chem. Phys. Lett.* **105**, 363 (1984).
50. C. W. Bauschlicher, S. R. Langhoff, P. R. Taylor and H. Partridge, *Chem. Phys. Lett.* **126**, 436 (1986).
51. C. W. Bauschlicher, S. R. Langhoff, P. R. Taylor, N. C. Handy and P. J. Knowles, *J. Chem. Phys.* **85**, 1469 (1986).
52. C. W. Bauschlicher, S. R. Langhoff, H. Partridge and P. R. Taylor, *J. Chem. Phys.* **85**, 3407 (1986).
53. C. W. Bauschlicher and P. R. Taylor, *J. Chem. Phys.* **85**, 6510 (1986).
54. C. W. Bauschlicher and P. R. Taylor, *J. Chem. Phys.* **86**, 1420 (1987).
55. C. W. Bauschlicher and P. R. Taylor, *J. Chem. Phys.* **86**, 858 (1987).
56. C. W. Bauschlicher and P. R. Taylor, *Theor. Chim. Acta* **71**, 263 (1987).
57. C. W. Bauschlicher and P. R. Taylor, *J. Chem. Phys.* **86**, 5600 (1987).
58. C. W. Bauschlicher, *J. Chem. Phys.*, **86**, 5591 (1987).
59. C. W. Bauschlicher and S. R. Langhoff, *J. Chem. Phys.* **86**, 5595 (1987).
60. C. W. Bauschlicher and S. R. Langhoff, *Chem. Phys Lett.* **135**, 67 (1987).
61. S. R. Langhoff, C. W. Bauschlicher and P. R. Taylor, *J. Chem. Phys.* **86**, 6992 (1987).
62. C. W. Bauschlicher, H. Partridge, S. R. Langhoff, P. R. Taylor and S. P. Walch, *J. Chem. Phys.* **86**, 7007 (1987).
63. C. W. Bauschlicher and S. R. Langhoff, *J. Chem. Phys.* **87**, 4665 (1987).
64. C. W. Bauschlicher and S. R. Langhoff, *Theor. Chim. Acta* **73**, 43 (1988).
65. C. W. Bauschlicher, *J. Phys. Chem.* **92**, 3020 (1988).
66. C. W. Bauschlicher, S. R. Langhoff and P. R. Taylor, *J. Chem. Phys.* **88**, 2540 (1988).
67. C. W. Bauschlicher, S. R. Langhoff, H. Partridge and D. P. Chong, *J. Chem. Phys.* **89**, 2985 (1988).
68. C. W. Bauschlicher and S. R. Langhoff, *J. Chem. Phys.* **89**, 2116 (1988).
69. C. W. Bauschlicher and S. R. Langhoff, *J. Chem. Phys.* **89**, 4246 (1988).
70. P. R. Taylor, A. Komornicki and D. A. Dixon, *J. Am. Chem. Soc.* **111**, 1259 (1989).
71. C. W. Bauschlicher and S. R. Langhoff, *J. Chem. Phys.* **87**, 2919 (1987).
72. S. J. Cole and R. J. Bartlett, *J. Chem. Phys.* **86**, 873 (1987).
73. J. N. Murrell, N. M. R. Hassian and B. Hudson, *Mol. Phys.* **60**, 1343 (1987).
74. K. Jankowski, R. Becherer, P. Scharf, H. Schiffer and R. Ahlrichs, *J. Chem. Phys.* **82**, 1413 (1985).
75. M. R. A. Blomberg and P. E. M. Siegbahn, *Chem. Phys. Lett.* **81**, 4 (1981).

76. B. H. Lengsfield, A. D. McLean, M. Yoshimine and B. Liu, *J. Chem. Phys.* **79**, 1891 (1983).
77. T. H. Dunning and P. J. Hay, in *Methods of Electronic Structure Theory*, H. F. Schaefer Ed., Plenum Press, New York, 1977, p. 1.
78. R. C. Raffenetti, *J. Chem. Phys.* **58**, 4452 (1973).
79. I. Shavitt, B. J. Rosenberg and S. Palalikit, *Int. J. Quantum Chem. Symp.* **10**, 33 (1976).
80. J. Almlöf and P. R. Taylor, *J. Chem. Phys.* **86**, 4070 (1987).
81. S. F. Boys and F. Bernardi, *Mol. Phys.* **19**, 553 (1970).
82. B. Liu and A. D. McLean, *J. Chem. Phys.* **59**, 4557 (1973).
83. S. R. Langhoff, C. W. Bauschlicher and P. R. Taylor, *J. Chem. Phys.* **88**, 5715 (1988).
84. J. Almlöf, T. U. Helgaker and P. R. Taylor, *J. Phys. Chem.* **92**, 3029 (1988).
85. C. W. Bauschlicher, *Chem. Phys. Lett.* **142**, 71 (1987).
86. T. H. Dunning, *J. Chem. Phys.* **90**, 1007 (1989).
87. R. Ahlrichs, K. Jankowski and J. Wasilewski, *Chem. Phys.* **111**, 263 (1987).
88. C. W. Bauschlicher, S. R. Langhoff and P. R. Taylor, *J. Chem. Phys.* **87**, 387 (1987).
89. H.-J. Werner and E.-A. Reinsch, in *Proceedings of the 5th European Seminar on Computational Methods in Quantum Chemistry*, P. Th. van Duijnen and W. C. Nieuwpoort, Eds., Max-Planck-Institut für Astrophysik, Munich, 1981, p. 206.
90. A. D. McLean, P. R. Bunker, R. M. Escribano and P. Jensen, *J. Chem. Phys.* **87**, 2166 (1987).
91. N. C. Handy, Y. Yamaguchi and H. F. Schaefer, *J. Chem. Phys.* **84**, 4481 (1986).
92. R. D. Cowan and D. C. Griffin, *J. Opt. Soc. Am.* **66**, 1010 (1976).
93. R. L. Martin, *J. Phys. Chem.* **87**, 750 (1983).
94. M. Krauss and W. J. Stevens, *Ann. Rev. Phys. Chem.* **35**, 5357 (1984).
95. W. D. Allen and H. F. Schaefer, *Chem. Phys.* **108**, 243 (1986).
96. J. Berkowitz, J. P. Greene, H. Cho and B. Ruscic, *J. Chem. Phys.* **86**, 1235 (1987).
97. K. Balasubramanian and A. D. McLean, *J. Chem. Phys.* **85**, 5117 (1986).
98. H. Basch, W. J. Stevens and M. Krauss, *Chem. Phys. Lett.* **109**, 212 (1984).
99. K. K. Sunil and K. D. Jordan, *J. Phys. Chem.* **92**, 2774 (1988).
100. M. Leleyter and P. Joyce, *J. Phys. B*, **13**, 2165 (1980).
101. T. H. Upton, *J. Phys. Chem.* **90**, 754 (1986).
102. D. M. Cox, D. J. Trevor, R. L. Whetten, E. A. Rohlfing and A. Kaldor, *J. Chem. Phys.* **84**, 4651 (1986).
103. M. A. Douglas, R. H. Hauge and J. L. Margrave, *J. Phys. Chem.* **87**, 2945 (1983).
104. H. Abe and D. M. Kolb, *Ber. Bunsenges. Phys. Chem.* **87**, 523 (1983).
105. D. S. Ginter, M. L. Ginter and K. K. Innes, *Astrophys. J.* **139**, 365 (1963)
106. Z. Fu, G. W. Lemire, Y. M. Hamrick, S. Taylor, J. Shui and M. D. Morse, *J. Chem. Phys.* **88**, 3524 (1988).
107. M. F. Cai, T. P. Dzugan and V. E. Bondybey, *Chem. Phys. Lett.* **155**, 430 (1989).
108. K. P. Huber and G. Herzberg, *Constants of Diatomic Molecules*, Van Nostrand Reinhold, New York, 1979.
109. P. R. Taylor and H. Partridge, *J. Phys. Chem.* **91**, 6148 (1987).
110. A. B. Meinel, *J. Astrophys.* **11**, 555 (1950).
111. S. R. Langhoff, R. L. Jaffe, J. H. Yee and A. Dalgarno, *Geophysical Research Lett.* **10**, 896 (1983).

112. D. Yaron, K. Peterson and W. Klemperer, *J. Chem. Phys.* **88**, 4702 (1988).
113. H.-J. Werner and P. Rosmus, in *Comparison of Ab Initio Quantum Chemistry with Experiment*, R. Bartlett, Ed., Reidel, Boston, 1985, p. 267.
114. J. Oddershede, *Phys. Scripta*, **20**, 587 (1979).
115. K. R. German, T. H. Bergeman, E. M. Weinstock and R. N. Zare, *J. Chem. Phys.* **58**, 4304 (1973) and references therein.
116. J. Brzozowski, P. Erman and M. Lyyra, *Phys. Scripta* **17**, 507 (1978).
117. W. L. Dimpfl and J. L. Kinsey, *J. Quant. Spectrosc. Radiat. Transf.* **21**, 233 (1979).
118. K. R. German, *J. Chem. Phys.* **64**, 4065 (1976).
119. I. S. McDermid and J. B. Laudenslager, *J. Chem. Phys.* **76**, 1824 (1982).
120. J. M. O. Matos, P.-Å. Malmqvist and B. O. Roos, *J. Chem. Phys.* **86**, 5032 (1087).
121. W. Meyer and P. Rosmus, *J. Chem. Phys.* **63**, 2356 (1975).
122. L. G. Piper, *J. Chem. Phys.* **70**, 3417 (1979).
123. M. W. Chase, J. L. Curnutt, J. R. Downey, R. A. McDonald, A. N. Syverud and E. A. Valenzuela, *J. Phys. Chem. Ref. Data* **11**, 695 (1982).
124. W. R. Graham and H. Lew, *Can. J. Phys.* **56**, 85 (1978).
125. A. Hofzumahaus and F. Stuhl, *J. Chem. Phys.* **82**, 5519 (1985).
126. K. M. Ervin and P. B. Armentrout, *J. Chem. Phys.* **86**, 2659 (1987).
127. S. R. Langhoff, C. W. Bauschlicher and P. R. Taylor, *Chem. Phys. Lett.* **135**, 543 (1987).
128. B. Liu, private communication.
129. J. Almlöf, B. J. DeLeeuw, P. R. Taylor, C. W. Bauschlicher and P. E. M. Siegbahn, *Int. J. Quantum Chem. Symp.* **23**, 345 (1989).
130. C. Sneden and D. L. Lambert, *Astrophys. J.* **259**, 381 (1982).
131. C. W. Bauschlicher, S. R. Langhoff and P. R. Taylor, *Astrophys. J.* **332**, 531 (1988).
132. H. Hotop and W. C. Lineberger, *J. Phys. Chem. Ref. Data* **14**, 735 (1985).
133. J. Berkowitz, W. A. Chupka and T. A. Walter, *J. Chem. Phys.* **50**, 1497 (1968).
134. S. R. Langhoff, C. W. Bauschlicher and H. Partridge, *J. Chem. Phys.* **87**, 4716 (1987).
135. S. R. Langhoff and C. W. Bauschlicher, *J. Chem. Phys.* **88**, 329 (1988).
136. N. Duric, P. Erman and M. Larsson, *Phys. Scripta* **18**, 39 (1978).
137. J. R. Peterson and J. T. Moseley, *J. Chem. Phys.* **58**, 172 (1973).
138. A. W. Johnson and R. G. Fowler, *J. Chem. Phys.* **53**, 65 (1970).
139. D. C. Cartwright and P. J. Hay, *Astrophys. J.* **257**, 383 (1982).
140. M. R. Taherian and T. G. Slanger, *J. Chem. Phys.* **81**, 3814 (1984).
141. J. O. Arnold and R. W. Nicholls, *J. Quant. Spectrosc. Radiat. Transf.* **12**, 1435 (1972).
142. S. P. Davis, D. Shortenhaus, G. Stark, R. Engleman, J. G. Phillips and R. P. Hubbard, *Astrophys. J.* **303**, 892 (1986).
143. J. M. Zucconi and M. C. Festou, *Astron. Astrophys.* **150**, 180 (1985).
144. R. R. Treffers, *Astrophys. J.* **196**, 883 (1975).
145. S. R. Langhoff and C. W. Bauschlicher, *Astrophys. J.* **340**, 620 (1989).
146. J. Anketell and R. W. Nicholls, *Rep. Prog. Phys.* **33**, 269 (1970).
147. J. Berkowitz, W. A. Chupka and G. B. Kistiakowsky, *J. Chem. Phys.* **25**, 457 (1956).
148. I. M. Campbell and B. A. Thrush, *Proc. Roy. Soc. A* **296**, 201 (1967).

149. P. K. Carroll, *J. Chem. Phys.* **37**, 805 (1962).

150. H. Partridge, S. R. Langhoff, C. W. Bauschlicher and D. W. Schwenke, *J. Chem. Phys.* **88**, 3174 (1988).

151. P. K. Carroll and N. P. Sayers, *Proc. Phys. Soc. London A* **66**, 1138 (1953).

152. I. Nadler and S. Rosenwaks, *J. Chem. Phys.* **83**, 3932 (1985).

153. K. P. Huber and M. Vervloet, *J. Chem. Phys.* **89**, 5957 (1988).

154. S. R. Langhoff and C. W. Bauschlicher, *Ann. Rev. Phys. Chem.* **39**, 181 (1988).

155. C. E. Moore, *Atomic Energy Levels U.S. Natl. Bur. Stand.* (*U.S.*), circ. no. 467, 1949.

156. J. A. Gray, S. F. Rice and R. W. Field, *J. Chem. Phys.* **82**, 4717 (1985).

157. S. P. Walch, C. W. Bauschlicher and S. R. Langhoff, *J. Chem. Phys.* **83**, 5351 (1985).

158. A. D. McLean and B. Liu, *J. Chem. Phys.* **91**, 2348 (1989).

159. H. Partridge, *J. Chem. Phys.* **90**, 1043 (1989).

160. A. E. Stevens, C. S. Feigerle and W. C. Lineberger, *J. Chem. Phys.* **78**, 5420 (1983).

161. W. J. Balfour, B. Lindgren and S. O'Connor, *Phys. Scripta* **28**, 551 (1983).

162. S. P. Walch, *Chem. Phys. Lett.* **105**, 54 (1984).

163. C. W. Bauschlicher and S. R. Langhoff, *Chem. Phys. Lett.* **145**, 205 (1988).

164. See, for example, the volume of *Chem. Rev.* (and references therein) devoted to chemical dynamics; *Chem. Rev.* **87** (1987).

165. A. Lifshitz, M. Bidani and H. F. Carroll, *J. Chem. Phys.* **79**, 2742 (1983).

166. L. L. Lohr, *Chem. Phys. Lett.* **56**, 28 (1978) and references therein.

167. D. A. Dixon, R. M. Stevens and D. R. Herschbach, *Far. Disc. Chem. Soc.* **62**, 110 (1977).

168. C. W. Bauschlicher, S. P. Walch, S. R. Langhoff, P. R. Taylor and R. L. Jaffe, *J. Chem. Phys.*, **88**, 1743 (1988).

169. C. W. Bauschlicher, S. R. Langhoff, T. J. Lee and P. R. Taylor, *J. Chem. Phys.*, **90**, 4296 (1989).

170. M. J. Frisch, B. Liu, J. S. Binkley, H. F. Schaefer and W. H. Miller, *Chem. Phys. Lett.* **114**, 1 (1985).

171. R. Steckler, D. W. Schwenke, F. B. Brown and D. G. Truhlar, *Chem. Phys. Lett.* **121**, 475 (1985); D. W. Schwenke, R. Steckler, F. B. Brown and D. G. Truhlar, *J. Chem. Phys.* **84**, 5706 (1986); D. W. Schwenke, R. Steckler, F. B. Brown and D. G. Truhlar, *J. Chem. Phys.* **86**, 2443 (1987).

172 D. R. Garmer and J. B. Anderson, *J. Chem. Phys.* **89**, 3050 (1988).

173. G. E. Scuseria and H. F. Schaefer, *J. Chem. Phys.*, **88**, 7024 (1988).

174. H. F. Schaefer, *J. Phys. Chem.* **89**, 5336 (1985) and references therein.

175. D. M. Neumark, A. M. Wodtke, G. N. Robinson, C. C. Hayden and Y. T. Lee, *J. Chem. Phys.* **82**, 3045 (1985).

176. D. M. Neumark, A. M. Wodtke, G. N. Robinson, C. C. Hayden, K. Shobatake, R. K. Sparks, T. P. Schafer, Y. T. Lee, *J. Chem. Phys.* **82**, 3067 (1985).

177. B. C. Garrett, D. G. Truhlar, R. S. Grev and A. W. Magnuson, *J. Phys. Chem.* **84**, 1730 (1980).

178. M. Desouter-Lecomte, J. C. Leclerc and J. C. Lorquet, *J. Chem. Phys.* **66**, 4006 (1977).

179. R. Grice and D. R. Herschbach, *Mol. Phys.* **27**, 159 (1974).

180. L. R. Kahn, P. J. Hay and I. Shavitt, *J. Chem. Phys*, **61**, 3530 (1974).

181. B. H. Lengsfield, P. Saxe and D. R. Yarkony, *J. Chem. Phys.* **81**, 4549 (1984).
182. P. Saxe, B. H. Lengsfield and D. R. Yarkony, *Chem. Phys. Lett.* **113**, 159 (1985).
183. H.-J. Werner and W. Meyer, *J. Chem. Phys.* **74**, 5802 (1981).
184. W. Müller, J. Flesch and W. Meyer, *J. Chem. Phys.* **80**, 3297 (1984).
185. L. B. Knight, K. D. Johannessen, D. C. Cobranchi, E. A. Earl, D. Feller and E. R. Davidson, *J. Chem. Phys.* **87**, 885 (1987).

MOLECULE–SURFACE SCATTERING AND REACTION DYNAMICS

ANDREW E. DEPRISTO AND ABDELKADER KARA

Department of Chemistry
Iowa State University
Ames, Iowa

CONTENTS

I. INTRODUCTION

Over the past decade, we have witnessed a phenomenal growth in the quantity and quality of both experimental and theoretical studies of the microscopic dynamics of molecule–surface scattering.* While much of this work has

*A number of excellent reviews of the experimental (Bernasek 1980; Barker and Auerbach 1984; Goodman 1984; Comsa and David 1985; Cardillo 1985; Ceyer 1988; Rettner 1988) and theoretical (Tully 1981; Barker and Auerbach 1984; Gerber 1987; Gadzuk 1988; Brenner and Garrison 1989) literature are available. An excellent book is also strongly recommended (Bortolani et al 1988) which consists of lectures given at the 1988 Trieste Summer School by a number of experts in gas–surface systems. The depth and breadth of this book is truly impressive. A less detailed, textbook-style monograph by Zangwill (1988) is also strongly recommended. Finally, a large number of texts treat catalysis and chemisorption phenomena at an elementary level (e.g., simple kinetic models). We cannot recommend any one in particular, except for historical reading, but instead suggest two lucid discussions of the relationship of surface reaction dynamics and kinetics to real catalytic processes that are found in Stoltze (1987) and Norskov and Stoltze (1987).

focused on cases in which the incident molecule remains intact, a substantial fraction has addressed the dynamics of elementary reactions in these complex systems. With the expectation that the latter effort will continue to grow in size and importance, we have undertaken to review the recent conceptual, theoretical, and computational developments in the treatment of reactions involving molecular gases and solid surfaces. We emphasize the microscopic understanding provided by atomic- and electronic-level theories and computer simulations. In particular, we try to address two questions:

What can happen when a molecule collides with a surface?

How does the underlying microscopic potential energy surface (PES) govern the event?

The study of molecule–surface interactions is very old because of its importance in a myriad of applications (Somorjai 1981; Gasser 1985; Zangwill 1988). The first ideas can be found as far back as the turn of the century in relationship to heterogeneous catalysis; later work was concerned with aerodynamics; and, more recently, in this "age of silicon," many investigations have been driven by the needs of the microelectronics industry. Nevertheless, it is only in the last two decades that detailed information has been learned about the microscopic details. This has occurred for a number of reasons. First, the early theoretical work mainly used thermodynamics and kinetics, which of course provide no information on the microscopic dynamics (except for fanciful interpretations). The few dynamical treatments focused on simplistic one-dimensional models and qualitative descriptions (Lennard-Jones 1932). Second, the early experiments utilized polycrystalline surfaces and relatively high pressures, both of which lead to confusion about the state of the surface, or the impinging molecule, or both. These observations are not intended to be critical comments about the early work. The fact is that the necessary theoretical and experimental tools were simply unavailable to these early researchers.

The situation has changed dramatically due to a number of advances in both experiment and theory. For the former, these have included techniques for the preparation and maintenance in UHV of clean single-crystal surfaces with low concentrations of defects and steps: new and more accurate spectroscopies to investigate the adsorbates, with the literal alphabet soup of acronyms, such as high-resolution low-energy electron diffraction (LEED), He scattering, infrared (IR) adsorption, reflection high-energy electron diffraction (RHEED), ultraviolet photoelectron spectroscopy (UPS), x-ray photoelectron spectroscopy (XPS), temperature programmed desorption (TPD), and secondary ion mass spectrometry (SIMS); production of well-collimated, rotationally and vibrationally cold, nearly mono-energetic beams of gas molecules to collide with the surface; and laser-based detection of state- and velocity-selected scattering products. These have eliminated many of the ambiguities and averages inherent in previous methods.

Theoretical advances have included the tremendous increase in computational power; sophisticated *ab initio* and density-functional-based techniques capable of treating tens of heavy metal atoms using pseudopotentials; dynamical methods for treatment of the full multidimensional dynamics of a strongly-interacting small system interacting weakly with an infinite thermal system; qualitatively correct representations of interaction potentials; and new techniques for the solution of the time-dependent Schrödinger equation. In our opinion, increased computational power is the foundation of theoretical treatments in the same way that UHV technology is the foundation of experimental ones: other developments are certainly important, but the entire field of study would regress by decades without this single development. Computational advances have transformed

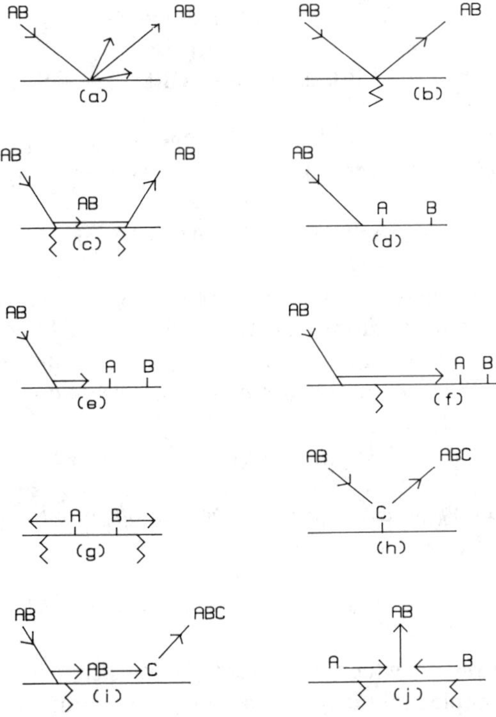

Figure 1. Schematic of the different types of dynamical processes involved in molecule–surface systems. The jagged line indicates energy exchange with the surface. Further details are contained in the text.

simulations from model studies of "Lennard-Jonesian" materials to realistic treatments of chemical systems.

To acquaint the reader with the subject, we have depicted in Fig. 1 a number of possible outcomes of the collision between a diatomic molecule (AB) and a surface (S). As in all fields, a certain amount of jargon is unavoidable. We sketch the events in Fig. 1 to help in this introduction, using the notation that $\mathbf{k}$, ε, E are the initial (pre-scattering) translational momentum, internal energy, and total (translational plus internal) energy of the molecule, respectively, with primes denoting final (post-scattering) quantities. The z-axis is assumed to be normal to the surface pointing into the vacuum. The diagrams in Fig. 1 may be summarized as follows:

(a) The molecule scatters without exchanging energy with the surface (*direct elastic scattering*) and with the angles of $\mathbf{k}'$ and $\mathbf{k}$ from the surface normal either equal (*specular scattering*) or different (*non-specular scattering*); the latter is due to *diffraction* from a corrugated surface and to energy exchange between translational (T) to internal ro-vibrational (R, V) degrees of freedom:

$$\mathrm{AB}(\mathbf{k}, \varepsilon, E) + \mathrm{S} \rightarrow \mathrm{AB}(\mathbf{k}, \varepsilon', E) + \mathrm{S}$$

(b) As in (a) but here the molecule exchanges some energy with the surface (*direct inelastic scattering*):

$$\mathrm{AB}(\mathbf{k}, \varepsilon, E) + S \rightarrow \mathrm{AB}(\mathbf{k}', \varepsilon', E') + \mathrm{S}$$

(c) As in (a) but here the molecule exchanges so much energy with the surface that it is trapped on the surface for a long time (*adsorbed*, *AB*(*a*)) before returning to the gas phase (*desorption*) and is thus labeled *indirect scattering or trapping–desorption*:

$$\mathrm{AB}(\mathbf{k}, \varepsilon, E) + \mathrm{S} \rightarrow \mathrm{AB(a)} + \mathrm{S} \rightarrow \mathrm{AB}(\mathbf{k}', \varepsilon', E') + \mathrm{S}$$

(d) The incoming molecule breaks apart upon initial collision, which is labeled *direct dissociation*:

$$\mathrm{AB}(\mathbf{k}, \varepsilon, E) + \mathrm{S} \rightarrow \mathrm{A(a)} + \mathrm{B(a)} + \mathrm{S}$$

(e) As in (d) but in which T → V energy transfer forms a vibrationally excited complex that exists for a number of vibrational periods before the bond breaks, and labeled *indirect dissociation*:

$$\mathrm{AB}(\mathbf{k}, \varepsilon, E) + \mathrm{S} \rightarrow \mathrm{AB}(\mathrm{a}; \mathbf{k}', \varepsilon', E') + \mathrm{S} \rightarrow \mathrm{A(a)} + \mathrm{B(a)} + \mathrm{S}$$

(f) As in (d) but in which the molecule exists for many vibrational periods, equilibrates with the surface, and eventually breaks apart, and labeled *dissociation via a precursor*:

$$AB(\mathbf{k}, \varepsilon, E) + S \rightarrow AB(a) + S \rightarrow A(a) + B(a) + S$$

(g) After molecular adsorption in (c) or dissociative chemisorption in "d–e," the adsorbates can move on the surface (*diffusion*).

(h) The incident molecule can collide with one of the previously formed adsorbates and form a new chemical species (*Eley–Rideal mechanism*):

$$AB(\mathbf{k}, \varepsilon, E) + C(a) + S \rightarrow ABC(g) + S$$

(i) The incident molecule can adsorb either molecularly or dissociatively and then react with one of the previously formed adsorbates, form a new chemical species, and desorb (*Langmuir–Hinshelwood mechanism*):

$$AB(\mathbf{k}, \varepsilon, E) + C(a) + S \rightarrow AB(a) + C(a) + S \rightarrow ABC(g) + S$$

(j) The second part of (i), reaction between two adsorbed species, is labeled *recombinative* or *associative desorption*

(k) The incident molecule, or some part of it, can react with the surface or physically modify the surface (not illustrated in Fig. 1).

Some of the above categories are not especially distinct, as for example (e) and (f) which differ only in the amount of time the molecule spends on the surface before dissociating. Some also have further subcategories. For example, precursor dissociation can be divided into *extrinsic* and *intrinsic* precursors. The former occurs when the surface is already covered with adsorbed species, A(a) and B(a), while the latter occurs for the clean surface. In addition, for some of the reactions it is possible that the species may not be adsorbed, especially when the molecule has enough energy initially: For example, the reaction, $AB(\mathbf{k}, \varepsilon, E) + S \rightarrow A(a) + B(g) + S$, is feasible if the reaction exoergicity is sufficient to break the B—S bond. Although not mentioned explicitly, transfer of energy between the molecular and solid degrees of freedom may be quite significant for all the more complex processes (d)–(h). The relevant solid's degrees of freedom for low energy excitations involve motion of the nuclei (*phonons*), and for metals excitation and deexcitation of electrons with energies close to the Fermi level (*electron–hole pairs* or *e, h*).

Because dissociative chemisorption dynamics will be a central topic of this review, we mention a few details of this process here. When the dissociation occurs irrespective of the incident kinetic energy and angle, and internal energy, the process is labeled *nonactivated*. By contrast, when the dissociation increases with incident kinetic energy, it is generally labeled *activated*. The dissociation probability or "initial sticking coefficient" for nonprecursor dissociation, universally denoted as S_0, depends on all three components of translational momentum. Two limiting cases are: S_0 depends only on $\hat{z}\cdot\mathbf{k}$ (*normal energy scaling*) and on $|\mathbf{k}|$ (*total energy scaling*).

In this review, we will focus on the interaction, and especially the scattering, of small molecules with clean, perfectly periodic solid surfaces. We will not describe scattering from adsorbate covered or nonperiodic surfaces (Gerber 1987) since these areas are much more complex computationally or insufficiently advanced conceptually. We will completely ignore more complex scattering events alluded to in (k), such as ejection of surface atoms into the gas phase either directly by sputtering or indirectly by forming volatile molecular species via reaction between the gas molecule and the surface atoms. In high-energy sputtering, excitations of the solid must be expanded to include collective excitation of the free electrons in a metal (i.e., *plasmons*), electronic excitation of core electrons of the surface atoms, and the previously mentioned ejection of the surface atoms. Such sputtering has been treated by a number of research groups. The interested reader is referred to the work of Garrison and co-workers (Garrison and Winograd 1982; Garrison et al 1988) for an introduction and relevant references. To our knowledge, theoretical treatments of the dynamics of formation of volatile products such as in the reaction of F with Si to form SiF_n are not sufficiently advanced since the underlying PES is essentially unknown, and must be extremely complex to describe the many arrangements occurring in such processes.

We will also omit elastic and inelastic scattering (except in the context of dissociative chemisorption reactions) since these have been subjected to intensive theoretical studies by many workers for over a decade, and for certain simple problems, much longer. A significant literature is available for such processes, including a number of reviews and books (Wolken 1976; Goodman and Wachman 1976; Barker and Auerbach 1984; Gerber 1987; Halstead and Holloway 1988; Zangwill 1988). The recent comprehensive overview of theory by Gerber (1987) provides an introduction to a number of topics, along with a representative reference list.

We have intentionally left off this wide variety of topics in order to focus on the detailed coverage of the dynamics and interactions of dissociative chemisorption reactions. Our goals are to introduce the concepts in this area, to illustrate these with representative experimental measurements, to present the state-of-the-art theoretical methodology, and to indicate the understanding provided by applications of these theoretical methods. We will try to distinguish that which is well understood from that which is controversial or unknown. In short, this review is intended to take practicing theoretical chemists or physicists to the current research level in a substantial part of molecule–surface reaction dynamics. It will also serve to acquaint experimentalists with current theories and especially with the relationship between various measurements and microscopic models of the dynamics. While we hope that this will allow them to provide more sophisticated interpretations of their own experiments, at the very least it will provide an appreciation for the complex nature of the dynamics in these systems.

As in any review, we have tried to provide an accurate assessment of the current literature. Omission of exhaustive references to any individual is intentional. The explosive increase in papers published, including the more modern tendency to present very similar work in a multitude of journals, makes it nearly impossible (and unnecessary) to reference every article. We have tried to provide a representative reference list for every idea and concept. If we have overlooked some important work, we regret the oversight and would be happy to hear from the researchers.

II. OVERVIEW OF EXPERIMENTAL MEASUREMENTS

A one-dimensional representation of the molecule–surface PES is often used to provide a qualitative description of experimental data; a schematic is shown in Fig. 2 with all energies measured with respect to the binding energy of the gas-phase molecule. We will discuss reduced dimensional PES in considerable detail in Sections III and VI, but for now just note that a 1-D PES has the advantage of simplicity in introducing the general characteristics of data. In Fig. 2 a number of special features are labeled:

1. The molecular well for physisorption, W_p, and energy for desorption, E_d, with typical values in the range of 0.01–0.1 eV (e.g., H_2 and N_2 interactions with metal surfaces, respectively);
2. The barrier to chemisorption, E_c, with $E_c \gg 0$ for kinetically nonreactive systems, $E_c > 0$ for activated chemisorption, and $E_c < 0$ for nonactivated chemisorption;
3. The chemisorption well, W_c, with typical values in the range 1–10 eV.

We have also indicated a few rate constants relevant to the disposition of a molecular physisorbed species: k_d and k_c are the rate constants for desorption and chemisorption, respectively, while α is the physisorption probability for an incident gas-phase molecule.

In a direct mechanism for dissociative chemisorption, W_p well plays a minor role since the important quantity is E_c. For an indirect mechanism, W_p plays two important roles: (1) acceleration of the incoming molecule to

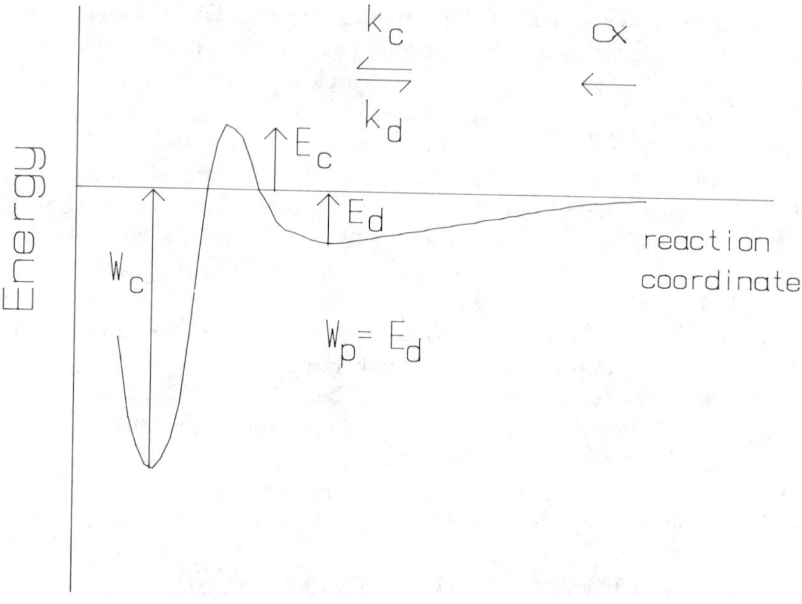

Figure 2. A one-dimensional representation of a molecule–surface interaction. The depth of the physisorption and chemisorption wells are W_p and W_c, respectively. The barrier to chemisorption is E_c and the barrier to desorption is E_d. The probability of physisorption is α, while the rate constants for desorption and chemisorption out of the physisorbed state are k_d and k_c, respectively.

allow for significant energy exchange with the lattice; and (2) stabilization of the trapped molecular species for a long enough time to allow significant energy exchange among the molecular degrees of freedom. For a precursor mechanism, W_p plays the same roles as in the indirect mechanism but the stabilization is long enough to allow for equilibration between the physisorbed molecule and the lattice. In the last case, a rate formulation using a steady-state assumption for physisorbed species yields the initial sticking coefficient for dissociation as:

$$S_0 = \alpha k_c/(k_c + k_d) \tag{2.1}$$

Figure 3. Shown is the sticking coefficient as a function of the initial kinetic energy (normal to the surface in the case of a and b) for the H_2/Ni(111) system. (a) data from Rendulic et al. (1989); (b) data from Robota et al. (1985) with the various incident angles shown as crosses and circles; (c) data from Hayward and Taylor (1986) at various incident angles as labeled on the figure.

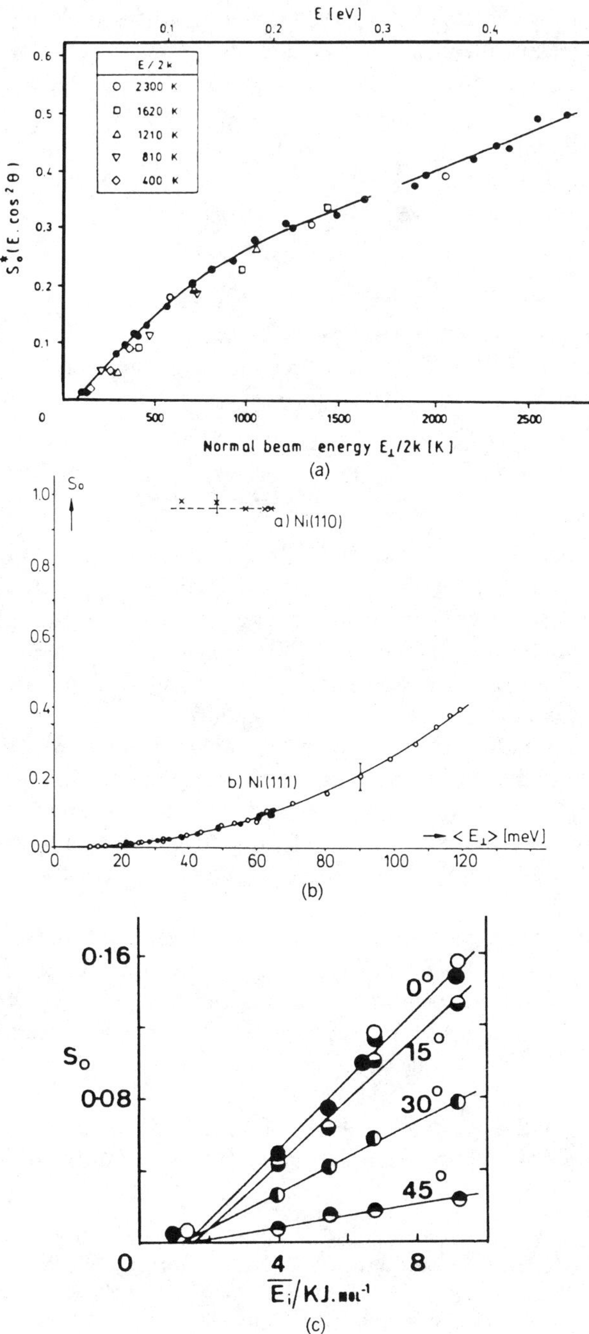
E [eV]
0.1
0.2
0.3
0.4
E / 2k
2300 K
1620 K
1210 K
810 K
400 K
$S_o^*(E.\cos^2\theta)$
Normal beam energy $E_\perp/2k$ [K]
So
a) Ni(110)
b) Ni(111)
< $E_\perp$ > [meV]
0°
15°
30°
45°
S_o
$\overline{E_i}$/KJ.mol^{-1}

(a)

(b)

(c)

Since increasing the incident kinetic energy, E_i, of the molecule decreases α, S_0 will decrease with increasing kinetic energy in a precursor mechanism. This is one signature of a precursor mechanism often looked for experimentally.*

To derive another, consider the simple Arrhenius rate forms:

$$k_d = \nu_d \exp(-E_d/kT) \tag{2.2a}$$

$$k_c = \nu_c \exp(-(E_d + E_c)/kT) \tag{2.2b}$$

where T is the surface temperature. (The two frequency factors are often of the same order of magnitude unless passage over the chemisorption barrier is very restrictive sterically, but this is of no importance here.) Using Eqs. (2.2a–b) in Eq. (2.1) yields

$$S_0 = \alpha[1 + (\nu_d/\nu_c)\exp(E_c)/kT]^{-1} \tag{2.3}$$

The dependence of S_0 on T follows from Eq. (2.3). For activated chemisorption (i.e., $E_c > 0$), S_0 increases with increasing T; this behavior is not unique since it can also occur for activated reactions in both direct and indirect mechanisms. By contrast, for nonactivated chemisorption (i.e., $E_c < 0$), S_0 decreases with increasing T; this is a unique signature for such a process that is often looked for experimentally.

We should note that for either activated or unactivated dissociative chemisorption, increasing the kinetic energy enough will lead to direct dissociation. Thus, at high kinetic energy, one should expect S_0 to increase with E_i. Of course, the value at which S_0 stops decreasing and starts increasing with E_i will depend upon the value of E_c and the detailed dynamics of the dissociation.

We limit the presentation of experimental data to those experiments utilizing supersonic molecular beam expansions in combination with UHV techniques. Supersonic beam experiments control the incident kinetic energy and angle of the molecule to a high degree, and thus are generally easier to interpret at the microscopic dynamical level than those experiments utilizing effusive beams.

We begin with H_2 scattering from various Ni (surfaces) which are among the most studied systems. Some of the main features of the dissociation are

*We should point out that more complex one-dimensional PESs are possible with, for example, two different molecularly physisorbed species separated by a barrier from each other and also by a barrier from the chemisorbed state (Grunze et al. 1984; Whitman et al. 1986). In such cases, the bahavior of S_0 for precursor mechanism chemisorption may not be so transparent.

well-known, but one still can find many discrepancies among the findings and hence the conclusions of different groups. The data on these systems will serve to illustrate both the accuracy and difficulty of the experimental measurements in this field.

Consider the dissociative chemisorption of H_2 on the close-packed Ni(111) surface. Three different set of results for S_0 are presented in Figs. 3a–c from Rendulic et al. (1989), Robota et al (1985), and Hayward and Taylor (1986), respectively. The values of S_0 are in general agreement when one notes that the data in Figs. 3b–3c correspond to $E_i < 0.1$ eV, which is the small linearly increasing leftmost part of the curve in Fig. 3a. Indeed, the data in Figs. 3a and 3c are in excellent agreement while that in Fig. 3b displays a slightly different shape. It is quite likely that such small disagreements in small sticking coefficients at low beam energy are due in part to the fact that each group used a different crystal, containing different amounts of steps and defects, and perhaps in part to the uncertainties inherent in the different techniques used to determine the dissociation probability. The general features that all the groups agreed on are: (1) the dissociation probability increases with incident kinetic energy, implying that the dissociation is activated; (2) there is no isotope effect, implying that tunneling is unimportant; (3) S_0 is a function of the normal energy; and (4) there is no evidence of a precursor state, implying that the dissociation is direct.

For a startling contrast, we consider the H_2/Ni(100) system. Two different set of results for S_0 are presented in Figs. 4a–b from Rendulic et al. (1989) and Hamza and Madix (1985), respectively. The values of S_0 are in qualitative and quantitative disagreement. The results of Rendulic et al. are much smaller, decrease more quickly with incident angle than would hold for normal energy scaling, display evidence for a precursor at low kinetic energy (note the kinetic energy dependence), and did not display an isotope effect between H_2 and D_2. The results for Hamza and Madix are much larger, obey normal energy scaling, do not exhibit any precursor type behavior, and exhibit an isotope effect. About the only feature the two studies do find in common is that the dissociation is activated.

A similar situation holds for the dissociation of H_2 on Ni(110). Two different set of results for S_0 are presented in Fig. 5 from Rendulic et al. (1989) and in Fig. 3 from Robota et al. (1985). In the range of 20–60 meV, the results of the former decrease from ≈ 0.5 to ≈ 0.4 while those from the latter are nearly constant at ≈ 0.96. The former also decrease with increasing T. It is not surprising that Rendulic concludes that a precursor mechanism is operative in this energy range while Robota et al. conclude that non-activated direct chemisorption occurs. The increase in S_0 at higher E_i in Fig. 5 also indicates a possible direct activated dissociative channel. Thus, these two experiments disagree completely even as to the basic mechanism.

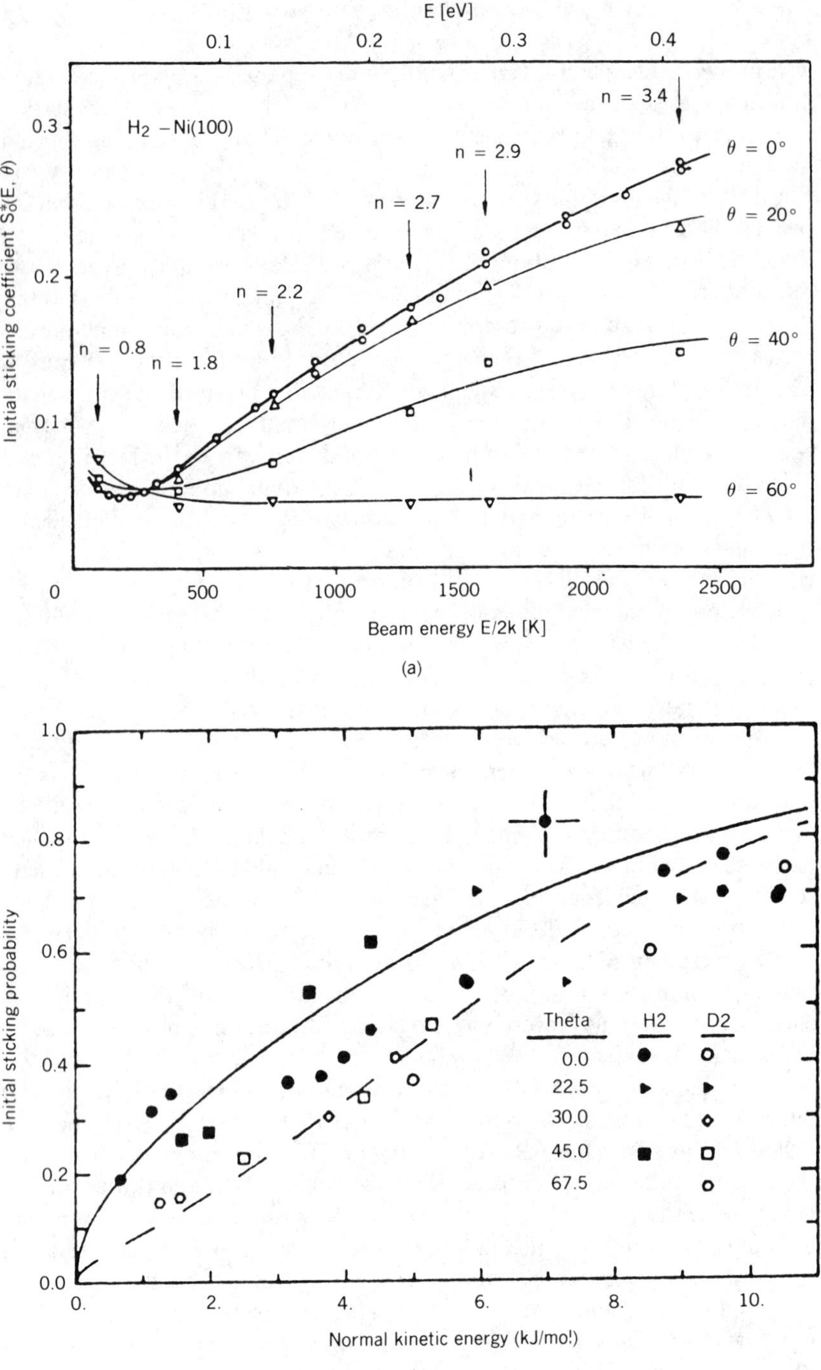
E [eV]
0.1
0.2
0.3
0.4
H_2 – Ni(100)
n = 3.4
n = 2.9
n = 2.7
n = 2.2
n = 1.8
n = 0.8
θ = 0°
θ = 20°
θ = 40°
θ = 60°
Initial sticking coefficient $S_0^*(E, \theta)$
Beam energy E/2k [K]
(a)
Theta
H2
D2
0.0
22.5
30.0
45.0
67.5
Initial sticking probability
Normal kinetic energy (kJ/mol)
(b)

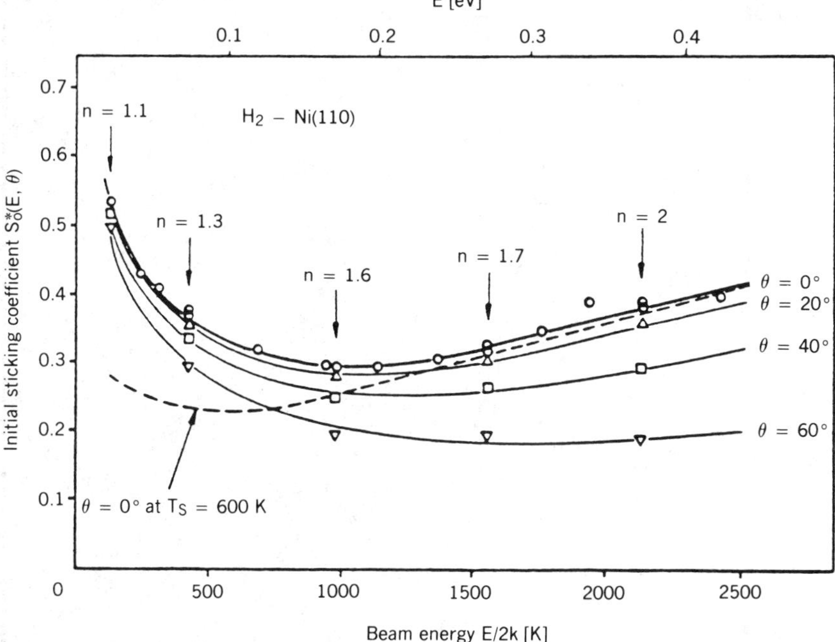

Figure 5. Same as Fig. 4a except for the H_2/Ni(110) system (Rendulic et al. 1989).

Finally, to close the case of H_2 dissociative chemisorption on Ni(surfaces), and to give the reader an idea of what may be the origin of the discrepancies, we briefly mention the effect of steps, surface defects, and adsorbates on the dissociation probability of H_2. After introducing defects on Pt(111), Poelsema et al. (1985) found that the dissociation probability is enhanced considerably. But this is not always the case since, in another experiment, Rendulic (1988) studied the effect of oxygen adsorption on the dissociation probability of H_2 on Ni(111) and Ni(110). Fig. 6 shows the variation of S_0 as a function of oxygen coverage. Note that at very low coverages, oxygen acts as a promoter in the H_2/Ni(111) system and an inhibitor in the H_2/Ni(110) system. We conclude that defects or adsorbates, even at low coverages, can affect dissociative sticking considerably and in as yet unknown ways. The simple arguments for promotion (e.g., defects act like nonactivated sites) and

---

Figure 4. Initial sticking coefficient for various incident angles, as labeled on the figure for the H_2/Ni(100) system: (a) as a function of the initial kinetic energy (Rendulic et al. 1989); (b) as a function of the initial normal kinetic energy (Hamza and Madix 1985). Ignore the specification of n on a.

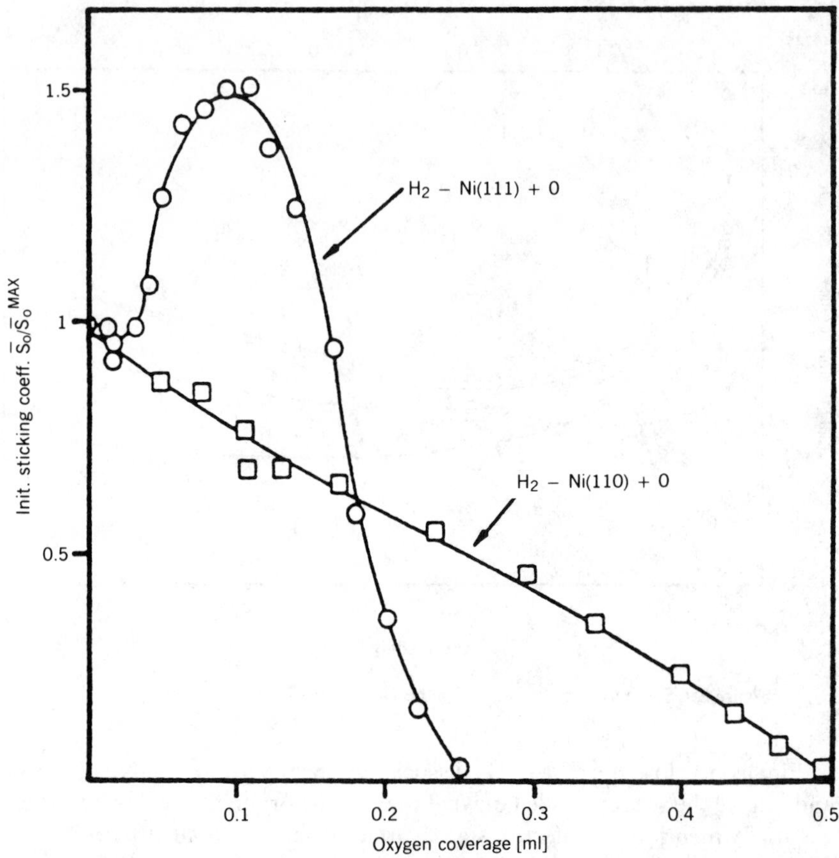

Figure 6. Initial sticking coefficient for H_2/Ni(111) and Ni(110) as a function of 0-atom converge; reprinted with permission from Rendulic (1988).

poisoning (e.g., adsorbates occupy and hence block nonactivated sites) are not sophisticated enough to describe the face sensitivity evident in Fig. 6. Much work, both experimental and theoretical, needs to be done in this area.

Another system that has been subject to detailed study is the dissociative chemisorption of CH_4 on metal surfaces. We focus on the investigations by Rettner et al. (1985a, 1986) on CH_4/W(110) and by Lee et al. (1987) on CH_4/Ni(111). Since the data on S_0 pertain to different systems, the numerical

---

Figure 7. Initial sticking coefficient as a function of the initial normal kinetic energy for: (a) CH_4 and CD_4 on W(110) from Rettner et al (1985a); and (b) CH_4 and CD_4 on Ni(111), from Lee et al. (1987), with permission.

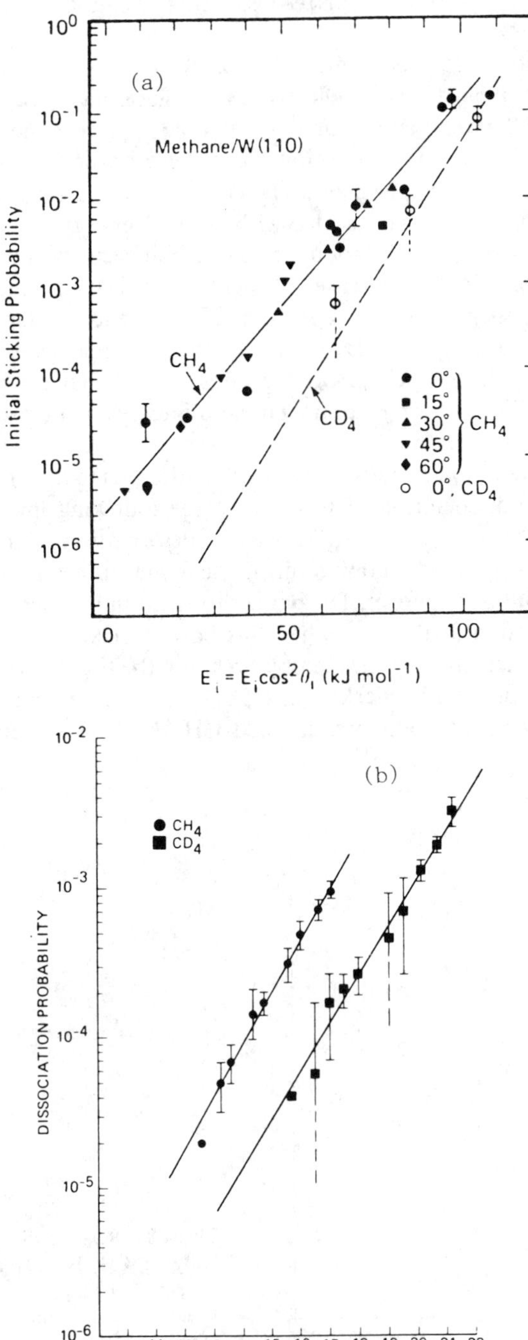

(a)
Methane/W(110)
Initial Sticking Probability
10^0
10^{-1}
10^{-2}
10^{-3}
10^{-4}
10^{-5}
10^{-6}
CH_4
CD_4
● 0°
■ 15°
▲ 30°
▼ 45°
◆ 60°
CH_4
○ 0°, CD_4
0
50
100
$E_\perp = E_i \cos^2\theta_i$ (kJ mol^{-1})
(b)
● CH_4
■ CD_4
DISSOCIATION PROBABILITY
10^{-2}
10^{-3}
10^{-4}
10^{-5}
10^{-6}
11 12 13 14 15 16 17 18 19 20 21 22
$E_\perp = E_i\,COS^2\theta_i$ Kcal/mole

values need not agree, of course. For both systems, it was found that vibrational energy is at least as effective as translational energy in promoting dissociation. However, neither group was able to determine which modes were responsible for this promotion since the vibrational modes were all excited at once in the experiment. The effect of translational energy on S_0 is shown in Fig. 7. First, S_0 scales with normal energy for both systems. Second, there is remarkably good agreement between the two sets of data regarding the exponential increase of S_0 with E_i, indicating a highly activated process. Third, there is a strong isotope effect in each system. In fact, the dissociation probability for CD_4 is smaller by about one order of magnitude as compared to that for CH_4. Both groups argued that this isotope effect cannot be attributed to zero-point energy differences but rather is due to tunneling of H atoms.

Lee et al. (1987) suggested a further explanation of the detailed mechanism involved in the dissociation of CH_4 along the following lines. When CH_4 collides with a surface, it experiences a deformation, perhaps from a tetrahedral to a pyramidal configuration, near the surface in order to allow a strong C—Ni bond to form. The barrier to H-atom loss becomes narrower as the C—Ni bond forms, and when this barrier becomes narrow enough, the H atom tunnels through it, thereby breaking the C—H bond. To further investigate this picture, Beckerle et al. (1987) performed an experiment where an Ar gas is incident upon preadsorbed CH_4/Ni(111) and found that the

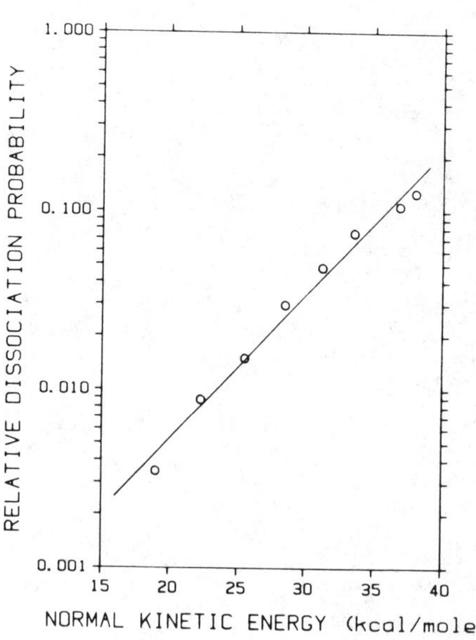

Figure 8. Relative dissociation probability of CH_4/Ni(111) induced by Ar impact as a function of the Ar initial kinetic energy normal to the surface, reprinted with permission from Beckerle et al. (1987).

dissociation probability varies as in Fig. 8 over a limited range of beam energies and angles. They concluded that the impact of the rare gas with the adsorbed molecule produces the aformentioned deformation of CH_4. For more details, the reader is invited to consult the excellent review by Ceyer (1988) on the subject. We would caution the reader that these interpretations of either the tunneling mechanism or the more detailed deformation-tunneling mechanism are founded upon extremely simple low-dimensional PES models. Perhaps it is our theoretical bias, but we would be more convinced by corroborating detailed simulations on realistic multidimensional PESs.

As a last example of dissociative chemisorption, we consider the case of N_2 on metal surfaces. It is well-known that dissociation of N_2 is the rate-determining step in ammonia formation, which has led to a large number of investigations.* Here, we present only representative molecular beam studies involving the W(110) surface (Pfnür et al. 1986), the W(100) surface (Rettner et al. 1988) and the Fe(111) surface (Rettner and Stein 1987). The purpose is to illustrate a few dynamical features that were not present in the previous examples, not to provide an exhaustive coverage of the molecular beam scattering literature on dissociative chemisorption of N_2.

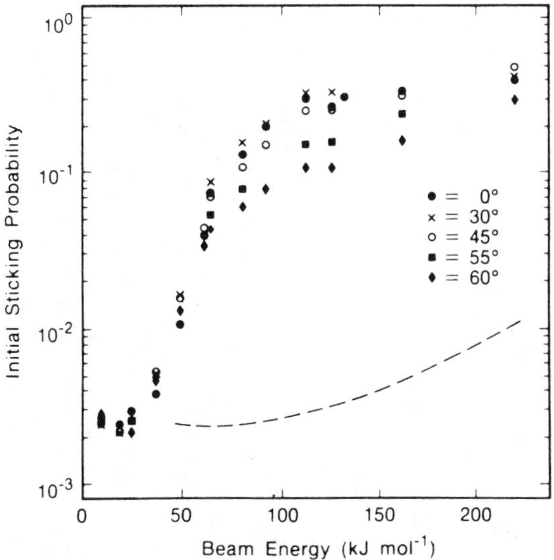

Figure 9. Initial sticking coefficient as a function of the initial kinetic energy in the N_2/W(110) system for various incident angles, from Pfnür et al. (1986) with permission.

*For an excellent review and a lucid discussion of the relationship between fundamental surface science studies and industrial catalytic synthesis of NH_3, see Stoltze (1987).

Results for S_0 in the N_2/W(100) system are shown in Fig. 9 as a function of E_i not $E_i \cos^2 \theta_i$. These data clearly demonstrate that S_0 scales with the total energy, which is especially intriguing since there is no indication of a precursor mechanism. Two possible explanations were advanced (Pfnür et al. 1986) to explain this behavior: (1) substantial corrugation of the surface leads to a strong coupling of the parallel and perpendicular translational momenta of the N_2; and (2) multiple collisions of the N_2 with the surface induces scrambling of the momentum components. These mechanisms imply a diffuse angular distribution that cannot be reconciled with the measurement shown in Fig. 10 for the angular distribution along the specular direction. Instead, this is an example of indirect dissociation which will be discussed in more detail in Sections III and VI.

Results for S_0 in the N_2/W(100) system are shown in Fig. 11. These data clearly exhibit the two signatures of an unactivated precursor mechanism: (1) S_0 decreases when E_i increases, at low values of E_i; and (2) S_0 decreases when the surface temperature increases. The N_2/W(100) system is one of the few that presents such an unambiguous interpretation. In addition, this

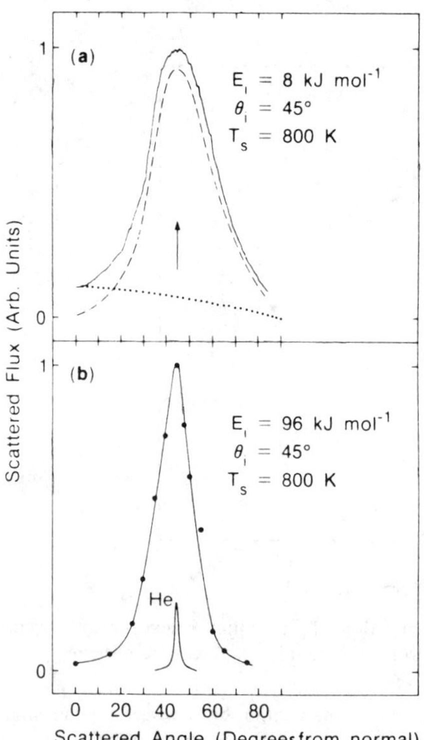

Figure 10. Angular distributions of scattered N_2 from W(110) at different initial kinetic energies as shown on the figure. The He specular peak is shown for comparison. The data are from Pfnür et al. (1986).

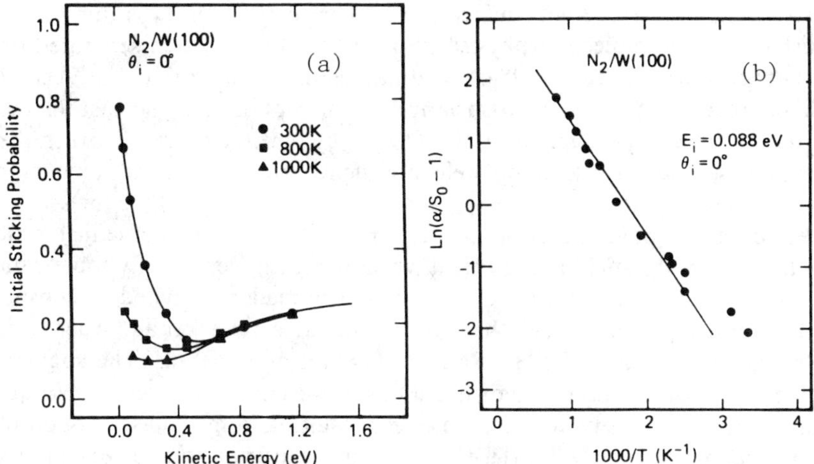

Figure 11. (a) Initial sticking coefficient as a function of the initial kinetic energy in the N_2/W(100) system for various surface temperatures; (b) variation of the sticking coefficient with the surface temperature. All data from Rettner et al. (1988).

system obeys a total energy scaling. We should remark that the leveling off of S_0 at large kinetic energy could signify a change from a precursor to a direct (or indirect) dissociation.

Results for S_0 in the N_2/Fe(111) system are shown in Fig. 12. These data are quite peculiar, exhibiting a strong increase of S_0 with increasing E_i but

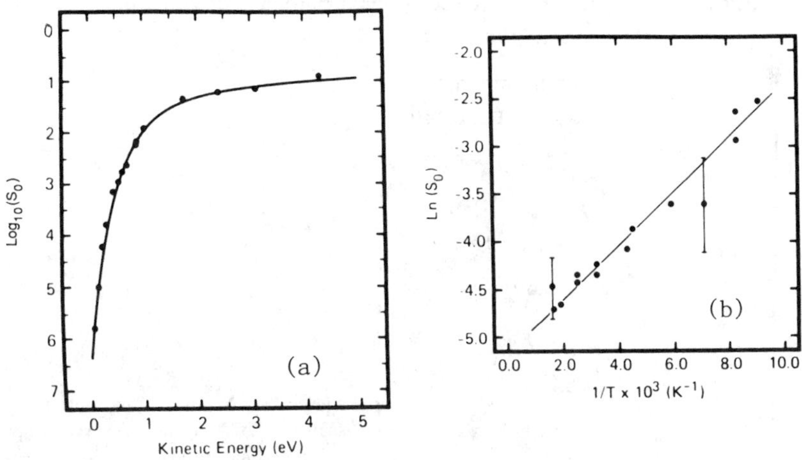

Figure 12. Same as Fig. 11 except for the N_2/Fe(111), from Rettner and Stein (1987) with permission.

also a strong decrease with increasing T. The increase of S_0 is due to the existence of two molecular physisorption states in this system separated by a barrier (Grunze et al. 1984; Whitman et al. 1986). Only the second physisorbed state leads to dissociative chemisorption. An increase in the initial kinetic energy helps the molecule to overcome the barrier between the first and second physisorption wells and hence to adsorb in the precursor state to dissociation.

Since the above description of experimental data focused almost exclusively on the sticking coefficient, we now present very briefly some data on the analysis of the scattered molecules. The reader is invited to consult a number of excellent reviews that exist on the subject (Barker and Auerbach, 1984; Comsa and David 1985; Rettner 1988) for more details. The angular, velocity, and internal energy distributions are usually measured. These are compared, respectively, to the predictions resulting from the assumption of equilibrium of the gas and surface: (1) the angular distribution is proportional to $\cos\theta$; (2) the velocity and internal energies are described by a boltzmann distribution.

NO scattering from the Ag(111) surface is the prototype for nonreactive scattering. Studies on this system have been reviewed by Rettner (1988) and some general behavior has emerged. Fig. 13 shows that the angular distribution for different final rotational energies is not of the form $\cos\theta$ but instead is slightly shifted downward from and broadened around the specular direction. This is consistent with transfer of small amounts of initial normal kinetic energy to phonons, and (due to the large shift for $J = 29.5$) also consistent with a slight $T \rightarrow R$ energy exchange. Fig. 14 shows that the translational energy of the scattered molecules for particular final rotational

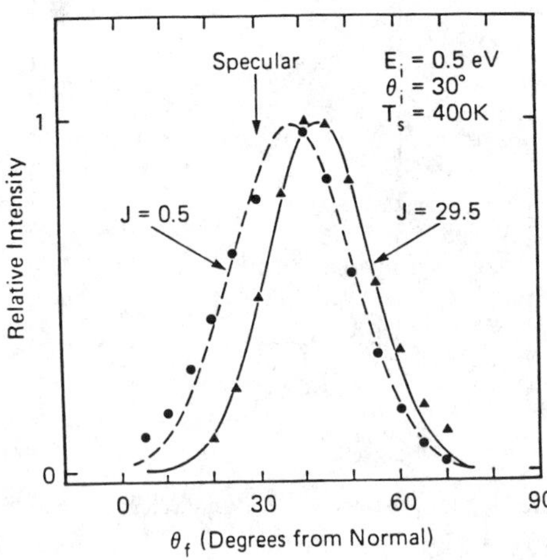

Figure 13. Angular distributions of scattered NO from Ag(111) for two different final rotational states $J = 0.5$ and $J = 29.5$. Reprinted with permission from Rettner (1988).

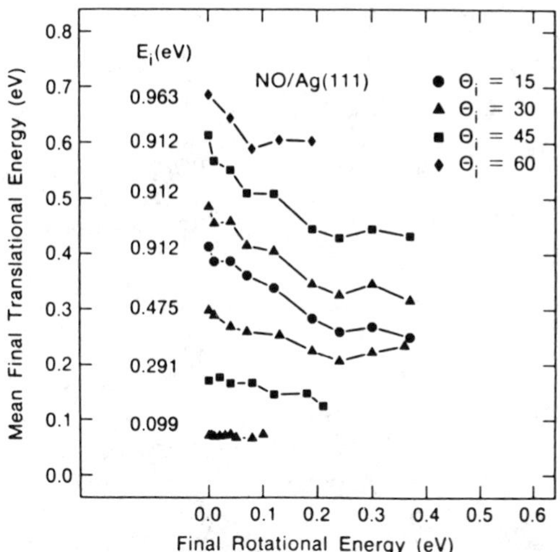

Figure 14. Mean translational energy, for scattered NO from Ag(111) as a function of final rotational energy for various incident kinetic energies and angles, from Rettner (1988) with permission.

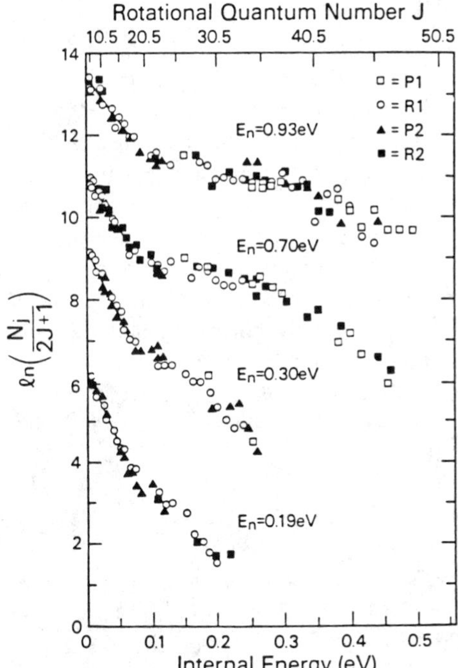

Figure 15. Rotational distributions of scattered NO from Ag(111) at different initial normal kinetic energies, from Kleyn et al. (1982) with permission.

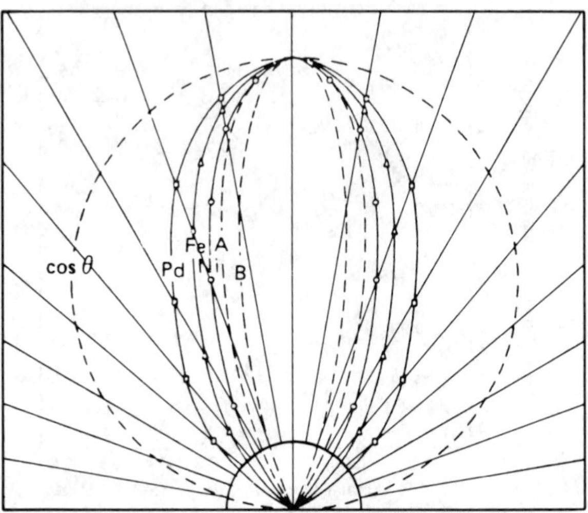

Figure 16. Angular distributions of desorbed H_2 from various metal surfaces as labeled on the figure. The data are from Comsa and David (1985). The A and B curves are theoretical predictions from Van Willigen (1968).

energies is not constant, independent of E_i, as would be the case for a Boltzmann velocity distribution. Finally, Fig. 15 shows that the rotational energy distribution is distinctly non-Boltzmann at higher incident normal kinetic energies and higher final rotational states. Such deviations are the norm for direct and indirect scattering (Barker and Auerbach 1984).

It is also the case that measurements of gas-phase molecular distributions are used to analyze the products of recombinative desorption. In this case, as indicated in Fig. 16, the angular distribution is observed to peak along the normal to the surface, being proportional to $\cos^n \theta$ with $n > 2$ and sometimes up to 12. Highly excited vibrational states are often found (Kubiak et al. 1984; Brown and Bernasek 1985; Mantell et al. 1986; Karikorpi et al. 1987; Harris et al. 1988b). We will have much more to say about such distributions in Section VI. The review by Comsa and David (1985) deals in detail with this subject.

III. POTENTIAL ENERGY SURFACES

Now that we have presented an overview of the type of experimental information to be explained on a microscopic basis, it is time to focus on the description of the interactions between the gas molecule and the solid surface in more detail. We leave the description of the dynamical techniques for Section IV, since the accuracy and topology of the PES often control the accuracy of the results of all dynamical simulation. In any case, dynamical simulations cannot be performed without some type of PES.

In order to keep the notation as simple as possible, we consider a diatomic molecule interacting with a solid with only a single type of atom from here on. The generalization of a PES to polyatomics is generally rather difficult (and pretty much unknown), while the treatment of more complex solids is generally straightforward (because we have limited the problem to low-energy processes). We denote the positions of the gas molecule's atomic nuclei by $(\mathbf{X}_1, \mathbf{X}_2)$ and the solid's atomic nuclei by $(\mathbf{Y}_i,\ i = 1, \ldots, N_S)$ or $\{\mathbf{Y}_i\}$. The equilibrium positions of the solid's atoms are $\{\mathbf{Y}_i^{(eq)}\}$. The masses are m_1, m_2 and m for the gas atoms and solid's atoms, respectively. The former are also located by the center of mass, $\mathbf{X} = (m_1\mathbf{X}_1 + m_2\mathbf{X}_2)/(m_1 + m_2)$, and relative, $\mathbf{r} = (\mathbf{X}_2 - \mathbf{X}_1)$, coordinates, with the equilibrium value of the bond length, r, denoted as $r^{(eq)}$. Associated with these coordinates are the total and reduced molecular masses, $M = (m_1 + m_2)$ and $\mu = m_1 m_2/(m_1 + m_2)$.

First, we focus on the basic physics of the interaction as described in a simple but very concise and useful model (Norskov et al. 1981; Norskov and Besenbacher 1987). At far distances (long-range) from the surface, the molecule–surface interaction is a van der Waal's attraction, of the form

$$V_{LR}(\mathbf{X}, \mathbf{r}, \{\mathbf{Y}_i\}) = -C_3[1 + a_3 P_2(\hat{r}\cdot\hat{z})]/Z^3 \tag{3.1}$$

where C_3 and a_3 depend upon the perpendicular and parallel polarizabilities of the molecule and the electronic response of the solid. The distance is Z $(\equiv \hat{z}\cdot\mathbf{X})$ as measured from the surface plane. The coefficient of the anisotropic term is generally $\leqslant 0.1$. As Z decreases further, a repulsive term arises from the extra kinetic energy repulsion due to the Pauli exclusion principle operating between the solid's and molecule's electrons, leading to the form:

$$V(\mathbf{X}, \mathbf{r}, \{\mathbf{Y}_i\}) = V_{SR}(\mathbf{X}, \mathbf{r}, \{\mathbf{Y}_i\}) - C_3[1 + a_3 P_2(\hat{r}\cdot\hat{Z})]/Z^3 \tag{3.2}$$

For a nonreactive system, V_{SR} is generally strongly dependent upon the orientation of the molecule, but only weakly dependent upon the bond length r and solid atom's positions. (The full AB–S PES, $V_{AB,S}$, is then just $V + V_{AB}(r)$ where the latter is the interaction in the isolated molecule.) The effect of solid atom displacement in V_{SR} is typically repulsive in the separation between each gas atom and each nearby solid atom. Because of the weak dependence of V on r, the variation of the bond length from $r^{(eq)}$ on the full AB–S PES is to greatly increase the energy irrespective of the height of the molecule above the surface, thereby leading to very small coupling of r and $\mathbf{X}$. These features are just the molecule–surface analog of gas-phase nonreactive PES with simple mechanical vibrational processes.

As emphasized by Norskov and co-workers, for reactive systems another effect occurs. As Z decreases, the affinity level (LUMO) of the molecule begins to feel an image-charge type attraction, causing a shifting of its energy as

$$E_a(Z) = E_a(\infty) - 1/4Z \tag{3.3}$$

As the distance Z gets smaller, the affinity level gets pulled below the Fermi level of the solid, and the molecular orbital fills. If this orbital is antibonding, then the molecular bond may be weakened sufficiently to rupture, thereby leading to dissociative chemisorption. Since filling of an antibonding level leads to an increase in the bond length, the PES will display a strong and

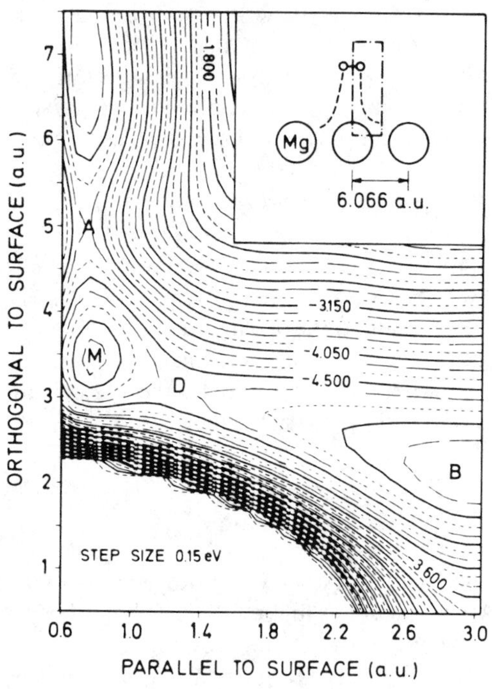

Figure 17. Contour plot of a two-dimensional cut in the H_2/Mg(100) PES for the configuration shown on the top right of the figure. The energies are relative to those of the free atoms. A is a barrier to molecular chemisorption. M is a molecular chemisorption well. D is a barrier to dissociation, and B is an atomic chemisorption well. The distance from the surface is measured from the first atomic layer, and the distance parallel to the surface is measured from the top site towards the bridge site. The figure is from Norskov et al. (1981) with permission.

nonmonotonic dependence upon r at small Z. When the Pauli repulsion increases more quickly than E_a decreases, a barrier to dissociation results.

More quantitative evidence for the charge transfer mechanism comes from self-consistent calculation of H_2 on Mg (Norskov et al 1981; Norskov and Besenbacher 1987). Calculated interaction energies are shown in the contour plot in Fig. 17. The vertical axis is Z and the horizonal axis is r. The energies in eV are calculated relative to those of the free H atoms. One immediately sees the various possible characteristics of a reactive PES:

1. A weak physisorption well at far distances from the surface ($Z \approx 7$ bohr here) with $r \approx r^{(eq)}$;
2. A barrier to molecular chemisorption closer to the surface ($Z \approx 5$ bohr here) with $r \approx r^{(eq)}$ and denoted by point A;
3. A more strongly-bound molecular chemisorption well at still closer distances ($Z \approx 3.5$ bohr here) with $r \geqslant r^{(eq)}$, but with considerable anharmonicity in r, and denoted by point M;
4. A barrier to dissociative chemisorption just slightly closer to the surface ($Z \approx 3$ bohr here) with $r > r^{(eq)}$ and denoted by point D;
5. A strongly bound atomic chemisorption well at still closer distances ($Z \approx 2\frac{1}{4}$ bohr here) with $r \gg r^{(eq)}$, and denoted by point B.

Not all of these features will appear for every system. For example, a kinetically nonreactive system can be thought of as having a very large value of either A or D. Of more relevance for molecule–surface reaction dynamics is the fact that all of these features may not appear in every geometry of a particular molecule–surface system. For example, if the H_2 bond was perpendicular to the surface, feature M would either disappear or be much smaller; the barrier D would disappear since point B would lie considerably above (≈ 2.3–2.5 eV) the reactant H_2 molecule. In this configuration, the PES would appear similar to that of a thermodynamically nonreactive system.

A brief mention of nomenclature is necessary here. First, note that the barrier at A is described as being in the entrance channel. This is merely a convenient name for an activation barrier in which the molecular bond length is "nearly unchanged" from its gas-phase value, $(r - r^{(eq)}) \ll r^{(eq)}$. By the same token, an exit channel barrier is one in which the molecular bond length is "significantly stretched" from its gas-phase value, $(r - r^{(eq)}) \approx r^{(eq)}$. These are subjective criteria, useful mainly as limiting cases. Realistic PES will almost certainly fall in between each limit.

To illustrate the charge transfer corresponding to the various marked points in this figure, we show the one-electron density of states in Fig. 18. This clearly shows that the anti-bonding $2\sigma_u^*$ level is not-filled in the physisorbed (P) and A points on the PES; is slightly filled early in the M

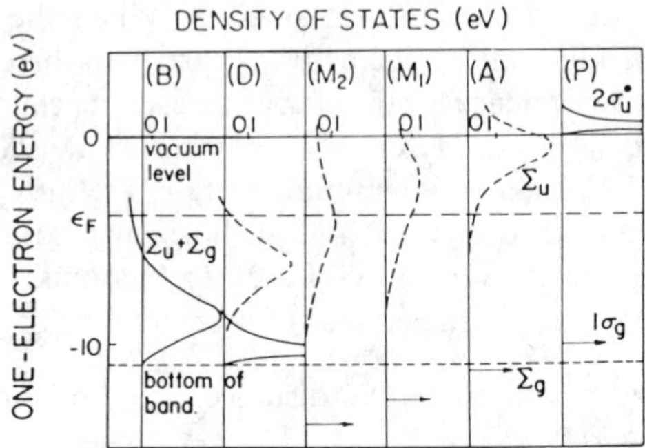

Figure 18. One-electron density of states for H_2 approaching Mg(100) as in Fig. 17. The figure is reprinted with permission from Norskov and Besenbacher (1987).

well (denoted M_1) and nearly half-filled late in the M well (denoted M_2); and is nearly fully filled by the dissociation barrier (D). This provides more quantitative evidence for the importance of negative ion formation in dissociative chemisorption.

We can use the contour plot to make a few general points about molecule–surface interactions. The points in the contour plot are often mapped onto a one-dimensional reaction path form, as in Fig. 2, and shown in Fig. 19 for the two-dimensional PES of Fig. 17. This figure illustrates a few important features in the contour plot concisely, and thus is often used to discuss real systems. We argue strongly against this practice since such conciseness is associated with a lack of detail. For example, in the contour plot, it is apparent that point D is associated with $r > r^{(eq)}$, while in Fig. 19,

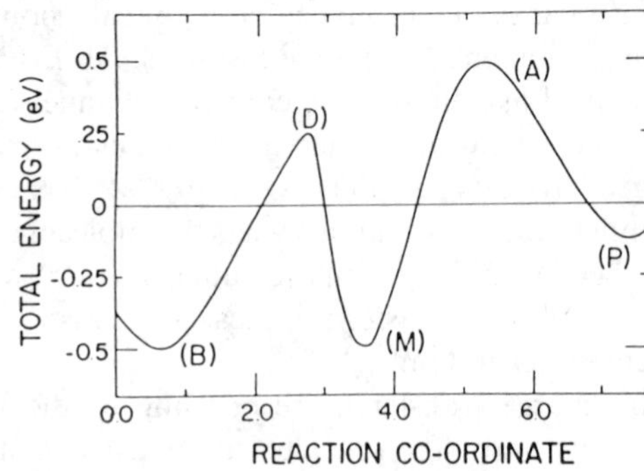

Figure 19. One-dimensional representation of the PES in Fig. 17. The figure is from Norskov and Besenbacher (1987).

this is obscured and could just as easily arise from a barrier at small Z with $r \approx r^{(eq)}$. Such different topologies of the PES will have important distinguishing characteristics in the dynamics.

Along the same lines, it is worthwhile mentioning that even the two-dimensional contour plot obscures some important features of the true multidimensional PES. In order to illustrate this fact, consider the following argument. Some of the features in the PES can be probed by experiments: the weakly physisorbed, molecularly and atomically chemisorbed species by surface spectroscopies; and the activation barriers by molecular beam scattering experiments. Thus, much of the PES can be determined and the reader may wonder why one does not simply join these individual regions together and "find" the PES!

The reason is that such a PES contour plot describes a dissociative chemisorption process for a particular location and orientation of the molecule above a particular geometry of the surface atoms. In real systems, all of the PES features will vary with orientation of the molecule $\hat{r}$, position of the molecule in the surface unit cell (X, Y), and displacement of the surface atoms $\{\mathbf{Y}_i - \mathbf{Y}_i^{(eq)}\}$. These other variables may complicate the problem tremendously, and may even control the particular dynamical process. For example, the atomically and molecularly chemisorbed species may be stable at different positions in the surface unit cell. Spectroscopies probes of these species would then determine two completely different regions of the PES without any means for connecting them.

It is apparent that computations of full PESs are needed. Considerable progress has been made in this area, and there are now becoming available *ab initio* and density-functional calculations of at least a few points on a full PES (Norskov et al. 1981; Fischer and Whitten 1984; Siegbahn et al. 1984; Panas et al. 1987; Yang and Whitten 1988; Siegbahn et al. 1988; Upton et al. 1988). However, the accuracy of the detailed calculations is still not sufficient for chemical accuracy, $\leqslant 1$ kcal/mole for instance, and the computational expense still prohibits calculation of enough points to map out the full multidimensional PES. The use of a reaction path Hamiltonian formalism (Miller et al. 1980) would reduce the number of required points but may not be particularly appropriate since the depth of the physisorption well can often reach 0.2–0.4 eV. This accelerates the molecule and thus prevents following the reaction path even at extremely low initial kinetic energies. For example, on the contour plot in Fig. 17, the well M of nearly 0.4 eV will cause acceleration along Z and not along r, even though the latter is close to the reaction coordinate near point D.

Two important concepts provided by recent *ab initio* and density-functional approaches are worth mentioning. The first is associated with the electronic mechanism for reducing the barriers to chemisorption. In transition metals, it has been found from local density calculations (Harris and Andersson 1985) that the presence of holes in the d-band lowers the activation barrier, A. The basic mechanism is rather simple: the increased Pauli repulsion as H_2 approaches the metal surface eventually leads to filling of the d-band at the expense of the s-band, allowing for stronger bonding between the s-orbitals of the metal and those of H. This increased bonding competes with the Pauli repulsion and inhibits a further increase in the activation barrier. The d-electrons have also been shown to be important in *ab initio* calculations of the interaction between H_2 and metal surfaces (Siegbahn et al. 1984; Panas et al. 1987; Yang and Whitten, 1988).

The second concept is associated with the use of clusters to model the interactions with an infinite surface. It has been known for a number of years that cluster models of the interactions between molecules and surfaces often do not provide interaction energies that converge quickly with increasing cluster size (Upton and Goddard 1981). Very recently, it has been suggested (Panas et al. 1988) that the problem involves the fact that bonding between an adsorbate and a cluster should involve orbital states of the cluster with certain symmetry in order to mimic the infinite solid. These states may be higher in energy than the ground state of the cluster, but the interaction energy should be determined with respect to this excited-state energy. Such states will presumably be much closer in energy to the ground state in the infinite system. Thus, in order to determine the binding energy with the surface, one must prepare a cluster wavefunction of the proper symmetry even if it is an excited state. A similar idea has been used in a study of the binding in the O_2/Ag(110) system (Upton et al. 1988). Whether such a rule holds in general needs to be determined by further study. However, we do note that preparation of such a wavefunction of proper symmetry may only be straightforward when the molecule reacts in a high-symmetry arrangement. We will have more to say about such low-symmetry sites in the dynamics section.

Even with considerable progress in both the understanding and computation of interaction energies, only a few energies can be provided on the PES. In the dynamics, the molecule can rotate and translate parallel to the surface, and the surface atoms can move, thereby distorting the PES. When the molecule rotates, the various barriers and even the gross topology of the PES changes. Similarly, the PES at different locations in the surface unit cell will be different. The dynamics in these other degrees of freedom may play a central role, and thus one must include them. Thus, it is necessary to develop computationally efficient global representations of PES that can be used to construct the full PES using all available information, experimental as well as theoretical. Ideally, the former will be especially focused on stable reactants and products, while the latter will provide some points in the activation barrier regions. Such a synthesis plays a major role in this field.

We now consider the most common method of construction of these global PESs, based upon a modified four-body LEPS form that was initially developed by McCreery and Wolken (1977). This was later modified and quantified to be more accurate for metal surfaces (Lee and DePristo 1987; Kara and DePristo 1988) and extended using embedded atom ideas (Truong et al. 1989a).

Consider a diatomic, AB, interacting with a surface, S. The basic idea is to utilize valence bond theory for the atom–surface interactions, V_{AB} and V_{BS}, along with V_{AB} to construct $V_{AB,S}$. For each atom of the diatomic, we associate a single electron. Since association of one electron with each body in a three-body system allows only one bond, and since the solid can bind both atoms simultaneously, two valence electrons are associated with the solid. Physically, this reflects the ability of the infinite solid to donate and receive many electrons. The use of two electrons for the solid body and two for the diatomic leads to a four-body LEPS potential (Eyring et al. 1944) that is convenient mathematically, but contains nonphysical bonds between the two electrons in the solid. These are eliminated, based upon the rule that each electron can only interact with an electron on a different body, yielding the modified four-body LEPS form. One may also view this as an empirical parametrized form with a few parameters that have well-controlled effects on the global PES.

The explicit form is,

$$V_{AB,S} = Q_{AS} + Q_{BS} + Q_{AB} - [J_{AB}(J_{AB} - J_{AS} - J_{BS}) + (J_{AS} + J_{BS})^2]^{1/2} \quad (3.4)$$

where Q and J are coulomb and exchange integrals, respectively, for each constituent. One important feature of this form is the non-additivity of the interaction potentials. For example, the AB interaction alone is given by

$$V_{AB} = Q_{AB} - |J_{AB}| \quad (3.5)$$

which evidently does not determine Q_{AB} and J_{AB} individually. Another important feature is the incorporation of all four asymptotic limits: AB(g) + S; A(a) + B(g) + S; A(g) + B(a) + S; A(a) + B(a) + S. For example, note that if the terms with A–S and B–S vanish, then $V_{AB,S} = Q_{AB} - |J_{AB}| = V_{AB}$, and if the terms with A–B vanish, then $V_{AB,S} = (Q_{AS} - |J_{AS}|) + (Q_{BS} - |J_{BS}|)$. We should emphasize that it is the precise division into Q and J for each interaction in Eq (3.4) that will control the topology and energies of the full PES for the reaction. This will be clear later, but first we must describe each type of term in Eq. (3.4).

Consider the A–B interaction, since this is the simplest:

$$Q_{AB} + J_{AB} = V_{AB} = D_{AB}\{\exp[-2\alpha_{AB}(r - r_{AB})] - 2\exp[-\alpha_{AB}(r - r_{AB})]\} \quad (3.6a)$$

$$Q_{AB} - J_{AB} = \tfrac{1}{2}[(1 - \Delta_{AB})/(1 + \Delta_{AB})] = D_{AB}\{\exp[-2\alpha_{AB}(r - r_{AB})] + 2\exp[-\alpha_{AB}(r - r_{AB})]\} \quad (3.6b)$$

The A–B interaction potential is represented by the Morse potential in Eq. (3.6a) with the parameters D_{AB}, α_{AB}, and r_{AB} characterizing the bond energy, range parameter, and equilibrium bond length ($= r^{(eq)}$ for the A–B bond), respectively, for the A–B molecule. These are specified by the A–B binding curve (Huber and Herzberg 1979). The A–B antibonding potential in Eq. (3.6b) is represented by the anti-Morse form. The Sato parameter, Δ_{AB}, is unspecified as yet. However, it clearly controls the division of the Morse- anti-Morse forms into Q_{AB} and J_{AB}. If one were to take the idea seriously that Eq. (3.6b) represents some excited state of the AB molecule, then Δ_{AB} could be determined from spectroscopic information about such an excited state. However, this is not appropriate since the "excited state" is really not one of the isolated molecule but one of the molecule–surface system. This was demonstrated especially clearly for the O_2/Ag(110) system (Lin and Garrison 1984) and was also emphasized in earlier work by Tully (1980b).

Now consider the A–S interaction (with the analogous equation for B–S derived by simply changing A to B everywhere). This interaction is more complicated than the A–B one, given by:

$$\begin{aligned} Q_{AS} + J_{AS} = V_{AS} &= D_{AH}\{\exp[-2\alpha_{AH}(r_{As} - R_{AH})] - 2\exp[-\alpha_{AH}(r_{As} - R_{AH})]\} \\ &\quad + \sum D_{AS}\{\exp[-2\alpha_{AS}(R_{A\beta} - R_{AS})] - 2\exp[-\alpha_{AS}(R_{A\beta} - R_{AS})]\} \end{aligned} \quad (3.7a)$$

$$\begin{aligned} Q_{AS} - J_{AS} = \tfrac{1}{2}[(1 - \Delta_{AS})/(1 + \Delta_{AS})] &= [D_{AH}\{\exp[-2\alpha_{AH}(r_{As} - R_{AH})] + 2\exp[-\alpha_{AH}(r_{As} - R_{AH})]\} \\ &\quad + \sum D_{AS}\{\exp[-2\alpha_{AS}(R_{A\beta} - R_{AS})] + 2\exp[-\alpha_{AS}(R_{A\beta} - R_{AS})]\}] \end{aligned} \quad (3.7b)$$

where

$$R_{A\beta} = |\mathbf{X}_A - \mathbf{Y}_\beta| \quad (3.8)$$

is the distance between gas atom A and solid atom β. The summation extends over all the atoms in the solid. The two general terms in Eq. (3.7) are required in order to accurately describe the bonding involving metals, as will now be explained.

First consider the terms in D_{AH}. *These represent the interaction between* A *and the valence electrons of the metal.* This is modeled by the interaction of A with jellium whose density is provided by the valence electrons of the metal at the position of A. The parameters are thus:

1. D_{AH} = strength of the interaction between atom A and jellium
2. α_{AH} = range of the above interaction
3. $R_{AH} = (3/4\pi n_0)^{1/3}$ where n_0 is the density at the minimum of the atom–jellium binding curve

These are determined from the self-consistent-field local density (SCF-LD) functional calculations (Puska et al. 1981) on the embedding energy of an atom in jellium. The values as a function of jellium densities are represented by a Morse-like form. (It is not strictly a Morse potential since density is the variable instead of distance.) The variable, r_{As}, depends upon the density of the solid at the position of A:

$$r_{As} = (3/4\pi n(\mathbf{X}_A))^{1/3} \tag{3.9}$$

The additional assumption that the density of the solid is well-represented by the sum over the individual atomic densities is usually made, leading to:

$$n(\mathbf{X}_A) = \sum n(R_{A\beta}) \tag{3.10}$$

where $n(R_{A\beta})$ is the atomic density of the solid atom β at the position of the gas atom A and again the summation extends over all the atoms of the solid. Since differentiation of the PES is needed, simple forms have been used for the atomic densities for each shell, where only the outer or valence electrons need be considered. For example, in W, the $6s^2$ and $5d^4$ have been included as (Kara and DePristo 1988a):

$$n(R_{A\beta}) = 2n_s(R_{A\beta}) + 4n_d(R_{A\beta}) \tag{3.11a}$$

where

$$n_s(x) = s_1[x - s_2]^{s_4} \exp(-s_3 x) \tag{3.11b}$$

$$n_d(x) = d_1 \exp(-s_2 x^{s_3}) \tag{3.11c}$$

The various parameters $\{s_k\}, \{d_k\}$ are determined by fitting to known atomic densities. As a matter of practicality, we should point out two features: (1) a fitting procedure may not be necessary since the densities can be easily represented quite accurately by a one-dimensional spline; and (2) the inclusion of inner-core electrons does not change the relevant densities at the separations, $R_{A\beta}$, of importance to low-energy collisions.

Next consider the terms in D_{AS}. *These represent the interaction between atom* A *and the localized electrons and nuclear charges of the metal.* These are described by a two-body interaction between A and each solid atom, β. In Eq. (3.7) these interactions are represented by Morse potentials for convenience. The parameters are defined below:

1. D_{AS} = strength of the localized two-body interaction between atom A and the atoms of the solid
2. α_{AS} = range of the above interaction
3. R_{AS} = position of the minimum of the two-body interaction

These parameters must be determined from information on the atom–surface interaction potential, which may come from either sufficient experimental or theoretical information, such as the binding energy, height, and frequency for different sites of adsorption, or the full binding curves above each site.

Assume that the binding curves for the asymptotic fragments are available and have been used to determine the relevant parameters: $(D_{AB}, \alpha_{AB}, r_{AB})$, $(D_{AS}, \alpha_{AS}, R_{AS})$ and $(D_{BS}, \alpha_{BS}, R_{BS})$. The parameters for the interaction with jellium, $(D_{AH}, \alpha_{AH}, R_{AH})$ and $(D_{BH}, \alpha_{BH}, R_{BH})$, are not adjustable, being determined by the SCF-LD values. Thus, the remaining variables in the PES, the so-called Sato parameters, Δ_{AB}, Δ_{AS}, and Δ_{BS}, are undetermined and available for flexible representation of the full molecule–solid reactive PES. We consider the effect of these on the PES later. At this point, we do want to emphasize that the basic physics and chemistry—: (1) interactions with localized and delocalized metal electrons; and (2) nonadditive chemical bonding—are correct. We should also note that the representation of an interaction in metals in terms of an embedding function (in jellium) plus two-body terms is identical in spirit to the embedded atom method (EAM) (Daw and Baskes 1984, 1988). The distinction here is that we do not use the EAM for the A–B interaction and explicitly incorporated nonadditive energies via the LEPS prescription, both of which are important for the accurate representation of the reactive PES.

We use the N_2/W(110) system as an example of the construction of the modified four-body LEPS PES. Since the molecule is homonuclear, the parameters for the $A = B = N$ atoms are identical. We use the explicit subscripts for the relevant bodies, but note that the original work (Kara and DePristo 1988a, 1988b, 1988c) used a simpler notation which is also followed

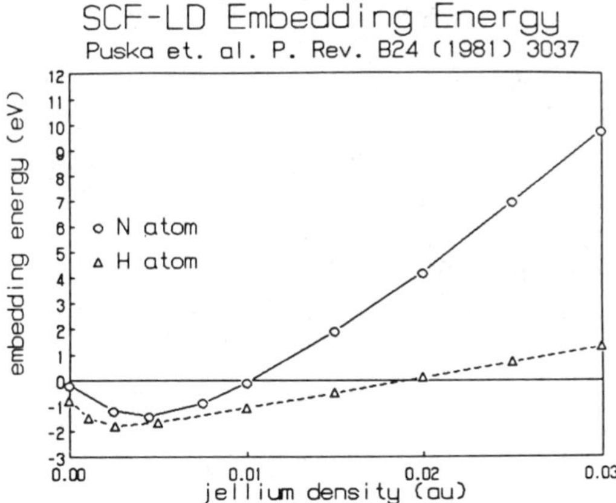

Figure 20. Embedding energy as a function of jellium density for two different atoms. The data are from Puska et al (1981).

in the figures: (1) $\Delta_{NW} \equiv \Delta_{GS}$; (2) $\Delta_{NN} \equiv \Delta_{GG}$; (3) $D_{NW} \equiv D_2$; (4) $\alpha_{NW} \equiv \alpha_2$; (5) $R_{NW} \equiv R_2$. As the first part of the interaction potentials, we specify $(D_{NN}, \alpha_{NN}, R_{NN})$ to be (9.9 eV, 1.42 bohr^{-1}, 2.06 bohr) from the N_2 spectroscopic information (Huber and Herzberg 1979).

For the N–W potential, we need both the homogeneous and two-body terms. Typical results of the embedding energy for an atom in jellium are shown in Fig. 20, based on SCF–LD calculations (Puska et al. 1981). For the N atom, the Morse-like parameters modeling the embedding energy are $(D_{AH}, \alpha_{AH}, R_{AH}) = (1.4\,\text{eV}, 0.78938\,\text{bohr}^{-1}, 3.7575\,\text{bohr})$, with the last number corresponding to $n_0 = 0.0045\,\text{bohr}^{-3}$. From the atomic Hartree–Fock calculations for the W-atom (McClean and McClean 1981), the 6s and 5d electron densities per electron were accurately represented over the important distance range 2 bohr $< r <$ 10 bohr by the forms of Eqs. (3.11b–c) with the numbers given below (in atomic units):

$$n_s(r) = 0.1398[r - 1.6992]^{1.1043} \exp(-1.3967r)$$

$$n_d(r) = 0.078513 \exp(-0.76027r^{1.5282})$$

This specifies the homogeneous part of the N—W potential.

The two-body part of the N—W potential was determined by fitting the full N—W potential (given the homogeneous part specified above) to the theoretically predicted values shown in Fig. 21. There is a strong dependence upon the binding site, at least between the atop vs. other sites. The lowest minimum is in very good agreement with the experimental value of 6.73 eV

(Ho et al. 1980; Tamm and Schmidt 1971). This yields estimates of the other N—W potential parameters as $(D_{NW}, \alpha_{NW}, R_{NW}) = (6.78\,\text{eV},\ 1.089\,\text{bohr}^{-1},\ 2.354\,\text{bohr})$. Note the dominance of the two-body contribution to the N—W bond as compared to the homogeneous energy contribution, since the former has a minimum of $-1.4\,\text{eV}$. Note also the small value of R_{NW} and the moderate value if α_{NW}, which combine to make the two-body terms generally attractive (except close to the surface on an atop site) and relatively short-ranged. These parameter values can change by a few percent and still provide an excellent fit to the calculated potential in Fig. 21. This will be important in adjustment of $V_{NN,W}$.

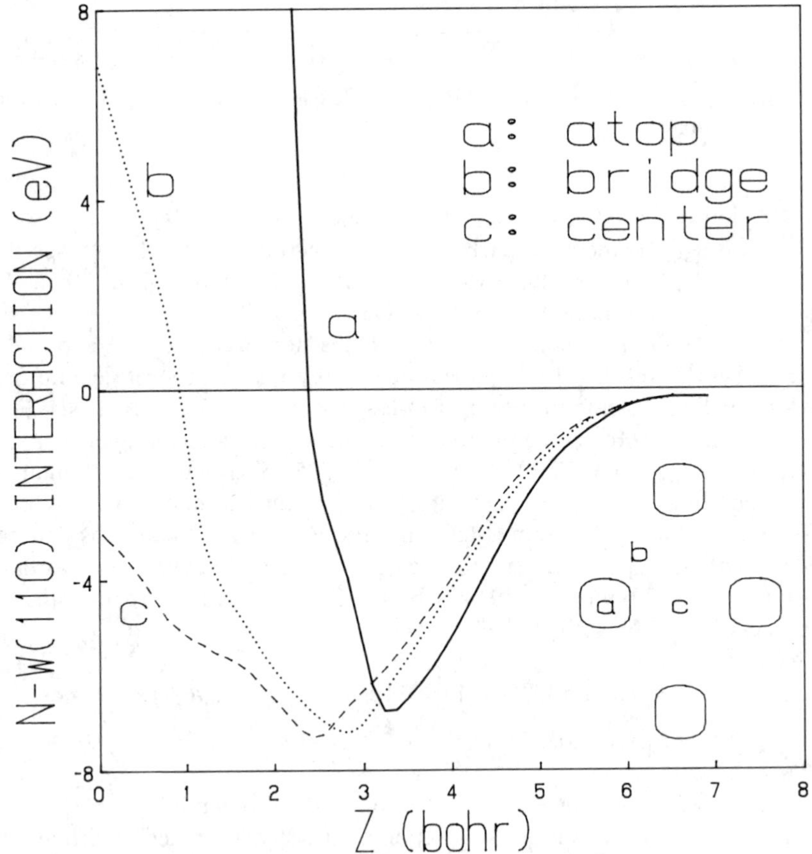

Figure 21. N–W(110) interaction as a function of the height of the N above the (rigid) W(110) surface for different sites. The plot is from Kara and DePristo (1988a) based upon theoretical calculations using the corrected effective medium one-body method (Kress and DePristo 1987).

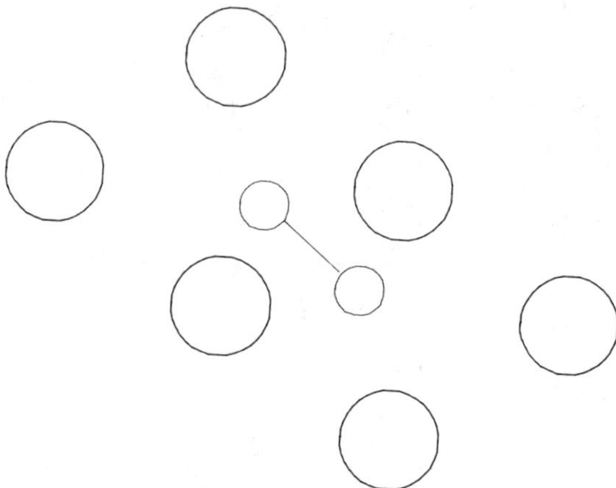

Figure 22. Configuration of the N_2/W(110) system with only a few W atoms shown for clarity. The figure is reprinted with permission from Kara and DePristo (1988a).

We illustrate the variety of topologies for the full NN–W PES that can be generated by varying mainly the Sato parameters, Δ_{NW} and Δ_{NN}, and slightly the two-body parameters, $(D_{NW}, \alpha_{NW}, R_{NW})$. The representative configuration is used in which a N_2 molecule approaches a rigid W(110) surface with the N_2 molecular axis parallel to the surface, the center of mass above the bridge site, and the two atoms pointing to opposite center sites, as shown in Fig. 22. While the actual dissociation occurs over a wide variety of locations in the unit cell as will be shown in Section VI, the orientation of the N_2 is always close to parallel when the dissociation occurs. For each set of parameters, a PES contour plot is made in the usual coordinates: bond length and height. For clarity, we show only for critical features from each: the molecular well depth in the entrance channel (point P in Fig. 19); the barrier energy (point D in Fig. 19); and the coordinates of the barrier location in both height and bond length. For the N_2/W(110) system, this parameter variation never yielded two barriers such as A and D in Fig. 17 since $(D_{NW}, \alpha_{NW}, R_{NW})$ were only allowed to vary slightly around the values specified above. (A different situation exists for the H_2/Ni(100) system discussed later.)

First, consider variation of the two Sato parameters with the results shown in Fig. 23. The curves correspond to fixed Δ_{NW} as a function of Δ_{NN} ($\equiv \Delta_{GG}$). In panel a, note that as either of the Sato parameters approaches 1, the barrier energy increases and the molecular well depth decreases. This is easy to understand since as either $\Delta \rightarrow -1$, the antibonding interaction increases to ∞ and it is the separation between the bonding and antibonding

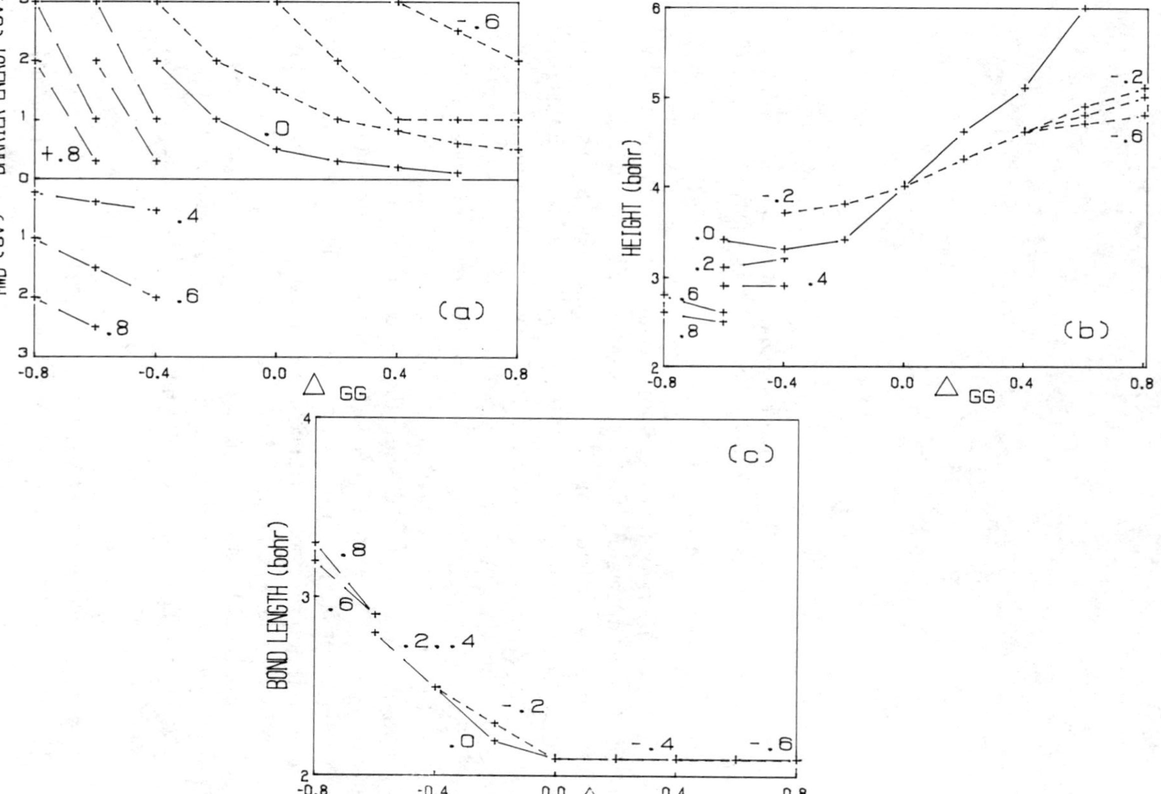

Figure 23. (a) Activation barrier and molecular well depth as function of the N–N and N–W Sato parameters; (b) same as (a) except for the height of the barrier location above the surface plane; (c) same as (a) except for the N–N bond length at the barrier location. The figure is reprinted with permission from Kara and DePristo (1988a).

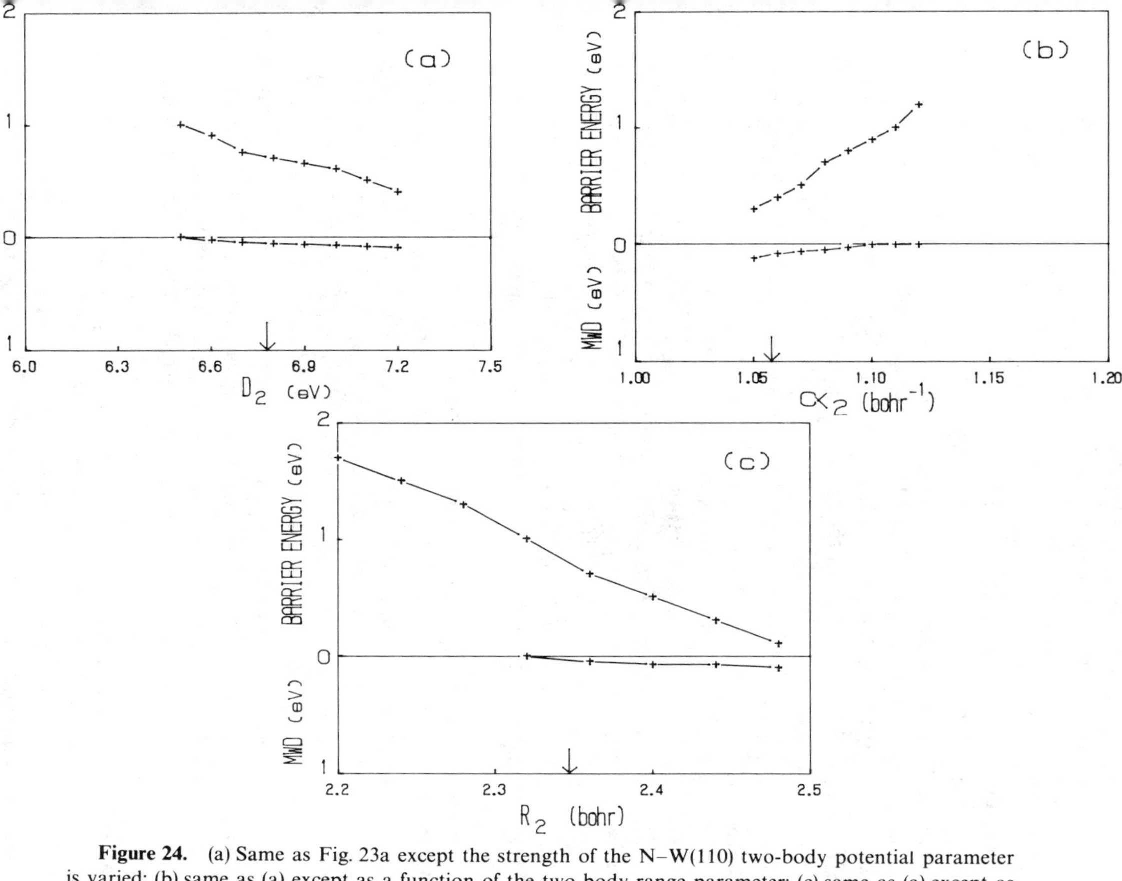

Figure 24. (a) Same as Fig. 23a except the strength of the N–**W**(110) two-body potential parameter is varied; (b) same as (a) except as a function of the two-body range parameter; (c) same as (a) except as a function of the two-body interaction minimum. The figure is reprinted with permission from Kara and DePristo (1988a).

interaction that controls the size of the dissociation barrier. Also note two important points: (1) For any reasonably sized barrier of 0–1 eV, there are an infinite set of possible Sato parameters. (2) The molecular well depth can take on values from small ($\leqslant 0.1$ eV) to large ($\geqslant 1.0$ eV). Thus, the parametrized PES is quite flexible with regard to the relevant energies.

Next, consider the location of the barrier as shown in panels b and c. Some of the curves extend over a limited range because the barriers exist only over a small range. It is possible to place the barrier in either the entrance channel or the exit channel by variation of the Sato parameters.

To conclude this investigation of the variation of $V_{NN,W}$ with parameters, we show the effect of slight changes in the two-body parameters on the barrier height and molecular well depth in Fig. 24. These changes have a very small effect on the location of the barrier, and thus this is not shown. By adjusting $(D_{NW}, \alpha_{NW}, R_{NW})$ it is possible to change substantially the size of the activation barrier without varying the topology of the PES or the molecular well depth, providing a way to fine-tune the PES.

Using experimental information on the physisorption well depth of ≈ 0.27 eV (Yates et al. 1976) and by comparison of dynamical simulations with molecular-beam scattering data on $S_0(E_i, \theta_i)$ (Pfnür et al. 1986), the values of $(\Delta_{NN}, \Delta_{NW}) = (-0.369, 0.241)$ were found for $(D_{NW}, \alpha_{NW}, R_{NW}) =$ (6.18 eV, 1.05 bohr^{-1}, 2.32 bohr). The latter values differ very slightly from those found by fitting the N–W calculated interaction potential in Fig. 21.

Next, we consider a LEPS PES for the H_2/Ni(100) system (Kara and DePristo, 1989), which is based upon recent ab initio calculations (Siegbahn et al. 1988) along with dynamically adjusted Sato parameters. Since the H—Ni binding energy is ≈ 2.7 eV while D_{AH} is ≈ 1.8 eV, the two-body terms do not dominate for this system. Contour plots in Fig. 25 demonstrate that there are small activation barriers of ≈ 0.035 eV and ≈ 0.045 eV in the entrance channel of the bridge$\rightarrow$center and atop$\rightarrow$center PES, respectively. The dependence of the entrance channel barrier on the orientation of H_2 has not been determined quantitatively, but the dependence on r is quite weak. The variation with position in the unit cell is also weak. In addition, for the atop$\rightarrow$center dissociation there is also a barrier in the exit channel. The exit channel barrier obviously has a strong dependence upon bond length. Thus, the PES is quite complex.

For completeness, we consider other descriptions of the PES. The first two are modifications of the above four-body LEPS approach. In the original articles (McCreery and Wolken 1977) and in much later work by a number of other groups (Tantardini and Simonetta 1981, 1982; Marcusson et al. 1981; Jackson and Metiu 1987; Chiang and Jackson 1987), the surface was assumed to be rigid. Then, it is unnecessary to consider atomic positions and densities

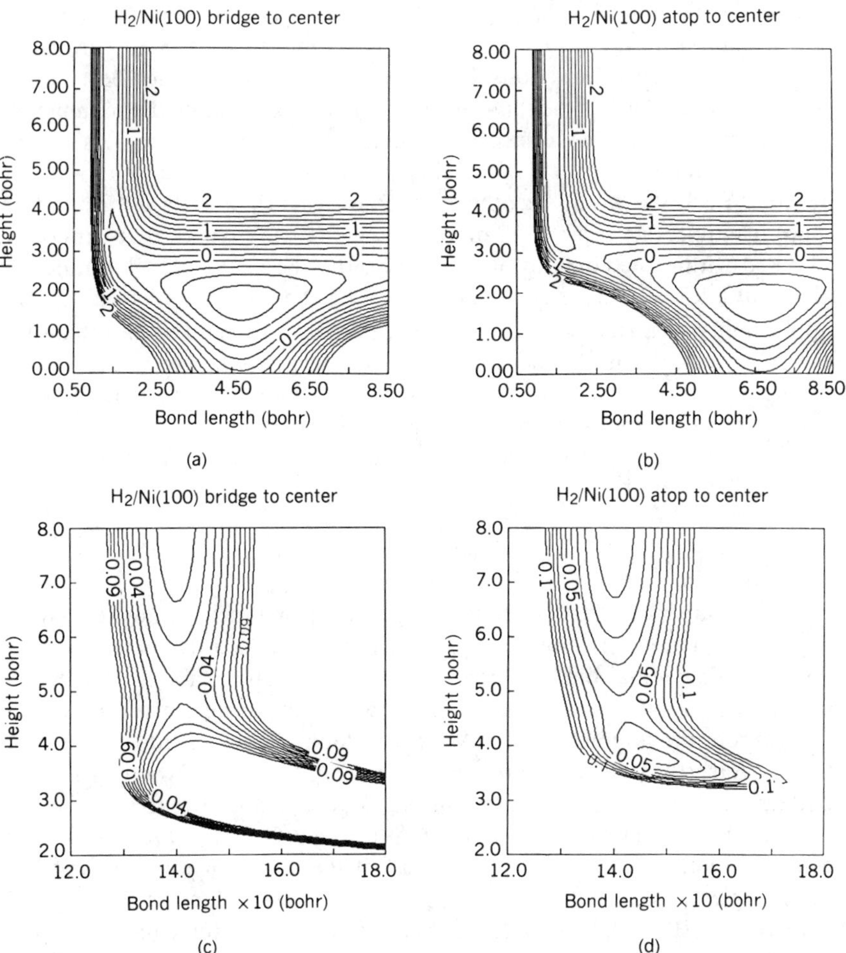

Figure 25. Two-dimensional contour plots for a LEPS potential of H_2/Ni(100) based upon the ab initio calculations of Siegbahn et al. (1988). The data are from Kara and DePristo (1989).

of the solid. One may simply replace Eqs. (3.7a–b) by

$$Q_{AS} + J_{AS} = D_{AS}\{\exp[-2\alpha_{AS}(Z_A - Z_{AS})] - 2\exp[-\alpha_{AS}(Z_{A\beta} - Z_{AS})]\} \tag{3.12a}$$

$$\begin{aligned} Q_{AS} - J_{AS} &= \tfrac{1}{2}[(1-\Delta_{AS})/(1+\Delta_{AS})] \\ &= D_{AS}\{\exp[-2\alpha_{AS}(Z_A - Z_{AS})] + 2\exp[-\alpha_{AS}(Z_A - Z_{AS})]\} \end{aligned} \tag{3.12b}$$

where the parameters D_{AS}, α_{AS}, and Z_{AS} are (periodic) functions of the horizontal position of atom A in the surface unit cell. (The Sato parameter, Δ_{AS}, could also be a function of this position but this has not been done.) The A–B interaction is the same as in Eq. (3.6). To determine the parameters, one again needs atom–surface binding curves at different sites. As before, the Sato parameters provide flexibility in representing the molecule–surface reactive PES. We would recommend against this rigid surface formulation since it requires an *a priori* assumption about the dynamics that cannot be checked within the confines of the model. It is better to use a formulation that can allow atomic motion and then hold the atoms fixed to check the applicability of a rigid surface. This is especially so since the movement of the surface may not just be important for energy exchange but also may provide a variation of the PES topology (e.g., barrier heights) with slight displacements of the positions of the atoms of the solid.

The second modification involves replacement of the antibonding A–B term in Eq. (3.6b) by

$$Q_{AB} - J_{AB} = V_{AB}^{(a)} \tag{3.13}$$

where the superscript (a) symbolizes a general new function, not related by any simple parametrization to V_{AB}. This provides added flexibility over the form in Eq. (3.6b), but requires further information about the AB–S interaction to determine the form, since the antibonding term must reflect the appropriate interaction of A with B in the presence of the surface. For the O_2/Ag(110) system, the ordered coverage of O(a) into $(N \times 1)$ overlayer patterns and the negative charge on the O atom led to a representation of $V_{AB}^{(a)}$ as a simple electrostatic interaction (Lin and Garrison 1984). A further generalization uses $V_{AS}^{(a)}$ and $V_{BS}^{(a)}$ instead of the anti-Morse forms, and has been implemented (without the D_{AH} and D_{BH} terms) for the CO/Pt(111) system (Tully 1980b). We would not recommend against these more complex models of the PES but would caution that much of the information available on atom–surface and molecule–surface interactions would not allow for the determination of this complex form. One must be especially careful in using electrostatic models for antibonding states of the A–S interaction because the electron shielding length on metals is short; a long-range electrostatic model may very well be replacing a through-metal interaction with a through-space one.

A different construction of the PES, based closely on the electron transfer mechanism, has been presented (Gadzuk and Holloway 1986; Gadzuk 1988; Holloway 1989). The two configurations are (AB, S) and (AB^-, S), described by $H_{1,1}$ and $H_{2,2}$ which are the interaction energies in the appropriate configuration. With $H_{1,2}$ as the mixing between these configurations, the adiabatic PES is just the lowest eigenvalue of the 2 × 2 Hamiltonian matrix, yielding:

$$V_{AB,S} = \tfrac{1}{2}\{H_{1,1} + H_{2,2} - [(H_{1,1} - H_{2,2})^2 + H_{1,2}^2]^{1/2}\} \tag{3.14}$$

This has been useful in constructing a number of distinct PESs for use in dynamical simulations (Gadzuk 1987a, 1987b, 1987c; Halstead and Holloway 1988; Harris et al. 1988a, 1988b; Holloway 1989). Since representative PESs will be shown in some detail in the section on dynamical results, we shall not dwell on this here.

Before leaving this section, we should mention the application of simple bond-order conservation models (Shustorovich 1985, 1986, 1987, 1988) to molecule–surface PESs. In this model, one assumes that each gas atom–surface atom bond is a simple Morse potential, yielding the total interaction between A–S as:

$$V_{AS} = \sum D_{AS}(x_{A\beta}^2 - 2x_{A\beta}) \tag{3.15a}$$

where the bond order is defined by

$$x_{A\beta} = \exp[-\alpha_{AS}(R_{A\beta} - R_{AS})] \tag{3.15b}$$

Analogous terms are defined for the B–S and A–B interactions and bond orders, $X_{B\beta}$ and x_{AB}, respectively. These potentials are directly analogous to those used in the four-body LEPS form but without the homogeneous embedding energy for the A–S and B–S interactions, and without the combining rule (and Sato parameters) to provide the separate Q and J for the full AB–S interaction. Instead, in the bond-order conservation model, two different assumptions are made. The first is that the number of solid atoms to include in the summation is restricted to nearest neighbors, a meaningful concept only for particular high-symmetry sites. The second, and most important, is that the total bond order is conserved:

$$\sum(x_{A\beta} + x_{B\beta}) + x_{AB} = 1 \tag{3.15c}$$

This leads to nonadditive energies.

From the above assumptions, it is possible to generate a relationship between the atomic chemisorption energies (B in Fig. 17) and the activation barrier for chemisorption:

$$\Delta E_{AB,S}^* = D_{AB} - (Q_A + Q_B) + Q_A Q_B/(Q_A + Q_B) \tag{3.16}$$

TABLE I
Experimental Values for Dissociative Chemisorption (kcal/mol)

A	Surface	Q_A	D_{AA}	$\Delta E^*_{AB,S}$	$\Delta E^*_{AB,S}$ − Eq. (3.15)
H	Fe(111)	62	103	≈0	10
	Ni(111)	63		≈2	8.5
	Ni(110)	62		≈0	10
	Cu(100)	58		≈5	16
N	W(110)	155	228	≈10	−4.5
	Fe(110)	138		≈8	21
	Fe(100)	140		≈2.5	18
	Fe(111)	139		≈ −0.8	19.5
O	Pt(111)	85	119	≈1	−8.5

where Q_A and Q_B are the heats of chemisorption of atoms A and B, respectively. There are a number of conceptual difficulties not addressed in such an approach, such as whether the barrier is A or D, what happens in low-symmetry arrangements, and even why pair potentials should be used for bonding on metal surfaces. However, we shall not discuss these problems in detail, but instead produce (Table I) for dissociative chemisorption of homonuclear diatomics (with all energies in kcal/mol). This is identical to Table II in Shustorovich (1988) except that we have illustrated the predictions of Eq. (3.15) rather than test the linearity of a relationship between activation barrier and atomic heat of chemisorption. We believe this clearly indicates the difficulty of any theoretical approach attempting to predict activation barriers, a property of a transient molecular species, from the properties of stable reactants and products.

IV. DYNAMICAL THEORIES

We begin with the general quantum mechanical approach to solution of the dynamics, namely the time-dependent Schrödinger equation for all the degrees of freedom:

$$i\hbar\partial\Phi/\partial t = H\Phi \tag{4.1a}$$

$$= [T(\mathbf{X}, \mathbf{r}, \{\mathbf{Y}_i\}) + V_{AB,S}(\mathbf{X}, r, \{\mathbf{Y}_i\}) + V_{SS}(\{\mathbf{Y}_i\})]\Phi \tag{4.1b}$$

T is the total nuclear kinetic energy operator of the gas and solid's atoms and $V_{SS}(\{\mathbf{Y}_i\})$ is the interaction among the solid's atoms. We have not included electronic degrees of freedom, notably the electron–hole pair excitations of the solid, since these will be discussed later. Eq. (4.1) has the formal solution

$$\Phi(t';\mathbf{X}',\mathbf{r}',\{\mathbf{Y}_i'\}) = \int \langle \mathbf{X}',\mathbf{r}',\{\mathbf{Y}_i'\}|\exp[-iH(t'-t)/\hbar]|\mathbf{X},\mathbf{r},\{\mathbf{Y}_i\}\rangle \times \Phi(t;\mathbf{X},\mathbf{r},\{\mathbf{Y}_i\})d\mathbf{X}\,d\mathbf{r}\,d\{\mathbf{Y}_i\} \tag{4.2}$$

The solution of a quantum mechanical equation with this many degrees of freedom is extremely difficult but, in contrast to the situation just a few years ago, it may not be implausible or out of the question.

To simplify the notation, we use the symbol q to signify all degrees of freedom in Eq. (4.2), which then becomes:

$$\langle t';q'|\Phi\rangle = \int \langle q'|\exp[-iH[t'-t]/\hbar)|q\rangle\langle t;q|\Phi\rangle dq \tag{4.3}$$

The necessary quantity is the propagator

$$U(t',q';t,q) = \langle q'|\exp[-iH(t'-t)/\hbar]|q\rangle \tag{4.4}$$

For multidimensional problems, the only practical method to evaluate such a quantity is via path integrals (Feynman and Hibbs 1965; Suzuki 1986; Makri and Miller 1987a, 1988; Doll et al. 1988a, 1988b). Most simply, this can be viewed as breaking the time interval into small parts:

$$\delta t = (t'-t)/N \tag{4.5a}$$

and rewriting the exponential operator in the form

$$\exp(-iH(t'-t)/\hbar) = [\exp(-iH\delta t/\hbar)]^N \tag{4.5b}$$

Substitution of Eq. (4.5b) into Eq. (4.4) and insertion of $N-1$ complete sets of states, $\{|q_i\rangle\langle q_i|, i=1,\ldots,N-1\}$, leads to the path-integral representation of the propagator:

$$U(t',q';t,q) = \int \langle q'|\exp(-iH\delta t/\hbar)|q_1\rangle\langle q_1|\exp(-iH\delta t/\hbar)|q_2\rangle\cdots \times \langle q_{N-1}|\exp(-iH\delta t/\hbar)|q\rangle dq_1\cdots dq_{N-1} \tag{4.5c}$$

For N large enough, each short-time propagator, $\langle q_i|\exp(-iH\delta t/\hbar)|q_{i+1}\rangle$, can be approximated via a simple formula such as:

$$\langle q_i|\exp(-iH\delta t/\hbar)|q_{i+1}\rangle \approx \langle q_i|\exp(-iT\delta t/\hbar)\exp(-iV\delta t/\hbar)|q_{i+1}\rangle \quad (4.6a)$$

or the more complicated form (Fleck et al. 1976; Wahnstrom and Metiu 1987):

$$\langle q_i|\exp(-iH\delta t/\hbar)|q_{i+1}\rangle \approx \langle q_i|\exp(-iV\delta t/2\hbar)\exp(-iT\delta t/\hbar) \times \exp(-iV\delta t/2\hbar)|q_{i+1}\rangle \quad (4.6b)$$

These two forms are accurate to different orders in δt with Eq. (4.6b) being of order $(\delta t)^3$ which is one order higher than Eq. (4.6a). In the limit of $\delta t \to 0$ (i.e., $N \to \infty$), both forms yield the same result for the full propagator in Eq. (4.5c). ($V = V_{AB,S} + V_{SS}$ in these equations.)

The important point is that the multidimensional integrals over $\{q_i, i = 1, \ldots, N-1\}$ must be performed via Monte Carlo (MC) methods. Even rather quantum-like degrees of freedom for nuclear motion are unlikely to require more than 10–20 intermediate paths or expansion states (depending upon V and of course the time difference δt). Thus, a quantum system with N degrees of freedom is isomorphic to a classical system with $\approx 10N$ degrees of freedom. This would appear to allow for evaluation of even $N = 100$ since classical MC simulations of fluids with many thousands of particles are now routine. The bottleneck is the computation of an oscillatory function via Monte Carlo methods since $\exp(iF)$ does not provide a positive definite sampling function. A possible solution may be to add a small imaginary time conponent $t -> t - i\hbar\beta$, and to evaluate Eq. (4.5c) via either newly developed stationary phase quantum Monte Carlo (QMC) methods (Makri and Miller 1987a, 1988; Doll et al. 1988a, b) or via analytic continuation methods (DePristo et al. 1989). At this time, it is not clear whether one or both of these approaches will be successful for the size of problems involved in molecule–surface reaction dynamics (e.g., 2 gas atoms and perhaps 4–12 surface atoms, at the minimum, yielding 18–45 degrees of freedom), but there is a cause for optimism.

For typical solids, it is likely that the motion of the atoms of the solid can be treated classically, except perhaps at very low surface temperatures. This greatly lowers the number of degrees of freedom in Eq. (4.2), which becomes:

$$\Phi(t + \Delta t; \mathbf{X}', \mathbf{r}') = \int \langle \mathbf{X}', \mathbf{r}'|\exp[-iH(t)\Delta t/\hbar]|\mathbf{X}, \mathbf{r}\rangle \Phi(t; \mathbf{X}, \mathbf{r})\, d\mathbf{X}\, d\mathbf{r} \quad (4.7a)$$

with

$$H(t) = [T(\mathbf{X}, \mathbf{r}\}) + V_{AB,S}(\mathbf{X}, \mathbf{r}, \{\mathbf{Y}_i(t)\})] \quad (4.7b)$$

T is the kinetic energy operator of only the gas atoms now, and the AB–S interaction potential is shown as an explicit function of t to indicate that the atoms of the solid move classically. The time-evolution of the quantum system is coupled to the classical equations of motion of the atoms of the solid

$$md^2\mathbf{Y}_j/dt^2 = -\nabla_{Y_j}\langle\Phi(t;\mathbf{X},\mathbf{r})|V_{\mathrm{AB,S}}(\mathbf{X},\mathbf{r},\{\mathbf{Y}_i\})|\Phi(t;\mathbf{X},\mathbf{r})\rangle - \nabla_{Y_j}V_{\mathrm{SS}}(\{\mathbf{Y}_i\}) \tag{4.8}$$

where m is the mass of the solid's atoms (which are assumed the same for simplicity). In Eq. (4.7) we have neglected variations of $H(t)$ over the time Δt since the lattice atoms move much more slowly than the impinging gas molecule.

The separation in Eqs. (4.7–4.8) and the use of the averaged potential in Eq. (4.8) follow from the general time-dependent self-consistent field (TDSCF) approach (Gerber et al. 1982; Gerber 1987; Makri and Miller 1987b). In this theory, a full wavefunction for two sets of coordinates, X and Y, is approximated via

$$\phi(t;X,Y) = \Phi(t;X)\Theta(t;Y) \tag{4.9}$$

The original TDSE

$$[T(X) + T(Y) + V(X,Y)]\phi(t;X,Y) = i\hbar\partial\phi(t;X,Y)/\partial t \tag{4.10}$$

is replaced by the set of equations:

$$[T(X) + \langle\Theta(t;Y)|V(X,Y)|\Theta(t;Y)\rangle]\Phi(t;X) = i\hbar\partial\Phi(t;X)/\partial t \tag{4.11}$$

$$[T(Y) + \langle\Phi(t;X)|V(X,Y)|\Phi(t;X)\rangle]\Theta(t;Y) = i\hbar\partial\Theta(t;Y)/\partial t \tag{4.12}$$

Equations (4.11–4.12) are the best variational solution of Eq. (4.10) with the simple product wavefunction in Eq. (4.9). The molecule–surface scattering description in Eqs. (4.7–4.8) results from:

1. the identification of $X \to (\mathbf{X},\mathbf{r})$ and $Y \to (\mathbf{Y}_i\}$;
2. the additional assumption that the wavefunction for $\{\mathbf{Y}_i\}$ is peaked around the classical value, $\{\mathbf{Y}_i(t)\}$, thereby replacing the TDSE in Eq. (4.12) by Hamilton's equations.

An important feature is that the TDSCF solution conserves energy, provided the original TDSE conserves energy. For the molecule–surface system, this means that energy exchange between the gas molecule and the solid is described correctly on the average.

This approximation illustrates the power of the time-dependent solution of the Schrödinger equation since the calculation of $\{\mathbf{Y}_i(t)\}$ will generally involve a relatively small amount of extra work over that needed to evolve $\Phi(t; \mathbf{X}, \mathbf{r})$ on a rigid surface. (For many purposes, the motion of the solid can be mimicked by a generalized Langevin equation approach (Zwanzig 1960; Adelman and Doll 1977; Adelman 1980; Tully 1980a; Lucchese and Tully 1983, 1984; DePristo 1984; Lee and DePristo 1986b; Xu et al. 1988; DePristo and Metiu 1989), which only requires a small number of Newtonian equations.) Even the integration over $\mathbf{X}$ and $\mathbf{r}$ in Eq. (4.8) is not particularly tedious since the wavefunction is already known on the grid of $(\mathbf{X}, \mathbf{r})$ values.

There is one computationally time-consuming feature, however. For a non-zero surface temperature, the initial values of $\{\mathbf{Y}_i\}$ and $\{d\mathbf{Y}_i/dt\}$ must be sampled. This implies that a considerable number of $\Phi(t; \mathbf{X}, \mathbf{r})$ must be propagated and then averaged incoherently (or coherently) to determine the full wavefunction.

It is worthwhile to reiterate that the quantum problem still involves both the translations and internal motion of the gas molecule, or 6 degrees of freedom. Solution of the six-dimensional TDSE is still a very difficult matter (unless the TDQMC approaches become practical). There are a multitude of further approximations that can be made if the scattering does not involve breaking the molecular bond, which are equivalent to the plethora of methods developed for gas-phase inelastic scattering. We will not consider these further here, but will simply refer the interested reader to the excellent review by Gerber (1987).

If reactions can take place, the most promising approach appears to involve direct (non-MC) solution of the TDSE for $\Phi(t; \mathbf{X}, \mathbf{r})$, a recent development (Kosloff and Kosloff 1983a, 1983b; Kosloff and Cerjan 1984). A review is available (Gerber et al. 1986), with developments and applications to surface scattering having been performed by Mowrey and Kouri (1986), Hellsing and Metiu (1986), Mowrey et al. (1987), Jackson and Metiu (1987), Chiang and Jackson (1987), and Halstead and Holloway (1988).

To illustrate the method, we consider Eq. (4.3) again but in this case where the dimensionality of q is small, say 1–6. Letting $t' - t = \Delta t$, and using the more accurate short-time propagator in Eq. (4.6b), we rewrite Eq. (4.3) as

$$\langle t + \Delta t; q' | \Phi \rangle = \int \langle q' | \exp(-iV\Delta t/2\hbar) \exp[-iT\Delta t/\hbar) \exp(-iV\Delta t/2\hbar) | q \rangle \langle t; q | \Phi \rangle dq \tag{4.13}$$

Since this is a low-dimensional integral, a MC evaluation is not necessary and maybe not even most appropriate. Instead, we use the fact that T is a local operator in momentum space and V is a local operator in coordinate space to insert a complete set of momentum states, $|p\rangle\langle p|$, into Eq. (4.13), yielding

$$\langle t+\Delta t; q'|\Phi\rangle = \int \langle q'|\exp(-iV\Delta t/2\hbar)|q'\rangle\langle q'|p\rangle\langle p|\exp(-iT\Delta t/\hbar)|p\rangle \langle p|q\rangle\langle q|\exp(-iV\Delta t/2\hbar)|q\rangle\langle t;q|\Phi\rangle\, dp\, dq \tag{4.14}$$

The transformations between coordinate and momentum space are accomplished via the fast Fourier transform (FFT) algorithm. This transforms the coordinate space evaluation of $\exp(-iV\Delta t/2\hbar)\langle t|\Phi\rangle$ into momentum space; the effect of $\exp(-iT\Delta t/\hbar)$ is determined in this space to yield the momentum space function; the momentum space function is transformed back to a coordinate space function via another FFT; and finally, the new coordinate space function is further evolved by the local operator $\exp(-iV\Delta t/2\hbar)$ to yield the coordinate space wavefunction at time $t+\Delta t$. The analogous technique can be accomplished with the simpler formula of Eq. (4.6a), but this does not reduce the number of FFT evaluations per time step, and thus is of no utility. This is clear when one notes that the values of $\langle q'|\exp(-iV\Delta t/2\hbar)|q'\rangle$ and $\langle p|\exp(-iT\Delta t/\hbar)|p\rangle$ are evaluated at the grid points (in the FFT) just once at the beginning of the calculation and stored. Essentially all the work is involved in the FFT computation.

The propagation procedure proceeds in the following manner. Start with a wavepacket, $\Phi(t=0;\mathbf{X},\mathbf{r})$, which is the product of an internal-state eigenfunction and a localized translational packet at large Z with an average velocity directed towards the surface. The wavepacket at time $t+\Delta t$ is then generated via Eq. (4.14) with $q=(\mathbf{X},\mathbf{r})$. Next, use $\Phi(\Delta t;\mathbf{X},\mathbf{r})$ to generate the packet at $2\Delta t$ via Eq. (4.14). This procedure is repeated many times until the final wavefunction, $\Phi(t\rightarrow\infty;\mathbf{X},\mathbf{r})$, is obtained. From this function, all scattering information is extracted via projection onto plane waves and appropriate eigenstates.

A number of advantages can be attributed to a time-dependent wavepacket approach. First, it allows computation of scattering information for one specific initial internal state of the reactant molecule. By contrast, time-independent methods generally calculate the values for all internal states at once, even though this information is often not desired. And, this scattering information is obtainable at a number of translational energies in a single calculation since the initial translational packet is composed of large numbers of initial plane waves. Second, the method is applicable to a nonrigid surface, as specifically indicated above. Third, the method is applicable to nonperiodic surfaces.

The major limitation is computational: for each time step, the computational time is proportional to $N_g \ln N_g$ where N_g is the number of grid points used in the evaluation of the FFT. Assuming for simplicity an equivalent and low number of grid points of $2^4 = 16$ in each degree of freedom, then in M degrees of freedom we have $N_g = 16^M$ which limits treatments to $M \leqslant 3$ on current supercomputers unless only a single calculation must be performed, in which case $M = 4$ may be feasible. However, even a rigid rotor-rigid surface collision entails $M = 5$ (unless further simplifying assumptions are made). The FFT approach is only applicable at present to atom–surface scattering unless the molecular internal coordinates, **r**, are not important in the dynamics (Karikorpi et al. 1987; Halstead and Holloway 1988).* However, in that case it should provide much more capabilities than the standard time-independent scattering theory (Goodman and Wachman 1976). This argument is based upon the assumption that the number of time steps is not too large, which limits the FFT approach to direct scattering.

The most direct treatment of reactive scattering for molecule–surface systems ignores completely the quantum mechanical nature of the molecular internal and translational motions. In this case, the wavefunction $\Phi(t; \mathbf{X}, \mathbf{r})$ is assumed to be strongly peaked around the classical values $\mathbf{X}(t)$ and $\mathbf{r}(t)$ yielding the equations of motion:

$$\mu d^2\mathbf{r}/dt^2 = -\nabla_r V_{\mathrm{AB,S}}(\mathbf{X}, \mathbf{r}, \{\mathbf{Y}_i\}) \tag{4.15a}$$

$$M d^2\mathbf{X}/dt^2 = -\nabla_X V_{\mathrm{AB,S}}(\mathbf{X}, \mathbf{r}, \{\mathbf{Y}_i\}) \tag{4.15b}$$

$$m d^2\mathbf{Y}_j/dt^2 = -\nabla_{Y_j} V_{\mathrm{AB,S}}(\mathbf{X}, \mathbf{r}, \{\mathbf{Y}_i\}) - \nabla_{Y_j} V_{\mathrm{SS}}(\{\mathbf{Y}_i\}) \tag{4.15c}$$

where μ is the reduced mass and M is the total mass of the gas molecule. These are rather standard molecular dynamics simulation equations and can be used directly, providing enough computer time is available to treat the large number of atoms of the solid needed to model a surface.

At this point, we have presented the global dynamical methods which, in principle, require knowledge of the entire AB–S or S–S interactions. There are a number of methods that involve only local dynamics, in the sense of using only localized properties of the full PES. As such, they will be much less dependent upon the large (and unknown) regions of configuration space that must be predicted by any PES form. If only small enough regions are needed, then these might even be provided by high-quality *ab initio* or density-functional calculations. This can be a mixed blessing since one must

*The recent combination of internal basis set techniques with the FFT for translational motion may allow for solution of the TDSE in five *special* dimensions, three translations and two rotations (Mowrey and Kouri 1986, Mowrey et al. 1987).

make *a priori* assumptions about which regions are most important without a full multidimensional dynamical simulation as a guide. In the worst case, this will yield dynamical results that are completely incorrect and inappropriate for the system under study.

The most common assumption is one of a reaction path in hyperspace (Miller et al. 1980). A saddle point on the PES is found and the steepest descent path (in mass-weighted coordinates) from this saddle point to reactants and products is defined as the reaction path. The information needed, except for the path and the energies along it, is the local quadratic PES for motion perpendicular to the path. The reaction-path Hamiltonian is only a weakly local method since it can be viewed as an approximation to the full PES and since it is possible to use any of the previously defined global-dynamical methods with this potential. However, it is local because the approximate PES restricts motion to lie around the reaction path. The utility of a reaction-path formalism involves convenient approximations to the dynamics which can be made with the formalism as a starting point.

To describe these approximations, we provide the treatment of dynamics within a correlation function formalism (Yamamoto 1960; Miller et al. 1983). Let the reaction-path coordinate be denoted by u and the other coordinates by U. Define the saddle point on the path as $u=0$. Then the thermally averaged rate constant for transition from $u<0$ to $u>0$ is given by

$$k(T)=\frac{1}{Q_{\mathrm{r}}}\int_{0}^{\infty} c_{\mathrm{ff}}(t)dt \tag{4.16a}$$

The flux–flux correlation function is defined by:

$$C_{\mathrm{ff}}(t)=\mathrm{Tr}\{\exp(iHt_{\mathrm{c}}^{*}/\hbar)\hat{F}\exp(-iHt_{\mathrm{c}}/\hbar)\hat{F}\} \tag{4.16b}$$

$$t_{\mathrm{c}}=t-i\beta\hbar/2 \tag{4.16c}$$

$$\hat{F}=(2m_{\mathrm{u}})^{-1}(\hat{p}_{\mathrm{u}}\delta(\hat{u})+\delta(\hat{u})\hat{p}_{\mathrm{u}}) \tag{4.16d}$$

where β is the usual $1/kT, \hat{F}$ is the flux operator for passage from $u<0$ to $u>0, \hat{p}_{\mathrm{u}}$ is the momentum operator conjugate to the reaction coordinate, and m_{u} is the mass associated with the reaction coordinate motion. Q_{r} is the reactant partition function with the restriction that $u<0$, which is important in a pure condensed-phase process such as diffusion of an atom on a surface (Wahnstrom and Metiu 1988). The trace operation in Eq. (4.16a) is performed over all degrees of freedom of the system except that of the reaction coordinate.

Equation (4.16) is exactly equivalent to solving the full TDSE, finding the transition probability at a particular initial system energy E, and then thermally averaging over all initial E. However, evaluation of Eq. (4.16) can be much more efficient. Equation (4.16) also forms the basis for convenient approximations to the dynamics, the most important of which is classical canonical transition-state theory (CTST). This approximation is based upon three assumptions: (1) all operators can be replaced by their classical variables; (2) only values of $p_u > 0$ contribute to the rate; (3) only the zero-time limit of Eq. (4.16a) is needed. This transforms Eq. (4.16) into:

$$k_{TST}(T) = (kT/h)(Q^{\ddagger}/Q_r)\exp(-\beta V^{\ddagger}) \tag{4.17}$$

where $Q^{\ddagger}$ and $V^{\ddagger}$ are the partition function (including all degrees of freedom except u) and barrier height, respectively, at the saddle point, i.e., the transition state. In the more accurate and modern versions of this theory, the location of the saddle point is varied to provide a minimum value for k_{TST}, termed canonical variational TST (CVTST) and, in addition, tunneling corrections are explicitly incorporated in order to incorporate some quantum mechanical features (Lauderdale and Truhlar 1985, 1986; Truhlar et al. 1986). These corrections lead to the form

$$k(T) = \Gamma(T)\, k_{TST}(T) \tag{4.18}$$

where $\Gamma(T)$ is a ubiquitous factor including all such corrections. Applications of this formula can utilize some or all corrections, ranging from simple CTST to very sophisticated CVTST with least-action paths for tunneling.

In all cases, TST provides a relatively simple method for the prediction of the rate constant. However, one must always keep in mind the assumption of TST, especially the short-time and positive momentum criteria, which prohibit recrossing of the saddle point. The fundamental idea that the dynamical process follows a reaction path must also be critically examined.

V. INCORPORATION OF LOW-ENERGY EXCITATIONS OF THE SOLID

Many of the methods in Section IV require explicit evaluation of the classical time-evolution of the atoms of the solid. Whenever any such calculation is attempted for a nonrigid classical surface, however, the treatment of the motion of the solid is not generally included simply by integration of Newton's equations. This is because the collision process, and $V_{AB,S}$, are quite strongly dependent upon the exact location (and hence motion) of only a small subset of the entire atoms of the solid, called the primary zone. These primary zone atoms are strongly disturbed by the collision, while the remainder of the solid remains near-equilibrium at temperature T. The role of the remainder of the solid is to allow for dissipation of the collision energy into the bulk and to provide energy from the bulk into the primary zone atoms via thermal fluctuations.

A basic introduction to the current generalized Langevin equation (GLE) formalism for gas–surface processes can be found in a number of articles (Adelman and Doll 1977; Adelman 1980; Tully 1980a; DePristo 1984, 1989). We will simply state the assumptions and the relevant equations. The basic physical idea is simple: the colliding molecule interacts strongly with a few nearby solid's atoms (i.e., the primary zone mentioned above) which then interact less strongly with their own neighbors and so forth (i.e., the secondary zone). Eventually, the energy exchange between the gas molecule and the primary zone's atoms in the initial collision is dissipated throughout the entire infinite solid.

Mathematically, this implies that while the primary zone's atoms may feel very anharmonic S–S forces and feel the force from the AB molecule, the secondary zone can be described by simple harmonic S–S restoring forces and should be relatively independent of the forces due to the AB molecule. Solving for the motion of the harmonic secondary atoms and substitution of this result into the equations of motion of the primary zone atoms yields a GLE for the latter:

$$md^2Y/dt^2 = -\nabla_Y V_{AB,S}(\mathbf{X}, \mathbf{r}, Y, Y_Q^{(eq)}) - \nabla_Y V_{SS}(Y) + m\underline{\underline{M}}(0)Y - m\underline{\underline{M}}(t)Y(0) - \int m\underline{\underline{M}}(t-t')dY(t')/dt'\,dt' + F(t) \tag{5.1}$$

The notation Y signifies all the coordinates of $\{\mathbf{Y}_i\}$ which are included in the primary zone, while the coordinates Y_Q are those remaining from $\{\mathbf{Y}_i\}$ that are not included in Y. The memory function and random force are defined by:

$$\underline{\underline{M}}(t) = \underline{\underline{\Omega}}_{PQ}^2 \cos(\underline{\underline{\Omega}}_{QQ}t)\underline{\underline{\Omega}}_{QQ}^{-2}\underline{\underline{\Omega}}_{QP}^2 \tag{5.2a}$$

$$F(t) = -m\underline{\underline{\Omega}}_{PQ}^2 \cos(\underline{\underline{\Omega}}_{QQ}t)Y_Q(0) - m\underline{\underline{\Omega}}_{PQ}^2 \sin(\underline{\underline{\Omega}}_{QQ}t)\underline{\underline{\Omega}}_{QQ}^{-1}dY_Q(0)/dt \tag{5.2b}$$

where P signifies a projection onto the primary zone and Q signifies a projection on the secondary zone. (Such projections simply keep track of the locations of the atoms.) The frequency matrix of the original solid is $\underline{\underline{\Omega}}$. Note that the interaction $V_{AB,S}$ depends upon the full set $\{\mathbf{Y}_i\}$ but with the secondary zone at equilibrium, i.e., $\{Y, Y_Q^{(eq)}\}$.

The memory kernel in Eq. (5.2a) involves the response of the full many-body system and thus retains memory of previous velocities. This differs from a standard Langevin equation which replaces the memory function by a δ function, leading to a temporally local friction. The random force, $F(t)$, and the memory kernel are related via the fluctuation–dissipation theorem:

$$\langle F(t)F(0)^{\mathrm{T}} \rangle = mkT\underline{\underline{M}}(t) \tag{5.3}$$

where the $\langle\,\rangle$ indicate an average over initial conditions of the secondary atoms. The derivation of Eq. (5.3) uses the equilibrium properties of the positions and velocities of the secondary zone atoms.

Equations (5.1–5.3) are equivalent to the original set of molecular dynamics equations in Eq. (4.15c) in their ability to mimic an infinite solid. Indeed, the GLE equations are no easier to solve in their present form. The advantage of the GLE approach occurs because it is possible to approximate the memory kernel and random force (Tully 1980a; DePristo 1984, 1989; Diestler and Riley 1985, 1987) to provide a reasonably accurate description of both the short-time and long-time (actually long-wavelength) response of the primary zone atoms to an external perturbation (e.g., a collision).

One expects a decaying and oscillating function on physical grounds, which can be represented by

$$\underline{\underline{M}}(t) = \underline{\underline{M}}_0^{1/2} \exp(-\underline{\underline{\gamma}}t)[\cos(\underline{\underline{\omega}}_1 t) + \tfrac{1}{2}\underline{\underline{\gamma}}\underline{\underline{\omega}}_1^{-1} \sin(\underline{\underline{\omega}}_1 t)]\underline{\underline{M}}_0^{1/2} \tag{5.4}$$

This is the multidimensional generalization of the position autocorrelation function of a Brownian oscillator and provides such a function with a number of unknown parameters. Using this approximate memory function allows replacement of the GLE in Eq. (5.1) by

$$md^2 Y/dt^2 = -\nabla_Y V_{\mathrm{AB,S}}(\mathbf{X}, \mathbf{r}, \{\mathbf{Y}_i\}) - \nabla_Y V_{\mathrm{SS}}(Y) + m\underline{\underline{M}}_0^{1/2}\underline{\underline{\omega}}_0 s \tag{5.5a}$$

$$md^2 s/dt^2 = m\underline{\underline{\omega}}_0\underline{\underline{M}}_0^{1/2} Y - m\underline{\underline{\omega}}_0^2 s - m\underline{\underline{\gamma}} ds/dt + f(t) \tag{5.5b}$$

where $f(t)$ is a Gaussian white-noise random force obeying

$$\langle f(t)f(0)^{\mathrm{T}} \rangle = 2mkT\underline{\underline{\gamma}}\delta(t) \tag{5.5c}$$

$\delta(t)$ is a Dirac delta function. The fictitious particles obeying the equations of motion of s are commonly referred to as ghost atoms, and thus this formulation is referred to as the GLE–ghost atom method.

One can determine these parameter matrices by two fundamentally different methods. The first, due to Tully (1980a), solves Eq. (5.5) at $T = 0$ without the AB–S interaction and with harmonic primary zone interactions;

and then uses this solution to find the phonon density of states. The parameters are adjusted to achieve agreement with the experimental results. The second, due to DePristo (1984), assumes a microscopic interaction model for the forces among the atoms of the solid, finds $\underline{\underline{\Omega}}$, and substitutes this into Eq. (5.2a) to define the exact $\underline{\underline{M}}(t)$; then Eq. (5.4) is forced to agree with the exact $\underline{\underline{M}}(t)$ at short times. In both methods, the low-frequency density of states is determined by the relationship $\underline{\underline{\gamma}} = \underline{\underline{1}}\pi\omega_D/6$ (Adelman and Doll 1977). The two methods will provide very similar parameters if an accurate frequency matrix, $\underline{\underline{\Omega}}$, is available.

The choice of methods is a matter of convenience. Both will capture the essential features of the GLE, namely frictional energy loss from the primary atoms to the secondary atoms and thermal energy transfer from the secondary atoms to the primary atoms. Both will provide a reasonable description of the bulk and surface phonon density of states of the solid. Neither will provide the exact time-dependent response of the solid due to the limited number of parameters used to describe the memory function.

To emphasize the physics again for the GLE–ghost atom technique, we note that the forces among the primary zone atoms are general and anharmonic. The coupling between the primary and ghost atoms is incorporated by the terms $\underline{\underline{M}}_0^{1/2}\underline{\underline{\omega}}_0 s$ and $\underline{\underline{\omega}}_0\underline{\underline{M}}_0^{1/2} Y$. The frictional force, $-m\underline{\underline{\gamma}}\,ds/dt$, and the gaussian random force are related via the local fluctuation dissipation theorem in Eq. (5.5c). This relationship ensures that in the long-time limit, the positions and velocities for the primary and ghost atoms become thermally distributed at temperature T.

It is worthwhile to make a few points about the GLE formalism. First, it is not possible to implement an (atomic-based) GLE in any practical and systematic way if the primary–secondary or secondary–secondary interactions are anharmonic. Second, a complication arises within the GLE formalism due to the localized nature of the primary zone atoms. In a molecule–surface collision, the initial localized interaction specifies a set of primary zone atoms as illustrated for a diatomic interacting with a BCC(110) surface in Fig. 26. (There is nothing fundamental about this surface or number of primary zone atoms; the same argument holds irrespective of the lattice and primary zone size). When the molecule moves outside of these 13 atoms, a new set of primary zone atoms must be defined, a process termed switching of the primary zone (Tully 1980a). There is no *a priori* method to consistently define new primary zone atoms since there is no information on the flow of energy from the original primary zone into the specific secondary zone atoms (i.e., the exact $\underline{\underline{M}}(t)$ is no longer available). The current practice is to assume that the motion of the molecule across the surface is slow compared to thermalization of the surface atoms. Then the new primary zone atoms are reinitialized from a thermal distribution. This method will break down if the

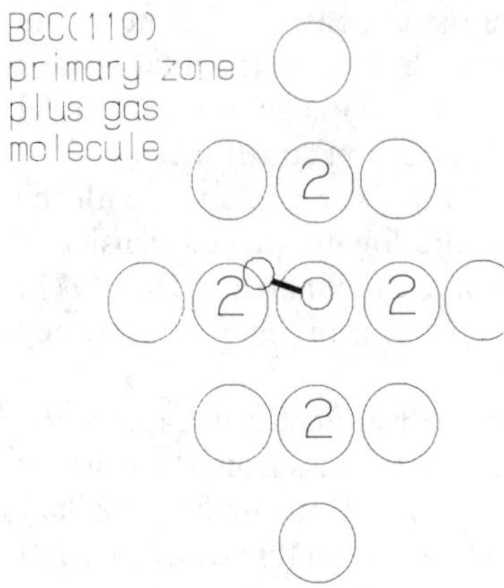

Figure 26. Diatomic molecule and primary zone atoms used for GLE simulations of dissociative chemisorption on a BCC (110) surface. The atoms labeled by "2" are in the second layer.

motion of the molecule is very fast or the distortion of the lattice is sufficiently great to inhibit thermalization on a fast enough time scale.

Third, we note that the collision defines a time scale over which the response of the solid is required, namely the collision time t_c (i.e., the duration of a strong interaction). Another time scale occurs in the GLE–ghost atom equations for the rate of energy dissipation. The rate of damping is determined by $|\underline{\gamma}| = |\underline{1}|\pi\omega_D/6$ while the energy exchange between primary and ghost atom subsystems is controlled by $|\underline{\omega}_0 \underline{M}_0^{1/2}|$. Typically, these two rates are very similar since they both reflect motions of the solid's atoms due to the forces between solid atoms. When $t_c\pi\omega_D/6 \ll 1$, the solid response is limited to the motion of the primary zone atoms, with negligible effects due to the remainder of the solid, i.e., the ghost atoms. When $t_c\pi\omega_D/6 \approx 1$, the response of the solid is truly a dynamical many-body phenomena since the energy exchange between the gas molecule and primary zone atoms is strongly influenced by the ghost atoms on the time scale of the collision duration. When $t_c\pi\omega_D/6 \gg 1$, the solid responds as a many-body equilibrium system, describable via either a classical Langevin treatment without memory or a temperature-dependent potential of mean force (Tully 1987). Note that $\pi\omega_D/6 = 1.5 \times 10^{-11}$ K-sec/θ_D where θ_D is the Debye temperature. A typical range is $100\,\text{K} < \theta_D < 400\,\text{K}$, yielding $3.8 \times 10^{-14}\,\text{sec} < \pi\omega_D/6 < 15.0 \times 10^{-14}\,\text{sec}$. A collision duration can vary from 10^{-15} to 10^{-11} sec, which allows for the three possibilities described above.

Fourth, the GLE is not a particularly simple method to implement. One must choose the parameter matrices for the memory function, which will differ for each material and surface face, a procedure that is intensive of human time. Changing the number of primary zone atoms requires redetermination of these parameters and major modifications of a computer code to implement the switching process. If a treatment of a particular solid and a surface face is to be the focus of investigation for an extended period of time, the investment in human time is definitely worthwhile. (If computational resources are limited to nonsupercomputers, it may be the only viable approach.) However, if a number of faces and materials are to be treated, still simpler methods combining LE and MD are desirable, although more expensive computationally (Berkowitz and McCammon 1982; Brooks and Karplus 1983; Lucchese and Tully 1983, 1984; Xu et al. 1988; Riley et al., 1988; DePristo and Metiu 1989).

Fifth, while little attention has been paid to the accuracy of the force model for the lattice, it is clear that simple primary zone interactions will not suffice as the dynamical simulations become increasingly more accurate. This will be especially true of simulations of reactions at finite surface coverages, and for processes that lead to reconstruction of the solid surface during the course of the reaction (Ladas et al. 1988; Sobyanin and Zhdanov 1987). In these cases, one must pay particular attention to an accurate microscopic description of V_{SS}, since the general argument that the major unknown is $V_{AB,S}$ cannot be correct.

Other interesting treatments of the solid motion have been developed in which the motion of the solid's atoms is described by quantum mechanics [Billing and Cacciatore 1985, 1986]. This has been done for a harmonic solid in the context of treatment of the motion of the molecule by classical mechanics and use of a TDSCF formalism to couple the quantum and classical subsystems. The impetus for this approach is the fact that, if the entire solid is treated as a set of coupled harmonic oscillators, the quantum solution can be evaluated directly in an operator formalism. Then, the effect of solid atom motion can be incorporated as an added force on the gas molecule. Another advantage is the ability to treat the harmonic degrees of freedom of the solid and the harmonic electron–hole pair excitations on the same footing. The simplicity of such harmonic degrees of freedom can also be incorporated into the previously defined path-integral formalism in a simple manner to yield influence functionals (Feynman and Hibbs 1965).

For completeness, we should mention that inclusion of single-phonon events can be accomplished within a time-independent scattering theory simply by using a first-order Taylor series expansion of the interaction potential between the gas molecule and the solid surface:

$$V_{AB,S}(\mathbf{X}, \mathbf{r}, \{\mathbf{Y}_i\}) = V_{AB,S}(\mathbf{X}, \mathbf{r}, \{\mathbf{Y}_i^{(eq)}\}) + \Sigma \nabla_{Y_j} V_{AB,S}(\mathbf{X}, \mathbf{r}, \{\mathbf{Y}_i\}) \cdot (Y_j - Y_j^{(eq)}) \tag{5.6}$$

where the derivative is evaluated at $\{\mathbf{Y}_i^{(eq)}\}$. The full system wavefunction would then be expanded in translational, vibrational, and phonon states.

Since the latter are continuously distributed, this leads to a set of coupled integro-differential equations in the standard time-independent theory of scattering (Goodman and Wachman 1976). The effect of the phonons can then be included via a first-order perturbation treatment of the phonon coupling. Because of the expansion in a basis set in the molecular degrees of freedom, this approach is limited to nonreactive systems at present.

Next we turn to a brief mention of the treatment of the other low-energy excitation modes, namely electron–hole pair (denoted by e–h) creation and annihilation. In our opinion, these will not be very important in general for translational to e–h pair processes since the coupling to phonons is so much stronger (Billing 1987) as indicated in Fig. 27 by comparing the energy loss in the left and right panels. The exception may be atomic H and D (Kirson et al. 1984).

The major role for e–h pair processes is likely in vibrational excitation and deexcitation for weakly interacting molecule–surface systems. In this regard, it is important to distinguish e–h pair processes from the adiabatic coupling of molecular and electronic degrees of freedom. For example, if the molecular bond length varies as the molecule approaches a surface as in

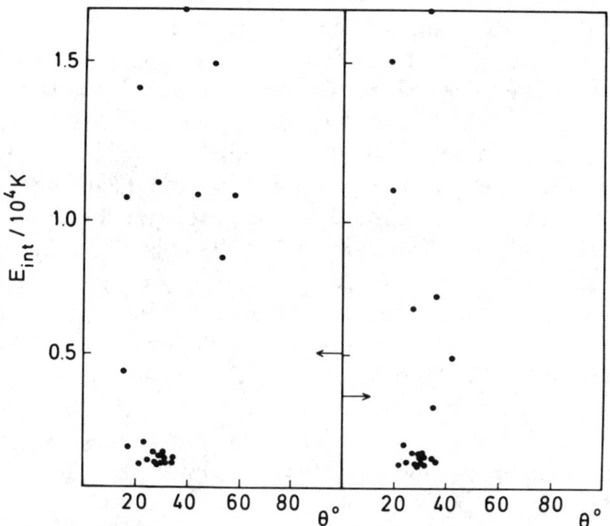

Figure 27. Energy transfer with the surface as a function of final scattering angle for CO/Pt(111). The initial kinetic energy, and vibrational and rotational states are 100 kJ/mol, 0, and 50, respectively. The left- and right-hand panels show the results, including only phonons and phonons plus electron–hole pair mechanisms, respectively. The figure is reprinted with permission from Billing (1987).

Figs. 17 and 25, exchange of translational and vibrational energy is probable. This process should be relatively independent of the surface temperature. By contrast, experimental data (Rettner et al. 1985b) for the NO/Ag(111) system have shown that NO is vibrationally excited with a probability of the form

$$P = f(E_i \cos^2 \theta_i) \exp(-E_a/kT) \quad (5.7)$$

which indicates that at $T = 0\,\mathrm{K}$ there is no vibrational excitation. The probabilities are quite small even at $T = 760\,\mathrm{K}$, increasing from 0.01 to 0.06 as $E_i \cos^2 \theta_i$ increases from 0.01 eV to 1.2 eV as shown in Fig. 28.

The treatment of such processes is provided by Newns (1986) along the lines developed by others (Persson and Persson 1980; Gadzuk 1983, 1984). The starting point is the Hamiltonian,

$$H = \varepsilon_a(t) n_a + \Sigma \varepsilon_k n_k + V(t) \Sigma \{c_a^\dagger c_k + h.c.\} + \omega_0 b^\dagger b + \tau n_a (b^\dagger + b) \quad (5.8)$$

$\varepsilon_a(t)$ is the energy level of the π^* level in NO and is dependent upon the position of the CM of NO from the surface, which provides the

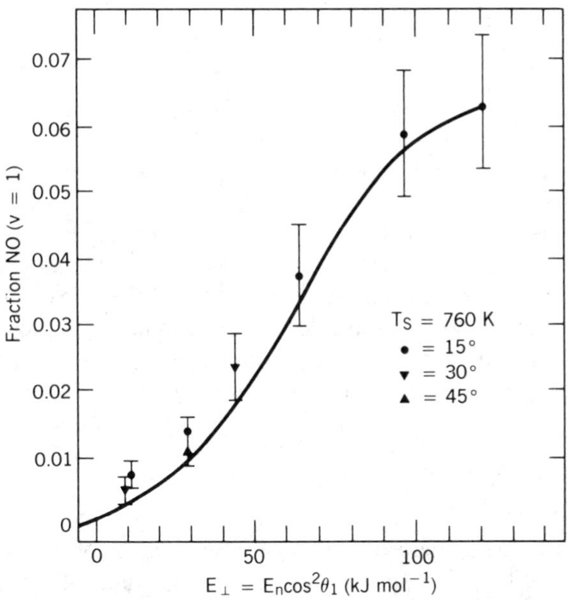

Figure 28. Vibrational excitation probability for the NO/Ag(111) system as a function of the initial normal kinetic energy. The curve is from Newns (1986) while the experimental data points are from Rettner et al. (1985b)

time-dependence via motion of the NO CM. $n_a = c_a^\dagger c_a$ is the occupation number operator for this level. ε_k and n_k refer to the electronic eigenstates of the metal's electrons. $V(t)$ is actually $V(Z(t)) = V^0 \exp(-\alpha Z(t))$, where $Z(t)$ is assumed to be known from motion on a PES. The frequencies of vibration in the NO and NO^- are assumed to be the same, ω_0. The operator b annihilates a vibrational quanta in either NO or NO^-. The electron–vibration coupling within the molecule is provided by the last term in Eq. (5.8). Thus, this equation contains two types of electron–vibration coupling, that between the e–h pairs of the solid and the π^* level of NO and that between vibrations of NO and NO^- (i.e., $n_a = 1$).

Solution of Eq. (5.8) for the operator b, and then determination of $n_a(t)$ by first-order perturbation theory yields a formula similar to Eq. (5.7) and the results in Fig. 28. A number of assumptions must be made to obtain quantitative agreement with experiment, including two critical ones. First, the energy of the π^* level as a function of the distance of the molecule from the surface is written in the image-dipole form:

$$\varepsilon_a(z) = \varepsilon_a(\infty) - 1/4(Z - Z_0) \tag{5.9}$$

Second, the lifetime broadening of the π^* level varies as

$$\Delta(Z) = \Delta^0 \exp(-2\alpha Z) \tag{5.10}$$

The parameter Z_0 controls the effect of the coupling with distance out into the vacuum, since it changes the energy gap between the excited and ground electronic level. Δ^0 controls the rate of deexcitation from the π^* level. The particular values are $Z_0 = -1.93$ Å and $\hbar/\Delta^0 = 6 \times 10^{-15}$ sec. The significant sensitivity to these values is in accord with a different approach based upon the use of an electronic friction, with the spatial extent of the electronic friction controlling the probability of energy dissipation to e–h pair (DePristo et al. 1986). Finally, a recent article by Billing (1987) is suggested for further reading since it provides a more collision-oriented approach to the treatment of e–h pair excitations for translational energy loss.

VI. APPLICATIONS

We now turn to an illustration of the theoretical results generated by the variety of previously described theories. Since most of the results that we shall show are for clean, perfect surfaces, it is important to indicate when this assumption is valid. Serri et al. (1983) developed a kinetic model for NO (molecular) chemisorption on Pt which incorporated adsorption, desorption, and diffusion of NO on terraces, diffusion from the terraces to the steps, and escape from the steps to the terraces. Adsorption and desorption directly from the steps was not included. The major conclusion was that, under the physically realistic condition of fast diffusion at low coverage and moderate to high temperature, the presence of steps plays a dominant role in thermal desorption. In dissociative chemisorption via either an intrinsic or extrinsic precursor, the presence of steps may also play a dominant role, as has been emphasized in a number of experimental studies (Comsa and David 1985; Lin and Somorjai 1984; Steinruck et al. 1986; Rendulic 1988), especially when the dissociative chemisorption probabilities are small.

We consider H_2 colliding with Ni and Cu surfaces for which the light mass of H_2 precludes significant energy transfer to the lattice. This indicates that for S_0 of substantial size, a direct mechanism should apply and thus S_0 and the dissociation probability, P_d, are equal. Since H_2 is so light, it is important to identify possible quantum mechanical effects, such as tunneling, zero-point energy, and diffraction. This has been accomplished in a number of articles (Muller 1987; Jackson and Metiu 1987; Chiang and Jackson 1987; Karikorpi et al. 1987; Halstead and Holloway 1988; Holloway 1989).

We start with an extremely simple model in which the H_2 orientation is fixed parallel to a rigid linear chain of Ni atoms (representing a row of Ni(100)) and is located above a Ni atom as shown in Fig. 29. Due to the

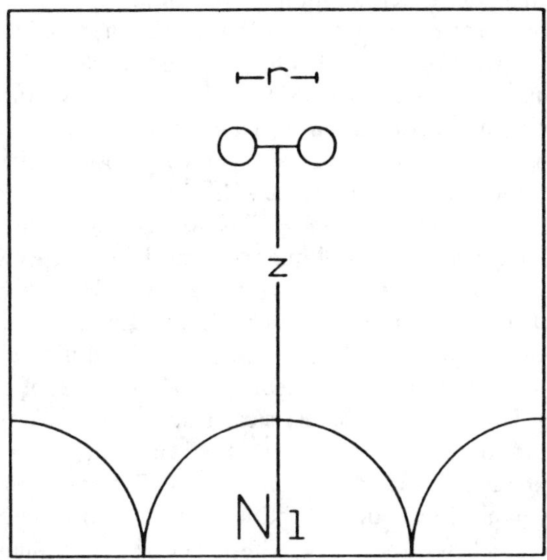

Figure 29. Geometry and coordinate system for H_2/Ni(100). The plot is from Chaing and Jackson (1987).

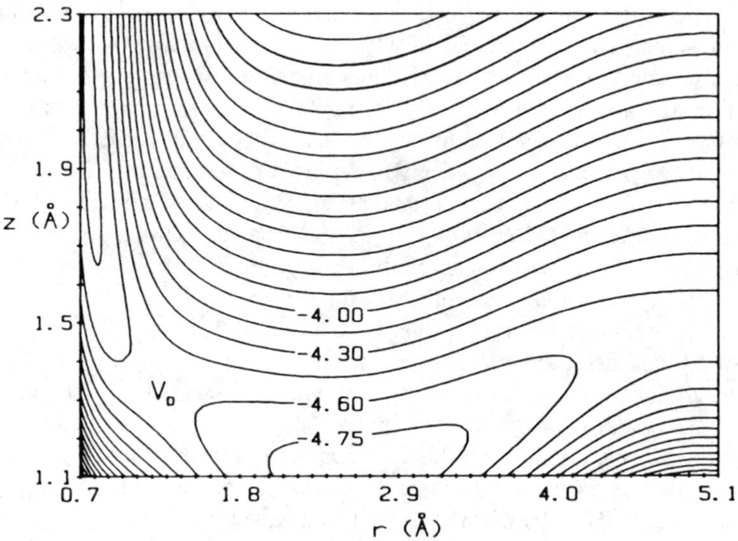

Figure 30. A contour plot of the PES for H_2/Ni(100) in the configuration of Fig. 29, from Chiang and Jackson (1987).

symmetry, only Z and r can vary, yielding a two-dimensional dynamical problem for which the exact solution of the time-dependent Schrödinger equation is feasible using the FFT algorithm as outlined in Section IV from Eq. (4.13) until above Eq. (4.15). This has been accomplished (Jackson and Metiu 1987; Chiang and Jackson 1987) on the PES shown in Fig. 30. This PES is a model for the dissociation over an atop atom towards the bridge sites on Ni(100), but this should not be taken too seriously since, as will be discussed later, the dissociation process likely does not even involve a symmetric site or orientation of H_2. This work is important because it provides a detailed investigation of quantum mechanical effects.

The results are shown in Fig. 31 for four isotopes of H with mass of 1, 2, 3, and 7, the last being fictitious but extremely useful in delimiting the importance of quantum effects. The classical and quantum values agree in general, with the former slightly exceeding the latter, except at the lowest kinetic energies. Even for H_2, the classical results are not terrible but are clearly not quantitative. It is interesting that the classical results are too large at low kinetic energies. This is due to two effects. The first is the occurrence of quantum mechanical reflection at the second atop position of the part of the wavepacket which surmounts V_0. This effect will be negligible on a real 3-D surface. The second, and more important, is the inability of the classical simulations to enforce zero-point energy restrictions throughout the

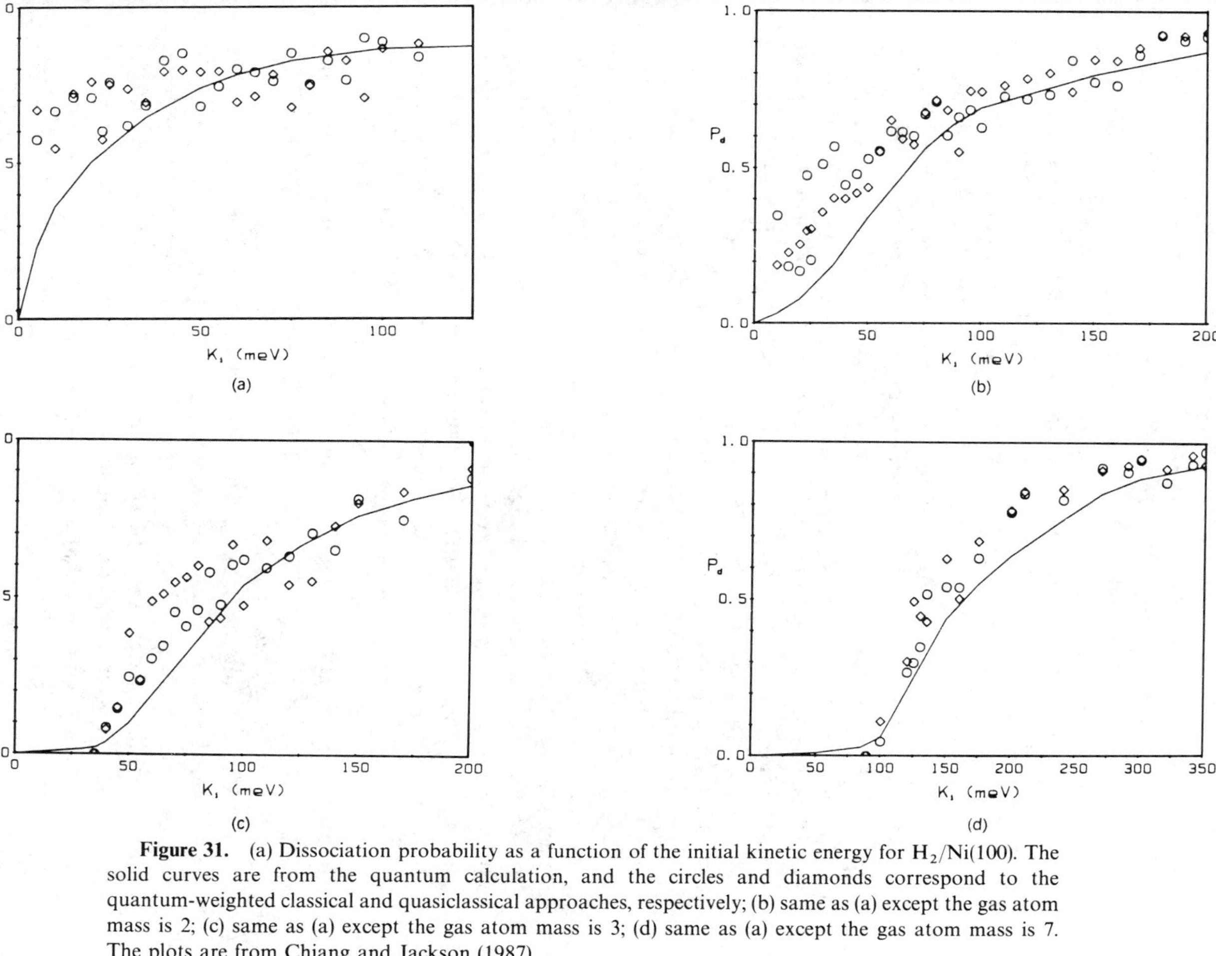

Figure 31. (a) Dissociation probability as a function of the initial kinetic energy for H_2/Ni(100). The solid curves are from the quantum calculation, and the circles and diamonds correspond to the quantum-weighted classical and quasiclassical approaches, respectively; (b) same as (a) except the gas atom mass is 2; (c) same as (a) except the gas atom mass is 3; (d) same as (a) except the gas atom mass is 7. The plots are from Chiang and Jackson (1987).

trajectory, thereby allowing the initial H_2 zero-point vibrational energy to be converted into motion along the reaction coordinate, which increases the classical S_0. Both effects will become less important for larger masses, as can be seen for the $m = 7$ results in Fig. 31.

It is interesting to note that the stretching of the H_2, without dissociation, as it approaches the surface can give rise to a trapped molecular physisorbed species (Muller 1987). Due to the lowering of the vibrational frequency as H_2 approaches to surface, some vibrational zero-point energy is released into translational energy which can then be transferred to a nonrigid surface. Loss of even a small fraction of this local translational energy will trap a molecule with low initial kinetic energy since the zero-point energy is typically much larger than the initial low kinetic energy. This is the vibrational analog of the Beeby mechanism for translations (Goodman and Wachman 1976): acceleration by the physisorption well provides a large local speed when the molecule hits the surface, and a small fractional loss of this local translational energy leads to trapping. The vibrational zero-point effect is independent of the well depth of the molecule–surface PES, except, of course, that a well must exist to support a physisorbed species.

It is also interesting that the shape of the P_d versus kinetic energy curves is not a step function even though only a single barrier occurs. This is easy to understand classically. Each trajectory is initiated with a different division between kinetic and potential vibrational energy (i.e., particular vibrational phase) as well as with the same translational kinetic energy. Thus, each trajectory oscillates in r as it decreases in Z. A trajectory which overshoots the reaction path and bounces off the hard wall will not dissociate because its motion will not lie along the reaction coordinate. Two schematic trajectories are shown in Fig. 32. The first trajectory is not reactive because of incorrect vibrational phase. At low kinetic energy, it will be most important to have precisely the correct phase to ensure motion over the minimum barrier. At higher kinetic energies, enough energy is available to surmount more that the minimum barrier, making the vibrational phase less important, and thus yielding larger P_d.

The above results demonstrate that even a single barrier located in the exit channel will not give rise to a step function change in S_0. They also demonstrate that even when the kinetic energy is not enough to surmount the minimum barrier, the ability of the PES to transform vibrational zero-point energy into motion along the reaction path can also lead to $S_0 > 0$. Thus, ascribing these two characteristics to tunneling without knowing the shape of the PES (Asscher et al. 1988) is not a good practice. When one factors in all the other variables affecting the reaction such as molecular orientation, position in the surface unit cell, and surface atom motion, it would be very premature indeed to identify a tunneling mechanism based

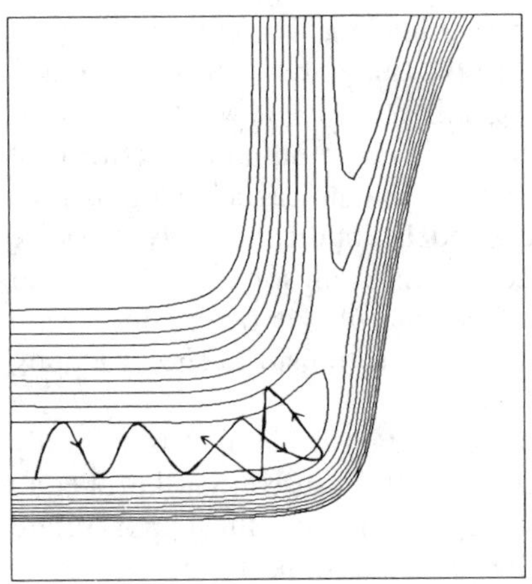

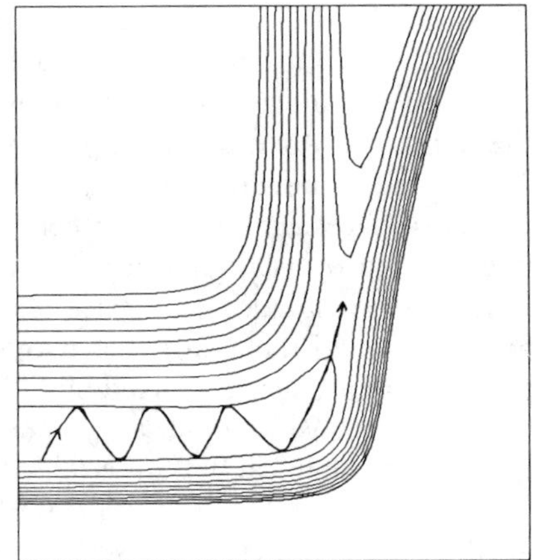

Figure 32. Illustrations of molecule–surface trajectories for different initial vibrational phase. (a) nonreactive; (b) reactive.

upon the variation of (experimental) S_0 with kinetic energy. Even the kinetic isotope effect may not be a good indicator since the effect of vibrational zero-point energy varies substantially with such a mass substitution.

In our opinion, the present evidence for tunneling in any dissociative chemisorption reaction is not overwhelming. Analyses have been made based upon one-dimensional tunneling formulas for the tunneling probability, such as the very simple inverse parabolic potential form:

$$P(E) = \exp[-(\pi L/\hbar)(m/V_0)^{1/2}(V_0 - E)] \tag{6.1}$$

Here L is the full width at half-maximum of the barrier, m is the effective mass of the tunneling particle, V_0 is the activation energy, and E is the energy along the reaction path. Indeed, disagreements have developed (Lo and Ehrlich 1986, 1988; Kay and Coltrin 1988) based upon such one-dimensional formulas. These are all moot points, in our opinion, for a number of reasons. First, it is not at all clear what energy to use in the formula since different PESs can transform different amounts of initial kinetic energy into energy along the reaction path. Second, more sophisticated multidimensional tunneling theories (Truhlar et al. 1986) demonstrate clearly that tunneling follows a least-action path and not a minimum-energy path, with the differences often being orders of magnitude in $P(E)$. Third, as we shall emphasize later, the use of restricted-dimensionality models without surface motion can significantly distort the dynamics. For example, if one assumes a one-dimensional potential, then it is obvious that a sharply increasing but nonstep function for S_0 versus E arises from tunneling. The converse is most definitely not true since a multitude of features can give rise to such a curve when the dynamics is truly multidimensional. We do not want to belabor the point, but do want to emphasize that the reader should be critical of simple models that do not include all the types of dynamical behavior possible in gas–surface scattering. Furthermore, models that completely separate internal and translational degrees of freedom in chemisorption (Brass and Ehrlich 1986) should also be viewed with a dose of skepticism since these do not allow for the natural behavior of the PES to transform energy among these different degrees of freedom.

We now consider models of higher dimensionality. The first is still fully quantum mechanical but a special form for the PES is used (Karikorpi et al. 1987; Halstead and Holloway 1988; Holloway 1989) that is most appropriate for the H_2/Ni and H_2/Cu systems. The basic assumptions are that a barrier to chemisorption exists in the entrance channel, much like point A in Fig. 17; this barrier is nearly independent of the molecular orientation and vibrational bond length; and all molecules that pass over this barrier dissociate (i.e. there is a sink in the PES). The justification for the orientation independence is the free rotational motion in the physisorbed species. The justification for the independence of bond length is the absence of a significantly stretched molecular bond at the position of the activation barrier A. The irreversible dissociative behavior after surmounting a point like A in Fig. 17 assumes that either no barrier or at most an extremely small barrier exists in the exit channel, unlike point D in that figure.

The distinguishing feature of this work, however, is that the activation barrier is allowed to vary with different positions of the H_2 in the surface unit cell. This is a clever approach which captures much of the influence of the corrugated surface. Moreover, under these approximations, the dynamical equations become only three-dimensional and, in fact, are identical to those of an atom scattering from a rigid, corrugated surface but with a potential with a sink. The TDSE is solved by the FFT procedure. Hence, it is possible to treat diffraction, tunneling, and dissociative chemisorption by accurate quantum techniques, at least on this simple type of PES.

Calculations of the diffraction intensities were performed for three different PESs. The first had the minimum activation barrier when the H_2 was located over a Ni atom (i.e., atop site), while the second had the minimum over a fourfold (i.e., center) site. The third did not allow for dissociation at all, since it did not have a barrier (i.e., the PES continually increased as Z decreased). The first is shown in Fig. 33a as a function of Z and position of H_2 center-of-mass along a line connecting atop and center sites. A different representation is shown in Figs. 33b–d, providing a graphic illustration of the energy dependence of the size of the hole. The PES with the minimum barrier over the center site was very similar.

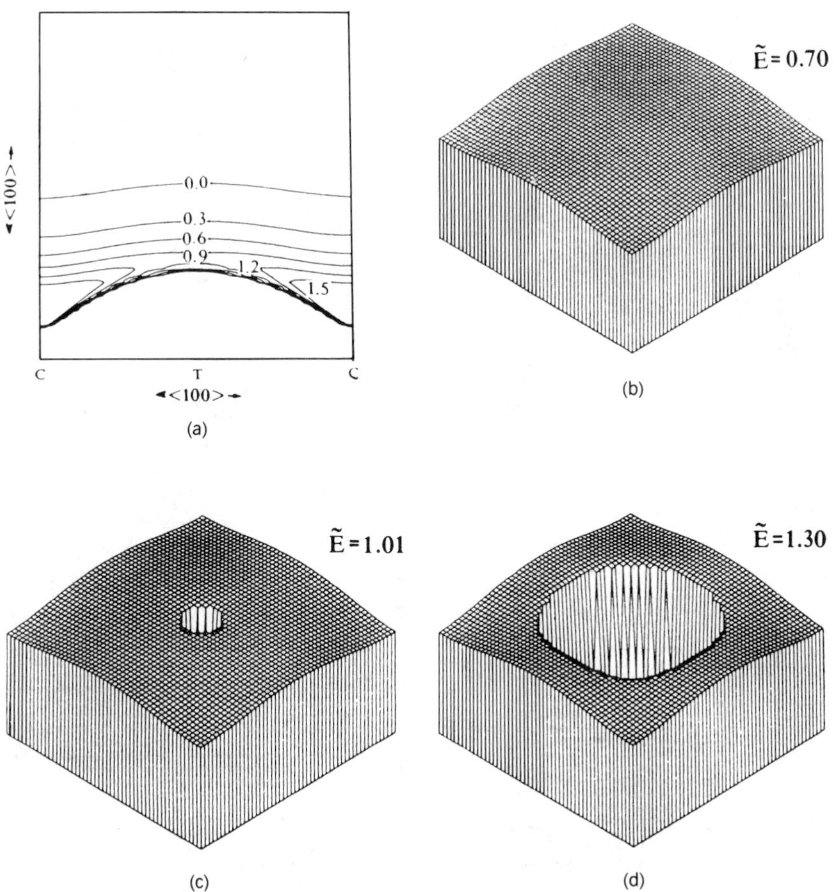

Figure 33. (a) Contour map of a model potential for an adsorbing H_2 with points C and T referring to center and atop sites, respectively; (b–d) surfaces of constant H_2 surface energy plotted across the unit cell. See the text for further description of the model. The figure is reprinted with permission from Karikorpi et al (1987).

Diffraction intensities are shown in Fig. 34 with the notation [00], [10], and [20] indicating the specular, first- and second-order diffracted beams, respectively. It is immediately apparent from the contrast between the three figures that the availability of the PES sink has a marked effect on all the beams. The specular beam does not go through a minimum at $\approx 350\,\text{meV}$ because as the energy is raised, dissociation becomes more probable, and thus more of the flux is channeled into this dissociation channel instead of into the reflected specular channel. Similarly, the peak values of the [10] and [20] beams are significantly reduced. In a nonreactive system, it is

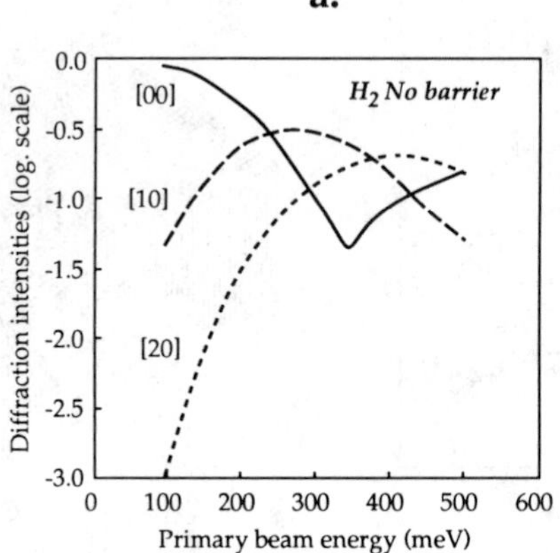

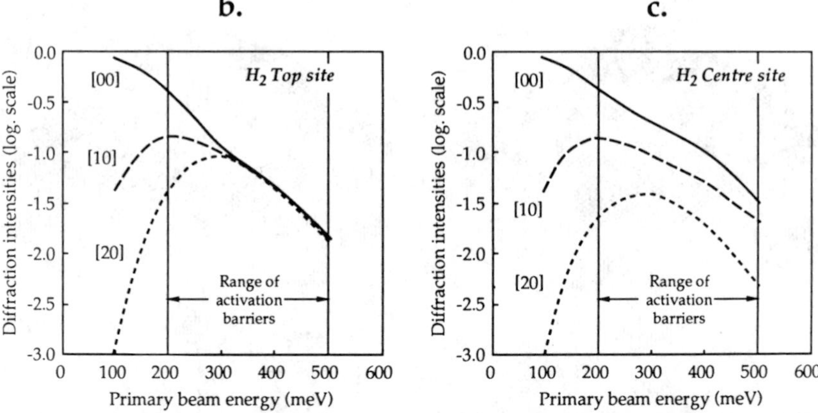

Figure 34. (a) H_2 intensities for the first three diffraction states as a function of initial energy for a primary beam at normal incidence to a potential having no reactive channel. The sharp minimum occurring in the specular beam at 370 meV arises as a consequence of destructive interference between molecules scattered at top and center sites; (b) same as (a) but here the minimum value of the activation barrier is above an atop site; (c) same as (a) but here the minimum barrier value is located above a center site. The figure is reprinted with permission from Halstead and Holloway (1986).

well-known that the corrugation of the potential increases with energy, thereby leading to larger diffracted intensities at high energy. In the reactive system, however, the competition for the flux at high energy by the dissociative channel significantly lessens the peak of the diffracted beams. Similar results were also found for D_2 scattering. The conclusion is that the energy dependence of the diffracted beams may contain significant information on the reactive PES, at least when the barrier lies in the entrance channel.

As a first test of some of the dynamical assumptions in the above model, we have performed GLE–ghost atom simulations using the PES described in Section III for H_2/Ni(100) and with a few contour plots shown in Fig. 25 for special configurations. The surface was allowed to move initially, but that yielded negligible changes compared to the rigid surface, and the results shown in Fig. 35 are for a rigid surface with all six degrees of freedom of the H_2 allowed to vary. First, note that there must be some dynamical features of orientation and bond length since S_0 does not become unity at $E_i > 0.1$ eV, which exceeds all the entrance channel barriers. On the other hand, S_0 is quite large (0.6–0.7) at such energies, and so most of the molecules do dissociate after surmounting the entrance channel barrier. Second, the angular distribution of scattered molecules is quite narrow, indicating small translational to rotational energy transfer. Nevertheless, the dependence of S_0 on initial rotational state at high kinetic energy is quite complex and by

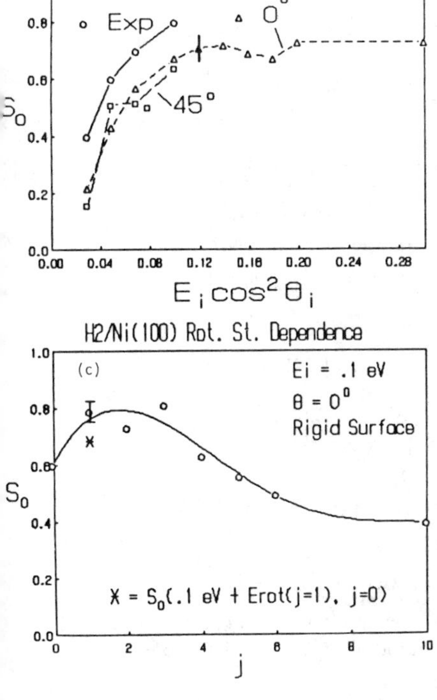

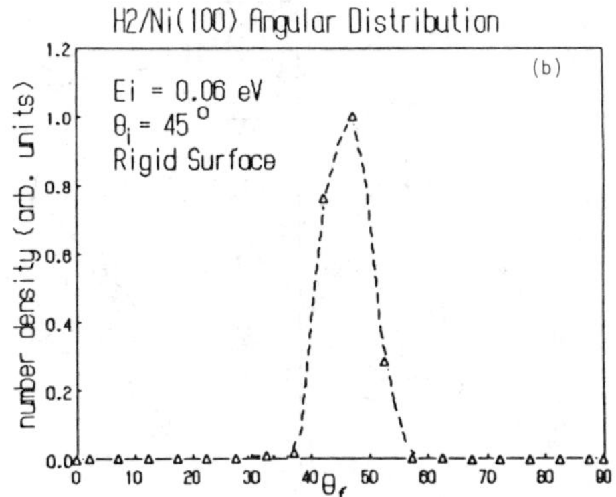

Figure 35. (a) Variation of the initial sticking coefficient with initial normal kinetic energy (circles represent experimental results from Hamza and Madix (1985) while triangles and squares represent classical GLE calculations at incident angles of 0° and 45°, respectively); (b) calculated number density of scattered H_2 vs. final polar angle; (c) calculated initial rotational state dependence of the dissociation probability. The plots are from Kara and DePristo (1989).

$j = 8$, S_0 decreases by 50% from the $j = 0$ result. These results show that the rotational dependence is quite substantial at high kinetic energies. However, we should emphasize that at low kinetic energies the rotational dependence may be much less important than the neglect of all quantum effects inherent in the classical simulation. Hence, the model of Karikorpi et al. and Halstead and Holloway may have real utility at low kinetic energy. We will discuss this in greater detail later in this section.

Further work will be needed to ascertain whether their conclusions about diffraction intensities remain valid even when rotations of the H_2 and motion of the surface atoms are included. This will require quantum mechanical calculations, not classical simulations. A recent method has been developed to treat simultaneous rotations and diffraction via a set of close-coupled TDSE (Mowrey and Kouri 1986; Mowrey et al. 1987) which are solve using the FFT procedure. This approach could be combined with a modification

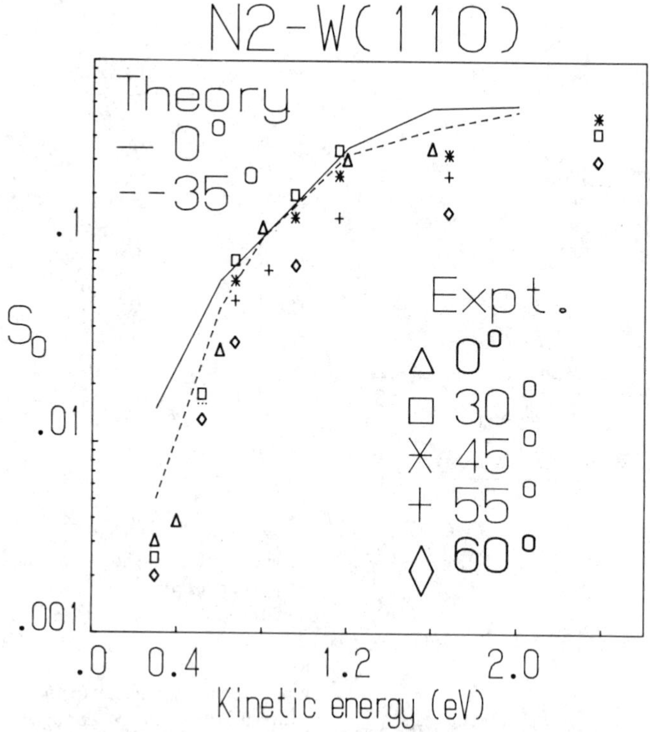

Figure 36. The initial sticking coefficient as a function of the initial kinetic energy for the N_2/W(110) system; experimental data are from Pfnür et al. (1986) and the theoretical curves are from Kara and DePristo (1988a).

of the PES used in the classical simulations to provide such a test. If a movable surface could be incorporated via the GLE–ghost atom method, these simulations would provide benchmarks of unquestioned importance to molecule–surface dynamics.

The location of the barrier in the entrance channel implies normal energy scaling, $S_0(E_i, \theta_i) = S_0(E_i \cos^2 \theta_i, 0)$. However, it is worthwhile to emphasize that such scaling behavior does not imply a barrier in the entrance channel. To see why, we note first that the shape of the PES shown in Figs. 17 and 25 arises from two general factors: (1) the molecular physisorption well is located further from the surface than the atomic chemisorption well; and (2) the bond must stretch in transforming the reactant molecule into the separated atomic adsorbates. Smooth contours drawn between these limits will yield a PES with a topology similar to those shown previously. Thus, most PESs should be able to transform translational motion normal to the surface into motion along the reaction path with considerable efficiency. This is the physical explanation of why many systems with exit channel barriers obey near-normal energy scaling, as was first discovered by Lee and DePristo (1986a, 1986b) and later emphasized by Kara-and DePristo (1988a).

A system that does not is N_2/W(110), rather showing a scaling with the total kinetic energy, $S_0(E_i, \theta_i) = S_0(E_i, 0)$ (Pfnür et al. 1986). It is also possible to duplicate this behavior using full GLE–ghost atom simulations on a four-body LEPS PES described in Section III (Kara and DePristo, 1988a, 1988b, 1988c) as shown in Fig. 36. This behavior was explained in terms of the topology of the PES shown in Fig. 37 which exhibits a very constricted activation barrier region. This enforces a much more stringent steric dependence on the molecule before reaction can occur, which does not allow for the indiscriminate transfer of translational to vibrational energy to be efficient at surmounting the barrier.

The scattered molecules are also interesting to analyze, and this is done in Fig. 38 for the angular distributions. The surprising features are the narrowness and peak at a slightly supraspecular angle, agreeing with the experimental data (Pfnür et al. 1986). The reason for these is that the molecules which scatter back into the gas phase almost invariably undergo only a single impact with the surface, because of an improper orientation during the initial collision. These scattered molecules do not sample the dissociative part of the PES but instead act just like the scattered molecules for inelastic scattering. Further evidence for this is seen in Fig. 39 for the distribution of rotational, vibrational, kinetic, and total energies for all the scattered molecules, each summed over all final angles. The surprising result is the negligible change in the vibrational energy even though E_i is large enough to excite N_2 $(n = 4)$. This demonstrates that the part of the PES which stretches the N_2 bond is not sampled by the scattered molecules,

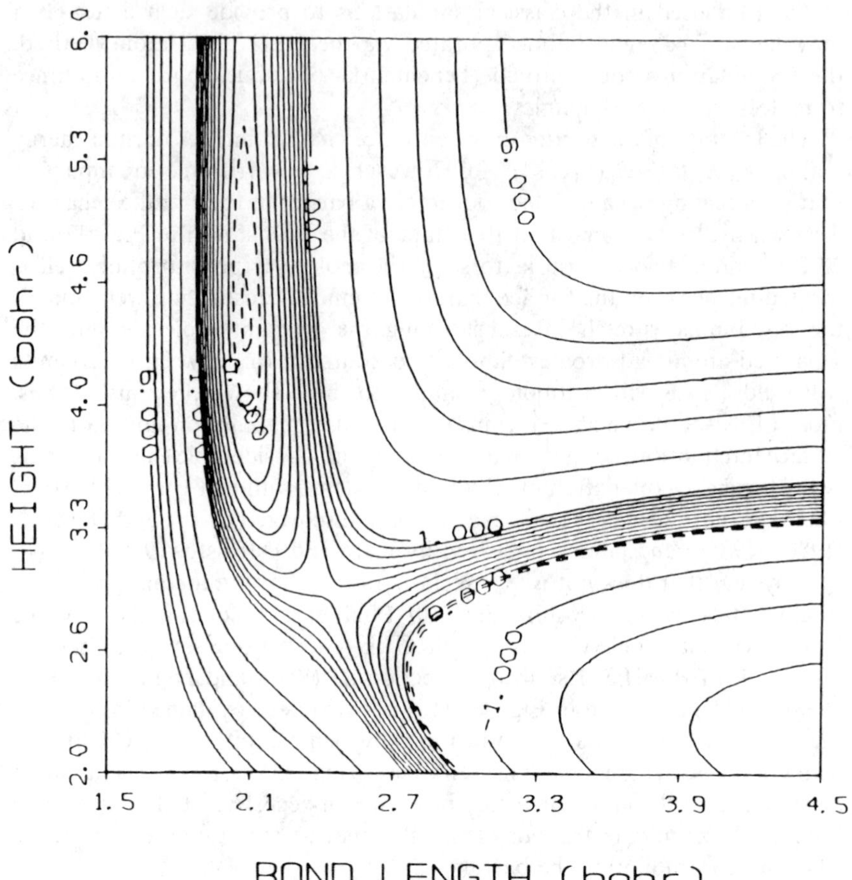

Figure 37. Contour plot of a two-dimensional cut in the N_2/W(110) PES for the configuration in Fig. 22 which yielded total energy scaling. The figure is reprinted with permission from Kara and DePristo (1988a).

corroborating the above idea. The scattering is not elastic however. The energy loss to W(110) is indicated by the average of the final total energy being less than the initial value by about 0.3 eV. By contrast, the change in E_i is about 0.5 eV, which demonstrates that about 0.2 eV is transferred from translational into rotational energy.

Another interesting feature of the dissociative chemisorption is the dependence upon initial vibrational and rotational excitation of the N_2. The results in Fig. 40 demonstrate that translational and vibrational energy are equally efficient at increasing S_0. This may be expected for a system in which

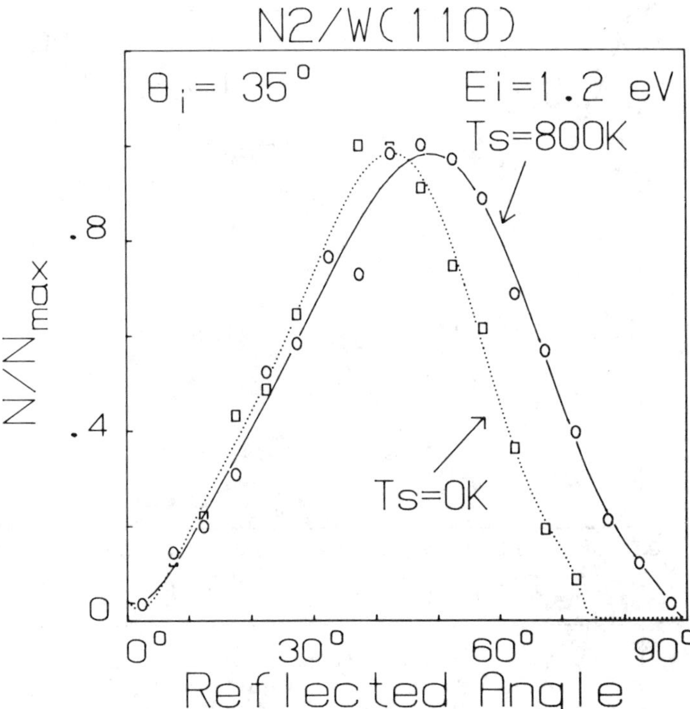

Figure 38. Number distribution of final angles for N_2 molecules scattered from the W(110) surface, from Kara and DePristo (1988a), with permission.

total energy scaling is found. By contrast, a complicated dependence upon rotational state is exhibited in Fig. 41. At each E_i there is a fact increase of S_0 at $j = 2$ followed by a nearly constant value. The increase is much larger than the variation of S_0 with E_i since the rotational energy of $j = 2$ is only ≈ 0.0015 eV. It appears that the dependence of S_0 on j is informative as related to the orientation dependence of S_0 for this system (Kara and DePristo 1988c).

In the N_2/W(110) system, S_0 remains less than unity even at kinetic energies well above those of the minimum barrier. By contrast, for the H_2/Ni(100) system considered previously, S_0 approaches unity at high kinetic energies. To understand this behavior more fully, we have determined the location of the molecule in the surface unit cell when dissociation occurred. It was possible to define such a position because the molecular bond length oscillated around its gas-phase value until increasing quickly and monotonically during the irreversible dissociation. The rapid increase in the

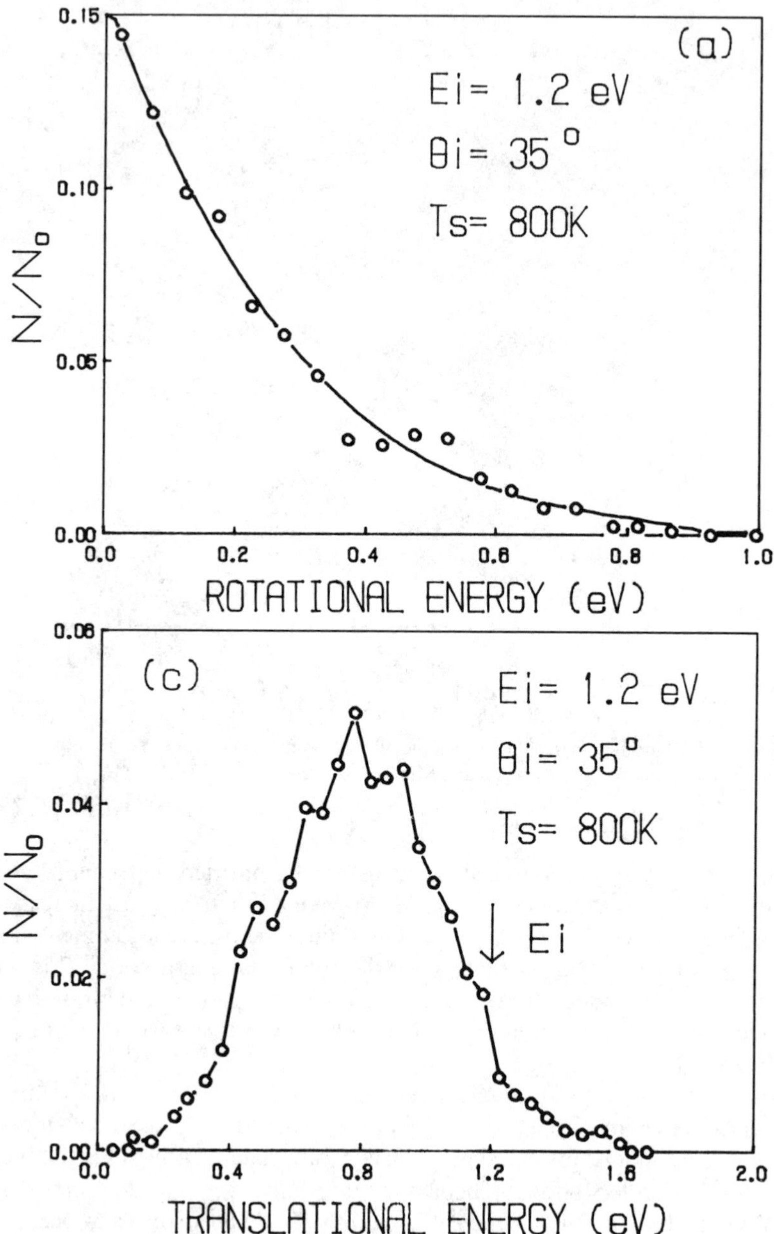

Figure 39. (a) Distribution of final rotational energy for N_2 molecules scattered from the W(110) surface; (b) same as (a) except for vibrational energy; (c) same as (a) except for kinetic energy; (d) same as (a) except for total molecular energy.

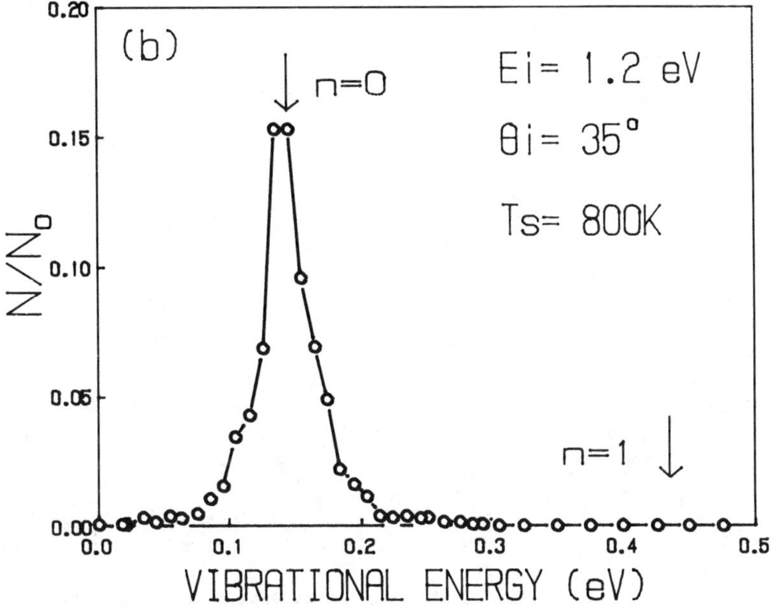

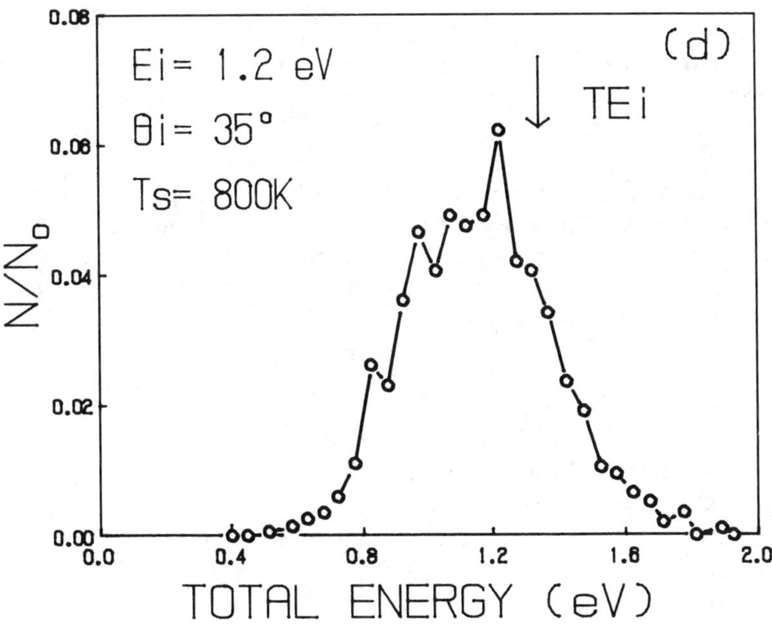

Figure 39. (*Continued*)

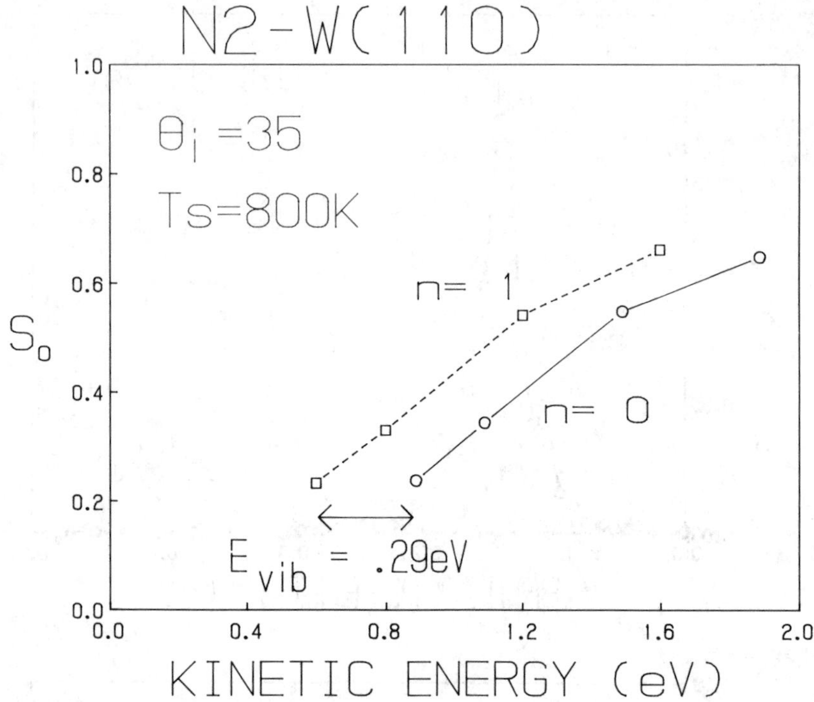

Figure 40. A comparison of the efficiency of vibrational and translational energies in the dissociative chemisorption of N_2 on W(110).

bond length occurred much more quickly than any translational motion of the molecule center of mass. (The rotation of the molecule occurred on the same time scale as the increase in bond length, which precluded a precise definition of the molecular orientation at dissociation.) Figs. 42 and 43 show the distribution of dissociation impacts for both systems at low and high kinetic energy. Three surprising features are apparent:

1. the retention of site selectivity in H_2/W(110), for $E_i = 0.14$ eV in Fig. 42a, even when $S_0 \approx 0.8$;
2. the absence of site selectivity in N_2/W(110), for the nonrigid surface at $E_i = 0.8$ eV in Fig. 43c, even when $S_0 \approx 0.15$;
3. the change from site-selective to non-site-selective behavior for the N_2/W(110) system at $E_i = 0.8$ eV, an effect which requires $E_i = 1.2$ eV in the rigid surface.

As to the reaction zone, note that although the shape of this zone is very

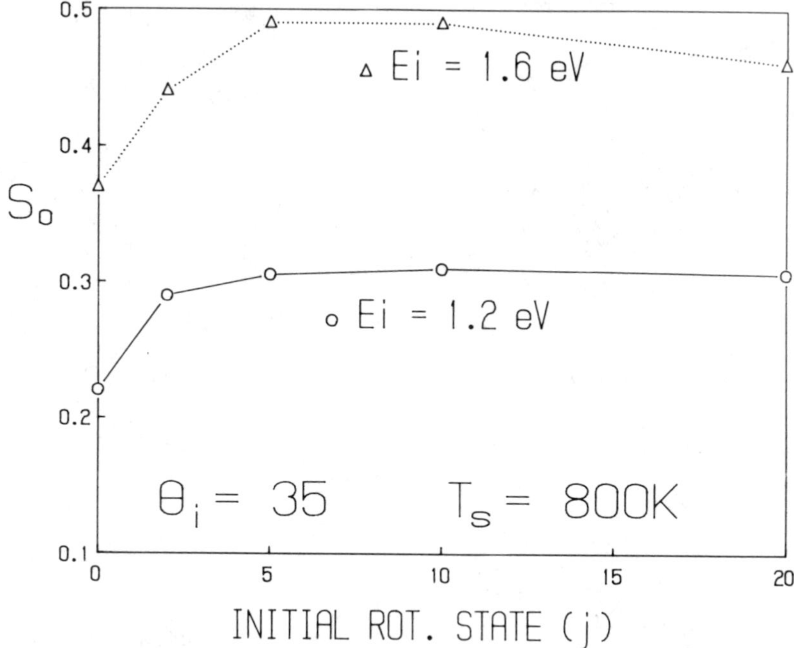

Figure 41. Initial rotational state dependence of the dissociative chemisorption probability of N_2 on W(110). The figure is reprinted with permission from Kara and DePristo (1988c).

distorted from the near-circular one shown in Figs. 33b–d, it could clearly be modeled along the same lines as done by Karikorpi et al. (1987) and Halstead and Holloway (1988) for that PES.

Rationalization of these results depends upon the influence of molecular orientation. In the N_2/W(110) case, molecular orientation is much more important than position in the surface unit cell, while in the H_2/Ni(100) case, the opposite relationship holds. This can be understood in terms of the variation of the underlying atom-surface PES and the size of the molecules. Rotation of the N_2 relative to the surface changes the N—W distance substantially on the scale that V_{NW} changes. By contrast, rotation of the H_2 relative to the surface changes the H—Ni distance only a small amount on the scale that V_{HNi} changes. Thus, rotation of the molecule is more important in the former case. In addition, location in the unit cell is more important in H_2/Ni(100) because the atop site does not bind H strongly enough to dissociate H_2 (e.g., atop, bridge, and fourfold sites bind H by $\approx 2.2, 2.5$, and 2.8 eV, respectively). By contrast, the relevant quantities for N—M are 5.8, 6.8, and 7.2 eV. In other words, the atom–surface PES is much more

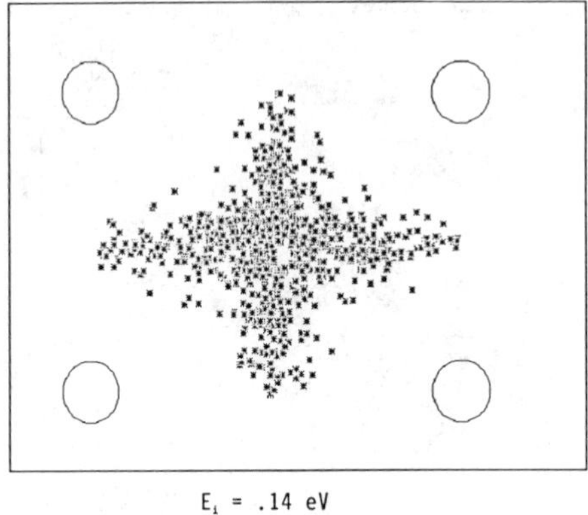

(a)

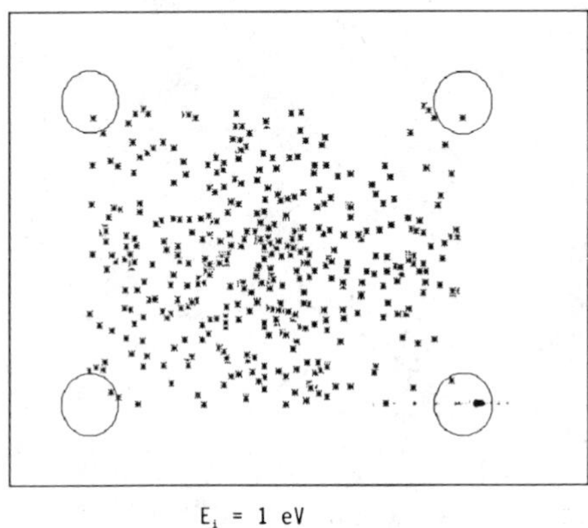

(b)

Figure 42. Location within the unit cell of the center of mass of the molecule at the occurrence of molecular dissociation in the H_2/Ni(100) system, for normal incidence and initial kinetic energies of 0.14 eV in (a) and 1 eV in (b). The plots are from Kara and DePristo (1989).

corrugated for the H—Ni(100) than N—W(110) systems. More work will need to be performed to ascertain the generality of these conclusions, but it is clear already that one must generally discuss these reactions in a much higher dimensionality than the simple one- or two-dimensional models. This lack of dominance by high-symmetry dissociation sites implies that *ab initio* and density-functional calculations of PES must sample a wide variety of sites in order to provide detailed information about the dissociation process.

Before leaving the discussion of dissociative chemisorption results, we note a few further points. First, nearly all simulations have either ignored the effect of the lattice or included it via simple harmonic motion of a primary zone coupled via the GLE to ghost atoms. A recent investigation (Agrawal et al. 1987) showed that for the H_2/Si(111) system the lattice potential plays

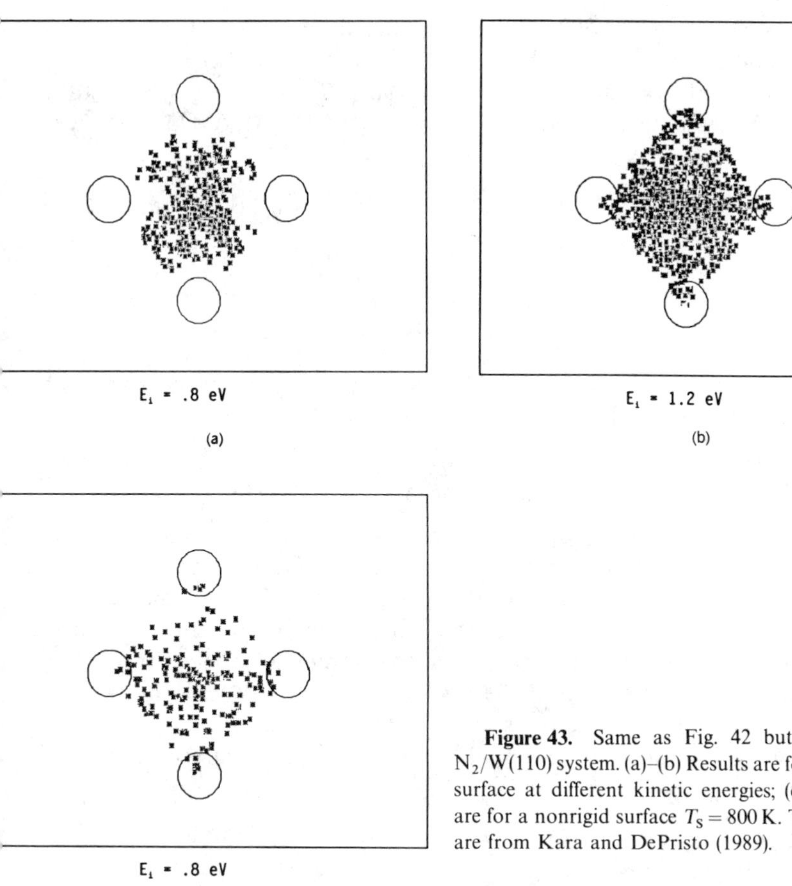

Figure 43. Same as Fig. 42 but for the $N_2/W(110)$ system. (a)–(b) Results are for a rigid surface at different kinetic energies; (c) results are for a nonrigid surface $T_S = 800$ K. The plots are from Kara and DePristo (1989).

a minimal role. This result should hold for other systems as long as the chemisorption process is not strongly coupled to the relaxation or reconstruction of the lattice. In the latter case [e.g., $H_2/W(100)$], a more quantitative treatment of the lattice will be necessary.

Second, we consider a recent statistical model for the dissociative chemisorption of H_2 (Hayward and Taylor 1988). This model does not depend upon a PES but instead makes two critical assumptions: (1) $S_0(E_i, \theta_i) = S_0(E_i, 0)f(\theta_i)$ where $f(\theta_i)$ is an arbitrary function; and (2) S_0 is independent of the initial molecular internal ro-vibrational energy. The first assumption is more restrictive than simple normal energy scaling and thus must be tested further. By contrast, the second is in general accord with PES-based models at the low temperatures 300–1200 K considered where only low rotational states would be populated (Halstead and Holloway, 1988; Lee and DePristo, 1986b, 1987a, 1987b; Kara and DePristo, 1989). Using these assumptions, it is easy to show that the number of dissociatively adsorbed molecules, N_{ads}, is

$$N_{ads} = P(2\pi MkT)^{-1/2}F(kT)^{-2}\int_0^\infty S_0(E_i,0)E_i\exp(-E_i/kT)dE_i \tag{6.2a}$$

$$F = \int_0^{\pi/2} f(\theta_i)\sin(2\theta_i)d\theta_i \tag{6.2b}$$

where P and T are the gas pressure and temperature, respectively. This result is compared to the TST equation:

$$N_{ads} = Kh^2P(2\pi MkT)^{-3/2}[Q^\ddagger/Q]\exp(-E_0/kT) \tag{6.3}$$

where K is the transmission coefficient, Q is the internal state reactant partition function of the reactants, and $Q^\ddagger$ is the partition function of the activated complex, excluding motion along the reaction coordinate. Q is evaluated under the rigid rotor approximation; vibrations are assumed to remain in the ground state. For the activated complex, all modes are assumed to be classical yielding $q^\ddagger \sim (kT)^q$ where $q = [n_v + \frac{1}{2}(n_r + n_t)]^\ddagger$ in obvious notation. Combining all these assumptions, and setting the two expressions for N_{ads} in Eqs. (6.2a) and (6.3) equal, yields the simple integral equation for $S_0(E_i,0)$:

$$\int_0^\infty S_0(E_i,0)E_i\exp(-E_i/kT)dE_i = B(kT)^q\exp(-E_0/kT) \tag{6.4}$$

where B is a constant. This equation can be solved easily by inverse Laplace transforms, yielding

$$S_0(E_i,0) = CE_i^{-1}(E_i - E_0)^{(q-1)} \tag{6.5}$$

Equation (6.5), with $q = 3$, is consistent with experimental results at low kinetic energy for the dissociative chemisorption of H_2 on Ni(111) (Hayward and Taylor 1986) and Fig. 3c. It will generally break down at high energies since $S_0 \rightarrow CE^{(q-2)}$ which approaches ∞, C, and O for $q > 2$, $q = 2$, and $q < 2$, respectively. Beyond that, the numerous assumptions and approximations inherent in such a model make it difficult to interpret the meaning of $q = 3$. It is certainly not possible to extract detailed information about any PES features using such a model.

Instead of this type of model, we suggest utilization of TST in the recently developed variational versions (Lauderdale and Truhlar 1985, 1986; Truhlar et al. 1986; Truong et al. 1989b). These can be either microcanonical (E_i) or canonical (T) forms. These also include multidimensional tunneling via least-action techniques as well as surface atom motion via embedded cluster models. They do require a PES, but in our opinion, it is preferable to at least indicate a PES rather than make a multitude of assumptions about the dynamics. Because of the speed of the VTST methods, large computing facilities would not be necessary, and a well-documented program exists for these calculations, which should be available by the time of publication of this review (from QCPE at the University of Indiana).

The third point is that most of the systems studied to date have an A—B bond energy which substantially exceeds either the A—S or B—S chemisorption energies. Thus, at low kinetic energies, the process $AB + S \rightarrow A(a) + B(a) + S$ is the only one that is energetically possible. It is not necessary for this inequality to hold: the I_2/Fe(100) is an example where $D_{I_2} = 1.55\,eV$ while the I—Fe chemisorption energy is 2.52 eV. This system was studied using classical trajectory techniques on a number of model PESs, with surface motion included via a single dissipative oscillator moving in the Z-direction only (Ron et al. 1987). The important conclusion is that the dissociative chemisorption process is not generally much more favored than the exchange process, $I_2(g) + Fe(100) \rightarrow I(g) + I(a) + Fe(100)$. It would be very interesting to determine the detailed information contained in the energy and angular distribution of the gaseous I atom. Such systems would be well worth studying in more detail.

Now, we consider a few other types of surfaces processes, but not in the same detail as dissociative chemisorption. The first is the Eley–Rideal type reaction, $A(g) + B(a) + S \rightarrow AB(g) + S$. This is the inverse of the exchange process mentioned above. All cases have considered systems in which D_{AB} exceeds the A—S and B—S interactions:

1. $O(g) + C(a) + Pt(111) \rightarrow CO(g) + Pt(111)$, $\Delta H \approx -6.1\,eV$ (Tully 1980b; Billing and Cacciatore 1985);
2. $H(g) + H(a) + W(100) \rightarrow H_2(g) + W(100)$, $\Delta H \approx -1\,eV$ (Elkowitz et al. 1976);
3. $H(g) + Cl(a) + S(model) \rightarrow HCl(g) + S(model)$, $\Delta H \approx -1.87\,eV$ (Shima and Baer 1985; Bear and Shima 1986; Ron and Baer 1988);
4. $CO(g) + O(a) + Pt(111) \rightarrow CO_2(g) + Pt(111)$, $\Delta H \approx -1.7\,eV$ (Billing and Cacciatore 1986).

The large amount of energy liberated in the process may show up in a number of places, including vibrations, rotations, and translations of the molecule;

phonons and e–h pairs of the solid; and defect formation and ejection of the surface atom. A realistic simulation of the CO/Pt(111) system used a GLE–ghost atom approach and found large vibrational excitation of the CO, but also noted that this would be significantly reduced by excitation of electron–hole pairs (Tully 1980b). Further study of these types of processes must await more accurate descriptions of e–h pair excitations.

Very-high-energy processes involving the dissociation of gas-phase species via $AB(g) + S \rightarrow A(g) + B(g) + S$ have treated using only the simplest PES or dynamics which did not allow all of the multiple rearrangement channels. The reader is referred to the work of Gerber and co-workers as reviewed in Gerber (1987). The major conclusions for the I_2/MgO system are: (1) dissociation occurs via a $T \rightarrow R$ energy transfer mechanism that excites the I_2 above its dissociation threshold; (2) the $T \rightarrow R$ process is so fast that it is unaffected by surface atom motion; (3) substantial energy transfer to the surface occurs in collisions that do not lead to dissociation.

The reader will note that we have been shifting considerations to processes in which the surface plays less and less of a role. Now we switch to processes that occur either totally or predominantly on the surface. Perhaps the simplest is diffusion, especially of atoms. While it has no direct connection with molecule–surface scattering, diffusion is important for a number of reasons. First, diffusion has been treated by a number of the most sophisticated theoretical dynamical methods outlined in Section IV, including: flux–flux correlation functions via both path integrals (with analytic continuation) (Jacquet and Miller 1985) and FFT propagation approaches (Wahnstrom et al. 1988; Haug et al. 1989); canonical VTST with tunneling corrections (Lauderdale and Truhlar 1985, 1986; Truong and Truhlar 1987; Rice et al. 1988); classical transition state theory with dynamical corrections and quantum modified potential (Valone et al. 1988).

Second, diffusion at low surface temperatures is typically a rare event because an adsorbed atom spends most of its time in the equilibrium position. Special tecnhiques for simulation of such rare events involve basically classical transition state theory plus short-time corrections (Keck 1967; Anderson 1973; Pechukas 1976; Montgomery et al. 1980; Grimmelmann et al. 1981; Doll and Voter 1987); the equivalent methods in quantum dynamics will involve short-time approximations to the quantum flux–flux correlation function. In the classical case, the idea is to start the system at thermal equilibrium on a dividing surface between reactants and products, sample trajectories from this distribution, and then integrate these for a short-time to determine those leading to products. In the quantum case, which is much less developed, a similar idea holds with replacement of the trajectory integration by short-time evolution of the flux–flux correlation function (Tromp and Miller 1986; Haug et al. 1989). These rare-event methods are very powerful, and should be invoked whenever processes occur with low probability due to the crossing of a barrier.

Third, diffusion is often an integral part of the dynamics in real systems under non-UHV and even under UHV conditions. These involve nonzero coverages of multiple adsorbates, nonequilibrium surface structures of adsorbates, and other complications. It is clearly important to understand the simplest case of single-atom diffusion on a surface since this will underlie the more complex situations. A number of important conclusions have been learned from studies of atomic diffusion: (1) lattice motion substantially increases the quantum mechanical diffusion constant as shown in Fig. 44 by

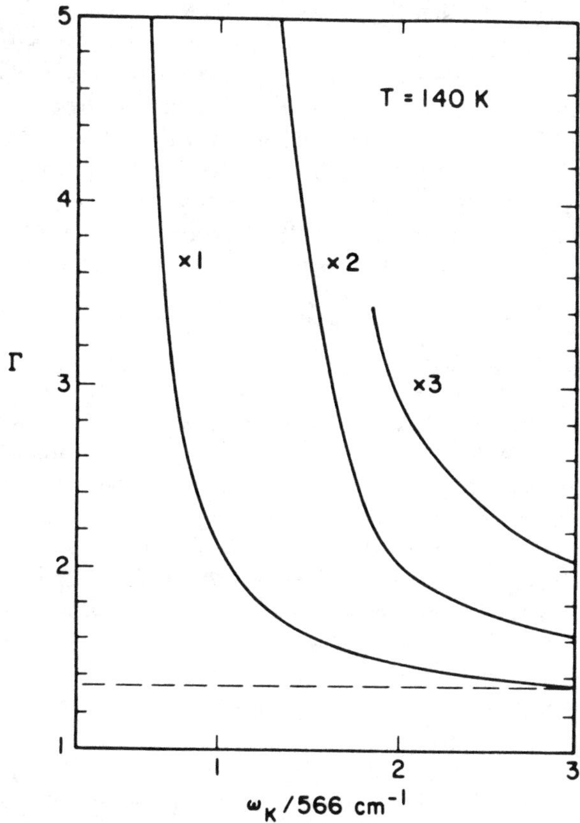

Figure 44. Correction factor to the transition-state theory rate as a function of the frequency of the surface phonon, for H-atom diffusion on a W(100) surface, at T = 300 K. The horizontal dashed line is for a moveable surface. The various curves correspond to different multiplicative factors of the strength of copling to the phonon. The plot is from Jaquet and Miller (1985).

providing a lower dynamical barrier in the low-temperature tunneling regime (Jacquet and Miller 1985; Rice et al. 1988; Lauderdale and Truhlar 1985, 1986; Valone et al. 1985); (2) motion perpendicular to the surface is strongly coupled to motion along the surface, leading to apparent diffusion even for a rigid surface (Wahnstrom et al. 1988; Haug et al. 1989). More details on diffusion studies can be found in a recent review (Doll and Voter 1987). We would caution the reader that nearly all diffusion studies have utilized the crudest pairwise additive potentials for both V_{AS} and V_{SS}. More accurate PES should be utilized in future realistic simulations.

Diffusion of two adsorbates into close proximity can lead to the recombinative or associative desorption process, A(a) + B(a) + S → AB(g) + S, which is the reverse of the dissociative chemisorption process. These two processes provide the same dynamical information, at the level of each individual trajectory, because of time-reversal invariance. However, because observables are averages over many trajectories, and the sampling of the reactants' phase space in the two processes is different, the observables may not be related by detailed balance. In other words, if nonequilibrium holds for the reactants in either process, then detailed balancing will not hold. The degree of deviation from detailed balancing in any system has been a point of some controversy (Balooch et al. 1974; Kubiak et al. 1984; Comsa and David 1985; Rendulic 1988). We have no intention of being drawn into this quagmire at this time since our concern is the interpretation of these processes in terms of features of the PES, and there is no doubt that the underlying PESs are the same.

The simplest dynamical model of associative desorption is based upon a one-dimensional PES with, of course, a single barrier in the exit channel for desorption (Van Wiligen 1968). (An exit channel barrier for desorption is equivalent to an entrance channel barrier for dissociative chemisorption.) Assuming an equilibrium distribution of adsorbates at the surface temperature T, the model predicts a Boltzmann distribution of velocities with the Z-component of velocity centered around $v^* = (2V_0/M)^{1/2}$ where V_0 is the barrier height. The angular distribution of desorbed molecules is then

$$N(\theta) = N(0)\frac{V_0 + kT\cos^2\theta}{(V_0 + kT)\cos\theta}\exp[-(V_0/kT)\tan^2\theta] \tag{6.6}$$

while the rotational and vibrational distributions are Boltzmann. The measured angular distributions tend to be broader than those calculated in this way as shown in Fig. 16 (Comsa and David 1985) while the vibrational state distribution can be distinctly non-Boltzmann by orders of magnitude, as for example in the H_2/Cu system (Kubiak et al. 1984). This should not be surprising since such a simple PES bears little resemblance to realistic PESs for the reverse dissociative chemisorption process.

The earliest multidimensional treatment of recombinative desorption was a classical trajectory simulation of the H(a) + H(a) + W(100, rigid) → H_2(g) + W(100, rigid) by McCreery and Wolken (1977). There were six degrees of freedom, but they did not use techniques for treating rare events. Thus, they had to construct an artificial initial distribution of reactants that allowed diffusion on a short enough time scale to allow two adsorbates to diffuse and recombine. The resulting distribution had a mean energy of over 3 eV, making their results inapplicable to a real system, but of course providing a striking example of why explicit use of rare-event formalisms is crucial in such studies.

We are aware of two applications of rare-event theory in the treatment of recombinative desorption. The first involves the H_2/Si(111) system (NoorBatcha et al. 1985; Raff et al. 1986) while the second involves the H_2/Cu system (Harris et al. 1988a, 1988b). The former work is strongly recommended for study, since it does treat a multidimensional problem by the rare-event formalism. However, here we use the latter as an example since it allows a close tie to the previous discussion of PESs for H_2/metal systems.

Two basic assumptions about the PES are made. First, a single barrier in the desorption entrance channel is assumed for every configuration of the molecule–surface system and r_{AB} is assumed to be significantly stretched at the location of this barrier. This is just like many of the systems described previously. The exceptions are the model of Karikorpi et al. (1987) and Halstead and Holloway (1988) in which only a barrier in the chemisorption entrance channel was used for the scattering of H_2 from Ni and the more complex two-barrier PESs for the H_2/Mg(100) system in Fig. 17 and the H_2/Ni(100) system in Fig. 25. Since the chemisorption entrance channel barrier in Fig. 25 is quite small compared to the chemisorption exist channel barrier, a first approximation would involve neglecting the former, which then yields a PES like that assumed in the work of Harris et al. Since Karikorpi et al. and Holloway and Halstead simply replace everything after the chemisorption entrance barrier with a sink, their PES model does not appear to be in accord with that of Harris et al. The second assumption is that the PES variation around the barrier is separable in the other H_2 coordinates: translations parallel to the surface, (X, Y), and orientations, $\hat{r}$. This is a rather rough approximation, as noted in the original article, but does have the nice feature of reducing the dimensionality of the initial sampling problem.

With these two assumptions, they ran many trajectories and determined the final angular, translational, and internal energy distributions. Their finding can be summarized as follows: (1) the vibrational and translational energies of the $H_2(g)$ significantly exceed kT; (2) the rotational energy of the H_2 is $\simeq kT$; (3) the angular distribution, $N(\theta)$, of the H_2 is strongly focused along a normal to the surface. The first and third findings follows from the curvature of the PES in the region of the chemisorption exit channel barrier. As emphasized previously, for dissociative chemisorption this curvature transforms incident translational energy into motion along the reaction coordinate (e.g., partial vibration). The reverse process starting from the barrier will clearly transform energy into both vibrations and translations. The second finding results from the reduced dimensionality assumptions of the model since there is no mechanism for significant rotational excitation (Harris et al. 1988b).

These findings are in general qualitative agreement with experimental data discussed in Section II with the exception of the experimental result that the translational energy distribution is nearly independent of θ. The theoretical results show that the average translational energy increases with θ. There is no satisfactory explanation for this discrepancy. It is possible that inclusion of rotational motion, and especially strong coupling between T, R, and V near the barrier region, would make a difference by allowing for more rotational energy at large θ, and thus less translational energy. In addition, the variation of the PES shape in the unit cell may also be important in this regard since it will give rise to different correlations between θ and translation/vibrational energy. In this regard, we note that a recent model does include different local sites for desorption, but does not allow for $T \rightarrow R$, V energy exchange (Ertl et al. 1988). Such a model is also inadequate to explain all experimental findings. We suspect that answers to these discrepancies must await simulations of the desorption including full degrees of freedom.

Before leaving the topic of recombinative desorption, we do want to make one observation about the N_2/W(110) system. As noted for the H_2/Ni(100) system, in recombinative desorption the incipient molecule will be formed at the activation barrier and then will move into the entrance channel region, thereby transforming vibrational energy to translational energy and a desportion pattern that peaks around the normal to the surface. This argument holds irrespective of the narrowness of the activation barrier region and thus there is no contradiction between desorption intensities peaked around the normal and total energy scaling of S_0.

VII. SUMMARY

It would not be worthwhile (or possible) to summarize all the information presented previously in just a few paragraphs. We do want to emphasize that the theoretical treatment of molecule–surface reaction dynamics has matured

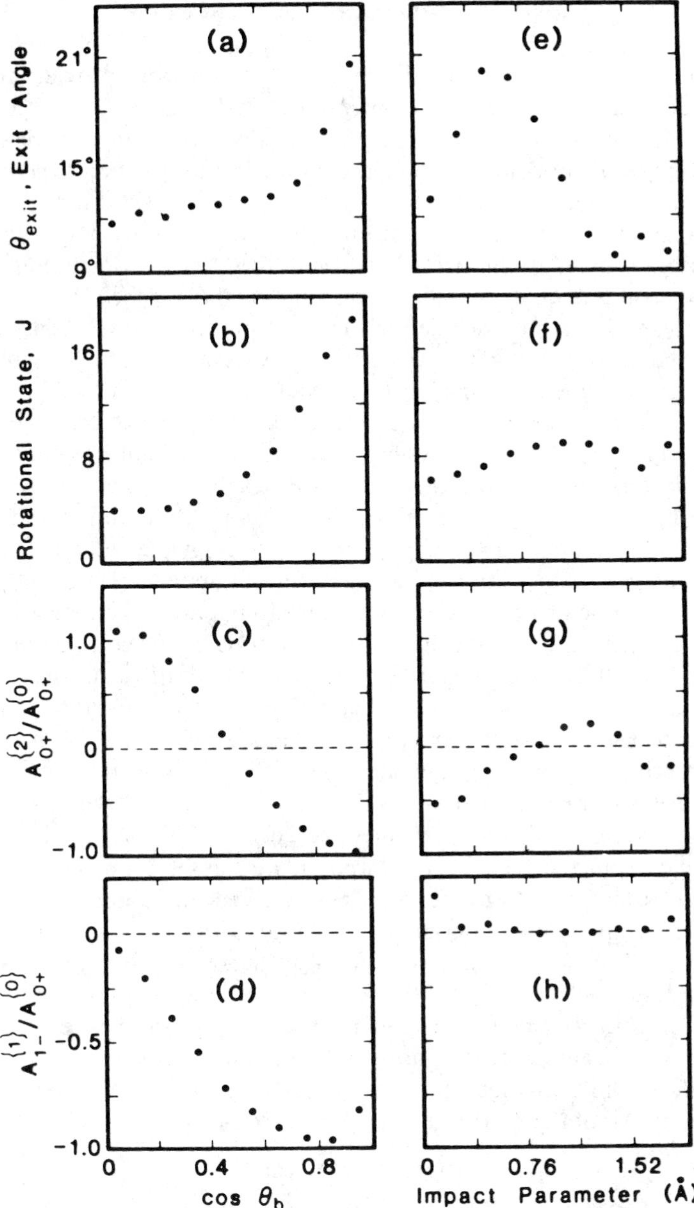

Figure 45. Calculated parameters of the scattered molecules vs. initial conditions for the N_2/Ag(111) system with $E_i = 0.3\,\text{eV}$, $T_S = 90\,\text{K}$, $T_{rot}(\text{init}) = 0\,\text{K}$, $\theta_i = 0^0$. The ratios $A^{(2)}_{0+}/A^{(0)}_{0+}$ measure the orientation of the final angular momentum vector. The initial orientation of the N_2 is described by θ_b where where $\cos(\theta_b) = 0$ corresponds to side-on approach geometry and $\cos(\theta_b) = 1.0$ corresponds to end-on approach geometry. The impact parameter describes the initial aiming point in the unit cell with $b = 0.0\,\text{Å}$ corresponding to an atop site and $b = 1.89\,\text{Å}$ a center site. The plots are from Kummel et al. (1988).

considerably in the past decade. Simulations can now provide detailed understanding of complex multidimensional dynamical processes in terms of the underlying potential energy hyperspace. It is no longer necessary to invoke simplistic one- or two-dimensional models to interpret experimental data only qualitatively.

We cannot resist a brief conjecture about the future. From the experimental viewpoint, a major problem is the inability to select the orientation of the incident molecule or the initial aiming point at the surface. This limitation means that even the most detailed calculations must be averaged considerably before comparison with data. An exciting new development using polarized laser detection methods may help alleviate this problem. In particular, detailed experimental studies of the angular momentum vector, j, have been presented for the N_2/Ag(111) system (Kummel et al. 1988). The microscopic trajectory analysis is shown in Fig. 45, with the lower two rows of panels measuring the angular momentum orientation θ_b providing the initial molecular orientation, and the impact parameter giving the initial aiming point. Analysis of this figure leads to the following points: (1) from (a) versus (e), the exit or scattering angle is determined largely by the location of the collision in the surface unit cell, or in other words by the corrugation of the surface; and (2) from (b) versus (f), (c) versus (g) and (d) versus (h), the final magnitude and direction of j are determined mainly by the molecular orientation geometry during the collision. Thus, experimental measurements now offer the real possibility of separating the effects of the surface corrugation and molecular orientation, or in the authors' words to "differentiate between the two hidden initial conditions in a gas–surface collision."

Second, from the theoretical viewpoint, we foresee a number of nearly exact quantum claculations that will provide benchmarks for comparison to more approximate theories, especially those of classical simulations. In addition, we expect that the recent trend to investigate collisions involving large molecules (Kwong et al. 1989; Xu et al. 1988; Lee 1989) will accelerate. Finally, we speculate that a new trend will develop involving the simulation of much more complicated surfaces, including bimetallics and oxides, those with defects and steps, and those with ordered and disordered coverages of adsorbates. All of these areas are truly in their infancy.

Acknowledgment

This work has been supported by the National Science foundation, division of Chemical Physics. Partial support by the Petroleum Research Foundation administered by the American Chemical Society has also been received.

References

Adelman, S. A. and J. D. Doll (1977). *Acc. Chem. Res.* **10**, 378.

Adelman, S. A. (1980). *Adv. Chem. Phys.* **44**, 143

Agrawal, P. M., L. M. Raff, and D. L. Thompson (1987). *Surf. Sci.* **188**, 402.

Anderson, J. B. (1973). *J. Chem. Phys.* **58**, 4684.

Asscher, M., O.M. Becker, G. Haase, and R. Kosloff (1988). *Surf. Sci.* **206**, L880.

Baer, M. (1984). *J. Chem. Phys.* **81**, 4526.

Baer, M. and Y. Shima (1986). *Chem. Phys. Lett.* **125**, 490.

Balooch, M., M. J. Cardillo, D. R. Miller, and R. E. Stickney (1974) *Surf. Sci.* **46**, 358.

Barker, J. A. and D. J. Auerbach (1984). *Surf. Sci. Reports* **4**, 1.

Beckerle, J. D., Q. Y. Yang, A. D. Johnson, and S. T. Ceyer (1987). *J. Chem. Phys.* **86**, 7236.

Berkowitz, M. and J. A. McCammon (1982). *Chem. Phys. Lett.* **90**, 215.

Bernasek, S. L. (1980). *Adv. Chem. Phys.* **41**, 477.

Billing, G. D. (1987). *Chem. Phys.* **116**, 269.

Billing, G. D. and M. Cacciatore (1985). *Chem. Phys. Lett.* **113**, 23.

Billing, G. D. and M. Cacciatore (1986). *Chem. Phys.* **103**, 137.

Bortolani, A., N. March, and M. Tosi (1988). *Interactions of Atoms and Molecules with Solid Surfaces*, (Plenum, New York).

Brass, S. G. and G. Ehrlich (1986). *Phys. Rev. Lett.* **57**, 2532.

Brenner, D. W. and B. J. Garrison (1989). *Adv. Chem. Phys.* **66**, 281.

Brooks, C. L., III and M. Karplus (1983). *J. Chem. Phys.* **79**, 6312.

Brown, L. S. and S. L. Bernasek (1985). *J. Chem. Phys.* **82**, 2110.

Cardillo, M. J. (1985). *Langmuir* **1**, 4.

Ceyer, S. T. (1988). *Ann. Rev. Phys. Chem.* **39**, 479.

Chiang, C.-M. and B. Jackson (1987). *J. Chem. phys.* **87**, 5497.

Comsa, G. and R. David (1985). *Surf. Sci. Reports* **5**, 145.

Daw, M. S. and M. I. Baskes (1984). *Phys. Rev.* **B29**, 6443.

Daw, M. S. and M. I. Baskes (1988). *Proc. Mater. Res. Soc.* (in press).

DePristo, A. E. (1984). *Surf. Sci,* **141**, 40.

DePristo, A. E. (1989). In *Interactions of Atoms and Molecules with Solid Surfaces*, eds. A. Bortolani, N. March, and M. Tosi, (Plenum, New York).

DePristo, A. E. and H. Metiu (1989). *J. Chem. Phys.* **90**, 1229.

DePristo, A. E., C.-Y. Lee, and J. M. Huston (1986). *Surf. Sci.* **169**, 451.

DePristo, A. E., K. Haug, and H. Metiu (1989). *Chem. Phys. Lett.* **155**, 376.

Diestler, D. J. and M. E. Riely (1985). *J. Chem. Phys.* **89**, 5753.

Diestler, D. J. and M. E. Riley (1987). *J. Chem. Phys.* **86**, 4885.

Doll, J. D. and A. Voter (1987). *Ann. Rev. Phys. Chem.* **38**, 413.

Doll, J. D., D. L. Freeman and M. J. Gillan (1988a). *Chem. Phys. Lett.* **143**, 277.

Doll, J. D., T. L. Beck, and D. L. Freeman (1988b). *J. Chem. Phys.* **89**, 5753.

Elkowitz, A. B., J. H. McCreery, and G. Wolken (1976). *Chem. Phys.* **17**, 423.

Ertl, A., L. Rohner, and H. Wilsch (1988). *Surf. Sci.* **199**, 537.

Eyring, H., J. Walter, and G. E. Kimball (1944). *Quantum Chemistry* (Wiley, New York).

Feynman, R. P. and A. R. Hibbs (1965). *Quantum Mechanics and Path Integrals*, (McGraw-Hill, New York).

Fischer, C. R. and J. L. Whitten (1984). *J. Chem. Phys.* **30**, 6821.

Fleck, J. A., Jr., J. R. Morris, and M. D. Feit (1976). *Applied Phys.* **10**, 124.

Gadzuk, J. W. (1983). *J. Chem. Phys.* **79**, 6341.

Gadzuk, J. W. (1984). *J. Chem. Phys.* **81**, 2828.

Gadzuk, J. W. (1987a). *J. Chem. Phys.* **86**, 5196.

Gadzuk, J. W. (1987b). *Surf. Sci.* **184**, 483.

Gadzuk, J. W. (9187c). *Chem. Phys. Lett.* **136**, 402.

Gadzuk, J. W. (1988). *Ann. Rev. Phys. Chem.* **39**, 395.

Gadzuk, J. W. and S. Holloway (1986). *J. Chem. Phys.* **84**, 3502.

Garrison, B. J. and N. Winograd (1982). *Science* **216**, 805.

Garrison, B. J., N., Winograd, D. M. Deaven, C. T. Reimann, D. Y. Lo, T. A. Tombrello, D. E., Jr. Harrison, and M. H. Shapiro (1988). *Phys. Rev.* **B37**, 7197.

Gasser, R. P. H. (1985). *An Introduction to Chemisorption and Catalysis by Metals*, (Clarendon Press, Oxford, England).

Gerber, R. B. (1987). *Chemical Reviews* **87**, 29.

Gerber, R. B., V. Buch, and M. A. Ratner (1982). *J. Chem. Phys.* **77**, 3022.

Gerber, R. B., R. Kosloff, and M. Berman (1986). *Computer Physics Reports* **5**, 59.

Goodman, D. W. (1984). *Acc. Chem. Res.* **17**, 194.

Goodman, F. O. and H. Y. Wachman (1976). *Dynamics of Gas Surface Scattering*, (Academic, New York).

Grimmelmann, E. K., J. C. Tully, and E. Helfand (1981). *J. Chem. Phys.* **74**, 5300.

Grunze, M., M., Golze, J. Fuhler, M. Neumann, and E. Schwarz (1984). In *Proceedings of the Eight International Congress on Catalysis* (Verlag-Chemie, Weinheim).

Halstead, D. and S. Holloway (1988). *J. Chem. Phys.* **88**, 7197.

Hamza, A. V. and R. J. Madix (1985). *J. Phys. Chem.* **89**, 5381.

Harris, J. and S. Andersson (1985). *Phys. Rev. Lett.* **55**, 1583.

Harris, J., T. Rahman, and K. Yang (1988a). *Surf. Sci.* **198**, L312.

Harris, J., S. Holloway, T. Rahman, and K. Yang (1988b). *J. Chem. Phys.* **89**, 4427.

Haug, K., G. Wahnstrom, and H. Metiu (1989). *J. Chem. Phys.* **90**, 540.

Hayward, D. O. and A. O. Taylor (1986). *Chem. Phys. Lett.* **124**, 264.

Hayward, D. O. and A. O. Taylor (1988). *Chem. Phys. Lett.* **146**, 221.

Hellsing, B. and H. Metiu (1986). *Chem. Phys. Lett.* **127**, 45.

Ho, W., R. F. Willis, and E. W. Plummer (1980). *Surf. Sci.* **95**, 171.

Holloway, S. (1989). In *Interactions of Atoms and Molecules with Solid Surfaces*, eds. A. Bortolani, N. March, and M. Tosi (Plenum, New York).

Huber, K. P. and G. Herzberg (1979). *Constants of Diatomic Molecules* (Van Nostrand, New York).

Jackson, B. and H. Metiu (1987). *J. Chem. Phys.* **86**, 1026.

Jacquet, R. and W. H. Miller (1985). *J. Phys. Chem.* **89**, 2139.

Kara, A. and A. E. DePristo (1988a). *Surf. Sci.* **193**, 437.

Kara, A. and A. E. DePristo (1988b). *J. Chem. Phys.* **88**, 2033.

Kara, A. and A. E. DePristo (1988c). *J. Chem. Phys.* **88**, 5240.

Kara, A. and A. E. DePristo (1989). To be pubished.

Karikorpi, M., S. Holloway, N. Henriksen, and J. K. Norskov (1987). *Surf. Sci.* **179**, L41.

Kay, B. D. and M. E. Coltrin (1988). *Surf. Sci.* **198**, L375.

Keck, J. C. (1967). *Adv. Chem. Phys.* **13**, 85.

Kirson, Z., R. B. Gerber, A. Nitzan, and M. A. Ratner (1984). *Surf. Sci.* **137**, 527.

Kleyn, A. W., A. C. Luntz, and D. J. Auerbach (1982). *Surf. Sci.* **117**, 33.

Kosloff, R., and C. Cerjan (1984). *J. Chem. Phys.* **1**, 3722.

Kosloff, R., and D. Kosloff (1983a). *J. Chem. Phys.* **79**, 1823.

Kosloff, D., and R. Kosloff (1983b). *J. Com. Phys.* **52**, 35.

Kress, J. D. and A. E. DePristo (1987). *J. Chem. Phys.* **87**, 4700.

Kubiak, G. D., G. O. Sitz, and R. N. Zare (1984). *J. Chem. Phys.* **81**, 6397.

Kummel, A. C., G. O. Sitz, R. N. Zare, and J. C. Tully (1988). *J. Chem. Phys.* **89**, 6947.

Kwong, D. W. J., N. DeLeon, and G. L. Haller (1988). *Chem. Phys. Lett.* **144**, 533.

Ladas, S., R. Imbihl, and G. Ertl (1988). *Surf. Sci.* **197**, 153.

Langhoff, S. R. and C. W., Jr. Bauschlicher (1988). *Ann. Rev. Phys. Chem.* **39**, 181.

Lauderdale, J. G. and D. G. Truhlar (1985). *Surf. Sci.* **164**, 558.

Lauderdale, J. G. and D. G. Truhlar (1986). *J. Chem. Phys.* **84**, 1843.

Lennard-Jones, J. E. (1932). Trans. Faraday Soc. **28**, 333.

Lee, C. Y. (1989). "A Theoretical Investigation of the Dissociative Chemisorption of Co_2 on the Ni(100) Surface" (preprint).

Lee, C. Y. and A. E. DePristo (1986a). *J. Chem. Phys.* **84**, 485.

Lee, C. Y. and A. E. DePristo (1986b). *J. Chem. Phys.* **85**, 4161.

Lee, C. Y. and A. E. DePristo (1987a). *J. Vac. Sci. Tech.* **A5**, 485.

Lee, C. Y. and A. E. DePristo (1987b). *J. Chem. Phys.* **87**, 1401.

Lee, M. B., Q. Y. Yang, and S. T. Ceyer (1987). *J. Chem. Phys.* **87**, 2724.

Lin, J.-H. and B. J. Garrison (1984). *J. Chem. Phys.* **80**, 2904.

Lin, T. H. and G. A. Somorjai (1984). *J. Chem. Phys.* **81**, 704.

Lo, T.-C. and G. Ehrlic (1986). *Surf. Sci.* **179**, L19.

Lo, T.-C. and G. Ehrlic (1988). *Surf. Sci.* **198**, L380.

Lucchese, R. and J. C. Tully (1983). *Surf. Sci.* **137**, 1570.

Lucchese, R. and J. C. Tully (1984). *J. Chem. Phys.* **80**, 3451.

Makri, N. and W. H. Miller (1987a). *Chem. Phys. Lett.* **139**, 10.

Makri, N. and W. H. Miller (1987b). *J. Chem. Phys.* **87**, 5781.

Makri, N. and W. H. Miller (1988). *J. Chem. Phys.* **89**, 2170.

Marcusson, P., Ch. Opitz and H. Muller (1981). *Surf. Sci.* **111**, L657.

Mantell, D. A., K. Kunimori, S. B. Ryali, G. L. haller, and J. B. Fenn (1986). *Surf. Sci.* **172**, 281.

McCreery, J. H. and G. Wolken (1977). *J. Chem. Phys.* **67**, 2551.

McLean, A. D. and R. S. McLean (1981). *At. Data Nucl. Data Tables* **26**, 197.

Miller, W. H., N. C. Handy, and J. E. Adams (1980). *J. Chem. Phys.* **72**, 99.

Miller, W. H., S. D. Schwartz, and J. W. Tromp (1983). *J. Chem. Phys.* **79**, 4889.

Montgomery, J. A., S. L. Holmgren, and D. Chandler (1980). *J. Chem. Phys.* **73**, 3688.

Mowrey, R. C. and D. J. Kouri (1986). *J. Chem. Phys.* **84**, 6466.

Mowrey, R. C., H. F. Bowen, and D. J. Kouri (1987). *J. Chem. Phys.* **86**, 2441.

Muller, J. E. (1987). *Phys. Rev. Lett.* **59**, 2943.

Newns, D. M. (1986). *Surf. Sci.* **171**, 600.

NoorBatcha, I., L. M. Raff, and D. L. Thompsay (1985). *J. Chem. Phys.* **83**, 1382.

Norskov, J. K. and F. Besenbacher (1987). *J. of the Less Commn Metals* **130**, 475.

Norskov, J. K. and P. Stoltze (1987). *Surf. Sci.* **189**, 91.

Norskov, J. K., A. Houmoller, P. Johansson, and B. I. Lundqvist (1981). *Phys. Rev. Lett.* **46**, 257.

Panas, I., P. Siegbahn, and U. Wahlgren (1987). *Chem. Phys.* **112**, 325.

Panas, I., J. Schule, P. Siegbahn, and U. Wahlgren (1988). *Chem. Phys. Lett.* **149**. 265.

Pechukas, P. (1976). In *Dynamics of Molecular Collisions, Part B*, ed. W. H. Miller, (Plenum, New York).

Presson, B. N. J and M. Persson (1980). *Solid State Commun.* **36**, 175.

Pfnür, H. E., C. T. Rettner, J. Lee, R. J. Madix, and D. J. Auerbach (1986). *J. Chem. Phys.* **85**, 7452.

Polesema, B., L. K. Verheij, and G. Comsa (1985). *Surf. Sci.* **152/153**, 496.

Puska, M. J., R. M. Nieminen, and I. Manninen (1981). *Phys. Rev.* **B24**, 3037.

Raff, L. M., I. NoorBatcha, and D. L. Thompson (1986). *J. Chem. Phys.* **85**, 3081.

Rendulic, K. D. (1988). *Appl. Phys.* **A47**, 55.

Rendulic, K. D., G. Anger, and A. Winkler (1989). *Surface Sci.* **208**, 404.

Rettner, C. T. (1988). *Vacuum* **38**, 295.

Rettner, C. T. and H. Stein (1987). *Phys. Rev. Lett.* **59**, 2768.

Rettner, C. T., H. E. Pfnür, and D. J. Auerbach (1985a). *Phys. Rev. Lett.* **54**, 2716.

Rettner, C. T., F. Fabre, J. Kimman, and D. J. Auerbach (1985b). *Phys. Rev. Lett.* **55**, 1904.

Rettner, C. T., H. E. Pfnür, and D. J. Auerbach (1986). *J. Chem. Phys.* **84**, 4136.

Rettner, C. T., H. Stein, and E. K. Schweizer (1988). *J. Chem. Phys.* **89**, 3337.

Rice, B. M., L. M. Raff, and D. L. Thompson (1988). *J. Chem. Phys.* **88**, 7221.

Riley, M. E., M. E. Coltrin, and D. J. Diestler (1988). *J. Chem. Phys.* **88**, 5934.

Robota, H. J., W. Vielhaber, M. C. Liu, J. Segner, and G. Etrl (1985). *Surf. Sci.* **155**, 101.

Ron, S. and M. Baer (1988). *Chem. Phys. Lett.* **146**, 265.

Ron, S., Z. B. Alfassi, and M. Baer (1987). *Chem. Phys. Lett.* **117**, 39.

Serri, J. A., J. C. Tully, and M. J. Cardillo (1983). *J. Chem. Phys.* **79**, 1530.

Shima, Y. and M. Baer (1985). *J. Chem. Phys.* **83**, 5250.

Shustorovich, E. M. (1985). *Surf. Sci.* **150**, L115.

Shustorovich, E. M. (1986). *Surf. Sci.* **175**, 561.

Shustorovich, E. M. (1987). *Surf. Sci.* **181**, L205.

Shustorovich, E. M. (1988). *Acc. Chem. Res.* **21**, 183.

Siegbahn, P., M. R. A. Blomberg, and C. W. Bauschlicher (1984). *J. Chem. Phys.* **81**, 2103.

Siegbahn, P., M. R. A. Blomberg, I. Panas, and U. Wahlgren (1988). *Theor. Chim. Acta.* **75**, 143.

Sobyanin, V. and V. P. Zhdanov (1987). *Suf. Sci.* **181**, L163.

Somorjai, G. A (1981). *Chemistry in Two Dimensions* (Cornell University Press, Ithaca, New York).

Steinruck, H. P., M. P. D'Evelyn, and R. J. Madix (1986). *Surf. Sci.* **172**, L561.

Stoltze, P. (1987). *Phys. Scripta* **36**, 824.

Suzuki, M. (1986). In *Quantum Monte Carlo Methods*, ed. Suzuki, M. (Springer-Verlag, New York).

Tann, P. W. and L. D. Schmidt (1971). *Surf. Sci.* **26**, 286.

Tantardini, G. F. and M. Simonetta (1981). *Surf. Sci.* **105**, 517.

Tantardini, G. F. and M. Simonetta (1982). *Chem. Phys. Lett.* **87**, 420.

Tromp, J. W. and W. H. Miler (1986). *J. Phys. Chem.* **90**, 3482.

Truhlar, D. G., A. D. Isaacson, and B. C. Garrett (1986). In *Theory of Chemical Reaction Dynamics*, ed. Baer, M., Vol. 4, 1 (CRC, Boca Raton, Florida).

Truong, T. N., D. G. Truhlar and B. C. Garrett (1989a). *J. Phys. Chem.* **93**, 8227.

Truong, T. N., G. C. Hancock and D. G. Truhlar (1989b). *Surf. Sci.* **214**, 523.

Tully, J. C. (1980a). *J. Chem. Phys.* **73**, 1975.

Tully, J. C. (1980b). *J. Chem. Phys.* **73**, 6333.

Tully, J. C. (1981). *Acc. Chem. Res.* **14**, 188.

Upton, T. H. and W. A. Goddard (1981). In *Chemistry and Physics of Solid Surfaces, Vol. III*, eds. Vandelow, R. and England, W. (CRC Press, Boca Raton, Florida).

Upton, T. H., P. Stevens, and R. J. Madix, (1988). *J. Chem. Phys.* **88**, 3988.

Valone, S. M., A. F. Voter, and J. D. Doll (1985). *Surf. Sci.* **155**, 687.

Van Willigen, W. (1968). *Phys Lett.* **A28**, 80.

Wahnstrom, G. and H. Metiu (1987) *Chem. Phys. Lett.* **134**, 531.

Wahnstrom, G. and H. Metiu (1988). *J. Phys. Chem.* **92**, 3240.

Wahnstrom, G., K, Haug, and H. Metiu (1988). *Chem. Phys. Lett.* **148**, 158.

Whitman, L. J., C. E. Bartosch, and W. Ho (1986). *J. Chem. Phys.* **85**, 3688.

Wolken, G. Jr. (1976). In *Dynamics of Molecular Collisions Part A*, ed. Myller, W. H. (Plenum, New York).

Xu, G.Q., S. L. Bernasek, and J. C. Tully (1988). *J. Chem. Phys.* **88**, 3376.

Yamamoto, T. (1960). *J. Chem. Phys.* **33**, 281.

Yang, H. and J. L. Whitten (1988). *J. Chem. Phys.* **89**, 5329.

Yates, J. T., R. Klein, and T. E. Madey (1976). *Surf. Sci.* **58**, 469.

Zangwill, A. (1988). *Physics at Surfaces* (Cambridge University Press, New Rochelle, New York).

Zwanzig, R. W. (1960). *J. Chem. Phys.* **32**, 1173.

PROBING ELECTROCHEMICAL INTERFACES WITH X-RAYS

HÉCTOR D. ABRUÑA

Department of Chemistry, Baker Laboratory
Cornell University
Ithaca, New York

CONTENTS

I. X-RAYS AND THEIR GENERATION

X-rays represent that region of the electromagnetic spectrum whose wavelengths range from about 0.01 to 100 angstroms. Because of their very short wavelengths and high penetration, x-rays are powerful probes of atomic structure. X-rays are traditionally produced by impinging an electron beam (at energies from about 20 to 50 keV) onto a target material such as copper, molybdenum, or tungsten. The sudden deceleration of the electron beam by the target material gives rise to a broad spectrum of emission termed *bremsstrahlung*. The wavelength in angstroms of the emitted x-rays is given by

$$\lambda(\text{Å}) = hc/E = 12{,}400/V \tag{1}$$

where lambda is in angstroms, and V is the accelerating voltage. The minimum wavelength of emission is obtained when all of the electron energy is converted to an x-ray photon. The intensity and wavelength distribution of this bremsstrahlung are both a function of the accelerating voltage, the current, and the target material. As the accelerating voltage is increased or a higher atomic number element is used as a target material, the emission distribution shifts to higher energies.

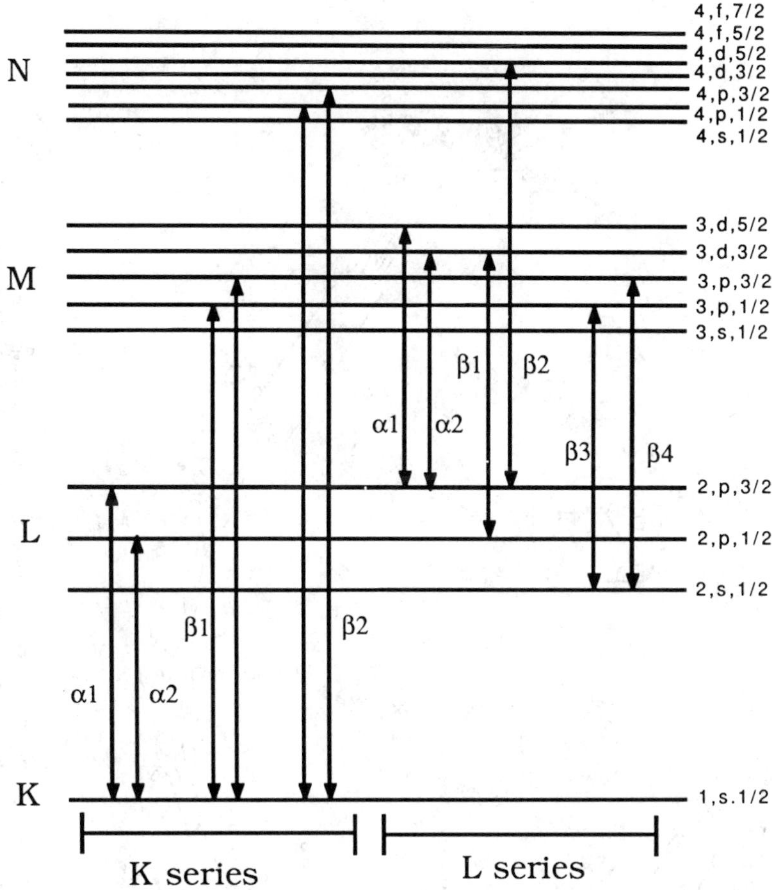

Figure 1. Energy-level diagram depicting part of the K and L series characteristic lines.

When the accelerating voltage reaches a threshold value (dependent on the nature of the target material), core electrons from the target material can be ejected. These vacancies are quickly filled by electrons in upper levels, resulting in the emission of x-ray photons of characteristic energies that depend on the natue of the target. The energies and intensities of characteristic lines depend on the nature of the core hold generated (e.g., K, L, or M shell vacancy) as well as the level from which the electron that fills the vacancy originates. Figure 1 shows a schematic of some of the more important x-ray emission lines. These characteristic lines are much more intense than the bremsstrahlung emission and are superimposed on the same as very sharp

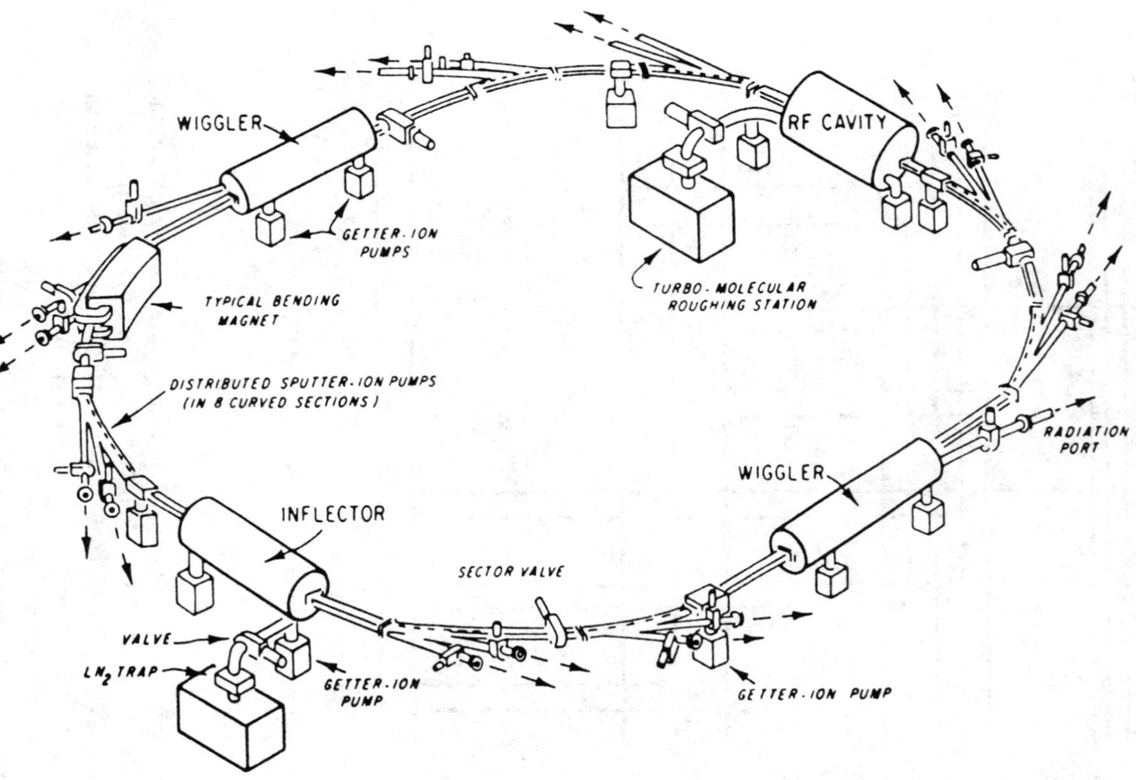

Figure 2. Schematic of a storage ring for use as a source of synchrotron radiation. (From Winick, H., and Doniach, S., eds., Synchrotron Radiation Research, Plenum, New York, 1980 with permission.)

emissions. The main difficulty with conventional x-ray sources is their low intensity, especially away from characteristic lines.

An alternative and the most generally employed source of x-rays for EXAFS experiments is radiation from synchroton sources[1] based on electron (or positron) storage rings.

II. SYNCHROTRON RADIATION AND ITS ORIGIN

The development of synchrotron radiation sources based on electron (or positron) storage rings has dramatically influenced the use of x-ray based techniques.[1] In its simplest form, a synchrotron source consists of a beam of electrons (positrons) orbiting at relativistic speeds in a storage ring (Fig. 2). The path of the electrons (positrons) is controlled by magnets, and they are maintained in orbit by constantly supplying energy through high-power RF generators. Synchrotron sources provide a continuum of photon energies at intensities that can be from 10^3 to 10^6 higher than those obtained with x-ray tubes, thus dramatically decreasing data acquisition times as well as making other experiments feasible.

The most attractive features of synchrotron radiation include:[2] (1) high intensity; (2) broad spectral range; (3) high polarization; (4) natural collimation; (5) small source-spot size; (6) stability; and (7) pulsed time structure.

Whenever a charged particle undergoes acceleration, electromagnetic waves are generated. An electron (or positron) in a circular orbit experiences an acceleration towards the center of the orbit and as a result emits radiation in an axis perpendicular to the motion. At relativistic speeds ($v \approx c$), the radiation pattern is highly peaked[3] and one can think of an orbiting searchlight in the shape of a thin slab as a good approximation to the radiation pattern. The cone of emitted radiation is characterized by an emission angle $1/\gamma$ where γ is the electron energy divided by its rest mass. Since the electron energy is typically of the order of $1\text{–}7 \times 10^9$ eV, whereas the rest mass of an electron is 5.1×10^5 eV, this results in a natural collimation effect which gives rise to very high fluxes on small targets.

The spectral distribution of synchrotron radiation is continuous and depends on a number of factors. Two that are particularly important are the electron energy (expressed in GeV; 10^9 eV) and the bending radius R (in meters) of the orbit. These are related by the critical energy E_c given by:

$$E_c(\text{keV}) = 2.21\text{E}^3/R$$

The critical energy represents the midpoint of the radiated power. That is, half of the radiated power is above and half below this energy. In general, useful fluxes are obtained at energies up to four times the critical energy, although in this region the output decreases dramatically. Figure 3 presents some flux curves for the Cornell High Energy Synchrotron Source (CHESS) operated at various electron beam energies.

The fact that the critical energy is inversely proportional to the bending radius is used in so-called insertion devices such as wiggler and undulator magnets and wavelength shifters.[4] These are magnetic structures that make the electron beam undergo sharp serpentine motions, thereby giving rise to a very short radius of curvature. The differences between these various insertion devices lies in the number of periods in the magnetic structure, the angular excursion of the electron beam (relative to $1/\gamma$) and the magnetic

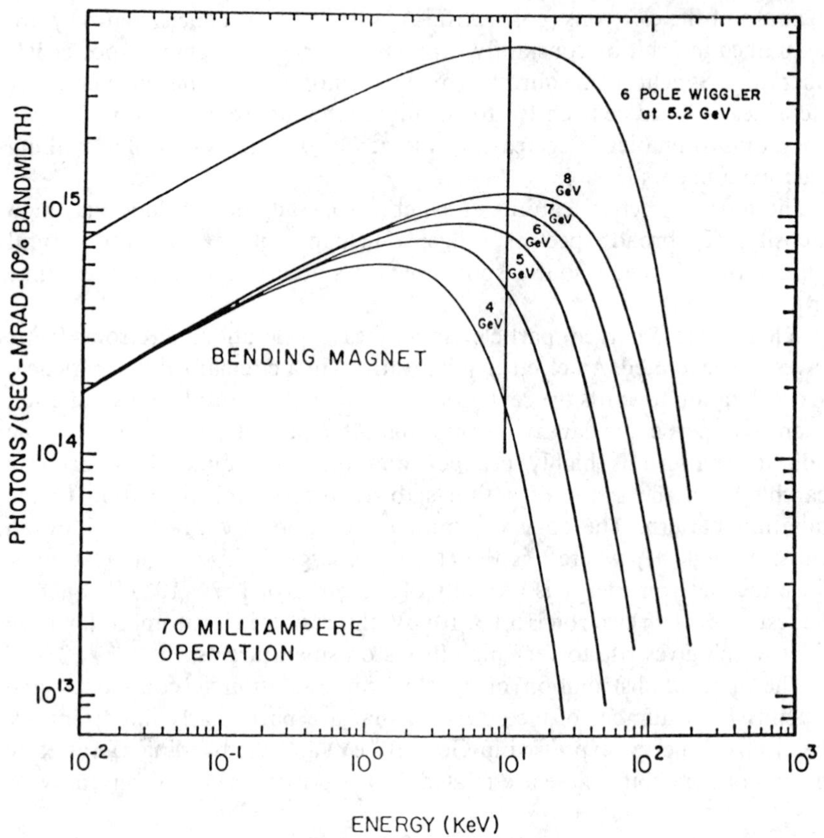

Figure 3. Photon flux as a function of energy for the Cornell High Energy Synchrotron Source (CHESS) operated at various accelerating voltages. The topmost curve is the radiation profile from the 6-pole wiggler magnet. (Figure courtesy of the Laboratory for Nuclear Studies at Cornell University.)

field. A wiggler is a magnetic device with a small number of periods in which the angular excursion is considerably greater than $1/\gamma$. The output from a wiggler is, to a good approximation, equal to that from a bending magnet with the same field multiplied by a factor equal to the number of poles. The topmost curve of Fig. 3 shows the output from the CHESS wiggler at an electron beam energy of 5.2 GeV. An undulator is a structure with many periods in which the angular excursion of the electron beam is smaller or of the same order as $1/\gamma$. Because of interference effects, the output from an undulator is highly peaked rather than continuous, and, in the limiting case, the output intensity is proportional to the square of the number of poles rather than directly proportional as for wigglers. Thus, at the characteristic peaks, undulators are exceedingly intense sources. A wavelength shifter is a device with a few poles but with a very high magnetic field. Its basic purpose is to shift the spectrum to higher energies.

Another very important property of synchrotron radiation is its very high degree of polarization. The radiation is predominantly polarized with the electric field vector parallel to the acceleration vector. Thus, in the plane of the orbit, the radiation is 100% plane-polarized. Elliptical polarization can be obtained by going away from the plane; however, intensities also decrease significantly.

The pulsed time structure, useful for kinetic studies, arises from the fact that in a storage ring the electrons are orbiting in groups or bunches. The specific beam energy, the number of bunches, and the circumference of the storage ring dictate the exact time structure. Although the time structure has not been employed in the study of electrochemical systems, it has been employed in time-resolved diffraction studies.

III. INTRODUCTION TO EXAFS AND X-RAY ABSORPTION SPECTROSCOPY

EXAFS or extended x-ray absorption fine structure refers to the modulations in the x-ray absorption coefficient beyond an absorption edge.[5] Such modulations can extend up to about 1000 eV beyond the edge and have a magnitude of typically less than 15% of the edge jump.

An x-ray absorption spectrum presents the absorbance of a sample (typically expressed as an absorption coefficient μ) as a function of the incident photon energy (Fig. 4). As the incident photon energy increases, there is a monotonic decrease (proportional to E^3) in the absorption coefficient. However, when the incident x-ray energy is enough to photoionize a core level electron, there is an abrupt increase in the absorption coefficient and this is termed an *absorption edge*. There are absorption edges that correspond to the various atomic shells and subshells so that a given atom will have

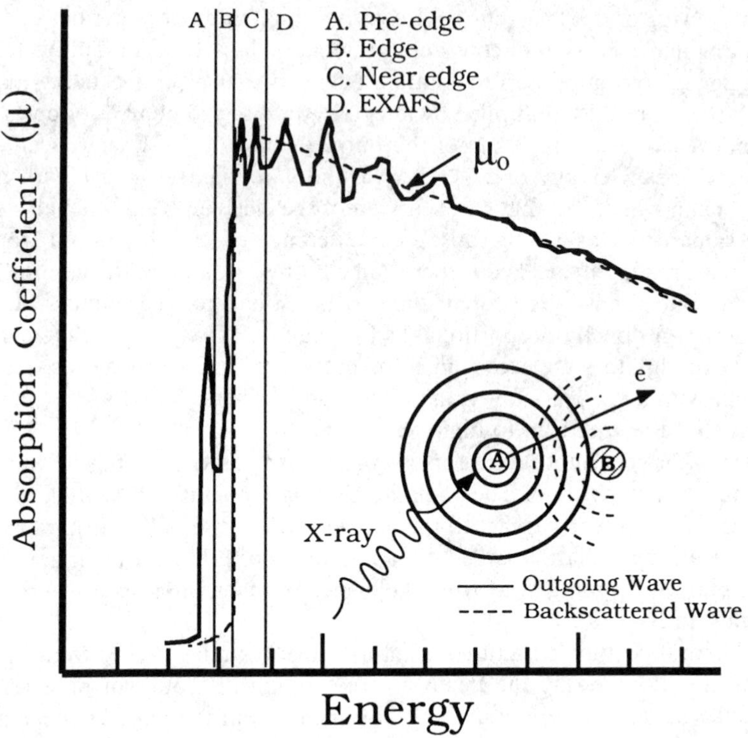

Figure 4. Depiction of the various regions in an x-ray absorption spectrum. Inset: Interference between outgoing (solid line) and backscattered (dashed line) waves.

one K absorption edge, three L edges, five M edges, and so forth, with the energies decreasing in the expected order K > L > M (see Fig. 1).

As the scan continues to higher energies beyond the edge, two different situations arise depending on whether or not the species under investigation has near neighbors (typically at 5 Å or closer). If there are no near neighbors, the absorption coefficient will again decrease in a monotonic fashion (proportional to E^3; dashed line labeled μ_0 in Fig. 4) until its next absorption edge or that of another element present in the sample is encountered.

In the presence of one or more near neighbors, there will be modulations in the absorption coefficient out to energies about 1000 eV beyond the edge. The modulations present at energies from about 40 eV to 1000 eV beyond the edge are termed EXAFS.

The phenomenon of EXAFS has been known since the 1930s through the work of Kronig,[6] who stated that the oscillations are due to the modification of the final state of the photoelectron by near neighbors. The absorption coefficient is a measure of the probability that a given x-ray photon will be absorbed and therefore depends on the initial and final states of the electon. The initial state corresponds to the localized electron in a core level. The final state is represented by the photoionized electron, which can be visualized as an outgoing photoelectron wave that originates at the center of the absorbing atom. In the presence of near neighbors, this photoelectron wave can be backscattered (inset in Fig. 4) so that the final state is given by the sum of the outgoing and backscattered waves. It is the interference between the outgoing and backscattered waves that gives rise to the EXAFS oscillations.

To a good approximation, the frequency of the EXAFS oscillations will depend on the distance between the absorber and its near neighbors, whereas the amplitude of the oscillations will depend on the numbers and type of neighbors as well as their distance from the absorber. From an analysis of the EXAFS, one can obtain information on near-neighbor distances, numbers, and type. A further advantage of EXAFS is that it can be applied to all forms of matter—solids, liquids, and gases—and that in the case of solids, single crystals are not required since EXAFS is only sensitive to short-range order (about 5 Å). In addition, one can focus on the environment around a particular element by employing x-ray energies around an absorption edge of the element of interest without interferences from other elements in the sample, except for those with very similar atomic number.

The simple description of EXAFS given above is based on the so-called single-electron, single-scattering formalism.[7] This treatment assumes that for sufficiently high energies of the photoelectrons, one can make the plane-wave approximation and in addition only single-backscattering events will be important. This is the reason why the EXAFS is typically considered for energies higher than 40 eV beyond the edge since in this energy region the above approximations hold well.

In addition to the EXAFS region, Figure 4 depicts three other regions: the pre-edge, edge, and near-edge regions, respectively. Below or near the edge, there can be absorption peaks due to excitations to bound states that can be so intense so as to dominate the edge region. Because the transitions are to localized states, the pre-edge region is rich in information pertaining to the energetic location of orbitals, site symmetry, and electronic configuration.

The position of the edge (in terms of energy) contains information concerning the effective charge of the absorbing atom. Thus, its location can be correlated with the oxidation state of the absorber in a way that is analogous to XPS measurements. Such shifts can be very diagnostic in the assignment of oxidation states and can be especially important in applied potential-dependent studies of electrochemical systems.

In the near-edge region, generally termed XANES (x-ray absorption near edge structure) of NEXAFS (near-edge x-ray absorption fine structure), when UV or soft x-rays are employed, the photoelectron wave has very small momentum and as a result the plane wave as well as the single-electron, single-scattering approximations are no longer valid. Instead one must consider a spherical photoelectron wave as well as the effects of multiple scattering. Because of multiple scattering, the photoelectron wave can sample much of the environment around the absorber, making this region of the spectrum very rich in structural information. However, the theoretical modeling is quite complex.

IV. THEORY OF EXAFS

The theoretical description of EXAFS given below is based on the single-scattering, short-range order formalism mentioned previously. The EXAFS can be expressed as the normalized modulation of the absorption coefficient as a function of energy:

$$\chi(E) = [\mu(E) - \mu_0(E)]/\mu_0(E) \tag{2}$$

Here $\mu(E)$ is the total absorption at energy E and $\mu_0(E)$ is the smooth atom-like absorption coefficient (dashed line in Fig. 4). In order to be able to extract structural information from the EXAFS, the data are transformed to wave vector (k) by using the relationship

$$k = \{[2m(h\nu - E_0)]/\hbar^2\}^{1/2} \tag{3}$$

where E_0 is defined as the threshold energy, which is typically close but not necessarily congruent with the energy at the absorption edge.

In wave-vector form, the EXAFS can be expressed as a summation over the various coordination (near neighbor) shells and is given by

$$\chi(k) = \sum_j \frac{1}{kr_j^2} N_j F_j(k) S_i(k) \exp^{-2\sigma_j^2 k^2} \exp^{-2r_j/\lambda(k)} \sin(2kr_j + \phi_j(k)) \tag{4}$$

where k represents the wave vector, r_j is the absorber–backscatterer distance, and N_j is the number of scatterers of type j with backscattering amplitude $F_j(k)$. The product of these last two terms gives the maximum amplitude. There are also amplitude reduction factors. $S_i(k)$ takes into account many-body effects such as electron shake-up and shake-off processes, whereas the term $\exp^{-2\sigma j^2 k^2}$ (known as the Debye–Waller factor) accounts for thermal vibration and static disorder. Finally, the term $\exp^{-2kj/\lambda(k)}$ takes into account inelastic scattering effects where $\lambda(k)$ is the mean free path of the photoelectron. $\phi_j(k)$ represents the total phase shift of the photoelectron.

Equation (4) can be divided into two main terms that correspond to amplitude and frequency, respectively.

A. Amplitude Term

The amplitude term

$$\frac{1}{kr_j^2} N_j F_j(k) S_i(k) \exp^{-2\sigma_j^2 k^2} \exp^{-2r_j/\lambda(k)} \tag{5}$$

can be subdivided into two main components: a maximum amplitude term and an amplitude reduction factor.

For a given shell, the maximum amplitude is given by the product of the number (N_j) of the j type of scatterer atom times its backscattering amplitude $F_j(k)$. This maximum amplitude is then reduced by a series of amplitude reduction factors which are considered below.

1. Many-Body Effects

The $S_i(k)$ term takes into account amplitude reduction due to many-body effects and includes losses in the photoelectron energy due to electron shake-up (excitation to upper localized levels of other electrons in the absorber), shake-off (ionization of low-binding-energy electrons in the absorber), and plasmon excitation processes. Such energy losses change the energy of the outgoing photoelectron wave so that it no longer has the appropriate wave vector for constructive interference, thus resulting in a reduction of the amplitude.

2. Thermal Vibrations and Static Disorder

Photoionization (and therefore EXAFS) takes place on a time scale that is very short relative to atomic motions, so the experiment samples an average configuration of the neighbors around the absorber. Thus, one needs to consider the effects of thermal vibration and static disorder, both of which will have the effect of reducing the EXAFS amplitude. These effects are considered in the so-called Debye–Waller factor which represents the mean-square relative displacement along the absorber–backscatterer direction and is given by

$$\exp^{-2\sigma_j^2 k^2} \tag{6}$$

This can be separated into static disorder and thermal vibrational components:

$$\sigma_j^2 = \sigma_{\text{vib}}^2 + \sigma_{\text{stat}}^2 \tag{7}$$

It is generally assumed that the disorder can be represented by a symmetric gaussian-type pair distribution function and that the thermal vibration will be harmonic in nature. Experimentally, one can only measure a total sigma. However, the two contributions can be separated by performing a temperature-dependence study of sigma, since the vibrational component will be inversely proportional to temperature whereas the static disorder term will be independent of temperature. Alternatively, the two contributions can be determined by having an a-priori knowledge of σ_{vib} from vibrational spectroscopy. Whereas there is little one can do to overcome the effects of static disorder, the effects of thermal vibration can be significantly decreased by performing experiments at low temperatures, and in fact many solid samples are typically run at liquid nitrogen temperatures in order to minimize such effects. In general, failure to consider the effects of thermal vibration and static disorder can result in large errors in the determination of coordination numbers (number of neighbors) and interatomic distances.[8,9]

3. *Inelastic Losses*

Photoelectrons that experience inelastic losses will not have the appropriate wave vector to contribute to the interference process. Such losses are taken into account by an exponential damping factor:

$$\exp^{-2r_j/\lambda(k)} \tag{8}$$

where r_j is the interatomic distance between the absorber and backscatterer and $\lambda(k)$ is the electron mean free path. This damping term limits the range of photoelectrons in the energy region of interest, and this is in part responsible for the short-range description of the EXAFS phenomenon.

B. Oscillatory Term

The frequency term or oscillatory part of the EXAFS takes into account the relative phases (i.e., phase shifts) between the outgoing and backscattered waves as well as the interatomic distance between absorber and scatterer. The outgoing photoelectron will experience the absorbing atom's phase shift $\delta_i(k)$ on its outward trajectory, the near neighbor's phase shift $\alpha_s(k)$ upon scattering, and the absorbing atom's phase shift once again upon returning. There is in addition a $2kr$ term which represents twice the interatomic distance between absorber and scatterer. Thus, the oscillatory part of the EXAFS is given by

$$\sin[2kr + 2\delta_i(k) + \alpha_s(k)] \tag{9}$$

Since the accuracy of the determination of interatomic distances depends largely on the appropriate determination of the relative phases, a great deal of attention has been given to this aspect. The problem arises from the fact that when the outgoing photoelectron wave is backscattered by near neighbors, it is the neighbor's electron cloud and not its nucleus that is largely responsible for the scattering. As a result, one needs to correct for this effect (through the use of phase shifts) since it is the internuclear distance that is the desired parameter. The correction can be achieved by ab initio calculation of the phases involved, or alternatively they can be determined experimentally through the use of model compounds. A more thorough discussion of phase correction will be given further on.

C. Data Analysis

The purpose behind the analysis of EXAFS data is to be able to extract information related to interatomic distances, numbers and types of backscattering neighbors. In order to accomplish this, there are a number of steps involved in the data analysis. These include:

1. Background subtraction and normalization
2. Conversion to wave-vector form
3. *K* weighing
4. Fourier transforming and filtering
5. Fitting for phase
6. Fitting for amplitude

1. Background Subtraction and Normalization

The first step in the analysis is the background subtraction. This consists of separating the modulation in the absorption coefficient from the smooth atom-like absorption (that is, the absorption that would be observed for an isolated atom). However, since the latter is generally not available, it is usually assumed that the smooth part of $\mu(E)$ at energies well beyond the absorption edge is a good approximation to $\mu_0(E)$. The observed EXAFS oscillations need to be normalized to a single atom value, and this is generally done by normalizing the data to the edge jump intensity.

2. Conversion to Wave-Vector Form

In order to extract structural information, the EXAFS must be expressed in terms of wave vector k. However, this requires a value for the threshold

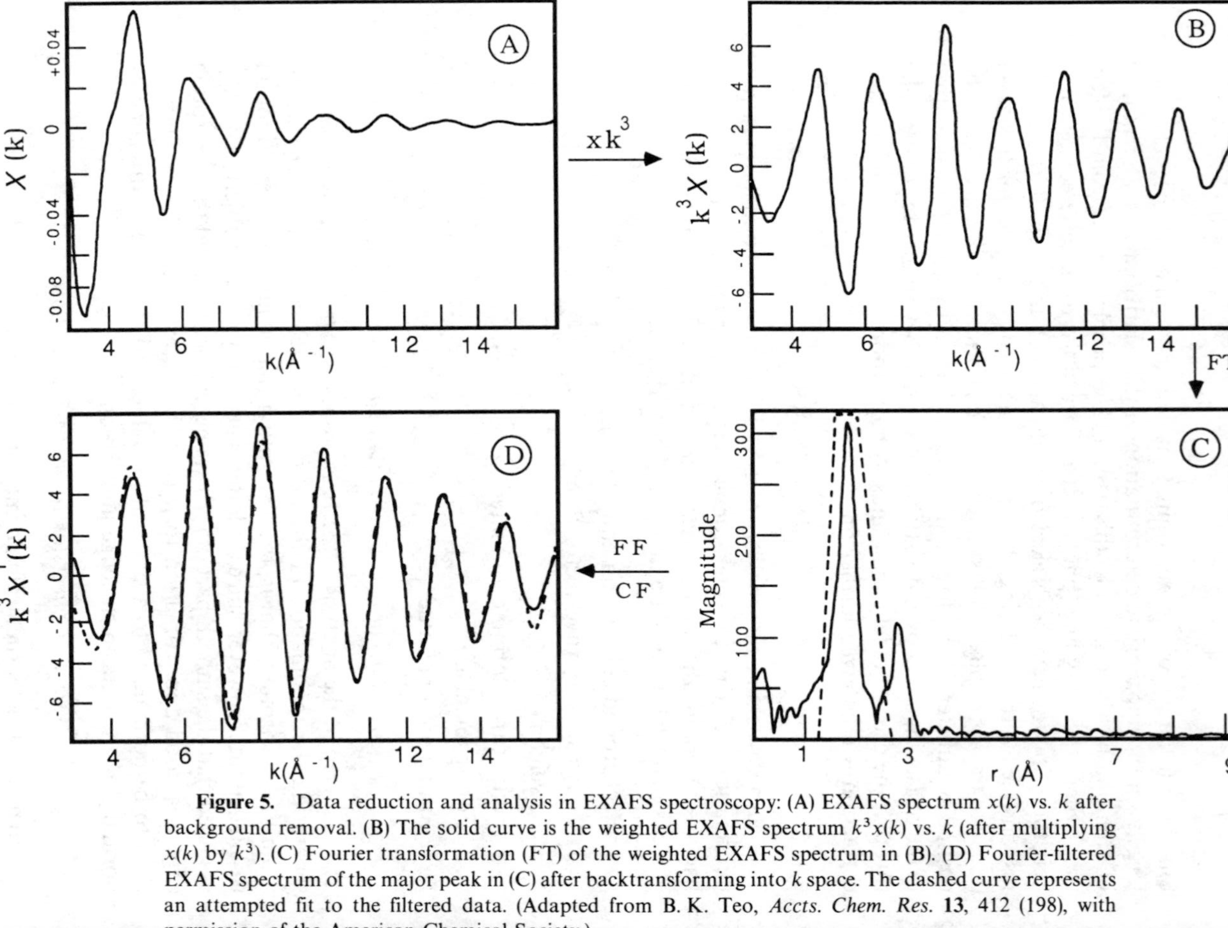

Figure 5. Data reduction and analysis in EXAFS spectroscopy: (A) EXAFS spectrum $x(k)$ vs. k after background removal. (B) The solid curve is the weighted EXAFS spectrum $k^3x(k)$ vs. k (after multiplying $x(k)$ by k^3). (C) Fourier transformation (FT) of the weighted EXAFS spectrum in (B). (D) Fourier-filtered EXAFS spectrum of the major peak in (C) after backtransforming into k space. The dashed curve represents an attempted fit to the filtered data. (Adapted from B. K. Teo, *Accts. Chem. Res.* **13**, 412 (198), with permission of the American Chemical Society.)

energy E_0. The choice is important because of its effect on the phase of the EXAFS oscillations, especially at low k values. The difficulty in determining E_0 arises from the fact that there is no way of identifying an edge feature with E_0. A procedure proposed by Beni and Lee[10] is to have E_0 as an adjustable parameter in the data analysis, and its value is changed until the observed phase shifts are in good agreement with theoretical values. When good model compounds are available (vide infra), the use of a fixed value for E_0 works well.[11] However, in many cases it is difficult to assess a priori whether a given material is a good model compound. Figure 5A depicts a generic EXAFS data set after background subtraction, normalization, and conversion to wave-vector form.

3. *k Weighing*

Once the data has been transformed to wave-vector form, it is generally multiplied by some power of k; typically k^2 or k^3. Such a factor cancels the $1/k$ factor in Eq. (4) as well as the $1/k^2$ dependence of the backscattering amplitude at large values of k. k weighing is important because it prevents the large amplitude oscillations (typically present at low k) from dominating over the smaller ones (typically at high k). This is critical since the determination of interatomic distances depends on the frequency and not the amplitude of the oscillations. Figure 5B depicts multiplication of the data in Fig. 5A by k^3.

4. *Fourier Transforming and Filtering*

Examination of the EXAFS formulation in wave-vector form reveals that it consists of a sum of sinusoids with phase and amplitude. Sayers, Lytle, and Stern[7a] were the first to recognize that a Fourier transform of the EXAFS from wave-vector form yields a function that is qualitatively similar to a radial distribution function and is given by

$$\phi(r) = \frac{1}{\sqrt{2\pi}} \int_{k_{\min}}^{k_{\max}} k^n \chi(k) \exp^{(2ikr)}\, dk \tag{10}$$

Such a function exhibits peaks (Fig. 5C) that correspond to interatomic distances between the central atom and the individual coordination shells but are shifted to smaller values (recall the distance correction mentioned above). This finding was a major breakthrough in the analysis of EXAFS data since it allowed ready visualization. However, because of the shift to shorter distances and the effects of truncation of the Fourier transform, such an approach is generally not employed for accurate distance determination. However it allows for the use of Fourier filtering techniques which make possible the isolation of individual coordination shells. For example, the dotted line Fig. 5C represents a Fourier filtering window that isolates the first coordination shell. After Fourier filtering, the data is back-transformed to k space (Fig. 5D) where it is fitted for amplitude and phase. The basic principle behind the curve-fitting analysis is to employ a parametrized function that will model the observed EXAFS and the various parameters are adjusted until the fit is optimized.

5. *Fitting for Phase*

Accurate distance determinations depend critically on the accurate determination of phase shifts. There are two general approaches to this problem: theoretical and empirical determination. The main approaches to the theoretical calculation of phase shifts are based on the Hartree–Fock (HF)[7e,12] and Hartree–Fock–Slater (HFS)[13,14] methods. However, both of these are too involved for general use. Teo and Lee[15] used the theoretical approach of Lee and Beni[7f] to calculate and tabulate theoretical phase shifts for the majority of elements. Use of these theoretical phase shifts requires the use of an adjustable E_0 in the data analysis (vide supra). Most recently, McKale and co-workers[16] performed ab initio calculations of amplitude and phase functions using a curved wave formalism for the range of k values $2 \leqslant k \leqslant 20$.

The second, and more commonly employed, approach is the empirical one based on the use of model compounds and the concept of phase transferability. This approach consists of employing a compound of known structure (e.g., by x-ray diffraction) that has the same absorber/backscatterer combination as that of the material of interest. The EXAFS spectrum of the known compound is obtained and the oscillatory part of the EXAFS is fitted. Since r is known in this case, the phase shift can be determined. Implicit in this treatment is the applicability of phase transferability, which states that for a given absorber/backscatterer combination, the phase shifts can be transferred to any compound with the same combination of atoms without regards to chemical effects such as ionicity or covalency of the bonds involved. This is based on the idea that at sufficiently high kinetic energies for the photoelectron (e.g., about 50 eV above threshold), the EXAFS scattering processes are largely dominated by core electrons and thus the measured phase shifts are insensitive to chemical effects. Thus, determination of the phase shift for an absorber/backscatterer pair in a system of known r allows for the determination of the distance in an unknown having the same atom pair. This was conclusively demonstrated by Citrin, Eisenberger and Kincaid[17] on a study of germanium compounds.

With good quality data and appropriately determined phase shifts, distances determined by EXAFS are typically good to ± 0.01 Å and sometimes better in favorable cases.

6. *Fitting for Amplitude*

Fitting for amplitude is employed in order to determine the types and numbers of backscattering atoms around a given absorber. In the absence of any information as to the probable nature of the backscatterer, identification is difficult, especially among atoms that have similar atomic number, because backscattering amplitudes are not a very strong function of atomic number. However, for the case of a heavy atom backscatterer, there is typically a resonance in the backscattering amplitude so that differentiation between light and heavy backscatterers can be readily made.

It should be mentioned that when a peak from a Fourier transform is filtered and back-transformed to k space, the envelope represents the backscattering amplitude for the near neighbor involved and this can be used in the identification of the backscattering atom.

If the identity of the backscatterer is known, the interest is in determining the number of near neighbors. In this case, one needs to compare the amplitude of the EXAFS of the material of interest (unknown) to that for a compound of known coordination number and structure. However, unlike transferability of phase, which is generally regarded as an excellent approximation, the transferability of amplitude is not. This is because there are many factors that affect the amplitude, and except for the case of model compounds with structure very similar to that of the unknown, these will not necessarily (and often will not) be the same. As a result, determination of coordination numbers (near neighbors) is usually no better than $\pm 20\%$.

V. SURFACE EXAFS AND POLARIZATION STUDIES

EXAFS is fundamentally a bulk technique due to the significant penetration of high-energy x-rays. The technique can be surface-sensitive if one knows a priori that the specific element of interest is present only at the surface. Alternatively, one can employ detection techniques or experimental geometries such that the detected signal arises predominantly from the surface or near-surface region.[18] Such techniques include electron detection (e.g. Auger, partial or total electron yield) and operating at angles of incidence that are below the critical angle of the particular material so that only an evanescent wave penetrates the substrate, resulting in a very shallow sampling depth. These aspects will be discussed further in the experimental section. In addition, there have been a number of reviews of surface EXAFS, with Citrin's being the most comprehensive.[18e]

For studies on single crystals, surface EXAFS offers an additional experimental handle and this refers to the polarization dependence of the signal.[19] As mentioned previously, synchroton radiation is highly polarized with the plane of polarization lying in the plane of orbit. Since only those bonds whose interatomic vector has a projection on the plane of polarizaton of the x-ray will contribute to the observed EXAFS, polarization dependence studies can provide a wealth of information on the geometric disposition of scatterers. Such studies are also extremely valuable in surface EXAFS studies of adsorbed layers since they allow for the determination of the site of adsorption as well as the structure and geometry of the adsorbed layer.[20,21] This is a significant enhancement over the information content of a conventional EXAFS experiment. Polarization-dependent surface EXAFS measurements have provided some of the best-defined characterizations of adsorbate structures.

VI. EXPERIMENTAL ASPECTS

A. Synchrotron Sources

There are a number of eperimental factors to be considered when performing an in situ surface EXAFS experiment in an electrochemical system. First and foremost is having access to a synchrotron source (for the reasons previously mentioned) with significant flux in the hard x-ray region (above 10 keV). In the United States such facilities include:

1. Cornell High Energy Synchrotron Source (CHESS)
2. Stanford Synchrotron Radiation Laboratory (SSRL)
3. National Synchrotron Light Source (NSLS) at Brookhaven National Laboratory
4. Advanced Photon Source (APS) to be built at the Argonne National Laboratory

In addition to having access to a synchrotron, one needs to pay close attention to detection schemes and the design of specialized equipment. These aspects are discussed below.

B. Detection

The mode of detection is usually dictated by the concentration of the species of interest, the nature of the sample, and the experiment. All of thse aspects have been considered in great detail by Lee et al.[5e], so only some of the most important ones will be covered here. In general, the measurement of any parameter that can be related to the absorption coefficient can be employed in a detection scheme; the most common ones are discussed below.

1. *Transmission*

For concentrated or bulk samples, a transmission experiment is both the simplest and most effective. It involves measuring the x-ray intensities incident and transmitted through a thin and uniform film of the material of interest. Careful analysis of signal-to-noise ratio considerations indicates that optimal results are obtained when the sample thickness is of the order of 2.5 absorption lengths. Since in transmission experiments Beer's Law applies, the data are usually plotted as $\ln(I/I_0)$ versus E. The x-ray intensities are typically measured using ionization chambers in conjunction with high-gain electrometers. Figure 6 depicts a typical transmission experimental set-up.

2. *Fluorescence*

For dilute samples, where absorption of the x-ray beam by the element of interest would be very low, a transmission geometry cannot be employed. Instead, fluorescence detection is the method of choice.[22,23] Fluorescence can be employed as a detection method because the characteristic x-ray fluorescence intensity depends on the number of core holes generated, which in turn depends on the absorption coefficient. Fluorescence detection is much more sensitive than transmission because one is measuring the signal over an essentially constant and typically low background. In general, the incident and fluorescent beams impinge and leave the surface at 45 degrees with the x-ray beam and the detector being at 90 degrees. The detector can be either an ionization chamber or a solid-state detector. The former is much simpler to implement, whereas the latter gives the best resolution. A filter, to minimize the contributions from elastic and Compton scattering, and soller slits are typically placed in front of the detector. The filter material is chosen so as to have an absorption edge that falls between the excitation energy employed and the energy of the characteristic x-ray photon emission from the element

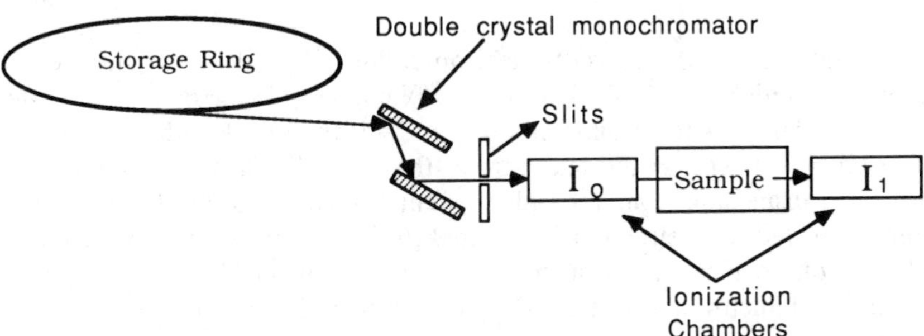

Figure 6. Schematic diagram of a transmission EXAFS experimental set-up. I_0 and I_1 refer to the incident and transmitted intensities, respectively.

of interest. Thus, the filter is generally made from the $Z-1$ or $Z-2$ element where Z represents the atomic numbr of the element of interest. For example, for the detection of CuK_α radiation at 8.04 KeV, a nickel filter would be employed. In this way the characteristic fluorescence is only slightly attenuated whereas both the elastic and Compton intensities are greatly reduced. However, there is the problem that often the K_β emision from the filter material is energetically very close to that of the K_α emission from the element of interest and thus may increase the background.

A solid-state detector, either Si(Li) (lithium-drifted silicon) or intrinsic germanium, offers the ability to discriminate on the basis of energy. The resolution can be as good as 150 eV although it degrades somewhat with increasing detector area. The main drawback with a solid-state detector is its limited count rate of approximately 15,000 cps. Since the detector accepts a wide range of photon enrgies from which the region of interest is chosen (via a single-channel analyzer), it can take significant amounts of time to obtain adequate statistics. In addition, the cost of solid-state detectors and associated electronics is much higher than that of ion chambers.

3. *Reflection*

When the sample under study is a planar surface, one can take advantage of x-ray optics to enhance surface sensitivity.[24] The most important aspect is specular or mirror reflection, and this is due to the fact that at x-ray energies the index of refraction of matter is slightly less than 1 and is given by

$$n = 1 - \delta - i\beta$$

$$\delta = \frac{1}{2\pi}\left(\frac{e^2}{mc^2}\right)\left(\frac{N_0\rho}{A}\right)[z + \Delta f']\lambda^2 \qquad (11)$$

$$\beta = \frac{\lambda\mu}{4\pi}$$

where $(e^2/(mc^2))$ is the classical electron radius, $(N_0\rho/A)$ is the number of atoms per unit volume, N_0 is Avogadro's number, ρ is the density, A is the atomic weight, z is the atomic number, and λ is the wavelength of the x-ray. The term $[z + \Delta f']$ is the real part of the scattering factor (including the so-called dispersion term f') and is essentially equal to Z. The imaginary part of the index of refraction β is related to absorption, where μ is the linear absorption coefficient. Considering an x-ray beam incident on a smooth surface and Snell's Law, one obtains that the critical angle for total reflection is given by:

$$\theta_{\text{crit}} = \sqrt{2\delta} \qquad (12)$$

Delta is of the order of 10^{-5} and θ_{crit} is typically of the order of a few milliradians. As long as the beam is incident below this critical angle, it is totally reflected and only an evanescent wave penetrates the substrate. This has two very important consequences. First, the penetration depth is of the order of 20 Å and thus one can significantly discriminate in favor of a surface-contained material. Compton and elastic scattering are also minimized. In addition, the reflection enhances the local intensity by as much as a factor of 4 as well as the effective "path length." All of these factors combined enhance the surface sensitivty of the technique and when combined with solid-state fluorescence detection, submonolayer amounts of material can be detected.[23]

Using a reflection geometry, one can employ the technique known as REFLEXAFS[25], which consists of measuring the ratio of the reflected and incident intensities as a function of energy. Although an EXAFS spectrum can be obtained from such a measurement, the process is somewhat involved since the reflectivity is a complex function of the angle of incidence, the refractive index, and energy.

Heald and co-workers[26] have made a careful comparison of REFLEXAFS versus measurements at grazing incidence with fluorescence detection. They conclude that in general, the latter offers enhanced sensitivity for studies of monolayers. However, the REFLEXAFS technique can be applied in a dispersive arrangement (*vide infra*) allowing for faster data acquisition and the possibility of performing kinetic studies on the millisecond time scale.

4. *Electron Yield*

Electron yield—Auger, partial, or total—can be employed as a means of detection since again these techniques depend on the generation of core holes. Because of the very small mean free paths of electrons, electron yield detection is very well suited for surface EXAFS measurements. However, due to this very same reason, in-situ studies of electrochemical interfaces are precluded. Details of electron yield EXAFS have been discussed by a number of authors.[18]

5. *Dispersive Arrangements*

In all the experimental techniques described up to this point, spectra were obtained by monitoring one of the above mentioned parametes as the incident energy was scanned by the use of monochromator crystals. This conventional mode of operation suffers from the fact that only a very narrow range of wavelengths is employed at a given time, and thus requires a significant amount of time to obtain a complete spectrum. In addition, it precludes real-time kinetic studies of all but the slowest of reactions. An alternative is to employ a dispersive arrangement[27] (Fig. 7) where by the use of focusing

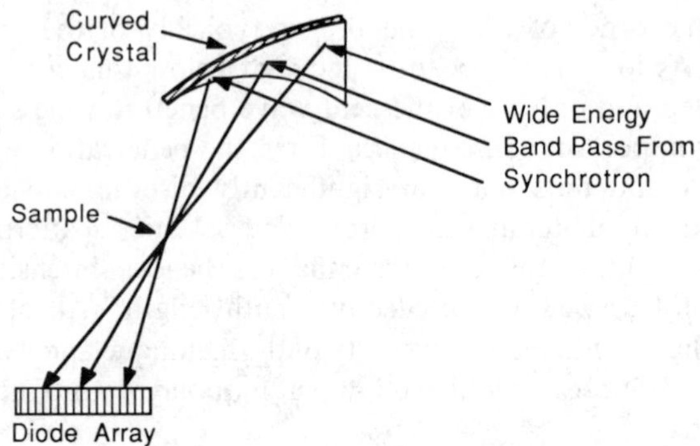

Figure 7. Schematic depiction of a dispersive EXAFS set-up.

optics (using a bent crystal), a range of energies can be brought to focus on a small spot. The exact energy spread will depend on the specific optical elements employed but a range of 500 to 600 eV represents a realistic value. Coupling this with a photodiode array (Fig. 7), allows for the simultaneous use of the full range of wavelengths; thus, a spectrum can be obtained in periods as short as milliseconds rather than minutes. This is of great significance because a number of relevant dynamic processes take place on this time scale. The applicaton of this approach to electrochemical studies will be discussed in a later section.

C. Electrochemical Cells

A number of cell designs have been employed in EXAFS studies of electrochemical systems. Of these, two general types can be identified depending on whether a transmission or a fluorescence mode of detection is employed. In a transmission mode, cells should be designed so as to minimize absorption losses due to the window material, elecrolyte, and the electrode itself. As a result, the windows are typically made of thin films (25 μm) of low absorbing materials such as polyethylene and polyimide (Kapton). The electrolyte layer thickness is typically small, and electrodes are generally metal films evaporated on a thin polymer film or small particles dispersed in a low Z matrix. Carbon can be employed in a variety of forms and shapes because of its low absorption.

Figure 8 shows the transmission cell employed by O'Grady and co-workers[28] in the study of the nickel oxide electrode. Heineman and co-workers[29] have employed a cell with reticulated vitreous carbon as an electrode material in x-ray spectroelectrochemical studies of various transition metal complexes.

When fluoresence detection is employed, the most important consideration

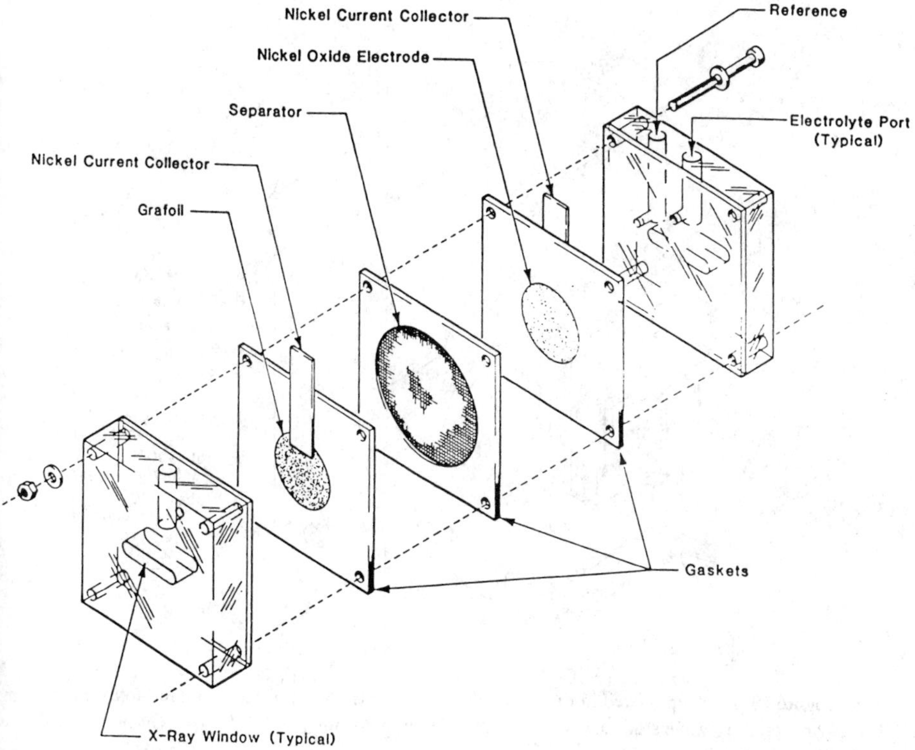

Figure 8. *In situ* transmission EXAFS cell for the study of Ni oxide electrodes. (From McBreen, J., O'Grady, W. E., Pandya, K. I. et al., *Langmuir* **3**, 428 (1986) with permission.)

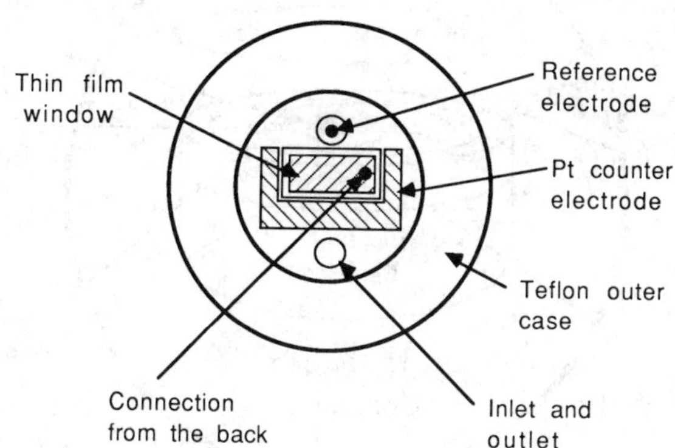

Figure 9. *In situ* cell for performing EXAFS studies on passivated iron films. (From Long, G. G., Kruger, J., and Kuriyama, M., in *Passivity of Metals and Semiconductors*, M. Froment, ed., Amsterdam, Elsevier, 1983, p. 139, with permission.)

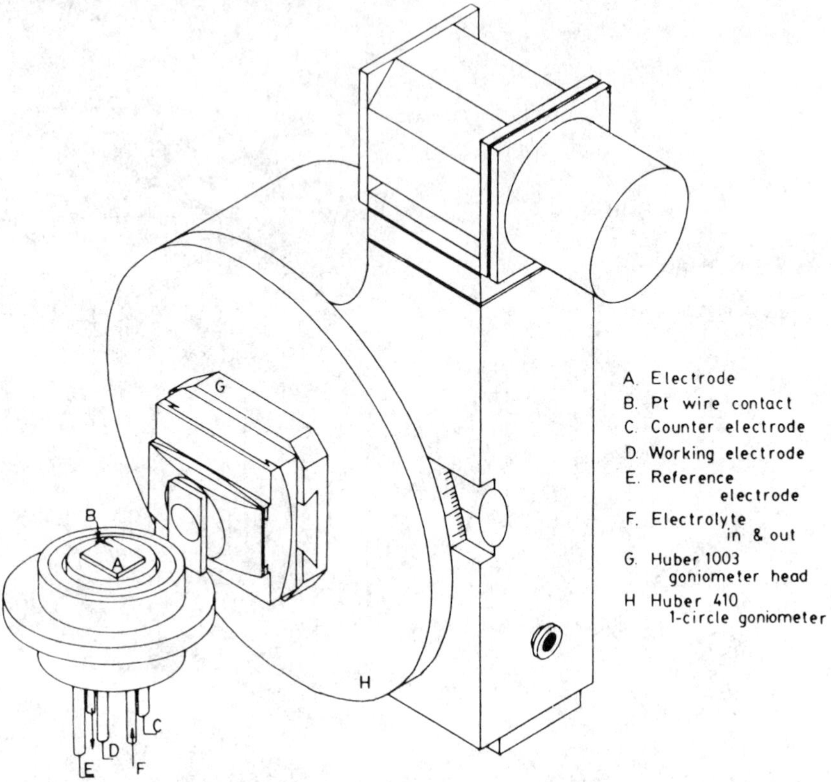

G
B
A
C
D
E
F
H
A. Electrode
B. Pt wire contact
C. Counter electrode
D. Working electrode
E. Reference electrode
F. Electrolyte in & out
G. Huber 1003 goniometer head
H. Huber 410 1-circle goniometer

Figure 10. Cell employed in the study of underpotentially deposited metal monolayers and in x-ray standing-wave studies. (From White, J. H., and Abruña, H. D., *J. Phys. Chem.* **92**, 7131 (1988), with permission.)

X-Ray Beam
10 μM Tefzel Window
Pt-disk working electrode
Pt counter electrode
Teflon Cell Body
Reference electrode port
Electrolyte ports
Huber 410 Goniometer
Viton O-ring

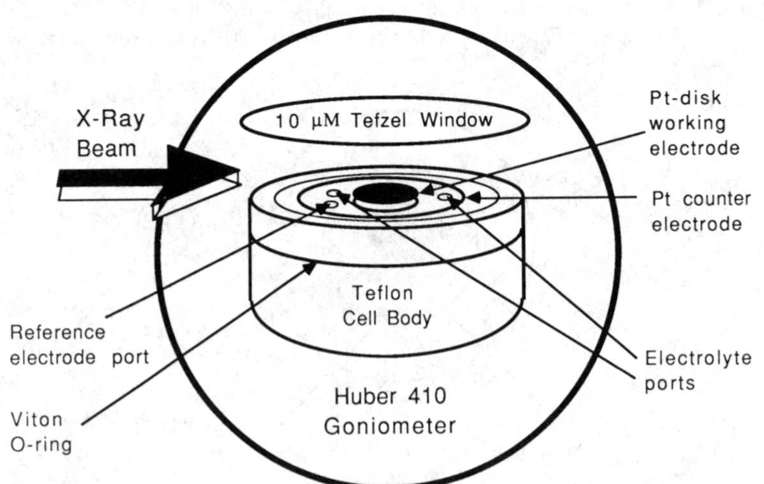

Figure 11. Cell employed in the study of electropolymerized films on electrodes. (From Albarelli, M. J., White, J. H., et al., *J. Electroanal. Chem.* **248**, 77 (1988), with permission.)

is to have a very thin window material as well as a thin layer of electrolyte. As a result, very thin films (6 μm) of polyethylene or polyimide are typically employed as windows. In addition, the cell configuration is of the thin layer type where a thin layer of electrolyte (10 μm) is trapped between the electrode and the window. An advantage of this cell configuration is that conventional bulk electrodes can be easily employed since transparency is not required. Figures 9, 10, and 11 show three different cells employed in the study of anodic films[30], underpotentially deposited (UPD) metal monolayers,[31] and chemically modified electrodes,[32] respectively.

VII. EXAFS STUDIES OF ELECTROCHEMICAL SYSTEMS

The discussion of EXAFS studies of electrochemical systems will be divided into:

(A). Oxide films
(B). Batteries and fuel cells
(C). Monolayers
(D). Spectroelectrochemistry
(E). Adsorption.

A. Oxide Films

Because of their relevance to corrosion phenomena, the study of passive films on electrode surfaces is an area of great fundamental, technological, and practical relevance. Despite decades of intensive investigation, there still exists a great deal of controversy as to the exact structural nature of passive films, especially when they are formed in the presence or absence of glass-forming additives such as chromium.

One of the main sources of controversy arises from the fact that many of the structural studies performed have been on dried films, and, as correctly pointed out by O'Grady,[33] the structure of dehydrated films can be significantly different from that of hydrated ones.

The use of surface EXAFS in the study of passive films represents a natural application of the technique, and in fact, the studies by Kruger and co-workers[30,34–36] on the passive film on iron represent the first reported. In their first studies, they employed vacuum-deposited iron films on glass slides and subsequently oxidized the films in either nitrite or chromate solution. They obtained the EXAFS spectra for the oxidized films, employing a photocathode ionization chamber (detecting the emitted electron current) and compared these with spectra for γ-FeO(OH), γ–Fe_2O_3, and Fe_3O_4. Although these studies were not in-situ, they did not require evacuation of the samples and therefore represent an intermediate situation between dehydrated films and in-situ experiments. They obtained spectra for Fe, Fe_3O_4, and the nitrite and chromate generated passive films. They noticed that the near edge region for the nitrite generated film showed evidence of an enhancement, similar to that observed for Fe_3O_4, indicative of an increase in the density of available final states with p-character. Such an enhancement is absent in the chromate formed films. These results point to a more covalent bonding in the chromate vs. the nitrite passivated films.

Upon Fourier transforming of the data, two peaks corresponding to Fe—O and Fe—Fe distances were obtained. The peaks in the Fourier transform of the chromate generated film were much less well-resolved than those for the nitrite films and this was ascribed to the presence of a glassy structure associated with the chromium. From a comparison of the edge jump for Fe and Cr, the authors estimate that the films had about 12% Cr.

They have also employed an in-situ cell for carrying out these experiments. Again they studied nitrite and chromate passivated films. The results obtained in this case were significantly different from the ex-situ measurements with the spectral features for both nitrite and chromate passivated films being quite similar. This surprising similarity underscores the importance of in-situ measurements.

Most recently,[37] they established the presence of Cr(VI) in oxide scales formed in Fe/Cr alloys oxidized at 600° C. In this study, they compared the chromium near-edge region of the oxidized film (Fig. 12b) with those of Cr_2O_3, CrO_3 and metallic chromium (Fig. 12a). The presence of Cr(VI) was clearly established by the appearance of the very characteristic sharp feature at about 6001 eV. Although these studies were ex situ, they demonstrate the utility of using the near-edge region (XANES) in the determination of oxidation state.

Hoffman and co-workers[38] have carried out a series of studies on the passive films on iron, with particular attention to cell design. They have employed a so-called bag[38] cell that allows for the in-situ passivation or cathodic protection of the iron films which were deposited onto gold films deposited on melinex (polymer film with excellent adhesive properties). In addition, they employed a setup in which the working electrode is partially immersed in solution and continuously rotated. In this way, they could expose the electrode to the x-ray beam with ostensibly only a very thin film of electrolyte. Under these conditions, they were able to obtain spectra of the film as prepared, a cathodically protected film as well as a film passivated in borate solution at 1.3 V. From an analysis of their data, they concluded that the passive film had an Fe—O coordination with 6 near neighbors at a distance of 2 ± 0.1 Å. The approach followed by these authors appears most appropriate since they were able to reduce the deposited films

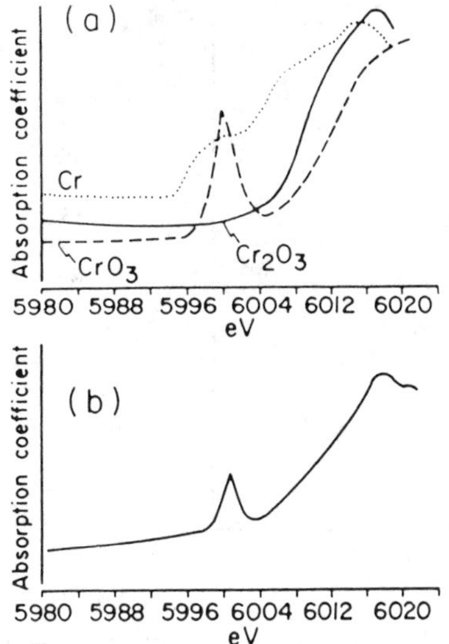

Figure 12. Cr K edge spectra for: (a) Cr, CrO_3, and Cr_2O_3; (b) oxide formed on a Fe 25% Cr film by oxidation in oxygen at 600°C (From Long, G. G., Kruger, J. and Tanaka, D., *J. Electrochem. Soc.* **134**, 264 (1987), with permission.)

to the metallic state and subsequently oxidize them. It would be most interesting to ascertain how the structure of the passive film varies through sequential reduction/passivation cycles.

Forty and co-workers[39] have used electron yield EXAFS to study (ex-situ) anodic films on aluminum prepared in tartrate and phosphoric acid electrolytes. The Al—O distance is different for the two film preparations. They conclude from these studies that in the tartrate formed film 80% of the aluminum ions are in octahedral sites and 20% in tetrahedral sites whereas for the films formed in phosphoric acid all of the aluminum ions appear to be located at tetrahedral sites. In addition, immersion of the phosphoric acid generated films into water at 85° C for 4 hr. gave rise to dramatic changes in the structure which were ascribed to hydration. (Fig. 13) No such changes were noted for the films generated in tartrate electrolyte.

These investigators have also studied[40] the passive films formed on iron and iron-chromium alloys upon immersion in sodium nitrite solution. Studies were carried out in a wet environment (which they term *in-situ*) as well as after dehydration. For a FeCr alloy (13% Cr) they find that the structure of the wet film is analogous to that of γ–FeOOH but with a higher degree of disorder, consistent with the Mossbauer results of O'Grady.[33] Upon dehydration, the structure transforms to one that is closer to that of γ–Fe_2O_3 but with reduced long-range order. In addition, they looked at the chromium EXAFS and found that the local structure around chromium in the passive films was similar to that of Cr_2O_3. They concluded that the presence of

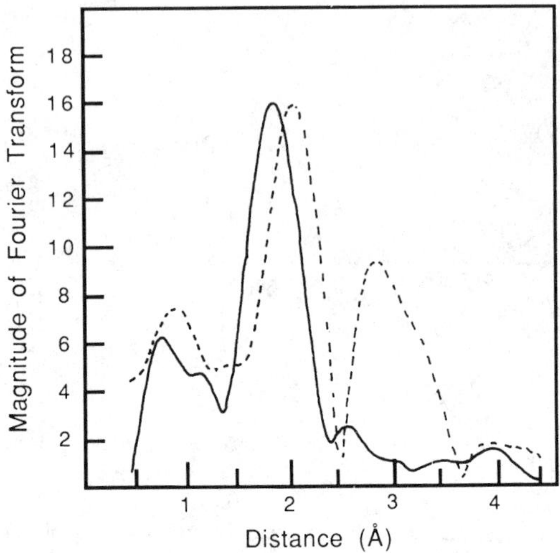

Figure 13. Fourier transform for aluminum oxide film prepared in phosphoric acid (——) immediately after preparation (– – –) after immersion in water at 85°C for 4 hr. (Adapted from El-Mashri, S. M., Jones, R. G and Forty, A. J., *Phil. Mag. A.*, **48** 665 (1983), with permission.)

chromium in alloys stabilized the γ–Fe_2O_3-like layers against dehydration, thus forming a glasy-like structure which enhances the stability of the passive film. These results are in good agreement with those of Kruger and co-workers mentioned previously.

Froment and co-workers[41–42] have employed REFLEXAFS (vide supra) for studying passive films on iron and nickel. Their early studies were concerned with demonstrating the applicability of the REFLEXAFS technique to electrochemical systems. Most recently,[42] they have used this technique to study the structure of passive films on Ni and on Ni–Mo alloy electrodes. For the Ni electrodes, they performed studies after reduction at −700 mV (vs. saturated mercurous sulphate electrode) as well as in the passive (+300 mV) and transpassive (+800 mV) regions. The Fourier transforms for the films in the passive region have a Ni—O peak at a distance that corresponds closely to that in bulk nickel oxide. However, no Ni–Ni interactions were observed. These investigators interpreted these results as consistent with a model that postulates an amorphous hydrated polymeric oxide.[43]

B. Batteries and Fuel Cells

The performance and lifetime of fuel cells and batteries can often be determined by the structural features (and their change) of the various components. Structural studies of these systems are generally very difficult because the materials involved are often noncrystalline so that x-ray diffraction measurements are of very limited use. Because of its ability to provide short-range order, EXAFS is ideally suited to the in situ study of such systems and a number of applications have been reported.[44]

Linford and co-workers[45] studied the changes in the structure of the electrolyte in the system: Metal/CuI: sulphonium iodide(5.5:1)/Metal as a function of discharge, since they had previously observed long induction periods in attaining a steady voltage. They examined the EXAFS around the copper K edge and found that the starting electrolyte had a structure essentially identical to that of γ–CuI. After 2,000 hours of discharge, the spectrum was significantly changed and this was ascribed to the incorporation of sulfur from the electrolyte.

O'Grady and McBreen[28] performed an extensive study of the nickel oxide electrode employing the cell shown in Fig. 8 in a transmission mode. The study of nickel oxide is complicated by the numerous species present and their interconversion. They found that the as-prepared β–$Ni(OH)_2$ has the same structure within the x–y plane as that determined by x-ray diffraction experiments but with a significant degree of disorder along the c-axis. Oxidation to the trivalent state results in contraction of the Ni—O and Ni—Ni distances along the x–y plane. Re-reduction of this material yields a structure similar to that of the freshly prepared $Ni(OH)_2$. Repeated oxidation–reduction cycles resulted in an increased disorder which is believed to be responsible for facilitating the electrochemical oxidation to the trivalent state.

Boudart and co-workers[46] recently reported on a study in which they monitored x-ray absorption spectra around the Pt L_{III} edge during the electrochemical reduction of oxygen at electrocatalysts containing 1 nm clusters of platinum dispersed on carbon. Changes in the threshold peak area and edge position in the spectra suggest that the oxidation state of the metal decreases linearly with the applied potential in the range from 1010 to 900 mV (vs NHE), which is similar to the behavior of bulk platinum electrodes. However, they also found that the overall oxidation state of the clusters was consistent with a surface stoichiometry that was closer to $PtO_{0.5}$ rather than PtO which is the one proposed for the surface of bulk electrodes.

O'Grady and Koningsberger[47] have studied metal–carbon interactions in carbon-supported platinum catalysts in fuel cells and concluded that there are two types of Pt–C interactions.

Van Wingerden and co-workers[48a] obtained the EXAFS spectra around the Co K edge for 5, 10, 15, 20-tetra-(*p*-chloro-phenyl)porphyrinatocobalt(III) supported on Norit BRX and heated in nitrogen to various temperatures up to 850° C. This study was geared to an understanding of the enhanced catalytic effect towards oxygen reduction that accompanies heat treatment.

They found that upon adsorption or after heating up to 550° C (where the catalytic activity is maximal), the cobalt chelate retains its square planar configuration. Further heating causes some decomposition, and at the highest temperatures metallic cobalt is obtained. Similar results were obtained by O'Grady and co-workers.[48b]

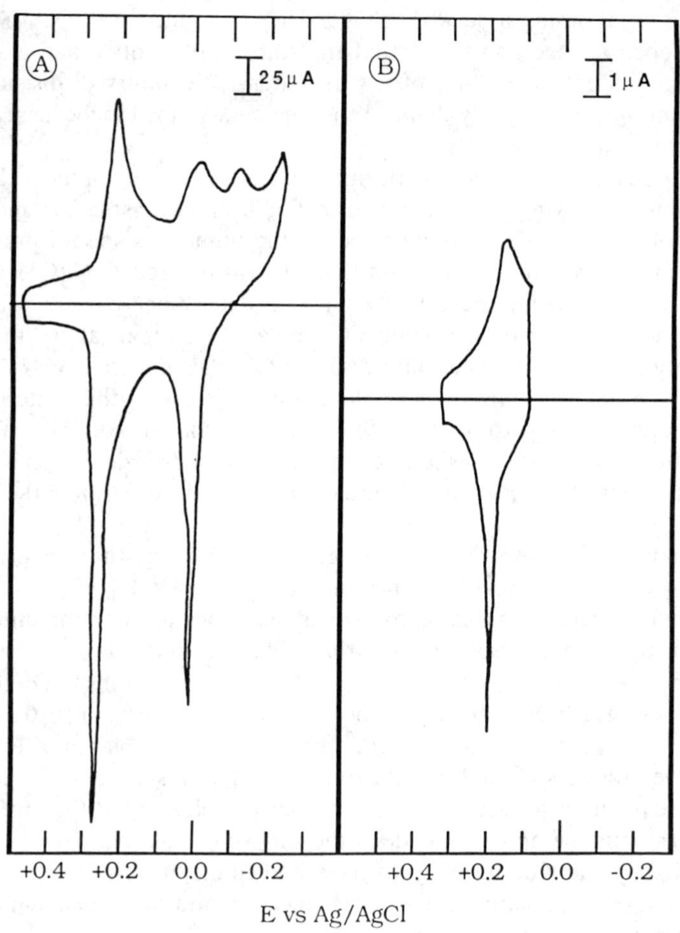

Figure 14. Voltammetric scans for the underpotential deposition of copper on (A) an epitaxial film of gold (111 orientation) on mica, and on (B) a bulk Pt(111) single-crystal electrode coated with a layer of iodine. Experimental conditions: (A) 1 M H_2SO_4 containing 5×10^{-5} M Cu^{+2}, sweep rate 1 mV/sec. (B) 0.1 M H_2SO_4 containing 5×10^{-5} M Cu^{+2}, sweep rate 1 mV/sec. (From Abruña, H. D., White, J. H., et al., *J. Phys. Chem.* **92**, 7045 (1988), with permission.)

C. Monolayers

The study of electrochemically deposited monolayers poses the strictest experimental constraints because of the very low signals involved. However, these studies can provide much detail on interfacial structure as well as on the effects of solvent and supporting electrolyte ions.

An especially attractive way of preparing metllic monolayers on electrodes is by a process known as underpotential deposition (UPD).[49] This refers to the deposition of metallic layers on an electrode of a different material. The first monolayer is deposited at a potential that is less negative (typically by several hundred millivolts) than the expected thermodynamic potential; hence the term underpotential deposition. This occurs over a somewhat narrow range of potentials where the coverage varies from zero to a monolayer. A distinct advantage of this approach is that it allows for precise control of the surface coverage from a fraction of a monolayer up to a full monolayer. Since subsequent electrodeposition (bulk deposition) will require a significantly different potential, very reproducible monolayer coverages can be routinely obtained. Thus, this represents a unique family of systems with which to probe electrochemical interfacial structure in situ.

A number of UPD systems have been investigated[50] including Cu/Au(111),[51a,b] Ag/Au(111),[51c] Pb/Ag(111)[31] and Cu/Pt(111)/I.[50,52] The first three systems involved the use of epitaxially deposited metal films on mica as electrodes which gives rise to electrodes with well-defined single-crystalline structures.[53] In the last case, a bulk platinum single crystal was employed. Because of the single-crystalline nature of the electrodes, polarization-dependence studies could be used to ascertain surface structure.

In order to minimize background scattering, thin layer cells were employed in all of these studies, and although very slow sweep rates had to be employed, well defined voltammetric responses were obtained. For example, Figs. 14A and B show voltammograms for the underpotential deposition of copper on an epitaxial film of gold (111 orientation) on mica and on a bulk Pt(111) single-crystal electrode that had been pretreated with a layer of adsorbed iodine. The very well-defined voltammetric features are indicative of well-ordered surfaces and good potential control.

1. *Cu UPD on a Au*(111) *Electrode*

This is one of the best characterized systems to date[50,51a,b] In these studies, spectra were obtained by monitoring the characteristic CuK_α fluorescence intensity at 8.04 KeV. Spectra were obtained with the polarization of the x-ray beam being either perpendicular (Fig. 15A) or parallel (Fig. 15B) to the plane of the electrode. In both cases, a number of well-defined oscillations were observed in addition to a sharply defined edge. From a qualitative

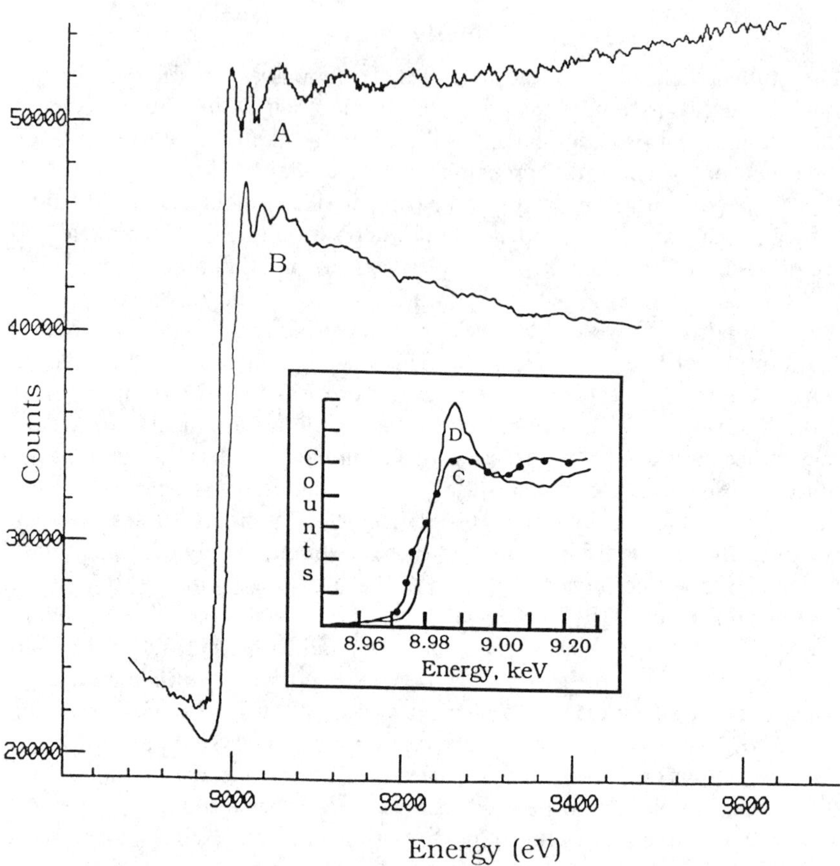

Figure 15. Fluorescence-detected (*in situ*) x-ray absorption spectrum for an underpotentially deposited (UPD) monolayer of copper on a gold (111) electrode with the plane of polarization of the x-ray beam being perpendicular (A) or parallel (B) to the plane of the electrode. Inset: Edge region of the x-ray absorption spectrum for a copper UPD monolayer before (C) and after (D) stripping. (From Abruña, H. D., White, J. H., et al., *J. Phys. Chem.* **92**, 7045 (1988), with permission.)

comparison of the spectra for copper foil and $CuAu_3$, it is clear that spectra for the UPD monolayer resemble the latter to a much greater extent, pointing to the strong influence exerted by the gold substrate in the UPD monolayer.

That the signal arises from the monolayer of copper on the gold surface can be demonstrated by adjusting the potential to a value (+0.50 V) where the monolayer is oxidized to Cu^{+2} and dissolved (stripped) into the thin layer of electrolyte. The spectrum of the edge region taken under these conditions is compared to that of the UPD layer in Figs. 15C and D. There are two noticeable differences. First, the edge position for the Cu^{+2} ions in solution (i.e., after oxidation and stripping of the monolayer) (Fig. 15C) is shifted to higher energy by about 2 eV relative to the value of the upd layer (Fig. 15D) and this is consistent with the higher oxidation state of Cu^{+2}. In addition, the presence of the very characteristic "white-line" for Cu^{+2} and its absence in the spectrum of the UPD layer further corroborate spectral assignments. It should be mentioned that the reason we are able to detect the Cu^{+2} ions in solution stems from the fact that upon oxidation and stripping, the concentration of Cu^{+2} within the thin layer volume is of the order of 1×10^{-3} M (as opposed to the bulk solution concentration of 10 μM) and therefore, detectable.

From analysis of the data, a number of salient features can be pointed out. First, the copper atoms appear to be located at three-fold hollow sites (i.e., 3 gold near neighbors) on the gold (111) surface with copper near neighbors. The Au/Cu and Cu/Cu distances obtained are 2.58 and 2.91 ± 0.03 Å, respectively. This last number is very similar to the Au/Au distance in the (111) direction, suggesting a commensurate structure. Most surprising,

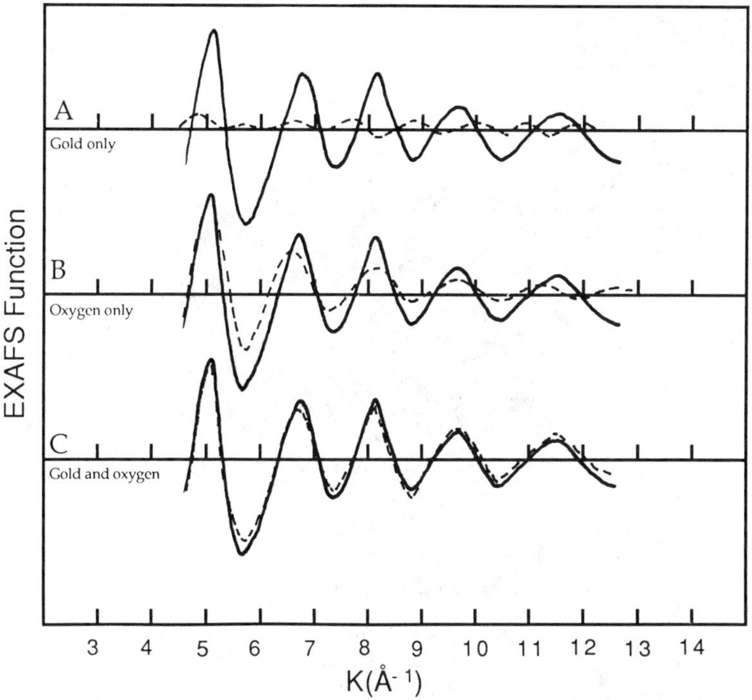

Figure 16. Experimental (——) EXAFS and fits (---) using gold (A), oxygen (B), or gold and oxygen (C) as backscatters. (Data corresponds to the spectrum in Figure 15A.)

however, was the presence of oxygen as a scatterer at a distance of 2.08 ± 0.02 Å. From analysis and fitting of the data, we obtain that the surface copper atoms are bonded to an oxygen from either water or sulfate anion from the electrolyte. That there might be water or sulfate in contact with the copper layer is not surprising, however, such interactions generally have very large Debye–Waller factors so that typically no EXAFS oscillations (or heavily damped oscillations) are observed. The fact that the presence of oxygen (from water or electrolyte) at a very well-defined distance is observed is indicative of a significant interaction and underscores the importance of in-situ studies.

Scattering by oxygen was only detected when the plane of polarization was normal to the electrode surface, suggesting that the oxygen is present atop the copper layer. The strong scattering by oxygen is most apparent when considering the fits to the Fourier filtered data. Figure 16 shows the Fourier-filtered experimental data and the fits with only gold (Fig. 16A) or oxygen (Fig. 16B) or both (Fig. 16C) as backscatterers. Whereas in the first two cases these were significant deviations in different parts of the spectrum, the fit with both gold and oxygen as backscatterers is excellent over the entire range of k values.

A pictorial representation of this system is shown in Fig. 17 where the

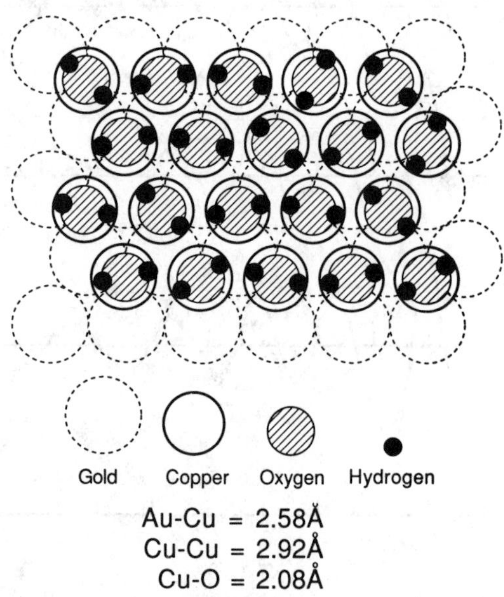

Figure 17. Structure of a copper UPD monolayer on a gold(111) electrode surface with water as the source of oxygen.

source of oxygen is presented as water. However, it should be mentioned that from the EXAFS experiment one cannot rule out sulfate anions as the source of oxygen. In fact, experiments by Kolb and co-workers[54] indicate that sulfate may be present since at the potential for monolayer deposition the electrode is positive of the potential of zero charge so that sulfate would be present to counterbalance the surface charge.

2. *Ag UPD on Au*(111)

Studies of Ag on Au(111)[51] yield very similar results in terms of the structue of the deposited monolayer, with the silver atoms being bonded to three surface gold atoms and located at three-fold hollow sites forming a

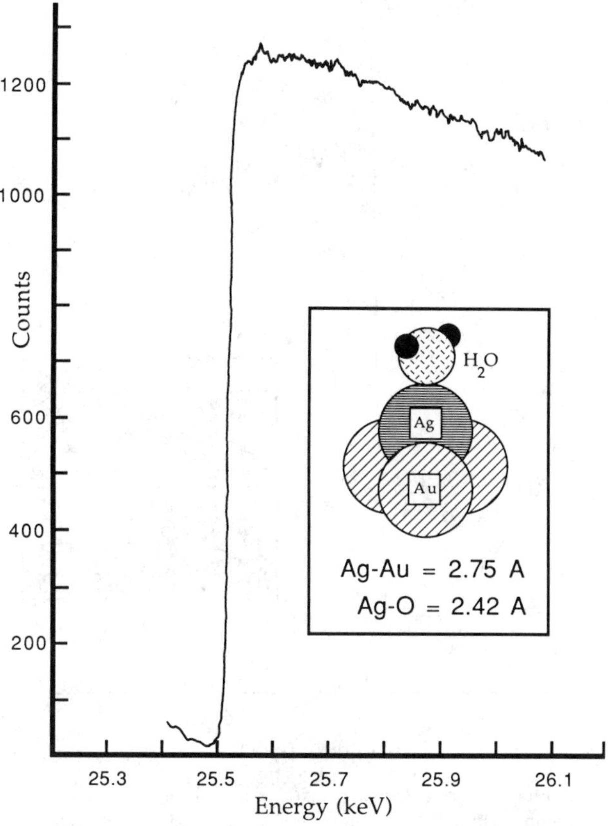

Figure 18. Fluorescence-detected (*in situ*) x-ray absorption spectrum for an underpotentially deposited (UPD) monolayer of silver on a gold(111) electrode. Inset: Proposed structure.

commensurate layer. Again, strong interaction by oxygen from water or electrolyte (perchlorate) was present.

Figure 18 shows an EXAFS spectrum for Ag UPD on Au with the plane of polarization perpendicular to the electrode surface. Only very shallow modulations were observed, and this was ascribed to a large Debye–Waller factor. The inset to Fig. 18 depicts the structure of the UPD monolayer which is analogous to the one obtained for copper on gold.

3. *Pb UPD on Ag*(111)

Melroy and co-workers[31] have obtained the EXAFS spectrum of Pb underpotentially deposited on a silver (111) electrode. In this case, no Pb/Ag

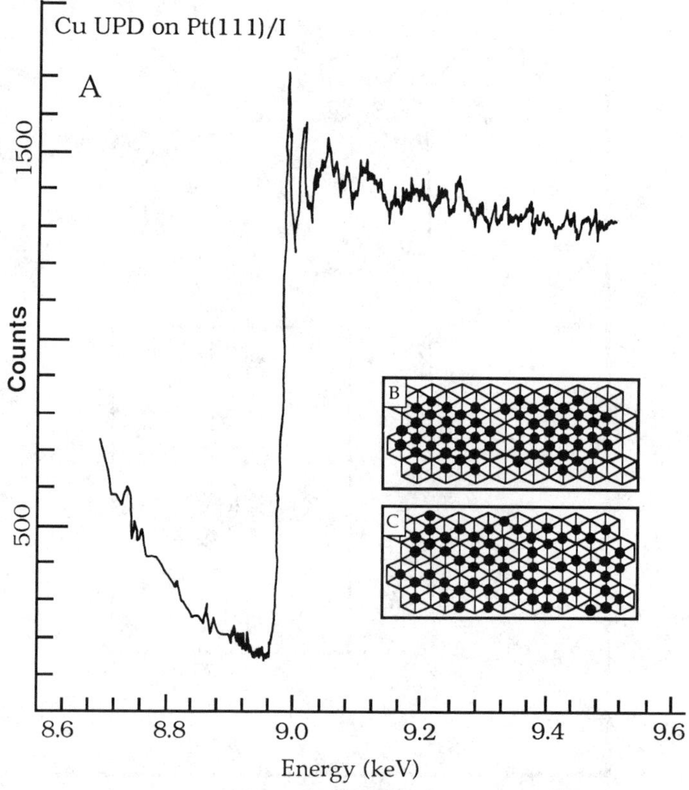

Figure 19. (A) Fluorescence-detected (*in situ*) x-ray absorption spectrum for an UPD half-monolayer of coppr on a Pt(111) single-crystal electrode. Inset: Depiction of models involving clustering (B) and random decoration (C). (From Abruña, H. D., White, J. H., et al., *J. Phys. Chem.* **92**, 7045 (1988), with permission.)

scattering was observed, and this was ascribed to the large Debye-Waller factor for the lead as well as to the presence of an incommensurate layer. However, data analysis as well as comparison of the edge region of spectra for the lead UPD, lead foil, lead acetate, and lead oxide indicated the presence of oxygen from either water or acetate (from electrolyte) as a back-scatterer.

They were also able to perform a potential dependence study of the lead oxygen distance. They find that it increases from 2.33 ± 0.02 to 2.38 ± 0.02 Å upon changing the potential from −0.53 to −1.0 V versus Ag/AgCl. This is consistent with the negatively charged electrode repelling a negatively charged or strongly dipolar adsorbate.

In all of these studies, it is clear that scattering from water and/or electrolyte plays a crucial role, once again underscoring the importance of *in-situ* measurements.

4. *Cu UPD on Pt*(111)/I

Most recently, we[52] have studied the structure of a half-monolayer of copper underpotentially deposited on a platinum (111) bulk single-crystal electrode annealed in iodine vapor. The spectrum, shown in Fig. 19A, exhibits five well-defined oscillations in addition to a sharply defined edge. It should be mentioned that in this experiment, the plane of polarization of the x-ray beam was parallel to the electrode surface so that it was most sensitive to in-plane scattering of copper by other copper neighbors. From analysis of the data, we determined a Cu—Cu distance of 2.85 ± 0.02 Å which is very close to the Pt—Pt distance in the (111) direction and suggests that the copper atoms are present at three-fold hollow sites and that they form a commensurate layer with the platinum substrate. More important, however, was the finding that the average number of Cu near neighbors was six. This strongly suggests that at half-monolayer coverage, the surface is better represented by one that contains large clusters (Fig. 19B) rather than by one that is randomly decorated with copper atoms or covered with a lattice with a large interatomic spacing (Fig. 19C). This is significant since it is a direct experimental documentation of a mechanism where monolayer formation involves nucleation and growth rather than random deposition with subsequent coalescence.

We also found that heating of the sample resulted in the loss of the pronounced features of the fine structure. This seems to indicate that the structure initially formed is a metastable one and that the stability is derived from the clustering.

D. Spectroelectrochemistry

EXAFS and XANES techniques have been applied in the more traditional type of spectroelectrochemical experiments where a thin layer cell configuration is employed. Drawing from extensive experience in the related UV–VIS measurements, Heineman, in collaboration with Elder, was the first to report on an *in-situ* EXAFS spectroelectrochemistry experiment.[55,56] Their first cell design employed gold minigrid electrodes similar to those typically employed in traditional UV–VIS experiments. They studied the ferro–ferricyanide couple in each oxidation state by monitoring the region about the iron K edge using fluorescence detection. From analysis of their data, they were able to determine that for Fe(II) there are 7.4 carbon atoms at 1.97 ± 0.01 Å, whereas for Fe(III) there are 6.8 carbon near neighbors at a distance of 1.94 ± 0.01 Å. Since, as mentioned previously, coordination number determination is usually no better that 20%, the numbers they find are in good agreement with the known value of 6. More interesting is the fact that they observed a contraction of the Fe—C bond upon oxidation, a finding that is contrary to results based on crystallographic studies. This points to the importance of *in-situ* measurements since by the applied potential one can precisely control the oxidation state of the species being studied. The determination of metal/ligand bond distances in solution and their oxidation state dependence are critical to the application of electron transfer

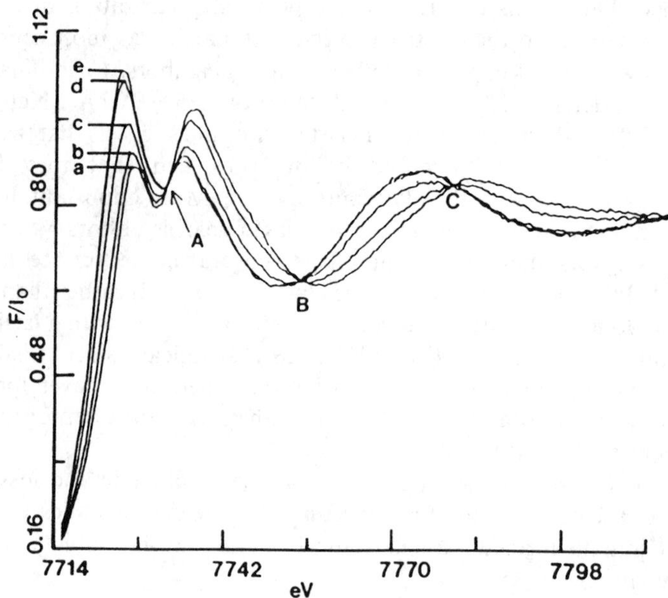

Figure 20. X-ray absorption spectra of 10 mM cobalt sepulchrate in 1 M sodium acetate at applied potential values of: (a) −0.30 V, (b) −0.58 V, (c) −0.60 V, and (d) −0.62 V vs. Ag/AgCl. (From Dewald, H. D., Watkins, J. W., et. al., *Anal. Chem.* **58**, 2968 (1986), with permission.)

theories, since such changes can contribute significantly to the energy of activation through the so-called inner sphere reorganizational energy term.

These authors have also developed a cell that employs reticulated vitreous carbon as a working electrode; they found that such a design allows for much faster electrolysis. Using such a cell, they have studied the $[Ru(NH_3)_6]^{+3/+2}$ couple, a cobalt(III/II) sepulchrate, as well as the Fe—C distance in cytochrome *C*. For the case of the cobalt sepulchrate, they were able to obtain spectra at five different applied potentials, and as shown in Fig. 20, very well-defined isosbestic points were obtained.

Most recently, they have developed a cell configuration for the study of modified electrodes[57] that employs, as a working electrode, colloidal graphite deposited onto kapton tape (typical window material). Such an arrangement minimizes attenuation due to the electrolyte solution. They coated the working electrode with a thin film of Nafion (a perfluoro sulfonate ionomer from E. I. DuPont de Nemours, Inc.) and incorporated $[Cu(2,9\text{-dimethy-}1,10\text{-phenanthroline})_2]^{+1}$ by ion exchange. They were able to obtain the EXAFS spectra around the copper K edge for the complex in both the Cu(I) and Cu(II) oxidation states.

Although the field of chemically modified electrodes has seen a tremendous growth in the recent past, there is very limited information on the structure of these layers. In an effort to bridge this gap, we have performed some *in-situ* EXAFS measurements on chemically modified electrodes.[32] We have studied films of $[M(\text{v-bpy})_3]^{2+}$ (v-bpy is 4-vinyl,4′methyl-2,2′bipyridine; M = Ru,Os) and $[Os(\text{v-bpy})_2(\text{phen})]^{+2}$ electropolymerized onto a platinum electrode and in contact with an acetonitrile/0.1M TBAP (tetra *n*-butyl ammonium perchlorate) solution and under potential control. We have found that for electropolymerized films of $[Ru(\text{v-bpy})_3]^{2+}$, the spectra obtained for electrodes modified with 5 or more equivalent monolayers are indistinguishable from the spectrum obtained for the bulk material. (It should be mentioned that a monolayer of $[Ru(\text{v-bpy})_3]^{+2}$ represents about 5.4×10^{13} molecules/cm^2 which is about 5% of a metal monolayer. This is mentioned since it is the metal centers that give rise to the characteristic fluorescence employed in the detection.) Upon fitting of the data for phase and amplitude, we obtain a Ru—N distance of 2.01 Å and a coordination number of six. These correlate very well with the known values of 2.056 Å and six, respectively. In addition, changes in oxidation state could be monitored by the shift in the position of the edge. For example, upon oxidation of the polymer film (at + 1.60 V) from Ru(II) to Ru(III), the edge position shifts to higher energy by about 1.5 eV. Thus, one can determine the oxidation state of the metal inside a polymer film on an electrode surface.

Similar results were obtained for the osmium complexes in terms of the local structure around the metal center and changes in the position of the edge as a function of oxidation state. In addition, we were able to correlate changes in the near-edge features with the coordination environment around the osmium metal center.

These results indicate that the structure of electroactive polymer films and the oxidation state of the metal center can be obtained at relatively low coverages, and this should have important implications in trying to identify the structure of reactive intermediates in electrocatalytic reactions at chemically modified electrodes.

Tourillon and co-workers[58-66] have also reported on a number of spectroelectrochemical studies, especially of electrodeposition of metals, particularly copper, on electrodes modified with poly 3-methylthiophene. What sets his experiments apart is the use of a dispersive approach (Fig. 7). As mentioned previously, in such a set-up, focusing optics are employed so as to have a range of energies (as wide as 500 eV) come to a tight focal spot at the sample. The beam then impinges a photodiode array so that all energies are monitored at once. The net result is to significantly decrease data acquisition times so that spectra can obtained in times as short as a few milliseconds. Thus, this opens up tremendous possibilities in terms of kinetic and dynamic studies. One of the more impressive results using this approach

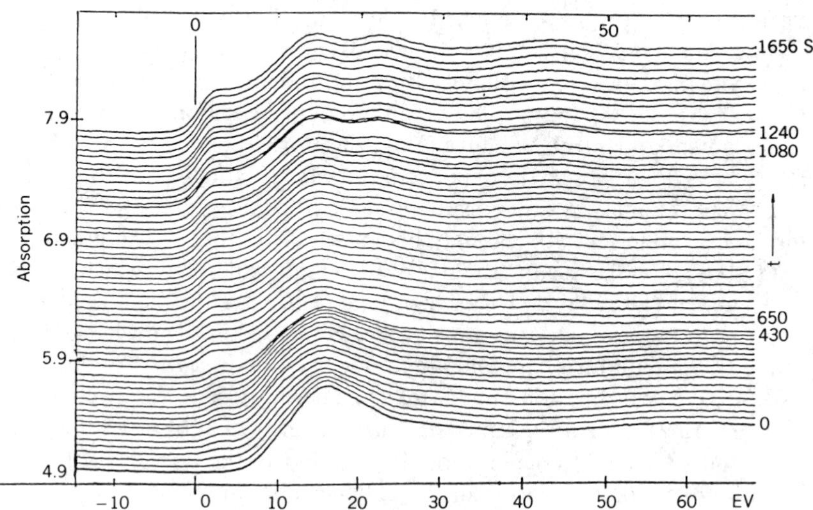

Figure 21. *In situ* measurements of the time evolution of the Cu *K*-edge when a platinum electrode coated with a polymeric film of polymethylthiophene is cathodically polarized in an aqueous solution containing 50 mM $CuCl_2$ (From Dartyge, E., Fontaine, A., et al. *Jnl. de Phys. Colloque. C8*, **47**, 607 (1986), with permission.)

is shown in Fig. 21, which shows spectra obtained around the copper K edge for a poly 3-methylthiophene film (on a platinum electrode) doped with Cu^{+2} ions. The potential of the electrode is stepped so as to reduce the Cu^{+2} ions to Cu^{+1} and subsequently to Cu^0. The spectra shown in Fig. 21 were taken at 7-second intervals, and the transitions from Cu^{+2} to Cu^{+1} and then to metallic copper are clearly evident. These authors have also employed this technique for the study of other metallic inclusions into poly 3-methylthiophene films, including Ir, Au, and Pt. This type of arrangement could open up new exciting possibilities in terms of kinetic studies.

Most recently, and in collaboration with McBreen and O'Grady,[67] these investigators have carried out a time-resolved XANES study of the nickel oxide electrode. They were able to study α–$Ni(OH)_2$, and β–$Ni(OH)_2$ with and without added $Co(OH)_2$. They observed a continuous shift of the edge position to higher energies consistent with the $Ni^{+2} \rightarrow Ni^{+3}$ electrochemical reaction. In addition, there were changes in the near-edge features that were related to changes in the Ni—O local environment. Figure 22 shows spectra at various potentials for α–$Ni(OH)_2$ during anodic (a) and cathodic (c) sweeps.

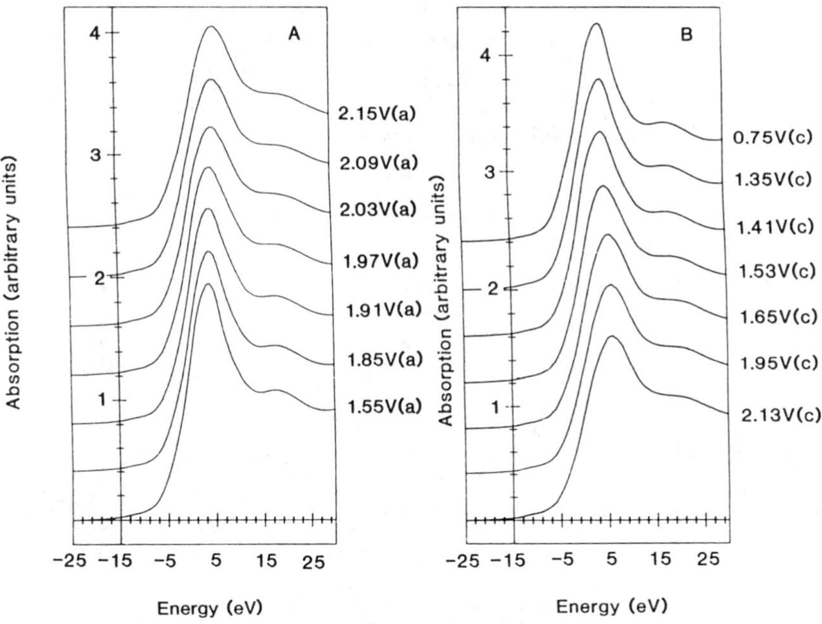

Figure 22. Near-edge spectra for α-$Ni(OH)_2$ at the indicated potentials and during (A) anodic (a) and (B) cathodic (c) sweeps. (From McBreen, J., O'Grady, W. E., et al., *J. Phys. Chem.* **93**, 6308 (1989) with permission.)

Tourillon and co-workers have also carried out extensive ex-situ studies on the structure of conducting polymers following the features around the carbon edge.[68]

E. Adsorption

Since the magnitude of the edge jump in an x-ray absorption spectrum is proportional to the number of absorbers, its potential dependence can be employed to measure potential/surface-concentration (electrosorption) isotherms in-situ. We have carried out such a study on iodide adsorption on a Pt(111) electrode surface.[69] The resulting isotherm, shown in Fig. 23, displayed two plateaus separated by a transition centered at about −0.10 V vs. Ag/AgCl. The iodide coverage at the most negative potential studied (−1.0 V) went to an essentially negligible value while that at the positive potential limit (+0.50 V) increased rapidly, suggesting an interfacial accumulation of absorbers.

The results at the most negative potentials can be interpreted in terms of a model where adsorbed iodine is converted to adsorbed iodide, which subsequently shows the behavior expected for a specifically adsorbed anion. That is, there is a marked dependence of the coverage on potential.

The plateaulike features in the isotherm may be explained in thermodynamic terms. For example, we attribute the flatness in the region between −0.50 and −0.10 V to the absence of potential dependence of coverage of a neutral adsorbate (iodine). On the other hand, between −0.1 and +0.3 V, further uptake involves adsorption of iodide. The potential

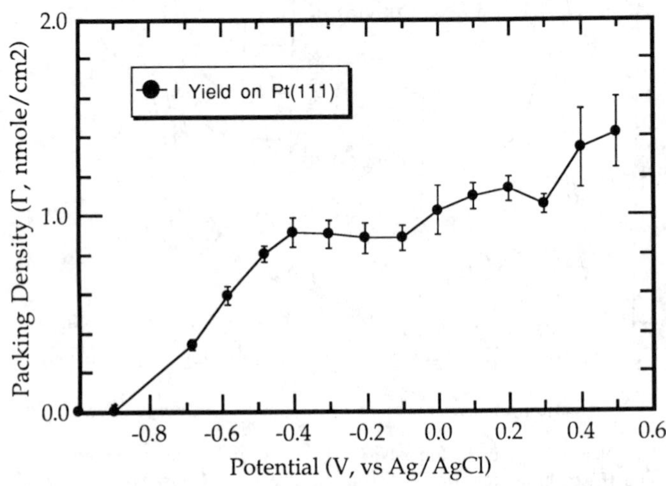

Figure 23. Electrosorption isotherm of iodine on a Pt(111) elecrode.

dependence here is due to interaction of the anion with an increasingly positive surface charge. The attainment of a saturation coverage is due to closest packing of iodine species as well as to possible lateral electrostatic interactions. The features in the electrosorption isotherm present at intermediate potentials may also be attributed to the formation of stable structures. Hubbard and co-workers[70] have shown that a transition from a mixed $\sqrt{3} \times \sqrt{3}$ and $\sqrt{7} \times \sqrt{7}$ structure to one that is purely $\sqrt{7} \times \sqrt{7}$ occurs between -0.2 and 0.0 V and is accompanied by a change in coverage from 3/9 to 4/9. A similar transition is observed here and is assumed, because of the similarity in the packing density change over the transition, to be due to the same structural transition. At the most positive potentials, we believe that the increase in the iodine/iodide concentration is produced by Faradaic charge flow, followed by association with the adsorbed iodine layer at that potential.

VIII. X-RAY STANDING WAVES

A. Introduction

The x-ray standing-wave technique represents an extremely sensitive tool for determining the position of impurity atoms within a crystal or adsorbed onto crystal surfaces.[71,72] This technique is based on the x-ray standing-wave fields that arise as a result of the interference between the coherently related incident and Bragg-diffracted beams from a perfect crystal, and is described by the theory of dynamical diffraction of x-rays.[71,73]

There are two general theories that describe observed intensities in x-ray diffraction; these are the kinematical and dynamical theories. The kinematical approach, which is the better known and most commonly employed, treats the scattering from each volume element independently of each other except for the incoherent power losses associated with the beam reaching and leaving a particular volume element.

The alternate theory, called the dynamical theory of x-ray diffraction,[71,73] takes into account all wave interactions within a crystal and must be used when considering diffraction from perfect or nearly perfect crystals. The dynamical approach represents a general theory to account for observed intensities in x-ray diffraction, and the kinematical approach can be considered one of its limiting cases. For small diffracting volumes and weak reflections, both theories predict the same results. However, when considering perfect or nearly perfect crystals and strong reflections, the kinematical approach gives an incorrect account of the diffracted intensities, and the dynamical approach must be employed.

One of the most dramatic illustrations of the need to employ the dynamical

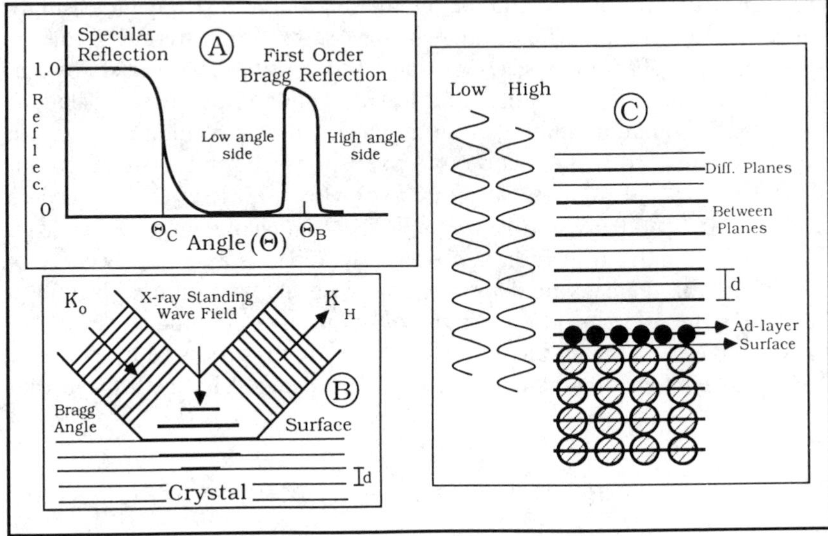

Figure 24. (A) Depiction of specular and first-order Bragg reflection from a surface. (B) Generation of an x-ray standing-wave field. (C) Movement of an x-ray standing-wave field in the $-\mathrm{H}$ direction upon advancing the angle of incidence across a Bragg reflection. (From Abruña, H. D., White J. H., et al., *J. Phys. Chem.* **92**, 7045 (1988), with permission.)

treatment to the above-mentioned conditions is the so-called Borrmann effect, or anomalous transmission.[74]

As mentioned above, the XSW field arises from the interference between the coherently related incident and Bragg-diffracted beams from the surface of a perfect crystal. In the vicinity of a Bragg reflection (Fig. 24A–B), an incident plane wave (with wave vector k_0) and a reflected wave (with wave vector k_H) interfere to generate a standing wave with a periodicity equivalent to that of the (h, k, l) diffracting planes. The ratio of the electric field amplitudes of the reflected and incident waves is given by

$$\frac{E_H}{E_0} = -|b|^{1/2}\frac{|P|}{P}\frac{(F_H F_{\bar{H}})^{\frac{1}{2}}}{F_{\bar{H}}}[\Delta\Theta \pm (\Delta\Theta^2 - 1)^{1/2}] \tag{13}$$

where $b = \gamma_0/\gamma_H$ and γ are the direction cosines for k_0 and k_H for a symmetrically cut $(b = -1)$ centrosymmetric crystal so that $F_H = F_{\bar{H}}$. P is the polarization factor and is 1 for σ polarization and $\cos 2\Theta_B$ for π polarization. F_H and $F_{\bar{H}}$ are the structure factors for the (h, k, l) and $(\bar{h}, \bar{k}, \bar{l})$ planes, respectively, and $\Delta\Theta$ represents the angular deviation from the Bragg angle Θ_B $(\Delta\Theta = \Theta - \Theta_B)$.

For $\Theta_B < \pi/2$ and for centrosymmetric crystals (e.g., Si, Ge) the reflectivity is given by

$$R = \left|\frac{E_H}{E_0}\right|^2 = |\Delta\Theta \pm (\Delta\Theta^2 - 1)^{1/2}|^2 \tag{14}$$

The electric field intensity of the standing-wave field at a point r is given by

$$I = |\bar{E}_{tot}|^2; \quad \bar{E}_{tot} = \exp(2\pi i v t)\exp(-2\pi i \bar{K}_0 \cdot \bar{r})[\bar{E}_0 + \bar{E}_H \exp(-2\pi i \bar{H} \cdot \bar{r})] \tag{15}$$

$$= |E_0|^2 \exp(-\mu_z Z)[1 + (E_H/E_0)^2 + 2P(E_H/E_0)\cos(v - 2\pi \bar{H} \cdot \bar{r})] \tag{16}$$

where the $(-\mu_z Z)$ factor takes into account normal and anomalous absorption and extinction.

The standing wave develops not only in the diffracting crystal, but also extends well beyond its surface. Estimates of this coherence length range to values as large as 1000 Å from the interface.[75] The nodal and antinodal planes of the standing wave are parallel to the diffracting planes, and the nodal wavelength corresponds to the d-spacing of the difracting planes. As the angle of incidence is advanced through the strong Bragg reflection, the relative phase between the incident and reflected plane waves (at a fixed point in the crystal) changes by π. Due to this phase change, the antinodal planes of the standing-wave field move in the $-H$ direction normal to the diffraction planes by 1/2 of a d-spacing, from a position halfway between the (h, k, l) diffracting planes (low-angle side of the Bragg reflection) to a position that coincides with them (high-angle side of the Bragg reflection) (see Fig. 24C). Thus, the standing wave can be made to sample an adsorbate or overlayer at varying positions above the substrate interface.

For an atomic overlayer positioned parallel to the diffracting planes, the nodal and antinodal planes of the standing wave will pass through the atom plane as the angle is advanced. Using an incident-beam energy at or beyond the absorption edge of the atoms in the overlayer, the fluorescence emission yield will be modulated in a characteristic fashion as the substrate is rocked in angle. The yield can be expressed as an integral that incorporates a distribution function $f(x)$ for the absorber:

$$\text{yield} = \int_{-d/2}^{+d/2} I f(x)\,dx$$

$$Y(z, \Theta) = \int I(z, \Theta) f(z)\,dz \tag{17}$$

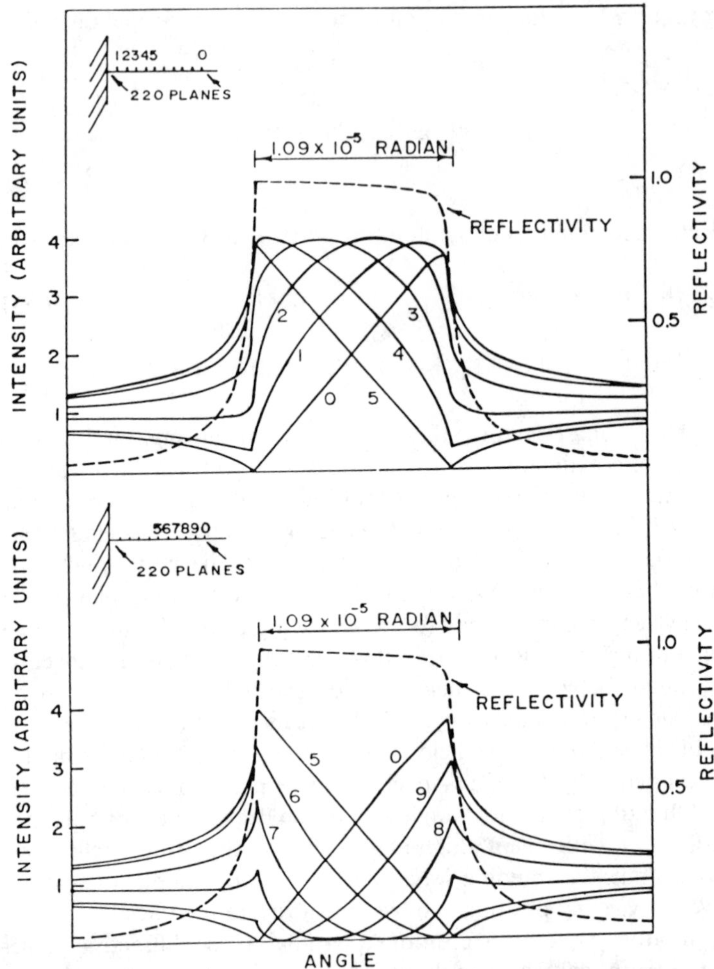

Figure 25. X-ray field intensities at extended Ge(220) lattice positions (0–9) for a perfectly collimated incident x-ray beam. An atomic ad-layer whose center falls on one of these positions would have its characteristic fluorescence intensity modulated in the same fashion. The dashed curve represents the Bragg reflectivity profile. (From M. J. Bedzyk, Ph.D. Thesis, SUNY Albany, 1982.)

Figure 25 depicts the angular dependence of the electric field and fluorescence yield for an adsorbate layer located at varying positions with respect to the diffracting planes. The phase and amplitude of this modulation (or so-called coherent position and coherent fraction) are a measure of the mean position $\langle Z \rangle$ and width $\sqrt{\langle Z \rangle^2}$ of the distribution of atoms in the overlayer. The coherent fraction (f_c) is defined as the mth coefficient of the Fourier series representation of the distribution of absorbers, and is incorporated into the yield equation as

$$Y(z, \Theta) = |E_0|^2\{(1 + R)(1 - f_c) + [(1 + R + 2P\sqrt{R}\cos(v - 2\pi\mathbf{H}\cdot\mathbf{r})]f_c\} \quad (18)$$

The coherent fraction (f_c) represents the fraction of absorbers occupying a distribution centered at a given coherent position. Since the Z scale of the XSW is modulo-d, if several coherent positions are possible, a single measurement will not be sufficient to unambiguously assign the positions of the absorbers. Thus, measurements in the specular reflection region (mostly applicable to layered synthetic microstructures (vide infra) or mirrors) or high-order Bragg measurements must be performed to allow such an assignment. The specular measurement is particularly useful in cases where changes in position across one or more d-spacings occur, since the periodicity of the standing wave in this regime is of the order of several hundred angstroms.

The Z scale of the standing wave field in the Bragg case is mod-d and points in the direction normal to the diffraction planes. Standing wave measurements of $\langle Z \rangle$ and $\sqrt{\langle Z \rangle^2}$ can be accurate to within 1% and 2% of the d-spacing, respectively.[76] Golovchenko and co-workers[77] have applied this technique to the study of surface adsorbates, monitoring the angular dependence of the fluorescence yield of bromine chemisorbed onto a Si(111) crystal. It should be mentioned that these experiments were performed while the crystal was covered by a thin film of methanol, pointing to the feasibility of performing experiments at the solid–liquid interface. Bedzyk and Materlik[78] used the angular dependence of the fluorescence yield of bromine to relate its position relative to the Ge(111) and Ge(333) diffracting planes. In addition, they demonstrated the feasibility of using higher-order reflections for determining the thermal vibration amplitude of the bromine adsorbate.

One of the problems associated with the implementation of the standing-wave technique is the fact that it requires the use of perfect or nearly perfect crystals. This presents a problem, especially for relatively soft materials such as copper, gold, silver, and platinum, which are not only very difficult to grow in such high quality, but are also very difficult to maintain in that state. Thus, most experiments have been performed on silicon or germanium single crystals. In addition, although the characteristic modulo-d length scale of a few angstroms of single crystals in these measurements is perfect for determining bond lengths between atom layers at single-crystal surfaces, it is inappropriate for a structural determination of systems extending over several tens of angstroms (e.g., ionic distributions at charged surfaces).

An alternative to the use of perfect crystals is the use of layered synthetic

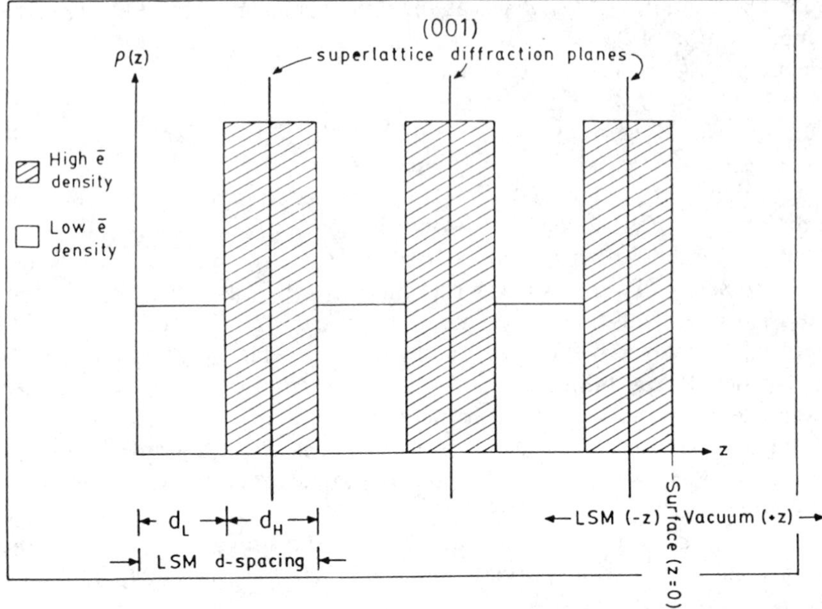

Figure 26. Density profile of an ideal (sharp interfaces) LSM as a function of depth. The (001) superlattice diffraction planes are centered in the layers of high-electron density material.

microstructures (LSMs).[79] LSMs are depth-periodic structures consisting of alternating layers of high and low electron density materials (such as tungsten and silicon, or platinum and carbon), and are of high enough quality to be considered superlattice crystals (Fig. 26). LSMs provide several advantages over natural crystals for x-ray standing-wave experiments:

1. LSMs can be produced with fundamental *d*-spacings ranging from 15 to 200 Å, as compared to a few angstroms for natural crystals. These large *d*-spacings give rise to long-period standing waves that are optimally suited to investigate surface systems with a large periodic scale.
2. The experimenter can choose to synthesize LSMs from a wide range of materials; there is only a limited choice of natural crystals. Even more important, the experimenter can choose the material to be used as the multilayer's top surface.
3. Experimental reflection curves from LSMs compare well with predictions from dynamical diffraction theory, and peak reflectivities are as high as 80%. Therefore, a well-defined standing wave can be produced (Fig. 27).

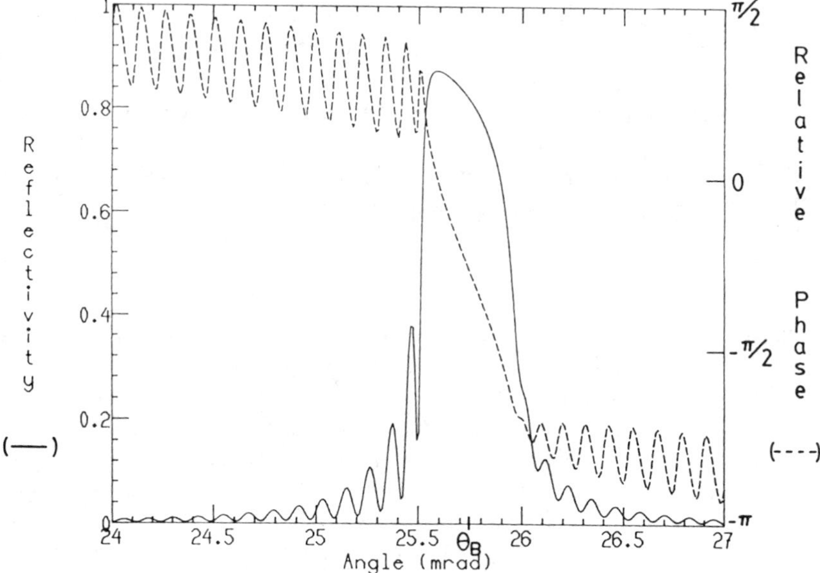

Figure 27. Theoretical reflectivity (solid line) for the Bragg reflection of a W/Si LSM (d-spacing = 25 Å) for 9.8-keV incident x-rays. Also shown is the relative phase (dashed line) of the incident and diffracted beams. This phase changes by π as the LSM is rocked across the strong Bragg diffraction.

4. Due to the low number of layer pairs that effect Bragg diffraction, LSMs have a rather large energy bandpass, and Bragg reflection angular widths of the order of milliradians rather than microradians, as in natural crystals, (Fig. 27). This last point is important as it considerably simplifies experimental design.

In addition to being good diffracting structures, LSMs possess surfaces of mirror quality, making them excellent x-ray reflectors as well. Therefore, they can also be used as substrates for generating total external or specular reflection XSWs.

Since the refractive index of matter for x-rays is less than unity, when x-rays are incident on a mirror surface at an angle smaller than the critical angle, the angle of refraction becomes imaginary. This means that all the incident-wave energy is going into the reflected wave, and total external or specular reflection is taking place. Under this specular reflection condition, the refracted wave propagates parallel to the surface with nearly the free-space wavelength, and an exponentially damped evanescent wave penetrates into the medium. At the same time, the incident and specularly reflected plane waves interfere to form a standing wave in the more optically dense medium above the mirror surface. The angular behavior of the specular standing wave can be described as follows. At $\theta = 0$ a node is at the mirror surface and the first antinode is at infinity. As the angle of incidence is increased, the first antinode moves inward, in a direction normal to the surface, until at the critical angle it coincides with the mirror surface. The trailing antinodes follow behind with a periodic spacing given by

$$D = \lambda/(2 \sin \theta) \tag{19}$$

Because the penetration depth of the evanescent wave is very small, x-ray reflection in the total reflection region is very sensitive to the physical and chemical conditions of the surface layers, and much attention has been given to this method as a structure probe.[80] However, little or no attention has been paid to the standing wave created above the reflecting surface and its possible applications as a long-period (hundreds of angstroms) structural probe. Such a probe would be ideal to study a system in which a plane of heavy atoms is displaced outward from a mirror surface by a low-z material tens of angstroms thick. Total-external-reflection standing waves could also be employed in the structural investigation, on an atomic scale, of ionic distributions at a charged surface, as these distributions are believed to extend several hundreds of angstroms.

B. Experimental Aspects

One of the main difficulties with x-ray standing-wave measurements is that this technique is experimentally very demanding. Although the experimental set-up is not particularly complex, alignment of the sample relative to the beam is critical. An XSW experiment typically consists of monitoring some signal proportional to the standing-wave electric field intensity as the angle of incidence is scanned across a strong Bragg reflection. A typical experimental setup is shown in Fig. 28 and consists of a collimated beam (defined by a pair of slits) whose intensity I_0 is monitored with an ion chamber, a sample stage, a reflected beam monitor I_R, and a detector at 90° relative to the x-ray beam. Of particular importance in this experiment is the angular resolution of the sample stage, since a typical reflection width for a single crystal will be of the order of tens of microradians and a few milliradians for LSMs. In both cases, however, high angular resolution is required in order to have a well-resolved reflectivity profile. In addition, when measuring fluorescence from an adsorbate layer, care must be taken to accurately subtract background radiation.

Typically, the signal from the fluorescence detector is passed through a pulse height analyzer to discriminate against unwanted contributions to the

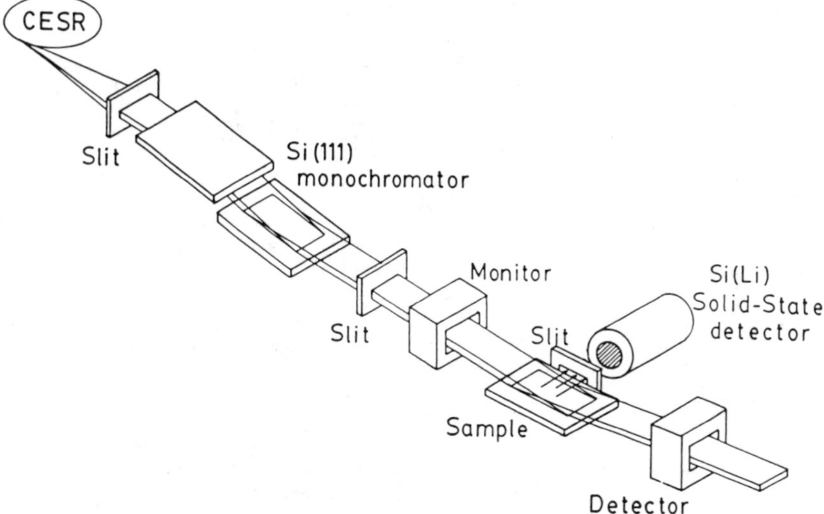

Figure 28. Experimental set-up for x-ray standing-wave measurements on an LSM.

signal. With an energy-dispersive semiconductor detector, an inelastic x-ray energy spectrum can be obtained with a multichannel pulse height analyzer for each angle of incidence. Subsequent analysis permits undesirable contributions and background to be subtracted from the total signal.

In the case of electrochemical experiments, cell design for XSW measurements is of critical importance. In general, experiments are performed in a thin-layer configuration and cell materials typically include teflon and Kel-F. Two cell designs previously employed in such measurements are shown in Figs. 10 and 29.

C. Analysis of Data and Interpretation of Results

Analysis of x-ray standing-wave data is based on a fit of the data (reflectivity and fluorescence yield) to those predicted from theory. However, the data must first be treated to extract yields corrected for background and other contributions. In general, the fluorescence yield is recorded as a function of angle of incidence. An energy dispersed spectrum for each angle is recorded in digital memory. Fitting of the desired emission line to an assumed functional form (usually a combination of Gaussian functions) and subtraction of an extrapolated polynomial background serve to render the data in a form suitable for reconstruction of the fluorescence yield as a function of the angle of the incident radiation.

The electric field intensity at a given point in the crystal must either be

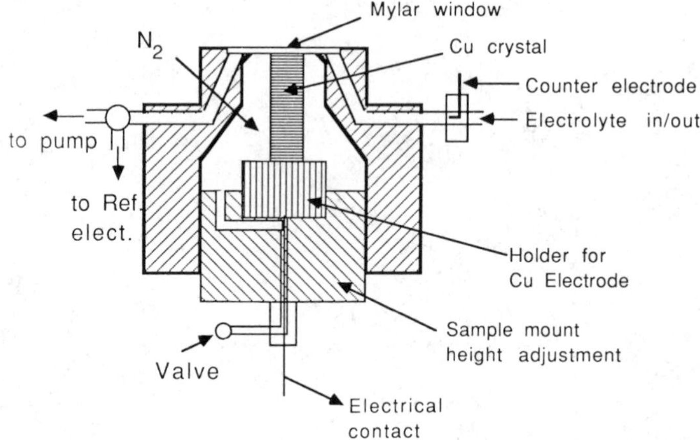

Figure 29. Electrochemical cell for *in situ* x-ray standing-wave measurements (From Materlik, G., Schmah, M. et al., Ber. Bunsenges. Phys. Chem. **91**, 292 (1987).

calculated form dynamical theory (i.e., wavefield amplitude from the dispersion equations) or from an optical theory approach. The latter approach is generally based on a stratified medium formalism in which the medium is divided into parallel slabs.[79,81] The continuity of the tangential components of the electric and magnetic fields at each of the resulting interfaces is the essential requirement invoked to obtain a recursion relation (containing Fresnel coefficients) describing the *e*-field amplitude raio for the reflected and incident beams. Such a treatment is applicable to the total external reflection condition as well as to diffraction.

The layered medium approach is particularly well suited for analysis of standing-wave electric fields in multilayered structures.[79] The recursion relation employed generally has the following form:

$$R_{j,j+1} = a_j^2 \left[\frac{R_{j+1,j+2} + F_{j,j+1}}{1 + R_{j+1,j+2} F_{j,j+1}} \right] = \frac{E_J^R(0)}{E_j(0)} \tag{20}$$

where $R_{j,j+1}$ = ratio of reflected E-field amplitude to incident E-field amplitude for jth layer

$R_{j+1,j+2}$ = ratio of reflected wave amplitude to incident wave amplitude for $(j+1)$th layer

$F_j, j+1$ = Fresnel coefficient for jth layer

a_j = complex amplitude factor at the $j, j+1$ interface

The reflectivity at a given interface is the squared modulus of $R_{j,j+1}$. The total reflectivity at a given angle of a structure consisting of n layers is

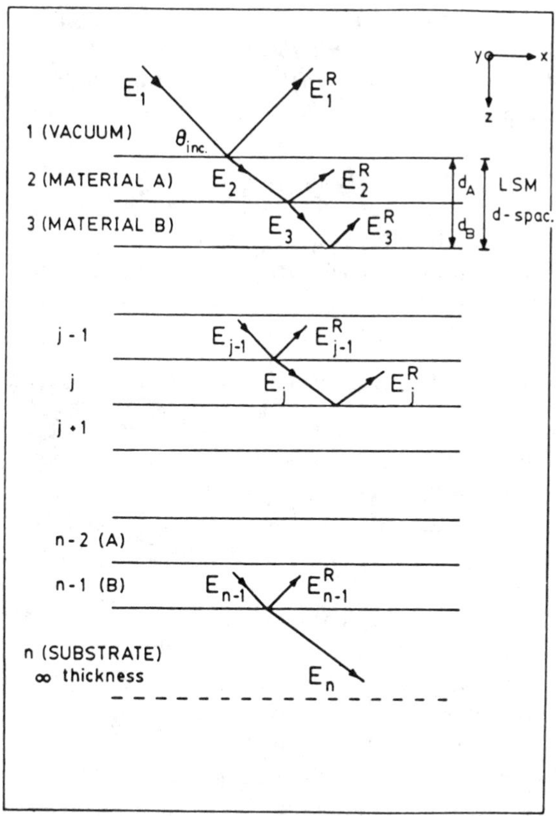

Figure 30. Cross-section of a stratified interface identifying the nomenclature used in describing the incident (E_j) and reflected (E_j^R) electric fields as well as the layer numbering.

obtained by applying the recursion relation $n - 1$ times from the substrate (or from the extinction length) to the topmost layer (Fig. 30).

The reflectivity in the total external reflection region is that of the topmost layer. It should be noted that in this region, the electric field intensity is a continuous function of angle of incidence. At the surface of the multilayer, the electric field intensity is given by[82]

$$I(\Theta) = |E_0|^2[1 + R_{1,2}|^2 + 2|R_{1,2}|\cos(v)] \tag{21}$$

In general, the E-field intensity in the total external reflection region can be determined from

$$I(z) = |E_0|^2[1 + |R_{1,2}|^2 + 2|R_{1,2}|\cos(v - 4\pi \sin\Theta_{\text{inc}} z/\lambda)] \tag{22}$$

Thus, one would anticipate a continuously periodic standing wave (of period $D = \lambda/2 \sin \Theta$) starting with an infinitely long periodicity at zero angle of incidence. This type of measurement has recently been employed in the examination of phase transitions in trilayers of cadmium and zinc stearates on a W/Si LSM.[83]

The coherent position of an absorber is obtained by fitting the observed fluorescence yields to equations of the form of Eq. (17) or (22). Incorporation of coherent fraction into these equations allows a fit to be obtained in terms of three parameters: normalized coverage of absorber, coherent position, and coherent fraction. In Bragg-diffraction standing-wave experiments, the coherent position is extracted modulo-d.

D. X-ray Standing-Wave Studies at Electrochemical Interfaces

Due to the experimental difficulties involved, there have been only three reports of XSW measurements at electrochemical interfaces. Materlik and co-workers have studied the underpotential deposition of thallium on single-crystal copper electrodes under both ex-situ[84] and in-situ[85] conditions. In addition, they report results from studies in the absence and presence of small amounts of oxygen.

In the ex-situ studies, the thallium layer was electrodeposited, and the electrode was subsequently removed from solution and placed inside a helium-filled box where the XSW experiments were carried out.

For the in-situ studies, an electrochemical cell was designed to hold the nearly perfect copper crystal in contact with a thin layer (20 to 50 μm) of electrolyte. Figures 29 and 31 show the cells employed in the in- and ex-situ experiments, respectively. In addition, Fig. 31 shows the voltammetric traces obtained for the deposition of T1 in the presence and absence of oxygen. In the experiments, they simultaneously monitored the reflectivity and the T1 fluorescence intensity. Figure 32 shows the results for the ex-situ study.

From an analysis of their data, they were able to determine that for the ex-situ case and in the absence of oxygen, the thallium atoms are located at two-fold sites at a mean distance of 2.67 ± 0.03 Å. For the in-situ case and again in the absence of oxygen, the data are consistent with the thallium atoms being at 2.58 ± 0.02 Å, but at three-fold sites.

In the presence of oxygen, there is a significant contraction of the mean distance of the thallium to 2.27 ± 0.04 Å. This is ascribed to a surface reconstruction of the copper induced by the adsorbed oxygen, which results in an inward shift of the copper surface atoms by about 0.3 Å. This is consistent with low-energy, ion-scattering studies. In general, these studies are of great significance since they demonstrate the applicability of the x-ray standing-wave technique to the in-situ study of electrochemical interfaces, even employing single crystals.

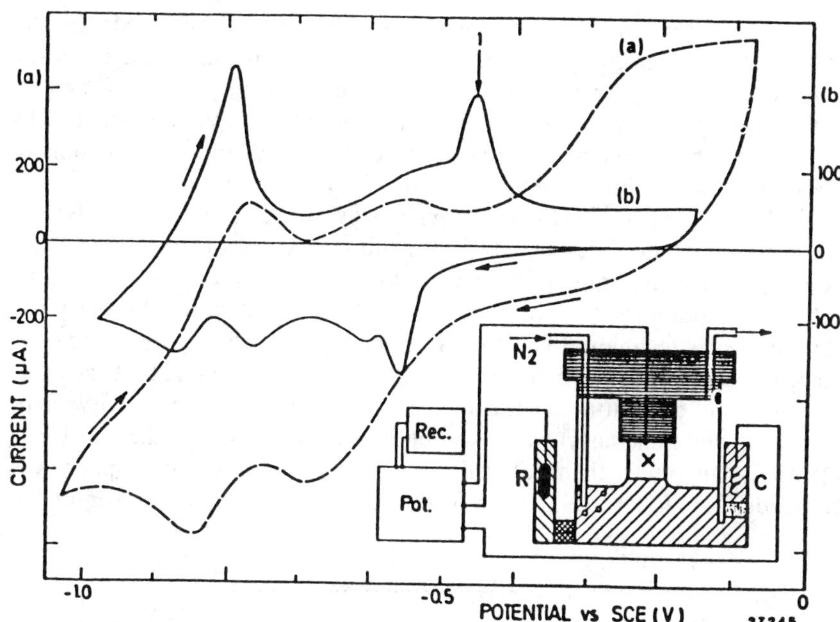

Figure 31. Voltammograms for T1 deposition onto a copper single crystal in the presence (a) and absence (b) of traces of oxygen. Inset: Electrochemical cell. (From Materlik, G., Schmah, M., et al., Ber. Bunsenges. Phys. Chem. **91**, 292 (1987).

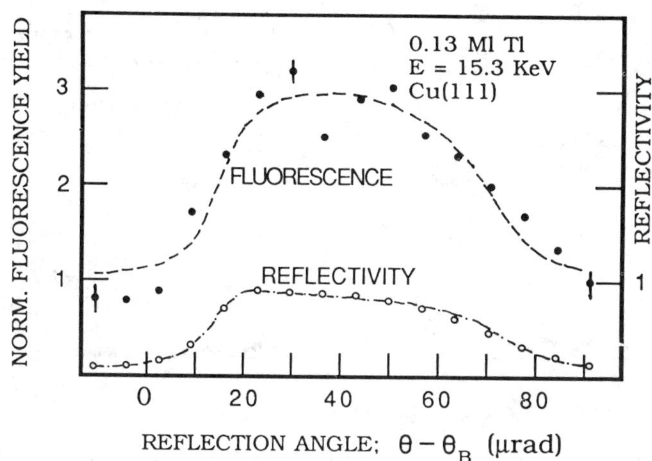

Figure 32. Angular dependence of the Cu(111) reflectivity and the normalized Tl_L fluorescence yield. ○, ● experiment – –,–·– least squares fit. Materlik, G., Zegenhagen, J., and Uelhoff, W., *Phys. Rev. B.* 32, 5502 (1985), with permission.)

We have performed some experiments on the use of LSMs in the investigation of electrochemical interfaces.[86] The system that we have studied involves the adsorption of iodide onto a platinum/carbon LSM followed by the electrodeposition of a layer of copper. The LSM samples consisted of 15 platinum/carbon layer pairs with each layer having 26 and 30 angstroms of platinum and carbon, respectively, with platinum as the outermost layer. 9.2 keV radiation from the Cornell High Energy Synchrotron Source (CHESS) was used to excite L-level and K-level fluorescence from the iodide and copper, respectively. Initially, the LSM was contacted with a 35-mM aqueous solution of sodium iodide for 15 minutes. It was then studied by the x-ray standing-wave technique. The characteristic iodine L fluorescence could be detected, and its angular dependence was indicative of the fact that the layer was on top of the platinum surface layer. A well-developed reflectivity curve (collected simultaneously) was also obtained. Following this, the LSM was placed in an electrochemical cell and half a monolayer of copper was electrodeposited. The LSM (now with half a monolayer of copper and a

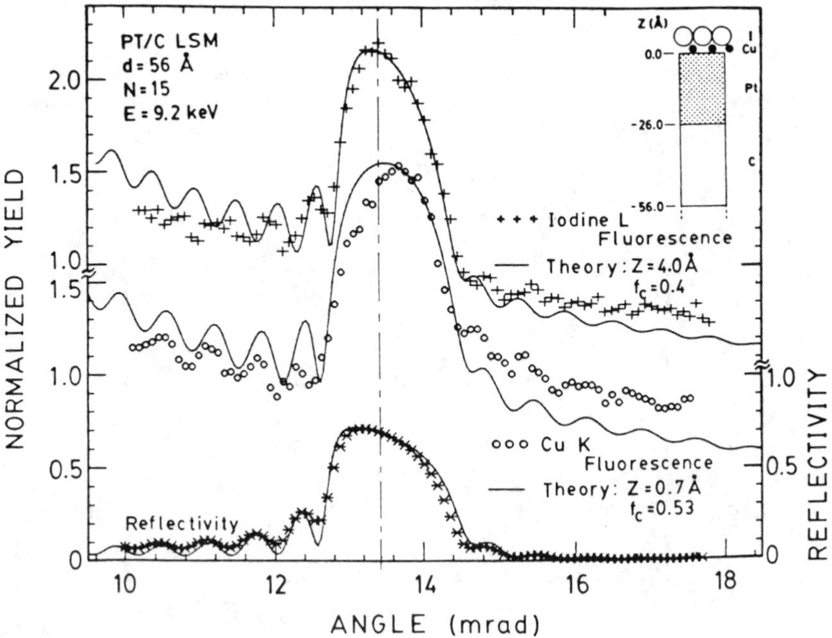

Figure 33. Experimental results and least squares fit of data (solid lines) for a Pt/C LSM covered with an electrodeposited layer of copper and an adsorbed layer of iodine. Topmost curve: I_L fluorescence. Middle curve: copper K_α fluorescence. Bottom curve: reflectivity. (From Abruña, H. D., White, J. H., et al., *J. Phys. Chem.* **92**, 7045 (1988), with permission.)

monolayer of iodide) was again analyzed by the x-ray standing-wave technique. Since the incident x-ray energy (9.2 keV) was capable of exciting fluorescence from both the copper and iodide, the fluorescence intensities of both elements (as well as the reflectivity) were obtained simultaneously. The results presented in Fig. 33 show the reflectivity curve and the modulation of the iodide and copper fluorescence intensities. The most important feature is the noticeable phase difference between the iodide and copper modulations, i.e., the location of the iodide and copper fluorescence maxima, with the copper maximum being to the right of the iodine maximum. Since the antinodes move inward as the angle increases, the order in which these maxima occur can be unambiguously interpreted as meaning that the copper layer is closer than the iodide layer to the surface of the platinum. Since the iodide had been previously deposited on the platinum, this represents unequivocal evidence of the displacement of the iodide layer by the electrodeposited copper. Similar findings based on Auger intensities and LEED patterns have been previously reported by Hubbard and co-workers.[87] In addition, from an analysis of the copper fluorescnce intensity, we were able to determine that the electrodeposited layer had a significant degree of coherence (53%) with the underlying substrate (Fig. 33).

Most recently, we have used the XSW technique to study the potential dependence of the adsorption of iodide onto a Pt/C LSM (layered synthetic microstructure) consisting of 200 layer pairs of carbon (14 Å) and platinum (26 Å), respectively). In this experiment, we simultaneously monitored the reflectivity across the first-order Bragg reflection as well as the characteristic iodine fluorescence intensity (Fig. 34). From an analysis of the data, we were able to determine that at −0.90 V essentially no iodide is present at the Pt/solution interface. As the potential is made progressively positive, the iodide fluorescence intensity increases and the peak maximum shifts to lower angles up to a potential of +0.40 V. At +0.49 V, the fluorescence intensity decreases and the peak maximum again moves towards higher angles. Also note the difference in the tails of each line, which seems to indicate a lack of coherence for the scatterers at the more positive potentials, in contrast to the more negative potential (−0.1 V) where coherence is observed in the XSW fluorescence yield due to the limited extent of the I^- diffuse layer. Qualitatively, these data appear to indicate that at −0.90 V there is no iodide adsorbed and that the coverage increases as the potential is made progressively positive. The movement of the peak maximum to lower angles indicates that, on average, the interfacial iodine density is moving away from the electrode surface.

The initial fits determined using a distribution consisting of a step at the electrode's surface to describe the I ad-layer formed there, and an exponential tail extending out from this step to describe the I^- diffuse layer, seem to

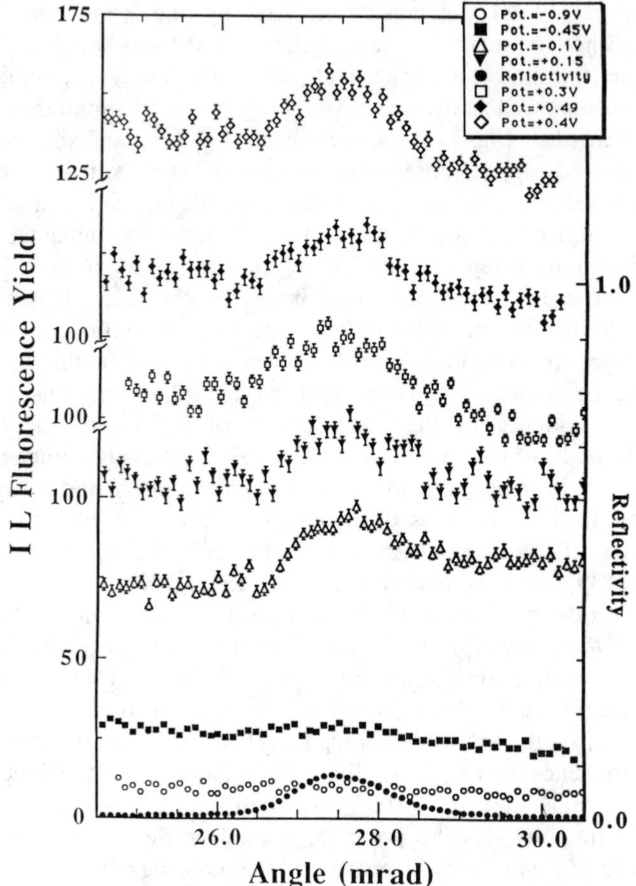

Figure 34. Normalized iodine fluorescence intensity vs. applied potential and reflectivity profile for a Pt/C LSM immersed in an aqueous solution containing 50 μM NaI/0.1 M Na_2SO_2. (From Abruña, H. D., White, J. H., et al., *J. Phys. Chem.* **92**, 7045 (1988), with permission.)

indicate that coherence is largely destroyed for the more positive potentials due to the extension of the diffuse tail across many d-spacings, where the observed offset in the tails indicates a coherent distribution of scatterers. It should be noted that the extension of iodide across many d-spacings requires that an E-field weighted fluorescence yield must be used, and this will, in general, lead to a rather complicated expression for the yield, which must be used in fitting the data.

As mentioned in the introduction, one of the limitations of the x-ray standing-wave technique is that it requires the use of perfect crystals.

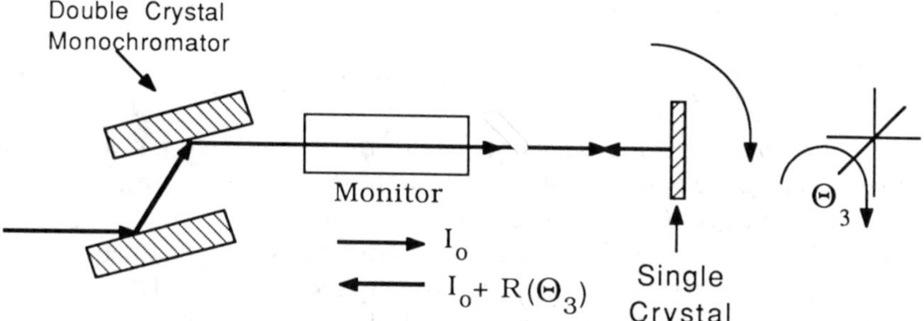

Figure 35. Experimental set-up for back-reflection x-ray standing-wave measurements.

Although LSMs can be employed, the layers within an LSM are amorphous and therefore not well-defined. It would be most desirable to be able to perform XSW studies on single-crystal surfaces without having to resort to perfect or nearly perfect crystals.

An alternative is offered by employing a backscattering geometry on a single crystal rather than the conventional near-grazing incidence. Such an experimental arrangement is depicted in Fig. 35. The dynamical theory of x-ray diffraction predicts that in a backscattering geometry the reflection widths for a single-crystal surface will be of the order of milliradians as opposed to microradians under conventional near-grazing incidence. This means that single crystals with a larger mosaic spread (such as those encountred in typical good quality single crystals) can be employed in such measurements.

In fact, very recent studies[88] have shown the feasibility of this approach to XSW studies, although all of these studies reported to date have been under UHV. However, the applicability to electrochemical interfaces is clear.

IX. SURFACE DIFFRACTION

A. Introduction

X-ray diffraction represents one of the most powerful methods to study the structure of crystalline materials, and diffraction methods can be employed in the study of two- and three-dimensional structural details. In this section, some of the basic concepts of diffraction are covered, but the interested reader is referred to the extensive literature on the subject for a more in-depth treatment.[89]

A crystal has long-range translational order and is composed of a periodic array of identical structural units termed unit cells. The unit cells in turn make up the crystalline lattice. Each atomic position in a crystal is described by a vector

$$\mathbf{R}_{xyz} = x\mathbf{a} + y\mathbf{b} + z\mathbf{c} \tag{23}$$

where $\mathbf{a}, \mathbf{b}$, and $\mathbf{c}$ are the lattice vectors of the crystal and x, y, and z are integers. There are two general approaches to the description of diffraction, the Bragg and Von Laue, respectively.

The simplest way to view diffraction is from the Bragg treatment where the atoms in a crystal occupy positions on sets of parallel lattice planes spaced at a characteristic distance d, (the d-spacing). In the Bragg treatment, when x-rays are incident on a set of planes (Fig. 36), a peak in the intensity of the scattered radiation will occur when reflections from successive planes interfere constructively. In order for this condition to be satisfied, the path length difference (equal to $2d \sin \theta$) must equal an integral number of wavelengths. Thus we obtain the well-known Bragg's law which states that

$$n\lambda = 2d \sin \theta \tag{24}$$

where n is an integer (order of the reflection), λ is the wavelength of the x-ray, and θ is the angle of incidence.

When discussing x-ray diffraction, it is convenient to define a scattering vector $\mathbf{Q} = k_0 - k_{\mathrm{H}}$ in terms of the incident (k_0) and scattered (k_{H}) wave vectors. Since the scattering is elastic, both k_0 and k_{H} have the same magnitude which is equal to $2\pi/\lambda$. Thus the magnitude of the scattering vector is $Q = (4\pi/\lambda) \sin \theta$.

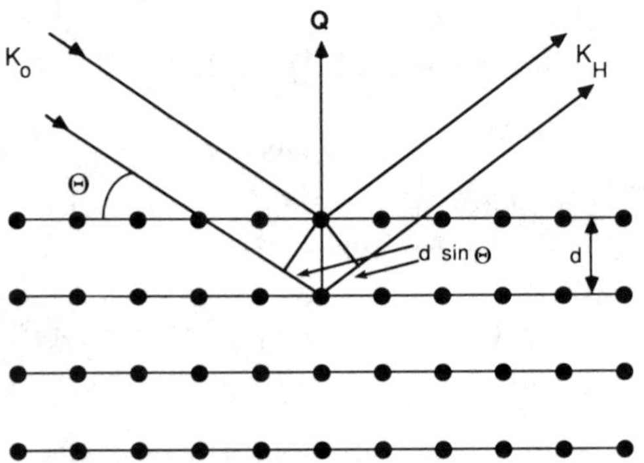

Figure 36. Bragg reflection from a set of lattice planes with interplanar distance d. K_0 and K_H refer to incident and reflected beams, respectively. The scattering vector (Q) is also shown.

This leads to an alternate expression of Bragg's law as

$$Q = 2\pi/d \tag{25}$$

Although Bragg's treatment of diffraction gives us an intuitive sense, experimentally one measures scattered intensities, and thus to gain an understanding of these processes it is best to consider an alternate treatment, which is that of Von Laue. Using the Born approximation, the scattering amplitude (A) can be calculated by summing the contribution from each electron:

$$A = \frac{e^2}{mc^2 R} \int \exp^{iQ\cdot r} \rho(r) d^3 r \tag{26}$$

where $\rho(r)$ is the electron density, $e^2/mc^2 R$ is the scattering from a single electron (neglecting the polarization factor $(1 + \frac{1}{2}\cos^2\theta)$), and R is the distance between the sample and the detector. The total electron density is obtained by summing the electron density of each atom in the crystal. Thus, the expression for the scattered amplitude becomes

$$A = \frac{e^2}{mc^2 R} \left(\int \exp^{iQ\cdot r} \rho(r) d^3 r \right) \left(\sum_{xyz} \exp^{iQ\cdot R_{xyz}} \right) \tag{27}$$

From this one can define the atomic form factor $f(Q)$ as the Fourier transform of the electron density, and the structure factor $F(Q)$ as

$$F(Q) = \sum \exp^{(iQ\cdot R_{xyz})} \tag{28}$$

The amplitude can thus be expressed as

$$A = \frac{e^2}{mc^2 R} f(Q) F(Q) \tag{29}$$

The scattered intensity (I) is given by the square of the amplitude and is equal to

$$I = |A|^2 = \left(\frac{e^2}{mc^2 R} \right)^2 |f(Q)|^2 \, |F(Q)|^2 \tag{30}$$

In order to determine the conditions necessary for diffraction to occur, the scattering vector **Q** must be expressed in terms of the reciprocal lattice vectors of the crystal:

$$\mathbf{Q} = q_a\mathbf{a}' + q_b\mathbf{b}' + q_c\mathbf{c}' \tag{31}$$

Where $\mathbf{a}'$, $\mathbf{b}'$ and $\mathbf{c}'$ are the reciprocal lattice vectors and are given by

$$\mathbf{a}' = \frac{b \times \mathbf{c}}{V}, \qquad \mathbf{b}' = \frac{c \times \mathbf{a}}{V}, \qquad \mathbf{c}' = \frac{a \times b}{V} \tag{32}$$

where V is the unit cell volume and is given by $(a \cdot b \times c)$.

The structure factor can then be expressed as

$$|F(Q)|^2 = |F(q_a, q_b, q_c)|^2 = \left| \sum_{x=1}^{N_a} \sum_{y=1}^{N_b} \sum_{z=1}^{N_c} \exp^{i(xq_a + yq_b + zq_c)} \right|^2 \tag{33}$$

$$= \frac{\sin^2(\frac{1}{2}N_a q_a)}{\sin^2(\frac{1}{2}q_a)} \frac{\sin^2(\frac{1}{2}N_b q_b)}{\sin^2(\frac{1}{2}q_b)} \frac{\sin^2(\frac{1}{2}N_c q_c)}{\sin^2(\frac{1}{2}q_c)} \tag{34}$$

where N_a, N_b, and N_c are the total number of unit cells in the a, b, and c directions, respectively.

When N_a, N_b, and N_c are large, this function is sharply peaked (at Bragg points) when the three Laue conditions are satisfied:

$$q_a = 2\pi h, \qquad q_b = 2\pi k, \qquad q_c = 2\pi l \tag{35}$$

where h, k, l are integers. Under these conditions, the intensity is $(N_a N_b N_c)^2$. This treatment (Von Laue's) describes the diffracted intensity in a three-dimensional crystal and can be used as a point of departure to determine what happens when we have reduced dimensionality, that is, what happens in going from three to two dimensions.

Consider a set of crystal planes with a given d-spacing and the corresponding Bragg points in the reciprocal lattice. If we progressively increase the d-spacing of the planes (by pulling them further apart), the Bragg points in reciprocal space will get concomitantly closer. In the limit of an infinite d-spacing, that is, having a single layer of atoms, instead of having discrete Bragg points in reciprocal space we will have a continuous line of diffracted intensity termed a Bragg rod.

We can also use the formalism presented previously to treat diffraction from a two-dimensional solid. In essence, we consider the diffracted intensity that would arise from a crystal that is only one layer thick, since we only need to satisfy two of the Laue conditions. In this case, the expression for the diffracted intensity reduces to

$$|F(q_a, q_b, q_c)|^2 = \frac{\sin^2 \frac{1}{2}N_a q_a}{\sin^2 \frac{1}{2}q_a} \frac{\sin^2 \frac{1}{2}N_b q_b}{\sin^2 \frac{1}{2}q_b} \tag{36}$$

The diffracted intensity in this case will be independent of the q_c vector.

As mentioned previously, a crystal will diffract x-rays with an intensity proportional to the square of the structure factor and is described by Eq. (33). The abrupt termination of the lattice at a sharp boundary (i.e., a surface) causes two-dimensional diffraction features termed crystal truncation rods (CTRs).[90] Measurements of CTRs can provide a wealth of information on surface roughness and may be useful in the determination of crystallographic phase information.[90,91]

Again, we take as a point of departure diffraction from a three-dimensional crystal and relax one of the Laue conditions ($q_c \neq 2\pi l$). The expression for the diffracted intensity reduces to

$$|F(2\pi h, 2\pi k, q_c)|^2 = N_a^2 N_b^2 \frac{\sin^2 \frac{1}{2}q_c N_c}{\sin^2 \frac{1}{2}q_c} \tag{37}$$

For large N_c the numerator on the last term on the right-hand side of Eq. (37) reduces to $\frac{1}{2}$ so that the structure factor becomes

$$|F(2\pi h, 2\pi k, q_c)|^2 = N_a^2 N_b^2 \frac{1}{2 \sin^2 \frac{1}{2}q_c} \tag{38}$$

Thus, in this case the diffracted intensity will consist of diffuse streaks that connect Bragg points, and these are the CTRs. It should be noted that while the scattering intensity from a CTR is comparable in magnitude to that for an isolated monolayer, it has a $(2\sin^2 \frac{1}{2}q_c)^{-1}$ dependence whereas the intensity from a monolayer is independent of q_c. This characteristic dependence of the scattering intensity with q_c can be used to obtain information on interface roughness. For an ideally truncated surface, the intensity will decay as $(2\sin^2 \frac{1}{2}q_c)^{-1}$ whereas it will decrease more sharply in the presence of interface roughness. Another point to be considered is that the ratio of intensities for a Bragg reflection to a CTR is of the order 10^5 so that intensities need to be measured over a rather wide dynamic range.

Figure 37 presents data of Robinson for tungsten surfaces. The dashed lines represent the decay of the diffracted intensity predicted by Eq. (38) and clearly there are significant discrepancies with the experimental values (open circles). If a roughness factor is introduced (Robinson used a partial occupancy model), the data can be fit quite well (solid lines).

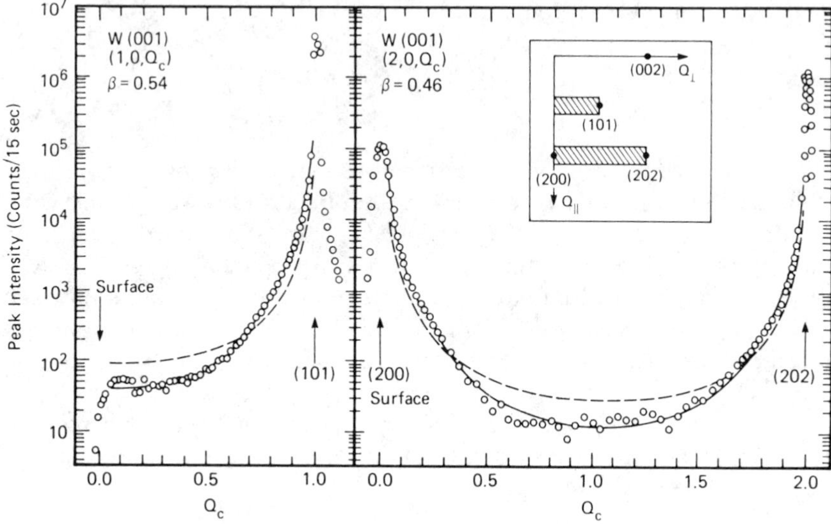

Figure 37. Crystal truncation rods (CTRs) for W(100) in ultrahigh vacuum. Open circles, experimental data. Solid and dashed lines are, respectively, fits with and without consideration of surface roughness. Inset: Reciprocal space diagram. (From Robinson, I. K., *Phys. Rev. B*. **33**, 3830, 1966, with permission.)

B. Experiment Aspects

A variety of experimental approaches and geometries can be employed in the study of electrochemical interfaces with diffraction techniques. The most applicable method is often dictated by the nature of the sample under study. For bulk materials or sufficiently thick films (such that the x-ray beam samples only that layer), conventional diffraction experiments can be peformed, and in fact, a number of in-situ x-ray diffraction studies of this type on electrochemical interfaces have been reported.[92]

In the case of thin films or monolayers, two different techniques can be employed; these are the total external-reflection Bragg diffraction (TERBD) technique introduced by Eisenberger and Marra[93] and the previously mentioned technique based on crystal truncation rods, introduced by Robinson,[90] which can provide in-plane structural information or information on interfacial roughness, respectively.

1. *Total External Reflection Bragg Diffraction*

This technique (sometimes referred to as *grazing incidence x-ray scattering*) essentially involves conventional Bragg diffraction under conditions of total external reflection.

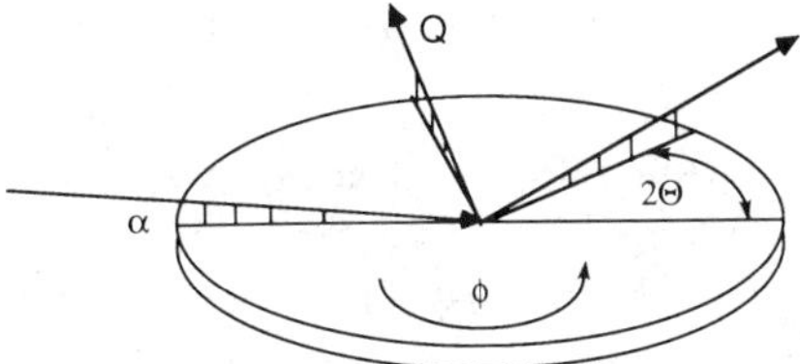

Figure 38. Scattering geometry in grazing-incidence surface diffraction.

In this geometry (Fig. 38), the angle of incidence, α, is small so that the scattering vector lies predominantly on the plane of the surface and as a result, this technique is most sensitive to in-plane scattering. For this same reason, however, this technique does not provide information on atomic correlations normal to the surface plane.

As originally described by Eisenberger and Marra,[93b] the angle of incidence is kept below the critical angle for the material under study so that the x-ray beam undergoes total external reflection. As mentioned previously, this has two very important consequences. First of all, since only an evanescent wave penetrates the substrate, the sampling depth is very shallow and of the order of 10 to 20 Å.

In addition, there is an enhancement in the reflected intensity given by

$$I_0 = I_i\{1 + R + 2\sqrt{R}\cos(v)\} \tag{39}$$

where I_i and I_0 are the incident and reflected intensities, respectively, and R is the reflectivity. When $\theta_{\text{incident}} = \theta_{\text{crit}}$: $v = 0$ and

$$I_0 = 4I_{\text{i}} \tag{40}$$

or

$$I_0/I_{\text{i}} = 4 \tag{41}$$

Thus a very significant enhancement can be obtained. In addition and again because of the grazing incidence geometry, the incident beam can have a large "footprint" so that it samples a large number of scatterers.

The kinematic treatment is generally used to interpret surface diffraction data. However, one must remember that the total external reflection effect is dynamial in nature so that the kinematic treatment is strictly not applicable. However, as has been pointed out by Vineyard,[94] a simple distorted wave approximation can be used, and this can be quite adequately treated in a kinematical approach. This last point greatly simplifies data interpretation.

2. *Crystal Truncation Rods*

As mentioned previously, CTRs arise as a result of the abrupt termination of a crystal lattice, and the diffuse diffracted intensity connects Bragg points in reciprocal space. In this case, the scattering vector is normal to the surface, and as a result, this technique is very sensitive to surface and interface roughness but not to in-plane atomic correlations. Thus, it yields information that is complementary to that obtained by grazing incidence diffraction. The most important feature of CTR is the characteristic decay of the scattered intensity described by Eq. (38). For surfaces that are not perfectly terminated (i.e., rough) the intensity will decay faster than predicted by this equation, and this can be used as a measure of root-mean-square surface roughness.

3. *Instrumental Aspects*

There are a number of experimental aspects that need to be considered when performing a surface diffraction experiment. Only some of these will be considered here; the interested reader is referred to recent comprehensive reviews.[95,96]

As with EXAFS and XSW experiments, the use of synchrotron radiation greatly facilitates surface diffraction experiments. Since diffraction experiments benefit greatly from an enhancement in the x-ray flux density (photons/cm^2 sec), a toroidal focusing mirror is often employed in order to focus the incoming beam (which is typically 6×2 mm) to a tight spot. The other optical elements present are similar to those employed in surface EXAFS and XSW experiments (e.g., Fig. 28).

In order to eliminate the loss in the scattering cross-section (by $\cos^2 2\theta$) that arises when the electric field is polarized in the plane of the scattered x-rays, a vertical scattering geometry is often employed. A four-circle diffractometer is generally employed in order to allow access to all points in reciprocal space.[97] Figure 39 shows a schematic drawing, and the various angles ϕ, χ, θ, and 2θ are defined there as well.

As mentioned previously, in a TERBD experiment, α is typically small so that the scattering vector lies largely on the plane of the surface and thus is most sensitive to in-plane atom correlations. For CTR, on the other hand, the scattering vector is normal to the surface so that it provides much information on atomic correlations in this direction, and thus is very sensitive to surface and interface roughness as mentioned previously.

Depending on which angles are varied and which remain fixed, various types of scans can be employed. A radial or Q scan corresponds to measuring the intensity along a radial scattering vector at a fixed angle ϕ. Experimentally, it involves moving the detector by an angle $\Delta 2\theta$ and the θ circle of the diffractometer by $\Delta\theta = (\Delta 2\theta)/2$. This is also commonly referred to as a theta,

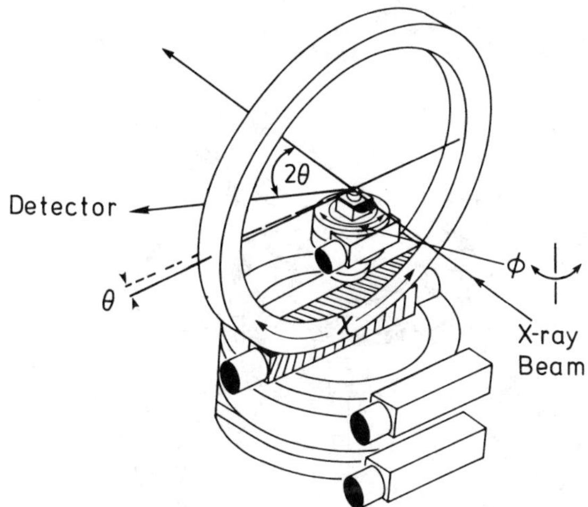

Figure 39. Schematic depiction of a four-circle diffractometer.

two-theta scan. The width of the diffraction peak obtained in a radial scan can be used to determine domain size.

An azimuthal, also known as a ϕ or rocking scan, involves measuring the intensity along an azimuthal arc at constant Q. Experimentally, it involves rotating the ϕ circle while maintaining θ and 2θ constant. These measurements are usually employed in determining the symmetry of an overlayer.

The third type of scan is a rod-scan which measures the intensity perpendicular to the surface. Experimentally, it involves scanning χ in the chi circle. However, in order to keep Q constant, some adjustments in θ and 2θ also need to be made. As mentioned before, these measurements give much information on surface and interface roughness. A θ–2θ scan is usually performed at each χ setting to obtain an integrated intensity measurement.

In addition, an α scan is often used to determine the angle of incidence that will give rise to the maximum scattering enhancement.

4. *Electrochemical Cells*

The requirements for an electrochemical cell to be employed in surface diffraction experiments are quite similar to those for surface EXAFS and XSW measurements. Thus, very similar cell designs have been employed. In order to minimize background scattering by the electrolyte, its thickness must be kept as thin as possible. This is because at the low angles of incidence typically employed, the effective path length can be quite long. Figure 40 shows a schematic of the cell employed by Samant et al.[98]

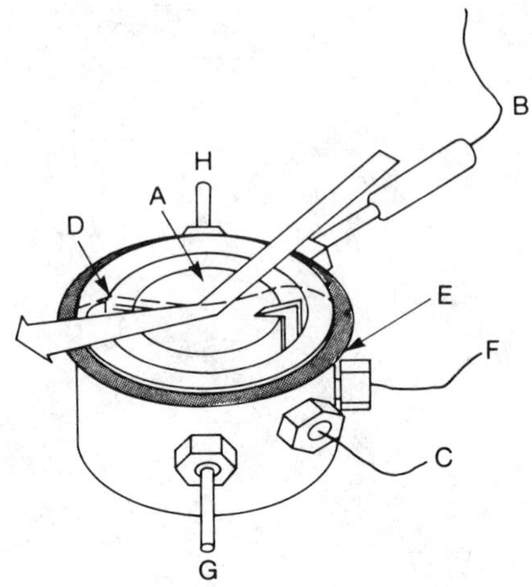

Figure 40. Electrochemical cell for *in situ* grazing-incidence x-ray scattering experiments. A, silver(111) electrode; B, Ag/AgCl reference electrode; C, Pt counter electrode; D, polypropylene window; E, O-ring; F, contact to Ag electrode; G, solution inlet; H, solution outlet. (From Samant, M. G., Toney, M. F., et al., *Phys. Rev. B.* **38** 10962 (1988), with permission.)

Scintillators are typically employed in the measurement of diffracted intensities. In some cases, the use of soller slits or an analyzer crystal is desirable in order to maximize resolution.

C. Examples

Although only a limited number of examples of surface diffraction at electrochemical interfaces currently exists, a wealth of information has emerged from them. The discussion will begin with studies performed at single-crystal surfaces followed by studies on polycrystalline elecrodes.

1. *Pb on Ag(111)*

The most thoroughly characterized surface diffraction study at a single crystal surface is that of underpotentially deposited lead on a silver(111) single crystal electrode (epitaxial film on mica).[98–101] These investigators began by measuring the CTR of the silver substrate since, unless these were clearly observed, it would be unlikely that diffraction from an electrodeposited monolayer would be detected.

Figure 41 shows radial (a), azimuthal (b), and CTR (c) scans in the (100)s direction for a Ag(111) electrode in contact with a sodium acetate solution. From the radial and azimuthal scans (Fig. 3.41a, b respectively), it is clear that the silver surface is of high quality.

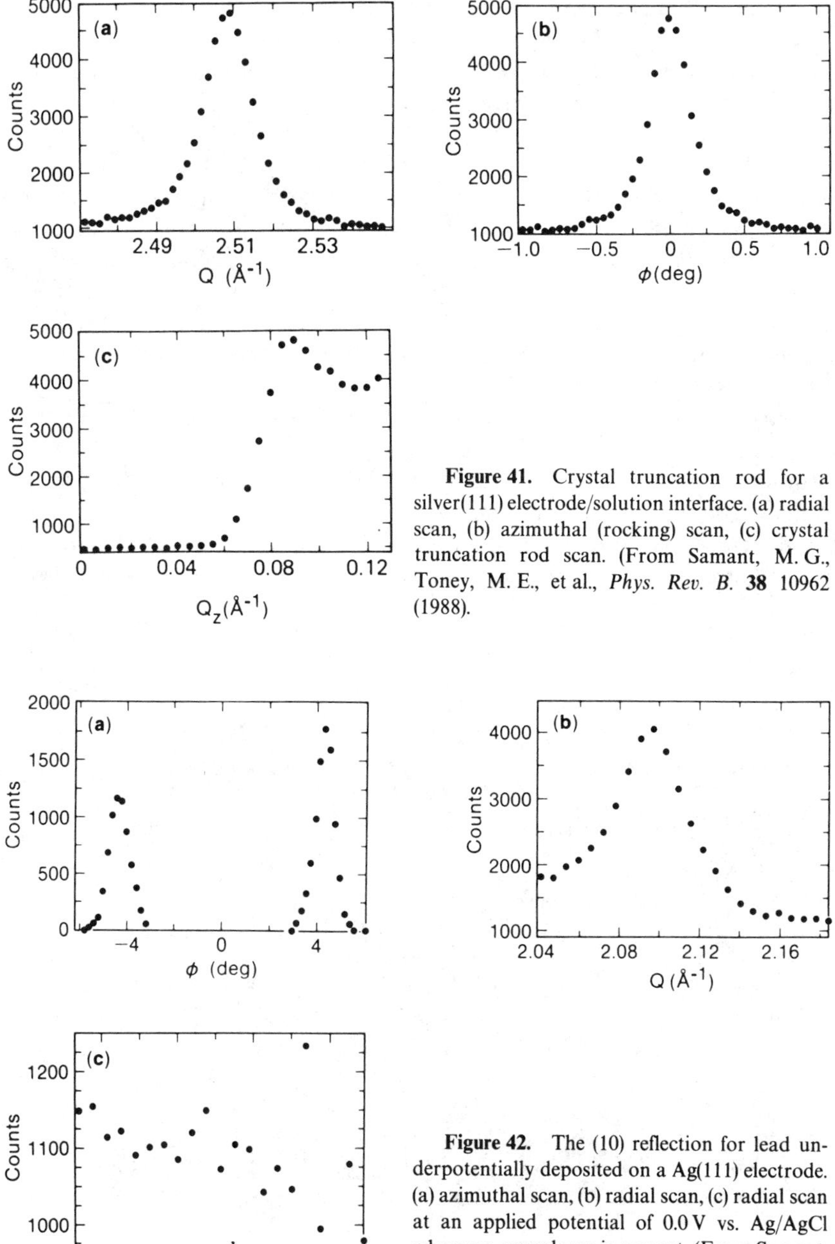

Figure 41. Crystal truncation rod for a silver(111) electrode/solution interface. (a) radial scan, (b) azimuthal (rocking) scan, (c) crystal truncation rod scan. (From Samant, M. G., Toney, M. E., et al., *Phys. Rev. B.* **38** 10962 (1988).

Figure 42. The (10) reflection for lead underpotentially deposited on a Ag(111) electrode. (a) azimuthal scan, (b) radial scan, (c) radial scan at an applied potential of 0.0 V vs. Ag/AgCl where no monolayer is present. (From Samant, M. G., Toney, M. F. et al., *J. Phys. Chem.* **92**, 220 (1988), with permission.)

Having characterized the substrate, they then investigated the structure of a monolayer of lead underpotentially deposited on the silver(111) electrode surface. Figures 42a and b, respectively, show radial and azimuthal scans from a lead monolayer underpotentially deposited on a Ag(111) electrode surface at a potential of -0.40 V vs. Ag/AgCl. That the diffraction arises from the lead monolayer was conclusively demonstrated by the fact that at an applied potential of 0.0 V vs. Ag/AgCl, where the monolayer is stripped, no diffraction peaks were observed (Fig. 42c). For the deposited monolayer, diffraction peaks spaced at 60° intervals were observed, consistent with the anticipated hexagonal lead overlayer. From these measurements, it was concluded that the electrodeposited lead forms a hexagonal monolayer that is incommensurate with the silver substrate and that at this potential, the Pb–Pb distance is 3.45 Å.

Another important feature is that the (10) reflections are rotated $\pm 4.5°$ from the silver $(\bar{2}11)$ direction ($\phi = 0$) which indicates that there exists a rotational epitaxy of the lead ad-layer relative to the silver substrate. This is pictorially presented in Figure 43.

Most recently, these same authors[100,101] have studied the potential dependence of the Pb–Pb distance in the monolayer and have found that it

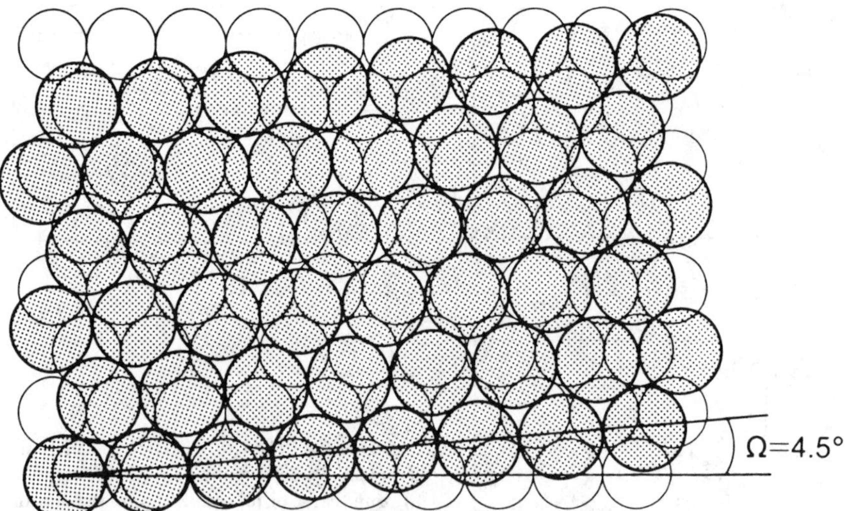

Figure 43. Depiction of the 4.5° of rotation epitaxy for a lead monolayer (shaded circles) underpotentially deposited on a silver(111) electrode (open circles). (From Samant, M. G., Toney, M. F. et al., *J. Phys. Chem.* **92**, 220 (1988), with permission.)

decreases (monolayer compression) with increasingly negative potential. At an applied potential of -0.55 V where bulk deposition commences, the compression is 2.8% relative to the bulk value. Furthermore, they have studied the effects of further deposition and have found that because of the compressive strain, the additional lead is not epitaxially deposited.

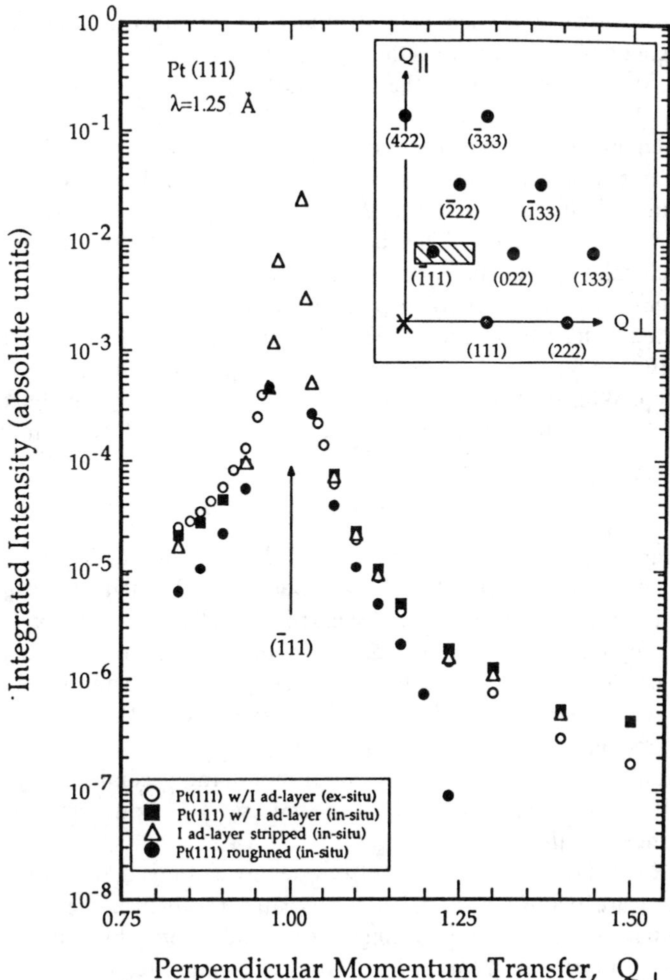

Figure 44. Crystal truncation rod for a Pt(111) electrode with an iodine monlayer: ○ = ex situ; ■ = in situ; △ = after electrochemical stripping of the iodine monolayer; ● = after electrochemical roughening.

2. *Surface Roughening of Pt(111)/I CTR*

We have recently been able to perform an in-situ surface diffraction study of a Pt(111) electrode covered with an iodine ad-layer. In this study, we were able to monitor the $\{\bar{1}11\}$ crystal truncation rod for the electrode in contact with air (i.e., ex-situ), as well as in contact with an aqueous sulfate electrolyte solution (i.e., in-situ). Furthermore we were able to monitor, in-situ, the features of this rod after electrochemically stripping the iodine ad-layer as well as after electrochemically roughening the electrode surface. The data are shown in Fig. 44, which also depicts the reciprocal lattice map for this particular surface. As can be ascertained from Fig. 44, the profiles for the electrode examined in-situ and ex-situ are virtually superimposed (the small shifts could be due to the fact that the electrode was annealed in iodine vapor between experiments). In addition, electrochemical stripping of the iodine ad-layer had only a small effect. This is to be contrasted with the profile after electrochemical roughening by potentiostatting at $+0.90\,V$ for 15 minutes. In this case, there are very significant deviations which can be unambiguously associated with a roughening of the surface. We are in the process of analyzing and fitting these data to determine the root-mean-square roughness after each step. What is clear, however, is that we can employ this technique to study, in-situ, ordered ad-lattices and surface roughening induced by electrochemical perturbation.

3. *Studies at Non-Single-Crystal Surfaces*

Fleischmann and co-workers have carried out a number of in-situ x-ray diffraction studies at roughened electrodes.[103–108] Their experiments are different from those previously discussed in that instead of a synchrotron source they employ a conventional x-ray tube ($\sim 1.5\,kW$) with a position-sensitive proportional counter for the measurement of the diffracted intensity. They have carried out experiments in both transmission (Laue) and reflection (Bragg) modes.

In addition, they employ a potential modulation technique (typically at 0.004 Hz) in order to subtract background and a 15-point smoothing filter to enhance signal-to-noise ratio.

Due to the fact that roughened electrodes are employed, the diffractograms measured are two-dimensional powder patterns which can be used in the determination of the *d*-spacings from ad-layers but cannot yield surface-orientational information. However, data acquisition is greatly simplified since only one type of scan (radial) needs to be performed.

The first system studied by these investigators was the adsorption of iodine at various coverages onto papyex (a large-area graphite material in which about half of the graphite basal planes lie within a few degrees of parallel

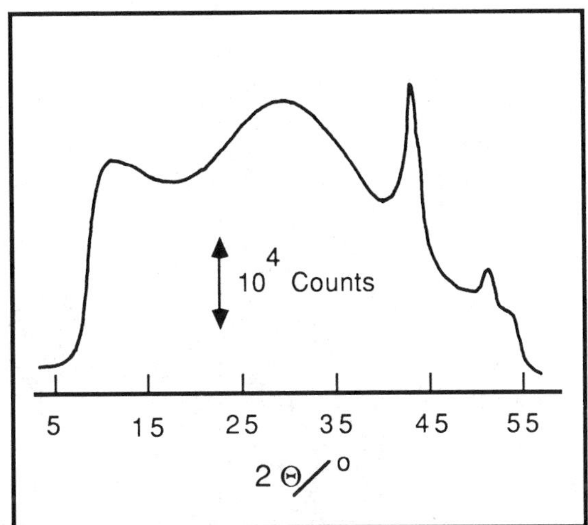

Figure 45. *In situ* x-ray diffractogram for an aged $Ni(OH)_2$ electrode held at +0.100 V vs. SCE in 1 M KOH. (From Fleishmann, M., Oliver, A., and Robinson, J., *Electrochimica Acta* **31**, 899 (1986), with permission.)

to the surface).[103] Although the experiments were not electrochemical in nature, the graphite was in contact with I_2/KI solutions. From these studies, they determined that the iodine formed a hexagonally closed-packed monolayer of adsorbed I_2 molecules. From the coverage dependence, they also determined that the adsorbate was present in islands.

These investigators have also studied structural changes in the nickel–hydroxide as well as nickel hydroxide–oxyhydroxide electrode system. As mentioned in the EXAFS section (vide supra), this system is rather complex because of interconversion of various species. For example, the as-deposited α nickel hydroxide ages (by dehydration) in concentrated KOH solution to yield β nickel hydroxide. Fleischmann et al. were able to study both aged and unaged films. For example, Fig. 45 shows the difference diffractogram for an aged $Ni(OH)_2$ electrode held at +0.100 V vs SCE in 1 M KOH. The broad peak in the middle arises from scattering by water, while the two sharper features are due to (111) and (200) diffractions from the nickel substrate. No detectable diffraction from the $Ni(OH)_2$ layer is apparent. However, upon taking a difference diffractogram (by subtracting diffractograms at +0.10 and +0.40 V) (Fig. 46A), a number of diffraction peaks are observed. This exemplifies the advantage of employing a potential modulation technique. Figure 46B gives assignments to the various features.

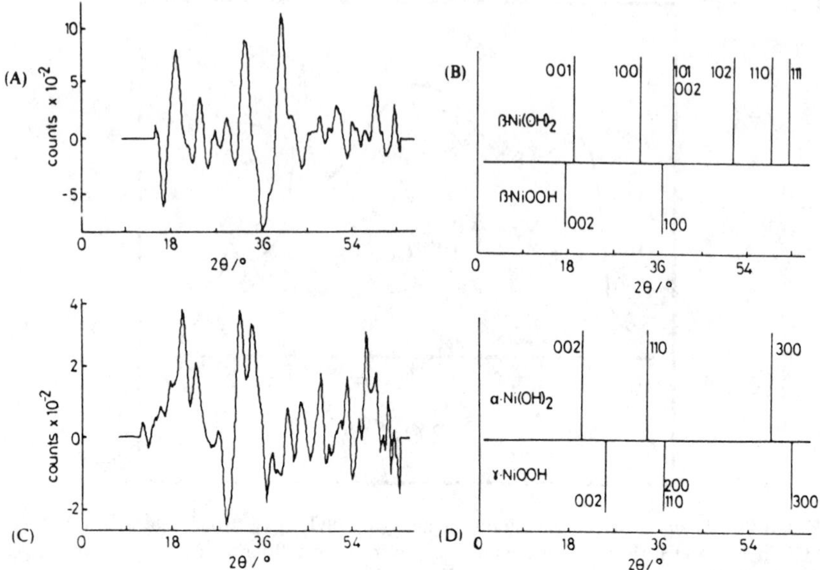

Figure 46. Difference diffractograms (obtained at +0.10 and +0.40 V vs SCE) for the nickel hydroxide/oxyhydroxide system for aged (A) and unaged (C) electrodes. (B) and (D) indicate the assignment of the features in A and C respectively. (From Fleishmann, M., Oliver, A., and Robinson, J., *Electrochimica Acta* **31**, 899 (1986), with permission.)

Figures 46C and D present analogous measurements for unaged electrodes. All the results are consistent with the structural changes that accompany aging as well as oxidation or reduction of the surface films.

These investigators have also studied the underpotential deposition of T1 and Pb on silver and gold electrodes.[107] They obtained diffractograms at regions that corresponded to bulk (Fig. 47A) and UPD (Fig. 47B) T1 deposition on Ag, respectively as well as in the double-layer region (Fig. 47C). In the case of T1 deposition within the UPD regime (where actually two monolayers are deposited), an enhancement of the Ag(111) diffraction peak was observed. This was attributed to the formation of a commensurate first monolayer of T1 on Ag(111), which requires the layer to be remarkably compressed given the significant differerences in atomic radii. Two additional and less well-defined features were observed and attributed to diffraction from a second layer of T1 that is incommensurate with the first. At potentials where bulk T1 deposition occurs, the T1(101) bulk reflection is clearly observed (Fig. 47A), suggesting that the deposited bulk T1 is highly ordered.

In the deposition of lead, they determined that the Pb atoms form an incommensurate layer. In addition, they noticed differences when using Ag

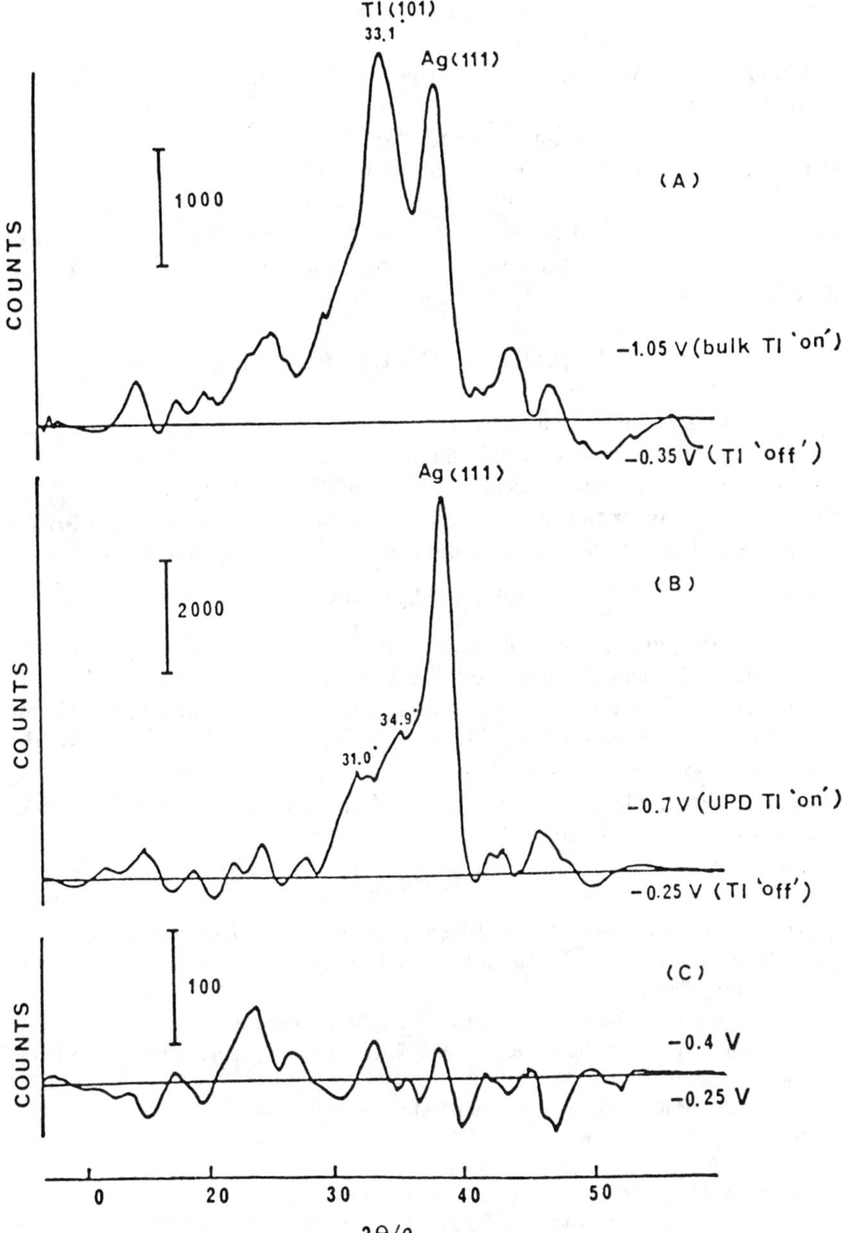

Figure 47. Difference diffractograms from roughened silver electrodes. (A) Potential modulation between −0.35 and −1.05 V (bulk T1 deposition). (B) Potential modulation between −0.25 and −0.70 V (underpotential deposition of two T1 monolayers). (C) Potential modulation between −0.25 V and 0.40 V (double-layer region). (From Fleischmann, M., and Mao, B. W., *J. Electroanal. Chem.* **247**, 297 (1988), with permission.)

vs. Au electrodes, and these were ascribed to differences in the double-layer properties of both metals.

Most recently, Fleischmann and Mao[106,108] have investigated the reconstruction of platinum surfaces in the presence of adsorbates. They note that the presence of CO or weakly adsorbed hydrogen causes a reconstruction with an enhancement of the reflection due to the (110) faces. Surprisingly, however, the presence of strongly adsorbed hydrogen did not result in a surface reconstruction.

X. CONCLUSIONS AND FUTURE DIRECTIONS

The use of x-rays is providing a rare glimpse of the in-situ structure of electrochemical interfaces of both fundamental and technological importance, and as these experiments become more widespread, a wide range of phenomena will be explored. I am certain that these studies will help provide the basis for a better understanding and control of electrochemical reactivity.

Acknowledgements

Our work was generously supported by the Materials Chemistry Initiative of the National Science Foundation, the Materials Science Center at Cornell University, the Office of Naval Research, the Army Research Office, Dow Chemical Co., and Xerox Corp. Special thanks to to Dr. James H. White and Mark Bommarito as well as to Dr. Michael J. Bedzyk (CHESS), Michael Albarelli, David Acevedo, Dr. Martin McMillan, and Dr. Ben Ocko (Brookhaven National Laboratories).

References

1. H. Winick and S. Doniach, eds., '*Synchrotron Radiation Research*,' Plenum, New York, 1980.
2. H. Winick in '*Synchrotron Radiation Research*, H. Winick, and S. Doniach, eds., Plenum, New York, 1980, p. 11.
3. D. H. Tomboulian and P. Hartman, *Phys. Rev.* **102**, 1423 (1956).
4. For an introductory discussion see: H. Winick, G. Brown, K. Halbach, J. Harris, *Physics Today* **34**, (May 1981), p. 50.
5. There have been numerous reviews of EXAFS over the last ten years. A selected number of leading references is listed below.
   a. E. A. Stern, *Sci. Am.* **234**(4), 96, 1976.
   b. P. Eisenberger, B. M. Kincaid, *Science* **200**, 1441 (1978).
   c. S. P. Cramer, K. O. Hodgson, *Prog. Inorg. Chem.* **25**, 1 (1979).
   d. B. K. Teo, *Accts. Chem. Res.* **13**, 412 (1980).
   e. P. A. Lee, P. H. Citrin, P. Eisenberger, B. M. Kincaid, *Rev. Mod. Phys.* **53**, 769 (1981).
   f. B. K. Teo, D. C. Joy, eds., *EXAFS Spectroscopy; Techniques and Applications*, Plenum, New York, 1981.
   g. A. Bianconi, L. Inoccia and S. Stippich, eds., *EXAFS and Near Edge Structure*, Springer-Verlag, Berlin, 1983.

h. K. O. Hodgson, B. Hedman and J. E. Penner-Hahn, eds., *EXAFS and Near Edge Structure III*, Springer-Verlag, Berlin, 1984.
i. B. K. Teo, *EXAFS: Basic Principles and Data Analysis*, Springer-Verlag, Berlin, 1986.

6. R. deL Kronig, *Z. Phys.* **70**, 317 (1931); **75**, 191, 468 (1932).
7. a. D. E. Sayers, E. A. Stern, F. W. Lytle, *Phys. Rev. Lett.* **27**, 1204 (1971).
   b. E. A. Stern, *Phys. Rev. B* **10**, 3027 (1974).
   c. E. A. Stern, D. E. Sayers, and F. W. Lytle, *Phys. Rev. B* **11**, 4836 (1975).
   d. C. A. Ashby and S. Doniach, *Phys. Rev. B* **11**, 1279 (1975).
   e. P. A. Lee, and J. B. Pendry, *Phys. Rev. B* **11**, 2795 (1975).
   f. P. A. Lee, and G. Beni, *Phys. Rev. B* **15**, 2862 (1977).
8. P. Eisenberger and G. S. Brown, *Solid State Commun.* **29**, 481 (1979).
9. a. T. M. Hayes, J. W. Allen, J. Tauc, B. G. Giessen, and J. J. Hauser, *Phys. Rev. Lett.* **40**, 1282 (1978).
   b. T. M. Hayes, J. B. Boyce, and J. L. Beeby, *J. Phys. C* **11**, 2931 (1978).
10. P. A. Lee and G. Beni, *Phys. Rev. B* **15**, 2862 (1977).
11. a. T. M. Hayes, P. N. Sen and S. H. Hunter, *J. Phys. C* **9**, 4357 (1976).
   b. S. P. Cramer, J. H. Dawson, K. O. Hodgson and L. P. Hager, *J. Am. Chem. Soc.* **100**, 7282 (1978).
   c. S. P. Cramer, W. O. Gillum, K. O. Hodgson, L. E. Mortenson, E. I. Stiefel, J. R. Chisnell, J. W. Brill and V. K. Shah, *J. Am. Chem. Soc.* **100**, 3814 (1978).
   d. S. P. Cramer, K. O. Hodgson, E. I. Stiefel and W. E. Newton, *J. Am. Chem. Soc.* **100**, 2748 (1978).
   e. T. Tullius, P. Frank and K. O. Hodgson, *Proc. Nat. Acad. Sci. U.S.A.* **75**, 4069 (1978).
12. R. F. Pettifer and P. W. McMillan, *Philos. Mag.* **35**, 871 (1977).
13. C. A. Ashby and S. Doniach, *Phys. Rev. B* **11**, 1279 (1975).
14. P. Lagarde, *Phys. Rev. B* **13**, 741 (1976).
15. B. K. Teo and P. A. Lee, *J. Am. Chem. Soc.* **101**, 2815 (1979).
16. A. G. McKale, B. W. Veal, A. P. Paulikas, S-K. Chan and G. S. Knapp, *J. Am. Chem. Soc.* **110**, 3763 (1988).
17. P. H. Citrin, P. Eisenberger and B. M. Kincaid, *Phys. Rev. Lett.* **36**, 1346 (1976).
18. a. E. A. Stern, *J. Vac. Sci. Tech.* **14**, 461 (1977).
   b. U. Landman and D. L. Adams, *J. Vac. Sci. Tech.* **14**, 466 (1977).
   c. J. Stohr, in *Emission and Scattering Techniques: Studies of Inorganic Molecules, Solids, and Surfaces*, P. Day, ed., D. Reidel Publ., Holland, 1981.
   d. J. Hasse, *Applied Phys. A* **38**, 181 (1985).
   e. P. H. Citrin, *Jnl. de Phys. Colloque C8* **47**, 437 (1986).
19. G. Beni and M. Platzman, *Phys. Rev. B* **14**, 1514 (1976).
20. a. F. Sette, C. T. Chen, J. E. Rowe and P. H. Citrin, *Phys. Rev. Lett.* **59**, 311 (1987).
   b. F. Sette, T. Hashizume, F. Comin, A. A. MacDowell and P. H. Citrin, *Phys. Rev. Lett.* **61**, 1384 (1988).
21. P. A. Lee, *Phys. Rev. B* **13**, 5261 (1976).
22. a. J. Jaklevic, J. A. Kirby, M. P. Klein, A. S. Robertson, A. S. Brown and P. Eisenberger, *Solid State Commun.* **23**, 679 (1977).
   b. J. B. Hastings, P. Eisenberger, B. Lengler and M. L. Perlman, *Phys. Rev. Lett.* **43**, 1807 (1979).
23. S. M. Heald, E. Keller and E. A. Stern, *Phys. Lett.* **103A**, 155 (1984).

24. a. R. W. James, *The Optical Principles of the Diffraction of X-rays*, Oxbow Press, Woodbridge, Connecticut, 1982.
b. see also: D. H. Bilderback, *SPIE Proc.* **315**, 90 (1982).

25. a. G. Martens and P. Rabe, *Phys. Stat. Solidi A* **58**, 415 (1980).
b. J. Goulon, C. Goulon-Ginet, R. Cortes and J. M. Dubois, *J. Phys.* **43**, 539 (1982).
c. S. J. Gurman and R. Fox, *Phil Mag. B* **54**, L45 (1986).

26. a. S. M. Heald, J. M. Tranquada and H. Chen, *Jnl. de Phys. Collogue C8*, **47**, 825 (1986).
b. E. A. Stern, E. Keller, O. Petipierre, L. E. Bouldin, S. M. Heald and J. Tranquada, in *EXAFS and Near Edge Structure III*, K. O. Hodgson, B. Hedman and J. E. Penner-Hahn, eds., Springer-Verlag, Berlin, 1984, p. 261.

27. a. E. Dartyge, A. Fontaine, A. Jucha and D. Sayers in *EXAFS and Near Edge Structure III*, K. O. Hodgson, B. Hedman and J. E. Penner-Hahn, eds., Springer-Verlag, Berlin, 1984, p. 472.
b. A. M. Flank, A. Fontaine, A. Jucha, M. Lemmonier and C. Williams in *EXAFS and Near Edge Structure*, A. Bianconi, L. Inoccia and S. Stippich, eds., Springer-Verlag, Berlin, 1983, p. 405.
c. D. E. Sayers, D. Bazin, H. Dexpert, A. Jucha, E. Dartyge, A. Fontaine and P. Lagarde, in *EXAFS and Near Edge Strucure*, A. Bianconi, L. Inoccia and S. Stippich, eds., Springer-Verlag, Berlin, 1983, p. 209.
d. H. Oyanagi, T. Matsushita, U. Kaminaga and H. Hashimoto, *Jnl. de Phys. Collogue C8* **47**, 139 (1986).
e. S. Saigo, H. Oyanagi, T. Matsushita, H. Hashimoto, N. Yoshida, M. Fujimoto and T. Nagamura, *Jnl. de Phys. Collogue C8*, **47**, 555 (1986).
f. J. Mimault, R. Cortes, E. Dartyge, A. Fontaine, A. Jucha and D. Sayers, in *EXAFS and Near Edge Structure III*, K. O. Hodgson, B. Hedman and J. E. Penner-Hahn, eds., Springer-Verlag, Berlin, 1984, p. 47.

28. J. McBreen, W. E. O'Grady, K. I. Pandya, R. W. Hoffman and D. E. Sayers, *Langmuir* **3**, 428 (1986).

29. H. D. Dewald, J. W. Watkins, R. C. Elder and W. R. Heineman *Anal. Chem.* **58**, 2968 (1986).

30. G. G. Long, J. Kruger and M. Kuriyama, in *Passivity of Metals and Semiconductors*, M. Froment, ed., Elsevier, Amsterdam, 1983, p. 139.

31. M. G. Samant, G. L. Borges, J. G. Gordon, O. R. Melroy and L. Blum, *J. Am. Chem. Soc.* **109**, 5970 (1987).

32. M. J. Albarelli, J. H. White, M. G. Bommarito, M. McMillan and H. D. Abruña, *J. Electroanal. Chem.* **248**, 77 (1988).

33. W. E. O'Grady, *J. Electrochem. Soc.* **127**, 555 (1980).

34. G. G. Long, J. Kruger, D. R. Black and M. Kuriyama, *J. Electrochem. Soc.* **130**, 240 (1983).

35. G. G. Long, J. Kruger, D. R. Black and M. Kuriyama, *J. Electroanal. Chem.* **150**, 603, (1983).

36. a. J. Kruger, G. G. Long, M. Kuriyama and A. I. Goldman, in *Passivity of Metals and Semiconductors*, M. Froment, ed., Elsevier, Amsterdam, 1983, p. 163.
b. J. Kruger and G. G. Long, *Proc. Electrochem. Soc.* **86–7**, 210, 1986.

37. G. G. Long, J. Kruger and D. Tanaka, *J. Electrochem. Soc.* **134**, 264 (1987).

38. a. M. E. Kordesch and R. W. Hoffman, *Nucl. Inst. Meth. Phys. Res.* **222**, 347 (1984).
b. R. W. Hoffman, in *Passivity of Metals and Semiconductors*; M. Froment, ed., Elsevier, Amsterdam, 1983, p. 147.

39. a. S. M. El-Mashri, R. G. Jones and A. J. Forty, *Phil. Mag. A* **48**, 665 (1983).
b. A. J. Forty, S. M. El-Mashri, R. G. Jones, R. Dupree and I. Farnon, *Surf. Interface Anal.* **9**, 383 (1986).

40. A. J. Forty, M. Kerkar, J. Robinson and M. Ward *Jnl. de Phys. Colloque C8* **47**, 1077 (1986).

41. a. L. Bosio, R. Cortes, A. Defrain, M. Froment and A. M. Lebrun, in *Passivity of Metals and Semiconductors*, M. Froment, ed., Elsevier, Amsterdam, 1983, p. 131.
b. L. Bosio, R. Cortes, A. Defrain and M. Froment, *J. Electroanal. Chem.* **180**, 265 (1984).
c. L. Bosio, R. Cortes and M. Froment, in *EXAFS and Near Edge Structure III*, K. O. Hodgson, B. Hedman and J. E. Penner-Hahn, eds., Springer-Verlag, Berlin, 1984, p. 484.

42. L. Bosio, R. Cortes, P. Delichere, M. Froment, and S. Joiret, *Surface Interface Anal.* **12**, 380 (1988).

43. O. J. Murphy, T. E. Poy and J. Bockris, *J. Electrochem. Soc.* **131**, 2785 (1984).

44. J. McBreen, J. E. O'Grady and K. I. Pandya, *J. Power Sources* **22**, 323 (1988).

45. R. G. Linford, P. G. Hall, C. Johnson and S. S. Hasnain, *Solid State Ionics* **14**, 199 (1984).

46. R. S. Weber Peuckert, R. A. Dallabetta and M. Boudart, *J. Electrochem. Soc.* **135**, 2535 (1988).

47. W. E. O'Grady and D. C. Koningsberger, *J. Chem. Phys.* (to be published).

48. a. B. Van Wingerden, J. A. R. Van Veen, and C. T. J. Mensch, *J. Chem. Soc. Farad. Trans. 1* **84**, 65 (1988).
b. J. McBreen, W. E. O'Grady, D. E. Sayers, C. Y. Yang and K. I. Pandya, *Proc. Electrochem. Soc.*, 1987, **87–12**, p. 182.

49. D. M. Kolb, in *Advances in Electrochemistry and Electrochemical Engineering, Vol. 11*, H. Gerischer and C. Tobias, eds., Pergamon Press, New York, 1978, p. 125.

50. H. D. Abruña, J. H. White, M. J. Albarelli, G. M. Bommarito, M. J. Bedzyk and M. McMillan, *J. Phys. Chem.* **92**, 7045 (1988).

51. a. L. Blum, H. D. Abruña, J. H. White, M. J. Albarelli, J. G. Gordon, G. Borges, M. Samant and O. R. Melroy *J. Chem. Phys.* **85**, 6732 (1986).
b. O. R. Melroy, M. G. Samant, G. C. Borges, J. G. Gordon, L. Blum, J. H. White, M. J. Albarelli, M. McMillan and H. D. Abruña, *Langmuir* **4**, 728 (1988).
c. J. H. White, M. J. Albarelli, H. D. Abruña, L. Blum, O. R. Melroy, M. Samant, G. Borges and J. G. Gordon, *J. Phys. Chem.* **92**, 4432 (1988).

52. J. H. White and H. D. Abruña, *J. Electroanal. Chem.* **274**, 185 (1989).

53. a. D. W. Pashley, *Phil. Mag.* **4**, 316 (1959).
b. E. Grunbaum, *Vacuum* **24**, 153 (1973).
c. K. Reichelt and H. O. Lutz, *J. Cryst. Growth* **10**, 103 (1971).

54. M. Zei, G. Qiao, G. Lehmpfhul and D. M. Kolb, *Ber. Bunsenges. Phys. Chem.* **91**, 349 (1987).

55. D. A. Smith, M. J. Heeg, W. R. Heineman and R. C. Elder, *J. Am. Chem. Soc.* **106**, 3053 (1984).

56. D. A. Smith, R. C. Elder and W. R. Heineman, *Anal. Chem.* **57**, 2361 (1985).

57. R. C. Elder, C. E. Lunte, A. F. M. M. Rahaman, J. R. Kirchoff, H. D. Dewald and W. R. Heineman, *J. Electroanal. Chem*, **240**, 361 (1988).

58. G. Tourillon, E. Dartyge, H. Dexpert, A. Fontaine, A. Jucha, P. Lagarde and D. E. Sayers, *J. Electroanal. Chem.* **178**, 357 (1984).

59. G. Tourillon, E. Dartyge, H. Dexpert, A. Fontaine, A. Jucha, P. Lagarde and D. E. Sayers, *Surf. Sci.* **156**, 536 (1985).

60. H. Dexpert, P. Lagarde and G. Tourillon, in *EXAFS and Near Edge Structure III*, K. O. Hodgson, B. Hedman and J. E. Penner-Hahn, eds., Springer-Verlag, Berlin, 1984, p. 400.

61. G. Tourillon, E. Dartyge, A. Fontaine and A. Jucha, *Phys. Rev. Lett.* **57**, 603 (1986).

62. E. Dartyge, A. Fontaine, G. Tourillon and A. Jucha, *Jnl. de Phys. Colloque C8* **47**, 607 (1986).

63. E. Dartyge, C. Depautex, J. M. Dubuisson, A. Fontaine, A. Jucha, P. Leboucher and G. Tourillon, *Nucl. Inst. Meth. Phys. Res.* **A246**, 452 (1986).

64. E. Dartyge, A. Fontaine, G. Tourillon, R. Cortes and A. Jucha, *Phys. Lett.* **113A**, 384 (1986).
65. A. Fontaine, E. Dartyge, A. Jucha, and G. Tourillon, *Nucl. Inst. Meth. Phys. Res.* **A253**, 519 (1987).
66. G. Tourillon, H. Dexpert and P. Lagarde, *J. Electrochem. Soc.* **134**, 327 (1987).
67. J. McBreen, W. E. O'Grady, G. Tourillon, E. Dartyge, A. Fontaine and K. I. Pandya, *J. Phys. Chem.* **93**, 6308 (1989).
68. a. G. Tourillon, H. Dexpert and P. Lagarde, *J. Electrochem. Soc.* **134**, 327 (1987).
    b. G. Tourillon, S. Raaen, T. A. Skotheim, M. Sagurton, R. Garrett and G. P. Williams, *Surf. Sci.* **184**, L345 (1987).
    c. G. Tourillon, A. Fontaine, M. Sagurton, N. R. Garrett and G. Williams, *Jnl. de Phys. Colloque C8* **47**, 579 (1986).
    d. G. Tourillon, A. Fontaine, Y. Jugnet, T. M. Duc, W. Braun, J. Feldhaus and E. Holub-Krappe, *Phys. Rev. B.* **36**, 3483 (1987).
    e. G. Tourillon, A. Fontaine, R. Garrett, M. Sagurton, P. Xu and G. Williams, *Phys. Rev. B* **35**, 9863 (1987).
69. J. H. White and H. D. Abruña, *J. Phys. Chem.* **92**, 7131 (1988).
70. F. Lu, G. N. Salaita, H. Baltruschat and A. T. Hubbard, *J. Electroanal. Chem.* **222**, 305 (1987).
71. B. W. Batterman and H. Cole, *Rev. Mod. Phys.* **36**, 681 (1964).
72. B. W. Batterman, *Phys. Rev.* **133**, A759 (1964).
73. R. W. James, *The Optical Principles of the Diffraction of X-rays*, Oxbow Press, Woodbridge, Connecticut, 1982.
74. a. G. Borrmann, *Phys. Zeit*, **43**, 157 (1941).
    b. G. Borrmann, *Zeit. F. Phys.* **127**, 297 (1950).
75. T. W. Barbee and J. H. Underwood, *Optics Comm.* **48**, 161 (1983).
76. M. J. Bedzyk and G. Materlik, *Phys. Rev. B* **31**, 4110 (1985).
77. P. L. Cowan, J. A. Golovchenko and M. F. Robbins, *Phys. Rev. Lett.* **44**, 1680 (1980).
78. M. J. Bedzyk and G. Materlik, *Phys. Rev. B* **31**, 4110 (1985).
79. J. H. Underwood and T. W. Barbee, in *AIP Conf. Proc.* **75**, 170, D. T. Atwood and B. L. Henke, eds., American Institute of Physics, New York, 1981.
80. J. M. Bloch, M. Sansone, F. Rondeber, D. G. Peiffer, P. Pincus, M. W. Kim and P. M. Eisenberger, *Phys. Rev. Lett.* **54**, 1039 (1985).
81. L. G. Parratt, *Phys. Rev.* **95**, 359 (1954).
82. G. M. Bommarito, M. S. Thesis, Cornell University, 1987.
83. M. J. Bedzyk, D. H. Bilderback, G. M. Bommarito, M. Caffrey and J. J. Schildkraut, *Science* **241**, 1788 (1988).
84. G. Materlik, J. Zegenhagen and W. Uelhoff, *Phys. Rev. B* **32**, 5502 (1985).
85. G. Materlik, M. Schmah, J. Zegenhagen and W. Uelhoff, *Ber. Bunsenges. Phys. Chem.* **91**, 292 (1987).
86. M. J. Bedzyk, D. Bilderback J. H. White, H. D. Abruña and G. M. Bommarito, *J. Phys. Chem.* **90**, 4926 (1986).
87. a. J. L. Stickney, S. D. Rosasco and A. T. Hubbard, *J. Electrochem. Soc.* **131**, 260 (1984).
    b. J. L. Stickney, S. D. Rosasco, B. C. Schardt and A. T. Hubbard, *J. Phys. Chem.* **88**, 251 (1984).
88. a. D. P. Woodruff, D. L. Seymour, C. F. McConville, C. E. Riley, M. D. Crapper, N. P. Prince and R. G. Jones, *Phys. Rev. Lett.* **58**, 1480 (1984).

b. D. P. Woodruff, D. L. Seymour, C. F. McConville, C. E. Riley, M. D. Crapper, N. P. Prince and R. G. Jones, *Surface Science* **195**, 237 (1988).

89. a. B. E. Warren, *X-ray Diffraction*, Addison-Wesley, Reading, Massachusetts, 1969.
    b. R. W. James, *The Optical Principles of the Diffraction of X-rays*, Oxbow Press, Woodbridge, Connecticut, 1982.
    c. B. D. Cullity, *Elements of X-ray Diffraction*, Addison-Wesley, Reading, Massachusetts, 1978.
    d. J. M. Cowley, *Diffraction Physics*, second ed., North-Holland, Amsterdam, Netherlands, 1984.
90. I. K. Robinson, *Phys. Rev. B* **33**, 3830 (1986).
91. a. I. K., Robinson, W. K. Waskiewicz, R. T. Tung and J. Bohr, *Phys. Rev. Lett.* **57**, 2714 (1986).
    b. S. G. J. Mochrie, *Phys. Rev. Lett.* **59**, 304 (1987).
    c. D. Gibbs, B. M. Ocko, D. M. Zehner and S. G. J. Mochrie, *Phys. Rev. B* **38**, 7303 (1988).
    d. B. M. Ocko, and S. G. J. Mochrie, *Phys. Rev. B* **38**, 7378 (1988).
92. a. S. U. Falk, *J. Electrochem. Soc.* **107**, 661 (1960).
    b. A. J. Salkind, C. J. Venuto and S. U. Falk, *J. Electrochem. Soc.* **111**, 493 (1964).
    c. G. Nazri and R. H. Muller, *J. Electrochem. Soc.* **132**, 1385 (1985).
    d. R. R. Chianelli, J. C. Scanlon and B. M. L. Rao, *J. Electrochem. Soc.* **125**, 1563 (1978).
    e. J. R. Dahn, M. A. Py and R. R. Haering, *Can. J. Phys.* **60**, 307 (1982).
93. a. W. C. Marra, P. Eisenberger and A. Y. Cho, *J. Appl. Phys.* **50**, 6927 (1979).
    b. P. Eisenberger and W. C. Marra, *Phys. Rev. Lett.* **46**, 1081 (1981).
    c. W. C. Marra, P. H. Fuoss and P. E. Eisenberger, *Phys. Rev. Lett.* **49**, 1169 (1982).
94. G. H. Vineyard, *Phys. Rev. B* **26**, 4146 (1982).
95. D. P. Woodruff and T. A. Delchar, *Modern Techniques of Surface Diffraction*, Cambridge University Press, Cambridge, England, 1986.
96. I. K. Robinson, in *Handbook on Synchrotron Radiation*, D. Moncton and G. S. Brown, eds. North-Holland, Amsterdam, 1988.
97. W. R. Busing and H. A. Levy, *Acta Cryst.* **22**, 457 (1967).
98. M. G. Samant, M. F. Toney, G. L. Borges, L. Blum and O. R. Melroy, *Surf. Sci. Lett.* **193**, L29 (1988).
99. M. G. Samant, M. F. Toney, G. L. Borges, L. Blum and O. R. Melroy, *J. Phys. Chem.* **92**, 220 (1988).
100. O. R. Melroy, M. F. Toney, G. L. Borges, M. G. Samant, J. B. Kortright, P. N. Ross and G. Blum, *Phys. Rev. B* **38**, 10962 (1988).
101. O. R. Melroy, M. F. Toney, G. L. Borges, M. G. Samant, J. B. Kortright, P. N. Ross and L. Blum, *J. Electroanal. Chem.* **258**, 403 (1989).
102. G. M. Bommarito, B. M. Ocko, D. A. Acevedo and H. D. Abruña, (manuscript in preparation).
103. M. Fleischmann, P. J. Hendra and J. Robinson, *Nature* **288**, 152 (1980).
104. M. Fleischmann, P. Graves, I. Hill, A. Oliver and J. Robinson, *J. Electroanal. Chem.* **150**, 33 (1983).
105. M. Fleischmann, A. Oliver and J. Robinson, *Electrochimica Acta* **31**, 899 (1986).
106. M. Fleischmann and B. W. Mao, *J. Electroanal. Chem.* **229**, 125 (1987).
107. M. Fleischmann and B. W. Mao, *J. Electroanal. Chem.* **247**, 297 (1988).
108. M. Fleischmann and B. W. Mao, *J. Electroanal. Chem.* **247**, 311 (1988).

PERIODIC SURFACES OF PRESCRIBED MEAN CURVATURE

D. M. ANDERSON

Physical Chemistry
Chemical Center
Lund, Sweden

H. T. DAVIS AND L. E. SCRIVEN

Chemical Engineering and Materials Science Department
University of Minnesota
Minneapolis, Minnesota

J. C. C. NITSCHE

Department of Mathematics
University of Minnesota
Minneapolis, Minnesota

CONTENTS

I. INTRODUCTION

This research was stimulated by the need for mathematically well-characterized surfaces that divide space into two distinct, multiply connected, interwined subspaces—spatial structures that are called *bicontinuous*. Sponge, sandstone, apple, and many sinters are examples of relatively permanent though chaotic bicontinuous structures in the material realm. In these, one of the subspaces is occupied by a solid that is more or less deformable and the other, though it may be referred to as void, is occupied by a fluid. Certain lyotropic liquid crystalline states are probably also examples, one subspace being occupied by amphiphile molecules oriented and aggregated into sheetlike arrays that are ordered geometrically, the other subspace being occupied by solvent molecules. The characteristic scale of these structures is so small—5 to 50 nm—that the issue of their bicontinuity has not been settled, however (Scriven 1976; Fontell 1981; Longeley and McIntosh 1983; Bull and Lindman 1974). Related liquid crystalline states that contain two incompatible kinds of solvent molecules, e.g., hydrocarbon and water, present a further possibility in which one subspace is rich in the first solvent, the other in the second, and the surface between lies within a multiply connected startum rich in oriented surfactant molecules (Scriven 1976). If two distinct surfaces dividing hydrocarbon-rich from surfactant-rich and surfactant-rich from water-rich are envisioned, this microstructure could be considered tricontinuous.

Certain equilibrium "microemulsion" phases that contain comparable amounts of hydrocarbon and water as well as amphiphilic surfactant may be chaotic bicontinuous (or tricontinuous) structures, maintained in a permanent state of fluctuating disorder by thermal motions (Scriven 1977), for they are fluid and give no indication of geometric order but there is compelling evidence of multiple continuity (Lindman and Stilbs 1982; Auvray et al. 1984; Clarkson et al. 1985; Kaler et al. 1983; Kahlweit et al. 1987; Bodet et al. 1988). Here we concentrate on geometrically ordered bicontinuous structures. That these are indeed promising models for lyotropic liquid crystalline states known as cubic or viscous isotropic phases has been demonstrated in x-ray diffraction studies by Lindblom et al (1979) and Longeley and McIntosh (1983); Mariani et al. (1988) and others.

Other possible areas of application of the periodic surfaces, or of disordered relatives of these, include structure of superconductors in the intermediate state (Shal'nikov 1941), sintering kinetics (Hench and Ulrich 1984), fluid flow through porous media (e.g., Zick and Homsy 1982), the topology of spacetime at the scale of Planck length (Wheeler 1957), the structure of the prolamellar body in certain plastics (Gunning 1965), certain phase-segregated block copolymers (Thomas et al. 1986; Anderson and Thomas 1988)) semiconductor-based separation processes (Faulkner 1984), shape-selective catalysts (Haggin 1982), and Fermi surfaces in electron band theory (Pippard 1954; see also Andersson 1983).

While there are triply periodic minimal surfaces that are known to be free of self-intersections, to date there has been no calculation of a triply periodic surface of constant, *nonzero* mean curvature that is embedded in $\mathbf{R}^3$. We compute and display five families of such surfaces, where every surface in a given family has the same space group, the same Euler characteristic per lattice–fundamental region, and the same dual pair of triply-periodic graphs that define the connectivity of the two labyrinthine subvolumes created by the infinitely connected surface. Each family is comprised of two branches, corresponding to the two possible signs of the mean curvature, and a minimal surface. The branches have been tracked in mean curvature, and the surface areas and volume fractions recorded, with the relation $dA = 2HdV$ carefully checked to hold. The three families that contain the minimal surfaces P and D of Schwarz and the I–WP minimal surface of Schoen terminate at configurations that are close-packed spheres. However, one branch of the family that includes the F–RD minimal surface of Schoen, and both branches of the family that includes the Neovius surface C(P), contain self-intersecting solutions and terminate at self-intersecting spheres. On approach to the sphere limit, whether self-intersecting or close-packed, the gradual disappearance of small "neck" or "connector" regions between neighboring "spherelike" regions is in close analogy with the rotationally symmetric surfaces of Delauney, and to the doubly periodic surfaces of Lawson (1970). We give what we suspect are analytical values for the area of the F–RD minimal surface, and a possible limit on the magnitude of the mean curvature in such families is proposed and discussed. We also report that the I–WP and F–RD minimal surfaces each divide $\mathbf{R}^3$ into two subspaces of *unequal* volume fractions. Existence proofs for some of these surfaces have recently been formulated (Karcher 1987) by reducing the problem to a Plateau problem in $\mathbf{S}^3$, or by appealing to standard p.d.e. existence proofs using as initial surfaces assemblies of Delauney surface patches (Kapouleas 1987).

The numerical method is based on a new approach to the formulation of the Galerkin, or weak form, of the problem of prescribed—not necessarily constant—mean curvature. The Surface Divergence Theorem is applied directly to a vector-valued function that is the product of a scalar weighting function and vector field chosen to enforce the boundary conditions. This formulation applied in the context of the finite element method provides a robust algorithm for the computation of a surface with: (1) mean curvature as a prescribed function of position, and (2) contact angle against an arbitrary bounding body as a prescribed function of position or of arc length. A parametrization scheme for triply periodic surfaces is described that calls only for knowledge of the two "skeletal" graphs; this is demonstrated by the computation of the triply periodic minimal surface S′—S″ hypothesized by Schoen, who described only the skeletal graphs associated with the surface. The parametrization allows for easy calculation of the scattering function for various density profiles based on the solutions, as well as the areas and volume fractions. For the three minimal surfaces—P, D, and C(P)—whose areas and volume fractions are known analytically, the numerical results are in agreement with these values. Furthermore, we review the history of such surfaces, and clear up some inconsistencies in the literature over the D minimal surface.

II. MATHEMATICAL BACKGROUND

In 1865, the mathematician H. A. Schwarz published the first example of a minimal surface with full three-dimensional periodicity and free of self-intersections (or embeded in $\mathbf{R}^3$) (Schwarz 1865, 1890), a surface now known as Schwarz's Diamond surface (or D). The designation (Schoen 1970) refers to the branching of the two congruent, intertwined but disjoint labyrinths lying on opposite sides of the infinitely connected orientable surface. Each labyrinth contains in its interior a diamond-branched, symmetric graph of degree four, that is, a periodic array of nodes connected by edges in which four equivalent edges meet at the tetrahedral angle at each equivalent node. Some authors, e.g. Mackay (1985), give the name F to this surface since it is of face-centered cubic symmetry. In Fig. 1 are shown: (a) an extensive portion of this surface, together with the two graphs; (b) a lattice-fundamental region (a lattice-fundamental, or L-F, region is the region that exactly fills space when the translational symmetries of the Bravais lattice are applied); and (c) a primitive patch bounded by straight lines lying in the surface. (See color insert for Figs. 1a–b.) This primitive patch is the surface of minimal area that spans a circuit of four edges of a regular tetrahedron (it was also derived independently at about the same time by Riemann, 1892.) This was the first analytical solution to a general problem later named after Plateau, in which the surface of least area spanning a given closed loop is sought, representing the equilibrium configuration of a soap film spanning the frame. Schwarz's 1865 publication also mentioned that a periodic surface could be built up by rotating the patch through lines of 180° rotational symmetry. The surface is connected and free of self-intersections, even though it contains a network of edges of congruent regular tetrahedra, and regular tetrahedra themselves cannot fill space without self-intersection.

It can be shown that any solution to a Plateau problem must be a surface whose mean curvature H at each point is zero. Recall that $H = \frac{1}{2}[\kappa_1 + \kappa_2]$, where κ_1 and κ_2 are the principal curvatures (reciprocals of the principal

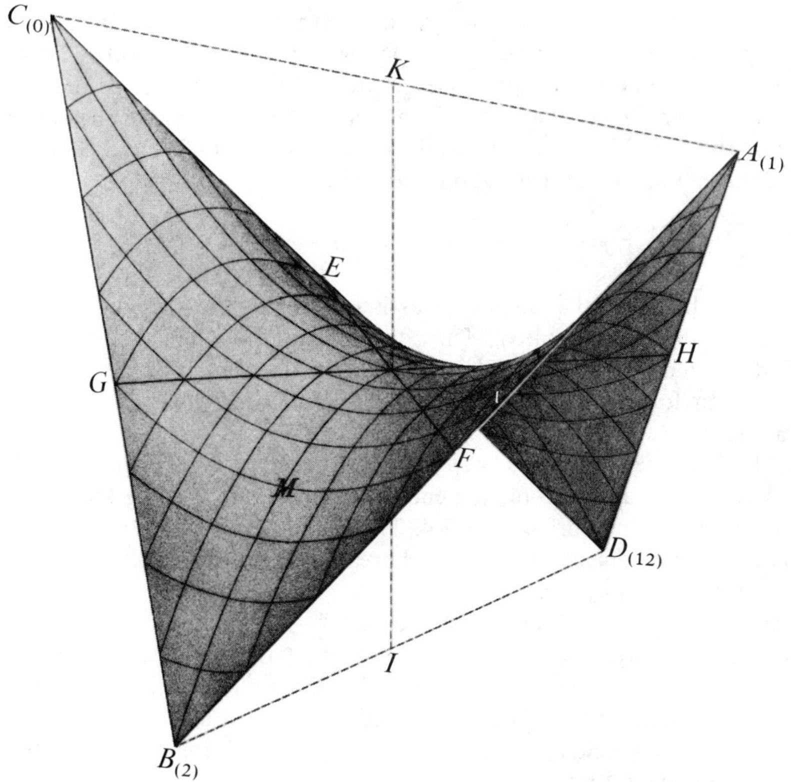

Figure 1c. A primitive patch of the D minimal surface bounded by straight lines that are axes of 180° rotational symmetry on the periodic surface. (Figure reprinted from Schwarz 1890.)

radii of curvature). A *minimal surface* is thus defined to be a surface of everywhere vanishing mean curvature [Lagrange (1761) derived a partial differential equation for a surface of least area that is equivalent to the condition that $H = 0$.] Each coordinate x_k of a minimal surface can be written as a harmonic function of certain surface coordinates (Weierstrass 1866; see also Bonnet 1853), i.e., $\partial^2 x_k/\partial u^2 + \partial^2 x_k/\partial v^2 = 0$. The reflection principle of harmonic functions then guarantees that a straight line lying in the surface is an axis of twofold rotational symmetry (Schwarz 1890). The maximum principle for harmonic functions requires that a minimal surface be wholly contained in the convex hull of its boundary curve. It can also be shown that, apart from the special case of a plane, the Gaussian curvature K is negative except at isolated points where it may be zero. (Nitsche, 1975, and Lawson, 1980, review these properties.) Mean and Gaussian curvature are the basic scalar invariants of the curvature dyadic, or second-rank tensors $\mathbf{b}(r)$, which describe the local state of surface curvature at points r on the surface. The curvature dyadic is the negative of the tangential gradient, or surface gradient ∇_S, of the unit normal vector $\mathbf{n}$ to one side of the surface (Weatherburn 1927). (In the classical treatment of the differential geometry of a surface, this operator is suppressed.) Thus

$$\mathbf{b}(r) = -\nabla_S \mathbf{n}, \quad H = 1/2\,\mathrm{tr}(\mathbf{b}) = -1/2 \nabla_S \cdot \mathbf{n} \tag{1}$$

Expecting the plane, the integral Gaussian curvature (or total curvature) of a minimal surface is negative. Thus, the integral Gaussian curvature of a periodic minimal surface must be infinite. By the Gauss–Bonnet theorem (the theorem for geodesic triangles is due to Gauss (1827) and a more general form is due to Bonnet (1848; see also Darboux 1894), $\iint_S K dA + \sum_i \int_{C_i} \kappa_g ds = 2\pi\chi$, the Euler characteristic χ of such a surface is infinite. But the Euler characteristic of a lattice–fundamental region is finite, and in the case of Schwarz's Diamond surface $\chi = -4$. The relation $\iint_S K dA = 2\pi\chi$ holds for every lattice–fundamental surface patch treated here because each is bounded by geodesic curves $\kappa_g = 0$.

Schoen [1970] has taken the relation $g = 1 - \chi/2$, which gives the genus g of a closed body in terms of its Euler characteristic, and applied it to a lattice–fundamental region of 17 of the 18 triply periodic minimal surfaces known (the eighteenth surface was added in proof in a footnote, and not discussed in any detail); thus he listed the value 3 for the genus of a lattice–fundamental region of the D surface, and of its conjugate surface P (see below). However, the correct formula for the genus of a surface with r closed boundary loops (r holes) is $g = 1 - (\chi + r)/2$. The Euler characteristic of an $n \times n \times n$ array of the L–F region of the P surface is $-4n^3$, and the number of boundary loops is $6n^2$, so that the genus per L–F region of the infinite P surface is in fact

$$g_{\text{L-F}} = \lim_{n \to \infty} [1 - (-4n^3 + 6n^2)/2]/n^3 = 2 \tag{2}$$

In fact, the value listed by Schoen for the genus of a L–F region must in each of the 17 cases be diminished by 1 to give the correct value for the genus per L–F region. The correct value can be found from the Euler characteristic listed in this paper by dividing this value by -2. The genus of an orientable surface is the maximal number of disjoint closed cuts that do not separate (disconnect) the surface. In the science of porous media, the average genus per unit volume of a dividing surface, whether periodic or not, provides a measure of the holyness of the medium (Rhines 1958; Pathak 1981).

Figure 1a. The Schwarz D or Diamond minimal surface. Four lattice-fundamental regions are shown, making the Euler characteristic − 16. The space group when considered as an oriented surface is F$\bar{4}$3*m*. The two graphs thread the two labyrinths created by the space-dividing surface.

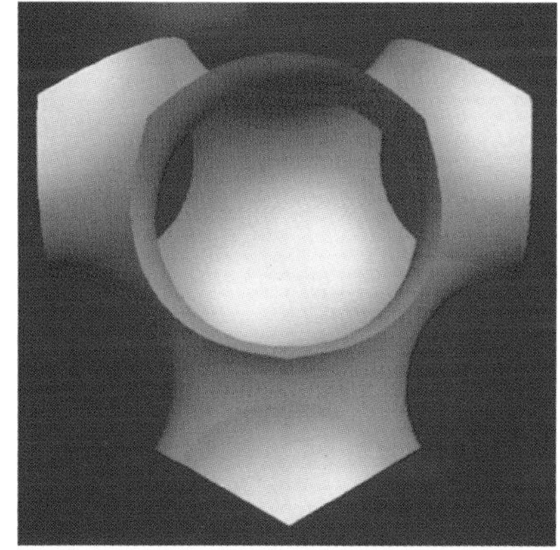

Figure 1b. A single lattice-fundamental region of the Schwarz D minimal surface, of Euler characteristic − 4.

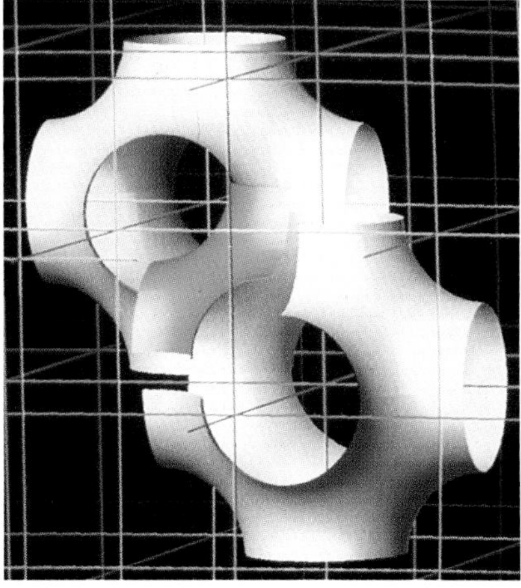

Figure 3a. The Schwarz P surface, the solution with $H^* = 0$, space group P*m*3*m*, Euler characteristic − 4 per lattice–fundamental region. The portion shown amounts to 1 7/8 unit cells.

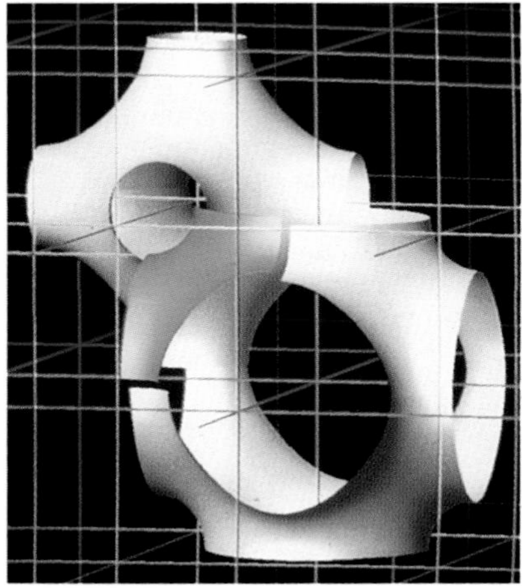

Figure 3b. The solution with $H^* = 1.0$, same space group and topological type as the Schwarz P minimal surface.

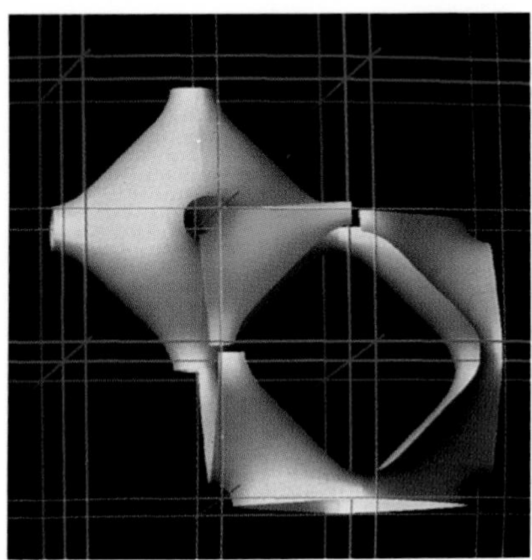

Figure 3c. The solution with $H^* = 1.7974$, same space group and topological type as the Schwarz minimal surface.

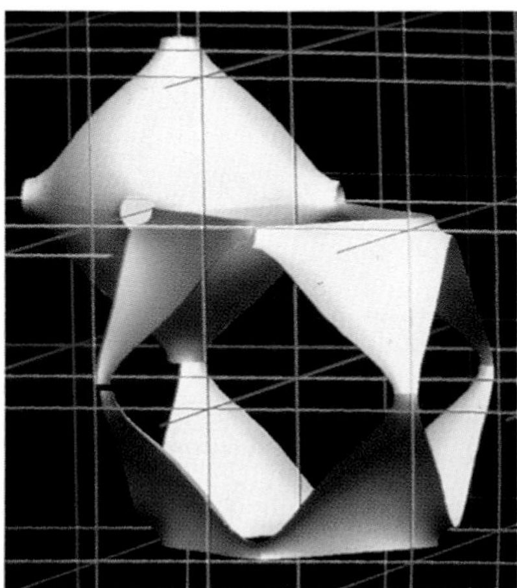

Figure 3d. Solution with $H^* = 2.0$, same space group and topological type as the Schwarz P minimal surface.

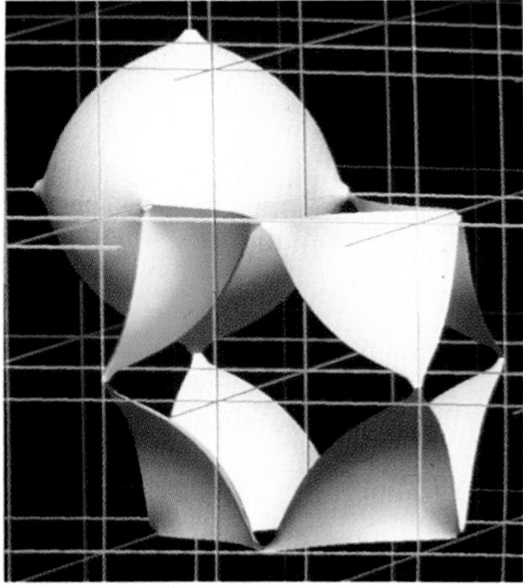

Figure 3e. Solution with $H^* = 2.133$, same space group and topological type as the Schwarz P minimal surface.

Figure 3h. A contact angle of 57.3° and a constant mean curvature of $H^* = 1.6$ have been prescribed.

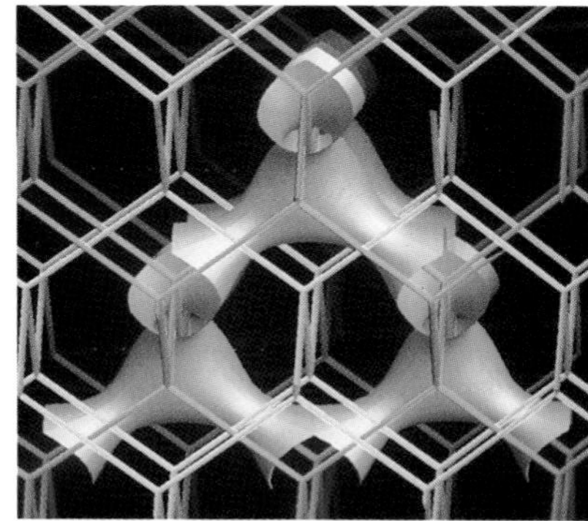

Figure 4b. Solution with $H^* = 3.8$, same space group and topological type is the Schwarz D minimal surface.

Figure 4c. Solution with $H^* = 4.836$, same space group and topological type is the Schwarz D minimal surface.

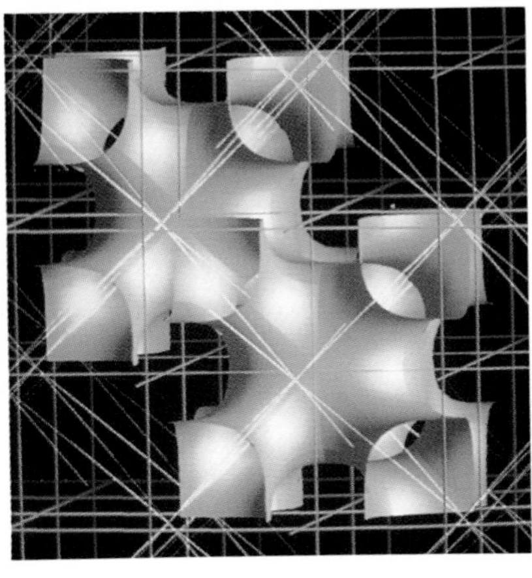

Figure 5a. The I–WP minimal surface, space group I$m3m$ and Euler characteristic -6 per lattice–fundamental region. Fifteen octants are shown.

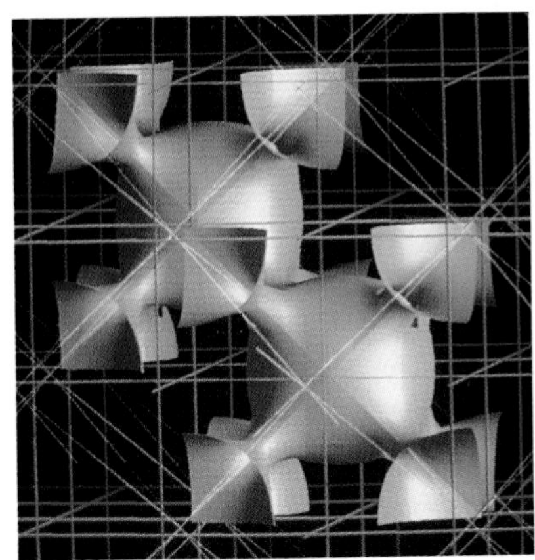

Figure 5b. Solution with $H^* = 2.0$, same space group and topological type as the I–WP minimal surface.

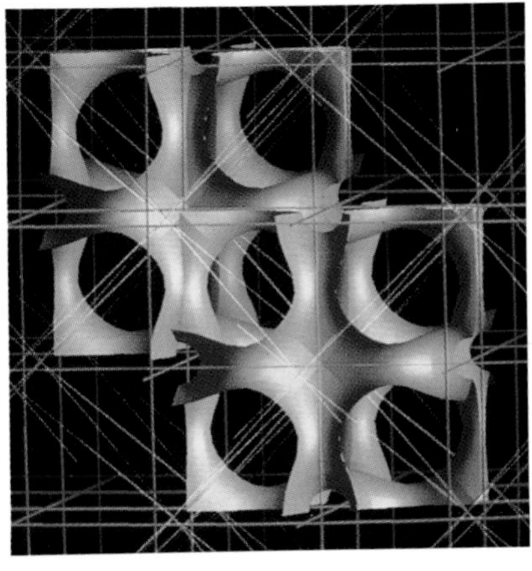

Figure 5c. Solution with $H^* = -3.55$, same space group and topological type as the I–WP minimal surface.

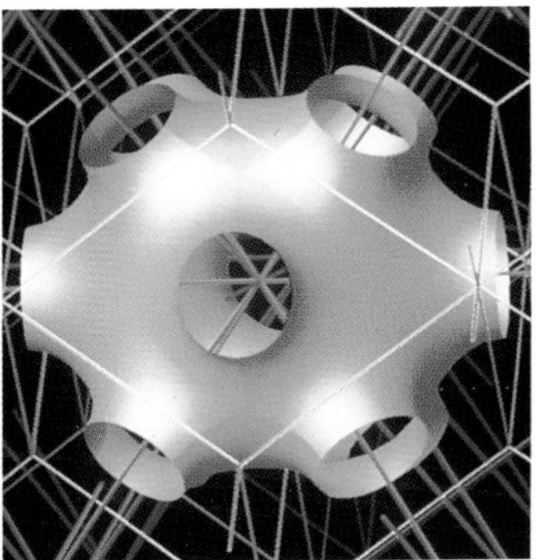

Figure 6a. The F–RD minimal surface. One lattice–fundamental region is shown. Space group is *Fm3m* and the Euler characteristic per lattice–fundamental region is -10. The twelve 'necks' are orthogonal to the twelve faces of the rhombic dodecahedron outlined by the red skeletal graph.

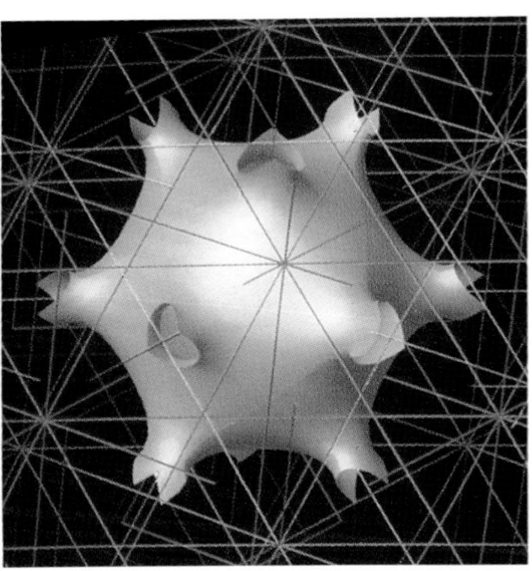

Figure 6b. Solution with $H^* = 1.86$, same space group and topological type as F–RD.

Figure 6c. Solution with $H^* = 2.5$, same space group and topological type as the F–RD minimal surface.

Figure 6d. Solution with $H^* = -2.4$, same space group and topological type as the F–RD minimal surface.

Figure 7a. The Neovius minimal surface C(P), space group Pm3m and Euler characteristic -16 per lattice–fundamental region. One unit cell is shown, which is also a lattice–fundamental region.

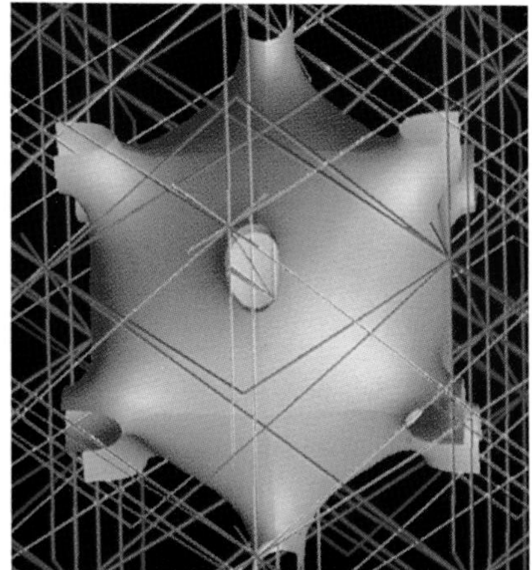

Figure 7b. Solution wth $H^* = 1.05$, same space group and topological type as the Neovius surface.

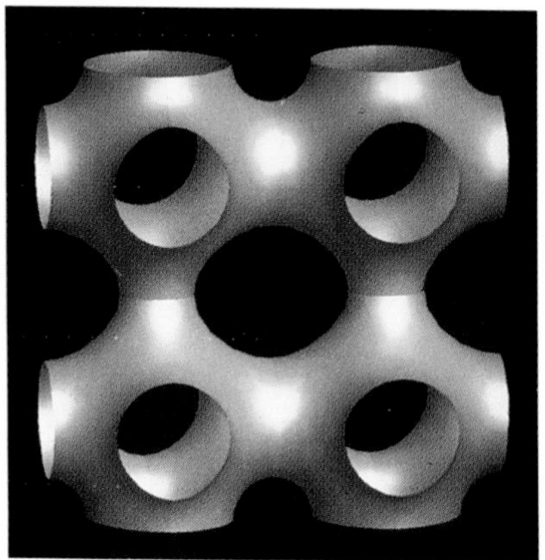

Figure 8. A periodic surface with the same topological type as the Schwarz P minimal surface, with sinusoidally varying mean curvature in the vertical direction. Two unit cells are shown side by side.

Schwarz made his discovery with the aid of Weierstrass's integral representation of minimal surfaces in terms of the following triplet of harmonic functions of the complex variable $u + iv$, where the domain is a complicated Riemann surface:

$$x(u, v) = \kappa Re \int_0^{u+iv} \frac{(1 - \sigma^2)e^{i\theta}}{\sqrt{R(\sigma)}} d\sigma \tag{3}$$

$$y(u, v) = \kappa Re \int_0^{u+iv} \frac{(1 + \sigma^2)e^{i\theta}}{\sqrt{R(\sigma)}} d\sigma \tag{4}$$

$$z(u, v) = \kappa Re \int_0^{u+iv} \frac{2\sigma e^{i\theta}}{\sqrt{R(\sigma)}} d\sigma \tag{5}$$

Schwarz was able to deduce that the diamond surface D could be obtained by taking $R(\sigma) = 1 - 14\sigma^4 + \sigma^8$, $\kappa = [2/K(1/2)]^{1/2} = 0.8389222985\ldots$, and $\theta = 0$. The three conjugate harmonic functions, generated by taking $\theta = \pi/2$, define the conjugate minimal surface to D also studied extensively by Schwarz (1890). This second surface is also periodic and free of self-intersections and is now known as Schwarz's Primitive or P surface in color insert, because each of the two labyrinths created by the surface contains a symmetric graph of degree six (six edges meet at each node) with cubes. The Bravais lattice of P is simple cubic, and its space group is Pm3m (no. 221 in the Crystallographic Table (1954)), Wyckoff notation n with 48 equivalent positions. The Bravais lattice of D is faced-centered cubic, and the space group is F$\bar{4}$3m (no. 216), Wyckoff notation i with 96 equivalent positions (see Schoen 1970, and Section IV of this chapter).

Neovius, a student of Schwarz, discovered another minimal surface, periodic and embedded in three dimensions (Neovius 1883). It is referred to simply as the Neovius surface, or as C(P) (Schoen 1970). Schoen has introduced the concept of complementary minimal surfaces, and his notation C(P) designates Neovius's surface as the complement of Schwarz's P surface. Two surfaces are said to be complementary if they contain the same straight lines and have the same space group. Despite these remarkable similarities, C(P) is considerably more complicated than P, the Euler characteristic of a lattice–fundamental region (which is also a unit cell) being -16 as opposed to -4 for P. The conjugate of Neovius's surface contains self-intersections (Neovius 1883). Neovius and Schwarz together discovered a total of five periodic minimal surfaces free of self-intersections, and no further examples were published until 1968. An example of a doubly periodic minimal surface is Scherk's surface (Scherk 1835), also studied by Schwarz (1890, vol. I, p. 147).

Schoen (1968, 1970) has proven or hypothesized an additional thirteen triply periodic embedded minimal surfaces. For one of these surfaces, which Schoen named the *gyroid*, an analytic solution with Weierstrass integrals was obtained. The surface is in fact related to both P and D by an associate transformation, in which the function $R(\sigma)$ and the Riemann surface in the Weierstrass representation are kept the same, and only θ is changed. The Gauss map, which maps each point of the surface to a point on the unit sphere representing the normal to the surface, is invariant under the associate (or Bonnet) transformation, and since the algebraic area of the Gauss map yields the Euler characteristic, the Euler characteristic of a corresponding region of the gyroid is the same as that of P and D.

Many of the remaining twelve examples discussed in Schoen's note must be considered conjectures. In many cases, physical models were built from plastic or with soap films. Although analytic representations for these surfaces have not yet been found, Schoen's contribution to a subject which had seen little progress in over 75 years was substantial. In this regard, it should be recalled that even in the case of rotationally symmetric surfaces, the constant-mean-curvature solutions (Delauney 1841) do not admit closed-form analytic representations (see also Kenmotsu 1980).

Schoen's system of notation for periodic minimal surfaces will serve equally well for periodic surfaces of constant, but not necessarily zero, mean curvature. Each surface is named after the skeletal graphs contained within the two labyrinthine regions into which $\mathbf{R}^3$ is partitioned by the surface. The two skeletal graphs are referred to as *dual skeletal graphs* and each has the same space group as the surface itself. If the surface contains straight lines, then these lines are two fold rotation axes whose action is to interchange the two sides of the surface; this implies that the two labyrinths—and therefore the two skeletal graphs—are congruent. The skeletal graph can then of course be called self-dual. The skeletal graphs of P, D, and C(P), for example, are all self-dual, and hence these surfaces can be specified by the name of just one graph. In the cases of P and D, this graph is symmetric (Schoen 1967, 1970). A graph is said to be symmetric if: (1) there exists a symmetry operation that is transitive on the edges that meet at a node; (2) there is a symmetry operation that is transitive on the nodes; (3) every vertex is joined to its z nearest neighbors (z is the degree of the graph); and (4) every vertex lies at the centroid of its z nearest neighbors. In the case of C(P), the graph is self-dual but not symmetric, having coordination symbol 12:4 (see Lines 1965); this graph is constructed by connecting the centroid of each cube to the twelve edge-midpoints, in a simple packing of cubes. Schoen chose to name the Neovius surface by its complementary relation to P rather than by this skeletal graph.

In cases where the minimal surface does not contain straight lines, it is possible that the skeletal graph is not self-dual (although it may be, as in the gyroid, where the two graphs are identical except that one is left-handed and the other right-handed). In cases where it is not, Schoen has given hyphenated names indicating both skeletal graphs. The three such cases which will be pursued in this work are called F–RD, I–WP, and S′–S″. F refers to the face-centered cubic graph, which is a symmetric graph of degree 12 obtained by connecting nearest neighbors in a face-centered cubic lattice; its dual graph is not symmetric, and is named RD because it consists of all the edges of space-filling assembly of rhombic dodecahedra (each rhombic dodecahedron defines a lattice–fundamental region in the face-centered cubic lattice). Alternatively, the RD graph can be constructed by joining the centroids of nearest-neighbor polyhedra in a space-filling assembly of regular octahedra and regular tetrahedra, indicating clearly that the coordination symbol of the graph is 8:4. The I graph is the body-centered graph, a symmetric graph of degree 8 obtained by connecting all nearest neighbors of a body-centered cubic lattice; its dual graph is not symmetric, and is named WP because a unit cell resembles the string of a wrapped package. The WP graph is constructed by joining the face centers to the four edge-midpoints for each cube in a simple packing, so the coordination symbol is 4:4:4 (the coordination symbol 4 is reserved for the symmetric graph of degree 4). The graph S′ is made by starting with parallel identical square tesselations, and joining corresponding edges in adjacent layers at their midpoints by edges which are perpendicular to the layers; its dual graph S″ starts with parallel identical square tesselations positioned halfway between the S′ layers and oriented 45° to the S′ squares, and edges perpendicular to the layers are erected at alternative vertices. Schoen's note (Schoen 1970) contains photographs of plastic models of F–RD and I–WP, but no visualizations of S′–S″ are given.

Thirteen of the eighteen examples introduced or reviewed by Schoen contain plane lines of curvature, though many of these do not contain straight lines (including the important examples F–RD and I–WP). A plane line of curvature is a curve on a surface which lies in a plane and whose tangent at each point is parallel to one of the principle directions of the curvature dyadic. By Joachimsthal's theorem (Joachimsthal 1846; see also Bonnet 1853), the surface meets this plane at a constant angle, which in all the periodic surfaces treated here is $\pi/2$. In these thirteen cases, Schoen has listed a kaleidescope cell—one of the seven convex polyhedra proven by Coxeter (see Coxeter 1963) to be the only generators of discrete groups of reflections—which orthogonally bounds a primitive patch of surface. The periodic surface can be obtained by repeated mirror reflections through these kaleidescope cell (or Coxeter cell) faces. For example, Schwarz showed that a primitive patch of the D surface is orthogonally bounded by a certain tetrahedron named "tetragonal disphenoid" by Coxeter. Thus D can be constructed by either solving a fixed-boundary problem and extending by repeated rotations, or by solving a free-boundary problem and extending by mirror reflections. The same is true of P and C(P), while others such as F–RD and I–WP that are not self-dual can be generated only by repeated reflections of a surface patch meeting orthogonally the faces of a Coxeter cell. The lattice–fundamental region for each periodic surface is composed of an integral number of Coxeter cells, and because the boundary curves are plane lines of curvature and the surface meets this plane at an angle of $\pi/2$, the boundary curves are geodesics, as stated above. This is thus a five-parameter family.

In his doctoral thesis, Meeks (1976) has given an existence proof of a family of triply periodic embedded minimal surfaces. Each surface corresponds to a placement of four antipodal pairs of points on the unit sphere.

Given fixed volume fractions, q and $1-q$, of the two subvolumes into which a given convex body is to be divided by a surface of a fixed topological type, the dividing surface of minimum area has constant mean curvature, and meets the boundary of the body orthogonally. A discussion has been given by Nitsche (1985). General existence proofs, in the context of geometric measure theory, can be found in Massari (1974) and Giusti (1984). However, these proofs do not provide partitioning surfaces of prescribed topological type. Results regarding interior and boundary regularity have recently been extended from area-minimizing surfaces to surfaces of stationary area (Grüter et al. 1981, 1986). Nitsche (1975) has reviewed the work on the problem of perturbing a minimal surface so as to have prescribed mean curvature, including the associated eigenvalue problem; however, very few constructive results exist for the case of orthogonality boundary conditions. Schoen (1970) has noted that the concept of dual skeletal graphs should apply to a constant-mean-curtature surface of the same topological type and space group as a periodic minimal surface. Recently, the Weierstrass representation has been extended to surfaces of nonzero mean curvature (Kenmotsu 1978), but to use his formulae to generate such a surface evidently requires its Gauss map, which in general is not known a priori. Recently, existence proofs have been given for certain triplyperiodic surfaces of constant, nonzero mean curvature (Karcher 1987). The conjugate surface to the minimal surface in $\mathbf{S}^3$ underlying such an H-surface is shown to be the solution to a well-posed Plateau problem in $\mathbf{S}^3$, the general existence of which has been proven. This is also the approach used by Lawson (1970) to prove the existence of doubly periodic surfaces of constant mean curvature. These existence proofs have not yet led to explicit calculations—to date there has been no calculation of a triply periodic embedded surface of constant, nonzero mean curvature reported in the literature.

In this chapter we report such surfaces, which we have computed with a new finite element formulation. Rivas (1972) has noted that finite-difference techniques are not well-suited for handling contact-angle boundary conditions. The finite element method has been shown to be an effective and versatile tool for the computation of minimal surfaces (Hinata et al. 1974; Wagner 1977), even when the surface does not admit singly valued orthogonal projections onto planes. The method we have developed can produce surfaces of nonuniform, nonzero mean curvature that is a prescribed function of position, and we report triply periodic examples of these as well. The method can produce a finite surface of prescribed mean curvature whose contact angle against a given boundary is a prescribed function of position; the obvious applications of this feature would be to fluid hydrostatics and hydrodynamics.

A rigorous mathematical existence proof for a periodic surface of small, nonzero constant mean curvature can be obtained with the methods of the theory of nonlinear elliptic differential equations. The resulting surface would be a perturbation of a known periodic minimal surface, but the intent of chapter is rather to exhibit numerical solutions that extend over wide ranges in mean curvature.

III. COMPUTATIONAL METHOD

In this section, we introduce the computational method in the form used for the surfaces exhibited in Section IV, i.e., where the prescribed mean curvature of the computed surface is everywhere constant and the boundary conditions are determined by two dual periodic graphs. We also give generalizations of the method for the computation for a surface of prescribed—not necessarily constant—mean curvature, with prescribed contact angle against surface. Generalization to the computation of space curves of prescribed curvature or geodesic curvature is available (Anderson 1986).

Given a pair of dual period skeletal graphs, G' and G'' with a certain space group, the first step is to identify the Coxeter cell capable of filling space by repeated applications of the mirror symmetries of the space group (see Coxeter 1963). There are only seven possibilities for this cell, each of which is either a tetrahedron, a rectangular parallelopiped, or a right prism.

Let C_1, $C_2, \ldots, C_f$ be the faces of C. This polyhedron C will contain at least a portion of one edge of each of the graphs, since the group of reflection symmetries that exactly fills space when applied to C also generates $G'(G'')$ when applied to the intersection of $G'(G'')$ with C. Call the edge-portions $l_1, l_2, \ldots, l_i \in G' \bigcap C$ and $m_1, m_2, \ldots, m_k \in G'' \cap C$. If i or $k > 1$, then $f > 4$ and in this case C is divided into tetrahedra $\{T_{pq}\}$; the tetrahedron T_{pq} is exactly the convex hull of the segments l_p and m_q. If l_p and m_q are parallel then T_{pq}

is not included. Let the faces of T_{pq} be T_{pqr}, $r = 1,2,3,4$. Any face in the entire collection $\{T_{pqr}\}$ that is not part of any C_n is shared by exactly two tetrahedra.

The unit interval $0 \leqslant u \leqslant 1$ is divided into i subintervals $\{I_p\}$, and l_p is parametrized by a linear mapping from I_p:

$$\mathbf{r} = \mathbf{a}_p + \mathbf{b}_p u, \quad u \in I_q \tag{6}$$

Similarly $O \leqslant v \leqslant 1$ is divided into k subintervals $\{J_q\}$ and m_q parametrized by a linear mapping from J_q:

$$\mathbf{r} = \mathbf{a}_q + \mathbf{b}_q v, \quad v \in J_q \tag{7}$$

The entire tetrahedron T_{pq} is now parametrized by sending a spine (Kistler 1983) from each point on l_p to each point on m_q (see Figure 2):

$$\begin{aligned}\mathbf{r}(u,v,w) &= (\mathbf{a}_p + \mathbf{b}_p u)(1-w) + (\mathbf{a}_q + \mathbf{b}_q v)w \\ &= \mathbf{a}_{pq}(u,v) + \mathbf{b}_{pq}(u,v)w, \quad u \in I_p, \quad v \in J_q, \quad 0 \leqslant w \leqslant 1\end{aligned} \tag{8}$$

The (u, v) computational domain D consists of the union of the rectangles $R_{pq} = I_p \times J_q$, one for each tetrahedron T_{pq}, with a pair of edges of two rectangles R_{pq} and $R_{p'q'}$ identified if T_{pq} and $T_{p'q'}$ share a face. In the cases of

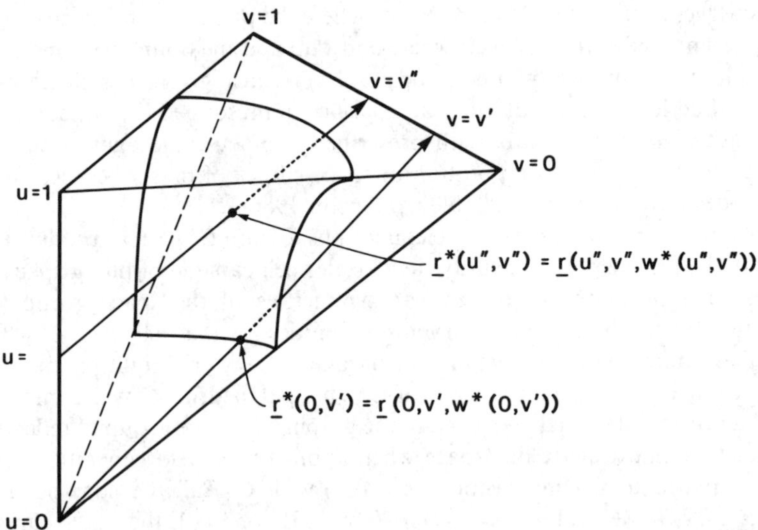

Figure 2. Parametrization of the primitive cell. One boundary spine and one interior spin are shown.

high symmetry examined in the next section, the identification of duplicated edges occurs automatically when the rectangles are assembled in the unit square. The parametrization of the surface patch S which spans C and is sufficient to generate the entire periodic surface by application of the group of reflections is defined by

$$\mathbf{r}^*(u,v) = \mathbf{a}_{pq}(u,v) + \mathbf{b}_{pq}(u,v)w^*(u,v) \tag{9}$$

The condition $0 < w^* < 1$ is necessary, though not sufficient, for the periodic surface to be free of self-intersections. The function $w^*(u,v)$ must be found to complete the characterization of the surface. Traditionally one solves the partial differential equation $\nabla_S \cdot \mathbf{n} = -2H$, which is a second-order p.d.e. for $w^*(u,v)$. A finite-difference or finite-element solution for $w^*(u,v)$ is sought using basis functions $\phi_i(u,v)$ and finding coefficients α_i such that

$$w^*(u,v) = \sum_{i=1}^{N} \alpha_i \phi_i(u,v) \tag{10}$$

A common way of generating residual equations from which α_i's are determined is the Galerkin method, which yields

$$\iint_S \phi_j [\nabla_S \cdot \mathbf{n} + 2H] dA = 0, \quad j = 1, \ldots, N \tag{11}$$

It is advantageous to replace Eq. (11) by an equivalent expression with lower differentiability requirements on the basis functions, in which the boundary conditions are automatically satisfied. The appropriate integration by parts formula is the Surface Divergence Theorem (SDT; Weatherburn 1927), which is an integral relation for a piecewise-differentiable vector-value function $\mathbf{F}$ defined on a surface:

$$\iint_S \nabla_S \cdot \mathbf{F} d\mathbf{A} = \int_{\partial S} \mathbf{m} \cdot \mathbf{F} ds - \iint_S 2H\mathbf{n} \cdot \mathbf{F} d\mathbf{A} \tag{12}$$

Here $\mathbf{F}$ is an arbitrary C^1 vector-valued function defined on a surface S with normal $\mathbf{n}$ and area element $d\mathbf{A}$, ∇_S is the surface divergence operator on S, H is the surface mean curvature and ds and $\mathbf{m}$ are the differential arc length and outward pointing unit tangent along the boundary ∂S of the surface. The outward pointing tangent is defined to be orthogonal to both the surface normal $\mathbf{n}$ and the unit tangent $\mathbf{t}$ to the boundary curve ∂S:

$$\mathbf{m} = \mathbf{t} \times \mathbf{n} = (d\mathbf{r}^*/ds) \times \mathbf{n}, \quad \mathbf{r}^* \in \partial S \tag{13}$$

However, if $\mathbf{F}$ is taken to be $\mathbf{n}\phi_j$, the result is Eq. (11), and so the SDT cannot be used to integrate Eq. (11) by parts since the SDT contains Eq. (11) as a special case.

The key to the present method is to choose the test functions $\mathbf{F}_j$ in terms of the fixed parametrization (Figure 6.2) and the weighting functions $\phi_j(u, v)$. An expedient choice for the test function $\mathbf{F}_j$ in the jth residual equation is

$$\mathbf{F}_j = \mathbf{M}\phi_j \tag{14}$$

where

$$\mathbf{M} = \partial \mathbf{r}/\partial w \tag{15}$$

In the present case $\mathbf{M} = \mathbf{b}_{pq}$. The scalar multiplication by the weighting function ϕ_j is done so that weighted residuals result on application of the SDT. Since each spine is parametrized by w, the vector $\mathbf{M}$ is tangent to the spine given by (u, v); in particular, if (u, v) is on the boundary of the computational domain, so that this spine runs along some C_n, then $\mathbf{M}$ will lie in the plane of that face. The orthogonality boundary conditions for the desired surface are equivalent to the condition that $\mathbf{m}\cdot\mathbf{F}_j(=\mathbf{m}\cdot\mathbf{M}\phi_j) = 0$, and the orthogonality boundary conditions are enforced by simply striking the boundary term from each residual equation.

To carry out a computation, H is prescribed and the surface operator ∇_S, the surface normal $\mathbf{n}$, and the area element data are expressed in terms of the solution vector and its first partial derivatives. The Jacobian of the parametrization is easily calculated to depend only on w:

$$\partial(x, y, z)/\partial(u, v, w)dudvdw = c_{pq}w(1 - w)dudvdw \tag{16}$$

where $c_{pq} = (\mathbf{b}_p \times \mathbf{b}_q)\cdot(\mathbf{a}_q - \mathbf{a}_p)$. The transformation of course is singular along l_p and m_q, where $w = 0$ or 1, but the surface lies inside the C and does not intersect these segments. The surface normal is computed from the surface basis vectors $\mathbf{r}^*_u = \partial\mathbf{r}^*/\partial u$, and $\mathbf{r}^*_v = \partial\mathbf{r}^*/\partial v$, where in differentiating $\mathbf{r}^*$ we encounter the partials $\partial w^*/\partial u$ and $\partial w^*/\partial v$:

$$\mathbf{n} = \mathbf{r}^*_u \times \mathbf{r}^*_v/|\mathbf{r}^*_u \times \mathbf{r}^*_v| \tag{17}$$

The surface operator is expressed in surface coordinates using a standard formula:

$$\nabla_S = \mathbf{s}_1\partial/\partial u + \mathbf{s}_2\partial/\partial v \tag{18}$$

or

$$\nabla_S = \frac{(G\mathbf{r}^*_u - F\mathbf{r}^*_v)}{EG - F^2}\partial/\partial u + \frac{(E\mathbf{r}^*_v - F\mathbf{r}^*_u)}{EG - F^2}\partial/\partial v \tag{19}$$

where $E = \mathbf{r}_u^* \cdot \mathbf{r}_u^*$, $F = \mathbf{r}_u^* \cdot \mathbf{r}_v^*$, and $G = \mathbf{r}_v^* \cdot \mathbf{r}_v^*$ are the fundamental magnitudes of first order. The vectors $\mathbf{s}_1$ and $\mathbf{s}_2$ are the surface reciprocal basis vectors. The area element is of the form $dA = (EG - F^2)^{1/2} du dv$ with our parametrization.

In the finite element method (Courant 1943; Turner et al. 1956), the basis functions are of compact support—that is, each ϕ_j is nonzero only on a small rectangle in D. Since the quantities that we want to calculate from the solutions, such as area, volume, and other surface or volume integrals such as the scattering function, do not call for explicit knowledge of the second or high partial derivatives, the most economical choice is bilinear basis functions, which are piecewise first differentiable. A method for estimating the Gaussian curvature from the solution is described later in this section.

The Galerkin-weighted residual equations are obtained by inserting the test function $\mathbf{F}_j$ into the SDT, with the boundary term deleted:

$$0 = R_j = \iint_S (\nabla_S \cdot \mathbf{F}_j + 2H\mathbf{n} \cdot \mathbf{F}_j) dA, \quad j = 1, \ldots, J \tag{20}$$

The integrals are evaluated by numerical quadrature, with the local mean curvature H prescribed at each Gauss point. The equations can be solved for the α_i's by Newton iteration.

Mean curvature is a dimensional quantity, with dimensions of inverse length. One definition of a dimensionless measure is the product $H^* = H \cdot a$ of mean curvature times the lattice parameter a. Thus, the solutions presented can be scaled to any size lattice. Corresponding areas and volumes are found by multiplying the dimensionless quantities by a^2 and a^3, respectively.

A good initial estimate of the solution is necessary, and this often requires considerable trial and error. Techniques for generating an estimate for the first solution in a family have been given (Anderson 1986); in the families treated here, the minimal surface was found first, since the nonlinearities in the term containing H generally slow convergence. Once a single solution is available, first-order continuation (Silliman 1979) in parameter H^* can then used to examine a full range of constant mean curvature surfaces with the same space group and skeletal graphs as the known surface. In this continuation scheme, the change in the nodal values resulting from a small increment in H is predicted by solving for $\partial/\partial H$ in the matrix equation:

$$(\partial R_j/\partial \alpha_k)(\partial \alpha_k/\partial H) + (\partial R_j/\partial H), \quad j = 1, \ldots, J \tag{21}$$

where the inverse of the Jacobian matrix $[\partial R_j/\partial \alpha_k]$ is known from the final Newton iteration at the old value of H. The column vector on the right-hand side of Eq. (21) has already been calculated as one of the contributions [after multiplication by $2H$; see Eq. (20)] in the residual R_j:

$$\partial R_j/\partial H = \iint_S \mathbf{n}\cdot\mathbf{F}_j dA, \quad j=1,\ldots,J \tag{22}$$

For each surface of nonzero H^*, the sign convention for H^* reverses when the (orientable) surface is viewed from the other side. For the cases where the skeletal graph is self-dual, this corresponds to the two choices for the unit cell, one in which the centroid is a node of G' and one where it is a node of G''. Thus in these cases we need only substitute positive values for H^* in the computation, and use this transformation to display unit cells with H^* negative. In cases where G' and G'' are not congruent, both positive and negative values of H^* must be explicitly substituted.

It is a simple matter to compute the area of each computed surface, as well as the volume fractions of the two labyrinths created by the partitioning surface:

$$\begin{aligned} A(H) &= \iint_S dA = \iint_D (EG - F^2)^{1/2}\, du\, dv \\ V(H) &= \iiint_{w<w^*} dV = \sum_{R_{pq}} \iiint_O^{w^*(u,v)} c_{pq} w(1-w)\, dw\, du\, dv \\ &= \sum_p \sum_q c_{pq} \int_{J_q} \int_{I_p} (w^{*2}/2 - w^{*3}/3)\, du\, dv \end{aligned} \tag{23}$$

This latter integral gives the volume in C lying on the same side of the surface as the skeletal graph G'. The relation

$$dA = 2H^* dV \quad \text{or} \quad A'(H) = 2HV'(H) \tag{24}$$

applies because each surface in a given family hits the faces of C orthogonally, and provides a check on the numerical accuracy of the constant-mean-curvature surfaces generated. [Gauss (1827, vol. V, p. 67; see also Nitsche 1975, p. 91) derived a formula for the first variation of a surface which is easily reduced to Eq. (24) when H is constant; the authors do not known where Eq. (24) was first explicitly stated.]

If H is specified as a function $H(u, v)$ of the surface coordinates, then this function is evaluated at each Gauss point P_m, where the value $H(P_m)$ is needed in those residual equations R_j satisfying $\phi_j(P_m) \neq 0$. Clearly a more rapidly-varying function will require a finer grid. If H is a specified function of spatial position $H(r)$, then this is rewritten, via Eqs. (9) and (10) as a function of u, v, and the solution $w^*(u, v)$. Examples of this are given in

Section IV. In principle, it is possible to prescribe H as a function of position *and slope* by including in the prescribed function the values of $\partial w^*/\partial u$ and $\partial w^*/\partial v$ already calculated at the Gauss points, though this has not been tried. When H is prescribed as a function of position and/or slope, then the Jacobian matrix should include the dependence of this prescribed H on the nodal values; however, if this function is slowly-varying and the user is willing to sacrifice quadratic convergence, this contribution to the Jacobian can be neglected or approximated.

The method is easily generalized to allow specification of the contact angle γ between S and the surface ∂C of a known body C along $\partial S = S \cap \partial C$; the precise definition of this contact angle is the angle between the surface normal $\mathbf{n}$ and the normal $\mathbf{N}$ to ∂C, both taken at $\mathbf{r}^*(u,v) \in \partial S$. Since the surface ∂C is known, its normal $\mathbf{N}$ at $\mathbf{r}^*(u,v)$ presumably can be written (except perhaps on a set of measure zero) as a function of position in space, and ultimately of u, v, and $w^*(u,v)$ using Eqs. 9 and 10. In cases where C is a polyhedron, $\mathbf{N}$ is a constant independent of u and v along faces of the polyhedron.

The parametrization $\mathbf{r}(u,v,w)$ of C must always be defined in such a way that the spines whose (u,v) coordinates are on the boundary ∂D of the computation domain lie entirely on the boundary ∂C of the body C. By the work of Kistler and Scriven (1983), it is known that there is a tremendous amount of versatility when spines are used. In the general case of a spine parametrization, the vector field $\mathbf{F}$ is still defined via Eqs. (14) and (15), recognizing that $\mathbf{M}$ is now a function of u, v, and w^*. In addition to this tangent vector ∂C along the (u,v) spine, we define another vector $\mathbf{M}^+$ also tangent to ∂C but orthogonal to the spine:

$$\mathbf{M}^+ = \mathbf{M} \times \mathbf{N} \tag{25}$$

and then

$$\mathbf{F}_j^+ = \mathbf{M}^+ \phi_j \tag{26}$$

We take $\mathbf{N}$ to be normalized so that

$$\mathbf{F}_j = \mathbf{N} \times \mathbf{F}_j^+ \tag{27}$$

Combining this with Eq. (13), we have that

$$\mathbf{F}_j \cdot \mathbf{m} = (\mathbf{N} \times \mathbf{F}_j^+) \cdot (\mathbf{t} \times \mathbf{n}) \tag{28}$$

By a simple vector identity this can be rewritten

$$\mathbf{F}_j \cdot \mathbf{m} = (\mathbf{N} \cdot \mathbf{t})(\mathbf{F}_j^+ \cdot \mathbf{n}) - (\mathbf{N} \cdot \mathbf{n})(\mathbf{F}_j^+ \cdot \mathbf{t}) = -\cos\gamma(\mathbf{F}_j^+ \cdot \mathbf{t}) \tag{29}$$

where the contact angle γ is specified between the normalized vectors $\mathbf{n}$ and $\mathbf{N}$, and the dot product of the normal and tangent vectors $\mathbf{N}$ and $\mathbf{t}$ to ∂C is zero. Another simplification occurs when we notice that the differential arc length ds in the expression $\mathbf{t} = d\mathbf{r}^*/ds$ cancels with the ds in the integrand of line integral of the SDT:

$$\begin{aligned} \mathbf{F}_j \cdot \mathbf{m} ds &= -\cos\gamma(\mathbf{F}_j^+ \cdot d\mathbf{r}^*/ds)ds \\ &= -\cos\gamma \mathbf{F}_j^+ \cdot d\mathbf{r}^* \end{aligned} \tag{30}$$

We specify the contact angle γ at each point of the boundary ∂S by inserting Eq. (30) into the SDT Eq. (12). In many cases the parametrization and the boundary ∂D of the computational domain are such that the line integral can be performed analytically between nodes. For example, if C is a polyhedron parametrized as in the past section, then each $\mathbf{M}_j$ (and therefore $\mathbf{M}_j^+$) is independent of w, since only straight spines are needed; furthermore $\mathbf{r}^*$ is linear in u and v (as a function $\mathbf{r}^*(u, v)$!) since the representation, Eqs. (8) and (9), with bilinear basis functions is piecewise linear along the polygonal boundary ∂D. Thus, if the contact angle is to be everywhere constant, the integration over a boundary grid line is easily performed analytically, as is done in an example in Section IV. If the contact angle has a relatively simple analytic form as a function of u and v or of arc length s, again the analytic integration should be possible. One can imagine, on the other hand, situations where the contact angle prescription is sufficiently complicated (particularly a function of the spatial coordinates on ∂S) that the integration would have to be performed by numerical Gaussian quadrature.

In the expansion in bilinear basis functions, the Gaussian curvature is concentrated in the solid angles at nodes, where four surface elements meet. The following scheme is a sensible method of estimating K at each node, and gives an accurate value of the integral Gaussian curvature over any region comprised of a nontrivial number of elements.

The integral of the Gaussian curvature over an area of surface S is exactly the algebraic (or signed) area of the image on the unit sphere of the spherical map, which for every point on S assigns a point on the unit sphere corresponding to the direction of the normal on S. The area of the spherical image is taken to be negative if the orientation of a closed loop is reversed with respect to the orientation on S, and positive if the sense is preserved. If this rule is applied to a very small surface patch S, over which the Gaussian curvature is approximately constant, then an approximation to the average pointwise Gaussian curvature is

$$K = (\text{signed area of spherical image})/(\text{area of } S) \tag{31}$$

The normals to a surface defined with bilinear basis functions are most accurately calculated at the midpoints of elements; in this calculation only first-partial derivatives are required. Thus, if we are interested in the value of K at an interior node N_j, a simple calculation at the midpoints of each of the four elements meeting at N_j supplies four points on the unit sphere. Girard's theorem then provides a convenient means of evaluating the area of the spherical patch bounded by geodesics that form a circuit through these four points. By choosing geodesics to join consecutive points, we also ensure that neighboring quadrilaterals join up exactly, without gaps or overlap; this is very important for the calculation of the total integral Gaussian curvature over a region of surface. To apply Girard's theorem, we will need the angle between two geodesics g_1 and g_2, where g_1 passes through points on the unit sphere corresponding to the unit vectors $\mathbf{n}_i$ and $\mathbf{n}_j$, and g_2 passes through the endpoints of $\mathbf{n}_i$ and $\mathbf{n}_k$. But this angle is just the dihedral angle between the planes whose sections on the sphere are g_1 and g_2. The unit normal to the plane containing $\mathbf{n}_i$ and $\mathbf{n}_j$ is

$$\mathbf{n}_{ij} = \mathbf{n}_i \times \mathbf{n}_j / |\mathbf{n}_i \times \mathbf{n}_j| \tag{32}$$

and similarly for $\mathbf{n}_{ik}$. Then the angle $\bar{\omega}$ between g_1 and g_2 is given by

$$\cos \bar{\omega} = \mathbf{n}_{ij} \cdot \mathbf{n}_{ik} \tag{33}$$

After calculating the four angles of the geodesic quadrilateral, its area is given by Girard's theorem as the sum of the four angles, minus 2π. A convenient test for the sign of this integral Gaussian curvature is to check the sense of the loop $\mathbf{n}_1 \to \mathbf{n}_2 \to \mathbf{n}_3 \to \mathbf{n}_4$, by

$$\operatorname{sign} K = \operatorname{sign}\{[(\mathbf{n}_2 - \mathbf{n}_1) \times (\mathbf{n}_3 - \mathbf{n}_2)] \cdot \mathbf{n}_2\} \tag{34}$$

The resulting value of the integral Gaussian curvature is then divided by the area of the surface element, given by one term in Eq. (15).

To estimate K at a node that lies on the boundary of D, we have used a method based on Joachimsthal's theorem (1846). Whenever the boundary conditions require a fixed contact angle against a plane (or against any surface, such as a sphere, on which every point is an umbilic), then the curve of this intersection is a line of curvature. In the triply-periodic surfaces of this paper, whether of constant mean curvature or not, the boundary curves on the primitive patch S are thus plane lines of curvature. A scheme to estimate $|\kappa_1|$, the magnitude of the curvature of this plane curve at a given point, is to circumscribe a circle about the triangle determined by three consecutive points centered about the given point. If a, b and c are the sides of this triangle and $s = (a + b + c)/2$, then the radius of curvature $R = 1/|\kappa_1|$ is given by Heron's formula:

$$R = \frac{abc}{4\sqrt{s(s-a)(s-b)(s-c)}} \tag{35}$$

There are then only two choices for the second principal curvature, because the sum of the curvatures equals $2H^*$. To distinguish between the two possibilities $\kappa_2 = 2H^* - |\kappa_1|$ and $\kappa_2 = 2H^* + |\kappa_1|$, the magnitude of κ_2 is estimated using the above formula applied to the images of three points that begin at the given node and move directly into D.

IV. RESULTS

A. The P Family

We now describe a family of triply periodic embedded H-surfaces that each have space group Pm3m and divide space into two labyrinthine regions, which are threaded by two fixed dual symmetric P graphs of degree six. The solution for $H = 0$ is the Schwarz P minimal surface, and the limiting configuration on each of the two branches consists of simple cubic close-packed spheres. Except for these limiting configurations, the Euler characteristic of a unit cell of surface, which is also a lattice–fundamental region, is -4, and the number of boundary loops is $r = 6$.

The Coxeter cell C for each member of this family is a quadrirectangular tetrahedron, e.g., given by $0 \leqslant z \leqslant x < y \leqslant 1/2$. Among the edges of C are representative edge-portions of each of the two skeletal graphs G' and $G'': l_1$: $\{(0, u/2, 0) | 0 \leqslant u \leqslant 1\} \in G'$ and m_1: $\{(1/2, 1/2, v/2) | \leqslant v \leqslant 1\} \in G''$. Thus the vector field $\mathbf{M}$ is simply $\mathbf{M}(u, v) = (1/2, 1/2 - u/2, v/2)$. The generic residual equation is now determined, using the equations of the previous section. The constant c_{pq} in the Jacobian, Eq. (9), is simply 1/4.

The computational domain is the unit square in u and v, and this was divided into a 15×15 mesh; i.e., 225 elements, and $16 \times 16 = 256$ nodes, so 256 basis functions and 256 residual equations. The Jacobian matrix was banded with a total bandwidth of 35. The first solution computed was the minimal surface, for which the initial estimate was an hyperbolic paraboloid. The nonlinear system of residual equations was solved by Newton iteration on a Cyber 124, each iteration using about 1 second cpu time. For nearly all the surfaces calculated, the mesh was an even mesh over the entire unit square. However, for the surfaces just near the close-packed spheres (CPS) limit, the nodes were evenly spaced in the u-direction but placed as follows in the v-direction: $v = 0, 1/60, 1/30, 0.05, 0.075, 0.1, 0.15, 0.2, 0.3, 0.4, 0.5, 0.6, 0.7,$

0.8, 0.9, and 1.0. In the case $H^* = 0$ the straight lines on the surface should satisfy $w = 1/2$ on nodes where $u = v$, and this was satisfied to 13 digits (the accuracy of the Cyber 124 with single precision) for every such node; this remarkable accuracy was in fact observed for all three minimal surfaces treated containing straight lines, namely P, D, and C(P). Also for these surfaces, the volume fraction calculated was 1/2 to 13 significant figures.

The computer graphics were prepared using a Cray-1S with the graphics package MOVIE.BYU (Christianson 1981). In this program, surfaces are represented by polygons. Limitations with the system at the University of Minnesota set an upper limit of 8,000 polygons. To meet this limitation, the 15×15 mesh was thinned to 8×8. A front clipping plane was used to reduce clutter and to eliminate nodes of the skeletal graphs that would appear with the wrong coordination number, where the polygon limitation did not allow of the edges coincident at these nodes to be included.

In the case of P family (Figure 3, see color insert for Figures 3a–3e), it was found best to exhibit an amount of surface that corresponds to $1\frac{7}{8}$ unit cells. There are two obvious choices possible for the unit cell; if a node of the G′ (blue) skeletal graph is designated as the origin, with the principal directions of the lattice coinciding with the coordinate directions, then the unit cube $\{(x, y, z) | -1/2 < x, y, z < 1/2\}$ centered at this origin could serve as the unit cell; alternatively the cube $\{(x, y, z) | 0 < x, y, z < 1\}$ centered at a node of the G″ (red) graph could be chosen. We have elected to exhibit both cells, which share the common octant $\{(x, y, z) | 0 < x, y, z < 1/2\}$. Thus, we have a conception of both the "solid" and the "void" regions, as it were, as well as a common octant showing how these join. Another octant of surface, different from this shared octant, has been exploded and colored blue-green; this is done not only to better define an octant of surface, but also because exploding this octant helps reveal the shared octant that would otherwise be obstructed.

Figure 3a through 3e are color photographs of the resulting surfaces for five selected values of H^*. Contour plots of the Gaussian curvature for the surfaces shown in Figures 6.3a–6.3d are shown in Figure 6.3a′–6.3d′. Because the P skeletal graph is self-dual, a plot for any nonzero value of H^* represents both the surface $H = H^*$ and $H = -H^*$, the sign depending on which side of the surface one is viewing it from, or equivalently, which of the two unit cells described in the previous paragraph is the object of interest and which is part of the "void." The volume fraction V of that portion of a unit cell lying on the same side of the surface as the blue skeletal graph is plotted versus H^* in Fig. 3f; likewise the sign convention for H^* is that we take the sign as seen from a node of the G′ skeletal graph. Figure 3g is a plot of A versus V, so that the slope dA/dV of the curve at any point is $2H^*$. In each graph, the endpoints represent simple cubic close-packed

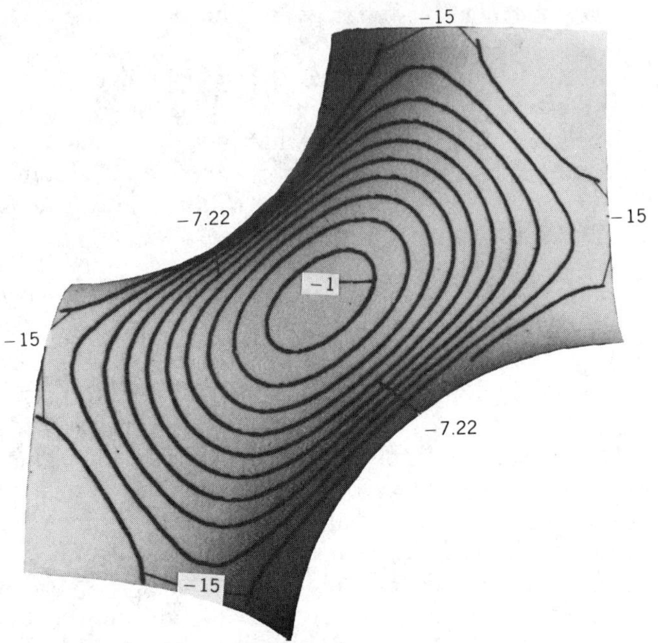

Figure 3a′. Contour plot of Gaussian curvature for the Schwarz P minimal surface, $H^* = 0$. The interval between contours is 1.56. One octant of a unit cell is shown.

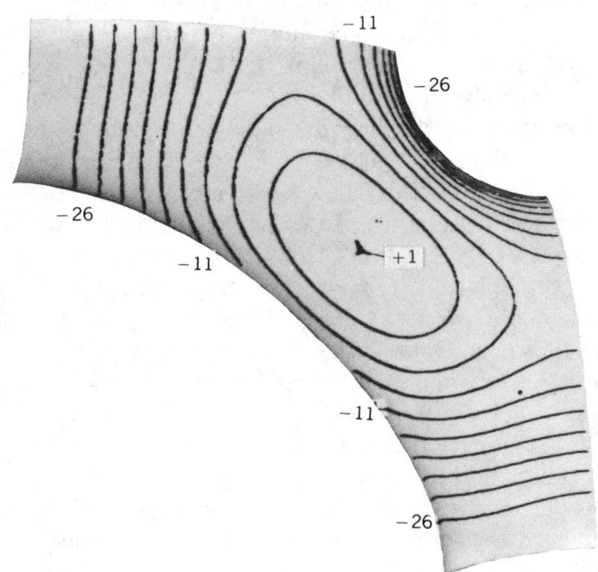

Figure 3b′. Contour plot of Gaussian curvature for an octant of the surface in Fig. 3b. Contour interval is 3.0.

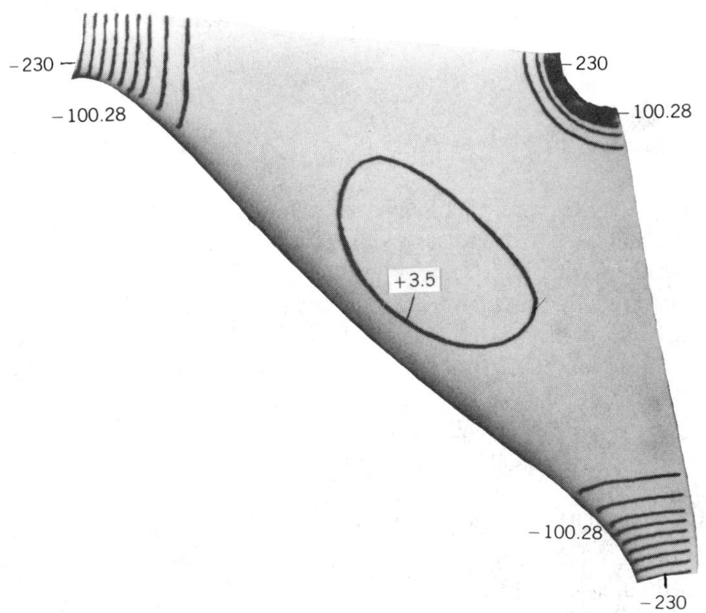

Figure 3c′. Contour plot of Gaussian curvature for an octant of the surface in Fig. 3c. Contour interval is 11.44.

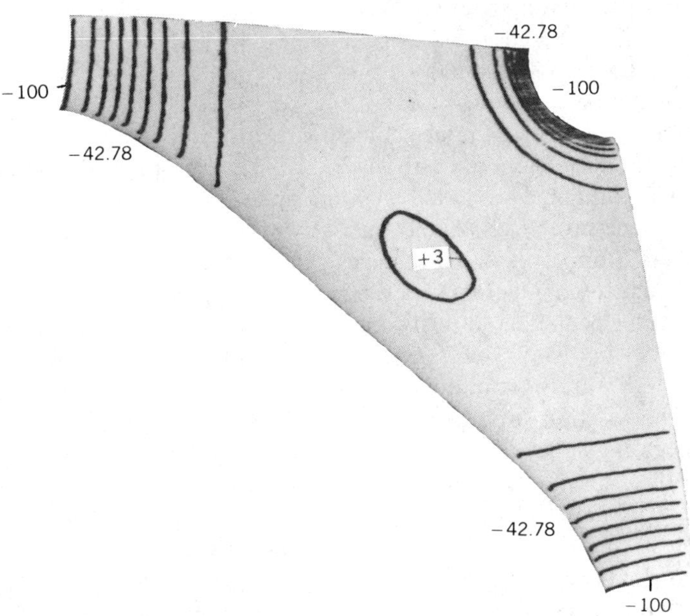

Figure 3d′. Contour plot of Gaussian curvature for an octant of the surface in Fig. 3d. Contour interval is 25.94.

spheres, where

$$A = \pi, \quad \text{and} \quad V = \pi/6 \quad \text{or} \quad 1 - \pi/6, \quad H^* = 2 \quad \text{or} \quad -2 \tag{36}$$

The point $H^* = 0$ represents Schwarz's primitive minimal surface, the only case for which the two choices of unit cell are congruent by virtue of the fact that this minimal surface contains straight lines.

Throughout this chapter we will continue to reserve the word "branch" for a set of surfaces that constitute all of the H-surfaces attainable from a certain close packing of spheres through a succession of H-surfaces having strictly the same sign of H. In the present family, there are exactly two branches satisfying this definition, so that they can be called simply the positive-H branch and the negative-H branch. Each branch will be broken down into three subbranches for discussion, those subbranches in the positive-H branch being called A, B, and C, and those in the negative-H branch being A′, B′, and C′ in the self-dual cases and D, E, and F in the non-self-dual cases (I–WP and F–RD).

The contour plot indicating the distribution of Gaussian curvature over an octant of the minimal surface is shown in Fig. 3a′. The values of K at the nodes were computed by the method described in the previous section. While it is true that there exists (Schwarz 1890) a simple, closed-form analytic formula for K, the coordinates in that formula—those of the Weierstrass representation [Eqs. (3)–(5)]—are not the curvilinear coordinates used in the present representation; thus, only in special cases can we determine what point on the computed surface corresponds to a value of K calculated with the analytic formula. Where it was possible to compare, the values of K computed compared very well with the analytical values, and the integral Gaussian curvature over the octant was within 1% of the value required by the Gauss–Bonnet theorem, namely $-\pi$ (the value $\chi = -4$ for a unit cell corresponds to an integral Gaussian curvature of $2\pi(-4) = -8\pi$). The point at the center of the octant, which lies on an axis of three-fold symmetry, is the only point with $K = 0$—there can be no points of strictly positive Gaussian curvature on a minimal surface.

On the solution subbranches labeled A and A′ in Fig. 3f, surface areas are decreasing as H^* increases in magnitude, and the volume fraction of the labyrinth of net positive curvature is monotonically decreasing; that is, on subbranch A where the net curvature is toward the G' graph, the volume fraction of this G' labyrinth decreases monotonically as H^* becomes more negative. Figure 3b (see color insert) represents both $H^* = 1.0$ and $H^* = -1.0$. The fact that the area is stationary at $H = 0$ is simply a consequence of the relation $dA = 2H^* dV$, which governs the entire family because of the orthogonality boundary conditions. The fact that the area is actually a local maximum within this family is discussed in Section V. Again, it should be emphasized that the term *minimal surface* refers to a property of area minimization under boundary conditions determined by a fixed boundary curve.

Figure 3b′ is the contour plot of the Gaussian curvature for the surface $H^* = 1.0$, and also for $H^* = -1.0$. In the cases of H^* nonzero, there is no analytic formula for K as in the case of the minimal surface. However, the integral Gaussian curvature was again within 1% of the required value of $-\pi$: the Euler characteristic of the surface remains -4 throughout the family. Note that regions of positive Gaussian curvature—synclastic regions—have grown out from the point on the threefold axis; the increase in integral Gaussian curvature from this region is countered by the more highly negative values of K along the boundary of the octant.

The relation $dA = 2H^* dV$ implies a relation between the first nonzero coefficients H^* in the expansions of A and V about $H^* = 0$. While the first variation of the area is zero at $H^* = 0$, the coefficient of H^{*2} is nonzero:

$$A(H^*) = A(0) - cH^{*2} + \cdots \tag{37}$$

So $dA = 2cH^* dH^*$, which must equal $2H^* dV$, giving $dV = -c dH^*$, or:

$$V(H^*) = V(0) - cH^* + \cdots \tag{38}$$

The value of $V(0)$ is of course 1/2, and the value of $A(0)$ is also known in terms of complete elliptic integrals of the first kind:

$$A(0) = 3K(1/2)/K'(1/2) = 2.3451068\cdots \tag{39}$$

The value of $A(0)$ computed with a 15×15 mesh was 2.34547, which is within 0.015% of this analytic value. Concerning the question of the analytical value for c, we mention that there is a simple formula, using a quantity that appeared in the Introduction as a normalization constant in the Weierstrass representation of the P and D minimal surfaces, that yields a value within 0.01% of the observed value. Recall the constant $\kappa = \sqrt{2}/K(1/2) = 0.8389222985\ldots$; if the representation is considered a dimensional equation, as is often conceptually useful, then this κ should be given units of inverse length. The observation is that the computed value of the constant c for the P family is, within the error of 0.03%, equal to $1/(2\kappa)^3 = 0.21171167\ldots$, and furthermore, the dimensions of both c and $1(2\kappa)^3$ are those of [length]^3. We suspect that this is indeed the exact value for the slope of the V–H^* curve in the P family.

The points labeled H_c^*, where A and A′ subbranches end and B and B′ begin, correspond to minima in area over the entire family. By the relation

$dA = 2H^*dV$, these also appear as cusps, with the V–A curve approaching the leaving with asymptotic slope $2H^*$. The magnitude of H^* at these points is $H_c^* = 1.7974 \pm 0.0002$, and the area of a unit cell is $A = 2.0025 \pm 0.0005$; the volume fraction of the side of greater curvature is $V = 0.2488 \pm 0.0002$. The computer graphic representing these two points is shown in Fig. 3c.

It is seen from Fig. 3c (see color insert) that three of the boundary curves to an octant of surface are very close to being linear near their midpoints, giving the surface a "squarish" appearance. This is discussed in Section V as being due to the proximity of the curve Ψ, describing the locus of points at which the Gaussian curvature vanishes, to these midpoints (Fig. 3c′). At some value of H^*, this contour Ψ separating the synclastic ($K > 0$) from the anticlastic ($K < 0$) regions makes tangential contact with each boundary curve, and does so at the midpoint of the curve by symmetry. Since Joachimsthal's theorem guarantees that the three curves are lines of curvature, and since $K = 0$ at a point implies that one of the principal curvatures is zero, it is true that each curve is asymptotically linear at its midpoint. For each branch where it was examined, the value of H^* at this midpoint was, within an error of about 0.15, equal to the value of H_c^* at which the minimum in area was found.

On subbranch B the magnitude of H^* is still increasing, but now the area is increasing with $|H^*|$, and the volume fraction of the labyrinth of net

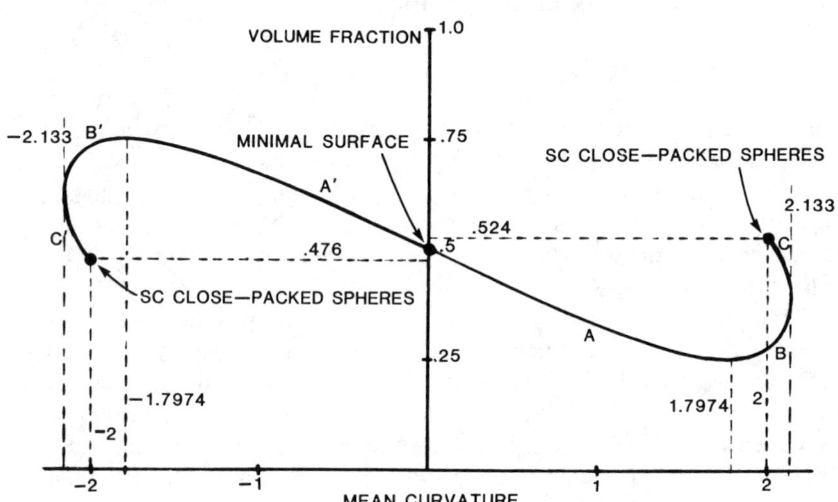

Figure 3f. Plot of volume fraction versus H^* for the P family.

positive curvature is increasing. Figure 3d (see color insert) is a graphic of the solution on this subbranch with $H^* = 2.0$, and Fig. 3d′ is the contour plot of K. At this value of H^*, the surface is largely synclastic, while the "neck" regions of negative Gaussian curvature exhibit some values of K less than -200. The fact that H^* is increasing in magnitude means that the graph of A versus V in Fig. 3g is concave upward, although this is slight because the increase in H^* is only from 1.7974 to 2.1335 ± 0.0005, the maximum value of H^* attained in this family. The endpoint of subbranch B(B′) corresponds to this maximum H^*, which appears as a turning point in Fig. 3f and as an inflection point in Fig. 3g. First-order continuation in H^* enabled us to get very close to this turning point, and the solution with $H^* = 2.133$ shown in Fig. 3e (see color insert).

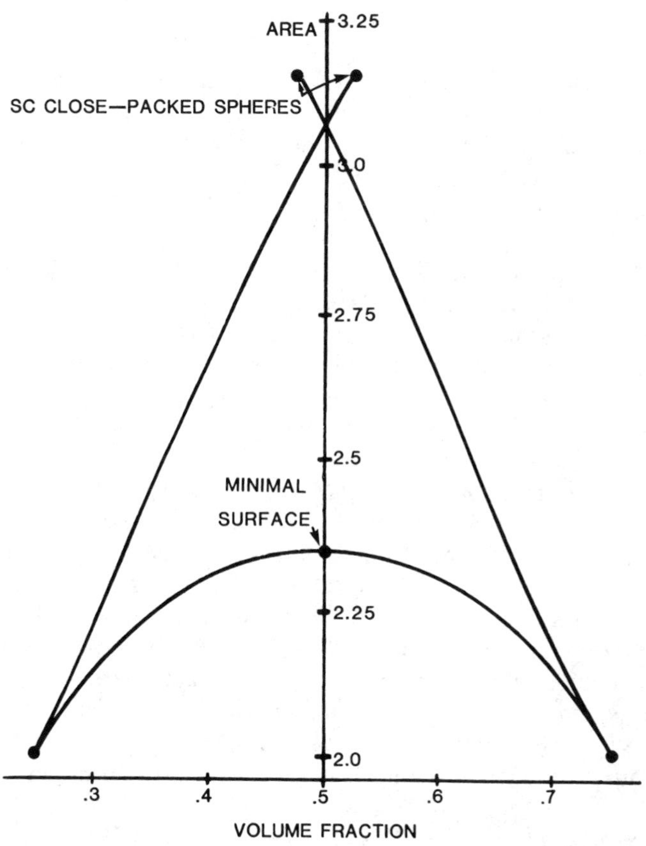

Figure 3g. Plot of area per unit cell versus volume fraction. The slope at each point is $2H^*$.

Past the turning point, on subbranch C or C′, the magnitude of H^* decreases from the maximum value of 2.1335 to the value 2 of the limiting CPS configuration. The curve in Fig. 3g is thus concave downward. Surfaces on this subbranch are very well described as spheres with small, nearly unduloidal necks connecting nearest neighbors. On approach to the CPS limit, the necks become increasingly smaller in both average radius and length. Because the neck length is shrinking while the lattice parameter is remaining constant, the average radius of the spherelike region in a cell must be increasing, and thus the area of a unit cell is increasing rather sharply as the C and C′ subbranches approach their CPS limits. The analogy with the pinching-off progression that occurs in the rotationally symmetric case is very close, and has been examined more closely (Anderson 1986). The asymptotic behavior as the CPS limit is approached has also been examined by a very different method in connection with existence proofs (Kapouleas, 1987).

In Fig. 3h is shown a cube of surface with $H^* = 1.6$, which has been computed to make a contact angle of 57.3° with the cube faces, using the equations of the previous section for prescribing contact angles. Obviously this surface would not be smooth if continued periodically, but is intended rather to demonstrate the generalized method.

B. The D Family

The starting point for this family is Schwarz's D minimal surface that is conjugate to the P minimal surface of the preceding family under the Weierstrass representation. The D minimal surface has been the object of investigation for many researchers in both the mathematical and the physical sciences, notably Schwarz (1890), Riemann (1892), Weierstrass (1903), Schoen (1970), Nitsche (1975), Longeley and McIntosh (1983), and Mackay (1985). We begin this discussion by clearing up certain inconsistencies in the literature. The space group of the minimal surface has been designated as Pn3m by some authors and F$\bar{4}$3 by others, resulting, for example, in differing values published for the normalized surface-to-volume ratio of a lattice–fundamental region of D (Mackay 1985). This apparent conflict will be resolved in this subsection, and the results of these authors will be shown to be in perfect agreement except for a difference with respect to treatment of the orientability of the minimal surface. F$\bar{4}$3m is the space group of each of the two symmetric skeletal graphs of degree four, and of the diamond-cubic CPS limiting configuration, with a packing fraction of 0.3401. Pn3m is the space group of the double-diamond packing found, for example, in cuprite (Bragg et al. 1965), or ice VII.

A simpler example of the effect of orientability on the space group of a minimal surface is provided by the P minimal surface. We saw above that the space group of the P surface is Pm3m, when the surface is viewed as an oriented surface with two distinct sides facing two distinct skeletal graphs. However, the space group of the surface, considered as unoriented, with its two sides and two associated labyrinths equivalent, is body-centered cubic, Im3m. This is because the point (1/2, 1/2, 1/2), for example, would be equivalent to the origin, since the two labyrinths in the case of $H = 0$ are congruent (this of course requires that H be identically zero, because of the change in sign of the mean curvature associated with crossing a surface). This has been pointed out by Fontell (1981), in the context of a physical problem. It does not seem to have been mentioned in the literature but it is true that the hexagonal faces of the truncated octahedron that encloses a lattice-fundamental region of the body-centered Bravais lattice cut the minimal surface along straight lines lying in the surface. Thus, in the case of the P minimal surface, there is a perfect analogue to the lattice-fundamental region proposed for the D minimal surface by Schwarz (his Tafel 2), Nitsche (Abbildung 27) (1975), Longeley and McIntosh (Fig. 3), and Mackay (Fig. 1). This L-F region for D consists of six replicas of the minimal surface patch that spans a circuit of four edges of a regular tetrahedron; the twelve edges that lie on the boundary line also on the faces of a cube which determines both an L-F region and a unit cell. In the case of P, however, the truncated octahedron which determines an L-F region severs the surface into patches that meet only in point contacts.

This explains why the values of normalized surface-to-volume are different for Mackay and Schoen. The value of 1.9193 given by Mackay is for an L-F region which is only half that of Schoen in both volume and area. Therefore it should be less, by a factor of $2^{1/3}$, than Schoen's analytic value $A/V^{2/3} = 2.4176538\ldots$, or $1.91889309\ldots$.

The volume determining the L-F region of the surface when considered oriented is composed of 24 replicas of the tatahedron known as the tetragonal disphenoid. As noted by Schoen (1970), this is the Coxeter cell C for a primitive patch of the D surface; note that this primitive patch is different from the one in Fig. 1b (see color insert), which is bounded by straight lines, and continued by rotations about these edges. The faces of C are isosceles triangles, and there are two right dihedral angles—these correspond under the conjugate transformation to the two right angles in the skew quadrilateral discussed above as lying in the P minimal surface. Giving the name U to the unit cell of Mackay cited above, the coordinates of the four vertices can be taken to be: $P_1(1/4, 1/4, 1/4)$, $P_2(-1/4, -1/4, -1/4)$, $P_3(1/4, 1/4, -3/4)$, and $P_4(3/4, -1/4, -1/4)$. The edge 1_1 connecting P_2 and P_3 is clearly parallel to a body diagonal of U, and in fact belongs to one of the skeletal graphs, say G'; a representative of the dual graph G'' is the edge m_1 connecting P_1 and P_4. Each face of this cell C is indeed a plane of mirror symmetry passing through a midplane of one of the six carpets T, and C contains a primitive patch S with a surface area amounting to exactly one-half of such a carpet. One can convert to the coordinates of the F$\bar{4}$3m unit cell by the transformation $(x', y', z') = 1/2(x + 1/4,\ y + 1/4,\ z + 1/4)$.

Figures 1a, 1b and 4a through 4c are the graphics representing solutions for selected values of $H^* = Ha$, where the lattice parameter a is that of the F$\bar{4}$3m space group. (See color insert for Figs. 1a, 1b, 4b and 4c.) Since the D skeletal graph is self-dual, the surfaces $H^* = h$ and $H^* = -h$ are identical except for interchange of the roles of the two skeletal graphs (see color section for blue and red skeletal graphs). Generally speaking, the progression of solutions found above in the P family from the minimal surface, through a minimum in area at H_c^*, past a turning point in H^* and to a CPS limit, is observed again the D family. One difference is that the value of H_c^* at the area minimum is greater than the value at the CPS limit in the D family, whereas it lies below the CPS value in the P family. This fact alone suggests that any link provided by the generalized Weierstrass representation (Kenmotsu 1978) between the P and D families would be limited, or at least more complicated than a simple rescaling.

The residual equations were solved for a 15×15 mesh on the Cyber 124 for values of H^* less than 4. For values greater than this, it was found

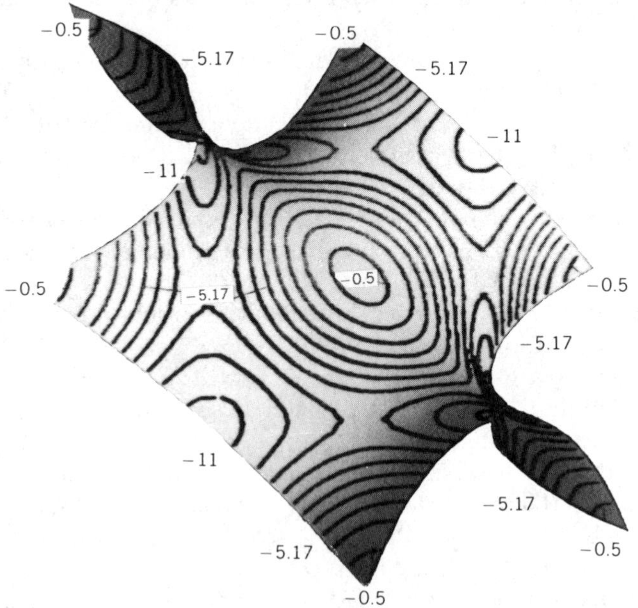

Figure 4a. Contour plot of Gaussian curvature for a portion of the Schwarz D minimal surface shown in Fig. 1a. The contour interval is 1.166.

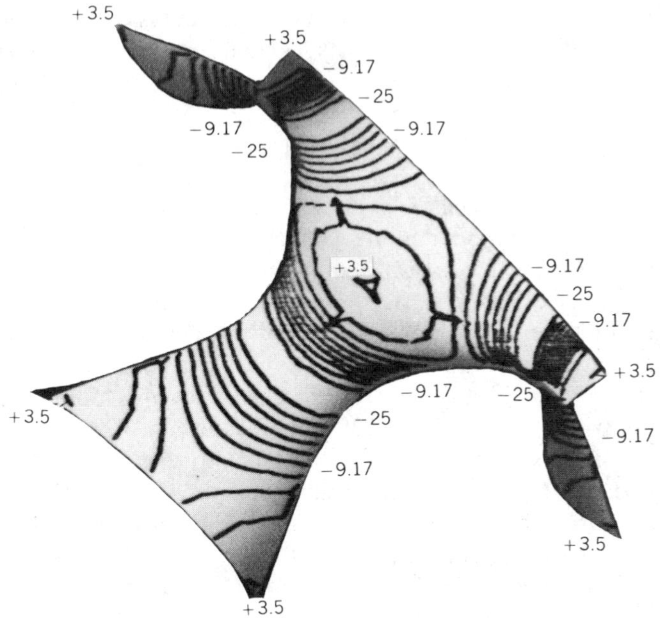

Figure 4b′. Contour plot of Gaussian curvature for a portion of the surface in Fig. 4b. Contour interval is 3.166.

necessary to increase the mesh to 20 × 20, and the equations were solved on a Cray-1S supercomputer. With a vectorized code, each iteration required approximately 0.18 second cpu time. A 20 × 20 mesh for a primitive patch is equivalent to 38,400 elements per unit cell.

In Fig. 1a (color insert) we have exhibited four L-F regions of the minimal surface; thus the surface area is equal to that in an $F\bar{4}3m$ unit cell, and the Euler characteristic is -16. The distribution of Gaussian curvature over a region made of six primitive patches is shown in Fig. 4a. With the lattice parameter of the P minimal surface, and of the D minimal surface, set equal to unity, the values of K for these two conjugate surfaces do not agree, as they would if the two surfaces were as calculated in the Weierstrass representation. The constant $\kappa = 0.8389222985\ldots$ in Eq. (1) was chosen to make the edge-length of the regular tetrahedron discussed above equal to unity, and with this some constant the lattice parameter of the P minimal surface is 2κ, not 1. In order to see the equality of K at corresponding points of the P and D minimal surfaces, it is necessary to rescale the values plotted in Figs. 3a′ and 4a according to these dimensions. The value of a principal radius of curvature at the midpoint of a fundamental patch (as in Fig. 1c) of the D surface, estimated numerically by Mackay (1985) at 0.29, can be evaluated analytically, using Schwarz's formula for the Gaussian curvature, as

$$1/[8(2-\sqrt{3})\mathrm{K}(7-4\sqrt{30}] = 0.296603823\ldots,$$

where K is the complete elliptical integral of the first kind.

On subbranches A and P′, the area of the surface decreases as H^* increases in magnitude, so that the area of the minimal surface is a local maximum, as in the P family. Also in analogy with the P family, the volume fraction of the labyrinth lying on the side of net positive curvature is decreasing as H^* increases in magnitude (see Fig. 4d). The coefficient of H^* in the expansion of V about $H^* = 0$, which is of course the coefficient of H^{*2} in the series for the area, is, within the error of 0.03%, equal to exactly 2/3 times the value for the P family. As in the case of the constant c for the P family, we cannot now prove this relation analytically. Figures 4b and 4b′ are for $H^* = \pm 3.8$.

At the end of subbranches A and A′, and beginning the subbranches B and B′, is the surface of minimum area at a value of $|H_c^*| = 4.836 \pm 0.001$, shown in Fig. 4c. The normalized surface-to-volume ratio for an L-F region at this minimum is 1.6527 ± 0.0005, which represents the smallest value for any surface in the five families treated here. It is interesting that this value,

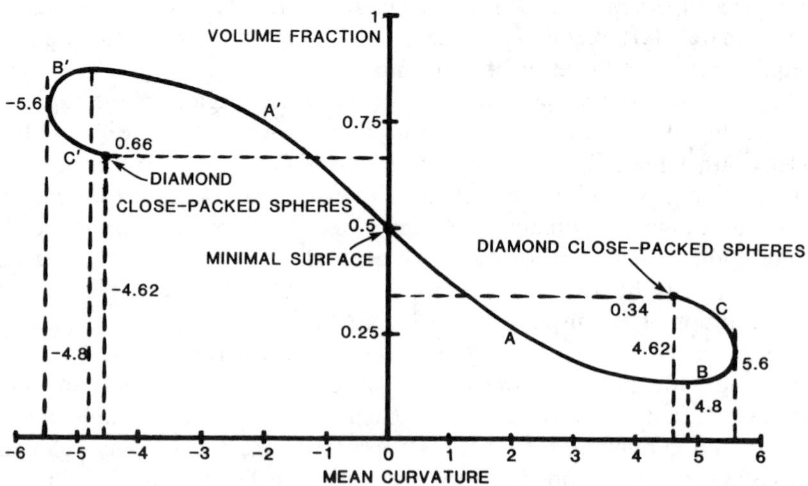

Figure 4d. Plot of volume fraction versus H^* for the D family.

which is also lower than any value previously listed for a triply-periodic minimal surface, occurs for a surface in the D family even though the P minimal surface has a value lower than that of the D minimal surface by 3%. In addition, the volume fraction ratio at this area minimum is 0.869195:0.130805, which represents the largest ratio of any surface (free of self-intersections) in the five families treated. This is not surprising since the packing fraction in the CPS limit is only 0.3401. Figure 4e is a plot of A versus V, in which the two minima in area appear as cusps.

On subbranches B and B′ we find surfaces consisting of spherelike regions lying inside the symmetry domains Schoen 1970, with nearly unduloidal necks piercing through each of the four faces of each symmetry domain and

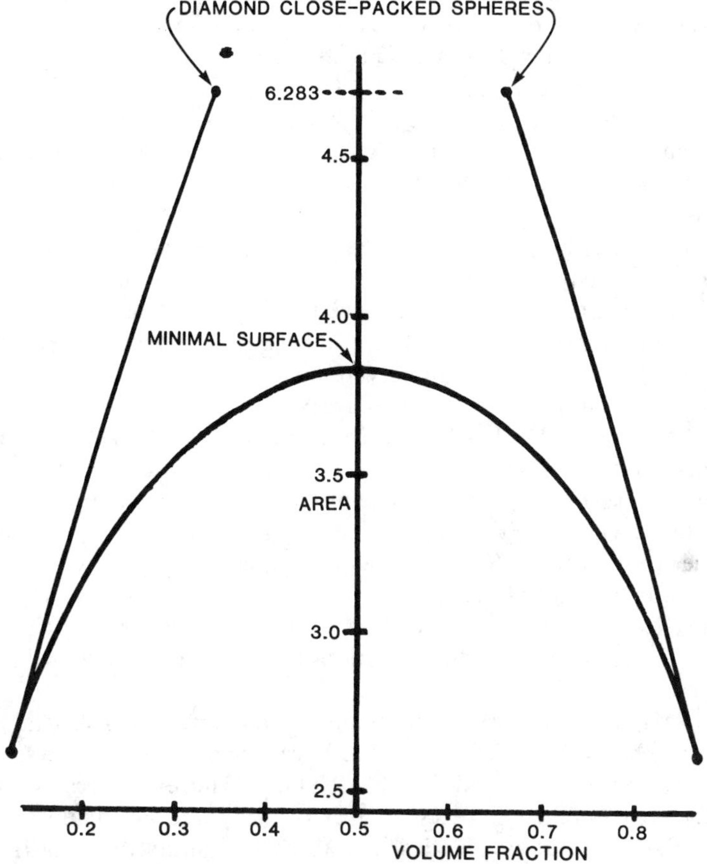

Figure 4e. Plot of area per unit cell versus volume fraction for the D family. The slope at each point is $2H^*$.

providing connections to the four neighboring spherelike regions. The magnitude of H^* then reaches a maximum of 5.6 ± 0.2, and past this turning point it begins to decrease, no subbranches C and C′. On C and C′ the area increases dramatically as the unduloidal necks pinch off, and H^* decreases slowly to its value of $8/(3)^{1/2}$ corresponding to the diamond–cubic CPS limiting configuration.

C. The I–WP Family

The minimal surface in this family was proposed by Schoen (1970) as the conjugate to a self-interesecting minimal surface partly studied by Stessmann (1934). Schoen also built plastic models of the surface, and identified the space group as $Im3m$ and the Coxeter cell as the quadrirectangular tetrahedron identical to that in the P family. Since the I graph is not self-dual—it and its dual WP were described in the Introduction—there cannot be straight lines lying in the minimal surface. This also means that the volume fractions of the two labyrinths need not be 1/2 (Mackay's [1985] claim that all the known triply periodic surfaces divide space into two congruent regions is thus mistaken), and Schoen did not give values for the volume fractions nor for the surface area. The Euler characteristic of an L-F region for any value of H^* is -6.

We have described the quadrirectangular tetrahedron C above, but the edges of C that represent the two skeletal graphs will be different from the P case. We take G' to be the symmetric I or body-centered graph, and G'' to be the nonsymmetric WP, or "wrapped package" graph, and take the sign convention to be that the mean curvature is positive if the net curvature is toward the I graph. Then the representative of G' is $1_1\colon\{(u/2, u/2, u/2) | 0 \leqslant u \leqslant 1\}$, and of G'' is $m_1\colon\{(v/2, 1/2, 0) | 0 \leqslant v \leqslant 1\}$. The cases of P, I–WP, and C(P) in fact correspond to the three ways of choosing opposing edge of the quadrirectangular tetrahedron. These two edges determine the parametrization in the usual way. From the minimal surface solution it is necessary to continue in both positive H^* and negative H^* directions explicitly, as opposed to the self-dual P and D cases. The subbranches on the positive H^* branch will be called A, B, and C, while those on the negative H^* branch will be called D, E, and F. The residual equations for a 15×15 uniform mesh were solved on a Cyber 124.

Figure 5a (see color insert) is the minimal surface, with the I skeletal graph is red and the WP graph in blue. We have elected to show the same $1\frac{7}{8}$ unit cells as were displayed for the P family. However, because of the body-centered symmetry, the two cubes (which share an octant) contain regions of surface that are identical, and this is true for all values of H^*. For the minimal surface, the volume fraction of the labyrinth containing the I graph is 0.53604 ± 0.0002, differing significantly from 1/2. The area of a unit

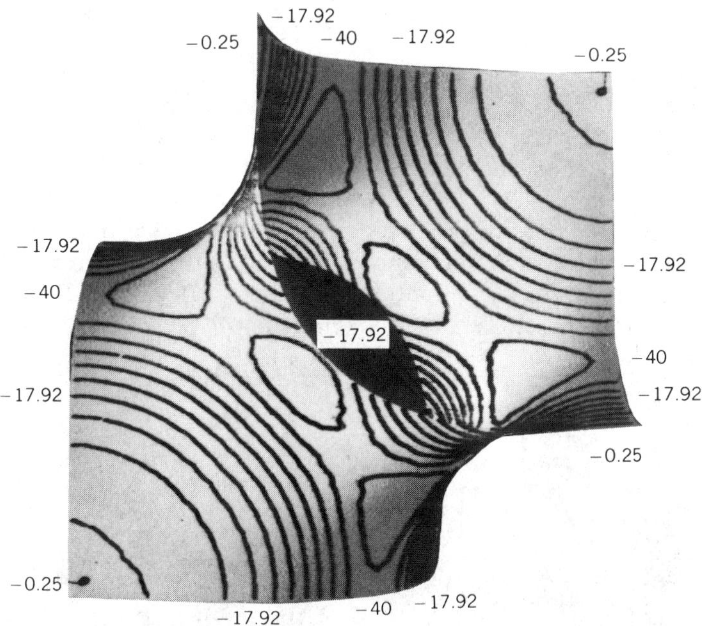

Figure 5a′. Contour plot of Gaussian curvature over an octant of the I–WP minimal surface in Fig. 5a. Contour interval is 4.42.

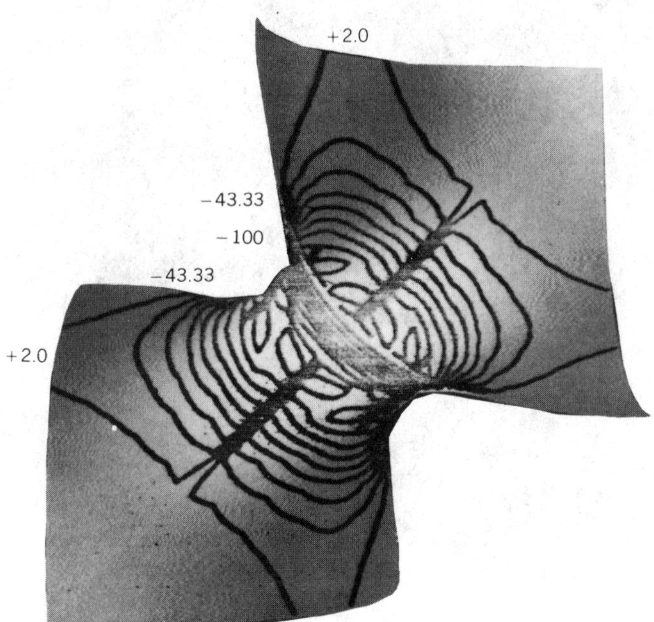

Figure 5b′. Contour plot of Gaussian curvature over an octant of the surface in Fig. 5b. Contour interval is 11.3.

cell of surface is 3.467 ± 0.003, and since an L–F region is one-half a unit cell, the value of $A/V^{2/3}$ for an L–F region is less than this by a factor of $2^{1/3}$, or 2.7521, which is 17.3% higher than the P value. Figure 5a′ shows the distribution of Gaussian curvature over an octant of unit cell. Each of the six corners of this patch is an isolated point of zero Gaussian curvature.

Continuing in the positive H^* direction, the volume fraction of the I labyrinth decreases along subbranch A (see Fig. 5d), and the area decreases. The constant c in Eqs. (37) and (38) is 0.1385 ± 0.0005, and this constant applies for approach to $H^* = 0$ from either above (subbranch A) or below (subbranch D).

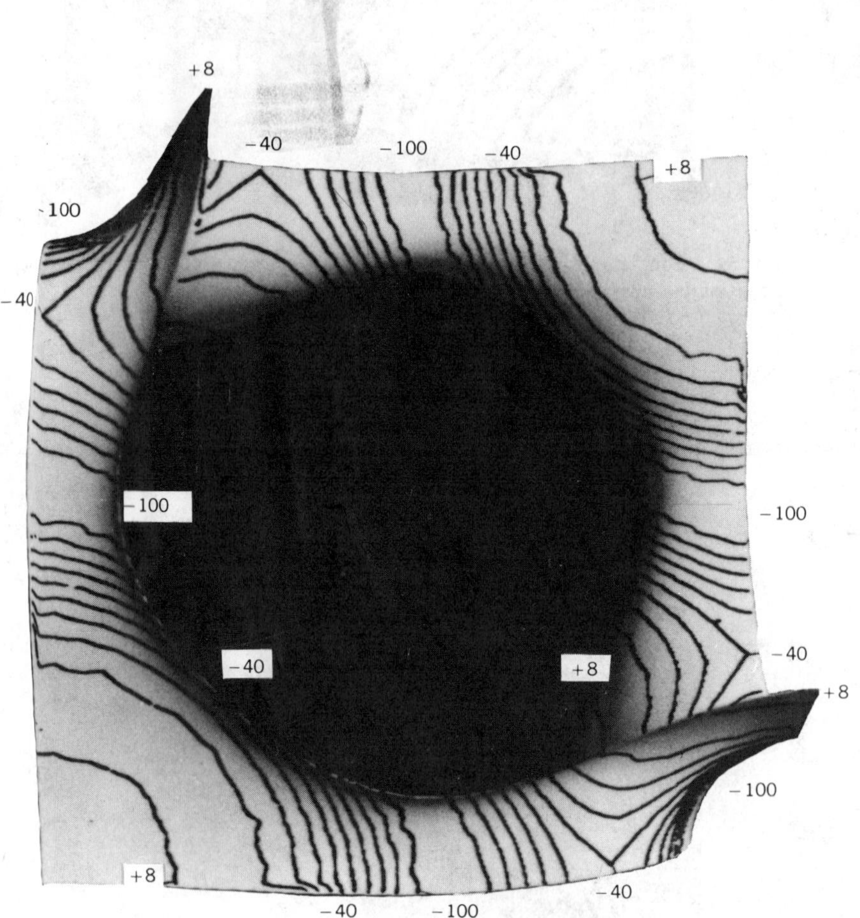

Figure 5c′. Contour plot of Guassian curvature over an octant of the surface in Fig. 5c. Contour interval is 12.0.

Subbranch A ends at the value $H_c^* = 1.940 \pm 0.005$ which corresponds to a local minimum in area and a global minimum in volume fraction. Figures 5b and 5b′ are for $H^* = 2.0$, and it is already evident at this value of H^* that the solution is approaching the BCC–CPS limiting configuration. (See color insert for Fig. 5b.) Subbranch B ends, and C begins at the turning point $H^* = 2.82 \pm 0.01$.

Subbranch D, on which the volume fraction of the I labyrinth is monotonically increasing, extends all the way to the value $H_c^* = -4.52 \pm 0.05$ corresponding to the global minimum in area, and the global maximum in volume fraction. The volume fraction ratio at this maximum is 0.857:0.143 (see Fig. 5d), almost as high as the maximum ratio in the D family, which is remarkable because the volume fraction ratio at the CPS limit on this branch is 0.6073:0.3927. The value of $A/V^{2/3}$ for an L-F region of this surface is 1.96 ± 0.01. Figures 5c and 5c′ are for the global maximum in volume fraction. (See color insert for Fig. 5c.) The volume fraction ratio at this maximum is 0.857:0.143 (see Fig. 5d), almost as high as the maximum ratio in the D family, which is remarkable because the volume fraction ratio at the CPS limit on this branch is 0.6073:0.3927. The value of $A/V^{2/3}$ for an L-F region of this surface is 1.96 ± 0.01. Figures 5c and 5c′ are for $H^* = -3.55$. Subbranch E ends at the turning point at $H^* = -5.28 \pm 0.05$. The CPS limit consists of spheres of radius $\frac{1}{4}$ centered at each face-midpoint, and each

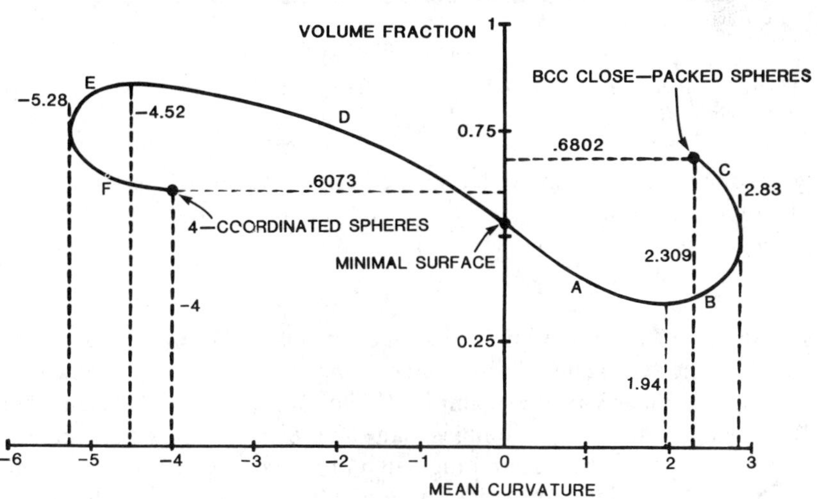

Figure 5d. Plot of volume fraction versus H^* for the I–WP family.

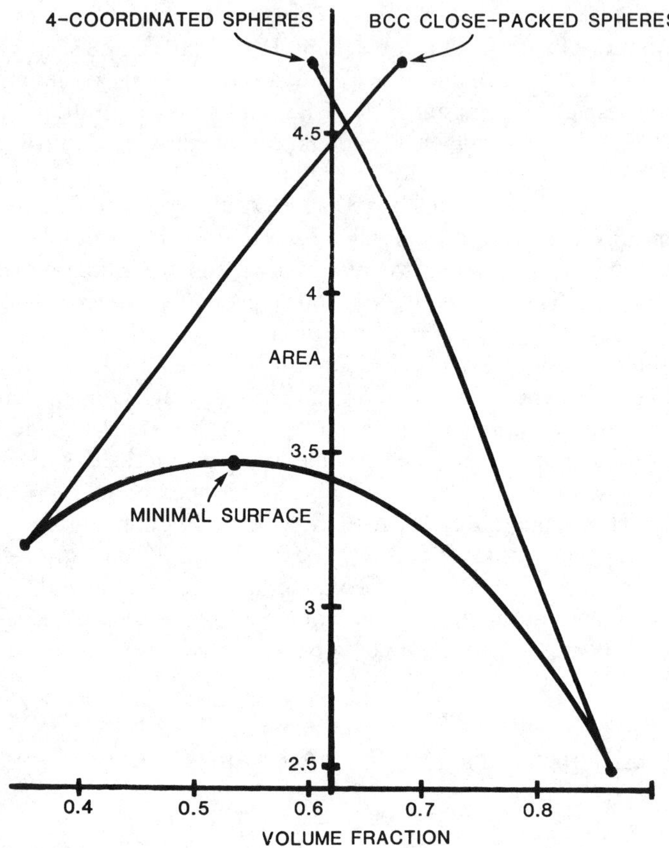

Figure 5e. Plot of area per unit cell versus volume fraction for the I–WP family. The slope at each point is $2H^*$.

edge-midpoint, of every unit cell. Figure 5d gives V versus H^*, and Figure 5e is the plot of A versus V.

D. The F–RD Family

As in the I–WP case, the minimal surface in this family was proposed by Schoen as the conjugate of a self-intersecting minimal surface studied by Stessmann (number V in Stessmann 1934); the conjugate to I–WP is number VI). Once again, the surface contains plane lines of curvature (corresponding to the straight edges that bound the patch of the conjugate surface) but no straight lines, and does not divide space 50:50. Schoen identified the space group as F$m3m$ and the Coxeter cell as a "trirectangular tetrahedron" (which is actually one-half of a tetragonal disphenoid), and incorrectly listed the genus of an L-F region as 6, meaning that its Euler characteristic is $\chi = 2 - 2g = -10$, which is the correct value for χ. This L-F region is determined by a rhombic dodecahedron, and the edges of this space-filing assembly of rhombic dodecahedra comprise the nonsymmetric RD skeletal graph that is dual to the symmetric F graph.

The origin of the coordinate system will be taken to be a node of the F graph, and the other three vertices of a Coxeter cell are $P_2(0, \frac{1}{4}, \frac{1}{4})$, $P_4(\frac{1}{4}, \frac{1}{4}, \frac{1}{4})$. The segment P_1P_2 is one-half an edge of the F graph, and P_3P_4 is an edge of the RD graph, and as usual these two representative edges determine the parametrization.

We make the sign convention that the curvature is with respect to the F graph, and for $H^* \geqslant 0$ we will display one L-F region, composed of 48 cells C. However, for the surface from the negative-H^* branch that is exhibited,

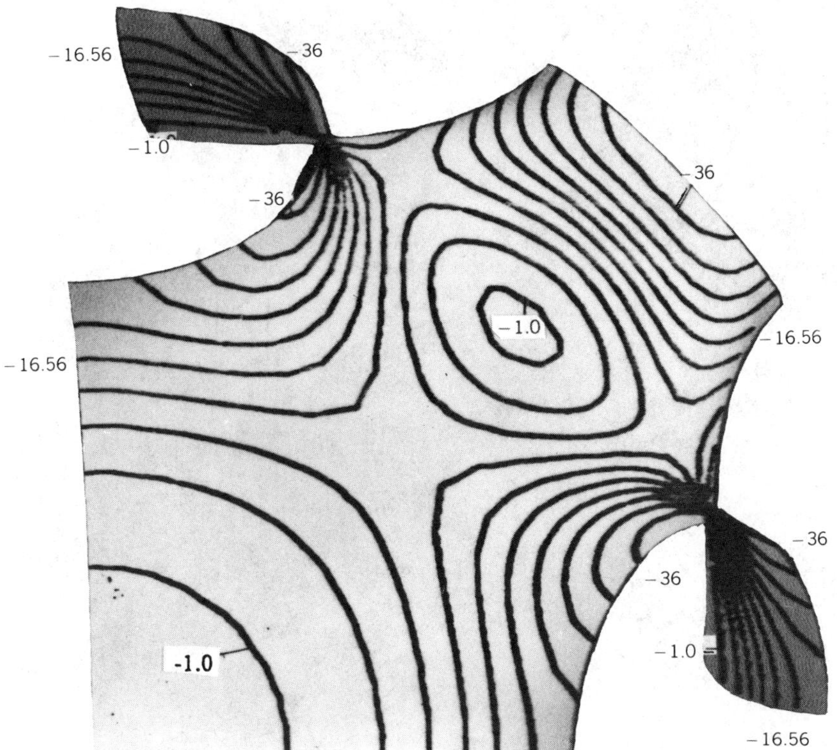

Figure 6a′. Contour plot of Gaussian curvature over a portion of the F–RD minimal surface. Contour interval is 3.889.

we have shown (Fig. 6b) a portion of surface made from 72 copies of C. The F graph is shown in blue and the RD graph as red. The residual equations were solved on a Cyber 124 using a 15 × 15 uniform mesh, or in some cases on a Cray-1S using a 20 × 20 mesh; no values for area or volume fraction have been published for the minimal surface to compare with.

The minimal surface in Figs. 6a (Gaussian contour plot in Figure 6a′) has volume fraction 0.5319 ± 0.0001. (See color insert for Fig. 6a.) The area computed for a unit cell was 4.773 ± 0.001, giving a volume of $A/V^{2/3}$ for an L-F region of 3.0065. This is in accord with the general trend of increasing values of $A/V^{2/3}$ as the Euler characteristic of the L-F region increases in

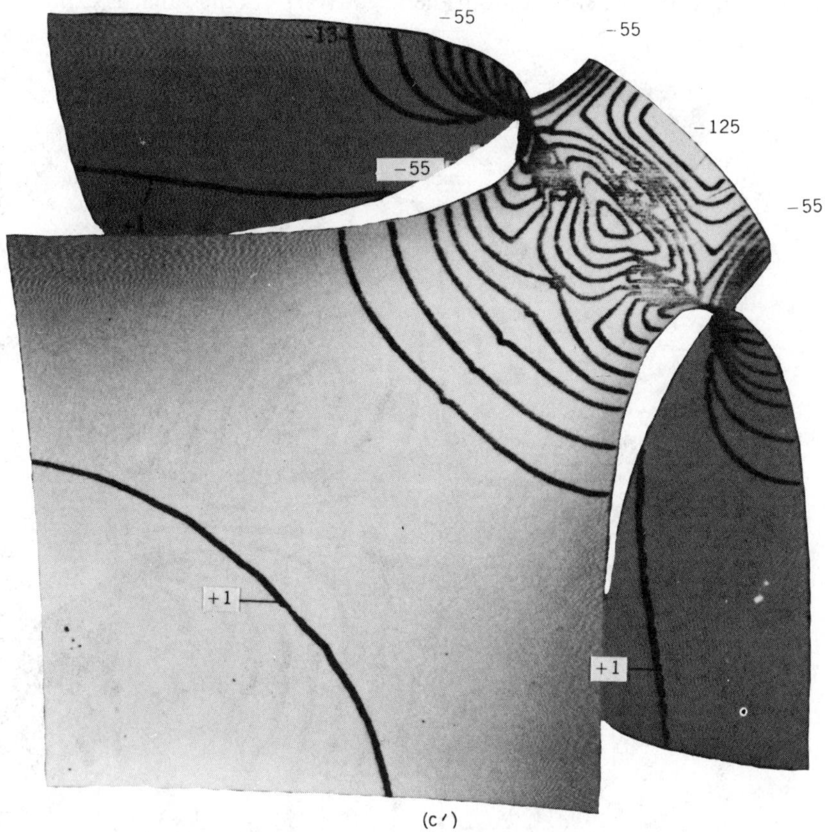

Figure 6.d′. Contour plot of Gaussian curvature over a portion of the F–RD minimal surface in Fig. 6c. Contour interval is 14.0.

magnitude. Figure 6a′ is the contour plot of K over a patch of surface corresponding to six Coxeter cells.

Subbranch A (see Figure 6e), on the positive-H^* branch, continues to a value of $H^* = 1.86 \pm 0.02$ (see Figure 6b, color insert) corresponding to the global minimum in area of 4.639 ± 0.001 (for a value of $A/V^{2/3} = 2.922$). There is some doubt involved in calling this a global minimum since the minimum in area on the negative-H^* branch is very close to this, namely 4.6395 ± 0.001. The volume fraction of the F labyrinth at $H^* = 1.86$ is 0.489 ± 0.0002. Since the FCC–CPS limit corresponds to $H^* = 2(2)^{1/2}$, the value of H^* at this area minimum is considerably less than the CPS value.

Convergence is comparatively easy to achieve on the first portion of subbranch B, in contrast with the I–WP case. Figure 6c (see color insert) is for $H^* = 2.5$, and in this figure it is quite clear that the solution branch is approaching the limiting FCC (12-coordinated) sphere-pack. The volume fraction at $H^* = 2.5$ is 0.4691 ± 0.0002, and the area is almost exactly that of the minimal surface, being 4.779 ± 0.001. The volume fraction first exceeds $\frac{1}{2}$ at $H^* = 2.67 \pm 0.02$.

Subbranch C begins at the turning point where H^* reaches the maximum of 2.92 ± 0.01. Using an unduloid as an initial estimate, it was possible to converge to a solution on this subbranch corresponding to $H^* = 2.89914$. This value is simply 1.025 times the value of H^* at the limit, $2(2)^{1/2}$, and an unduloid of mean curvature 2.05 was rescaled by $(2)^{1/2}$ for the initial guess. The area of the solution was 5.997 ± 0.005 and the volume fraction 0.6303 ± 0.005.

Self-intersecting solutions occur on the negative-H^* branch. The edges of the RD skeletal graph are all of equal length, and one can easily imagine a limiting configuration consisting of close-packed spheres of radius $3^{1/2}/8$, centered at each node of the RD graph whether 4- or 8-coordinated, but this is not found to be the limit. Rather, the branch seeks a limit that puts spheres of radius $3^{1/2}/4$ at 8-coordinated nodes only, and self-intersections must occur since nearest-neighbor, 8-coordinated nodes are only a distance $(2)^{1/2}/2$ apart. As seen in Fig. 6d ($H^* = -2.4$), for solutions on this branch before the point where self-intersections occur, the 4-coordinated nodes are centers of connector regions which join four neighboring spherelike regions. The contour plot of Gaussian curvature for a portion of the $H^* = -2.4$ surface is shown in Figure 6d′.

We have stated that self-intersections are not examined closely in this study, but one important point must be made about the self-intersecting solutions on this branch. If the volume of overlap is counted twice in each case, one can complete the area-H^*, volume-H^*, and area-volume curves for this branch all the way to the limiting configuration with the fundamental relation $dA = 2H^*dV$ obeyed throughout. For the limiting configuration, the

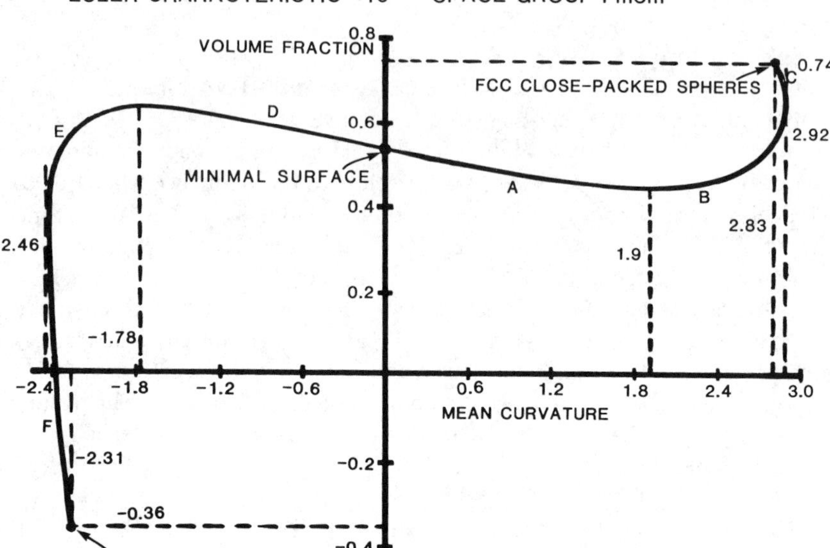

Figure 6e. Plot of area per unit cell versus H^* for the F–RD family.

sphere radius of $(3)^{1/2}/4$ gives a volume and area per sphere of $\pi(3)^{1/2}/16$ and $3\pi/4$, respectively; since there are four spheres associated with each unit cell, the equivalent volume fraction for the RD labyrinth is $\pi(3)^{1/2}/4 = 1.360349\ldots$, and the area per unit cell is $3\pi = 0.42477\ldots$. So we must assign a negative volume fraction to the F labyrinth, namely -0.360349. Applying the mean value theorem between this endpoint (where $H^* = -4/3^{1/2} = -2.309\ldots$) and the solution for $H^* = 2.4$ where the area is found to be 5.0048 and the volume fraction of the F layrinth is 0.54531 yields: $\Delta A/\Delta V = (3\pi - 5.0048)/(-0.360349 - 0.54531) = -4.8806 = 2H^*$. $H^* = -2.4403$ is an acceptable intermediate value because the most negative value obtained on the path joining these solutions is -2.51 ± 0.02. In Fig. 6f, showing area versus volume fraction, this is seen as a nearly linear portion of the curve ending at the limit just discussed, with the slope of the curve lying between 2×-2.51 and 2×-2.309. The maximum value for the volume fraction of the F labyrinth on this branch, namely 0.6272 ± 0.0002, is achieved at $H^* = -1.76 \pm 0.02$, well before self-intersections occur.

We suspect that the exact analytical value for the dimensionless area $A/V^{2/3}$ of an L-F region of the F-RD minimal surface is $3K(k)/K(k')$ where $k^2 = 8(3)^{1/2}/[21 + 4(3)^{1/2}]$. This value, 3.00535..., is approximately 0.05% lower than the value computed from the numerical solution. $A/V^{2/3} =$

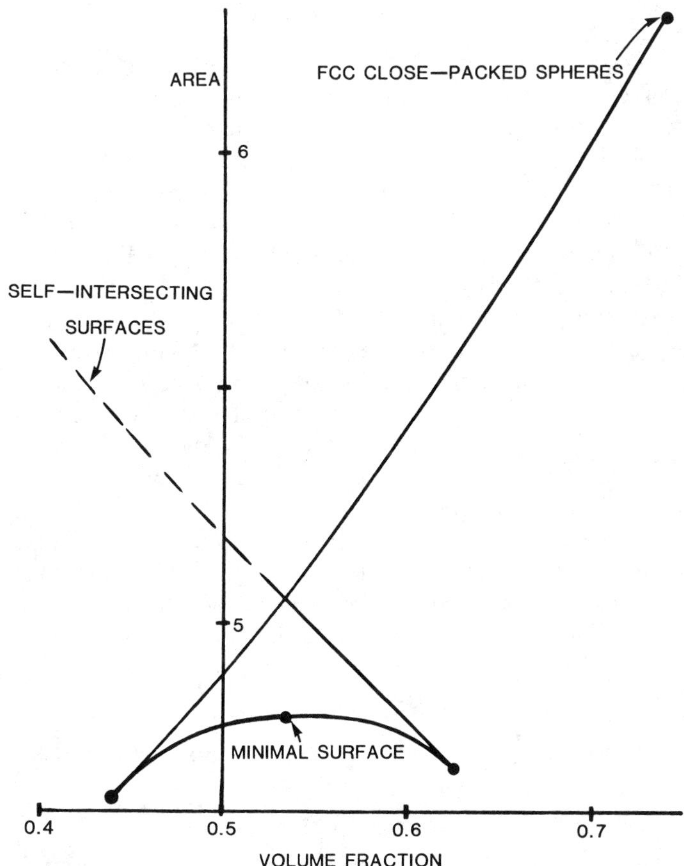

Figure 6f. Plot of area per unit cell versus volume fraction for the F–RD family. The slope at each point is $2H^*$.

$3K(k)/K(k')$ is due to Schoen (1970), who adapted a result of Schwarz (1890), and the work of Stessmann (1934) shows that the same formula applies to the F–RD minimal surface. Stessmann's analysis was not complete enough to yield the modulus k, but by Schwarz's work it is known that $k^2 = (a_2 - a_3)/(a_1 - a_3)$ where $\{a_1, a_2, a_3\}$ are roots (in descending order) to a cubic polynominal that is not explicitly known for the F–RD minimal surface. For the F–RD minimal surface, we suspect the roots to be $\{14, -7 + 4(3)^{1/2}, -7 - 4(3)^{1/2}\}$, where the latter two roots are roots to the polynomial $s^2 + 14s + 1$ that occurs in Stessmann's analysis, and the three roots satisfy $a_1 + a_2 + a_3 = 0$. This has been discussed in more detail (Anderson 1986).

E. C(P) Family

The minimal surface in this self-dual family is often called simply "Neovius' surface" after the pupil of Schwarz who discovered the surface and its Weierstrass representation (Neovius 1883). As mentioned in Section II.B, Schoen named the minimal surface C(P) to indicate that it is complementary to the P minimal surface, containing the same straight lines and having the same space group Pm3m. While the unit cell, which is also an L-F region, and the Coxeter cell are the same as for the P surface, the Euler characteristic per L-F region is -16 as opposed to -4 for the P family. The nonsymmetric, self-dual skeletal graph is constructed by connecting the centroid of every cube in a packing to the twelve edge-midpoints, so that the coordination symbol is 12:4; the two replicas of the graph are offset $(\frac{1}{2}, \frac{1}{2}, \frac{1}{2})$.

The edges of the quadrirectangular tetrahedron C that represent the skeletal graphs G′ and G″ are l_1:$\{(u/2, u/2, 0) | 0 \leqslant u \leqslant 1\}$ and m_1:$\{(v/2, 1/2, v/2) | 0 \leqslant v \leqslant 1\}$. For the minimal surface, the straight line segment that lies in that part S of the surface contained in C is the locus of $u = 1 - v, w = 1/2$, or in (x, y, z)-coordinates, $x = 1/4, y + z = 1/2$,exactly as for the P minimal surface in either coordinate system. However, the integral Gaussian curvature over this patch is four times as great in C(P), being $-2\pi/3$. To capture the regions of sharp curvature requires a fine mesh, and it was found necessary to use

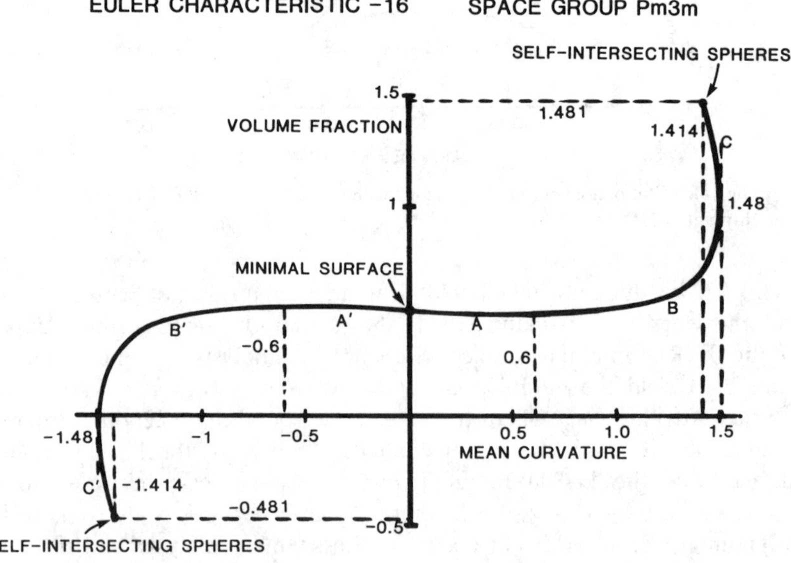

Figure 7c. Plot of volume fraction versus H^* for the C(P) family. Errors are much higher than in the previous for families.

a 20×20 mesh in all cases. Even with this refinement, the accuracy in area and volume fraction values was much lower than that in the preceding cases; for example, the computed area of the minimal surface was higher than the analytic value by about 0.1%.

Figure 7a (see color insert) shows the minimal surface, corresponding to the cube $-\frac{1}{2} \leqslant x, y \leqslant \frac{1}{2}, 0 \leqslant z \leqslant 1$; that is, two neighboring unit cells have been bisected and two adjoining half-cells exhibited. This brings out the noncircular nature of the "holes" surrounding four-coordinated nodes. The area of a unit cell of the minimal surface, which is also the value of $A/V^{2/3}$ for an L-F

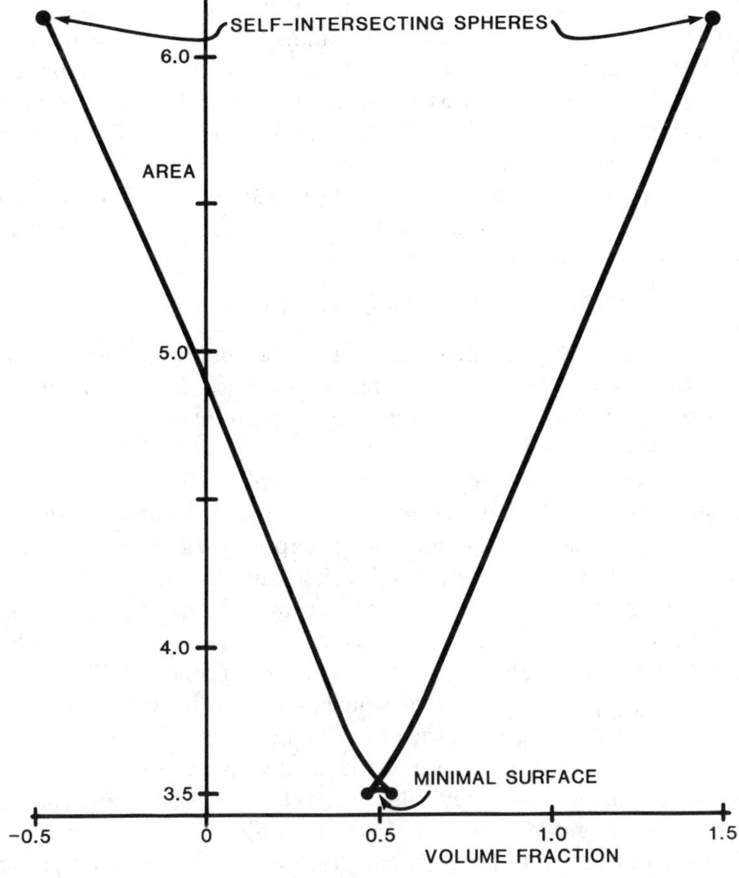

Figure 7d. Area per unit cell versus volume fraction for the C(P) family. The slope at each point is $2H^*$.

region, is known analytically to be $3K[(2/3)^{1/2}]/K[1/(3)^{1/2}] = 3.51048\ldots$ (Neovius 1883; Schoen 1970).

On subbranch A, the area, and the volume fraction of the labyrinth of positive curvature, decrease as always but only very slightly in this family (Figs. 7c and 7d). By the time the subbranch ends, at the area minimum corresponding to $H_0^* = 0.60 \pm 0.02$, the area has decreased only by about 0.2%, and the volume fraction is 0.480 ± 0.001. Figure 7b (see color insert) shows the surface with $H^* = 1.05$.

On subbranches B and B′ a maximum $|H^*|$ of 1.48 ± 0.01 is reached, with the area increasing very rapidly past $H^* = 1.4$. As in the RD branch of the previous family, we must again count intersection volumes twice, so that the relation $dA = 2H^* dV$ is observed all the way to the limit point. When this is done we find that the endpoint of the B(B′) branch, where the value $H^* = 1.48$ is reached, corresponds to a volume fraction just less than unity, and the area is approximately 50% greater than that of the minimal surface.

Subbranches C and C′ consist of self-intersecting solutions. The limiting configuration has $H^* = (2)^{1/2}$, $A = 2\pi$, and "pseudo"-volume fraction 1.48096.... Taking a point on the subbranch B—$H^* = 1.35$, $A = 3.7555$, $V = 0.59273$—and applying the mean-value theorem we obtain $\Delta A/\Delta V = 2.8458 = 2 \times 1.4229$, which is a good mean value for a path from $H^* = 1.35$ to a maximum of 1.48 and back to $(2)^{1/2}$.

F. Generalized P Surfaces

In this subsection, we discuss surfaces with the same topological type as ("homeomorphic to") the Schwarz P minimal surface, with mean curvature a prescribed periodic function of one spatial coordinate, say $H = f(y)$; the period of f will coincide with the period of the surface, or with twice that distance. Such a surface with f nonconstant is of lower symmetry than the P$m3m$ surfaces treated above because of the loss of invariance under the interchanges $y \leftrightarrow x$ and $y \leftrightarrow z$. The space group will in fact be P4/m2/m2/m, number 123 in the Crystallographic Tables, with 16 equivalent positions. This tetragonal space group is characterized by two lattice parameters, the unit cell having dimensions $a \times b \times a$. As in the previous cases, dimensional analysis shows that we can take $a = 1$ without loss of generality. The Coxeter cell C is a right prism the base of which is a right isosceles triangle; the mirror planes comprising the faces of the prism are $x = z, x = 0, z = 1/2, y = 0$ and $y = s$; the relation between s and the lattice parameter b is given next.

We give the name "six-armed cell" to the cell that is homeomorphic to a unit cell of the P minimal surface. If the length s of this six-armed cell is equal to the full period of the function $f(y)$ (or any integral multiple of it), then the six-armed cell is also a unit cell and $b = s$. If in addition $f(y)$ possesses mirror symmetry about the plane $y = b/2$, i.e., $f(y) = f(b - y)$, then this is a plane of mirror symmetry for the periodic surface. In these cases, this mirror plane allows us to divide the six-armed (unit) cell into sixteen equivalent portions. However, in the cases where the length s of the six-armed cell is only half the period of f, the crystallographic unit cell is actually of length $b = 2s$ and is of Euler characteristic -8. In these cases, each of the sixteen positions of the space group $P4/m2/m2/m$ corresponds to one-eighth of the six-armed cell, while in the first case with $b = s$ the portion corresponding to a position of the space group is half that, one-sixteenth of the six-armed cell. In order to write a single code to handle all these cases, we chose to divide the six-armed cell into eight equivalent portions, ignoring the fact that in special cases these may be further bisected.

The parametrization used was

$$\mathbf{r}(u,v,w) = \begin{cases} (2v, s, 1)(1-w) + (0, s-2us, 0)w & \text{if } 0 \leqslant u \leqslant 1/2 \\ (2v, s, 1/2)(1-w) + (0, 2us-s, 2u-1)w & \text{if } 0 \leqslant v \leqslant 1/4, \\ & 1/2 \leqslant u \leqslant 3/4 \\ (1/2, 3s/2 - 2vs, 1/2)(1-w) + (0, 2us-s, 2u-1)w & \text{if } 1/4 \leqslant 3/4, \\ & 1/2 \leqslant u \leqslant 3/4 \\ (2-2v, 0, 1/2)(1-w) + (0, 2us-s, 2u-1)w & \text{if } 3/4 \leqslant v \leqslant 1, \\ & 1/2 \leqslant u \leqslant 3/4 \end{cases} \tag{40}$$

The computational domain was divided into 1,401 elements, for 186 nodes. With this coarse grid, errors in area determinations were on the order of 0.5% (volume fraction errors were about 0.2%). With s set equal to unit, f was chosen to be sinusoidal with period 2: $f(y) = \alpha \sin \pi(y - \frac{1}{2})$. Thus the unit cell is actually homeomorphic to two unit cells of the P surface, so that $b = 2$. The amplitude α was varied up to a value of 0.6, and the result for 0.6 is shown in Fig. 8 (see color insert); two unit cells are shown side by side.

With f identically zero, s was varied from 0.98 to 1.04, to investigate expansions of the area and volume fractions in s about $s = 1$. Strictly speaking, there is a (removable) singularity in the curve of area of a unit cell at $s = 1$ because of the change in symmetry from tetragonal to cubic (when f is constant), so we choose to report the area of a six-armed cell. It was found that to whithin 0.5%, the value of $A/V^{2/3}$ remained constant at the value 2.3451... of the Schwarz surface. The equivalence of the two labyrinths when H is identically zero guarantees that the volume fractions remain exactly at $\frac{1}{2}$, at any value of s for which a solution exists; this held to within the error of 0.2% over the range examined.

We have computed surfaces with piecewise linear f, namely $f(y) = py + q$, $0 \leqslant y \leqslant s$, $f(y) = p(2s - y) + q, s \leqslant y \leqslant 2s$ with p and q small. Defining $H_{\text{ave}} = ps/2 + q$, we estimate the area per six-armed cell to be $A \cong$

$(2.3451 - 2.117H_{\text{ave}}^2 - 0.023p^2)s^{2/3}$. For the case $s = 1$, we can give estimates for the volume fraction V of the labyrinth containing the origin and of the average radius $R_0(R_s)$ of the hole in the plane $y = 0$ $(y = s)$; $V \equiv 0.5 - 0.2117H_{\text{ave}}$, $R_0 \equiv 0.25 + 0.095H_{\text{ave}} + 0.047p$, $R_s \equiv 0.25 + 0.95H_{\text{ave}} - 0.047p$.

G. S′–S″ Surface

In this subsection we demonstrate, by means of an example, the computation of a triply-periodic minimal surface from knowledge of only its two skeletal graphs. As mentioned in Section II.A, Schoen hypothesized a minimal surface that he named S′–S″, for "large-square graph, small-square graph," and described the skeletal graphs S′ and S″, but did not provide pictures or any other data on the surface. Contrary to the cases of I–WP and F–RD that were also conceived by Schoen, S′–S″ is not known to be related to any surface whose Weierstrass representation is skew pentagon, the five edges being normal to the five faces of the Coxeter cell C for S′–S″ described below, and while Schoen proved that only eight such pentagonal modules could generate surfaces with discrete noncubic symmetry groups, he did not elucidate these surfaces. The results in this subsection support, but of course do not prove, the existence of the minimal surfaces S′–S″.

As noted by Schoen, the space group for S′–S″ is tetragonal, P4/m 2/m 2/m, the same as that of the previous subsection, and again the Coxeter cell C is the isosceles right triangular prism with faces: $x = z, x = 0, z = 1/2, y = 0$ and $y = b/2$. The unit cell has been given dimensions $1 \times b \times 1$, and no change of symmetry would occur for $b = 1$, as opposed to the case in the previous subsection. There are two edge-portions of each of the skeletal graphs $G'(= \text{S}'')$ among the nine edges of C; call these $l_1 : \{(1/2 - u, b/2, 1/2) | 0 \leqslant u \leqslant 1/2\}, l_2 : \{(0, b - bu, 1/2) | 1/2 \leqslant u \leqslant 1\} \in G'$ and $m_1 : \{(0, b/2 - v, 0) | 0 \leqslant v \leqslant 1/2\}, m_2 : \{(v - 1/2, 0, v - 1/2) | 1/2 \leqslant v \leqslant 1\} \in G''$. The three pairs l_1m_1, l_1m_2, and l_2m_2 of opposing edge-portions determine spines that fill out C exactly, and the pair of parallel edge-portions l_2 and m_1 does not contribute any spines, of course. The three rectangles in the u–v unit square corresponding to the three active pairs determine an L-shaped computational domain, with the square $\{(u, v) | 1/2 \leqslant u \leqslant 1, 0 \leqslant v \leqslant 1/2\}$ corresponding to the inactive pair l_2m_1 omitted. We have thus determined the parametrization, and the residual equations, without ever having seen any portion of the surface.

An initial estimate was formulated by inventing a "composite" distance from an arbitrary point to each skeletal graph (details in Anderson 1986), and the surface determined by the locus of points equidistant in this measure from the two graphs was used successfully for the surface with $H^* = 1.0075$ and $b = 0.936$. From this converged solution, b and H^* were incremented to smaller values simultaneously, with convergence becoming steadily easier, until a minimal surface with $b = 0.64$ was found. The area of such a unit cell

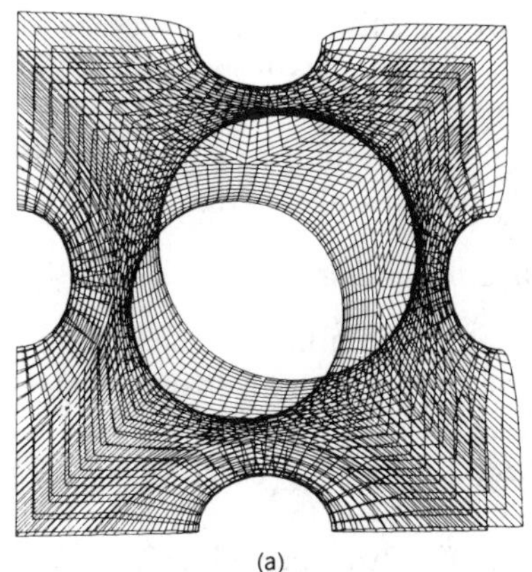

(a)

Figure 9a. Schoen's minimal surface S′–S″ with $b = 0.78$. One unit cell is shown, looking almost directly down the y-axis. The difference in the diameters of the two nearly-circular "holes" is due to the perspective.

was 2.0185; this area increased to 2.2640 when b was increased to 0.76, and within the error this represents a constant value of $A/V^{2/3}$ (equal to 2.718). We were not able to achieve convergence with $b = 0.80$, but this is not a proof that no minimal surface with $b \geqslant 0.80$ exists, of course. Figure 9a is a line drawing of a unit cell of the minimal surface with $b = 0.78$. Figure 9b is a line drawing of a unit cell with $b = 0.78$ and $H^* = 2.015$. The feature that best distinguishes S′–S″ from the P minimal surface has to do with the connectivity between unit cells lying at the same value of y; to get from the centroid $(0, 0, 0)$ of one unit cell to the centroid, say $(1, 0, 0)$, of a neighboring unit cell while remaining within the labyrinth containing G' requires a jig-jag path, say via the point $(1/2, 0, 1/2)$. On the other hand, the P minimal surface allows straight paths connecting nearest neighbor unit cells. The "holes" in the S′–S″ unit cell centered about $y = 0$ are very similar to those in the surface

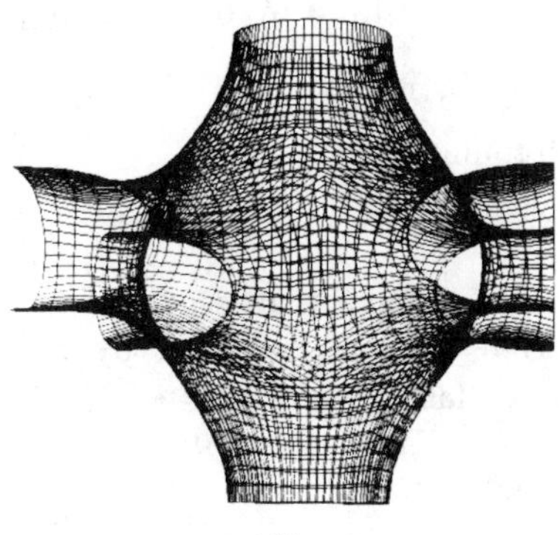

(b)

Figure 9b. Variant of the surface in Fig. 9a with $b = 0.936$ and $H^* = 1.0$. In this view the y-axis is vertical.

C(P); in both cases, at each edge-midpoint of the cube a "throat" comes in orthogonally to the two adjacent cube faces, with throats from each of the four cells that share this edge-midpoint joining smoothly. In the C(P) surface, this situation is repeated at each of the twelve edge-midpoints of the unit cell, whereas in S′–S″ it occurs only on the four edges parallel to the *y*-axis.

H. Scattering Function Calculations

At the time of this writing, the most immediate application for the surfaces discussed in this section is to provide models for viscous isotropic phase liquid crystals and certain phase-segregated polymer blends and block copolymers. In this subsection, we show how to calculate diffraction peak intensities from a class of model structures based on these surfaces. The method applies to scattering-density profiles (electron densities for x-ray scattering) determined by:

(A) Contrast between the two labyrinths (bulk scattering);

(B) Contrast between the surface—imagined as a thin layer—and two labyrinths of equal density (film scattering); or

(C) A combination of (A) and (B).

Case (C) involves three (electron) densities, one for each labyrinth and one for the surface layer. Because relative intensities are all that are usually measured, only one parameter is needed to allow for any combination of three densities. The scattering function calculated in case (B) is frequently referred to as the *structure factor* of the surface; for a minimal surface characterized by a self-dual skeletal graph, this will correspond to a different space group than that of (A) or (C), or of case (B) with nonzero mean curvature (see the D family, above).

A model structure of type AA) based on a triply periodic embedded surface S parametrized via the present method as $S\colon\{\mathbf{r}(u,v,w^*(u,v))|(u,v)\in D\}$, is in general described (for the purpose of relative intensities) by the following step-change density profile:

$$\rho_b(\mathbf{r}) = \begin{cases} 1 & \text{if } \mathbf{r} \in \text{labyrinth containing } G' \\ 0 & \text{if } \mathbf{r} \in \text{labyrinth containing } G'' \end{cases} \tag{41}$$

The portion of the volume inside the unit cell U for S that is part of the labyrinth containing the skeletal graph G' is exactly described by the condition $w < w^*$, and $w > w^*$ describes the G''-labyrinth, or:

$$\rho_b(\mathbf{r}(u,v,w(u,v))) = \begin{cases} 1 & \text{if } w < w^*(u,v) \\ 0 & \text{if } w > w^*(u,v) \end{cases} \tag{42}$$

so that

$$A_b(\mathbf{h}) = \iiint_U \rho_b(\mathbf{r}) e^{i\mathbf{h}\cdot\mathbf{r}} d^3\mathbf{r} = \iiint_{U:w<w^*} e^{i\mathbf{h}\cdot\mathbf{r}} d^3\mathbf{r} \tag{43}$$

In case (B), all of the scattering contrast is from a high-density region concentrated in a layer or "film" about the surface. The singularity can be considered the limiting value of the project of the excess density and the layer thickness; that is, a region $[\mathbf{r}^* + s\mathbf{n}, \mathbf{r}^* \in S, |s| \leqslant \tau/2)]$ of density $1/\tau$ in the limit as $\tau \downarrow 0$. The strength α can be assigned to the film contrast by multiplying the unit-strength singularity by α. The unit-strength singularity yields the structure factor A_f of the surface

$$A_f(\mathbf{h}) = \lim_{\tau \downarrow 0} \iiint_{[\mathbf{r} = \mathbf{r}^* + s\mathbf{n}, \mathbf{r}^* \in S \cap U, |s| \leqslant \tau/2]} (1/\tau) e^{i\mathbf{h}\cdot\mathbf{r}} d^3\mathbf{r} = \iint_{S \cap U} e^{i\mathbf{h}\cdot\mathbf{r}} dS \tag{44}$$

At zero wave vector, a unit-strength singularity at the surface, taken over a unit cell of surface, yields the area of the surface.

The amplitude in case (C) at a wave vector of **h** is the sum of the amplitudes of A_b and A_f from the bulk and film contrasts:

$$A(\mathbf{h}) = A_b(\mathbf{h}) + \alpha A_f(\mathbf{h}) \tag{45}$$

The integration is over the unit cell U, and this region is produced by applying the mirror symmetries of the space group (but of course *not* the translational symmetries) to the Coexeter cell C. However, since the dependence on **h** is through the dot product $\mathbf{h}\cdot\mathbf{r}$ it is equivalent to apply the mirror symmetries of the space group to the vector **h** and integrate over just the Coxeter cell for each permutation of **h**, and then add the resulting partial amplitudes.

Because of the perfect periodicity of these profiles, the intensities will be nonzero only at a discrete set of wave vectors. The Miller indices $(h\,k\,l)$ of the allowed reflections, as listed in the *International Tables of X-ray Crystallography* (1952) for a given space group and Wyckoff position, determine the wave vectors for which intensities must be evaluated. The wave vector that corresponds to the reflection with Miller indices $(h\,k\,l)$ is $\mathbf{h} = (2\pi h/a, 2\pi k/a, 2\pi l/a)$, where the lattice parameter a can be taken equal to unity here.

If the scattering densities on the A-rich and B-rich sides of the dividing surface are taken to be $\psi = 1 - \phi_A$ and $\psi = -\phi_A$, respectively, then the divergence theorem can be used to reduce the volume integration over the unit cell to a surface integral (Hosemann and Bagchi 1962):

$$\iiint e^{i\mathbf{q}\cdot\mathbf{r}} \psi(\mathbf{r})\, d^3\mathbf{r} = (i/q^2) \iint \mathbf{n}\cdot\mathbf{q}\, e^{i\mathbf{q}\cdot\mathbf{r}}\, dA \tag{46}$$

This surface integral can be performed analytically over each triangular patch in the finite element representation of the surface. For a space triangle with vertices $\mathbf{r}_1, \mathbf{r}_2$, and $\mathbf{r}_3$, this integral is

$$\{[\cos(a+b) - \cos(a+c)]/b(c-b) - [\cos a - \cos(a+c)]/bc\}M/q^2 \quad (47)$$

where $a = \mathbf{r}_1 \cdot \mathbf{q}, b = \mathbf{r}_2 \cdot \mathbf{q} - a, c = \mathbf{r}_3 \cdot \mathbf{q} - a$, and $M = |(\mathbf{r}_2 - \mathbf{r}_1) \times (\mathbf{r}_3 - \mathbf{r}_1)|$. The form factor calculated at reflections that are forbidden in the P$n3m$ space group (all ($hk0$) with $h+k$ odd) was in all cases less than 10^{-19}. The form factor calculated for a very small value (0.01) of H^* compared very well with the form factor of the minimal surface computed numerically by Mackay (1985); structure for the case of a very small H^* consists of a very thin shell bisected by the minimal surface.

Table I lists the two amplitudes for three members of the P family. For the minimal surface, the reflections that are forbidden in BCC symmetry are listed as zero amplitude, and in each case the computed amplitudes were less than 10^{-10}. For the I–WP family, there is no change in symmetry for $H^* = 0$, as seen in Table II. The structure factor for the D minimal surface is shown in Table III, and the last column in this table lists the values published by Mackay 1985a and 1985b; the results are in agreement.

When the amplitudes A_b and A_f have been calculated, the intensity $I(\mathbf{h})$ is calculated as $I = [A_b + \alpha A_f]^2$. Thus a listing of the two amplitudes for

TABLE I
Amplitudes for Three P Surfaces

	$H^*=0$		$H^*=-1.8$		$H^*=2$(CPS)	
(hkl)	Ab	Af	Ab	Af	Ab	Af
(000)	0.5	2.345107	0.75123	2.00264	0.52360	3.14159
(100)	0.16696	0	0.13022	−0.58291	0.15915	0
(110)	0	−0.44936	−0.04747	−0.20807	0.00392	−0.68158
(111)	−0.06057	0	−0.00837	0.57051	−0.04261	−0.43061
(200)	0	−0.54494	−0.01823	−0.15113	−0.03979	0
(210)	−0.0147	0	−0.01027	0.19855	−0.02041	0.30209
(211)	0	0.45618	0.02748	−0.13002	−0.00079	0.40312
(220)	0	−0.04256	0.00263	0.22521	0.01822	0.18147
(221)	0.01863	0	−0.00758	−0.30676	0.01768	0
(300)	−0.01849	0	0.00477	0.01301	0.01768	0
(310)	0	0.09885	0.00185	0.05618	0.01311	−0.15432
(311)	0.01583	0	−0.00608	−0.12509	0.00672	−0.25285
(222)	0	−0.40418	−0.01295	0.36989	0.00028	−0.28684
(320)	−0.00633	0	0.01065	−0.07833	−0.00501	−0.26224
(321)	0	−0.21779	−0.00703	−0.09816	−0.00853	−0.19387

TABLE II
Amplitudes for Three I–WP Surfaces

	$H^* = -3.55$		$H^* = 0$		$H^* = 20$	
(hkl)	Ab	Af	Ab	Af	Ab	Af
(000)	0.8477232	2.60518	0.536401	3.466965	0.3521078	3.197987
(110)	0.04481	−0.67059	0.10465	−0.30988	0.11261	0.10664
(200)	−0.08244	0.89956	−0.09021	−0.52743	−0.01901	−0.97923
(211)	0.00172	0.12524	−0.02475	−0.12206	−0.02093	−0.43372
(220)	−0.03122	−0.07605	0.00983	−0.28466	−0.00459	−0.12859
(310)	0.02583	−0.10469	−0.00127	0.68039	−0.02468	0.2102
(222)	0.00489	−0.51241	−0.00108	0.50945	−0.01796	0.06465
(321)	−0.00203	0.17153	0.00344	−0.03437	0.00541	0.24275

TABLE III
Amplitudes for D Minimal Surface (Mackay's Figures Included for Comparison)

(hkl)	Ab (this work)	Af (this work)	Af (Mackay)
(000)	0.50000	1.91928	1.9193
(111)	0.01347	0	0
(200)	0	0	0
(220)	0	0.4780	0.4775
(311)	0.00166	0	0
(222)	0	0.4840	0.4866
(400)	0	−0.2610	

TABLE IVa
Amplitudes for P Surface with $H^* = 1.8$, bilayer Arrangement

(hkl)	$H^* = 1.8$ Ab (bilayer)
(000)	0.50246
(110)	−0.09489
(200)	−0.03629
(211)	0.05498
(220)	0.00533
(310)	0.00366
(222)	−0.02595
(321)	−0.01407

each reflection permits easy calculation of the intensities for any number of choices of α; this simplicity is important in fitting the parameter α to experimental data.

A member of a family characterized by a self-dual graph can define a structure that has a different space group than that of each skeletal graph, by considering the region between the surfaces $H^* = h$ and $H^* = -h$ to be the region of unit density. We have seen that the transformation $(u, v, w) \rightarrow (1-u, 1-v, 1-w)$ takes a surface S with $H^* = -h$—positive with respect to the G' graph—into its counterpart with the same curvature toward the G'' graph. This situation is easily represented in the present parametrization by introducing the upper limit of integration $w''(u, v) = 1 - w^*(1-u, 1-v)$, and $e^{i\mathbf{h}_i \cdot \mathbf{r}}$ is integrated over $w \in [w^*, w'']$; the condition $w'' > w^*$ must of course be observed over the entire surface, and for the P and D families this holds over the branches A and A′, and portions of B and B′. This represents, for example, an idealized density profile for a binary amphiphile–water cubic phase in which two equivalent but disjoint layrinths of water are separated by a continuous bilayer of amphiphile. Table IVa

TABLE IVb
Amplitudes for D Surface with $H^* = 3.2$, Bilayer Arrangement

(hkl)	Ab (bilayer)
(000)	0.3460
(110)	0.02559
(111)	0.01845
(200)	−0.00484
(210)	0
(211)	−0.00372
(220)	−0.00008
(221)	0.00141
(222)	−0.00576
(300)	0
(310)	−0.00181
(311)	−0.00173
(320)	0
(321)	−0.00108
(322)	−0.00371
(330)	−0.00148
(331)	−0.00197
(332)	0.00285
(333)	0.00169
(400)	0.00201

Note: indices correspond to $Pn3m$ unit cell, in contrast with table 3.

shows the results of the calculation for the first eight allowed reflections for this case, based on the P surface with $H^* = 1.8$, and Table IVb is for a bilayer structure based on the D surface with $H^* = 3.2$.

V. CONCLUSIONS

A new approach to the computation of surfaces of prescribed mean curvature has been described and five new families of embedded, triply periodic surfaces of constant mean curvature tracked between periodic limiting configurations of sphere packings. These five families include as members all of the embedded triply periodic minimal surfaces that can be constructed from minimal surface patches spanning identical skew quadrilaterals, or from conjugates of such patches—for it was proven by Schoenfliess (1891; see also Schwarz 1890, vol. I, p. 221; Stessmann 1934) that exactly six periodic minimal surfaces can be constructed from minimal surface patches spanning quadrilateral boundaries, and of these surfaces and their conjugates (which are bounded by four plane lines of curvature) all but five are known to contain self-intersections. The representation of the surface in the new formulation allows for accurate calculation of scattering functions for a class of density profiles determined by the solutions.

Many important questions and conjectures remain unresolved. It is not known whether these solutions are the only embedded H-surfaces for the five dual pairs of skeletal graphs studied, for example. An important issue is whether or not there exists a bound on the mean curvature attainable in such families; for all of the branches studied here, and for the family of unduloids with a fixed repeat distance (Anderson 1986), the dimensionless mean curvature $H^\# = H\lambda$ is always less than π, where λ is the sphere diameter in the sphere-pack limit. It is possible that there exists an upper bound on $H^\#$ lower than π that depends on the coordination number, or the Euler characteristic. For the P, D, I, WP, F, and RD branches, the islands R^+ over which $K > 0$ coalesce wih neighboring R^+ regions at a critical mean curvature that is the same (to within an error in H^* of about 0.15) as the value H_c^* corresponding to the local minimum in surface area. We have given what we suspect to be the analytical value for the area of the F–RD minimal surfaces, and for the first nonzero coefficient in both the area and volume expansions about $H^* = 0$ in the P family.

The fact that a local maximum in area occurs at $H^* = 0$ in each family is not predicted by any known theorem. Schwarz (1890, vol. I, p. 150) showed that under the orthogonality boundary conditions, the second variation of the area is negative for a minimal surface bounded by the planes of a tetrahedron, but this only means that some normal perturbation which preserves the orthogonality boundary conditions decreases the area, and does not say anything specifically about those perturbations with constant mean curvature.

The I–WP and F–RD minimal surfaces have been shown to provide two counterexamples to a conjecture that has previously been made (Meeks 1978, p. 81, Conjecture 6): "A triply periodic minimal surface disconnects R^3 into two regions with asymptotically the same volume." The volume fraction of the labyrinth containing the symmetric skeletal graph is 0.5360 ± 0.0002 for the I–WP minimal surface and 0.5319 ± 0.0001 for F–RD.

In all of the solution branches investigated, we have observed qualitatively similar behavior: surface areas first decrease, moving away from the minimal surface until a minimum in area is reached at some H_c^*, which also corresponds to the point of most unequal volume fractions; then, past H_c^* a turning point in H^* occurs at some H_{max}^*, and past H_{max}^* areas increase dramatically until a limiting configuration is reached consisting of identical spheres, either close-packed or self-intersecting. Table V records some of the significant parameters observed on the branches examined. The branches are listed according to the coordination number in the sphere-pack limit; if two branches have the same coordination, the symmetric graph is listed first. Some trends are indicated by the table, namely that for branches with higher coordination number in the sphere limit, we observe: (1) lower values of $H_{max}^\#$, (2) lower values of $|V(0) - V(H_c^*)$, (3) lower values for the slope $dV/dH^\#|_{H^\#=0}$ of the V–$H^\#$ curve at $H = 0$; and (4) higher values of $A(H_c)/V(H_c)^{2/3}$ per lattice fundamental region. A heuristic argument can be given as to the source of the trends in Table V. Near the CPS limit on a given branch, the solution patch inside a single Coxeter cell is well-approximated by a patch of an unduloid (or Delauney surface), and in fact we have successfully used unduloids as initial estimates for solutions near the CPS limits of the P and F branches. In the case of unduloids, it can be shown (Anderson 1986) that $H^\#$ increases monontonically from $H^\# = 2$ at the sphere limit to the limiting value of $H^\# = \pi$ at the cylinder limit. However,

TABLE V
Parameters indicating Excursion in Mean Curvature, Volume Fraction, and Surface area

Branch	Coordination	$H^\#$max	$\lvert V(0) - V(H_c)\rvert$	$dV/dH^\# \lvert H^\# = 0$	$A(H_c)/[V(H_c)]^{2/3}$
D	4	2.42	0.369	0.327	1.653
WP	4	2.64	0.357	0.277	1.960
P	6	2.13	0.250	0.212	2.002
I	8	2.45	0.117	0.160	2.530
RD	8	(2.18)	0.095	0.083	2.922
F	12	2.07	0.093	0.102	2.922
C(P)	12	(2.09)	0.020	0.032	3.503

in each of the branches treated here, $H^{\#}$ increases from 2 on moving away from the sphere limit but must eventually start to decrease and head toward the value $H^{\#}=0$ of the minimal surface. It is not unexpected that a higher coordination number would impose greater constraints on a branch of solutions, so that the family of unduloids with its coordination of 2 would reach the highest value of $H^{\#}_{max}$, while those branches with coordination numbers of 12 never go higher than $H^{\#}=2.1$. In short, we are imposing constraints on the solutions, by fixing a higher coordination number, that tend to limit excursions in the mean curvature, in volume fraction ratios, and in surface areas.

Acknowledgments

Financial support for this work was provided by the Department of Energy, the National Science Foundation, and the Minnesota Supercomputer Institute.

Note added in proof: This paper was written and circulated for a rather lengthy period before being submitted for publication. As as result some very recent references are not represented, including some papers in which the results of this work were used in subsequent modeling. We first of all apologize for any recent works of relevance which may have been inadequately represented, particularly those of H. Karcher and coworkers. We then go on to mention references in which the surfaces computed herein, and in the coordinate system designed herein, have been used in the modeling of surfactant and block copolymer microstructures. In Anderson and Thomas (1988), the models, and the form factors calculated for them, were used in a thermodynamic treatment of the bicontinuous cubic morphology occuring in star biblock copolymers. In that paper, model based on interfacial surfaces of constant mean curvature were shown to yield lower free energies than alternative interconnected-rod and parallel surface models. In Anderson, Gruner and Leibler (1988), variations in bilayer thickness were calculated for models based on surfaces of constant mean curvature, and shown to be dramatically less than variations in mean curvature for models with constant bilayer thickness. In Thomas, Anderson, Henkee and Hoffman (1988), as well as in Anderson and Strom (1988), TEM micrographs were simulated from models based on surfaces of constant mean curvature. The results for the structure factors and surface areas calculated in this work were used in Radler, Rhadiman, DeVallera and Toprakcioglu (1989) for the analysis of neutron scattering contrast experiments on cubic phases. Approximate equations relating volume fraction, average bilayer thickness, and average mean curvature were derived in Anderson, Wennerstrom and Olsson (1989) and used in the description of so-called L_3 phases, and shown to be the same for both constant mean curvature and constant thickness models. And in Lindman et al. (1988) as well as in Anderson and Wennerstrom (submitted), the diffusion equation was solved numerically in the geometries determined by these H-surfaces to yield values of the geometric obstruction factor and compared with NMR self-diffusion results on cubic phases, L_3 phases, and microemulsions.

References

Anderson, D. M. 1986, "Studies in the Microemulsion," Ph.D. Thesis, University of Minnesota.

Anderson, D. M., Thomas, E. L., 1988, *Macromolecules* **21**, 3221–3230.

Andersson, A. 1983, *Angewandte Chemie* **22** (2), 69–170.

Auvray, L., Cotton, J., Ober, R., and Taupin, C. 1984, *J. Phys. Chem.* **88**, 4586–4589.

Bashkirov, N. M. 1959, *Soviet Physics, Crystallography* **4**, 442–447.

Bodet, J.-F., Bellare, J. R., Davis, H. T., Scriven, L. E., and Miller, W. G. 1988 *J. Phys. Chem.* **92**, 1898–1902.

Bonnet, O. 1848, *J. Ecole Polytechnique* **19**, 1–146.

Bonnet, O. 1853, *C.R. Acad. Sci. Paris* **37**, 529–532.

Bragg, Sir L., Claring bull, G. F., and Taylore, W. H. 1965, *Crystal Stuctures of Minerals*, Cornell University Press, New York.

Bull, T. and Lindman, B. 1974, *Mol. Cryst. Liq. Crystal* **28**, 155–160.

Christiansen, H. N., Stepheson, M. B., Nay, B. J., Ervin, D. G., and Hales, R. F. 1981, *NCGA '81 Conf. Proc.*, June 1981, 275–282.

Clarkson, M. T., Beaglehole, D., and Callaghan, P. T. 1985, *Phys. Rev. Lett.* **54**, 1722.

Coxeter, H. S. M. 1963, *Regular Polytopes*, Macmillan, New York.

Courant, R. 1943, *Bull. Am. Math. Soc.* **49**, 1–23.

Darboux, G. 1894, *Lecons sur la theorie generale des surfaces et les applications geometriques du calcul infitesimal*, Vol. III, Ch. VI, Gauthier-Villars, Paris.

Delauney, C. 1841, *J. Math. Pures. Appl. Ser. 1* **6**, 309–320.

Faulkner, L. R. 1984, "Chemical Microstructures on Electrodes," *Chemical and Engineering News*, Feb. 27, 1984, 28–45.

Fontell, K. 1981 *Mol. Cryst. Liq. Cryst.* **63**, 59–82.

Gauss, C. F. 1827, *Werke*.

Giusti, E. 1984, *Minimal Surfaces and Functions of Bounded Variation*, Birkhauser, Boston–Basel–Stuttgart.

Grüter, M., Hildebrandt, S., and Nitsche, J. C. C. 1981, *Manuscr. Math.* **35**, 387–410.

Grüter, M., Hildebrandt, S., and Nitsche, J. C. C. 1986 *Acta Math.* **156**, 119–152.

Gunning, B. E. S. 1965, *Protoplasma* **60**, 111–130.

Haggin, J. 1982, "Shape Selectivity Key to Designed Catalysts," *Chemical and Engineering News*, Dec. 13, 1982, p. 9.

Hench, L. L. and Ulrich, D. R., Eds. 1984, *Ultrastructure Processing of Ceramics, Glasses, and Composites*, Wiley-Interscience, New York.

Hinata, M., Shimasaki, M., and Kiyono, T. 1974, *Math. of Computation* **28** (125), 45–60.

Hosemann, R. and Bagchi, N. 1962, *Direct Analysis of Diffraction By Matter*, North-Holland, Amsterdam.

Hyde, S. T. 1986, *Infinite Minimal Surfaces and Crystal Structures*, Ph.D. Thesis, Monash University.

Hyde, S. T., Andersson, S., Ericsson, B., and Larsson, K. 1984, *Z. Krist.* **168**, 213–219.

Joachimsthal, F. 1846, *J. R. Angew. Math.* **30**, 347–350.

Kahlweit, M., Strey, R., Hasse, D., Kuneida, H., Schmeling, T., Faulhaber, B., Borkovec, M. J., Eicke, H.-F., Busse, G., Eggers, F., Funck, Th., Richmann, H., Magid, L. J., Soderman, O., Stilbs, P., Winkler, J., Dittrich, A., and Jahn, W. 1987, submitted to *J. Colloid and Interface Sci.*

Kaler, E. W., Bennett, K. E., Davis, H. T., and Scriven, L. E. 1983, *J. Chem. Phys.* **79**, 5673–5684.

Kapouleas, A. 1987, private communication.

Karcher, H. 1987, private communication.

Kenmotsu, K. 1978, "Generalized Weierstrass Formula for Surfaces of Prescribed Mean Curvature," in *Minimal Submanifolds and Geodesics*, Morio Obata Kaigai Publications, Tokyo, 73–76.

Kenmotsu, K. 1980, *Tohoku Math. Journ.* **32**, 147–153.

Kistler, S. 1983, "The Fluid Mechanics of Curtain Coating and Related Viscous' Free Surface Flows with Contact Lines," Ph.D. Thesis, University of Minnesota.

Kistler, S. and Scriven, L. E. 1983, "Coating Flow Computations," in *Computational Analysis of Polymer Processing*, Pearson, J. R. A., and Richardson, S. G., Eds., Applied Science Publishers Ltd., Barking, Essex, England, Ch. 8.

Lagrange, J. L. 1761, *Miscellanea Taurinensia* **2**, 173–195; *Oeuvres de Lagrange*, Vol. 1, Gauthier-Villas, Paris, 335–362.

Larsson, K. 1983, *Nature* **304**, 664.

Lawson, H. B. 1970, *Anal. Math.* **92**, 335.

Lawson, H. B. 1980, *Lectures on Minimal Submanifolds*. Vol. 1, Publish or Perish, Inc., Berkeley, California.

Lindblom, G., Larsson, K., Fontell, K., and Forsen, S. 1979, *J. Am. Chem. Soc.* **101** (19) 546–5470.

Lindman, B. and Stilbs, P. 1982, "Characterization of Microemulsion Structure Using Multicomponent Self-diffusion Data," in *Surfactants in Solution*, Mittal, K. L. and Lindman, B., Eds. L-103, 1651–1662.

Lines, L. 1965, *Solid Geometry*, Dover, New York.

Longeley, W. and McIntosh, J. 1983, *Nature* **303**, 612–614.

Mackay, A. L. 1985a, *Nature* **314**, 604–606; also *Physica* 1985b, **131B**, 300–306.

Mariani, P., Luzzati, Vl., and Delacroix, H., 1988, J. Molec. Biol. **204** (1) 165–189.

Massari, U. 1974, *Arch. Rat. Mech. Anal.* **55**, 357–382.

Meeks, W. 1978, "Lectures on Plateau's Problem," Escola de Geometria Differential, Universidade Federal do Ceara.

Meeks, W. 1976, Ph. D. Thesis, University of California, Berkeley.

Neovius, E. 1883, *Bestimmung Zweier Speciellen Periodischen Minimalflachen*, J. C. Frenckell and Sons, Helingfors.

Nitsche, J. C. C. 1975, *Vorlesungen uber Minimalflachen*, Springer-Verlag, Berlin.

Nitsche, J. C. C. 1985, *Arch. Rat. Mech. Anal.* **89**, 1–19.

Pathak, P. 1981, *Porous Media: Structure, Strength and Transport*, Ph.D. Thesis, University of Minnesota.

Pippard, A. B. 1954, *Adv. in Electronics* **6** 1; also *Proc. Roy. Soc.* **A224**, 273.

Plateau, J. A. F. 1873, *Statique Experimentale et Theorique des Liquides Soumis aux Forces Moleculaires*, 2 vols., Gauthier-Villars, Paris.

Rhines, F. N. 1958, "A New Viewpoint of Sintering," *Trans. MPA* (1058) Plansee pub. 91–101.

Riemann, B. 1892, *Gesammelte Mathematische Werke*, 2 vols., B. G. Teubner, Leipzig.

Rivas, A. P. 1972, "Meniscus Computations: Shapes of Some Technologically Important Liquid Surfaces," M. S. Thesis, University of Minnesota.

Scherk, H. F. 1835, *J. R. Angew. Math.* **13**, 185–208.

Schoen, A. H, 1967, *Not. Amer. Math. Soc.* **14**, 661.

Schoen, A. H. 1968, *Not. Amer. Math. Soc.* **15**, 727; also 1969 **16**, 519.

Schoen, A. H. 1970, "Infinite Periodic Minimal Surfaces without Self-intersections," *NASA Technical Note* TD-5541.

Schoenfliess, A. 1891, *C. R. Acad. Sci. Paris* **112**, 478–480; also 515–518.

Schwarz, H. A. 1865, *Monatsberichte der Koniglichen Akademie der Wissenscaften zu Berlin*, Jahrgang 1865, 149–153.

Schwarz, H. A. 1890, *Gesammelte Mathematische Abhandlungen*, Springer-Verlag, Berlin.

Scriven, L. E. 1976, *Nature* **263**, 123–125.

Scriven, L. E. 1977, "Equilibrium Bicontinuous Structures," in *Micellization, Solubilization, and Microemulsions*, Vol. 2, Mittal, K. L., Ed., Plenum Press, New York, 877–893.

Shal'nikov, A. 1941, *Zhurnal Eksperimental noi i teoretischeskoi fiziki* **11**, 202–210.

Silliman, W. J. 1979, *Viscous Film Flows With Contact Lines*, Ph.D. Thesis, University of Minnesota, Minneapolis.

Stessmann, B. 1934, *Math. Zeit.* **33**, 417–442.

Thomas, E. L., Alward, D. B., Kinning, D. J., Martin, D. C., Handlin, Jr., D. L., and Fetters, L. J. 1986, *Macromolecules* **19**, 2197.

Turner, M. R., Clough, R. W., Martin, H. C., and Topp, L. J. 1956, *J. Amer. Sci.* **23**, 805–823.

Wagner, H. J. 1977, *Computing* **19**, 35–58.

Weatherburn, C. E. 1927, *Differential Geometry of Three Dimensions*, Vol. 1, Cambridge University Press, Cambridge, England.

Weierstrass, K. 1866, *Monatsber. Berlin Akad*, Berlin.

Weierstrass, K. 1903, *Mathematische Werke*, 3 vols., Mayer and Muller, Berlin.

Wheeler, J. A. 1957, *Ann. Phys.* **2**, 604–614; also Misner, C. W. and Wheeler, J. A. 1957, *Ann. Phys.* **2**, 525–603.

Zick, A. A. and Homsy, G. M. 1982, *J. Fluid Mech.* **115**, 13–26.

International Tables for X-ray Crystallography, 1952, Vol. 1, N. S. M. Henry and K. Lonsdale, Eds. The Kynoch Press, Birmingham, England.

Added references:

Anderson, D. M. and Strom, P. 1988, in *polymer Association Structures: Liquid Crystals and Microemulsions*, ed. M. El-Nokaly, ACS Symposium Series, 204–224.

Anderson, D. M. and Thomas, E. L. 1988, *Macromolecules* **21**, 3221–3230.

Anderson, D. M., Wennerstrom, H. and Olsson, U. 1989, *J. Phys. Chem.* **93**, 4243–4253.

Anderson, D. M. and Wennerstrom, H. *J. Phys. Chem.* (submitted).

Anderson, D. M., Gruner, S. M. and Leibler, S. 1988, *Proc. Nat. Acad. Sci.* **85**, 5364–5368.

Lindman, B., Shinoda, K., Olsson, U., Anderson, D. M., Karlstrom, G. and Wennerstrom, H. 1988, Proceedings of 6th International Conference on Colloid and Interface Science in Hakone, Japan.

Radler, J. O., Rhadiman, S., DeVallera, A. and Toprakcioglu, C. 1989, *Physica B*, **156**, 398–401.

Thomas, E. L., Anderson, D. M., Henkee, C. S. and Hoffman, D. 1988, *Nature*, **334**, 598–601.

RECENT DEVELOPMENTS IN THE STUDY OF MONOLAYERS AT THE AIR–WATER INTERFACE

CHARLES M. KNOBLER

Department of Chemistry and Biochemistry
University of California
Los Angeles, CA 90024-1569 U.S.A.

CONTENTS

I. INTRODUCTION

A. Scope

The spreading of oil on the surface of water has been known since the eighteenth century and has been the subject of scientific investigation for over 100 years. Shortly before the turn of the century, Lord Rayleigh proposed[1] that films consisted of monolayers of spherical molecules; two decades later Langmuir and Harkins independently[2] recognized that the formation of a monolayer and its structure were associated with the amphiphilic and rod-like nature of the molecules. They proposed, for example, that in monolayers of fatty acids, the carboxylic head group is immersed in the water surface while the hydrocarbon tail remains above the surface.

There has been a continuous interest[3] in monolayers since this pioneering work, motivated both by a desire to understand their nature and by the possible relation of monolayers to other structures, such as bilayers and micelles, which are built up from amphiphiles. This research has relied mainly on the methods that originated with the early workers in the field, chief among them measurements of surface pressure–area isotherms. These techniques are primarily thermodynamic in nature and therefore provide no direct information about the microscopic nature of the monolayer. Moreover, since the monolayer cannot be seen, the character of surface phases and the transitions between them can only be inferred. The experiments are difficult and subject to systematic errors, and their interpretation has often been ambiguous.

The recent growth of interest in interfacial phenomena has enlarged and broadened the community of scientists engaged in studies of monolayers. New techniques employed in this research have enabled long-standing controversies to be resolved and have provided new insights into the microscopic structure of monolayer phases. In this paper, we will discuss much of this new work and describe the picture of monolayer structure that has evolved from it. Experiments, including simulations, will be stressed, but some comparisons will be made with theory. The discussion will be limited to monolayers at the air–water interface, which are sometimes called *Langmuir monolayers*, as distinct from *Langmuir–Blodgett films*, which are prepared by transferring Langmuir monolayers to solid supports.

The focus will be mainly on monolayers of fatty acids or alcohols and of phospholipids, because these are the simplest systems and have been the subject of the most recent research. These substances are similar in that their hydrophobic tails consist of long-chain hydrocarbons; there are two chains per phospholipid and one per acid or alcohol, as shown in Table I. The chain lengths typically range from 13 to 20 carbons, the lower limit being

TABLE I
Some Common Monolayer-Forming Amphiphiles

Saturated Fatty Acids	$CH_3(CH_2)_n-C(=O)OH$
n	
12	tetradecanoic (myristic) acid
13	pentadecanoic acid
14	hexadecanoic (palmitic) acid
16	octadecanoic (stearic) acid
18	eicosanoic (arachidic) acid
20	docosanoic (behenic) acid
Alcohols	$CH_3(CH_2)_mCH_2OH$
m	
16	1-octadecanol
19	1-heneicosanol
Phospholipids	$CH_3(CH_2)_n-C(=O)-O-CH_2$ $CH_3(CH_2)_n-C(=O)-O-{}^*CH$ $H_2C-O-P(=O)(O^-)-O-X$

	DMPC (dimyristoyl phosphatidylcholine)	DMPA (dimyristoyl phosphatidic acid)	DPPC (dipalmatoyl phosphatidylcholine)
n	12	12	14
X	$-CH_2CH_2\overset{+}{N}(CH_3)_3$	$-H$	$-CH_2CH_2\overset{+}{N}(CH_3)_3$

[Note asymmetric carbon marked by asterisk; molecules exist in enantiomeric forms]

dictated by the need to work with molecules that are not very soluble in the water subphase. Although both saturated and unsaturated chains have been studied extensively by classical methods, the newer experiments have dealt with the saturated molecules.

The hydrophilic moieties that constitute the head groups are usually ionizable. For example, the pK's of the fatty acids are of the order of 5 and that of phosphatidic acid is 2. Interactions between head groups may therefore depend on the pH and ionic strength of the subphase. Most studies on fatty acid monolayers have been performed at pH = 2, in order to keep the fraction of head groups ionized negligibly small.

B. Preliminaries

By far the most common experiment performed on Langmuir monolayers is the determination of surface pressure–area isotherms. The monolayer can be prepared by depositing a solution of the amphiphile in a volatile solvent on a clean water surface; the film spreads spontaneously as the solvent evaporates. The surface pressure π is defined as the difference between γ_0, the surface tension of pure water, and γ, the surface tension of the surface covered by the monolayer:

$$\pi = \gamma_0 - \gamma \tag{1}$$

A barrier can be slid across the water surface to change the area accessible to a fixed quantity of amphiphile (see Fig. 1); alternatively, the area per molecule may be decreased by successive additions of amphiphile to a surface of constant area.

A typical $\pi - A$ isotherm for a pure fatty acid or phospholipid is shown in Fig. 2. There are three regions of relatively steep slope separated by two distinct plateaus. The significance of these regions has generally been taken to be the following:

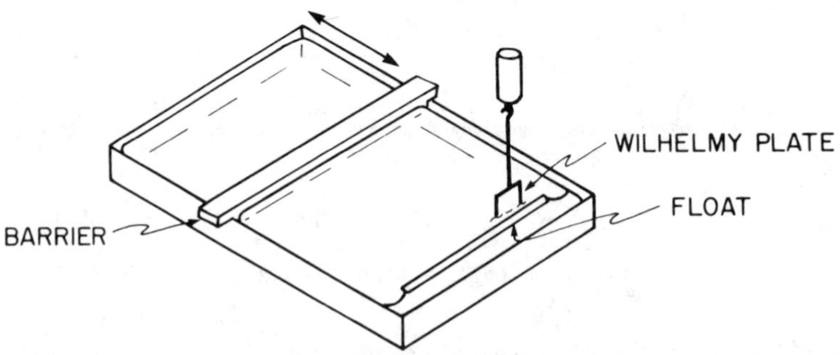

Figure 1. Schematic diagram of a Langmuir trough. The monolayer is deposited to the right of the barrier, and the barrier can be moved across the surface to change the area accessible to the monolayer. The surface pressure can be measured either by determining the force on a float that separates the monolayer from a clean water surface, or from the difference in the force exerted on the Wilhelmy plate when the plate is suspended in pure water and in water covered by the monolayer.

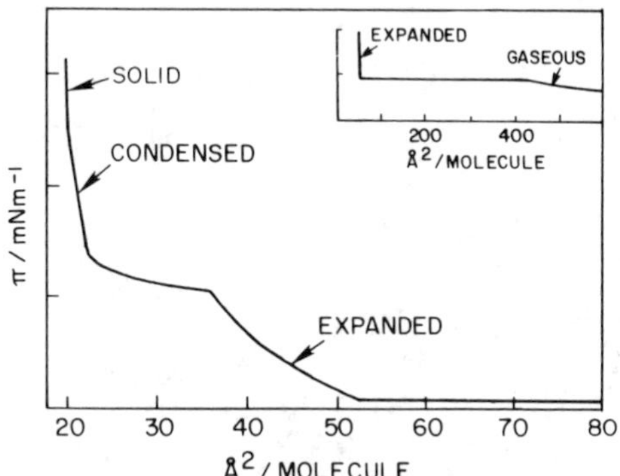

Figure 2. Surface pressure–area isotherm for fatty acids and phospholipids. The inset shows the behavior at large molecular areas. Four one-phase regions are identified: gaseous, liquid expanded, liquid condensed, and solid. This isotherm is typical of pentadecanoic acid at room temperature. For a higher-molecular weight acid, such as octadecanoic acid, there is no LE region at room temperature; the isotherm does not rise steeply until an area of about 25 Å^2.

1. At large areas, the monolayer is a two-dimensional gas.
2. The sharp break at the first plateau is the onset of condensation to a liquid, and the plateau represents two-phase coexistence. In the terminology of Adam and Harkins,[4] this liquid is called the *liquid-expanded (LE) phase.*
3. The LE one-phase region extends from the end of the first plateau to the start of the second. At this point, a phase transition takes place to another condensed phase, called the *liquid-condensed (LC) phase.* In the literature on phospholipids, the LE–LC transition region is called the *main* transition.

   The LE–LC transition region is lost at low temperature, as shown in the isotherms of pentadecanoic acid (Fig. 3). A regular progression is observed in the isotherms of an homologous series of amphiphiles, such as the saturated fatty acids. The isotherms are shifted to higher temperature with increasing chain length. In the case of the fatty acids, there is an 8–10 K shift per methylene group. Thus, isotherms of octadecanoic acid at room temperature correspond to those of pentadecanoic acid near 0°C and therefore do not show the LE–LC transition.
4. An additional feature, a change of slope or small kink, is sometimes evident at high pressure and may be interpreted as a transition from the LC phase to a solid phase. At smaller areas, the monolayer is a two-dimensional crystalline solid with a structure related to that of the three-dimensional crystal.

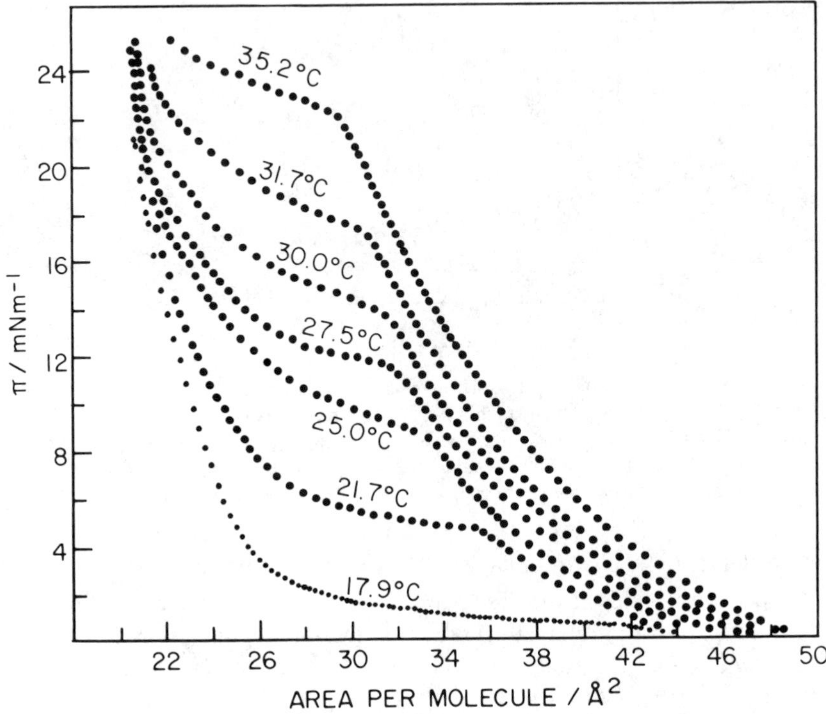

Figure 3. Isotherms of pentadecanoic acid (PDA). After Harkins and Boyd.[8]

5. The monolayer eventually undergoes *collapse* at high pressures, the transformation into a more stable three-dimensional phase. The pressure at which the monolayer and the bulk phase are in equilibrium is called the *equilibrium spreading pressure.* Since the bulk phase must be nucleated, collapse often does not occur until the pressure of the monolayer is well above the equilibrium spreading pressure.

Until recently, the characterization of the monolayer phases was based mainly on the shapes of the isotherms—the compressibility in each one-phase region and the packing densities of each phase, which can be taken as the zero-pressure intercepts of the steep portions of the isotherms. Since the head groups are constrained to be on the water surface and the van der Waals radii of the chains are known, these features of the isotherm can be matched to simple models for the monolayer phases.

The identification of the LE–G plateau as a two-phase region is consistent with the requirement[3] that at constant temperature and ordinary pressure the isotherm of a one-component monolayer system be horizontal at a first-order transition. Surface-potential measurements[3] in this part of the phase diagram show temporal fluctuations that are indicative of a heterogeneous monolayer; the fluctuations disappear when the monolayer is compressed into the LE one-phase region.[5,6]

The evidence for a first-order transition at the second plateau has been more ambiguous. As seen in Fig. 3, the isotherm is not horizontal, and this has sometimes been interpreted as an indication that the transition is second-order rather than first-order. On the other hand, fluctuations have been observed in the surface potential,[6] which is consistent with the coexistence of two phases.

C. A Case in Point: The G–LE Transition in Pentadecanoic Acid

It is well-known[7] that it is difficult to determine phase boundaries from isotherm measurements. A striking example of this problem is the G–LE transition in pentadecanoic acid (PDA). Harkins and Boyd[8] studied monolayers of PDA at a variety of temperatures and proposed that the G–LE two-phase region ended at a critical point at a temperature in excess of 35°C. The nature of this two-dimensional critical point was examined by Kim and Cannell,[9,10] who made isotherm measurements at closely spaced temperatures between 14.8 and 34.7°C. (Earlier, Hawkins and Benedek[11] carried out a similar investigation, but their failure to control the pH of the subphase left their results in doubt.)

These are very difficult experiments because the pressures in the G–LE region are very low and great care must therefore be taken to avoid even minute amounts of surface-active impurities, which can cause large drifts in the pressure and erroneously high pressure readings. Kim and Cannell employed a differential technique in which the pressure was measured with respect to that of a monolayer maintained at a fixed area in the two-phase region; their results are shown in Fig. 4.

By fitting the isotherms in the one-phase regions and locating the intersections with the constant-pressure two-phase regions, Kim and Cannell were able to locate the phase boundaries at each temperature, i.e., A_G and A_{LE}, the areas per molecule corresponding to the coexisting phases. These areas are given in Table II. They obtained the critical temperature by fitting the isothermal compressibility at the inflection point in supercritical isotherms and along the liquid side of the coexistence curve to a scaling equation of the form

$$\kappa_T = \Gamma |T - T_c|^{-\gamma} \tag{2}$$

in which Γ is a constant. The value $T_c = 26.27°\text{C}$ was determined by the requirement that the exponents γ and γ' (*not* surface tensions!), which describe

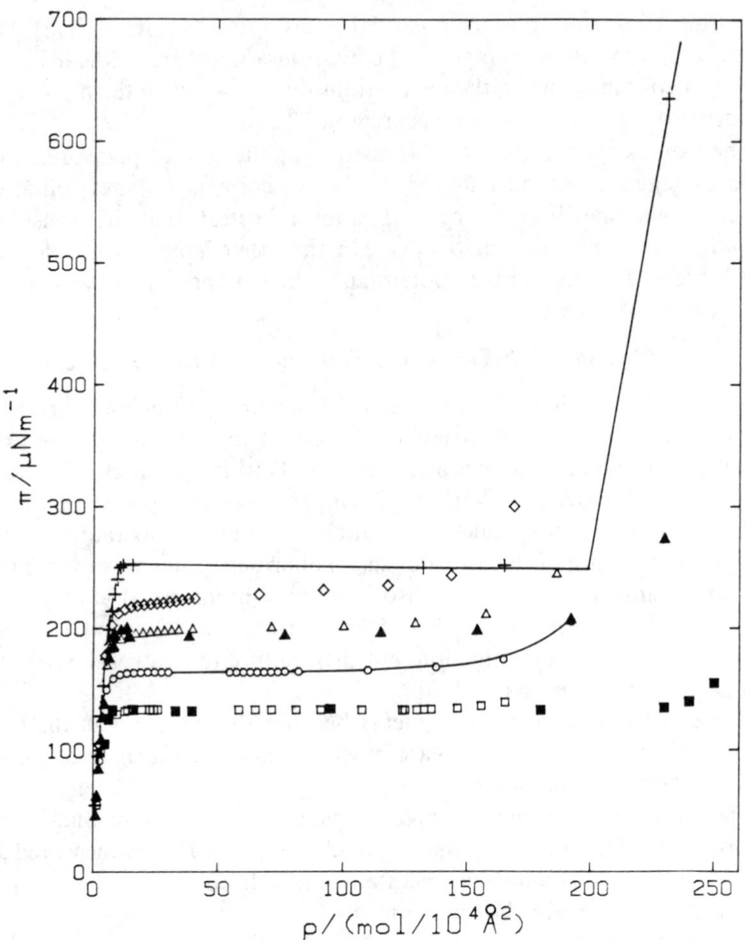

Figure 4. Isotherms of PDA in the G–LE region. The open symbols have been taken from the thesis of Kim[10] and the closed symbols and pluses from that of Pallas.[14] ■ 20°C; □ 20.03°C; ○ 24.97°C; ▲ 30°C; △ 30.10°C; ◇ 34.69°C; + 40°C. For ease of viewing, not all of Kim's or Pallas' isotherms are shown. The line drawn through the 40°C data is that of Pallas and Pethica,[13] that through the 24.97°C data is taken from that on the π-A curve of Kim and Cannell.[9]

the behavior above and below the critical temperature, be equal. A notable result of their analysis is that γ and β, the exponent that describes the shape of the coexistence curve, have the mean-field values 1 and 1/2, respectively, rather than the two-dimensional Ising values,[12] 7/4 and 1/8.

The G–LE coexistence region in pentadecanoic acid has been investigated more recently by Pallas and Pethica.[13,14] Like Kim and Cannell, they made

TABLE II
The LE–G Transition in PDA
Pressures and Molecular Areas at Coexistence

$T/°C$	$\pi/\mu Nm^{-1}$	$A_G/Å^2$	$A_{LE}/Å^2$	Reference
14.8	101.50	680	62.5	9
15	102±2	2000	20	14
20	132±2	1500	43	14
20.03	133.19	614	81.7	9
21.96	144.40	580	94.6	9
23.72	156.59	510	124	9
24.97	164.19	403	146	9
25	162±2	1300	45	14
25.63	169.82	394	160	9
30	192±2	1200	47	14
35	210	1000	60	41
40	252±2	850	51	14

extraordinarily careful measurements over a wide range of temperatures (15–40°C), but used a conventional (Wilhelmy plate) pressure-measuring technique rather than a differential method. As can be seen in Table II, at the temperatures at which they overlap, the pressures determined in the two sets of measurements are virtually identical. On the other hand, the densities of the coexisting phases are astonishingly different. Moreover, Pallas and Pethica believe that there is a horizontal slope in the isotherms even at 40°C and that Kim and Cannell's value of T_c must be in error by at least 14 K.

Pallas[14] suggests that the slope in Kim and Cannell's isotherms above 26.3°C is the result of a leaching out of surface-active impurities from the Teflon trough in which the measurements were made. The differences in the coexisting densities are attributed[13] to the effect of residual impurities in the sample of pentadecanoic acid used by Kim and Cannell; this does not seem likely.

When the data points obtained by the two groups are plotted on the same graph (Fig. 4), it is evident that they agree within the combined experimental error. A major part of the discrepancy stems from the interpretation of the data, the ways in which the isotherms have been drawn through the points. Kim and Cannell have required that all points in the two-phase region differ in pressure by no more than 0.02 μNm^{-1}. When this criterion is employed, the intersections of the one-phase and two-phase portions of the isotherm are not acute. This is consistent with the expectation that subcritical isotherms near T_c will be rather flat in one-phase regions close to the coexistence curve.

The sensitivity of the pressure measurement technique used by Pallas and Pethica was 0.2 μNm^{-1}. The precision of the measurements is much poorer, probably because they were made mostly by successive additions, and Pallas and Pethica include in the two-phase regions points that differ by as much as 3 μNm^{-1} from the mean pressure. In constructing the isotherms, they have chosen to have the one-phase and two-phase portions meet at an acute angle. This choice leads to coexisting densities markedly different from those reported by Kim and Cannell. Which of the interpretations in correct? As we shall see, other types of experiments allow us to choose between them.

II. NEW EXPERIMENTAL APPROACHES

Several new techniques have been developed by which it is possible to establish the boundaries between monolayer phases and the nature of the transitions between them. Other approaches have begun to provide direct information about the microscopic structures of monolayer phases. In this section, we will briefly describe these methods and their application to Langmuir monolayers.

A. Imaging Methods

The disagreements about the location of a critical point might easily be settled if the monolayer phases could be viewed directly. Early attempts[15] were made to observe the texture of monolayers by dark field microscopy, but the refractive index differences between the phases is small and only collapsed portions of the film are clearly discernible. It has only been recently that direct visual observations of Langmuir monolayers have been made.

The high sensitivity with which fluorescence can be detected has led to its widespread use in biological research. Fluorescent probes that can be bonded to specific substances or have high solubility in only one phase are able to provide optical contrast between different structures. This contrast difference can be employed in fluorescence microscopy to allow the texture of monolayers to be observed *in situ* on a water substrate.

McConnell,[16] Möhwald,[17] and their co-workers were the first to exploit fluorescence microscopy in the study of monolayers. A detailed description of the method has been given by Lösche and Möhwald.[18] The amphiphile is doped with a fluorescent probe, which is usually also an amphiphile. For example, the probe NBD–DPPE, which consists of the chromophore NBD (7-nitrobenz-2-oxa-1, 3-diazol-4-yl) bonded to dipalmitoylethanolamine, has been used in studies of phospholipid monolayers; NBD-hexadecylamine has been employed for measurements on fatty acids. Differences in the degree of fluorescence within the monolayer can reflect differences in density between phases or differences in solubility.

The fluorescence of a probe can also be quenched by interactions with its local environment. It has been observed[19] that the fluorescence intensity is not conserved when stearic acid (octadecanoic acid) doped with 12-NBD-stearic acid undergoes a transition from a condensed phase to the gaseous phase. NBD is known to be quenched in polar environments and the loss in fluorescence is thought to be associated with the change in chain orientation with respect to the water surface that occurs as a result of the phase transition.[20]

A schematic diagram of a fluorescence apparatus is shown in Fig. 5. Light from a mercury or xenon lamp or laser passes through filters and a dichroic mirror into the optics of the microscope. This exciting radiation passes through the objective and is focused on the monolayer, which has been prepared in a small trough that is mounted on the microscope stage. The dichroic mirror reflects at the wavelength of the chromophore

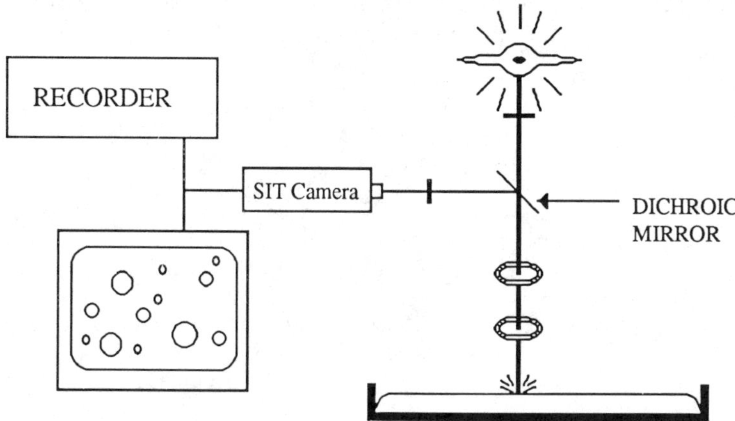

Figure 5. Schematic diagram of an apparatus for fluorescence microscopy of monolayers. The exciting light, a mercury or xenon arc or a laser, passes through a narrow-band filter into the microscope optics and is focused on the surface of a monolayer that contains a fluorescent probe. The fluorescence is captured by the microscope and separated from the exciting light by a dichroic mirror. It is reflected into a silicon-intensified-target television camera. The image can be viewed on a monitor or recorded. Some experiments[18] have used an inverted microscope in which the surface is viewed through the subphase.

fluorescence. The image is observed with a high-sensitivity television camera and displayed on a monitor, or recorded on tape, or both. Photographs are usually made of the monitor screen but the image can also be recorded directly on high-contrast film.

Unless measurements are made on a substance that is itself fluorescent, fluorescence studies require the addition of an impurity, the probe, to the monolayer. It can be expected that phase transitions in doped monolayers will not be as sharp as in pure systems. The effect of impurity can be investigated by varying the probe concentration. It is possible that a probe can be "line active," tending to concentrate at the interface between two surface phases. In such a case even very small concentrations of probe might have marked effects on the structure of the monolayer or on the dynamics of growth of a new phase. As discussed below, however, there is good evidence that these problems do not limit the utility of the fluorescence method.

McConnell and co-workers have demonstrated[21] that information about the orientation of molecules in a condensed monolayer phase can be obtained by examining the polarization of the fluorescence. The underlying principle is shown in Fig. 6. The monolayer is assumed to be made up of domains of tilted amphiphiles and, for simplicity, the polarizability tensor of the fluorescent probe bonded to some of the molecules is taken to be perpendicular to the chain axis. If p-polarized light is directed to the surface at an angle with respect to the surface normal, the fluorescence excited in the domains will not be identical. The contrast difference between the domains will be reversed if the horizontal component of the direction of the

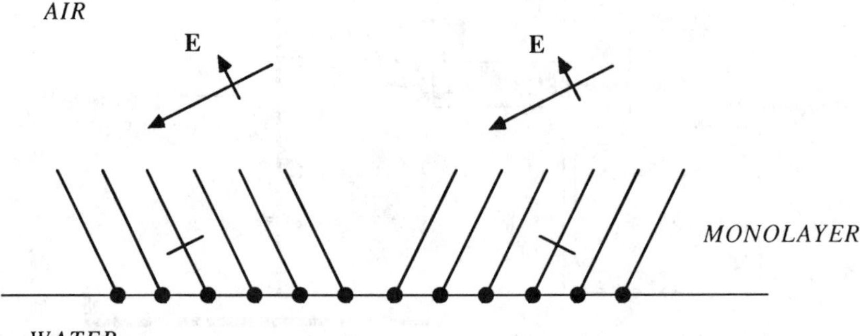

Figure 6. Determination of orientation from polarized fluorescence emission experiments. The orientation of the amphiphile chains in neighboring domains is indicated by the sloping parallel lines in the monolayer, and the orientation of the polarizability tensor of the fluorophore is represented by the line segment perpendicular to the chain axis. Long arrows show the direction of the incident light, and small arrows represent the direction of the electric field for p-polarized light. After Moy et al.[21]

illuminating light is changed by 180°. Other experimental arrangements can also be employed. The method has been used to examine the orientational order in a phospholid.[21]

B. Reflection and Diffraction Methods

1. *Ellipsometry*

Ellipsometry, a classical technique that is commonly employed in the study of thin films on solid supports, has long been used to investigate Langmuir monolayers.[3] The relative ease with which such measurements can now be carried out, and their high sensitivity and high spatial resolution, has led to renewed interest in the method. Rasing et al.[22] have recently described studies on PDA in which the phase retardation was measured as a function of the surface pressure. Fluctuations in the signal can be correlated with the texture of the monolayer, and estimates of the sizes of the domains and their character can be obtained from quantitative analysis of the data.

For an optically isotropic Langmuir monolayer, the phase retardation depends only on the thickness of the film. There is an additional contribution, however, if the refractive index of the monolayer is anisotropic. This feature of ellipsometry has recently been exploited by Bercegol et al.[23] to study the molecular alignment in monolayers of octadecanoic acid.

2. *X-Ray and Neutron Studies*

X-ray diffraction studies of monolayers provide information about the structure within the plane of the interface, and x-ray reflection measurements give information about the structure in the perpendicular direction. The scattering from monolayers is weak, so even with the high-intensity beams obtainable with synchrotron sources, measurements are confined to high-density monolayer phases.

In reflectivity experiments, the source and detector geometry are chosen so that the scattering vector **Q** is parallel to the *z*-axis (perpendicular to the plane of the monolayer), as shown in Fig. 7. The reflectivity $R(\mathbf{Q})$ is then given by[24]

$$R(\mathbf{Q}) = R_{\mathrm{F}}(\mathbf{Q}) \left| \rho_s^{-1} \int (\mathrm{d}\rho/\mathrm{d}z) \exp\left[iQz\right] \mathrm{d}z \right|^2 \tag{3}$$

which is valid in the large-**Q** limit when the reflectivity is low.

In this expression R_{F} is the Fresnel reflectivity of the water–air interface, i.e., the reflectivity of an infinitely sharp interface between homogeneous media, ρ_s is the scattering-length density of the subphase and $\rho(z)$ is that of the monolayer. In the measurements, the ratio $\mathrm{R}/\mathrm{R}_{\mathrm{F}}$ is determined as a function of **Q**.

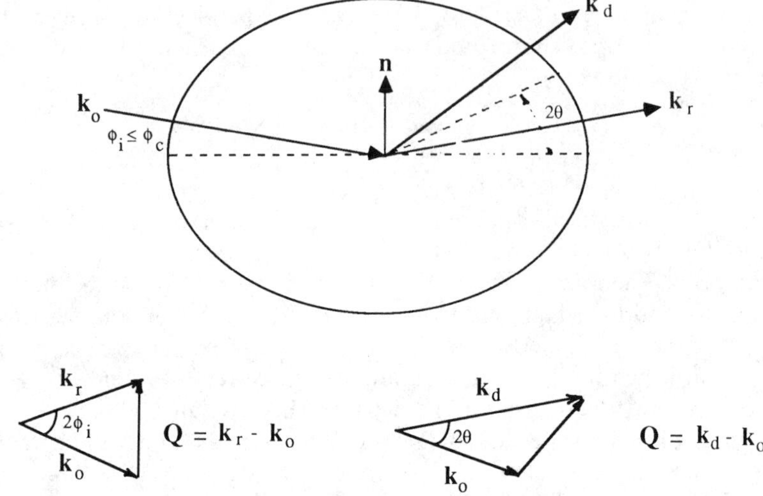

Figure 7. Scattering geometry: vertical scattering (reflectivity) and in-plane scattering (diffraction). The x-rays (or neutrons) with incident wave vector $\mathbf{k}_o$ strike the surface at a small angle of incidence ϕ_i which is less than the critical angle for total external reflection, ϕ_c. In reflectivity measurements, the vertical incidence and exit angles are equal so the difference between $\mathbf{k}_o$ and the reflected wave vector $\mathbf{k}_r$ is perpendicular to the scattering plane. In diffraction measurements, the difference $\mathbf{k}_o - \mathbf{k}_d$ lies essentially in the plane.

The scattering data $R(\mathbf{Q})$ cannot be inverted to give $\rho(z)$; instead the measured reflectivities are compared to those predicted from models of the interface. If a structure is assumed, ρ can be calculated by averaging over the contributions of each of the atoms:

$$\rho = \sum_j N_j b_j \tag{4}$$

where N_j is the number density of atom type j, and the scattering length b_j is the product of the atomic number of the atom and the scattering length of an electron, $Z_j r_e$.

For simplicity, monolayers are pictured as a series of homogeneous slabs—air, tail, head, and water—each with a characteristic average scattering-length density. Thus, there are at least four parameters: the thicknesses of the head and tail slabs and their densities. In some cases, a fifth parameter is added that characterizes the smearing out of the interfaces by surface roughness. The range of the parameters is constrained by a knowledge of the structure of the amphiphile, the chain length of the tail, and the chemical composition of the tail and head.

One feature of the reflectivity can be interpreted more easily, however. The reflectivity falls to a well-defined minimum when there is constructive interference between rays reflected from the water–amphiphile interface and the amphiphile–air interface. For a film on water, the minimum satisfies the condition[25]

$$Q_{\min}(l_t + l_h/2) = 3\pi/2 \tag{5}$$

where l_t and l_h are the lengths of the head and tail, respectively.

If the scattering vector is approximately parallel to the interface, a diffraction pattern can be observed that is characteristic of the in-plane structure of the interface.[25,26] (The background scattering of the subphase can be reduced by having the incident beam strike the surface at an angle slightly smaller than the critical angle for total reflection, thereby illuminating the sample only to a depth of about 100 Å.)

As in any diffraction study, three kinds of information can in principle be derived from such experiments: (1) the spacing of the diffracting planes from the diffraction peak positions, (2) the translational correlation length from the peak width, and (3) the structure factors from the integrated peak intensity. In a typical synchrotron measurement, the primary beam is defined by a slit 2mm wide and 70mm high. Since the angle of incidence is of the order of 0.1°, the surface area illuminated is about 1 cm^2. The patterns observed are therefore more likely to be powder patterns representing diffraction from many randomly oriented domains than the isolated peaks characteristic of single crystals.

Treatment of the experimental data requires that the diffraction peak be fitted to a function that characterizes the peak shape convoluted with the instrumental function; a background, which may depend on Q, must also be subtracted. The choices involved in these procedures are not unequivocal. Even with intense synchrotron sources, higher-order reflections are not usually observed, so the assignment of a structure is based on the position of only one peak, i.e., on only one lattice spacing. The interpretation of the structures is made plausible by knowledge of the molecular structure of the amphiphile and of the crystal structures of similar molecules.

Reflectivity measurements on monolayers have also been carried out with neutrons.[27,28] In general, x-ray sources are much more intense than neutron sources; even laboratory (rotating anode) x-ray sources are two orders of magnitude more intense than neutron sources. On the other hand, the scattering-length density for neutrons of an atom is that of its nucleus and, as a result, the contrast between different parts of the interface can be altered by isotopic substitution. For hydrocarbon chains, a significant change in contrast is easily accomplished by deuteration because the magnitude of the scattering length for protons is only about half that for deuterons. The lengths are also of different sign ($b_H < 0$) so, for example, the contribution of water to the reflectivity can be made negligible by using a mixture of H_2O and D_2O as the subphase.

C. Spectroscopic Methods

1. *Second-Harmonic Generation (SHG)*

When an intense laser beam strikes a medium with a structure that is not centrosymmetric, a second harmonic of the laser frequency is generated. A boundary between uniform media lacks inversion symmetry, and a second harmonic can therefore be generated at an interface. If the signal comes mainly from a monolayer on the surface, then the nonlinear susceptibility $\chi^{(2)}$ that is responsible for SHG has the form[29]

$$\chi^{(2)} = N_s \langle \alpha^{(2)} \rangle \tag{6}$$

where N_s is the surface density of the molecules and $\langle \alpha^{(2)} \rangle$ is the nonlinear polarizability averaged over the molecular orientation distribution.

In general, $\chi^{(2)}$ is a third-rank tensor and the relationship between the measured signal and the molecular orientation is complex. But if $\alpha^{(2)}$ is dominated by a single component along a molecular axis *and* the axis is randomly distributed in the azimuthal plane, there are only two nonvanishing components of $\chi^{(2)}$ and in this case $\langle \theta \rangle$, the average value of the polar angle between the axis and the surface normal, can be determined by comparing the polarization of the second harmonic signal with that of the incident radiation. This technique has been employed by Rasing et al.[29] to study PDA.

A related method that has been applied to Langmuir monolayers is sum-frequency generation (SFG) spectroscopy.[29b,30] The signal is generated by mixing an infrared wave of frequency ω_{IR} with a visible wave of frequency ω to yield an output ω_{SF}. When ω_{IR} matches a vibrational frequency, there is a resonant enhancement of the output. The nonlinear polarization that governs the SFG process can be written

$$\mathbf{P}^{(2)}(\omega_{SF} = \omega + \omega_{IR}) = \chi^{(2)} : \mathbf{E}(\omega)\mathbf{E}(\omega_{IR}) \tag{7}$$

where $\chi^{(2)}$, the sum of the resonant and nonresonant second-order, nonlinear susceptibilities, vanishes in a medium with inversion symmetry. Thus, like SHG, SFG is surface-specific; another feature common to the two techniques is the need to make approximations about the average value of the susceptibility in order to interpret the results.

2. *FT–IR Spectroscopy*

A key parameter in the structure of monolayer phases is the chain conformation of the amphiphile. Infrared spectroscopy, which is known to be an effective tool for determining conformation in bulk hydrocarbons, has recently been applied to Langmuir monolayers. Dluhy[31] investigated theoretically the possibility of studying monolayers at the air–water interface by external reflection infrared spectroscopy, a technique that had been applied to Langmuir–Blodgett films on reflective metal substrates. He found that the experiments were feasible and that finite values of the mean-square electric fields for both parallel and perpendicular polarizations are present at the air–water interface in contrast to the air–metal interface, where only the parallel component can be observed. Thus, in principle, the average orientation of the chains with respect to the surface can also be determined. (The term *chain orientation* can be defined unambiguously only when the chain is in some fixed conformation. If there is *gauche-trans*-disorder, one cannot speak of orientation separate from conformation.)

An apparatus for performing single-pass FT–IR external reflection measurements on monolayers has been described by Dluhy et al.[32] The beam from a commercial FT–IR spectrometer is diverted onto the surface of a Langmuir trough by a gold-coated mirror, where it is focused to a spot about 1 cm in diameter; another mirror directs the reflected beam to a high-sensitivity, liquid N_2-cooled detector. The optimum range for the angle of incidence is about 25°.

Conformational changes in hydrocarbon chains can be related to shifts in the CH_2 vibrations near 3000 cm^{-1}. In a typical measurement,[33] 4096 scans at a resolution of 8 cm^{-1} are summed, and absorption maxima can be located with a precision of 0.1 cm^{-1}.

D. Langmuir–Blodgett Films

Monolayers can be transferred to solid surfaces by passing the substrate through the interface, a layer being deposited at each passage. Most of the research on these Langmuir–Blodgett (LB) films has been focused on the properties of multilayers, but investigations have also been carried out on monolayers. Measurements on LB monolayer films may provide information about the structures of the Langmuir monolayers from which they have been prepared. There is no certainty, however, that structural features are retained when the film is transferred to the support or that the support has not induced significant changes.

There are several important methods for determining the structure of monolayers that can be employed only when the substrate is solid. For example, LB films can be studied by electron diffraction,[34,35] a technique that allows small regions of the film to be probed and therefore does not produce a powder-averaged diffraction pattern. With scanning electron microscopy, the textures of films shadowed with a metal coating can be directly observed.[34] Microscopic images of the films have also been obtained recently by scanning tunneling microscopy.[36]

A new imaging technique, surface-plasmon microscopy, has been developed by Rothenhäusler and Knoll and applied to Langmuir–Blodgett films.[37] Plasmon surface polaritons ("surface plasmons," or PSPs) are surface

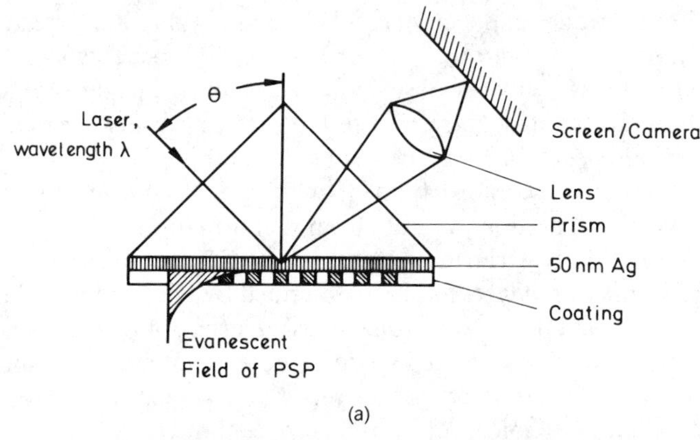

(a)

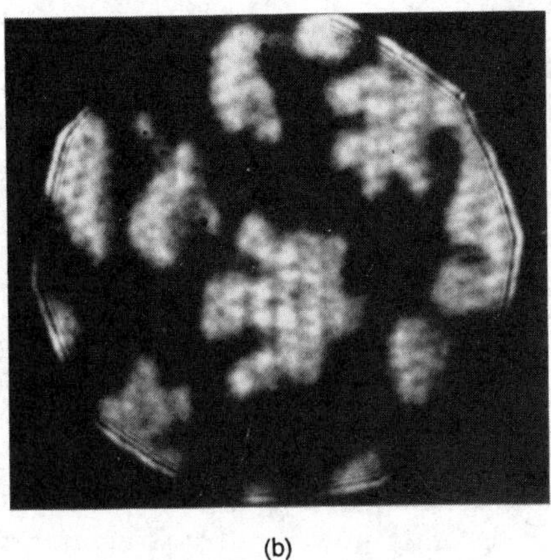

(b)

Figure 8. Surface-plasmon microscopy. (a) The coupling of the incident photons to the surface mode can be accomplished by using a prism. The evanescent field of the surface plasmon extends into the monolayer coating, and the light that is scattered or diffracted when the resonance condition is not fulfilled can be imaged by the lens. After Rothenhäusler and Knoll.[37] (b) SPM images of a monolayer of DMPA in the LE–LC region transferred to a solid support.

electromagnetic modes associated with a polarization charge wave propagating along a metal–dielectric interface. The field associated with surface plasmons decays exponentially in the direction perpendicular to the surface.

Photons can be converted into such modes by means of a plasmon coupler, which may be a grating or a prism, as shown in Fig. 8. The coupling is resonant and occurs whenever the momentum matching condition $\mathbf{k}_{ph} = \mathbf{k}_{sp}$ between the parallel component of the photon wave vector $\mathbf{k}_{ph}$ and the PSP wave vector $\mathbf{k}_{sp}$ is fulfilled. The value of $\mathbf{k}_{sp}$ depends on the optical thickness of the surface coating. In regions of the surface in which the resonance condition holds, the wave propagates and there is little leakage from the surface. But when the light enters regions that are off resonance, it is reflected, diffracted, or scattered out of the surface and can be coupled back through the prism and imaged by a lens (see Fig. 8).

An advantage of surface-plasmon microscopy is that great contrast between layers of different thicknesses that can be obtained in very thin films without the use of the probe molecules. It is a requirement of the method, however, that there be an underlying coating of a conductor such as silver that carries that surface plasmon field. There is a possibility that images can be obtained from the water/monolayer film that adheres to a vertical slide that has been partially drawn through the air–water interface.[38]

III. MONOLAYERS OF FATTY ACIDS

A. Phase Diagram

1. *Recent Isotherm Measurements*

As noted in Section I·C, isotherms of PDA in the G–LE transition region have been studied[9–11,13,14] with great care; the LE–LC transition has also been investigated. Pethica and co-workers[14,39,40] have performed systematic studies of the experimental factors that can affect the slope of π-A isotherms—purity of materials (amphiphile, spreading solvent, subphase, acid); materials from which the trough is constructed; pressure-measuring technique; humidity in the trough environment; rate of compression of the monolayer (stepwise and continuous); and the method by which the density of the monolayer is changed (compression/expansion vs. addition). They conclude that if careful attention is paid to the details of the experiment, the slope in isotherms of PDA and hexadecanoic (palmitic) acid at the LE–LC transition is horizontal and, therefore, that the LE–LC transition is clearly first-order.

Pegg and Morrison[41] have also found virtually flat isotherms at the LE–LC transition in PDA in the temperature range 23–33°C. They

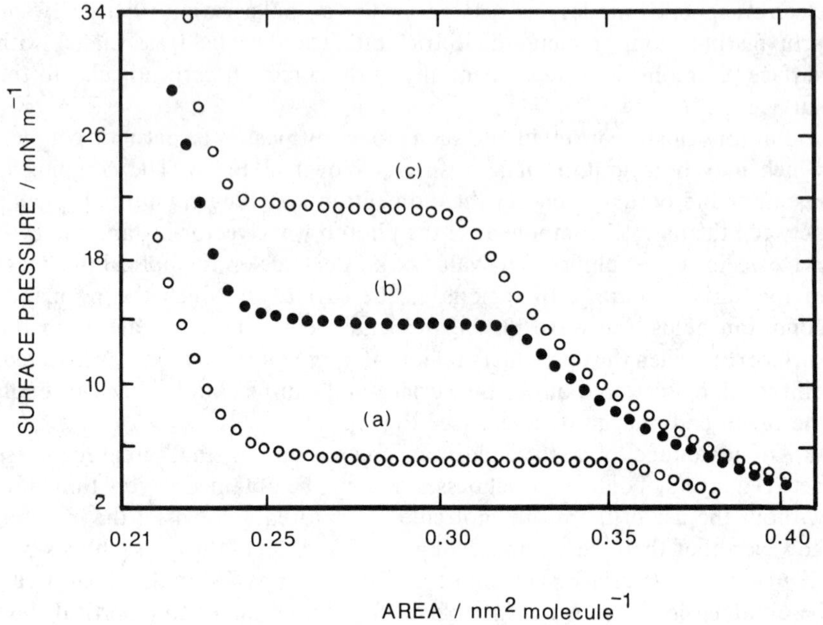

Figure 9. Isotherms of PDA in the LE–LC region.[41] (a) 22.7°C, (b) 29.2°C, (c) 33.0°C.

TABLE III
The LE–LC Transition in PDA
Pressures and Molecular Areas at Coexistence

T/°C	π/mNm^{-1}	A_{LE}'/Å^2	A_{LC}/Å^2	Reference
19	1.3	39		44
25	7.2±0.1	31	21.5	14
25	7.6±0.2	35		41
25	8.1	32		44
29	12.9	29		44
30	13.2±0 1	28	20	14

attribute[42] their success in obtaining horizontal plateaus mainly to a reduction in the surface-active impurities that are leached out of the trough or introduced from the air, and to control of the humidity; the purity of the amphiphile plays a lesser role. Their isotherms, which have been obtained by continuous expansion, are shown in Fig. 9, and the values of the pressure at the plateau are compared with those of other workers in Table III.

Extensive isotherm measurements on PDA have also been reported by Moore et al.[43] Their measurements were carried to high pressures, and they were able to observe and track the temperature dependence of the small kink that appears at areas of about 22 Å^2. These data are also summarized in Table III.

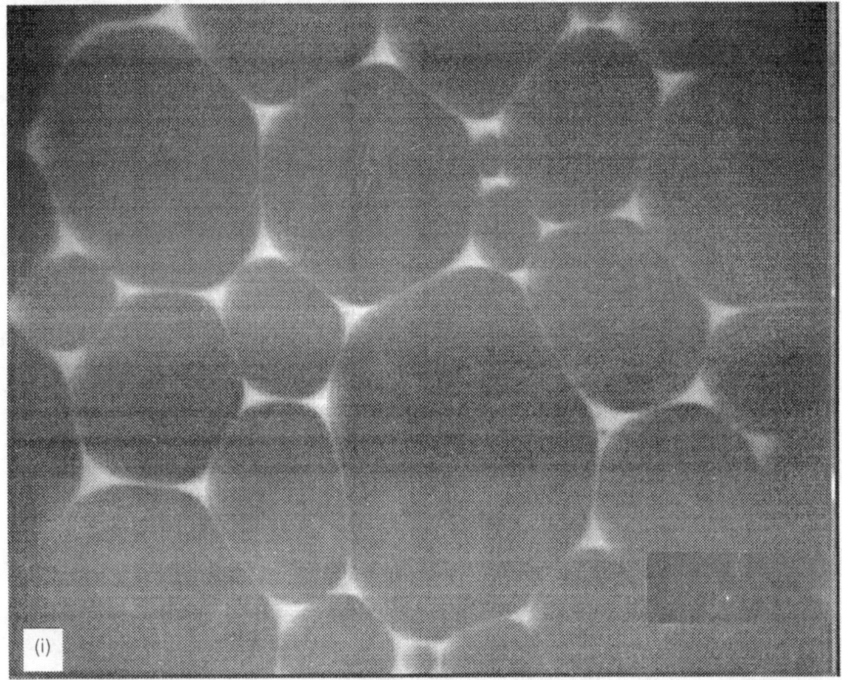

Figure 10. Fluorescence microscope images of PDA. The probe was NBD-hexadecylamine and the probe concentration was 1 mol%. (a) Isotherm at 20°C. The monolayer is prepared in the G–LE transition region (iii); bar = 100 μm. Expansion produces a foam structure (ii) that grows with time (i). With compression, (iii) is converted to an all-white phase and small dark islands appear at the LE–LC two-phase boundary (iv). The islands grow with further compression (v) and eventually deform and coalesce (vi). The granular appearance of the LE phase in (vi) is the result of demixing of the dye; at low dye concentrations this effect is not seen. (b) Measurements as a function of time after a quench from 20°C to 14°C at a fixed molecular area of 52 Å^2. Circular bubbles of the gas phase nucleate around the LC domains and grow with time.

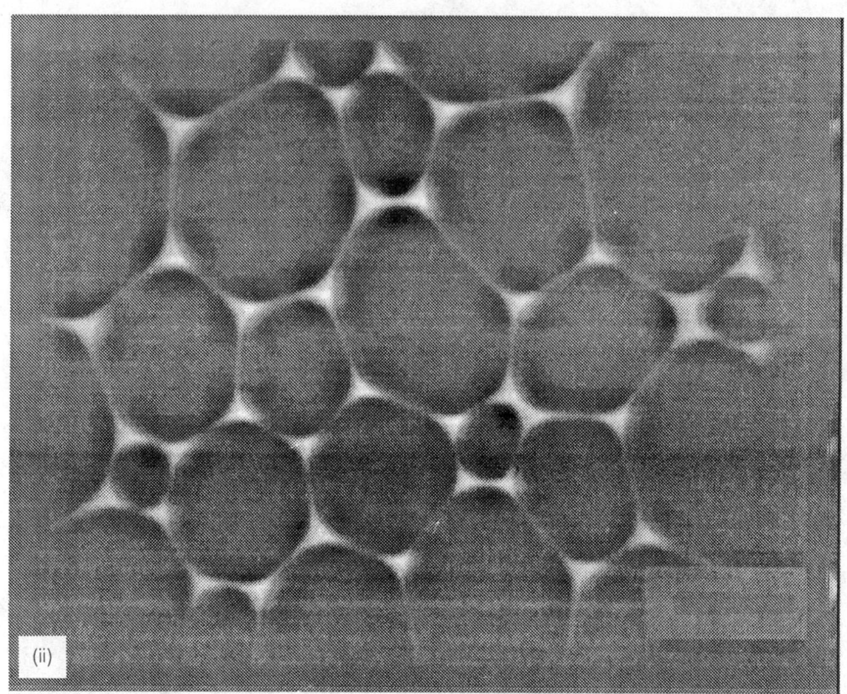

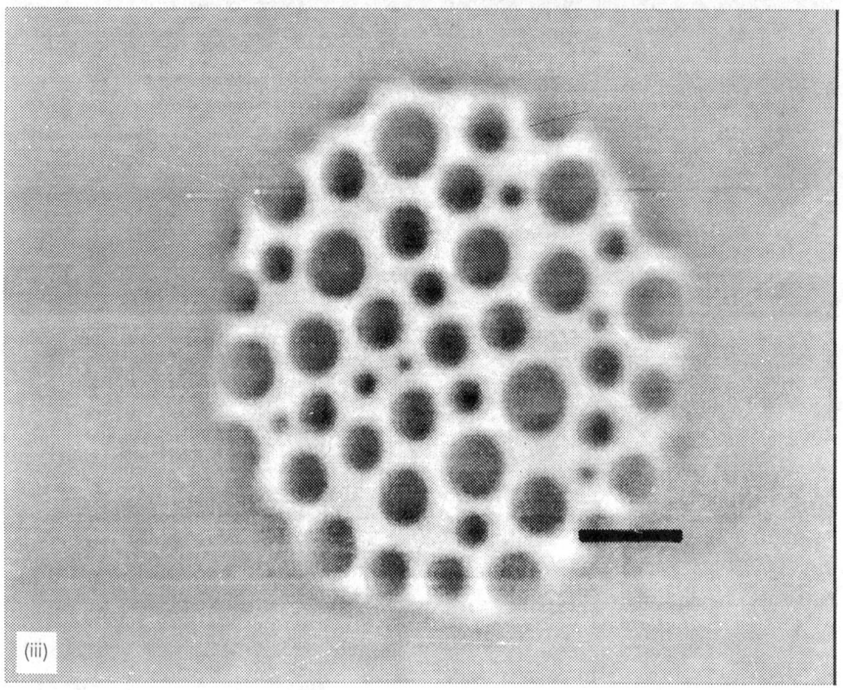

Figure 10(a) (*Continued*)

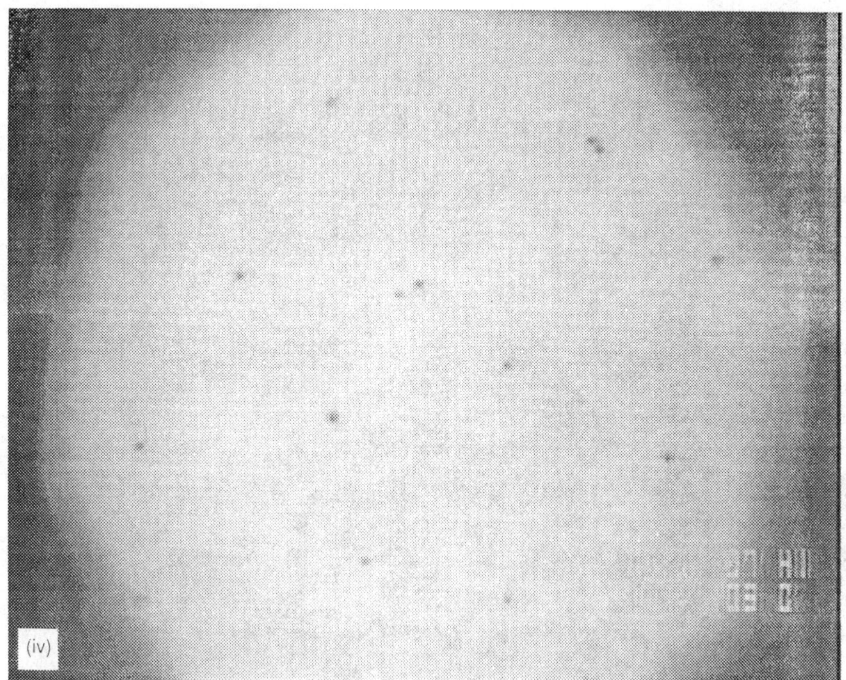

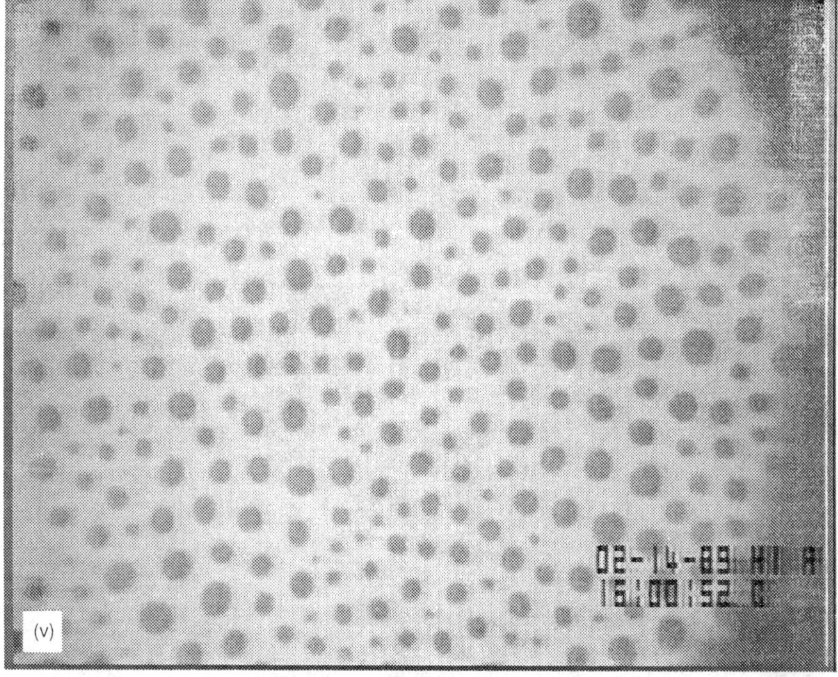

Figure 10(a) (*Continued*)

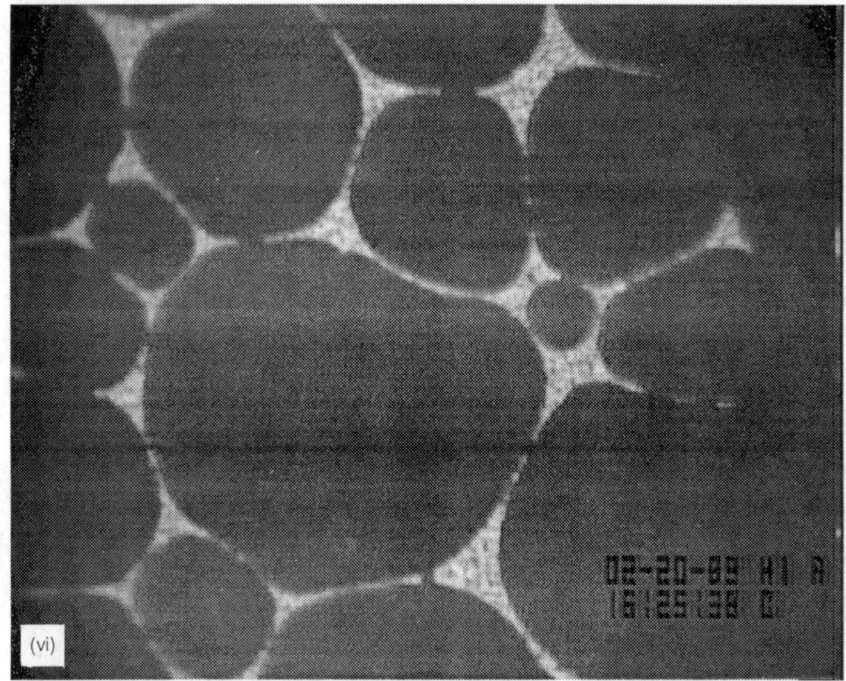

Figure 10(a) (*Continued*)

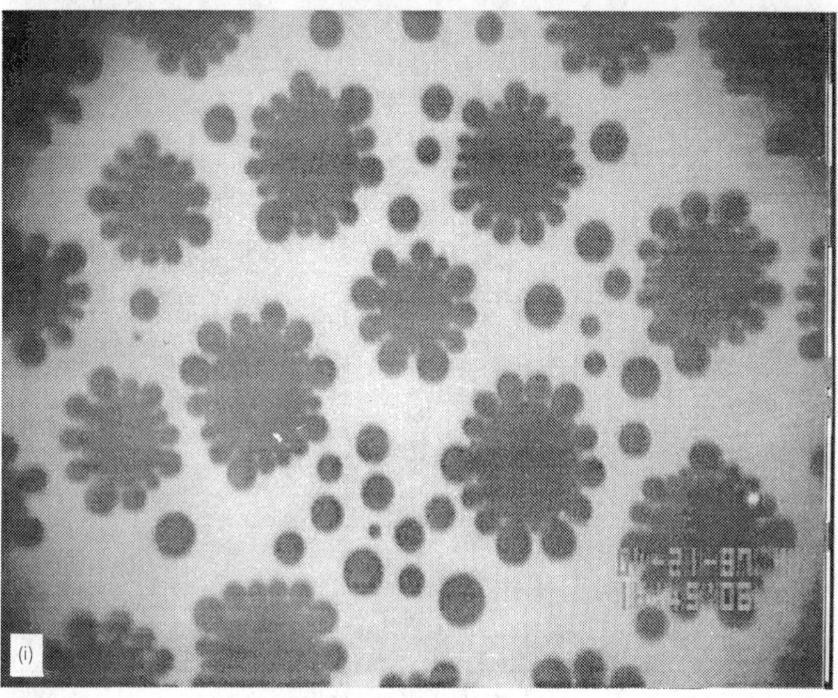

Figure 10(b) (*Continued*)

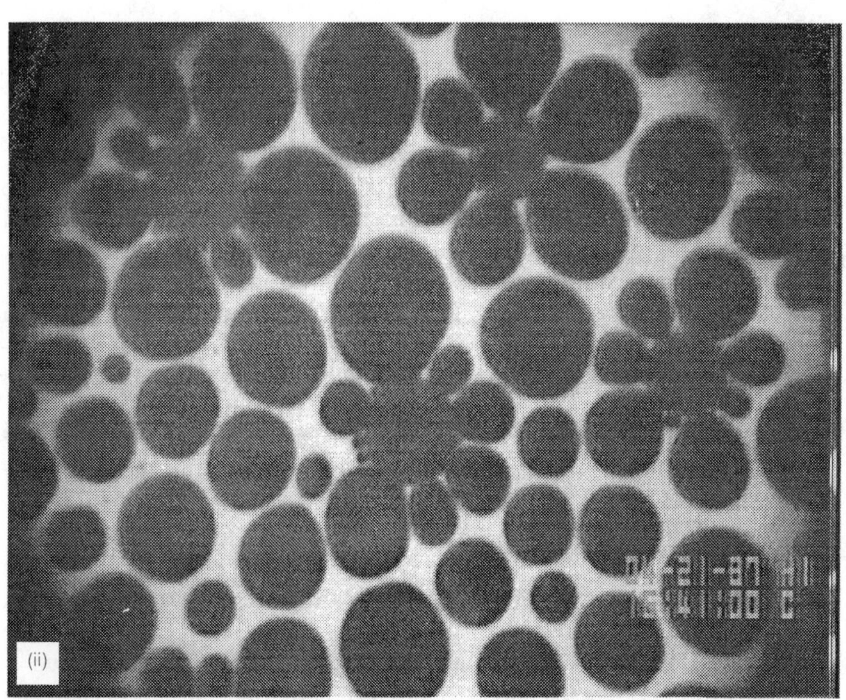
(ii)

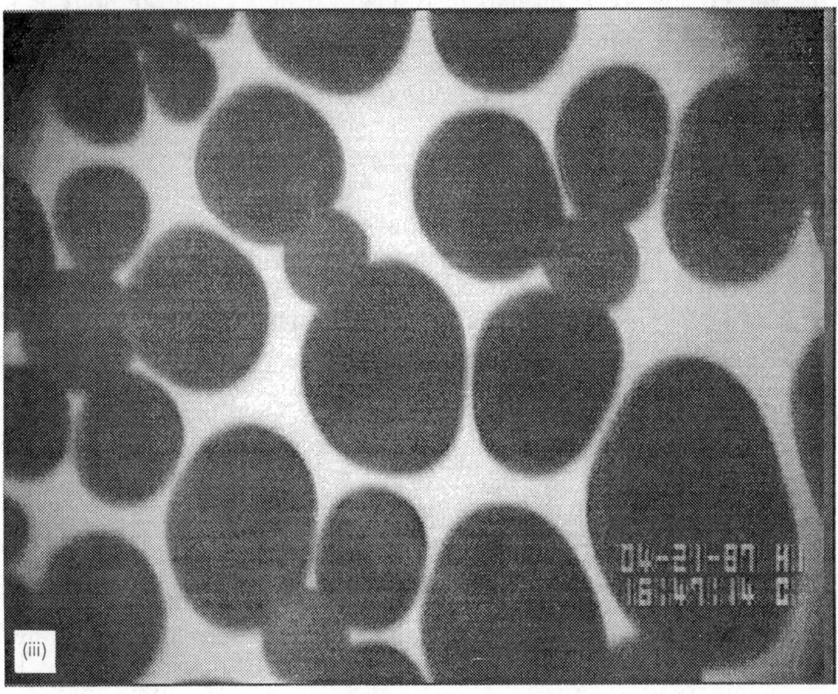
(iii)

2. *Fluorescence Measurements*

The phase behavior of octadecanoic (stearic) acid,[19] tetradecanoic (myristic) acid,[44] and PDA[43] has been examined by fluorescence microscopy. The observations are similar; the most complete measurements have been carried out on PDA. Figure 10 shows the progression of images when a PDA monolayer doped with a fluorescent probe is studied at 25°C.

At a molecular area that both Kim and Cannell, and Pallas and Pethica agree should be in the G–LE two-phase region, the film is clearly heterogeneous; there are circular dark domains, which can be associated with the gas phase, and a bright, continuous LE phase. If the monolayer is expanded, the bubbles of gas increase in size and the bright phase between them becomes increasingly thinner until the structure resembles a two-dimensional foam (see Section VII). When the expansion is stopped, the LE phase very slowly reorganizes into droplets.

If the initial two-phase system is compressed instead of expanded, the gas bubbles decrease in size and eventually disappear at a sharply defined molecular area, A_{LE}, leaving a homogeneous, uniformly bright field. The transition is reversible; circular droplets of gas reappear when the film is expanded to areas slightly larger than A_{LE}. The bright field persists when the compression is continued to areas smaller than A_{LE} until, abruptly at an area A'_{LE}, small dark spots become evident. These spots, the first traces of the LC phase, grow into circular domains as the compression is continued. Here again, the transition is reversible: the dark spots appear and vanish at the same molecular area.

At first, the LC domains maintain their circular shapes on compression and form a roughly hexagonal array. (Complex, dendritic shapes are observed[45] during compression in the LE–LC transition at higher temperatures. These will be discussed in Section VII.) The growing LC domains eventually deform and coalesce. If the probe concentration is large ($\approx 1\%$), small droplets of the probe appear to separate out of the LE phase as the amount of this phase diminishes. This effect is not seen at low probe concentrations. If a monolayer that consists almost entirely of LC phase is expanded, bright circular domains of the LE phase immediately appear within the dark regions, demonstrating that the probe is at least slightly soluble in the LC phase.

Further compression leads to a collapse of the film. The point of collapse is marked by the appearance throughout the monolayer of small, brightly fluorescent, faceted crystallites.

Coexistence curves can be traced out by determining the limits of the two-phase regions in isotherm measurements at different temperatures. Alternatively, molecular areas at coexistence can also be determined by observing the temperature at which a phase boundary is crossed in measurements at constant area. Supercooling is observed, however, and measurements in which a phase is lost upon heating are more reproducible than those in which a phase appears on cooling.

Direct measurements of A_{LC}, the molecular area that corresponds to the LC side of the LE–LC coexistence region, cannot be made unequivocally because of the proximity to collapse and because of the steepness of the coexistence curve. It can be determined, however, by application of the "lever rule." If a monolayer is prepared at a molecular area A located within the LE–LC two-phase region, then the mole fraction of the system that is in the LE phase is given by

$$x = (A - A_{LC})/(A'_{LE} - A_{LC}) \tag{8}$$

The fraction of the fluorescence microscope field that is bright corresponds to the "area fraction" Φ of the LE phase:

$$\Phi = xA'_{LE}/A \tag{9}$$

The area fraction, which can be determined by computer analysis of the microscope images,[44,46] is a linear function of the monolayer density. If measurements of Φ are made at a number of values of the density within the two-phase region, the intercepts A_{LC}^{-1} and $A_{LE}'^{-1}$ can be determined by least squares analysis.

The phase boundaries for PDA determined by fluorescence microscopy by Moore et al.[43] are plotted in Fig. 11. Also shown in the figure are the molecular areas A'_{LE} that correspond to the beginning of the LE–LC plateau determined by them from π–A isotherm measurements on pure PDA. Over the full temperature range of the measurements, there is a close coincidence between these areas and the values of A'_{LE} obtained in the fluorescence experiments.

There is no unequivocal method of estimating the value of A_{LC} from isotherm measurements. Pallas and Pethica[40] took A_{LC} to be the intersection of the LE–LC plateau and the extrapolation of the steep portion of the isotherm beyond the transition. The LC molecular area can be no smaller than that at which the kink is observed at high pressures. Values of this area at a number of temperatures are also plotted in Fig. 11, and they cluster around 20 Å^2, 1 to 2 Å^2 smaller than the values of A_{LC} determined from the fluorescence measurements.

Fluorescence studies at monolayer densities in the G–LE region show the presence of two phases even at temperatures close to 40°C, in agreement with Pallas and Pethica's interpretation of their isotherm measurements. The

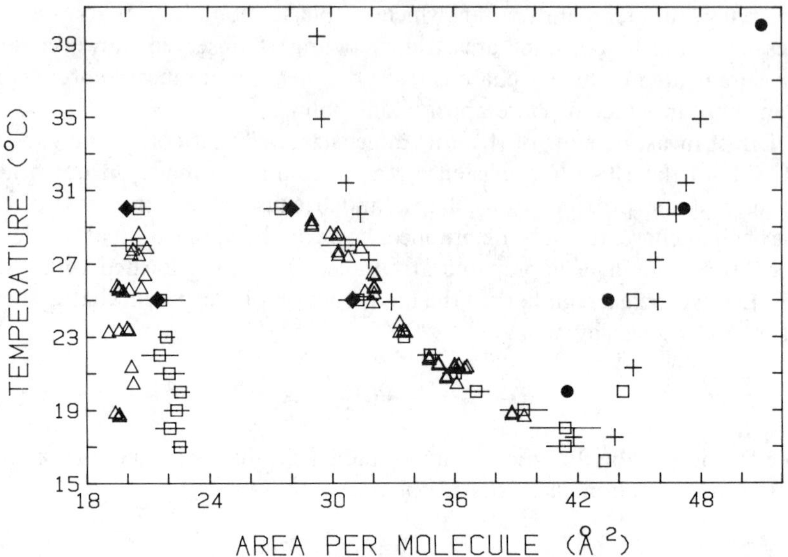

Figure 11. *T–A* diagram for PDA as determined from fluorescence and isotherm measurements, Moore et al.[43] □ fluorescence, △ isotherm; ● Pallas and Pethica;[13] + Harkins and Boyd;[8] ◆ Pallas and Pethica.[40] The fluorescence points with error bars have been obtained by the lever rule, while the others have been determined by observing the disappearance of a phase on heating. The isotherm points on the LC side of the LE–LC transition represent the location of the high-pressure kink in the isotherm.

values of A_{LE} and A'_{LE} obtained from isotherms are included in Fig. 11. The fluorescence measurements agree well with the Pallas and Pethica[13,14] and Harkins and Boyd[8] determinations of A_{LE}; there is reasonable agreement about the values[8,40] of A'_{LE}.

A triple point exists at which there is a equilibrium between the G, LE, and LC phases. At this point, $A_{LE} = A'_{LE}$. Harkins and Boyd[8] estimated that the coexistence curves meet at 17°C at an area of 44 Å^2 and Pallas and Pethica report the intersection at 38 ± 4 Å^2 and 17 ± 2°C. The triple-point temperature can be located directly in fluorescence measurements by determining the temperature at which a third phase appears when the temperature of a two-phase system is lowered at a fixed area.

The succession of images observed in an experiment in which a PDA monolayer in the LE–LC region is cooled just below the triple point is shown in Fig. 10b. The gas phase nucleates on the LC phase and is easily distinguished from it because the gas bubbles are markedly noncircular and highly compressible. If the temperature is now held constant, the area of the LE phase diminishes and the LC and G phases grow. The transformation rate becomes increasingly slow, however, and some LE phase persists even after two days. The highest temperature at which this three-phase region can be observed in the fluorescence studies is 17°C, in good agreement with the triple-point temperatures determined from isotherms.

The fluorescence and isotherm measurements are consistent with the generalized phase diagram for fatty acids shown schematically in Fig. 12. Presumably the G–LE coexistence region ends in a critical point but none of the measurements that have been carried out can fix T_c and define the asymptotic shape of the coexistence curve with any precision. The measurements show a clear narrowing of the range of LE–LC coexistence, which suggests that this may end at some sort of critical point as well, but here again experiments do not yet provide much direct information.

Suresh et al.[45] have argued that the formation of dendritic structures during compression of tetradecanoic acid monolayers at the LE–LC transition is related to the vanishing of the line tension between the phases with the approach to a critical point. They believe that for tetradecanoic acid T_c lies at about 31°C, which, if we apply the rule of thumb that there is a 10 K change in the phase diagram per CH_2, would place an LE–LC critical point for PDA at ~40°C. If the LC and LE phases do not have the same symmetry, e.g., if the LC phase is an ordered solid, then the coexistence region cannot

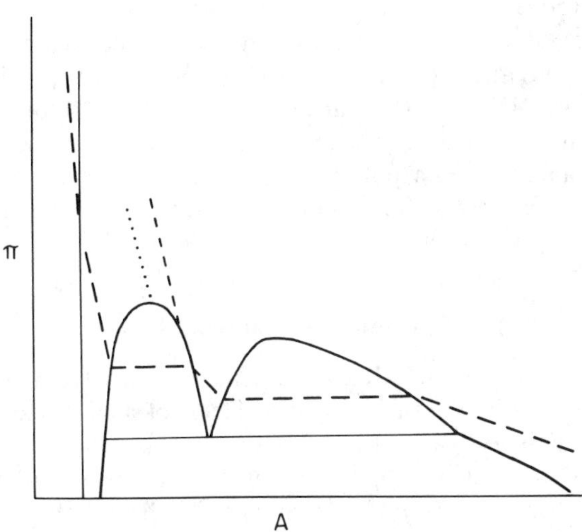

Figure 12. Generalized *T–A* phase diagram for fatty acids. A typical π–A isotherm is indicated by the dashed line. Two possible intersections of the line of second-order transitions with the LE–LC coexistence curve are shown. In the case of the dotted line, the intersection is a tricritical point.

end at an ordinary critical point without the intervention of a second-order transition. A line of second-order transitions can intersect the LE–LC coexistence curve below T_c, as shown in Fig. 12, in which case the critical point is an ordinary critical point. Many such phase diagrams are known. If the second-order line intersects the top of the curve, then the coexistence curve terminates at a tricritical point; this is the way in which the ^3He–^4He coexistence curve ends.[47]

Dervichian[48] observed that in plots of the area at which the LE–LC plateau in fatty acids begins against the temperature, the points for low-temperature isotherms appear to fall on a straight line while those at high temperature fall on another, steeper line; the lines intersect at an area of about 30 Å^2. He suggested that at this intersection the transition changed from first-order to second-order. Albrecht et al.[49] have proposed that the LE–LC region in phospholipids ends at a tricritical point, but this termination, which can be thought of as a rather special case of the intersection of the first- and second-order transitions, is unlikely to be general.

A second-order transition would produce no obvious change in fluorescence image, so the close correspondence found by Moore et al.[44] between the position of the abrupt change in slope in their isotherms and the start of the two-phase region in the fluorescence observations, rules out Dervichian's interpretation of the temperature dependence of the isotherms. If a second-order line intersects the coexistence curve, it does so at a temperature above 31°C and an area smaller than 28 Å^2.

Bercegol, et al.[23] have recently discovered an anisotropic solid phase that coexists with the LE phase in *pure* 12-NBD stearic acid. (At room temperature, the labeled acid, unlike stearic acid, exhibits an LE–LC transition.) Fluorescence microscope studies show that the phase forms needle-like domains, which polarization studies demonstrate are optically anisotropic. The corresponding dipole moment is perpendicular to the long axis.

B. Structures of Condensed Phases

As is often the case, systems that are amenable to study by one technique are difficult to study by another. The monolayers of single-chain amphiphiles that have been investigated by diffraction techniques are composed of relatively long molecules, which do not exhibit an LE phase at room temperature. Moreover, a number of the studies have been carried out on subphases containing divalent metal ions because these are often incorporated into Langmuir–Blodgett films. Such ions react chemically with carboxyl groups and cause a marked, strongly pH-dependent change in the shape of isotherms.[50] The relation between the measurements discussed in the previous section and some of the diffraction studies is therefore not as clear-cut as one would wish.

1. *Reflectivity Measurements*

X-ray reflectivity studies have been performed on eicosanoic (arachidic) acid,[25] octadecanoic acid,[51] and lysine-N-pentadecanoamide

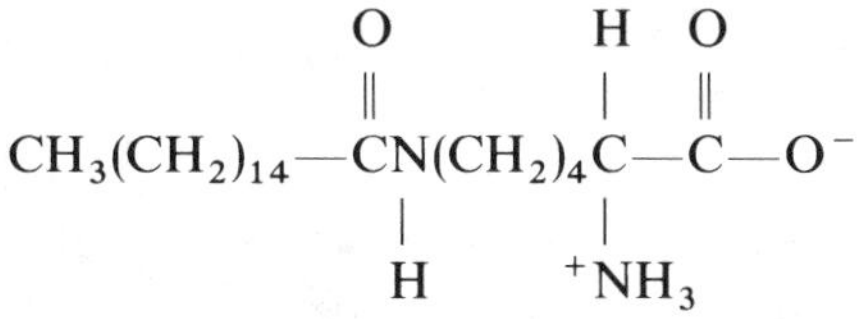

$$CH_3(CH_2)_{14}\!-\!\overset{\displaystyle O}{\overset{\|}{C}}\underset{\displaystyle H}{\underset{|}{N}}(CH_2)_4\overset{\displaystyle H}{\overset{|}{C}}\!-\!\overset{\displaystyle O}{\overset{\|}{C}}\!-\!O^-\quad (\text{with } {}^+NH_3 \text{ on the } \alpha\text{-carbon})$$

[palmitoyl-lysine];[52,53] docosanoic acid has been studied both with x-rays[27] and neutrons,[28] and preliminary neutron reflectivity measurements have been carried out on the ester

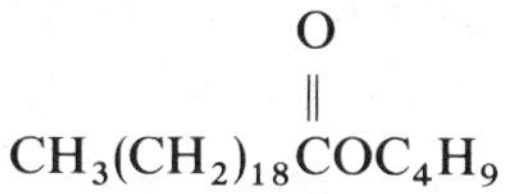

$$CH_3(CH_2)_{18}\overset{\displaystyle O}{\overset{\|}{C}}OC_4H_9$$

[*n*-butyl eicosanoate].[54] Most of the experiments were performed at pressures at which the monolayer density is very nearly close-packed.

The x-ray reflectivity of docosanoic acid monolayers spread on solutions of $LaCl_2$ and CsCl at pH 6.5, and at $2.75 \leqslant \text{pH} \leqslant 7.0$ on solutions of $CdCl_2$ was investigated by Grundy et al.[27] They analyzed their measurements in terms of a model in which there are two layers: (1) the hydrocarbon chain, and (2) the carboxyl group and the metal ion. There are eight adjustable parameters: the thicknesses of the layers; their scattering densities; three standard deviations to describe the roughness at boundaries between air/hydrocarbon, hydrocarbon/(COO + metal), and (COO + metal)/water; and the background.

In a parallel neutron reflectivity experiment,[28] measurements on fully deuterated docosanoic acid were performed as a function of pH on $CdCl_2$ solutions and on a metal-free substrate. Because the neutron scattering-length density of the head group and metal ion is little different from that of the 1:1 H_2O/D_2O mixture employed as a subphase, an analysis in terms of a single layer was possible. The precision of the neutron data was insufficient to allow the roughness parameters to be treated as adjustable and these were taken from the x-ray experiments, leaving only the thickness of the hydrocarbon-chain layer and its scattering-length density to be determined from the fit.

The neutron and x-ray reflectivity analyses are resonably consistent and indicate that the chain is fully extended and perpendicular to the surface at surface pressures above 16 mN m^{-1}. In measurements at low pH, where there are no Cd^{2+} ions associated with the monolayer, there is evidence of a chain tilt of perhaps 30° when the monolayer is expanded to twice the close-packed area. At this area, the data could be fit only by assuming that the monolayer is heterogeneous, with islands of a dense phase of dimension larger than 0.2 mm. Such heterogeneity has been observed in fluorescence studies of octadecanoic acid.[19]

Monolayers of lead octadecanoate at a molecular area of 19 Å^2 have been

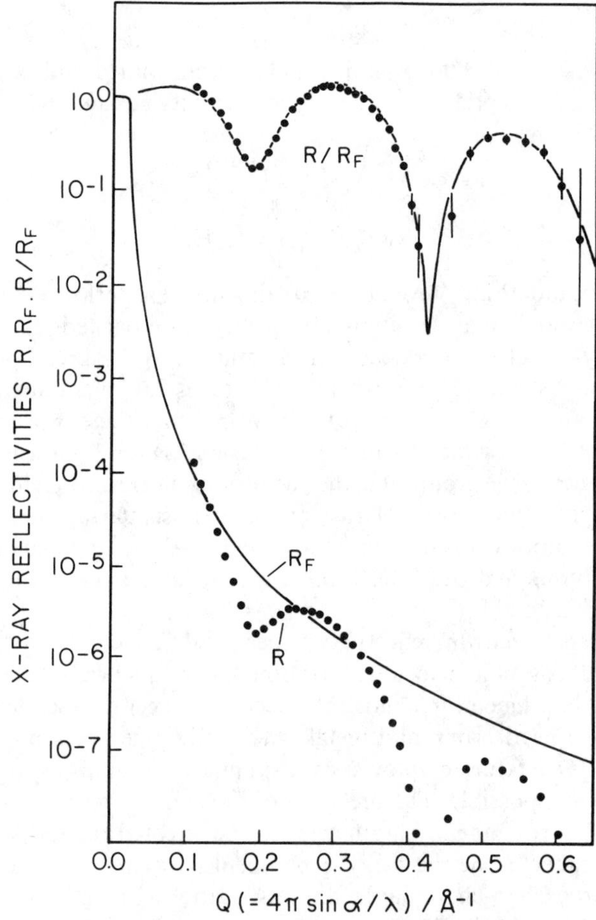

Figure 13. Reflectivity measurements on arachidic acid at $\pi = 30\,\text{mN m}^{-1}$. The subphase was 0.05 M $CaCl_2$ at pH 5.5. The smooth curve is the Fresnel reflectivity, R_F, which would be obtained if the interface were infinitely sharp. After Kjaer et al.[25]

examined by Bosio et al.[51] The monolayer was formed by depositing stearic acid on the surface of a 3×10^{-5} M solution of lead acetate buffered to a pH of 7.5. They give little detail about the slab model that has been used to interpret the measurements, but appear to have assumed values for the scattering densities and the length of the tail. The chains are found to be vertical. (Electron diffraction measurements on an LB monolayer of cadmium octadecanoate[35] show a dominant tilt angle of 8°, too small to be detected in the Langmuir film measurements.) An unusual feature is the observation that the reflectivity increases with time over a period of 10 hours. This change is attributed to chemical equilibria involving $C_{17}COO^-$, $C_{17}COOPb^+$, and $(C_{17}COO)_2Pb$.

The x-ray measurements that have been described were carried out with laboratory sources. The limitation on the precision with which the profile can be derived from the data is the dynamic range of the intensity measurement, which is determined by the ability to detect the signal above background. There is considerable advantage, then, to perform reflectivity measurements with synchrotron radiation.

Kjaer et al.[25] employed a synchrotron source in their investigation of arachidic acid. They were able to measure a reflectivity signal that was 10^{-8} of the direct beam intensity; for comparison, the measurements by Grundy et al.[27] were able to cover a range of only about 10^5. The measurements were carried out at a pressure of $30\,\mathrm{mN\,m^{-1}}$ ($A = 20\,\text{Å}^2$) on a 0.05 M $CaCl_2$ solution at pH 5.5. Their data are shown in Fig. 13; they were fitted to the model shown in Fig. 14, with a single roughness parameter. The tail thickness, 24.6 Å, corresponds closely to an all-*trans* chain perpendicular to the surface. In contrast, NEXAFS studies[55] of a compressed Langmuir–Blodgett film of calcium arachidate on Si show that the chains are titled at an angle of 30° from the perpendicular, which suggests that a change in orientation occurs when the film is transferred to the support. On the other hand, the NEXAFS experiments show the chains in cadmium arachidate to lie within 15° of the surface normal, while the chains in arachidic acid are disordered.

2. *In-Plane Scattering*

Several studies of the in-plane scattering of single-chain amphiphile monolayers have also been reported. Dutta et al.[56] measured the scattering of 1.24-Å synchrotron radiation from monolayers of octadecanoic acid on an 8×10^{-5} M lead acetate substrate. (They do not report intensity changes like those observed by Bosio et al.[51]) Measurements were carried out at "room temperature" for pressures in the range 18–25 $\mathrm{mN\,m^{-1}}$. They observed a single diffraction peak in all samples at $Q = 1.60\,\text{Å}^{-1}$, which corresponds to a d spacing of 3.93 Å. To determine the half-width of the peak, they fitted it both with a three-dimensional finite-size (Gaussian) structure factor and

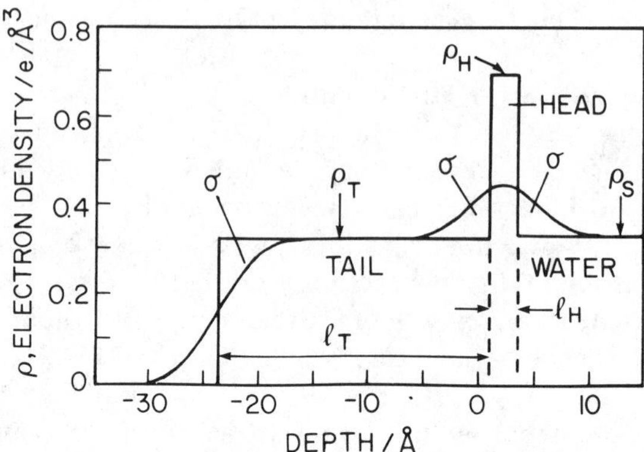

Figure 14. Interface profile for arachidic acid determined from the reflectivity data shown in Fig. 13: $\rho_T = 0.326\,Å^{-3}$, $\rho_H = 0.698\,Å^{-3}$, $\rho_S = 0.334\,Å^{-3}$, $\sigma = 3.0\,Å$, $l_T = 24.6\,Å$. $l_H = 2.47\,Å$. After Kjaer et al.[25]

with a structure factor appropriate to a finite two-dimensional solid, finding 250 and 280 Å, respectively.

If the peak is taken to be the first peak of a triangular structure, the area of the unit cell is 17.8 Å^2. Since the isotherms at the measured pressure give an average area of 19 Å^2 per chain, the unit cell is consistent with an array of vertical chains only if it is assumed that the monolayer does not uniformly cover the surface. The existence of such heterogeneity in compressed octadecanoic acid monolayers may be supported by dark-field electron micrographs of Langmuir–Blodgett films,[57] which show the presence of small irregular holes.

Room temperature measurements by Kjaer et al.[25] of arachidic acid on water at $\pi = 27\,\text{mN}\,\text{m}^{-1}$ also show a single diffraction peak. It lies at 1.52 Å^{-1}, which corresponds to a d spacing of 4.13 Å. The correlation length as determined from the half-width at half-maximum is said to be about 90 Å, but the procedure used to fit the peak is not described.

The most extensive diffraction study yet undertaken is that of the alcohol 1-heneicosanol ($C_{21}H_{43}OH$) by Barton et al.[58] Pressure-area isotherms for this alcohol have an unusual shape that changes irregularly with temperature, as shown in Fig. 15. Diffraction experiments were carried out as a function of pressure at several temperatures between 15 and 30°C and as a function of temperature (7–16°C) at a constant pressure of $30\,\text{mN}\,\text{m}^{-1}$. The variation of the diffraction peak position with pressure at two temperatures is shown in Fig. 16.

At high pressure the peak is located at about 1.51 Å^{-1}, a value consistent with the packing of the chains in a hexagonal lattice. There is a sharp change

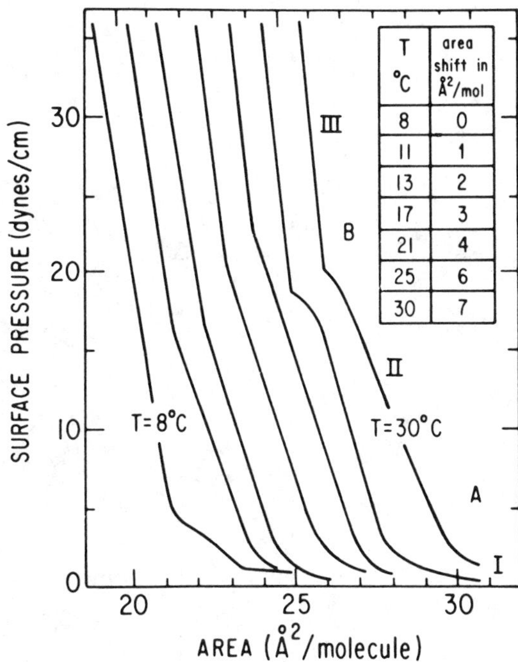

Figure 15. Pressure-area isotherms for 1-heneicosanol from 8°C to 30°C. For clarity, successive curves have been displaced by 1 Å^2 along the abscissa. From Barton et al.[58]

to smaller wave number as the pressure is lowered, the fall occurring at the kink in the isotherms. A parallel decrease is observed in the intensity of scattering. The translational correlation length, which was obtained by fitting the peak shapes to a Lorentzian, is roughly 50 Å and remains essentially constant.

The packing of hydrocarbon chains deduced from the diffraction measurements on single-chain amphiphiles is consistent with that determined in LB films of cadmium stearate by electron diffraction.[35] The diffraction pattern of the LB films has a six-fold symmetry consistent with a fully hexagonal symmetric structure. The d spacing is 4.20 ± 0.10 Å, essentially identical to that found in Langmuir monolayers. Scans of the diffracted intensity along a radial direction through the center of the Bragg spots show a broad intensity distribution that is a characteristic of translational disorder; the correlation length is only about 10 lattice spacings. In contrast, the narrow angular distribution of the spot intensity demonstrates that there is a high degree of bond orientational order, extending over more than 10^6 unit cells.

The behavior of the correlation length and intensity of the 1-heneicosanol monolayer at the kink in the isotherm is consistent with a first-order phase

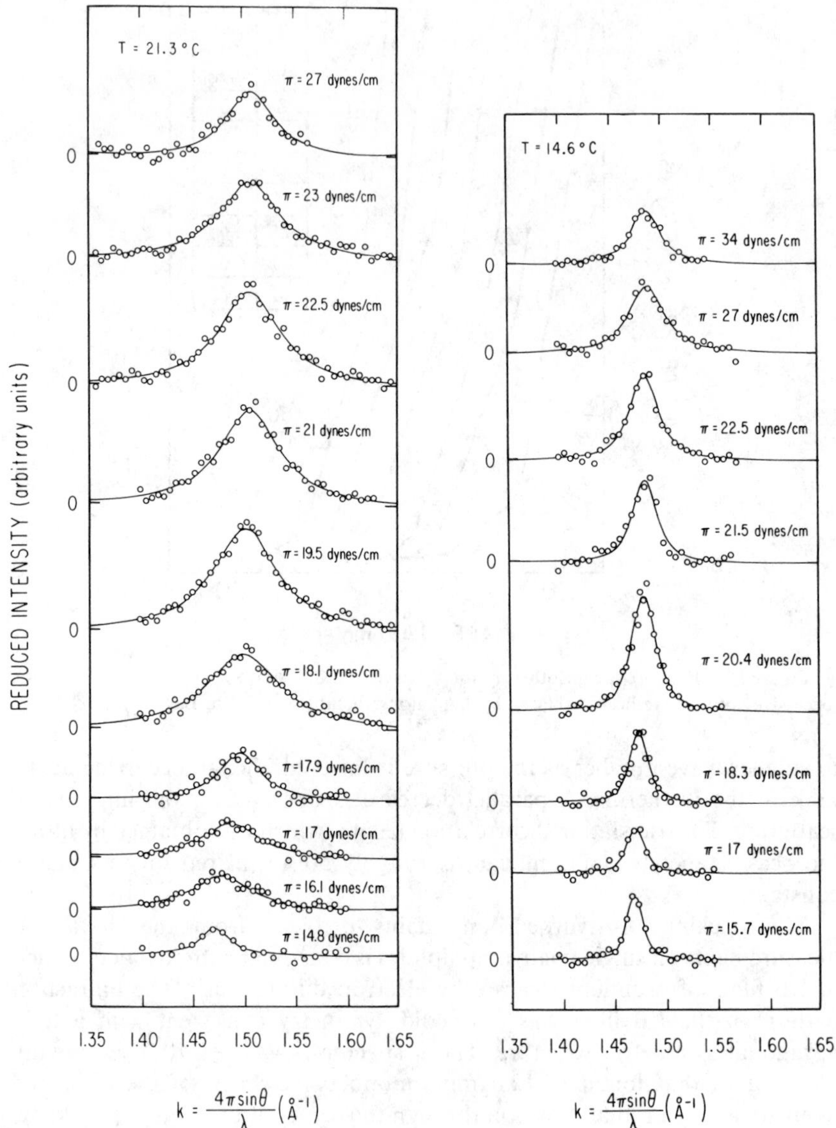

Figure 16. In-plane diffraction measurements for 1-heneicosanol as a function of pressure at 21.3°C and 14.6°C. From Barton et al.[58]

transition in which there is coexistence of a hexagonal solid with a uniformly scattering fluid phase. On the other hand, this interpretation would require that the peak position not change, which is at odds with the observations. Barton et al. suggest that the kink corresponds to the point at which increasing compression removes the last of the gauche conformations in the tails of the amphiphile. They argue that *gauche* conformations are most easily accommodated near the top of the monolayer, so that the packing near the head is not immediately affected by the continuous transition, and the width of the diffraction peak should therefore be unchanged. In a note added in proof, however, they report the results of molecular dynamics simulations[59] of PDA that show the conformations of *gauche* conformations to be smaller at the middle of the chain than at either end.

An unusual feature of the heneicosanol measurements was a second scattering peak, which appeared to arise at high pressure at temperatures below 16°C. It was attributed to a weakly first-order phase transition analogous to the rotator II-to-rotator I transition in lamellar crystalline n-alkanes with $n = 23, 25$. In the rotator II phase, rapid reorientation of the chain around its axis leads to a pseudohexagonal structure. When the chains can no longer reorient, the symmetry of the structure is reduced to a uniaxially distorted hexagonal structure.

In a subsequent paper, Lin et al.[60] report that the pseudohexagonal structure is not an equilibrium phase. They put forward the hypothesis that it arises because of the uniaxial compression of the film at the start of the experiments. Measurements made at temperatures between 10 and 20°C show that three peaks can be observed immediately after compression, one at 1.52 $Å^{-1}$, associated with the hexagonal phase, and two associated with the pseudohexagonal phase, a large peak at 1.485 $Å^{-1}$ and a very small one at 1.6 $Å^{-1}$. The 1.52 $Å^{-1}$ peak grows with time while the other peaks disappear. A fit of the time dependence of the areas of the 1.52 $Å^{-1}$ and 1.485 $Å^{-1}$ peaks to exponential functions gives a characteristic time for the decay of the unstable phase that varies from 2600 s at 10°C to 900 s at 20°C.

The monolayers of palmitoyl lysine studied by Grayer Wolf et al.[52] are a special case. A network of hydrogen bonds is formed at the interface by the glycine moieties in the amino acid lysine, and as a result the translational correlation length is 500 ± 100 Å. The in-plane scattering shows two peaks, which allows the two-dimensional powder pattern to be matched with a small number of structures for the monolayer. When the diffraction studies are combined with reflectivity measurements and interpreted in the light of plausible chain conformations, a choice can be made from among the structures. The chains are tilted and in an all *trans* conformation, and an essential feature of the structure is the hydrogen bonding between the secondary amide groups in the chains.

3. *The LE Phase*

The reflectivity and diffraction measurements that have just been discussed were made at temperatures below the range of stability of the LE phase. Some information about the structure of the LE phase has been obtained from other experiments. Second harmonic generation measurements[29] on PDA have been found to be sensitive to the orientation of the head group. If the chain is assumed to be rigid, the experiments can be interpreted in terms of chain tilt. At 25°C, the tilt angle of the chain with respect to the surface normal is nearly 90° at a molecular area of 45 $Å^2$ but falls sharply with compression to about 45° at 32 $Å^2$, where the LE–LC transition begins. The angle then changes linearly with area, reaching 30° at the end of the transition region. Measurements made between 20 and 30°C suggest that the tilt angle at the start of the LE–LC transition is independent of the temperature.

A strong CH_2 symmetric stretch signal at 2850 cm^{-1} and a broad background in the range 2930–2880 cm^{-1}, also attributable to the CH_2 stretches, appear in SFG spectra measured in the LE region of PDA.[30] The intensity of these features increases when the monolayer is expanded. If the chains were straight, there would be near inversion symmetry along them and the CH_2 modes could not be observed. The presence of the modes therefore is indicative of an increase in the fraction of *gauche* conformations along the chain as the area increases.

IV. PHOSPHOLIPID MONOLAYERS

A. Phase Diagram

Isotherms of phospholipids have the same features as those of fatty acids, but the terminology applied to the transitions and phases is somewhat different.[49] The LE–LC transition, which begins at a pressure denoted by π_c, is called the *main* transition and the phases involved are called the *fluid* (LE) and *gel* (LC) phases. The transition at higher density, which is marked by a kink in the isotherm, occurs at a pressure labeled π_s. A transition between the fluid phase and a gas is also observed, but this has not been the subject of recent investigations. As in the case of fatty acids, a corresponding states principle applies,[49] but for phospholipids the temperatures at which characteristic features of the diagram appear depend in a regular fashion on the chain length of the tail *and* on the nature of the head group.

There have been many interpretations of the failure to observe a horizontal slope in the isotherms at the main transition.[34,61–63] On the other hand, Pallas and Pethica[40] argue that the isotherm in the phopholipid DPPC *is* horizontal as long as the amphiphile is sufficiently pure and great care is taken in controlling the experimental conditions. This has not been confirmed by other investigators.

The first fluorescence microscope investigations of monolayers[16,17] were carried out on phospholipids. The progression of images on compression along an isotherm is identical to that observed in PDA, and it is clear from them that the main transition is first-order.

B. Structures of Condensed Phases

Möhwald, Als-Nielsen, and their co-workers have performed x-ray studies on DMPA[25,26] and on DPPC.[64,65] Their results have been summarized by Möhwald[65] and by Helm et al.[46] In-plane diffraction has been investigated for DMPA, for which $\pi_c = 10\,\mathrm{mN\,m^{-1}}$, $A_c = 80$ Å^2/molecule and $\pi_s = 40\,\mathrm{mN\,m^{-1}}$, $A_s \approx 40$ Å^2/molecule. Measurements were carried out at room temperature at pressures from about 12 to $60\,\mathrm{mN\,m^{-1}}$ (Fig. 17). No peak could be detected at the lowest pressure; a peak barely above the noise was observed at about $16\,\mathrm{mN\,m^{-1}}$, which corresponds to a molecular area of about 52 Å^2.

Below π_s, the peaks are weak and broad; they sharpen and intensify at higher pressure. The variation of the lattice spacing with pressure and the correlation range with molecular area are shown in Fig. 18. The lattice spacing, which is about 4 Å, decreases essentially linearly with pressure, although there may be a discontinuity or a slight change in slope at π_s. In contrast, the correlation range remains constant at about 10 lattice spacings throughout the main transition and then rises at π_s.

Electron diffraction studies[66] of Langmuir–Blodgett films of DMPA and fluorescence microscope studies on the monolayer[17,65] are helpful in interpreting the x-ray measurements. If one assumes that the gel and solid phases in the monolayer have the hexagonal structure found in the LB films, the lattice spacings correspond to the (1,0) planes and are consistent with a unit cell containing one alkyl chain. If the chains were tilted with respect to the surface, the hexagonal symmetry would be lost and the diffraction peak would be split; no splitting is observed.

A plot of the reflectivity of a DMPA monolayer at a pressure above π_s is presented without comment by Kjaer et al.[25] If one takes the position of the minimum from the graph and applies Eq. (5), it appears that the chains are vertical. The same group has performed reflectivity measurements on DPPC,[64] which they analyze in terms of a two-layer model in which there are five parameters (2 scattering-length densities, 2 lengths, and a roughness parameter). They find that at high pressure the chains are uniformly tilted 30° from the surface normal. The tilting of the chains in DPPC has also been demonstrated by fluorescence polarization.[21]

The sharp hexagonal electron diffraction pattern seen in the gel phase in

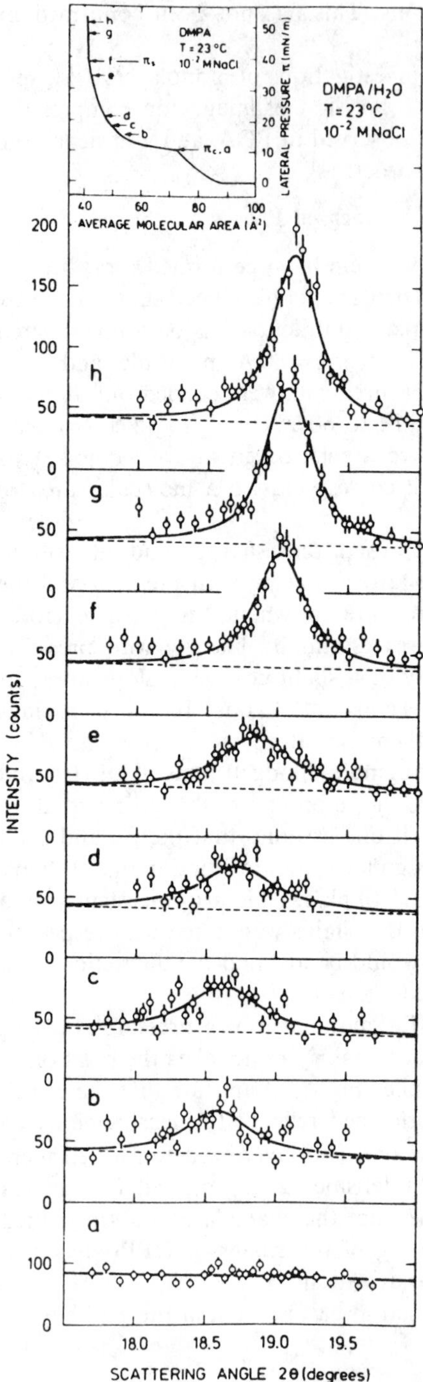

Figure 17. In-plane diffraction measurements as a function of surface pressure for DMPA. The inset shows the points along the isotherm at which the diffraction studies were made. From Helm et al.[46]

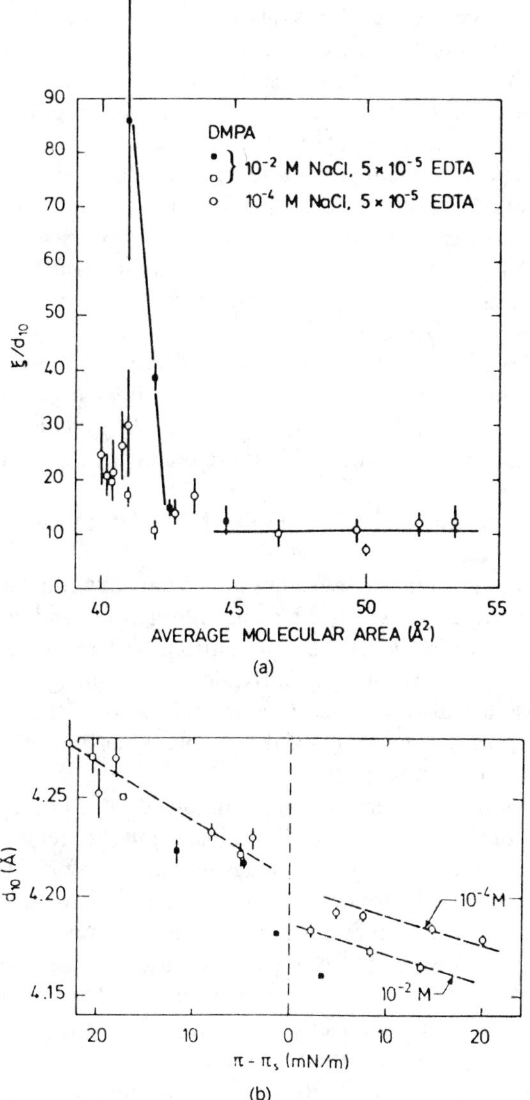

Figure 18. (a) Correlation range ξ in units of the lattice spacing d_{10} as a function of molecular area, and (b) lattice spacing as a function of pressure for DMPA. Derived from the diffraction data shown in Fig. 17. From Helm et al.[46] Figures 17 and 18 reproduced by copyright permission of the Rockefeller University Press.

LB films is evidence for long-range *orientational* order, which one assumes was present in the monolayer from which the LB film was prepared. On the other hand, the intrinsic linewidths of the x-ray peaks show that the *translational* order in the monolayer is short-range, roughly 40 Å.

Diffraction from the monolayer was observed only in a region where the isotherm is steep, so the failure to observe a constant lattice spacing during the main transition can be attributed to the compressibility of the gel phase. None of the measurements shows coexistence of two phases at the gel–solid transition. A sharp jump in correlation length and intensity is consistent with a first-order transition, but there is no abrupt change in lattice parameter and no evidence of hysteresis.

Although no diffraction peaks can be observed below π_c in DMPA, reflectivity measurements can be carried out. It is observed[25] that at a molecular area of 87 ± 7 Å^2 the film thickness is 40% smaller than it is above π_s, which corresponds to a tilt of 50° if the chains in the solid are vertical.

V. SIMULATIONS OF MONOLAYERS

Several research groups have begun molecular dynamics simulations of Langmuir monolayers in which the chain interactions and chain conformations are accurately modeled. (A few simulations with more simplified models have already been carried out.[67]) Full-scale simulations in which the details of the subphase are also included are complex, and in the two papers that have already appeared the scope of the problem has been reduced by using a structureless planar substrate.

The simulation by Bareman et al.[68] involved 90 chains, each of which was made up of 20 identical pseudoatoms meant to represent methylene groups and the terminal methyl. Changes in chain conformations, including bond-bending and chain torsional motions, were allowed, with potentials known to provide good representations of short-chain alkanes. Interchain interactions were computed from pairwise additive Lennard-Jones 12–6 potentials. A 9–3 potential, which is appropriate for the interaction of a L–J particle above the surface of a bulk medium, was taken for the interaction of each methylene with the surface, with a well depth 5 times deeper than that for the CH_2—CH_2 interaction. The simulations were carried out with periodic boundary conditions for three fixed areas, 21, 26, and 35 Å^2 per chain.

Two-dimensional cross-sections of instantaneous equilibrium MD configurations for each of the areas are shown in Fig. 19. In all three configurations, the systems appear to exist in two phases, one with disordered chains and one with more ordered chains. In no case are the chains perpendicular to the surface. Calculations of the density distribution of methylene groups along the chain and the distribution of average chain tilt

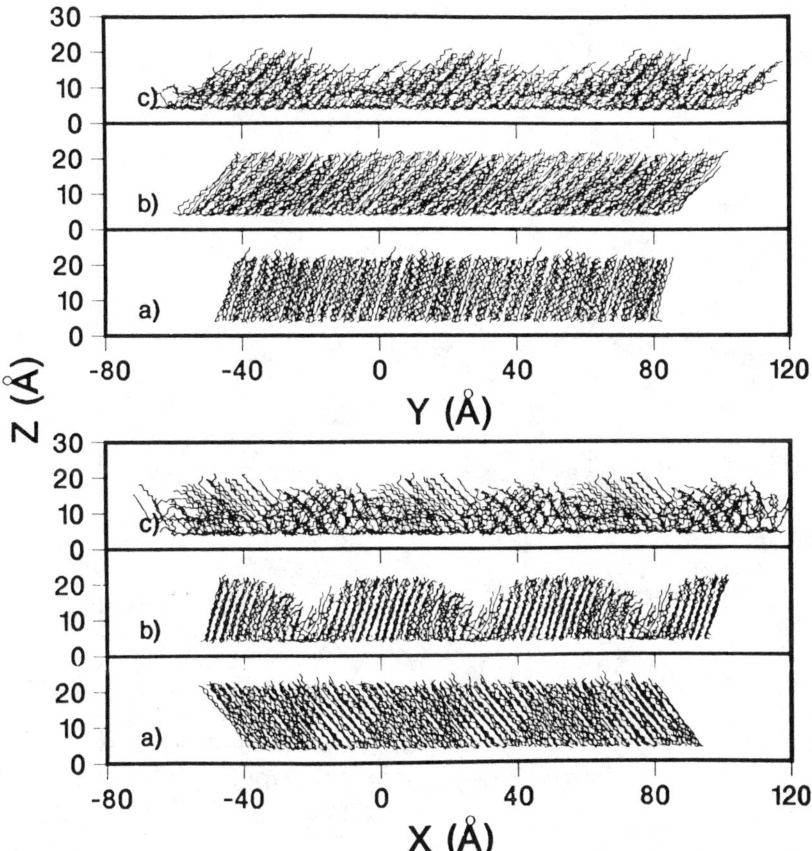

Figure 19. Orthogonal two-dimensional cross-sections of instantaneous MD configurations at three different areas from the simulation by Bareman et al.[68] (a) 21 Å^2, (b) 25 Å^2, (c) 35 Å^2.

angles show (Fig. 20) that a 21 Å^2 the density is essentially uniform along the chain and that the tilt angle is sharply peaked at 40° with respect to the surface normal.

Both distributions change markedly with area. At 35 Å^2 the peak in the angular distribution has shifted to 60° and the distribution has broadened to include angles greater than 90°, indicating that some chains are folded back to the surface. This is reflected as well in the chain density, which is now peaked close to the surface. The density profiles also show a layering of close-packed methylenes normal to the surface. Calculations by Bareman et al. suggest that the concentration of *gauche* conformers ranges from about one to two per chain as A varies from 21 to 35 Å^2; they are more likely to

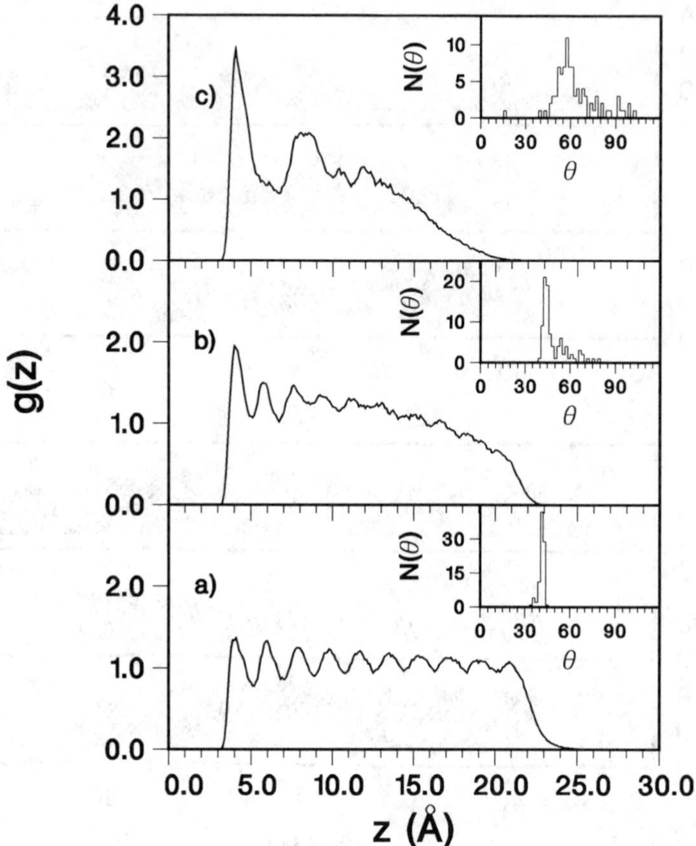

Figure 20. Density and chain-angle distributions at three different areas from the simulation by Bareman et al.[68]

be found near the ends than in the middle of the chain. For a chain of 20 carbons, an average of two *gauche* conformers per chain contributes about 5R to the entropy, a rather large effect in comparison to the energy cost, which is the order of RT/2 per conformer.

Harris and Rice[59] have specifically attempted to model PDA in their simulations. The amphiphiles are made up of pseudoatoms representing one methyl, 13 methylenes, and a COOH head, each of which has appropriate potential parameters and mass. Like Bareman et al. they accounted for intramolecular interactions with bending and torsional potentials appropriate for hydrocarbons. A Lennard-Jones 12–6 potential was used for intermolecular interactions and for interactions between pseudoatoms on the same chain separated by more than three bonds, with parameters that reproduce the internal energies and crystal structures of small alkanes. The well depths for the various interactions were: methylene–methylene 70.4K; methyl–methyl 90.7 K; headgroup–headgroup 151 K; methylene–water 289 K; headgroup–water 1492 K. Simulations were carried out at 300, 350, and 400 K for an MD cell containing 100 molecules. Both constant-pressure and constant-area runs were performed.

For all the temperatures studied, the simulations lead to crystalline monolayers with strong orientational ordering of the chains in a distorted hexagonal structure even in the absence of an applied pressure. The chains are tilted at an angle of 30° to the surface normal. Most of the *gauche* defects are found near the ends of the chains, as observed in crystals of *n*-alkanes, but even at 400 K the average number of *gauche* conformations per chain is only about one.

The relevance to this simulation to real systems is uncertain because the phase diagram of this simplified model, which shows no expanded state even at 400 K, is much different from the Langmuir monolayers it was meant to simulate. Harris and Rice believe that the replacement of the atomic structure of the substrate by a uniform continuum may be the weakest point of the simplified models.

VI. AN OVERALL PICTURE OF THE MONOLAYER PHASE DIAGRAM AND STRUCTURES OF THE PHASES

The results of the thermodynamic and structural experiments on both the fatty acids and the phospholipids can be combined to provide a picture of the phase diagram and the nature of the monolayer phases.

The general phase diagram sketched in Fig. 12 seems to be a correct representation of the phases and phase boundaries in both phospholipids and fatty acids and alcohols. There can no longer be any doubt that both the G–LE and LE–LC transitions are first-order. This has been demonstrated directly by fluorescence microscopy and by other methods as diverse as second harmonic generation, surface potential, and x-ray diffraction. In the case of PDA, there is excellent agreement between boundaries determined by isotherm measurements and those located visually. While shape instabilities are observed in two-phase regions of the phase diagram (see Section VII), there is no evidence for the existence of *equilibrium* hexagonal and striped phases that have been predicted[69] from dipolar models for monolayers.

It seems likely that the G–LE region ends in a critical point, but this has not yet been demonstrated by direct experiment. The detailed shape of the G–LE coexistence curve is still unknown. Although the LE–LC coexistence region in PDA has been reasonably well mapped, there have not yet been successful investigations of critical behavior. Isotherm measurements on oleic (hexadecenoic) acid suggest[3a] that the LE–LC transition is no longer present at room temperature. Moore[70] has performed fluorescence studies on oleic acid that confirm the loss of the transition, so it is clear that there *is* a termination to LE–LC coexistence.

A third transition, that between the LC phase and another condensed phase, seems evident in many isotherm measurements and appears to be confirmed by the diffraction studies on DMPA and PDA. It has not been observed by fluorescence microscopy (perhaps because there is no contrast between the two phases). Evidence for transitions between condensed phases in compressed monolayers is also found in surface shear modulus measurements.[71] These lead to the counter-intuitive conclusions, however, that the high pressure transition in $C_{20}H_{41}OH$ is pressure-induced *melting*, that in the next member of the homologous series, $C_{21}H_{43}OH$, there is reentrant behavior in which melting is followed by freezing as the pressure is increased along the isotherm, and that in the C_{22} alcohol a kink in the isotherm is associated with a solid–solid transformation. Rasing et al.[22] have argued that what is thought to be a LC–S transition in PDA is a nonequilibrium effect caused by the overcompression of LC domains that repel each other.

The microscopic picture of the structure of the LE phase that emerges from the recent measurements is that of a fluid in which the chains have an average tilt angle with respect to the surface normal that decreases continuously as the film is compressed,reaching 45° at the start of the LE–LC transition in PDA and in DMPA. Compression of the LE phase also induces a continuous change from *gauche* to *trans* conformations in the hydrocarbon tail. As shown by FTIR,[33] the CH_2 stretching frequency in DPPC at a molecular area of 100 Å^2 corresponds to that found in bulk suspensions. It decreases linearly with decreasing area and in the LC phase reaches a value comparable to that in pure bulk DPPC.

The LC or gel phase has hexagonal or pseudohexagonal long-range orientational order but only very limited translational order, about 10 or 20 lattice spacings. The chains of the fatty acids and DMPA are vertical, but those of DPPC are tilted at an angle of 30°. (The SHG studies on PDA also suggest a 30° orientation for PDA at the LC side of the LE–LC coexistence region, but this is less certain.) The difference in chain orientation between DMPA and DPPC is very likely related to the head groups. We have already noted the differences in chain tilt between calcium and cadmium arachidates in LB films. Despite the fact that there are two chains per head group in DMPA and only one in PDA, the LC phases of these two amphiphiles seem little different, suggesting that the structure is determined primarily by chain interactions and not by the head groups. There has been no evidence of crystalline faces in the fluorescence studies; the LC phase is easily deformed and can be seen to flow.

The phase formed on compression of the LC phase is also hexagonal and is orientationally ordered. The translational correlation length, while longer than that in the LC phase, is only of the order of 30 lattice spacings. From packing considerations, it is necessary that the chains be in the fully extended all-*trans* conformation.

Despite the major efforts and successes of the past few years, our understanding of the nature of the condensed phases remains fragmentary and, necessarily, the driving forces for the phase transitions cannot be well understood. Short-range translational order and long-range orientational order are hallmarks of a two-dimensional hexatic phase,[72] and it has been suggested[73] that the solid phase in a phospholipid monolayer might be a hexatic glass. But orientational order in the presence of positional disorder can also occur in structures consisting of domains with different chain tilt angles.[35] Helm et al.[46] have discussed these two possibilities. They point out that in principle they can be distinguished by line shape analysis of the

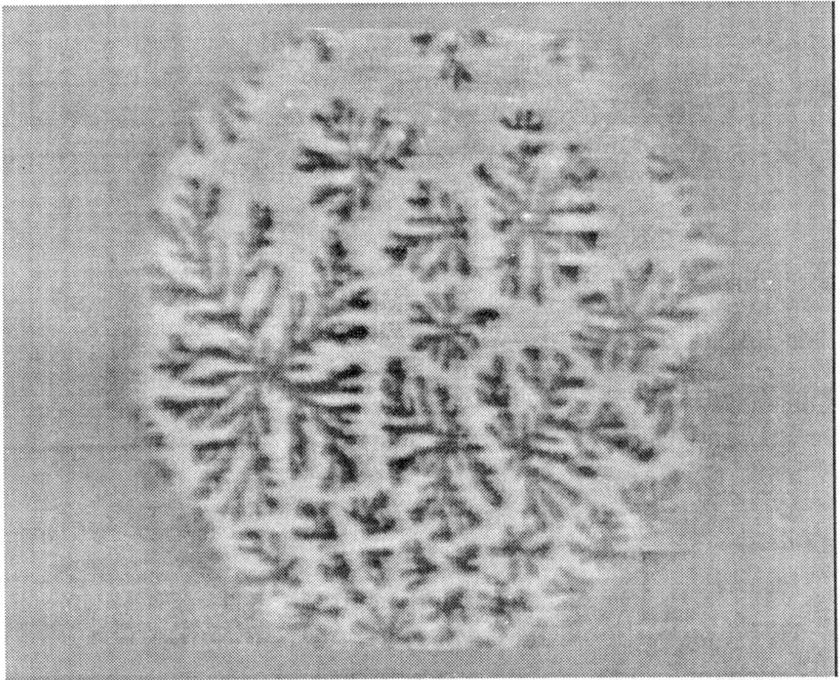

Figure 21. Dendritic structure observed immediately after formation of a PDA monolayer in the LE–LC two-phase region at 30.1°C and 22 Å^2.

diffraction peak, but this has not yet been done with sufficient confidence to be a definitive test.

Attempts are being made to develop phenomenological theories[74] of the monolayer phases, but the problem of dealing realistically with the many interactions has not been solved. Phase transitions in condensed monolayer phases have also been investigated with two-dimensional lattice models.[75,76] In such theories, the ordering processes are described in terms of order parameters whose precise physical significance need not be defined. They therefore do not lead to detailed physical descriptions of the monolayer phases.

VII. COMPLEX SHAPES OF DOMAINS

Although the focus of this chapter has been on the equilibrium properties of monolayer phases, we close with a mention of the complex shapes of domains that can arise as a new phase grows at the expense of another. Much of the research in this area has been the work of McConnell[77] and

Figure 22. Spiral structures observed in mixtures of DPPC with 2 mol% cholesterol. R. M. Weis and H. M. McConnell, unpublished.

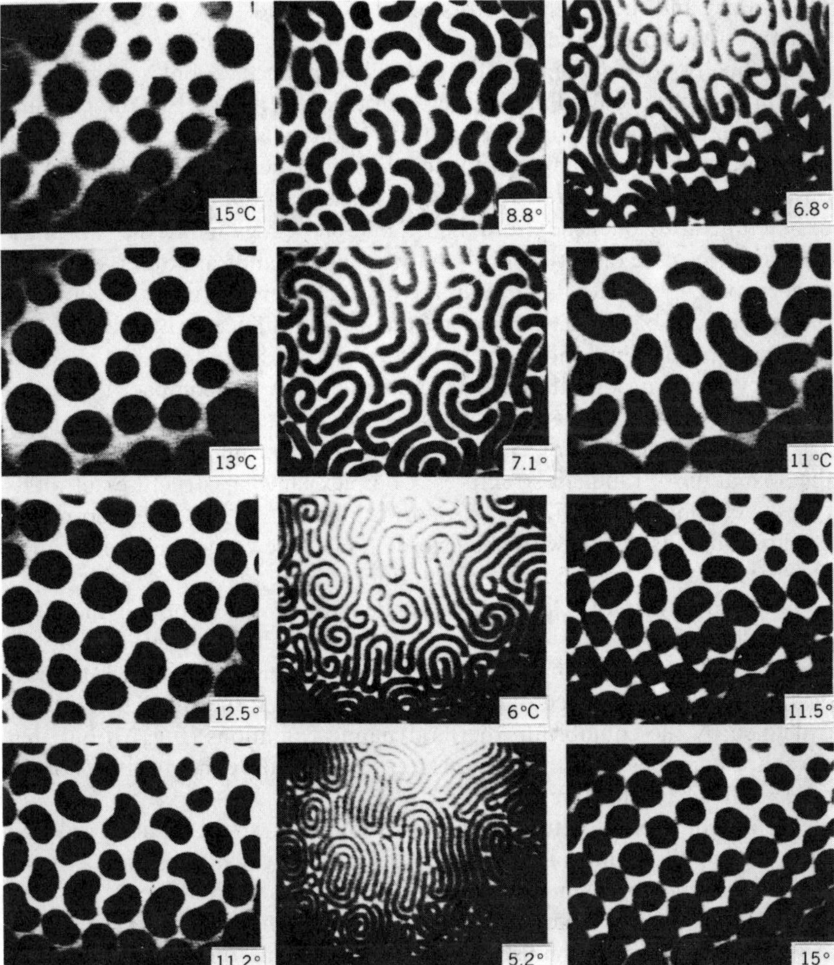

Figure 23.* Changes in lamellar structures with temperature in mixtures of DPPE with cholesterol. The sequence, during which the area remained constant at 54 Å^2 and the pH was 11.4, begins (upper left-hand corner) at 15°C. The monolayer is then cooled to 5.2°C and heated back to the starting temperature. It is seen that the lamellar thickness is a characteristic of each temperature but that the pattern is not reproduced. From Heckl and Möhwald.[83]

*Reprinted by permission of VCH Publishers, Inc., 220 East 23rd St., New York, N. Y., 10010 from: Heckl and Möhwald: Ber. Bunsenges. Physik. Chemie, 90, 1159 (1986), Figure 3.

Möhwald[65] and their collaborators on phospholipids, but complex domain structures have been observed in fatty acids[19,44,45] as well.

As already noted, when the LE phase is expanded, bubbles of gas grow and thin lamellae of fluid are left between them, forming a two-dimensional foam structure (Fig. 10a) in which the size of the cells and the distribution of the number of sides changes with time.[19] Dendritic structures (Fig. 21) are observed in the LE–LC transition region when the monolayer is compressed, or in temperature quenches from the LE one-phase region into the two-phase region. In the case of fatty acids,[45] only circular islands are observed at low temperature, even with rapid compression. The temperature threshold for the appearance of dendritic patterns is quite sharp. Such structures are also found in phospholipid monolayers.[78]

A key factor in formation of structures that are not compact is the line tension between the phases, which suppresses instabilities at an interface. The addition of cholesterol to a monolayer lowers the line tension and favors the formation of ramified structures. Complex lamellar structures of many types, including spirals (Fig. 22), are formed in the LE–LC transition region of phospholipids doped with cholesterol.[79,80] If the amphiphiles are chiral, the spirals have a unique sense of rotation.

The tendency to form lamellar structures can be understood as the result of dipole–dipole repulsions between the uniaxially ordered amphiphile molecules; theoretical analyses of the pattern formation based on this principle have been carried out.[81] It appears[82,83] that the lamellar thickness is defined by the equilibrium thermodynamic conditions (Fig. 23) but that the shape and density of the patterns are determined by the dynamics of the process by which they are formed.

Are any of these structures typical of those that would be observed in a pure amphiphile? The role played by the probe, which is essential to the fluorescence method, is not completely clear. It has been argued[84] that the formation of dendritic structures in phospholipids is the result of "constitutional supercooling," a mechanism that depends on the differential solubility of an impurity between two phases. This may not be the case: similar patterns have been observed in LB films by surface-plasmon microscopy,[85] for which no probe is added. The foam structures at the LE–G transition have also been attributed by some to the presence of the probe, but foams have also been observed[86] in monolayers composed solely of a labeled amphiphile.

Acknowledgments

The preparation of this chapter, and some of the work described in it, were supported by the National Science Foundation under grants CHE86-04038 and INT84-13698. I have benefitted from discussions with Silvère Akamatsu,

Bill Gelbart, Bill Hamilton, Mike Klein, Brian Moore, Ian Pegg, Francis Rondelez, and Greg Smith. Originals of figures were kindly provided by Mike Klein, Harden McConnell, Helmuth Möhwald, and Wolfgang Knoll. Faith Flagg, Daniella Fritter, and Brian Moore assisted me in preparing other figures and the manuscript.

References

1. Lord Rayleigh, *Phil. Mag.* **48**, 331 (1899).
2. I. Langmuir, *J. Am. Chem. Soc.* **39**, 1848 (1917) discusses the independent recognition of this key idea.
3. For a general discussion, see (a) G. L. Gaines, Jr., *Insoluble Monolayers at Liquid–Gas Interfaces*, Wiley-Interscience, New York, 1966; (b) A. W. Adamson, *Physical Chemistry of Surfaces*, 4th Ed., Wiley, New York, 1982, Ch. 4.
4. W. D. Harkins, *Physical Chemistry of Surface Films*, Reinhold, New York, 1952, Ch. 2.
5. M. W. Kim and D. S. Cannell, *Phys. Rev. A* **14**, 1299 (1976).
6. S. R. Middleton and B. A. Pethica, *Faraday Symp. Chem. Soc.* **16**, 109 (1981).
7. See, for example, M. E. Fisher, *J. Chem. Soc. Faraday Trans. II* **82**, 1824 (1986).
8. W. D. Harkins and E. Boyd, *J. Phys. Chem.* **45**, 20 (1941).
9. M. W. Kim and D. S. Cannell, *Phys. Rev. A* **13**, 411 (1976).
10. M. W. Kim, Ph. D. Thesis, University of California, Santa Barbara, 1975.
11. G. A. Hawkins and G. B. Benedek, *Phys. Rev. Lett.* **32**, 524 (1974).
12. H. E. Stanley, *Introduction to Phase Transitions and Critical Phenomena*, Oxford University Press, New York, 1971.
13. N. R. Pallas and B. A. Pethica, *J. Chem. Soc. Faraday Trans. I* **83**, 585 (1987).
14. N. R. Pallas, Ph. D. Thesis, Clarkson College of Technology, Potsdam, New York, 1983.
15. H. Zocher and F. Stiebel, *Z. Physik. Chem.* (*Leipzig*) **A147**, 401 (1930).
16. V. von Tscharner and H. M. McConnell, *Biophys. J.* **36**, 409 (1981).
17. M. Lösche, E. Sackmann, and H. Möhwald, *Ber. Bunsenges. Phys. Chem.* **87**, 848 (1983).
18. M. Lösche and H. Möhwald, *Rev. Sci. Instrum.* **55**, 1968 (1984).
19. B. G. Moore, C. M. Knobler, D. Broseta, and F. Rondelez, *J. Chem. Soc. Faraday II* **82**, 1753 (1986).
20. C. M. Knobler, *J. Chem. Soc. Faraday II* **82**, 1853 (1986).
21. V. T. Moy, D. J. Keller, H. E. Gaub, and H. M. McConnell, *J. Phys. Chem.* **90**, 3198 (1986).
22. Th. Rasing, H. Hsiung, Y. R. Shen and M. W. Kim, *Phys. Rev. A* **37**, 2732 (1988).
23. H. Bercegol, F. Gallet, D. Langevin, and J. Meunier, *J. Phys.* (*Paris*) **50**, 2277 (1989).
24. J. Als-Nielsen, *Z. Phys.* **B61**, 411 (1985).
25. K. Kjaer, J. Als-Nielsen, C. A. Helm, P. Tippmann-Krayer, and H. Möhwald, *Thin Solid Films* **159**, 17 (1988).
26. K. Kjaer, J. Als-Nielsen, C. A. Helm, L. A. Laxhuber, and H. Möhwald, *Phys. Rev. Lett.* **58**, 2224 (1987).
27. M. J. Grundy, R. M. Richardson, S. J. Roser, J. Penfold, and R. C. Ward, *Thin Solid Films* **159**, 43 (1988).

28. J. E. Bradley, E. M. Lee, R. K. Thomas, A. J. Willat, J. Penfold, R. C. Ward, D. P. Gregory, and W. Waschkowski, *Langmuir* **4**, 821 (1988).

29. (a) T. Rasing, Y. R. Shen, M. W. Kim, and S. Grubb, *Phys. Rev. Lett.* **55**, 2903 (1985); (b) Y. R. Shen, *Nature* **337**, 519 (1989).

30. P. Guyot-Sionnest, J. H. Hunt, and Y. R. Shen, *Phys. Rev. Lett.* **59**, 1597 (1987).

31. R. A. Dluhy, *J. Phys. Chem.* **90**, 1373 (1986).

32. R. A. Dluhy, M. L. Mitchell, T. Pettenski, and J. Beers, *Appl. Spectrosc.* **42**, 1289 (1988).

33. M. L. Mitchell and R. A. Dluhy, *J. Am. Chem. Soc.* **110**, 712 (1988).

34. A. Fischer, M. Lösche, H. Möhwald, and E. Sackmann, *J. Physique Lett.* **45**, L-785 (1984).

35. S. Garoff, H. W. Deckman, J. H. Dunsmuir, M. S. Alvarez, and J. M. Bloch, *J. Phys.* (*Paris*) **47**, 701 (1986).

36. D. P. E. Smith, A. Bryant, C. F. Quate, J. P. Rabe, Ch. Gerber, and J. D. Swalen, *Proc. Natl. Acad. Sci. USA* **84**, 969 (1987); C. A. Lang, J. K. H. Horber, T. W. Hansch, W. M. Heckel, and H. Möhwald, *J. Vac. Sci. Technol.* **A6**, 368 (1988).

37. B. Rothenhäusler and W. Knoll, *Nature* **332**, 615 (1988).

38. B. U. Rucha, J. Rabe, G. Schneider, and W. Knoll, *J. Colloid Interface Sci.* **114**, 1 (1986).

39. S. R. Middleton, M. Iwahashi, N. R. Pallas, and B. Pethica, *Proc. Roy. Soc.* (*London*) *A* **396**, 143 (1948).

40. N. R. Pallas and B. A. Pethica, *Langmuir* **1**, 509 (1985).

41. I. L. Pegg and G. Morrison, unpublished.

42. I. L. Pegg and G. Morrison, personal communication.

43. B. G. Moore, C. M. Knobler, S. Akamatsu, and F. Rondelez, *J. Phys. Chem*., in press.

44. F. Rondelez, J. F. Baret, K. A. Suresh, and C. M. Knobler, *Proc. 2nd Inter. Conf. on Physico-Chemical Hydrodynamics*, Huelva, Spain, 1987.

45. K. A. Suresh, J. Nittman, and F. Rondelez, *Europhys. Lett.* **66**, 437 (1988).

46. C. A. Helm, H. Möhwald, K. Kjaer, and J. Als-Nielsen, *Biophys. J.* **52**, 381 (1987).

47. C. M. Knobler and R. L. Scott, in *Phase Transitions and Critical Phenomena*, Vol. 9, Wiley, New York, 184, p. 164.

48. D. G. Dervichian, *J. Colloid Interface Sci.* **90**, 71 (1982).

49. O. Albrecht, H. Gruler, and E. Sackmann, *J. Phys.* (*Paris*) **39**, 301 (1978).

50. J. A. Spink and J. V. Saunders, *Trans. Faraday Soc.* **51**, 1154 (1955).

51. L. Bosio, J. J. Benattar, and F. Rieutord, *Rev. Phys. Appl.* **22**, 775 (1987).

52. S. Grayer Wolf, L. Leiserowitz, M. Lahav, M. Deutsch, K. Kjaer, and J. Als-Nielsen, *Nature* **328**, 63 (1987).

53. S. Grayer Wolf, E. M. Landau, M. Lahav, L. Leiserowitz, M. Deutsch, K. Kjaer, and J. Als-Nielsen, *Thin Solid Films* **159**, 29 (1988).

54. J. E. Bradley, E. M. Lee, R. K. Thomas, A. J. Willat, J. Penfold, R. C. Ward, D. P. Gregory, and W. Waschkowski, *Langmuir* **4**, 821 (1988).

55. J. P. Rabe, J. D. Swalen, D. A. Outka, and J. Stoehr, *Thin Solid Films* **159**, 275 (1988).

56. P. Dutta, J. B. Peng, B. Lin, J. B. Ketterson, M. Prakash, P. Georgopoulos, and S. Ehrlich, *Phys. Rev. Lett.* **58**, 2228 (1987).

57. N. Uyeda, T. Takenaka, K. Aoyama, M. Matsumoto, and Y. Fujiyoshi, *Nature* **327**, 319 (1987).

58. S. W. Barton, B. N. Thomas, E. B. Flom, S. A. Rice, B. Lin, J. B. Peng, J. B. Ketterson, and P. Dutta, *J. Chem. Phys.* **89**, 2257 (1988).

59. J. Harris and S. A. Rice, *J. Chem. Phys.* **89**, 5898 (1988).
60. B. Lin, J. B. Peng, J. B. Ketterson, P. Dutta, B. N. Thomas, J. Buontempo, and S. A. Rice, *J. Chem. Phys.* **90**, 2393 (1989).
61. A. Georgallas and D. A. Pink, *J. Colloid Interface Sci.* **89**, 107 (1981).
62. L. W. Horn and N. L. Gershfeld, *Biophys. J.* **18**, 301 (1977).
63. J. F. Nagle, *Ann. Rev. Phys. Chem.* **31**, 157 (1980).
64. C. A. Helm, H. Möhwald, K. Kjaer, and J. Als-Nielsen, *Europhys. Lett.* **4**, 697 (1987).
65. H. Möhwald, *Thin Solid Films* **159**, 1 (1988).
66. A. Fischer and E. Sackmann, *J. Phys. (Paris)* **45**, 517 (1984).
67. See, A. J. Kox, J. P. J. Michels, and A. F. Wiegel, *Nature* **287**, 317 (1980) and the references cited therein.
68. J. P. Bareman, G. Cardini, and M. L. Klein, *Phys. Rev. Lett.* **60**, 2152 (1988).
69. See, for example, D. Andelman, F. Brochard, and J. F. Joanny, *J. Chem. Phys.* **86**, 3673 (1987).
70. B. G. Moore, Ph.D. Thesis, University of California, Los Angeles, 1989.
71. B. M. Abraham, K. Miyano, J. B. Ketterson, and S. Q. Xu, *Phys. Rev. Lett.* **51**, 1975 (1983).
72. B. J. Halperin and D. R. Nelson, *Phys. Rev. Lett.* **41**, 121 (1978).
73. D. R. Nelson, M. Rubenstein, and F. Spaepen, *Phil. Mag.* **A46**, 105 (1982).
74. See, for example, Z.-G. Wang and S. A. Rice, *J. Chem. Phys.* **88**, 1290 (1988) and the references therein.
75. O. G. Mouritsen and M. J. Zuckermann, *Chem. Phys. Lett.* **135**, 294 (1987).
76. C. M. Roland, M. J. Zuckermann, and A. Georgallas, *J. Chem. Phys.* **86**, 5852 (1987).
77. See, H. M. McConnell and V. T. Moy, *J. Phys. Chem.* **92**, 4520 (1988) and the references therein.
78. A. Miller and H. Möhwald, *J. Chem. Phys.* **86**, 4258 (1987).
79. R. M. Weis and H. M. McConnell, *J. Phys. Chem.* **89**, 4453 (1985).
80. W. M. Heckl, M. Lösche, D. A. Cadenhead, and H. Möhwald, *Eur. Biophys. J.* **14**, 11 (1986).
81. D. J. Keller, H. M. McConnell, and V. T. Moy, *J. Phys. Chem.* **90**, 2311 (1986).
82. C. A. Helm and H. Möhwald, *J. Phys. Chem.* **92**, 1262 (1988).
83. W. M. Heckl and H. Möhwald, *Ber. Bunsenges. Phys. Chem.* **90**, 1159 (1986).
84. M. Lösche and H. Möhwald, *Eur. Biophys. J.* **11**, 35 (1984).
85. W. Hickel, D. Kamp, and W. Knoll, personal communication.
86. S. Akamatsu and F. Rondelez, personal communication.

Note added in proof: Recent x-ray studies of docosanoic acid monolayers demonstrate the existence of three condensed phases (H. Möhwald, private communication). Fluorescence measurements on esters (K. Stine and C. M. Knobler, unpublished) show evidence of a phase transitions between two LC phases. Thus, the generalized phase diagram discussed in Section VI is clearly oversimplified.

RECENT PROGRESS IN THE STATISTICAL MECHANICS OF INTERACTION SITE FLUIDS

P. A. MONSON[1,2] AND G. P. MORRISS[3]

[1] *Department of Chemical Engineering*
University of Massachusetts
Amherst, Massachusetts

[2] *Physical Chemistry Laboratory*
South Parks Road
Oxford, U.K.

[3] *Research School of Chemistry*
Australian National University
Canberra, Australia

CONTENTS

I. INTRODUCTION

A variety of approaches have been used in the development of models for the interactions between nonspherical molecules,[1,2] reflecting both the complexity of such interactions and the paucity of truly accurate quantitative information about them that could be used to assess the suitability of different models. Perhaps the most widely studied approach, especially over the last decade or so, is the interaction site formalism, in which the intermolecular pair potential is written as a sum of potentials between interaction sites on each molecule. We have

$$u(1,2) = \sum_{\alpha,\gamma} u_{\alpha\gamma}(r_{\alpha\gamma}) \tag{1.1}$$

where $u(1,2)$ is the overall intermolecular pair potential, a function of the positions and orientations of the two molecules (the notation 1 denotes both the position and orientation of molecule 1), and $u_{\alpha\gamma}(r_{\alpha\gamma})$ is the interaction potential between sites α and γ. The site–site potential has been modeled using a variety of different forms, including hard sphere and Lennard-Jones 12–6 potentials, and can be extended to the treatment of multipole–multipole interactions by the addition of Coulombic site–site interactions between discrete charge distributions placed in each molecule. The molecules can be considered as rigid arrays of interactions sites, or alternatively the effects of molecular flexibility can be incorporated. In principle, this formalism can be applied in a very wide range of problems from the properties of small nonspherical molecules to those of water, aqueous solutions, and polymers or large molecules of biological interest. For this reason, theoretical treatments of the statistical mechanics of interaction site systems are of considerable importance and can be expected to have an impact in a variety of contexts.

The purpose of this article is to describe some of the recent efforts in this area. The principal focus will be upon relatively small molecules in homogeneous fluid states, although applications to polymeric systems and inhomogeneous fluids will also be described. The subject at hand has been the focus of previous reviews,[3–6] and the present article is intended to be complementary to these, if written from a somewhat different perspective. We make a broader survey of the theoretical techniques available and focus on issues which we feel have received insufficient attention in previous discussions. We hope that the article will serve as a compact and critical guide to what might otherwise be seen as a rather bewildering and disjointed array of techniques and results.

To a large extent, theories of interaction site fluids, and molecular fluids in general, have their basis in the theoretical progress in the understanding of simple atomic fluids that was made in the 1950s and 1960s. Thus, there has been considerable attention devoted to extending integral equation and perturbation theories as well as cluster expansion techniques to fluids of anisotropic molecules. It turns out that such work is fraught with difficulties. First, it cannot be overstated that molecular anisotropy makes the statistical mechanics much more difficult to solve in a numerically tractable fashion. Second, although numerous ingenious theoretical strategies have been developed that render the problem more tractable, this is frequently at the cost of accuracy in the results. This balance between tractability and accuracy will be a major theme of the present discussion. Finally, in moving from simple atomic fluids to molecular fluids, a variety of new phenomena associated with the molecular anisotropy are encountered and new techniques must be developed in order to treat these phenomena. These include the calculation of the dielectric constant and Kirkwood orientational correlation parameters. Such quantities pose considerable problems since they are closely asociated with long-range orientational correlations in the system that have been found to be especially sensitive to weaknesses in approximate theories.

The scope of this article reflects both the interests and expertise of its authors as well as the constraint of keeping the article to a reasonable length.

We have chosen to focus almost entirely on the theoretical techniques themselves and on applications that test the accuracy of these techniques. Thus, we will generally refer to the rather large literature on the study of interaction site fluids via Monte Carlo and molecular dynamics simulations only as it impinges directly upon the theoretical developments. We have also taken advantage of the availability of review articles by Chandler[3,4] and Rossky[5] in limiting our discussion of applications such as the calculation of specific liquid structures and chemical equilibrium, which were described in detail in those articles.

II. Formal Results

In this section, we review some of the important formal results in the statistical mechanics of interaction site fluids. These results provide the basis for many of the approximate theories that will be described in Section III, and the calculation of correlation functions to describe the microscopic structure of fluids. We begin with a short review of the theory of the pair correlation function based upon cluster expansions. Although this material is featured in a number of other review articles, we have chosen to include a short account here so that the present article can be reasonably self-contained. Cluster expansion techniques have played an important part in the development of theories of interaction site fluids, and in order to fully grasp the significance of these developments, it is necessary to make contact with the results derived earlier for simple fluids. We will first describe the general cluster expansion theory for fluids, which is directly applicable to rigid nonspherical molecules by a simple addition of orientational coordinates. Next we will focus on the site–site correlation functions and describe the interaction site cluster expansion. After this, we review the calculation of thermodynamic properties from the correlation functions, and then we consider the calculation of the dielectric constant and the Kirkwood orientational correlation parameters.

A. From the Grand Partition Function to Pair Correlation Functions

The fundamental problem in classical equilibrium statistical mechanics is to evaluate the partition function. Once this is done, we can calculate all the thermodynamic quantities, as these are typically first and second partial derivatives of the partition function. Except for very simple model systems, this is an unsolved problem. In the theory of gases and liquids, the partition function is rarely mentioned. The reason for this is that the evaluation of the partition function can be replaced by the evaluation of the grand canonical correlation functions. Using this approach, and the assumption that the potential energy of the system can be written as a sum of pair potentials, the evaluation of the partition function is equivalent to the calculation of the two-particle correlation function $g_2(1,2)$. This correlation function gives structural information about the system directly, but also most thermodynamic quantities can be written as integrals involving $g_2(1,2)$. Thus, evaluating the correlation function has replaced the partition function as our central aim. The correlation function $g_2(1,2)$ will turn out to be the solution of two coupled equations, one of which is exact and the other approximate. Efforts have failed to solve the original problem exactly. However, the formalism has allowed us to set up approximate equations that are tractable.

Much of this section follows a review article by Stell[7] which, despite many more recent accounts, (including a particularly readable overview by Andersen[8]) remains the clearest and most thorough treatment. The language is that of graph theory, which we shall introduce as needed. Many of the theorems (or gaphical manipulations) fall under the general heading of topological reduction. Rather than attempting rigorous proofs, we will motivate the result as the details can be found in Stell's article. Also the use of functional differentiation[9] is helpful at various stages. It turns out to be just as easy to handle nonuniform systems as it is to handle uniform ones, so this generalization is introduced at the start. It can also be immediately generalized to apply to mixtures and to the case in which the potential energy includes n-body terms (in this case $g_n(1,\ldots,n)$ is the correlation function of interest). Finally, and mostly importantly, the development that we give illustrates the correct systematic approach to liquid state theory. It outlines the conceptual framework that is the basis of integral equations and their closure relations, and perturbations theories. It sets forth techniques with which to understand the approximations involved and to apply them in practical examples. As an example, the graphical methods employed enable us to obtain the prescription for the general term of cluster expansion without the use of combinatorial arguments, and the exact closure relation is obtained as a trivial consequence. Thus the basis for the development of approximate theories is revealed.

The grand canonical partition function for a system of volume V, temperature T, and chemical potential μ (the chemical potential and fugacity or activity z are related by $z = \exp[\beta\mu]$) is given by

$$\Xi(V,T,\mu) = \sum_{N=0}^{\infty} \frac{z^N}{N!} \int d1 \ldots dN \exp[-\beta V_N(1,\ldots,N)] \qquad (2.1.1)$$

where the notation $d\mathrm{K} = d\mathbf{r}_{\mathrm{K}}\, d\omega_{\mathrm{K}}/\Omega$ includes both the position coordinates $\mathbf{r}_{\mathrm{K}}$ and the orientational coordinates ω_{K} of particle K (note that $d\mathbf{r}_{\mathrm{K}} = V$ and $d\omega_{\mathrm{K}} = \Omega$). The function $V_N(1,\ldots,N)$ is the general N-body potential energy function. We use the subscript to denote that a function ϕ_M is a function of the coordinates (position and angular) of M particles. Using the assumption that the N-body potential can be written as a sum of single-particle and two-particle contributions only, that is

$$V_N(1,\ldots,N) = \sum_{i=1}^{N} \phi_1(i) + \sum_{i<j}^{N} \phi_2(i,j) \tag{2.1.2}$$

the grand canonical partition function becomes

$$\Xi = \sum_{N=0}^{\infty} \frac{1}{N!} \int d1\ldots dN \left(\prod_{i=1}^{N} z_1(i) \right) \left(\prod_{i<j}^{N} \exp[-\beta\phi_2(i,j)] \right) \tag{2.1.3}$$

where we have defined a position-dependent fugacity $z_1(i) = z\exp[-\beta\phi_1(i)]$. By a nonuniform system, we mean one where ϕ_1 is a nonconstant function of position, for example, surface effects may be included via a single-particle potential ϕ_1. Any semiclassical factors such as h^{-3N} (where h is Planck's constant) can also be absorbed into the definition of $z_1(i)$.

The power of the graphical expansion technique become apparent when we write down the first five terms in Eq. (2.1.3) as

Ξ = 1 + ● + ●- - -● + △ + ⊠ +

The notation is more concise and compact, but also we can write down a prescription for the whole infinite, sum.

$\Xi = 1 +$ the sum of all distinct simple graphs consisting of black z_1-circles and some or no e_2-bonds such that there is one e_2-bond joining pair of z_1-circles (2.1.4)

At this stage, we need some graphical definitions. With each *circle* we associate a function of position (and orientation) of one particle; in this expansion the function is $z_1(i)$. *White circles* are labeled, while *black circles* are unlabeled and an integral with respect to all coordinates is implied. With each *bond* we associate a function of the coordinates of two particles (both position and orientation); in this expansion the function is $e_2(i,j) = \exp[-\beta\phi_2(i,j)]$. Other examples would be the scalar separation or perhaps the dipole–dipole potential. A *graph* is a collection of circles and bonds such that each bond has a circle at each end. A *simple* graph has at most one bond between each pair of circles. With each graph we associate a factor s^{-1} where s is the symmetry number of the graph; that is, s is the number of equivalent ways in which the black circles can be labeled. Incorporating the symmetry number into the definition of the graph removes all cumbersome combinatorial factors from the graphical expansions. However, a major part of each graphical proof is to show that various graphical resummations do not lead to extra combinatorial factors.

As the separation between particles i and j becomes large, the function $e_2(i,j) \to 1$. If we define the Mayer f-function by $f_2(i,j) = e_2(i,j) - 1 = \exp[-\beta\phi_2(i,j)] - 1$, then $f_2(i,j) \to 0$ as the separation between i and $j \to \infty$. Replacing all e_2-bonds (- - -) in the expansion of the grand partition function by $(1+f)$-bonds (1 + ——) gives

$\Xi = 1 +$ the sum of all distinct simple graphs consisting of black z_1-circles some or no f-bonds (2.1.5)

This expansion contains all the graphs in Eq. (2.1.4) obtained by replacing e_2-bonds by 1, which is the subset of graphs with any number of black circles but no bonds. These are part of the class of *product* graphs. Writing out the first few terms in Eq. (2.1.5), we find that

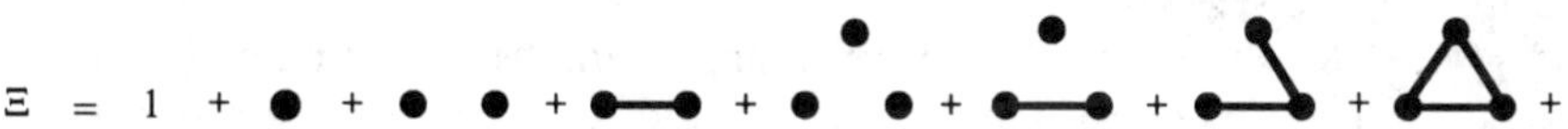

Ξ = 1 + ● + ● ● + ●—● + ● ● ● + ●—● ● + (two-bond chain) + (triangle) +

where the second, fourth, and fifth graphs are product graphs. This leads to the first theorem—the product theorem.

Theorem

Let G be a set of distinct connected graphs and let F be the set containing G plus all possible products (including multiple products) of all members of G, then

$$1 + F = \exp[G] \tag{2.1.6}$$

The proof is obtained by expanding the exponential. It is easy to see that all graphs in F are generated. The main difficulty is concerned with showing that the symmetry numbers of the product graphs are correct. Using the product theorem, it follows that

Ξ = exp [● + ●—● + (two-bond chain) + (triangle) +]

or

$\log \Xi =$ the sum of all distinct simple connected graphs consisting of black z_1-circles and some or no f-bonds. (2.1.7)

The grand canonical K-particle distribution function is defined to be

$$\rho_K(1,\ldots,K) = \frac{1}{\Xi} \sum_{N \geqslant K} \frac{z^N}{(N-K)!} \int d(K+1)\ldots.dN \exp[-\beta V_N(1,\ldots,N)] \quad (2.1.8)$$

Using the assumption of pairwise additive potentials and the previous definition of $z_1(i)$, we can write down the graphical expansion for ρ_K.

$\rho_K(1,\ldots,K) = (1/\Xi)$[the sum of all distinct simple connected graphs consisting of K white z_1-circles labeled $1,\ldots,K$ and some or no black z_1-circles with one e-bond joining each pair of z_1-circles] (2.1.9)

We can then follow the previous development to replace e-bonds by $(1+f)$-bonds. This gives a sum of connected and disconnected (product) graphs of f-bonds and z_1-circles. The disconnected graphs, whose factors contain no white z_1-circles, sum to cancel the factor Ξ in the denominator, so that

$\rho_K(1,\ldots,K) =$ the sum of all distinct simple graphs consisting of K white z_1-circles labeled $1,\ldots,K$ and some or no black z_1-circles with some or no f-bonds, such that there is at least one path from each black circle to a white circle (2.1.10)

Note that for $K \geqslant 2$ this graphical sum contains disconnected graphs that become connected if a $f_2(1,2)$ bond is added. For example the expansion for $\rho_2(1,2)$ contains

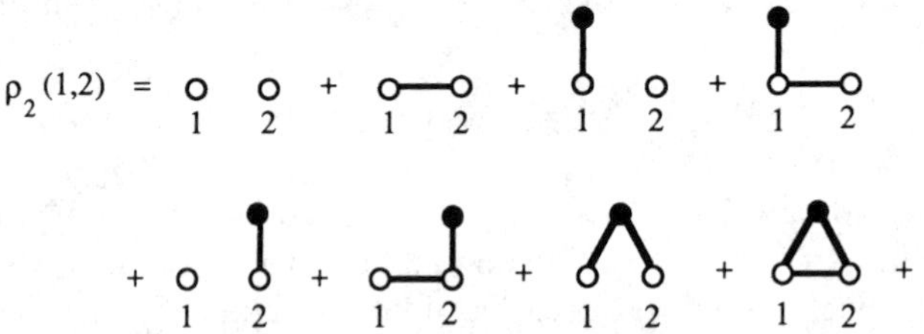

$\rho_2(1,2)$ = 1 2 + 1 2 + 1 2 + 1 2

+ 1 2 + 1 2 + 1 2 + 1 2 +

At this point, it is convenient to introduce the functional derivative. The K-particle distribution function can be written as a Kth order functional derivative, so together with a graphical interpretation of the action of a functional derivative, this is the easiest method of derivation. The *functional F*

$$F\{f\} = \int_a^b f(x)\,dx \quad (2.1.11)$$

depends upon the value of the function f at all the values of x in the interval $[a, b]$. Dividing the interval $a < x < b$ into n steps we may approximate the functional F by

$$F = \Delta x \sum_{m=0}^{n} f(x_m) = \sum_{m=0}^{n} f_m \Delta x \tag{2.1.12}$$

We may ask what happens to F as we vary the function f. In the discrete case, F is a function of all the f_m, so that

$$\delta F = \sum_{m=0}^{n} \frac{\partial F}{\partial f_m} \delta f_m \Delta x \tag{2.1.13}$$

Turning this sum back into an integral (by taking $n \to \infty$, and $\Delta x \to 0$ so that the product $n\Delta x$ remains fixed at $b - a$), we define the functional derivative of F with respect to the function f by

$$\delta F = \int \frac{\delta F}{\delta f(x)} \delta f(x)\, dx \tag{2.1.14}$$

This is convenient because the various correlation functions may be defined as functional derivatives. Armed with a graphical interpretation of functional differentiation, we can obtain graphical expansions for the correlation functions we need. If Γ is a set of graphs then the nth order functional derivative with respect to γ is written in graphical language as

$$\frac{\delta^n \Gamma}{\delta\gamma(1) \ldots . \delta\gamma(n)} = \begin{array}{l}\text{the sum of all distinct graphs obtained from } \Gamma \text{ by changing } n \\ \text{black } \gamma\text{-circles to white 1-circles and labeling them } 1, \ldots, n\end{array} \tag{2.1.15}$$

This is the interpretation of functional derivative with respect to the function associated with a circle (not a bond).

The advantage of functional differentiation is that the grand canonical k-particle distribution function can be written as

$$\rho_K(1, \ldots, K) = \frac{1}{\Xi} \prod_{i=1}^{K} z_1(i) \frac{\delta^K \Xi}{\delta z_1(1) \ldots . \delta z_1(K)} \tag{2.1.16}$$

The most compact graphical expansion we have is for the logarithm of Ξ, so it is more convenient to consider a distribution function based upon a functional derivative of $\log \Xi$. This leads naturally to the Ursell cluster functions $u_K(1,\ldots,K)$ defined by

$$u_K(1,\ldots,K)=\left(\prod_{i=1}^{K} z_1(i)\right)\frac{\delta^K \log \Xi}{\delta z_1(1)\ldots.\delta z_1(K)} \tag{2.1.17}$$

The Ursell cluster functions are usually defined by their relationship with the grand canonical distribution functions $\rho_K(1,\ldots,K)$

$$\rho_K(1,\ldots,K)=\sum\prod u_\alpha(i,\ldots,i_{n_\alpha}) \tag{2.1.18}$$

where this is a sum of products carried out over all partitions of the set $\{1,\ldots,K\}$. For $K=1$ we have trivially that

$$\rho_1(1)=u_1(1) \tag{2.1.19}$$

For $K=2$ all possible partitions of the set $\{1,2\}$ are $\{1,2\}$ and $\{1\}\{2\}$ so that

$$\rho_2(1,2)=u_2(1,2)+u_1(1)u_1(2) \tag{2.1.20}$$

Using our expansion for log Ξ it can be shown that

$\rho_1(1)=$ the sum of all distinct connected simple graphs consisting of one white z_1-circle labeled 1, some or no black z_1-circles, and f-bonds (2.1.21)

$\rho_2(1,2)=$ the sum of all distinct simple graphs consisting of 2 white z_1-circles labeled 1 and 2, some or no black z_1-circles, some or no f-bonds such that there is at least one path from each black circle to a white circle (2.1.22)

Notice that $\rho_2(1,2)$ contains disconnected graphs, but those which are disconnected can be connected by an f-bond between 1 and 2. All of the disconnected graphs arise from the product term $u_1(1)u_1(2)$.

If an *articulation* circle is removed, the graph falls into 2 or more disconnected pieces, at least one of which contains no white circles. Take an arbitrary graph in the expansion of $\rho_2(1,2)$, say

○ ○
1 2

for example. Then the expansion of $\rho_2(1,2)$ also contains the graphs

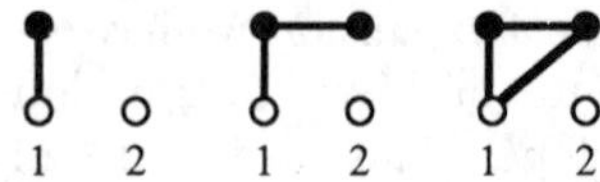

In each of these graphs the white z_1-circle labeled 1 is an articulation circle. These graphs, along with all others with an articulation 1 circle, can be resummed to change the z_1-circle labeled 1 into a ρ_1-circle. This process can be repeated for each articulation circle in a graph (both black circles and white circles), and hence $\rho_2(1,2)$ becomes

$\rho_2(1,2) =$ the sum of all distinct simple graphs consisting of 2 white ρ_1-circles, some or no black ρ_1-circles, some or no f-bonds such that there is a path from each black circle to a white circle, and the graphs are free of articulation circles (2.1.23)

This is called the *first replacement theorem*. The set of graphs with z_1-circles becomes the set of graphs with ρ_1-circles and no articulation circles.

Each graph in the expansion for $\rho_2(1,2)$ has two white ρ_1-circles labeled 1 and 2. Factoring these out, we may define a new 2-particle correlation function $g_2(1,2)$ by

$$\rho_2(1,2) = \rho_1(1)\rho_1(2)g_2(1,2) \tag{2.1.24}$$

For a system of spherical atoms, with no external field, $g_2(1,2)$ is the usual radial distribution function $g(r)$. The first graph in the expansion of $g_2(1,2)$ consists of two white 1-circles and no bond. This graph trivially has the value 1, so we can define $h_2(1,2)$ by

$$g_2(1,2) = h_2(1,2) + 1 \tag{2.1.25}$$

The graphical expansion for $h_2(1,2)$ is then

$h_2(1,2) =$ the sum of all distinct connected simple graphs consisting of 2 white 1-circles, some or no black ρ_1-circles, and at least one f-bond, such that the graphs are free of articulation circles (2.1.26)

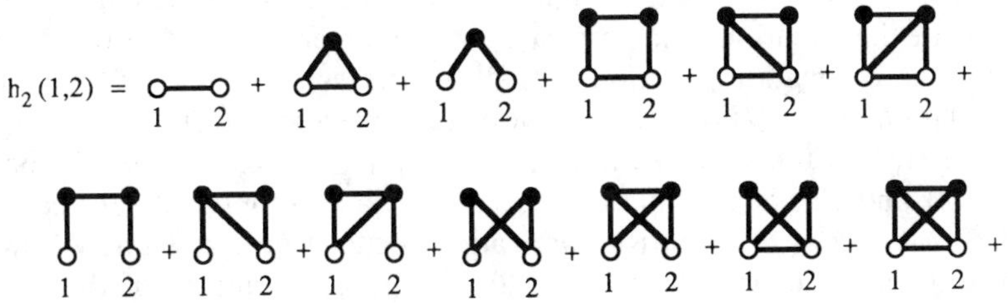

The next concept we need is the *nodal* circle. If all paths (constructed from a sequence of bonds) from circle 1 to circle 2 pass through the same black circle, then that black circle is called a nodal circle. If one nodal circle is removed, the graph falls into two disconnected pieces, both of which contain a white circle. A graph may have any number of nodal circles from zero upwards. As an example, the following graph with white circles

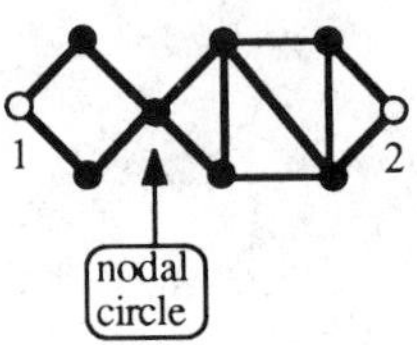

has a single nodal circle.

The graphical expansion of $h_2(1,2)$ in Eq. (2.1.26) can be written as the sum of two sets of graphs,

$$h_2(1,2) = c_2(1,2) + t_2(1,2) \tag{2.1.27}$$

where $c_2(1,2)$ is the set of graphs in $h_2(1,2)$ without nodal circles, and $t_2(1,2)$ is the set of graphs with nodal circles. It is easy to see that all the graphs in $t_2(1,2)$ can be constructed by joining members of $c_2(1,2)$ at nodal circles. So

$h_2(1,2)$ = ○—○ (1, 2) + ○—●—○ (1, 2) + ○—●—●—○ (1, 2) +

where here the bonds are c_2-bonds and the black circles are ρ_1-circles. This equation can be written as

$$h_2(1,2) = c_2(1,2) + \int d3\, c_2(1,3)\rho_1(3)h_2(3,2) \tag{2.1.28}$$

This is the Ornstein–Zernike equation. It is an exact integral equation relating the two 2-particle correlation functions $h_2(1,2)$ and $c_2(1,2)$. It is possible to motivate this equation form purely physical arguments; the idea is to interpret the total correlation function $h_2(1,2)$ as the sum of all possible direct correlations, thus $c_2(1,2)$ is termed the direct correlation function. We imagine that $h_2(1,2)$ is the sum of the direct correlation between 1 and 2 (that is $c_2(1,2)$), and all chains of direct correlations via a third, fourth etc., particle. The weakness of this heuristic derivation is that we do not know how to write down an expression for $c_2(1,2)$. The great advantage of the formal graphical derivation is that we have the graphical expansion for $c_2(1,2)$. $c_2(1,2)$ is simply the subset of graphs in the expansion for $h_2(1,2)$ with no nodal circles.

$c_2(1,2) =$ The sum of all distinct connected simple graphs consisting of 2 white 1-circles, some or no ρ_1-circles and, f-bonds such that the graphs are free of nodal circles and articulation circles (2.1.29)

We are now in position to derive a second equation—one that relates $c_2(1,2)$ to $h_2(1,2)$ and other known functions. This second equation is normally called a closure relation, and when combined with the Ornstein–Zernike equation, we have a closed system of equations to solve (two coupled equations in two unknowns h_2 and c_2). In principle, the graphical expansion of $c_2(1,2)$ in Eq. (2.1.29) is the exact closure relation, and if we could calculate all of the graphs we would have the exact solution. In practice, this has not been possible and all closure relations involve some type of approximation.

To generate the Percus–Yevick (PY) closure relation,[10] we consider the graphical expansion for $g_2(1,2)$ obtained from Eq. (2.1.26). Each graph in this expansion occurs both with and without a $f_2(1,2)$ bond, so we can factor out $(1+f_2(1,2)) = e_2(1,2)$. This gives

$$g_2(1,2) = e_2(1,2)\{1 + t_2(1,2) + E_2(1,2)\} \tag{2.1.30}$$

$E_2(1,2)$ is the set of graphs that have no nodal circles. The PY approximation is obtained by setting $E_2(1,2) = 0$. Straightforward algebra using Eq. (2.1.27) gives

$$c_2(1,2) = \{1 - \exp[\beta\phi_2(1,2)]\}g_2(1,2) \tag{2.1.31}$$

The second commonly used closure relation is the hypernetted chain approximation (HNC). This can be derived by taking the logarithm of both sides of Eq. (2.1.30).

$$\begin{aligned}\log[g_2(1,2)] + \beta\phi_2(1,2) &= \log[1 + t_2(1,2) + E_2(1,2)] \\ &= t_2(1,2) + B_2(1,2)\end{aligned} \tag{2.1.32}$$

We use the standard series expansion of $\log(1+x)$ to interpret the logarithm of a set of graphs. Essentially, we again obtain a set of nodal graphs plus a set of nodeless graphs and the HNC approximation is obtained by setting $B_2(1,2) = 0$. The set of graphs $B_2(1,2)$ is often referred to as the *bridge diagrams.*

$$c_2(1,2) = -\beta\phi_2(1,2) + h_2(1,2) - \log[g_2(1,2)] \qquad (2.1.33)$$

The derivations of the closure relations (2.1.31) and (2.1.33) are both approximate, but we include them in this section in order to complete the discussion of theoretical approaches to atomic fluids. These methods can be applied to molecular fluids directly without fundamental changes (as we will see in Section III), but integral equation approaches developed particularly for molecular systems often involve closure relations that are generalizations and extensions of the ideas presented here.

B. Interaction Site Formalism

The treatment described in the last section is appropriate for both atomic and molecular fluids, as the coordinate representation of each particle included the possibility of molecular orientation. This approach is convenient if the pair potential can be written as a function of the molecular coordinates $\mathbf{r}_K, \omega_K$. In many circumstances, however, the interaction between two molecules can best be written as the sum of pairs of site–site interactions, where typically the sites are the centers of the constituent atoms. If α and γ label sites on molecules 1 and 2, respectively, then the total interaction between the two molecules is given by

$$\phi_2(1,2) = \sum_{\alpha,\gamma} \phi(1^\alpha, 2^\gamma) = \sum_{\alpha,\gamma} u_{\alpha\gamma}(r_{\alpha\gamma}) \qquad (2.2.1)$$

In the remainder of this review, we will drop the subscript 2 on both the potential and the correlation function g. The subscripts $\alpha\gamma$ will denote a site–site correlation function rather than a center–center function, and we will use u for the potential rather than ϕ.

In the interaction site formalism, the natural pair correlation to consider is the site–site function $g_{\alpha\gamma}(r)$ which is a measure of the probability of finding a site γ a distance r from a site α on a different molecule, regardless of the orientations of the two molecules. This approach has been developed by Chandler,[11] Ladanyi and Chandler,[12] and by Cummings, Sullivan, and Gray.[13] The development we give here is loosely based on these works.

It is easy to obtain the site–site pair correlation function from the full molecular correlation function $g(1,2)$, as

$$g_{\alpha\gamma}(r) = \int_{r_{\alpha\gamma}\,\text{fixed}} d\omega_1\, d\omega_2\, g(1,2) \qquad (2.2.2)$$

where ω_i is now the normalized angular variable whose integral is unity. As before, we can define a site–site total correlation function $h_{\alpha\gamma}(r) = g_{\alpha\gamma}(\mathrm{r}) - 1$. We have already derived a graphical expansion for $h(1, 2)$, but it consists of f_2-bonds containing the total interaction between molecules 1 and 2. We can transform this expansion for $h(1, 2)$ into an expansion containing site–site f-bonds by considering a single f-bond:

$$f(1,2) = \exp\left(-\beta \sum_{\alpha\gamma} u_{\alpha\gamma}(r_{\alpha\gamma}) \right) - 1 = \prod_{\alpha\gamma} (1 + f_{\alpha\gamma}(r_{\alpha\gamma})) - 1 \qquad (2.2.3)$$

so each f-bond in the full molecular $h(1, 2)$ is replaced either a single site–site f-bond, or a product of site–site f-bonds. For a fluid of diatomic molecules, for example, each bond in a graph generates fifteen terms in the expansion of the site–site correlation functon $g_{\alpha\gamma}$.

$$\begin{aligned} f_2(1,2) = {} & f_{\alpha\alpha} + f_{\alpha\gamma} + f_{\gamma\alpha} + f_{\gamma\gamma} + f_{\alpha\alpha} f_{\alpha\gamma} + f_{\alpha\alpha} f_{\gamma\alpha} + f_{\alpha\alpha} f_{\gamma\gamma} \\ & + f_{\alpha\gamma} f_{\gamma\alpha} + f_{\alpha\gamma} f_{\gamma\gamma} + f_{\gamma\alpha} f_{\gamma\gamma} + f_{\alpha\alpha} f_{\alpha\gamma} f_{\gamma\alpha} + f_{\alpha\alpha} f_{\alpha\gamma} f_{\gamma\gamma} \\ & + f_{\alpha\alpha} f_{\gamma\alpha} f_{\gamma\gamma} + f_{\alpha\gamma} f_{\gamma\alpha} f_{\gamma\gamma} + f_{\alpha\alpha} f_{\alpha\gamma} f_{\gamma\alpha} f_{\gamma\gamma} \end{aligned} \qquad (2.2.4)$$

The symmetry number and power of density associated with each graph is unchanged by this bond replacement. The set of graphs obtained in the expansion depends upon the distance which is kept fixed in the angular integration in Eq. (2.2.2). If $r_{\alpha\gamma}$ is kept fixed then the first graph is simply two labeled circles with a single site–site f-bond between them. The second graph is more interesting as the site–site f-bond is between one of the labeled white circle 1^{α} and a black site γ circle. However, this black circle is part of molecule 2 and is fixed at a particular distance from 2^{α} by the chemical bond between α and γ. We represent this fixed bond length by a new function, an s_2-bond, which for rigid molecules is $\delta(|\mathbf{r}_i - \mathbf{r}_j| - L)$ where L is the length of the bond. The fifteen graphs obtained from the first graph in $h_2(1, 2)$ are

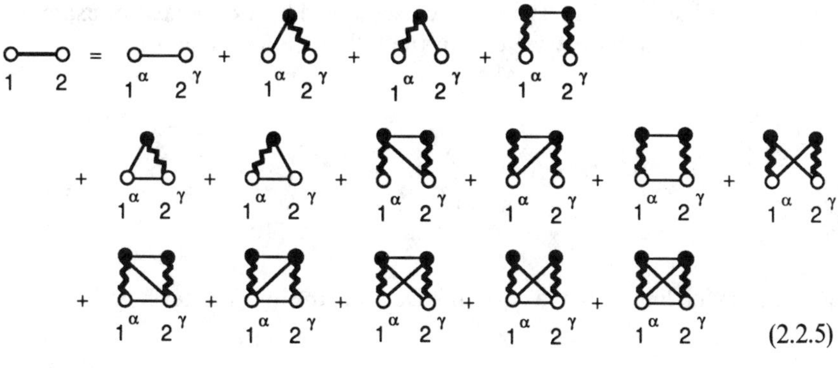

1 2 = 1^{α} 2^{γ} + 1^{α} 2^{γ} + 1^{α} 2^{γ} + 1^{α} 2^{γ}

+ 1^{α} 2^{γ} + 1^{α} 2^{γ} + 1^{α} 2^{γ} + 1^{α} 2^{γ} + 1^{α} 2^{γ} + 1^{α} 2^{γ}

+ 1^{α} 2^{γ} + 1^{α} 2^{γ} + 1^{α} 2^{γ} + 1^{α} 2^{γ} + 1^{α} 2^{γ} (2.2.5)

Continuing this process, the graphical expansion for $h_{\alpha\gamma}(r)$ is given by

$h_{\alpha\gamma}(r) =$ the sum of all distinct simple graphs consisting of black circles, one or more site–site **f**-bonds and some or no $\mathbf{s}_n$-bonds, such that at most one $\mathbf{s}_n$-bond is connected to each circle, no $\mathbf{s}_n$-bond exists between the two white circles, and the graphs are free of articulation circles (2.2.6)

The black circles in this expansion do not each have a factor of density associated with them. The power of density must be determined by considering the graph in $h(1,2)$, from which it is derived. For similar reasons, we drop the positional dependence of the density associated with black circles. In general, an m-site fluid would lead to $\mathbf{s}_2$- through to $\mathbf{s}_m$-bonds. The fixed distances implied by an $\mathbf{s}_m$-bond do not necessarily correspond to chemical bonds in the molecule.

Given the graphical expansion for the site–site total correlation function $h_{\alpha\gamma}(r)$ it is possible to generate on Ornstein–Zernike like equation. We will adopt the physical approach that the total correlation function is the sum of all possible direct site–site correlations, both intramolecular correlations via an $\mathbf{s}_m$-bond, or intermolecular correlations via a $c_{\alpha\gamma}(r)$. This defines a site–site direct correlation function $c_{\alpha\gamma}(r)$. From the expansion (2.2.5), it follows that at most one $\mathbf{s}_m$-bond is connected to each circle.

$h_{\alpha\gamma}(r) =$ the sum of all distinct simple chain graphs consisting of black circles, one or more $c_{\eta\nu}$-bonds, and some or no $\mathbf{s}_m$-bonds, such that at most one $\mathbf{s}_m$-bond is connected to each circle. The graph with an $\mathbf{s}_m$-bond between the two white circles is excluded. (2.2.7)

In this expansion, at the same order in density, we find the chain graph consisting of $mc_{\alpha\gamma}$-bonds, and all possible graphs obtained by replacing a circle (either black or white) with two circles joined by a $\mathbf{s}_m$-bond. For example, at zeroth order in density we obtain the following four graphs:

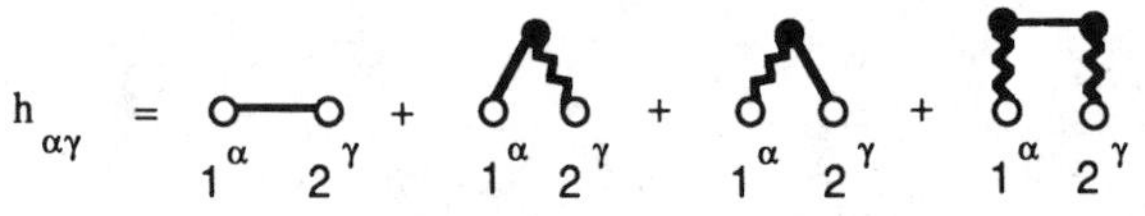

(2.2.8)

where each full line bond is a $c_{\eta\nu}$-bond. Defining the hypervertex function ω,

$$\omega_{\alpha\gamma}(r) = \delta_{\alpha\gamma}\delta(r) + (1 - \delta_{\alpha\gamma})\delta(r - L) \tag{2.2.9}$$

these graphs can be resumed to give

$h_{\alpha\gamma}$ = [graph: 1^α — 2^γ] + [graph: 1^α — — 2^γ] + ...

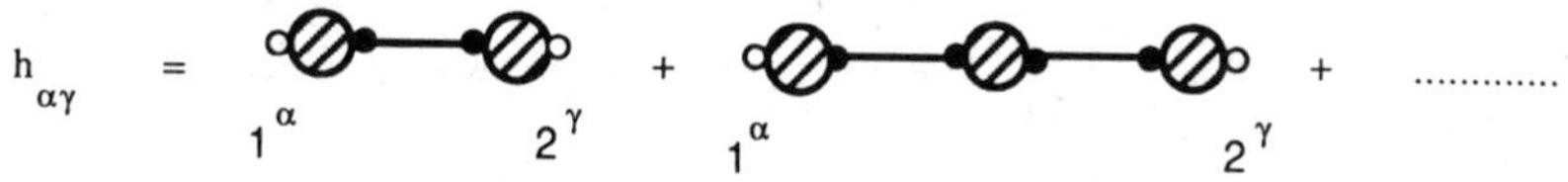

(2.2.10)

or in graph theory language

$h_{\alpha\gamma}(r) =$ the sum of all simple chain graphs consisting of ω-hypervertices and $c_{\eta\nu}$-bonds. The two unbonded circles on the terminal hypervertices are white circles. (2.2.11)

Equation (2.2.11) is the Ornstein–Zernike like equation for site–site correlation functions first derived by Chandler and Andersen.[14] It is more familiar when written as an integral equation.

$$H = WCW + \rho WCH \tag{2.2.12}$$

Each of the symbols H, W, and C is a matrix of site–site correlation functions. If these are real space functions, then a convolution is implied by the matrix multiplication, however it is more usual to write site–site Ornstein–Zernike (SSOZ) equation in k-space, where

$$\hat{x}_{\alpha\gamma}(k) = \int d\mathbf{r}\, x_{\alpha\gamma}(r) \exp[i\mathbf{k}\cdot\mathbf{r}] \tag{2.2.13}$$

for x, an element of matrix H or C. The elements of the W matrix are given by substituting Eq. (2.2.9) into Eq. (2.2.13). The result is

$$\hat{\omega}_{\alpha\gamma}(k) = \delta_{\alpha\gamma} + (1 - \delta_{\alpha\gamma})\frac{\sin kL_{\alpha\gamma}}{kL_{\alpha\gamma}} \tag{2.2.14}$$

The SSOZ equation is an exact integral equation relating the two site–site correlation functions $h_{\alpha\gamma}(r)$ and $c_{\alpha\gamma}(r)$. Strictly speaking, it is simply a definition of the site–site direct correlation function $c_{\alpha\gamma}(r)$. This fact is made clearest in the derivation by Høye and Stell.[15] Here they consider the Ornstein–Zernike equation for an equimolar mixture of two species α and γ. For each atom of type α, an atom of type γ is fixed a distance L away. This is essentially what would happen if a chemical reaction between α and γ went to completion. The term in the OZ equation that includes the total correlation function $h(r)$ becomes the sum of intramolecular correlations $\omega_{\alpha\gamma}$, and intermolecular correlations $h_{\alpha\gamma}$. Dividing the direct correlation term into density-dependent and density-independent terms leads naturally to the site–site direct correlation function $c_{\alpha\gamma}$.

The difficulty with the SSOZ equation is that although we have a well defined site–site direct correlation function $c_{\alpha\gamma}$, we can not easily obtain a graphical expansion for it that would be the basis of approximation schemes that are the site–site analogues of the PY and HNC approximations. Although we can clearly separate the expansion (2.2.5) into nodal and nodeless graphs, recombining the nodeless graphs in chains leads to graphs that do not occur in the original expansion. We will have more to say about this difficulty and its resolution in Section III.

C. Thermodynamic Properties from Site–Site Correlation Functions

In this section, we show the connection between the site–site correlation functions $g_{\alpha\gamma}(r)$ and the thermodynamic properties of the fluid. The definition of the $g_{\alpha\gamma}(r)$'s involves an integral over the orientations of the two molecules, so it is reasonable to expect that some structural information has been lost. This is indeed the case, but despite the pessimistic summary given by Gray[16] which is now somewhat out of date, most important properties are accessible from $g_{\alpha\gamma}(r)$.

The configurational energy is easy to calculate from $g_{\alpha\gamma}(r)$. It is given by

$$\frac{\beta U}{N} = \sum_{\alpha\gamma} \frac{\beta\rho}{2} \int d\mathbf{r}\, g_{\alpha\gamma}(r) u_{\alpha\gamma}(r) \tag{2.3.1}$$

We can separate the site–site interaction potential into two parts, a reference part $u_{\alpha\gamma}^{(0)}(r)$, and a perturbation $u_{\alpha\gamma}(r)$ which contains the attractions and other slowly varying forces. It can be shown that the Helmholtz free energy A is given by

$$\frac{\beta A}{N} = \frac{\beta A_0}{N} + \sum_{\alpha\gamma} \frac{\beta\rho}{2} \int_0^1 d\lambda \int d\mathbf{r}\, g_{\alpha\gamma}(r;\lambda) u_{\alpha\gamma}(r) \tag{2.3.2}$$

where A_0 is the Helmholtz free energy when the perturbation $u_{\alpha\gamma}(r)$ is turned off. The λ is called a charging parameter, which switches on the perturbation and $g_{\alpha\gamma}(r;\lambda)$ is the charging-parameter-dependent, site–site correlation function.

The usual compressibility theorem for site–site models is given by

$$kT\frac{\partial\rho}{\partial P}\bigg|_{V,T} = 1 + \rho\int d\mathbf{r}_{12}\int d\omega_1\int d\omega_2 h(1,2) = 1 + \rho\hat{h}_{\alpha\gamma}(0) \qquad (2.3.3)$$

where P is the pressure. The function $h_{\alpha\gamma}(0)$, we will show in the next section, is independent of *the site indices.*

In addition to the above expressions it is also possible to obtain expressions for the isothermal compressibility and other thermodynamic properties from the site–site direction correlation functions[17,18]. Such expressions require assumptions about the range of the site–site direct correlation functions and also, in some cases, are restricted to certain closures of the SSOZ equation. We will consider such routes to the thermodynamic properties in Section V.C.

The virial route to the pressure is based on an idea that originates with Green. It assumes that the isotropic volume expansion can be regarded as a scaling of the center of mass coordinates at fixed orientation and hence the virial pressure can written as[19]

$$\beta P = \rho - \frac{\beta\rho^2}{6}\int d\mathbf{r}_{12}\int d\omega_1\int d\omega_2 r_{12}\frac{\partial\phi(1,2)}{\partial r_{12}}g(1,2) \qquad (2.3.4)$$

In the site–site representation, the virial pressure involves higher spherical harmonic coefficients of the site–site pair correlation functions.[20] The difficulties involved in the standard virial route to the pressure in polymer systems have been well illustrated in recent work.[21]

D. Dielectric Constant and Kirkwood Parameters

In a series of papers, Høye and Stell[22] have shown that the value of the dielectric constant and the associated Kirkwood G-factors can be related to the long-range behavior of the site–site correlation function $h_{\alpha\gamma}(r)$. The relations obtained are exact connections between site–site correlation functions and system properties. In the following sections, we will develop these connections further by considering the results obtained from particular integral equations and closure relations.

It is convenient to develop a theory for these properties in k-space, and we will need to understand their small k behavior. It can be shown that for an arbitrary correlation function $h_{\alpha\gamma}$

$$\hat{h}_{\alpha\gamma}(k) = \frac{4\pi}{k}\int_0^\infty dr\, r\, h_{\alpha\gamma}(r)\sin kr \qquad (2.4.1)$$

If $h_{\alpha\gamma}(r)$ is analytic, and decays fast enough so that the integral exists, the Fourier transform is an even function of k. Further, $\sin(kr)$ has an expansion in odd powers of k and thus $\hat{h}_{\alpha\gamma}(k)$ will have a power series expansion in even powers of k:

$$\hat{h}_{\alpha\gamma}(k) = \hat{h}_{\alpha\gamma}(0) + k^2\hat{h}^{(2)}_{\alpha\gamma} + k^4\hat{h}^{(4)}_{\alpha\gamma} + \cdots \tag{2.4.2}$$

The site–site correlation function $h_{\alpha\gamma}(r)$ is defined as an integral of the full angle-dependent correlation function $h(1,2)$ with the distance between site α on molecule 1 and site γ on molecule 2 held fixed,

$$h_{\alpha\gamma}(r) = \iint_{r_{\alpha\gamma}=r} h(1,2)\, d\omega_1\, d\omega_2 \tag{2.4.3}$$

For a linear molecule, the vector separation between sites α and γ can be written as

$$\begin{aligned} \mathbf{r}_{\alpha\alpha} &= \mathbf{r} + d_\alpha\hat{\mathbf{s}}_2 - d_\alpha\hat{\mathbf{s}}_1, \quad \mathbf{r}_{\alpha\gamma} = \mathbf{r} - d_\gamma\hat{\mathbf{s}}_2 - d_\alpha\hat{\mathbf{s}}_1 \\ \mathbf{r}_{\gamma\alpha} &= \mathbf{r} + d_\alpha\hat{\mathbf{s}}_2 + d_\gamma\hat{\mathbf{s}}_1, \quad \mathbf{r}_{\gamma\gamma} = \mathbf{r} - d_\gamma\hat{\mathbf{s}}_2 + d_\gamma\hat{\mathbf{s}}_1 \end{aligned} \tag{2.4.4}$$

where $\mathbf{r}$ is the vector separation between the centers of the molecules, $\mathbf{s}_i$ is the orientation of dipole i, and d_α is the distance of site α from the center of the molecule. The Fourier transform of $h_{\alpha\gamma}(r)$ can be written as

$$\begin{aligned} \hat{h}_{\alpha\alpha}(k) &= \int d\mathbf{r}_{\alpha\alpha} h_{\alpha\alpha}(r_{\alpha\alpha}) \exp(i\mathbf{k}\cdot\mathbf{r}_{\alpha\alpha}) \\ &= \iiint d\mathbf{r}_{\alpha\alpha} h(1,2) \exp(i\mathbf{k}\cdot\mathbf{r}_{\alpha\alpha})\, d\omega_1\, d\omega_2 \end{aligned} \tag{2.4.5}$$

Using Eq. (2.4.4) to change variables from $\mathbf{r}_{\alpha\alpha}$ to $\mathbf{r}$, we find that

$$\hat{h}_{\alpha\alpha}(k) = \iint \hat{h}(1,2) \exp(i\mathbf{k}\cdot d_\alpha\hat{\mathbf{s}}_2) \exp(-i\mathbf{k}\cdot d_\alpha\hat{\mathbf{s}}_1)\, d\omega_1\, d\omega_2 \tag{2.4.6}$$

The first observation we can make from this expression is that for $k=0$, both the exponential terms are unity, so that $\hat{h}_{\alpha\alpha}(k)$ is independent of the site indices $\alpha\alpha$.

To derive a relation between the dielectric constant and the site–site correlation function, we begin by considering a fluid of linear polar molecules. For a dipolar fluid, it is convenient[23] to expand the full correlation function

$h(1,2)$ as

$$h(1,2) = h_S(r) + h_D(r)D(1,2) + h_\Delta(r)\Delta(1,2) + \text{higher-order terms} \tag{2.4.7}$$

where the angle-dependent functions $D(1,2)$ and $\Delta(1,2)$ are given by

$$D(1,2) = 3(\hat{\mathbf{r}}\cdot\hat{\mathbf{s}}_1)(\hat{\mathbf{r}}\cdot\hat{\mathbf{s}}_2) - \hat{\mathbf{s}}_1\cdot\hat{\mathbf{s}}_2 \quad \text{and} \quad \Delta(1,2) = \hat{\mathbf{s}}_1\cdot\hat{\mathbf{s}}_2 \tag{2.4.8}$$

$\mathbf{s}_i$ is the orientation of dipole i and $\hat{\mathbf{r}}$ is the direction of the vector from 1 to 2. The Fourier transform of $h(1,2)$ can be written as

$$\hat{h}(1,2) = \hat{h}_S(k) + \bar{h}_D(k)\hat{D}(1,2) + \hat{h}_\Delta(k)\Delta(1,2) + \cdots \tag{2.4.9}$$

where

$$\hat{D}(1,2) = 3(\hat{\mathbf{k}}\cdot\hat{\mathbf{s}}_1)(\hat{\mathbf{k}}\cdot\hat{\mathbf{s}}_2) - \hat{\mathbf{s}}_1\cdot\hat{\mathbf{s}}_2 \quad \text{and} \quad \hat{\Delta}(1,2) = \hat{\mathbf{s}}_1\cdot\hat{\mathbf{s}}_2 \tag{2.4.10}$$

Here $\hat{h}_s(k)$ and $\hat{h}_\Delta(k)$ are the usual Fourier transforms, while $\bar{h}_D(k)$ is a Hankel transform,

$$\bar{h}_D(k) = \int dr\, h_D(r)\, j_2(kr) \tag{2.4.11}$$

The long wavelength behavior of the expansion coefficients of $h(1,2)$ is known to be[24]

$$\rho\bar{h}_D(0) = -3y\left(\frac{\varepsilon-1}{3y}\right)^2 \frac{1}{\varepsilon} \tag{2.4.12}$$

$$\rho\hat{h}_D(0) = \frac{(\varepsilon-1)(2\varepsilon+1)}{3y\varepsilon} - 3 \tag{2.4.13}$$

where $y = 4\pi\rho\beta\mu^2/9 = 4\pi\rho\beta(qL)^2/9$ is the usual dimensionless measure of dipole strength. Following Høye and Stell, we expand the exponentials in Eq. (2.4.6) in terms of spherical Bessel functions j_n and Legendre polynomials P_n as

$$\exp(i\mathbf{k}d_\alpha\cdot\hat{\mathbf{s}}_2) = \sum_{n=0}^{\infty} (2n+1)i^n j_n(kd_\alpha)P_n(\hat{\mathbf{k}}\cdot\hat{\mathbf{s}}_2) \tag{2.4.14}$$

and it can be shown that

$$\hat{h}_{\alpha\alpha}(k) = j_0(kd_\alpha)j_0(-kd_\alpha)\hat{h}_S(k) - j_1(kd_\alpha)j_1(-kd_\alpha)[2\bar{h}_D(k) + \hat{h}_\Delta(k)] + \cdots \tag{2.4.15}$$

Similar equations can easily be obtained for the other site–site correlation functions. Combining these results, we have, as $k \to 0$, that

$$\begin{aligned}\Delta\hat{h} &= \hat{h}_{\alpha\alpha}(k) + \hat{h}_{\gamma\gamma}(k) - \hat{h}_{\alpha\gamma}(k) - \hat{h}_{\gamma\alpha}(k) \\ &= \tfrac{1}{9}k^2(d_\alpha + d_\gamma)^2[2\bar{h}_D(0) + \hat{h}_\Delta(0)] + O(k^4) \\ &= \frac{k^2L^2}{9p}\left[\frac{\varepsilon - 1}{y\varepsilon} - 3\right] + O(k^4)\end{aligned} \tag{2.4.16}$$

where $L = d_\alpha + d_\gamma$. A completely general analysis for molecules of arbitrary symmetry leads to the result

$$\Delta\hat{h} = \sum_{\alpha\gamma} q_\alpha q_\gamma \hat{h}_{\alpha\gamma} = \frac{k^2\mu^2}{9\rho}\left(\frac{\varepsilon - 1}{y\varepsilon} - 3\right) \quad \text{as} \quad k \to 0 \tag{2.4.17}$$

These two expressions connect the dielectric constant with the small k behavior of the Fourier transforms of site–site correlation functions. The next step is to determine how $\hat{h}_{\alpha\gamma}(k)$ behaves as $k \to 0$, from particular integral equations and closure relations.

III. APPROXIMATE THEORIES

As discussed earlier, our principal focus is the structure and thermodynamics of homogeneous fluids modeled with interaction site potentials. A wide variety of theoretical techniques have been used in this context. In reviewing the efforts in this area, one can choose from a number of different approaches. Perhaps the most natural is to divide theories into those based upon the explicit use of site–site correlation functions and those based upon the use of orientation-dependent molecular pair correlation function. Our review will proceed somewhat along these lines, especially in the context of integral equation theories. With respect to perturbation theories, however, we divide our discussion into a treatment of theories in which the perturbation is the anisotropy in the intermolecular forces, and, secondly, a treatment of theories in which the high temperature or λ expansion is used to calculate contributions of long-range forces to the free energy. Finally, we review some recent work in which cluster-expansion-based peturbation theories are extended to interaction site systems.[25]

A. The Molecular Ornstein–Zernike Equation

The Ornstein–Zernike equation can be written in a general form applicable to fluids of rigid nonspherical molecules

$$h(1,2) = c(1,2) + \rho \int c(1,3)h(2,3)\, d3 \tag{3.1.1}$$

where $h(1,2)$ is the total correlation function, $c(1,2)$ is the direct correlation function, and it is understood, for example, that 1 denotes both the position and orientation of molecule 1. As in the theory of simple fluids, we must introduce an independent relationship between $h(1,2)$ and $c(1,2)$ in order to construct a solution to Eq. (3.1.1). In the Percus–Yevick approximation, for example, we have

$$c(1,2) = y(1,2)f(1,2) \tag{3.1.2}$$

where $y(1,2)$ is defined through the relation

$$g(1,2) = y(1,2)[1 + f(1,2)] \tag{3.1.3}$$

and is called the background correlation function[26] or cavity distribution function.[27] A moment's reflection leads one to appreciate the futility of attempting to solve Eqs. (3.1.1) and (3.1.2) without some further simplifying measures. For linear molecules, the correlation functions will depend upon the separations of the centers of mass and three angles describing the orientation of the molecules relative to the vector joining the centers of mass, as illustrated in Fig. 1. The mere tabulation of such a function is a daunting problem in itself.

The most commonly used approach to the problem is to expand the correlation functions and their Fourier transforms in a series of orthogonal functions, usually the spherical harmonics. This approach was pioneered by Chen and Steele[28] in the case of the Percus–Yevick approximation for hard diatomic fluids. More recently, the approach has been generalized to arbitrary

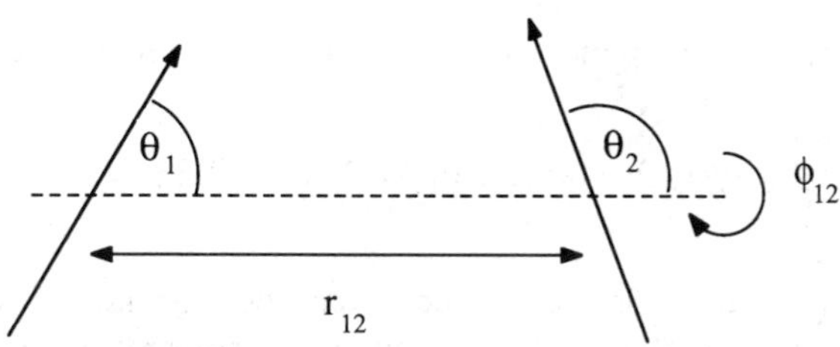

Figure 1

closures by Lado,[29,30] who extended the general treatment of the mean sperical approximation (MSA) for anisotropic potentials given by Blum.[31] It is worthwhile to point out that this approach as a whole has been applied far more often to intermolecular potentials of the generalized Stockmayer type than to interaction site systems (for a review see Gray and Gubbins[1]), although it is not our intention to review this work here. Neither do we set out to give a detailed treatment of the spherical harmonic expansion technique. Rather we seek to focus on issues such as the convergence of the spherical harmonic expansions which become especially important when applying these techniques to interaction site fluids. With this in mind, we will begin with a brief description of Lado's approach.

Lado begins his general treatment by writing the following exact expression for the direct correlation function

$$c(1,2) = h(1,2) - \ln y(1,2) + B(1,2) \tag{3.1.4}$$

where $B(1,2)$ is an unknown function which is the sum of the so-called *bridge diagrams*.[9] The HNC approximation results after the neglect of $B(1,2)$. The PY approximation results from linearization of the logarithm in addition to the neglect of $B(1,2)$. Introducing the function

$$\gamma(1,2) = h(1,2) - c(1,2) \tag{3.1.5}$$

we can rewrite the Ornstein–Zernike equation as

$$\gamma(1,2) = \rho \int c(1,3)[\gamma(2,3) + c(2,3)]\, d3 \tag{3.1.6}$$

Defining the Fourier transform of any one of these functions by, for example,

$$\hat{\gamma}(1,2) = \int \gamma(1,2) e^{i\mathbf{k}\cdot\mathbf{r}_{12}} d\mathbf{r}_{12} \tag{3.1.7}$$

we can rewrite the Ornstein–Zernike equation in Fourier space as

$$\hat{\gamma}(1,2) = \rho \int \hat{c}(1,3)[\hat{\gamma}(2,3) + \hat{c}(2,3)]\, d\omega_3 \tag{3.1.8}$$

where the integration is performed over the normalized orientational coordinates of molecule 3. By introducing the expansion in spherical harmonics of $\gamma(1,2)$, $c(1,2)$ and their Fourier transforms, viz.

$$\gamma(1,2) = 4\pi \sum_{l_1,l_2,m} \gamma_{l_1,l_2,m}(r_{12}) Y_{l_1,m}(\omega_1) Y_{l_2,-m}(\omega_2) \tag{3.1.9}$$

and

$$\hat{\gamma}(1,2) = 4\pi \sum_{l_1,l_2,m} \hat{\gamma}_{l_1,l_2,m}(k) Y_{l_1,m}(\omega'_1) Y_{l_2,-m}(\omega'_2) \tag{3.1.10}$$

we can rewrite Eq. (3.1.8) as a set of coupled algebraic equations involving the harmonic coefficients of the Fourier transforms of $c(1,2)$ and $\gamma(1,2)$. Notice that the orientations ω'_1 and ω'_2 are defined relative to the wave vector $\mathbf{k}$ so that the expansion coefficients of the Fourier transforms are not simply the Fourier transforms of the expansion coefficients in real space. To relate the coefficients in the two expansions, it is necessary to rewrite each set of expansion coefficients in terms of those in a spherical harmonic expansion in a space-fixed coordinate system. In the space-fixed coordinate system, the harmonic coefficients in real and Fourier space are mutual Hankel transforms.

In some respects, this approach is very attractive since, if the spherical harmonic expansions of the correlation functions are sufficiently rapidly convergent, the approximate solution of the Ornstein–Zernike equation for a molecular fluid can be placed upon essentially the same footing as that for a simple atomic fluid. The question of convergence of the spherical harmonic expansions turns out to be the key issue in determining the efficacy of the approach, so it is worthwhile to review briefly the available evidence on this question. Most of the work on this problem has concerned the spherical harmonic expansion of $g(1,2)$ for linear molecules. This work was pioneered by Streett and Tildesley,[32,33] who showed how it was possible to write the spherical harmonic expansion coefficients as ensemble averages obtainable from a Monte Carlo or molecular dynamics simulation via

$$g_{l_1,l_2,m}(r_{12}) = 4\pi g_{000}(r_{12}) \langle Y_{l_1,m}(\omega_1) Y_{l_2,-m}(\omega_2) \rangle \tag{3.1.11}$$

where the angled brackets denote an ensemble average. Streett and Tildesley looked at how the results obtained for $g(1,2)$ were affected by resumming various numbers of terms in the expansion, for fluids of hard diatomics and 12–6 diatomics. By examining the results for a variety of fixed relative orientations, they concluded that the convergence of the expansion was poor for intermolecular separations where the orientation dependence of the repulsive interactions is very large; for a homonuclear diatomic, this being $r_{12} < \sigma + L$ (where σ is the site diameter and L the bond length). Other investigators have obtained simulation results for $g(1,2)$ itself at selected relative orientations and reached similar conclusions.[34,35] Another manner in which the convergence can be quantified is through a consideration of the mean square deviation arising from various truncations of the series;[36] it can be shown that the expansion of a function in orthonormal basis with unit weight function leads to a minimization of this quantity. An important point to note is that for hard core site–site potentials $g(1,2)$ has a discontinuity at the distance of closest approach for each relative orientation and such behavior cannot, of course, be reproduced by a series expansion. To a large extent, the convergence problems of the spherical harmonic expansion of $g(1,2)$ are already evident in its low-density limit[26,36–38]

$$g(1,2) = \exp[-u(1,2)/kT] \tag{3.1.12}$$

Of course, the above considerations may not be relevant to the problem at hand, since in solving the OZ equation, the important functions are $\gamma(1,2)$, its Fourier transform and $B(1,2)$.[30] It is to be expected that $\gamma(1,2)$ will vary less quickly between different orientations and will be continuous even for hard core potentials. Thus, its expansion in spherical harmonics should be better behaved than that of $g(1,2)$. Computer simulation cannot be used to obtain $\gamma(1,2)$ but Lado has presented some evidence based upon his solution of the RHNC approximation for a hard diatomic fluid using a spherically averaged bridge function that the convergence is good. Nevertheless, the results he presents are, in our view, for a rather short diatomic bond length and may not be conclusive.

The approach has been applied to hard core interaction site systems by Chen and Steele[28] using the PY approximation and Lado[30] using the RHNC approximation with a spherically averaged bridge function. The results of Chen and Steele were compared with Monte Carlo results by Streett and Tildesley[32] for a moderate density and found to be in very close agreement. Lado[30] also compared his results with computer simulation data and found qualitatively good agreement, but there were substantial errors for the highest bond lengths considered at liquid-like densities. More recently, Lado[39] has applied the RHNC to the treatment of Lennard-Jones diatomics using an improved approximation for the bridge function based upon results for hard diatomics. The agreement with molecular dynamics results of Streett and Tildesley is very good.

In our view, there is a need for much more work in this area. For example, it would be very useful if more extensive studies with the PY approximation were made in order to investigate whether it is possible to obtain results for diatomics of comparable accuracy to those for hard spheres. Also, we do not believe that the influence of the convergence of the spherical harmonic expansions upon the accuracy of the solutions has been fully explored, and a more comprehensive study would be welcome. This approach has seemed unattractive in view of the investment of computer time and programming effort required. However, the computational limitations are becoming less significant, and if the issue of the convergence of the spherical harmonic expansions can be resolved in a more systematic fashion, then it appears likely that the approach may attract more interest. There are already clear indications of this with regard to other models of molecular shape effects.[40]

B. Ornstein–Zernike Like Equations for Site–Site Correlation Functions

As we discussed in Section II.B, site–site correlation functions provide a very useful formalism for describing the structure of fluids modeled with interaction site potentials. In this formalism, information equivalent to $g(1,2)$ is obtained from the set of site–site correlation functions and intramolecular correlation functions. For this reason, a great deal of effort has been put into the development of integral equation theories for these correlation functions. The seminal contribution in this area was made by Chandler and Andersen,[41] who sought to write an integral equation of the Ornstein–Zernike form in which the set of site–site total correlation functions were related to a set of site–site direct correlation functions. Their equation has the form

$$H = WCW + \rho WCH \tag{3.2.1}$$

where H and C are matrices with elements given by $H_{\alpha\gamma} = h_{\alpha\gamma}(r)$ and $C_{\alpha\gamma} = c_{\alpha\gamma}(r)$, the site–site total and direct correlation functions respectively. The elements of W are given by

$$W_{\alpha\gamma} = \delta_{\alpha\gamma}\delta(r) + s_{\alpha\gamma}(r) \tag{3.2.2}$$

where $s_{\alpha\gamma}(r)$ is the intramolecular correlation function which describes the distribution of sites within a single molecule. $\delta(r)$ and $\delta_{\alpha\gamma}$ are the Dirac and Kronecker delta functions, respectively. For molecules treated as a rigid array of interaction sites, we have

$$s_{\alpha\gamma}(r) = (1 - \delta_{\alpha\gamma})(4\pi L_{\alpha\gamma}^2)^{-1}\delta(r - L_{\alpha\gamma}) \tag{3.2.3}$$

The matrix products in Eq. (3.2.1) denote the operation of matrix multiplication together with convolution so that, for example,

$$(CH)_{\alpha\gamma} = \sum_{\beta} \int d\mathbf{r}' c_{\alpha\beta}(|\mathbf{r} - \mathbf{r}'|)h_{\beta\gamma}(\mathbf{r}') \tag{3.2.4}$$

Equation (3.2.1) can be rearranged to give

$$H = WC(I - \rho WC)^{-1}W \tag{3.2.5}$$

where I denotes the identity matrix. Approximate solutions of Eq. (3.2.1) can be obtained by assuming a closure relationship between the site–site total and direct correlation functions. The first such closure considered by Chandler and co-workers[41–43] in studies of hard sphere interaction site potentials was

$$c_{\alpha\gamma}(r) = 0 \qquad r > \sigma_{\alpha\gamma} \tag{3.2.6}$$

and

$$h_{\alpha\gamma}(r) = -1 \qquad r < \sigma_{\alpha\gamma} \tag{3.2.7}$$

This will be recognized as a transcription of the mean spherical approximation (MSA) or mean spherical model (MSM) to site–site language. Chandler and Andersen intended this closure of Eq. (3.2.1) for hard sphere sites to provide a theory for the reference system correlation functions in a perturbation theory for the free energy of interaction site fluids. For this reason, they referred to Eqs. (3.2.1), (3.2.6), and (3.2.7) as the reference interaction site model (RISM) equations. This terminology has been widely adopted in subsequent work. Other workers, notably Cummings and Stell,[17] prefer to use a different terminology. They refer to Eq. (3.2.1) as the site–site Ornstein–Zernike (SSOZ) equation, and to Eqs. (3.2.6) and (3.2.7) as the reference interaction site approximation (RISA). In doing so, they seek to distinguish the model describing the interactions in the system (hard sphere interaction sites) from the approximation used in the solution of the integral equation. In the remainder of this article, we adopt the latter approach or philosophy with a slight modification in that we refer to Eqs. (3.2.6) and (3.2.7) as the SSOZ–MSA theory.

Other closures have been used in the solution of the SSOZ equation. We may regard Eqs. (3.2.6) and (3.2.7) as a special case of a PY-like approximation for hard sphere potentials, so that more generally we could write

$$c_{\alpha\gamma}(r) = y_{\alpha\gamma}(r) f_{\alpha\gamma}(r) \tag{3.2.8}$$

where $y_{\alpha\gamma}(r) = g_{\alpha\gamma}(r)\exp[u_{\alpha\gamma}(r)/kT]$ and $f_{\alpha\gamma}(r)$ is the site–site f-function. We will call Eq. (3.2.8) the SSOZ–PY approximation. It was first applied to site–site Lennard-Jones potentials by Kojima and Arakawa[44] and subsequently by others.[45–47] Another possibility inspired by the hypernetted chain (HNC) approximation is

$$c_{\alpha\gamma}(r) = \exp[\gamma_{\alpha\gamma}(r) - \beta u_{\alpha\gamma}(r)] - \gamma_{\alpha\gamma}(r) - 1 \tag{3.2.9}$$

where $\gamma_{\alpha\gamma}(r) = h_{\alpha\gamma}(r) - c_{\alpha\gamma}(r)$. This closure was first used by Rossky and co-workers[48,49] in their work on interaction site models of polar fluids in which Coulombic site–site interactions arise. They refer to Eq. (3.2.1) together with Eq. (3.2.9) as the extended RISM equations. We prefer and will use the designation SSOZ–HNC for the approximation given by Eq. (3.2.9). Rossky[5] has recently reviewed the results obtained from applications of Eq. (3.2.9) to studies of the structure of polar fluids. Finally, the SSOZ–MSA approximation can be written more generally by replacing Eq. (3.2.6) by

$$c_{\alpha\gamma}(r) = -u_{\alpha\gamma}(r)/kT \quad r > \sigma_{\alpha\gamma} \tag{3.2.10}$$

Equation (3.2.10) was first applied by Morriss and Perram[50] in their analytic solution for dipolar hard diatomics (see Section IV for a discussion of the analytic solution of the SSOZ–MSA).

Considerable effort has been devoted to establishing the theoretical status of the SSOZ equation and the closure approximations that have been used to solve it. Equation (3.2.1) can be regarded as a defining equation for a site–site direct correlation function $c_{\alpha\gamma}(r)$, just as the OZ equation serves as a defining equation for the direct correlation function of an atomic fluid. In the case of an atomic fluid, the Ornstein–Zernike equation implies that the direct correlation function can be obtained by performing a resummation of those diagrams in the cluster expansion of the total correlation function that are free of nodal circles. It is tempting to assume that the site–site direct correlation function defined by Eq. (3.2.1) would also have this property; i.e. that $c_{\alpha\gamma}(r)$ would be the sum of those interaction site cluster diagrams in $h_{\alpha\gamma}(r)$ without nodal circles. Unfortunately, this turns out not to be the case, as was shown by Chandler and co-workers,[11,51] and as we shall now illustrate by considering the iterative solution of the SSOZ equation in the PY approximation in the limit of zero density. We begin with the diagrammatic expansion for $h_{\alpha\gamma}(r)$ in f-bonds and ρ-circles in Eq. (2.2.6). The first few terms corresponding to the low-density limit of this expansion are shown in Eq. (2.2.5). Equation (3.2.11) shows the diagrams appearing in the expansion of the cavity function $y_{\alpha\gamma}(r)$ at zeroth order in density.

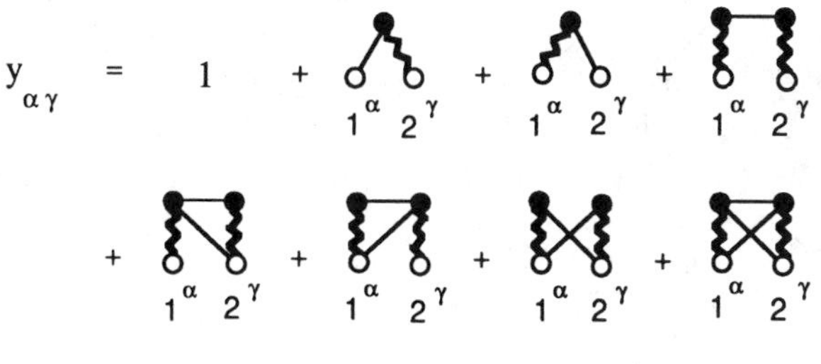

$$y_{\alpha\gamma} = 1 + \cdots \tag{3.2.11}$$

Now we can combine this expression for $y_{\alpha\gamma}(r)$ with the SSOZ–PY approximation of Eq. (3.2.8) to obtain an initial diagrammatic estimate for $c_{\alpha\gamma}(r)$. Substitution of this expression into the zero density SSOZ equation generates a diagrammatic expression for $h_{\alpha\gamma}(r)$ from which a new expression for $y_{\alpha\gamma}(r)$ may be obtained and so on. By continuing this iterative process we can build up the diagrammatic representation of $h_{\alpha\gamma}(r)$ at zero density. We find that two classes of diagram are obtained. The first is a subset of those in the exact result at zero order in density consisting of the first eleven diagrams in Eq. (2.2.5).

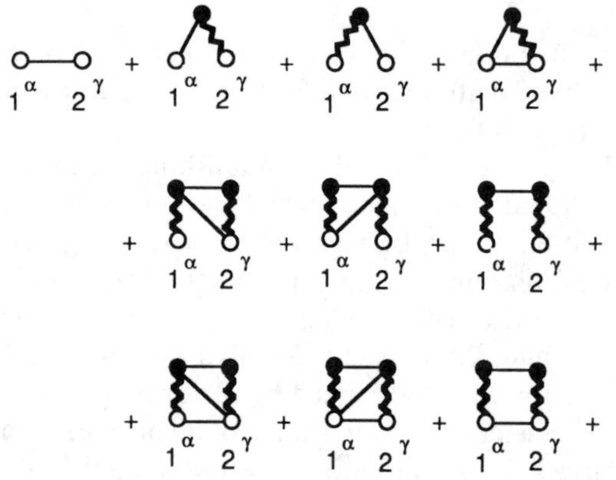

(3.2.12)

Notice that the diagrams neglected are similar in topology to those neglected at second order in density by the PY approximation for atomic fluids.[12] The second class of diagrams is of the type shown below.

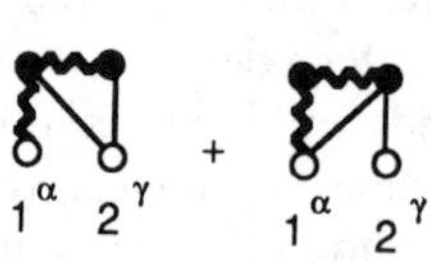

(3.2.13)

There are an infinite number of such diagrams at each order in density. These diagrams are unallowed since in each case one or more black circles are intersected by more than one *s*-bond. Similar results can be obtained by considering the SSOZ–HNC approximation, and a different analysis has been used[25] to show that the SSOZ–MSA also leads to the resummation of unallowed diagrams. Other features of the SSOZ equation associated with the inclusion of unallowed diagrams and neglect of allowed diagrams even at zeroth order in density have been discussed by Monson.[18] Another important feature of this approach involves the results obtained when auxiliary sites, which make no contribution to the intermolecular potential, are incorporated.[13] Such an approach is useful, for example, in calculating the center of mass correlation function in a diatomic fluid. It is found that the results obtained from the SSOZ–PY theory for the site–site and center of mass correlation functions depend upon the diameter of the auxiliary site.

There are other striking results that emerge from the use of the SSOZ equation with any of the closure equations given above to calculate the dielectric constant and the Kirkwood angular correlation parameters. We consider some of these in the next section.

C. The Dielectric Constant from the SSOZ Equation

The site–site direct correlation function $c_{\alpha\gamma}(r)$ often contains nonanalytic terms which come directly from the particular site–site interaction potential and the assumed closure relation. The most common such term comes from the site–site Coulomb interaction, in combination with any closure relation whose long-range form is $-\beta u(r)$. In general, the Fourier transform of $c_{\alpha\gamma}(r)$ is of the form

$$\hat{c}_{\alpha\gamma}(k) = -\frac{4\pi\beta q_\alpha q_\gamma}{k^2} + \hat{c}_{\alpha\gamma}(0) + k^2\hat{c}^{(2)}_{\alpha\gamma} + k^4\hat{c}^{(4)}_{\alpha\gamma} + \cdots \tag{3.3.1}$$

There is also the possibility of a nonanalytic term of $O(k^3)$ in the expansion of $\hat{c}_{\alpha\gamma}(k)$ due to the r^{-6} decay in the site–site Lennard-Jones interaction.

An expression for the dielectric constant from the SSOZ equation was first alluded to by Chandler.[52] The result for linear and tetrahedral molecules in the MSA closure was obtained by Morriss,[53] but the derivation we give here is more general and was obtained by Sullivan and Gray.[54] They consider the SSOZ equation in the following form given in (2.2.12). Note that when $k = 0$ all the elements of the matrix $\hat{W}(k)$ are equal to one. We can define a projection operator P by $P = \hat{W}(0)/n$, and the associated complementary operator Q by $Q = I - P$ (where I is the identity matrix). It is easy to show that Q and P are orthogonal. Operating on both sides of the SSOZ equation with Q gives

$$Q\hat{H}(k) = Q(k^2\hat{W}^{(2)} + k^4\hat{W}^{(4)} + \cdots)\hat{C}[\hat{W} + \rho\hat{H}] \tag{3.3.2}$$

as $Q\hat{W}(0) = nQP = 0$. The next step is to develop a general power series expansion for $Q\hat{H}(k)$. For polar fluids, we have shown that $\hat{C}$ has an expansion of the form

$$\hat{C}(k) = k^{-2}\hat{C}^{(-2)} + \hat{C}^{(0)} + k^2\hat{C}^{(2)} + \cdots \tag{3.3.3}$$

where the coefficient of k^{-2} has a particularly simple form; $\hat{C}^{(-2)} = -4\pi\beta q_\alpha q_\gamma$. The square bracketed term in Eq. (3.3.2) has an expansion of the form

$$\hat{W}(k) + \rho\hat{H}(k) = \hat{W}(0) + \rho\hat{H}(0) + k^2[\hat{W}^{(2)} + \rho\hat{H}^{(2)}] + k^4[\hat{W}^{(4)} + \rho\hat{H}^{(4)}] + O(k^6) \tag{3.3.4}$$

We have already seen that $\hat{W}(0)$ is matrix with each element equal to one; also we have shown (after Eq. (2.4.6)) that each $\hat{h}_{\alpha\gamma}(0)$ is independent of the site labels α and γ, so that

$$\hat{W}(0) + \rho\hat{H}(0) = (\hat{\omega}(0) + \rho\hat{h}(0))nP \tag{3.3.5}$$

Combining the results at Eqs (3.3.3), (3.3.4), and (3.3.5) gives

$$\hat{C}(k)[\hat{W}(k) + \rho\hat{H}(k)] = k^{-2}\{n(\hat{\omega}(0) + \rho\hat{h}(0))\hat{C}^{(-2)}P\} + k^0\{\hat{C}^{(-2)}(\hat{W}^{(2)} + \rho\hat{H}^{(2)}) + \hat{C}^{(0)}Pn(\hat{\omega}(0) + \rho\hat{h}(0))\} \tag{3.3.6}$$

It is straightforward to see that charge neutrality implies that the coefficient of k^{-2} is identically zero.

$$(\hat{C}^{(-2)}P)_{ik} = -4\pi\beta q_i \sum_j q_j = 0 \tag{3.3.7}$$

Substituting this into the RHS of Eq. (3.3.2) and equating the coefficients of k^0 gives

$$Q\hat{H}^{(2)} = Q\hat{W}^{(2)}\{\hat{C}^{(-2)}(\hat{W}^{(2)} + \rho\hat{H}^{(2)}) + n(\hat{\omega}(0)\,\rho\hat{h}(0))\hat{C}^{(0)}P\} \tag{3.3.8}$$

Operating on this equation from the RHS with Q removes the second term as $PQ = 0$. The result is

$$Q\hat{H}^{(2)}Q = Q\hat{W}^{(2)}\hat{C}^{(-2)}(\hat{W}^{(2)} + \rho\hat{H}^{(2)})Q \tag{3.3.9}$$

From Eq. (2.4.17) we see that the dielectric constant involves the combination of $\hat{h}_{\alpha\gamma}$ and the site charges q_α and q_γ. To obtain this combination, we operate on both sides of Eq. (3.3.9) with $\hat{C}^{(-2)}$, from both the left and right. Using the result that $Q\hat{C}^{(-2)} = \hat{C}^{(-2)}$, and that Q commutes with $\hat{C}^{(-2)}$, we find that

$$\hat{C}^{(-2)}\hat{H}^{(2)}\hat{C}^{(-2)} = \hat{C}^{(-2)}\hat{W}^{(2)}\hat{C}^{(-2)}(\hat{W}^{(2)} + \rho\hat{H}^{(2)})\hat{C}^{(-2)} \tag{3.3.10}$$

Substituting in the forms for $\hat{W}^{(2)}$ and $\hat{C}^{(-2)}$ it can be shown that

$$\Delta\hat{h} = \frac{-4\pi\beta M^2 k^2}{1 + 4\pi\rho\beta M} \tag{3.3.11}$$

where

$$M = \sum_{ij} q_i q_j \hat{\omega}_{ij}^{(2)} = \frac{\mu^2}{3} \tag{3.3.12}$$

so that

$$\Delta\hat{h} = \frac{k^2\mu^2}{3\rho}\left(\frac{-3y}{1 + 3y}\right) \tag{3.3.13}$$

Combining this with Eq. (2.4.17) we obtain the result that

$$\varepsilon_{\text{ssoz}} = 1 + 3y \tag{3.3.14}$$

for any closure relation whose long-range form is given by $-\beta u(r)$. This is a serious deficiency of the SSOZ equation and its closure. There have been differing schools of thought regarding the cause of this trivial result. Chandler has claimed that it is a deficiency of the SSOZ equation itself, whereas Stell has pointed out that the SSOZ equation can be easily (and rigorously) derived from the mixture OZ equation, so the assumed closure relation $c_{\alpha\gamma}(r) \sim -\beta u_{\alpha\gamma}(r)$ must be incorrect. Although the latter view is reasonable, no formal cluster series expansion for $c_{\alpha\gamma}(r)$ has been obtained, so it appears that the $c_{\alpha\gamma}(r)$ defined by the SSOZ equation is not a function amenable to simple closure approximations.

Evidently, there are serious theoretical deficiencies associated with the SSOZ equation and the site–site direct correlation function defined by it. Two rather different perspectives have emerged as a means of correcting these deficiencies. One approach, originated by Cummings and Stell,[17] is to abandon the link with the cluster expansions and regard the SSOZ equation strictly as a definition of a site–site direct correlation function. Cummings, Morriss and Stell[55] and Cummings and Stell[56] have shown that deficiencies in approximate solutions of the SSOZ equation can be interpreted in terms of the asymptotic behavior of the site–site direct correlation function for large values of r. For a number of interaction site systems, they have derived the exact asymptotic form of $c_{\alpha\gamma}(r)$ defined by the SSOZ equation. For example, they show that for a dipolar hard dumbbell,

$$\lim r \to \infty \quad c_{\alpha\gamma}(r) = -q_\alpha q_\beta A/kT \tag{3.3.15}$$

where the parameter A is related to the dielectric constant via

$$\varepsilon = (1 + 3yA)/[1 + 3y(A - 1)] \tag{3.3.16}$$

Thus, if we know the dielectric constant of the fluid, then Eq. (3.3.15) could be used to improve the results obtained from the SSOZ–MSA approximation, for example. This has been done for dipolar hard dumbbell fluid by Lee and Rasaiah[58] using Monte Carlo simulation results for the dielectric constant (Morriss[60]), and by Rossky, Pettitt and Stell.[59] However, this approach does not seem to have any value as a predictive tool. Moreover, in the case of some interaction site models, notably hard linear triatomics, the site–site direct correlation function is not a short-range function but in fact increases with increasing r.[57] This notwithstanding, the Cummings–Stell analysis remains an important contribution to our understanding of the SSOZ equation.

The G_2 parameter determines the magnitude of the Kerr constant as well as the depolarized light scattering intensities in fluids of linear nonpolar molecules. Høye and Stell[22] have developed general relations between G_l and the site–site correlation functions. Using these results, Sullivan and Gray showed that the G_2 parameter for any symmetric linear molecule is identically zero.[54]

D. The Chandler–Silbey–Ladanyi Equation

Chandler and his co-workers have taken a somewhat different point of view in seeking to improve upon the SSOZ equation.[51,61] They start from the perspective that the integral equation itself is flawed since all the tractable closures correspond to the resummation of unallowed diagrams in the interaction site cluster series for the site–site total correlation function. They have formulated an integral equation in which the direct correlation function does indeed correspond to the subset of diagrams in the interaction site cluster series in which there are no nodal circles. The key to their development is a grouping of the site–site total and direct correlation functions into four classes depending upon how the root points are intersected by **s**-bonds. They write

$$H = H_0 + H_l + H_r + H_b \tag{3.4.1}$$

$$C = C_0 + C_l + C_r + C_b \tag{3.4.2}$$

The subscript 0 denotes contributions from diagrams where the roots points in the diagrams are intersected only by **f**-bonds. The subscript l denotes contributions from diagrams where the left root point is intersected by an s-bond. The subscript r denotes contributions from diagrams where the right root point is intersected by an s-bond. The subscript b denotes contributions from diagrams where both root points are intersected by s-bonds. The zeroth order terms in the components of H and C are shown in Fig. 2. Chandler, Silbey, and Ladanyi have shown that with this decomposition, the components of the total and direct correlation functions are related by the integral equation

$$W + \rho H = (I - \rho\Omega C_0)^{-1}\Omega \tag{3.4.3}$$

where

$$\Omega = (I - \rho C_l)^{-1}(W + \rho C_b)(I - \rho C_r)^{-1} \tag{3.4.4}$$

They refer to Eq. (3.4.3) as the proper integral equation in view of the fact that the direct correlation function so defined does correspond to the sum of the nodeless diagrams in the interaction site cluster expansion of the total correlation function. Following Lupkowski and Monson,[25] we shall refer to Eq. (3.4.3) as the Chandler–Silbey–Ladanyi (CSL) equation. Interestingly, the component functions have a simple physical interpretation. The elements of H_0 correspond to the total correlation functions for pairs of sites at infinite dilution in the molecular solvent. The elements of the sum of H_0 and H_l (or H_r) matrices are the solute–solvent site–site total correlation functions for sites at infinite dilution in the molecular solvent.

Shortly after the work of Chandler et al.,[51,61] Rossky and Chiles[62] presented a slightly different but equivalent version of the CSL equation which makes the nature of the integral equation and its relationship to the SSOZ equation somewhat more transparent. To understand their approach, it is necessary to introduce the particle–particle representation of the site–site correlation functions (Cummings and Stell[17]). If we define

$$H^{(p)} = S/\rho + H \tag{3.4.5}$$

and

$$C^{(p)} = C + (I - W^{-1})/\rho \tag{3.4.6}$$

then the SSOZ equation may be written in the form

$$H^{(p)} = C^{(p)} + C^{(p)}\bar{\rho}H^{(p)} \tag{3.4.7}$$

where $\bar{\rho} = \rho I$. Written in this way, the SSOZ equation has the same form as the OZ equation for a mixture. Rossky and Chiles sought to preserve the

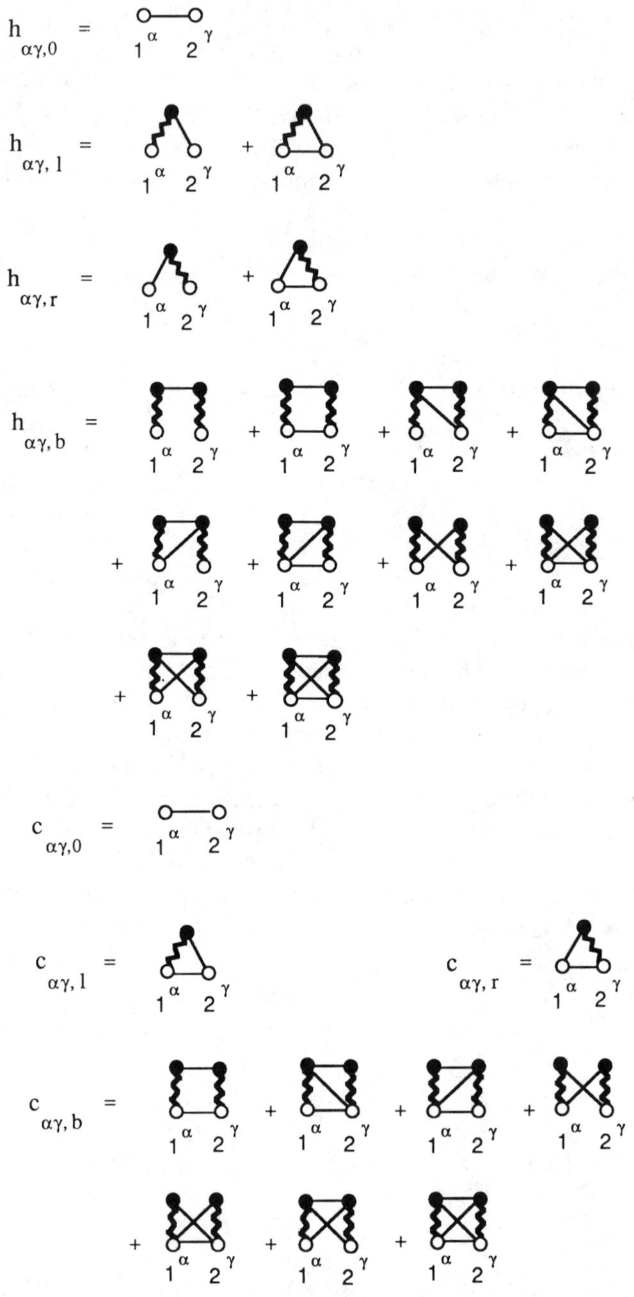

$h_{\alpha\gamma,0}$ = 1^{α} 2^{γ}

$h_{\alpha\gamma,1}$ = 1^{α} 2^{γ} + 1^{α} 2^{γ}

$h_{\alpha\gamma,r}$ = 1^{α} 2^{γ} + 1^{α} 2^{γ}

$h_{\alpha\gamma,b}$ = 1^{α} 2^{γ} + 1^{α} 2^{γ} + 1^{α} 2^{γ} + 1^{α} 2^{γ}

+ 1^{α} 2^{γ} + 1^{α} 2^{γ} + 1^{α} 2^{γ} + 1^{α} 2^{γ}

+ 1^{α} 2^{γ} + 1^{α} 2^{γ}

$c_{\alpha\gamma,0}$ = 1^{α} 2^{γ}

$c_{\alpha\gamma,1}$ = 1^{α} 2^{γ} $c_{\alpha\gamma,r}$ = 1^{α} 2^{γ}

$c_{\alpha\gamma,b}$ = 1^{α} 2^{γ} + 1^{α} 2^{γ} + 1^{α} 2^{γ} + 1^{α} 2^{γ}

+ 1^{α} 2^{γ} + 1^{α} 2^{γ} + 1^{α} 2^{γ}

Figure 2.

form of Eq. (3.4.7) but to redefine the matrices in such a way that in the second term, only those multiplications which will give rise to allowed terms in the interaction site cluster expansion are included. In this way, the direct correlation function thus defined would correspond to the sum of nodeless graphs in the interaction site cluster series for $H^{(p)}$. Rossky and Chiles make the same decomposition of the total and direct correlation functions as Chandler et al. but they use the subscripts *aa, sa, as, ss* instead of $0, l, r, b$ to denote the components of $H^{(p)}$ and $C^{(p)}$. The matrices that lead to a satisfactory equation structure are obtained by replacing each element in $H^{(p)}$ and $C^{(p)}$ by a 2×2 block with elements given by the *aa, sa, as,* and *ss* components. For example, we have

$$H_{\alpha\gamma} \to \begin{bmatrix} H_{\alpha_a\gamma_a} & H_{\alpha_a\gamma_s} \\ H_{\alpha_s\gamma_a} & H_{\alpha_s\gamma_s} \end{bmatrix} \tag{3.4.8}$$

The diagonal elements of the matrix $\bar{\rho}$ are replaced by 2×2 matrices of the form

$$\begin{bmatrix} \rho & \rho \\ \rho & 0 \end{bmatrix}$$

and the off-diagonal elements are replaced by 2×2 zero matrices. The result of this is an augmented matrix equation of dimension $2m \times 2m$ where m is the number of sites in the molecule. The direct correlation function defined by this equation can be shown to be equivalent to that in the CSL equation and corresponds to the sum of nodeless diagrams in the interaction site cluster series for the site–site total correlation function. From this point on, we will adopt the Rossky–Chiles formalism (RCF) in our remaining discussions of the CSL equation.

Chandler et al.[51,61] and Rossky and Chiles have shown how it is possible to develop closure approximations for the CSL equation. For example, the analogue of the PY approximation is

$$\begin{aligned} C_{\alpha_a\gamma_a} &= f_{\alpha\gamma}(1 + \gamma_{\alpha_a\gamma_a}) \\ C_{\alpha_a\gamma_s} &= f_{\alpha\gamma}\gamma_{\alpha_a\gamma_s} \\ C_{\alpha_s\gamma_a} &= C_{\gamma_a\alpha_s} \\ C_{\alpha_s\gamma_s} &= s_{\alpha\gamma} + f_{\alpha\gamma}\gamma_{\alpha_s\gamma_s} \end{aligned} \tag{3.4.9}$$

Rossky and Chiles refer to this as the ISF–PY approximation. They have applied this closure to a Lennard-Jones 12–6 diatomic model representative of liquid chlorine. They find that when compared with computer simulation results, the agreement is not as good as that obtained from the solution of the SSOZ–PY approximation. This represents something of a disappointment in view of the extra complexity of the CSL equation and the fact that it does have a firmer theoretical basis than the SSOZ equation. It is worthwhile to reflect upon the origin of this from the perspective of the interaction site cluster series. The key difference between the SSOZ–PY and ISF–PY approximations is that the ISF–PY does not resum the infinite set of unallowed diagrams which are included in the SSOZ–PY approximation. The allowed diagrams resummed to the two theories are the same. However, the allowed diagrams that are neglected in both approximations constitute a much larger contribution to the total correlation function than the diagrams neglected in the PY approximation to the OZ equation. The topology of the neglected diagrams is similar in each case; the diagrams are of the bridge and parallel types. However, since in the interaction site cluster series, diagrams of this topology can be built up from both **f**-bonds and **s**-bonds, such contributions can arise at zeroth order in density, and at all orders in density they make up a larger fraction of the diagrams present. On the other hand, in the molecular cluster series, diagrams of the bridge and parallel type do not appear until second order in density. The superior accuracy of the SSOZ–PY equation over the ISF–PY is perhaps a reflection of some cancellation of the unallowed diagrams summed in SSOZ–PY with the neglected allowed diagrams. However, when such a cancellation might occur seems totally unpredictable. Thus, from this point of view, the CSL equation does not appear to represent the advance that one might have hoped for from a more rigorous approach. However, we shall see later in Section III.G that the ideas which form the basis of this approach can be used with some success to develop perturbation theories for interaction site fluids.

E. Perturbation Theories—Molecular Anisotropy as the Perturbation

As we mentioned in the opening paragraph, thermodynamic perturbation theory has been used in two contexts in applications to interaction site fluids. In this section, we will describe efforts to treat the thermodynamics and structure of interaction site fluids in terms of a perturbation expansion where the reference system is a fluid in which the intermolecular forces are spherically symmetric. In developing thermodynamic perturbation theories, it is generally necessary to choose both a reference system and a function for describing the path between the reference fluid and the fluid of interest. The latter choice is usually made between the pair potential and its Boltzmann factor. Thus one writes either

$$u(1,2) = u_0(r_{12}) + \lambda \Delta u(1,2) \tag{3.5.1}$$

or

$$f(1,2) = f_0(r_{12}) + \lambda \Delta f(1,2) \tag{3.5.2}$$

where λ varies from 0 to 1. In the treatment of interaction site systems, the first of these can generally be ruled out since the anisotropy of the repulsive part of the intermolecular potential will make the perturbation term extremely large for certain configurations, although there is an important exception to this as will be seen later. On the other hand, the Mayer function is always a more well-behaved function. The most commonly used choice of reference system is to make $f_0(r_{12})$ equal to the orientation average of $f(1,2)$, i.e.

$$f_0(r_{12}) = \iint f(1,2)\, d\omega_1\, d\omega_2 \tag{3.5.3}$$

where the integrations are performed over the orientations normalized by the total orientation space for each molecule; 4π for linear molecules and $8\pi^2$ for nonlinear molecules. This choice of reference system can be partly justified in terms of the fact that to first order the Helmholtz free energy is given by

$$A = A_0 - \frac{\rho^2}{2kT} \iint y_0(r_{12})[\Delta f(1,2)]\, d1\, d2 + \cdots \tag{3.5.4}$$

and with this choice of the reference system the first-order term vanishes. Recent interest in the approach started with the work of Perram and White[63] and Smith[64], although their work is predated by some twenty years by that of Cook and Rowlinson,[65] who considered Eq. (3.5.3) as a means of generating an effective spherical potential for molecular fluids. At zeroth order, the pair distribution is obtained as

$$g(1,2) = y_0(r_{12})[1 + f(1,2)] + \cdots \tag{3.5.5}$$

Calculation of the second-order term in Eq. (3.5.4) and the first-order term in Eq. (3.5.5) requires knowledge of the triplet distribution function in the reference fluid[63,64] which is usually replaced by the Kirkwood superposition approximation. Following Smith,[64] we will refer to the approach as a whole as the reference averaged Mayer (RAM) function theory. Another choice of reference system based upon a division of the Mayer function is that of hard spheres with a diameter chosen so that the first-order term in the free energy vanishes. This gives rise to the so called blip function theory.[66,67]

There have been numerous applications of the RAM theory to both interaction site[68–71] as well as other models[72] of molecular fluids. There is clear evidence that the theory works best for purely repulsive intermolecular potentials[69] since under these circumstances the Mayer function always lies between -1 and zero. For strongly attractive potentials at low temperature, the theory is not at all successful. The other feature of the approach which should come as no surprise is that it is most suitable for systems with small anisotropy and increases in accuracy at low density. The theory has most often been applied to the calculation of the molecular correlation function and its spherical harmonic expansion coefficients.[68] However, it can also be adapted to the treatment of site–site correlation functions by choosing the origin of the coordinate system to be one of the interaction sites rather than the molecular center of mass.[68] The best results are obtained for the orientation average of $g(1,2)$.[69] Perhaps the most successful application has been in the context of calculating reference system correlation functions[73,74] for use in an extension of the Weeks–Chandler–Andersen perturbation theory to interaction site fluids, which we shall describe in Section III.F.

Another approach that has recently received considerable attention is to look for choices of reference system for which the perturbation to the intermolecular potential and that to the Mayer function are in one-to-one correspondence. Such a procedure was formalized by Lebowitz and Percus[75] who showed that if

$$\Delta u(1,2) = \operatorname{sgn}\,[u(1,2) - u_0(r_{12})] \tag{3.5.6}$$

where $\operatorname{sgn}(x)$ is the signum function

$$\begin{aligned} \operatorname{sgn}(x) &= -1 \quad x<0 \\ &= 0 \quad x=0 \\ &= 1 \quad x>0 \end{aligned} \tag{3.5.7}$$

then the perturbation expansion would be independent of whether the Mayer function or intermolecular potential was used to parametrize the perturbation. Lebowitz and Percus showed that the reference potential satisfying these requirements corresponds to taking the median value of the full potential or, equivalently, the Mayer function for a given intermolecular separation. Equation (3.5.5) turns out to be the limit as B becomes very large of the expression

$$\Delta u(1,2) = \sinh^{-1} B[u(1,2) - u_0(r_{12})] \tag{3.5.8}$$

which was used earlier by Shaw et al.[76] in a rather successful study of the equation of state of a diatomic 12–6 model of dense fluid nitrogen. It was this work which first stimulated interest in this type of approach. The method has since been applied quite widely and seems to give quite accurate thermodynamics for dense Lennard-Jones and hard sphere diatomics.[77] However, the structural predictions are rather poor.[78,79] A good critical discussion of the limitations of the approach has been given by MacGowan et al.[80]

Generally speaking, the theories described in this section have not been a great success except for mildly anisotropic systems. With hindsight, this might well have been anticipated. Perturbation theories for the correlation functions of fluids can only be expected to be accurate under two circumstances. The first of these is if the perturbative forces make no significant contribution to the fluid structure. Since it is the anisotropy of the intermolecular forces, especially arising from the molecular shape, which gives rise to the physically interesting contributions to the structure of molecular fluids, this criterion will not be satisfied except for mildly anisotropic molecules. The second circumstance is if there are methods for calculating and resumming the effects of the perturbative forces upon the fluid structure. Such methods are only feasible for much weaker perturbations to the fluid structure than those arising from molecular shape. Nevertheless, research in this area has been valuable, even if only in the sense that it is always useful to try the simplest approach first.

F. Perturbation Theories—Long Range Forces as the Perturbation

We turn now to a discussion of perturbation theories based upon extensions of the Barker–Henderson[81] and Weeks–Chandler–Anderson[82] theories to interaction site potentials. Such theories seek to treat the properties of the fluid as a perturbation about a reference fluid with anisotropic repulsive forces only. The theories have been formulated both explicitly in terms of division of the site–site potential into reference and perturbation potentials (Tildesley,[83] Lombardero et al.[84])

$$u_{\alpha\gamma}(r) = u_{\alpha\gamma}^{(0)}(r) + \lambda \Delta u_{\alpha\gamma}(r) \tag{3.6.1}$$

and in terms of a corresponding division of the total intermolecular potential[73]

$$u(1,2) = u_0(1,2) + \lambda \Delta u(1,2) \tag{3.6.2}$$

In the former case, the terms in the perturbation expansion of the Helmholtz free energy may be obtained in terms of site–site distribution functions. For example, to first order we have

$$A = A_0 + \frac{\rho^2}{2} \sum_{\alpha,\gamma} \int\int g_{\alpha\gamma}^{(0)}(r) \Delta u_{\alpha\gamma}(r) d\mathbf{r}_1^{(\alpha)} d\mathbf{r}_2^{(\gamma)} \tag{3.6.3}$$

Similarly, if a division is made of the full potential, then the terms in the perturbation expansion of the Helmholtz free energy are obtained in terms of the molecular distribution functions.

$$A = A_0 + \frac{\rho^2}{2} \int\int g_0(1,2) \Delta u(1,2) d1\, d2 \tag{3.6.4}$$

Most investigations in this area have been concerned with the Weeks–Chandler–Andersen division of the potential. For the molecular potential, the reference system is defined by

$$\begin{aligned} u_0(1,2) &= u(1,2) + \varepsilon(1,2) \quad & r_{12} < r_m(1,2) \\ &= 0 & r_{12} > r_m(1,2) \end{aligned} \tag{3.6.5}$$

and the perturbation by

$$\begin{aligned} \Delta u(1,2) &= -\varepsilon(1,2) \quad & r_{12} < r_m(1,2) \\ &= \quad u(1,2) & r_{12} > r_m(1,2) \end{aligned} \tag{3.6.6}$$

where $\varepsilon(1,2)$ and $r_m(1,2)$ are the well depth and the separation at which the potential is minimum for a fixed relative orientation of the molecules. The WCA potential can be similarly defined in the site–site formalism. There is a slight difference between the two reference systems since in the molecular frame both attractive and repulsive site–site interactions may contribute to a potential that is repulsive as a whole.

The free energy and distribution functions of the reference system are usually related to those of a hard sphere interaction site fluid (i.e., in the case of homonuclear diatomics this would be a hard dumbbell fluid) using generalizations of the blip function theory.[82] There are several empirical representations of the equation of state for hard sphere interaction site fluids (for a review see Boublik and Nezbeda[85]) which give good agreement with the results from Monte Carlo simulations (see, for example, Streett and Tildesley[86]) and from which the free energy may be obtained by integration.

The most difficult and least satisfactory part of the theories is the calculation of the cavity distribution functions for the hard sphere interaction site fluids. If the site–site formalism is being used, then the only suitable method available is the solution of the SSOZ–PY or RISM equation.[83] Results obtained in this way are reasonably accurate but are not as good as those obtained for hard spheres from the PY equation or semi-empirical corrections to it. A good semi-empirical representation of the site–site distribution functions for some classes of hard sphere interaction site fluids would be a welcome contribution in this area. If the molecular frame is used, then the method most often used is the RAM theory at zeroth order[73] in which

$$y(1,2) = y_0(r_{12}) \tag{3.6.7}$$

The evidence available suggests that the two approaches are about equally accurate,[69] although the approach based on site–site correlation functions is more readily generalized to the treatment of multipolar interactions as well as to the effect of the attractive forces upon the structure and free energy at moderate and low density. In addition to the efforts made at extending the WCA theory to interaction site fluids, the Barker–Henderson theory[81] has also been extended to these systems by Lombardero, Abascal, Lago and their co-workers.[84,87]

G. Cluster Perturbation Theory

As first-order perturbation theories based upon repulsive force reference fluids, the theories described in the previous section can be expected to give accurate results for dense fluids, and this turns out to be the case. Nonetheless, two important problems have not been addressed in the above treatments. The first is how treat structure-determining perturbations arising from the attractive forces at low density, and the second is how to treat the effects of electrostatic interactions arising from multipole moments in the molecules. In this section, we discuss some recent efforts to tackle these problems based upon cluster expansion techniques. Our treatment is based upon a recent paper by Lupkowski and Monson[25] in which the problem is discussed in some detail. The principal difficulty is that when the perturbative forces influence the fluid structure, it is necessary to incorporate higher-order terms in the perturbation expansion. The calculation of such terms generally requires knowledge of multibody correlation functions for the reference fluid, and there are no satisfactory and computationally convenient routes to these functions for interaction site systems. Efforts in this direction which illustrate the difficulties have been made by Sandler and co-workers[88,89] and Wojcik and Gubbins.[90]

An alternative to the calculation of higher-order terms in the λ expansion is provided by analysis of the interaction site cluster expansions of the correlation functions and Helmholtz free energy into contributions arising from the short-range and long-range site–site interactions. This leads to extensions to interaction site systems of theories such as the optimized cluster theory of Andersen and Chandler,[91,92] and the self consistent Γ-ordered theory of Stell and co-workers[93] developed for simple fluids. Such theories have been highly successful in the treatment of the Lennard-Jones equation of state at moderate and low density as well as in studies of electrolyte solutions. Lupkowski and Monson[25] have recently shown how such theories can be extended in a systematic and rigorous fashion to interaction site systems. The principal difficulty in doing this is the correct treatment of the effects of intermolecular correlations so as to avoid the resummation of unallowed diagrams in the approximations developed.

The starting point of their treatment is the interaction site cluster expansions of the Helmholtz free energy[27] and the site–site distribution functions.[12] They make a division of the site–site potential into short-range and long-range contributions as follows:

$$u_{\alpha\gamma}(r) = u^{(0)}_{\alpha\gamma}(r) + w_{\alpha\gamma}(r) \tag{3.7.1}$$

With this decomposition, the Mayer function can then be written as

$$f_{\alpha\gamma}(r) = f^{(0)}_{\alpha\gamma}(r) + [1 + f^{(0)}_{\alpha\gamma}(r)] \sum_{n=1}^{\infty} \frac{\phi^n_{\alpha\gamma}(r)}{n!} \tag{3.7.2}$$

where

$$\phi_{\alpha\gamma}(r) = -w_{\alpha\gamma}(r)/kT \tag{3.7.3}$$

and $f^{(0)}_{\alpha\gamma}(r)$ is the Mayer function for the reference fluid. When this expression is inserted into the interaction site cluster expansion for $h_{\alpha\gamma}(r)$ and the Helmholtz free energy, we have

$$h_{\alpha\gamma}(\mathbf{r}^{(\alpha)}_1, \mathbf{r}^{(\gamma)}_2) = \{\text{sum of all allowed interaction site diagrams with two white circles, any number of black circles, with zero or one } \mathbf{f}^{(0)} \text{ bond and any number of } \boldsymbol{\phi} \text{ bonds between any pair of circles, and with zero or more } \mathbf{s} \text{ bonds}\} \tag{3.7.4}$$

and

$$\mathscr{A} = A/VkT = V^{-1}\{\text{sum of all allowed interaction site diagrams with two or more black circles, with zero or one } \mathbf{f}^{(0)} \text{ bond and any number of } \boldsymbol{\phi} \text{ bonds between any pair of circles, and with zero or more } \mathbf{s} \text{ bonds}\} \tag{3.7.5}$$

The next step in the theory is to make a topological reduction so that the dependence upon the $\mathbf{f}^{(0)}$ bonds is replaced by a dependence upon $\mathbf{h}^{(0)}$ bonds. Lupkowski and Monson show that this topological reduction cannot be carried out unless the reference total correlation functions are decomposed in the manner described in Section III.B in connection with the development of the CSL integral equation. This result is analogous to the derivation of the CSL equation as a topological reduction of the interaction site cluster series for h. They do, however, show that there is subset of diagrams in Eq. (3.5.3) which can be written as a series of simple chains of $\boldsymbol{\phi}$, $\mathbf{h}^{(0)}$ and $\mathbf{s}^{(2)}$ bonds without making the decomposition of $h^{(0)}$. This set of diagrams is the so-called renormalized potential or chain sum which appears in the cluster perturbation theories for atomic fluids[91] and which may be readily evaluated via Fourier transform techniques. After carrying out the topological reductions, the following exact expression is obtained:

$$h_{\alpha\gamma}(\mathbf{r}_1^{(\alpha)}, \mathbf{r}_2^{(\gamma)}) = h_{\alpha\gamma}^{(0)}(\mathbf{r}_1^{(\alpha)}, \mathbf{r}_2^{(\gamma)}) + \{\text{sum of all allowed interaction site diagrams with two white circles, any number of black circles, with one or more } \boldsymbol{\phi} \text{ bonds and zero or one } \mathbf{h}_{aa}^{(0)}, \mathbf{h}_{sa}^{(0)}, \mathbf{h}_{as}^{(0)}, \text{ or } \mathbf{h}_{ss}^{(0)} \text{ bond between any pair of circles, and with zero or more } \mathbf{s} \text{ bonds and no reference articulation pair of circles}\} \tag{3.7.6}$$

The corresponding expression for the free energy is

$$\mathscr{A} = \mathscr{A}_0 + V^{-1}\{\text{sum of all allowed interaction site diagrams with two or more black circles, with zero or one } \mathbf{h}_{aa}^{(0)}, \mathbf{h}_{sa}^{(0)}, \mathbf{h}_{as}^{(0)} \text{ or } \mathbf{h}_{ss}^{(0)} \text{ bond and one or more } \phi \text{ bonds between any pair of circles and with zero or more } \mathbf{s} \text{ bonds, and no reference articulation pair or circles}\} \tag{3.7.7}$$

Two contributions to the second term on the right-hand side of Eq. (3.7.7) can be readily identified. The first consists of those diagrams with only two white circles joined by a $\boldsymbol{\phi}$ bond and either a $\mathbf{h}_{aa}^{(0)}$, $\mathbf{h}_{sa}^{(0)}$, $\mathbf{h}_{as}^{(0)}$ or $\mathbf{h}_{ss}^{(0)}$ bond. These diagrams can of course be written in terms of only a single $\mathbf{h}^{(0)}$ bond and have the same form as the first-order term in the λ expansion. In common with cluster perturbation theories for atomic fluids,[91] this contribution is called $\mathscr{A}_{\text{HTA}}$, where HTA denotes high-temperature approximation. A second contribution consists of the sum of the so called ring diagrams, $\mathscr{A}_{\text{RING}}$, and like the case of the chain sum this contribution can be written without explicitly making a decomposition of $\mathbf{h}^{(0)}$. Thus we have

$$\mathscr{A} \simeq \mathscr{A}_0 + \mathscr{A}_{\text{HTA}} + \mathscr{A}_{\text{RING}} \tag{3.7.8}$$

where $\mathscr{A}_{\text{HTA}}$ is essentially the same as the A_1 term in Eq. (3.6.3) apart from a factor of $(VkT)^{-1}$ and

$$\mathscr{A}_{\text{RING}} = -\tfrac{1}{2}(2\pi)^{-3} \int \{Tr[(\bar{\rho} + \bar{\rho}H_0\bar{\rho})\boldsymbol{\Phi}] + \ln\det[I - (\bar{\rho} + \bar{\rho}H_0\bar{\rho})\boldsymbol{\Phi}]\}\, d\mathbf{k} \tag{3.7.9}$$

In Eq. (3.7.9) H_0 is the matrix of particle–particle correlation functions for the reference system and has the same form as the matrix H used in the Rossky–Chiles[62] reformulation of the CSL integral equation. The matrix $\bar{\rho}$ is also the same as that used by Rossky and Chiles.

The renormalized potential obtained from the chain diagrams in Eq. (3.7.6) can be regarded as an approximation to the influence of the perturbation potential upon the fluid structure, and $\mathscr{A}_{\text{RING}}$ is the corresponding approximation to the free energy. This approximation is the extension of the random phase approximation (RPA) for simple fluids (Hansen and McDonald[9]) to interaction site systems. We would also point out that since the RPA sums diagrams in the interaction site cluster series for the correlation functions and free energy which are only simple chains or rings, it can be formulated without using the decomposition of the site–site total correlation functions described in Section III.B. In fact, this was done in the seminal paper of Chandler and Andersen.[41] As is well-known, however, the RPA gives rise to nonzero corrections to the total correlation function in regions where the repulsive forces should cause this to vanish. The solution to this problem can be obtained by an optimization procedure[91] leading to the optimized random phase approximation (ORPA). Lupkowski and Monson have shown that in order for this procedure to be applied in the interaction site case, the optimization must be done explicitly in terms of the four components $h_{aa}^{(0)}$, $h_{sa}^{(0)}$, $h_{as}^{(0)}$, and $h_{ss}^{(0)}$. If this is not done, then the approximations obtained will include contributions from unallowed diagrams. They give details of the optimization procedure and describe a diagrammatic interpretation of the results which is analogous to that given by Andersen et al.[92] for the simple fluid ORPA. The approximation is referred to as the ISF–ORPA in order to be consistent with the terminology used by Rossky and Chiles in their treatment of approximations to the CSL equation.

Lupkowski and Monson have also discussed the connection between the approximations described above and the integral equation approximations based upon the CSL equation described in Section III.B. They show that the ISF–ORPA is the same as a reference mean spherical approximation (RMSA) to the CSL integral equation, following the work of Madden[94] in connection with the corresponding approximations for simple fluids. This relationship provides the basis of the computational methods that have been used so far. The approach has been applied to the calculation of the influence of attractive forces upon the phase diagrams of Lennard-Jones 12–6 diatomic fluids[95] as well as to dipolar diatomic fluids,[96,97] with quite promising results, as we shall see in Section V.

IV. SOLUTION METHODS FOR INTERACTION SITE INTEGRAL EQUATIONS

In this section we consider some of the solution methods which have been used to solve the integral equations for the correlation functions in interaction site fluids. We begin by considering analytic solutions of the SSOZ–MSA equation which were developed by Morriss and coworkers[98]. We then focus on numerical solutions for arbitrary closure approximations; particularly, extensions of the method developed by Gillan[99] which have become standard. In this latter context the discussion is restricted to the SSOZ equation. However, the methodology described is readily extended to any set of coupled OZ-like equations. The numerical solution is a flexible and efficient method of solving site–site integral equations for polar and molecular fluids, and is only limited in its application by the computing hardware available. The analytic solution, on the other hand, is restricted to very simple systems, typically those with a very small number of distinct correlation functions. However, the analytic solution gives precise information about such properties as the dielectric constant, which is unavailable from the numerical solution. In this sense, the two techniques are complementary.

A. Analytic Solution for Diatomic Symmetric Molecules

It is perhaps surprising that it is possible to solve the SSOZ equation for a number of simple molecules. For diatomic symmetry molecules with hard sphere pair interactions, the SSOZ equation with PY closure has been solved analytically using a Weiner–Hopf technique introduced by Baxter.[100] We consider a diatomic molecule consisting of two fused hard spheres of diameter σ, with their centres a distance L apart. For a fluid composed of these molecules, each of the four correlation functions is the same by symmetry, and the SSOZ equation reduces to a scalar equation

$$\hat{h}(k) = (1 + \hat{w}(k))\hat{c}(k)(1 + \hat{w}(k) + 2\rho\hat{h}(k)) \tag{4.1.1}$$

where

$$\hat{w}(k) = \frac{\sin kL}{kL}$$

The hard sphere pair interaction implies that

$$h(r) = -1 \quad \text{for} \quad r \leqslant \sigma$$

and the PY approximation implies that

$$c(r) = 0, \quad \text{for} \quad r > \sigma$$

There is a class of molecular systems that can also be written as scalar SSOZ equations, and then solved using the same method. These molecules all have the same diameter hard sphere sites, and all intersite distances are equal (regardless of whether the two sites are bonded). Obvious examples are the 3-site equilateral triangle and the 4-site regular tetrahedron (the 4-site model is a possible model for methane or carbon tetrachloride). In general, if the molecule has n-sites, $\hat{w}(k)$ is replaced by $(n-1)\hat{w}(k)$ and 2ρ is replaced by $n\rho$. One of the conditions required for the solution is that $[1+(n-1)\hat{w}(k)]$ is strictly positive for all values of k. This is first violated when $n = 6$, so in principle a solution is possible for $n = 5$, however this molecule is geometrically impossible in three dimensions. It is rather interesting that the SSOZ equation can be written down for this class of molecules for any value of n, but its analytic solution recognizes a limitation that is almost in agreement with geometric considerations.

To begin the factorization, the SSOZ is rewritten in the form

$$[1 + \hat{w}(k) + 2\rho\hat{h}(k)][(1 + \hat{w}(k))^{-1} - 2\rho\hat{c}(k)] = 1 \tag{4.1.2}$$

where

$$\begin{aligned} \hat{h}(k) &= \frac{4\pi}{k}\int_0^\infty dr\, r\, h(r)\sin kr \\ &= 2\pi\int_{-\infty}^\infty dr\, J(|r|)e^{ikr} \end{aligned} \tag{4.1.3}$$

and the new function $J(r)$ is related to $h(r)$ by

$$J(r) = \int_r^\infty dt\, t\, h(t) \tag{4.1.4}$$

The integral in the Fourier transform of the direct correlation function $c(r)$ can be truncated at σ to give

$$\begin{aligned} \hat{c}(k) &= \frac{4\pi}{k}\int_0^\infty dr\, r\, c(r)\sin kr \\ &= 4\pi\int_0^\sigma dr\, S(r)\cos kr \end{aligned} \tag{4.1.5}$$

where the function $S(r)$ is

$$S(r) = \int_r^\sigma dt\, t\, c(t) \tag{4.1.6}$$

Consider the function

$$\hat{A}(k) = [1 + \hat{w}(k) + 2\rho\hat{h}(k)]^{-1} = (1 + \hat{w}(k))^{-1} - 2\rho\hat{c}(k) \tag{4.1.7}$$

When the fluid is in a disordered state, $h(r)$ approaches zero fast enough to ensure that $\hat{h}(k)$ is finite for all real values of k, thus $\hat{A}(k)$ has no zeros on the real axis. For such a fluid, it can be shown that $\hat{A}(k)$ can be factorized as

$$\hat{A}(k) = \hat{Q}(k)\hat{Q}(-k) \tag{4.1.8}$$

where

$$\hat{Q}(k) = 1 - 2\pi\rho \int_0^\infty dr\, Q(r) e^{ikr} \tag{4.1.9}$$

It now remains to determine the function $Q(r)$. It is straightforward to show that $Q(r) = 0$ for $r < 0$, and it can be shown that

$$2\pi\rho Q(r) = \sum_{n=-\infty}^{\infty}{}' \zeta_n e^{-i\lambda_n r} \quad \text{for} \quad r > \sigma \tag{4.1.10}$$

(the prime denotes that the $n = 0$ term is excluded from the summation). The parameters ζ_n are the residues of the poles of $(1 + \hat{w}(k))^{-1}$

$$\zeta_n = \frac{i}{\hat{w}'(k)\hat{Q}(-\lambda_n)} \tag{4.1.11}$$

and the ζ_n are density-dependent through the density dependence of $\hat{Q}(-\lambda_n)$. The λ_n are the positions of these poles, given by the solutions of the equation

$$1 + \hat{w}(\lambda_n) = 0 \tag{4.1.12}$$

If we extend the range of validity of this form of $Q(r)$ to $0 < r < \infty$, then $Q_0(r)$ is defined to be

$$2\pi\rho Q(r) = 2\pi\rho Q_0(r) + \sum_{n=-\infty}^{\infty}{}' \zeta_n e^{-i\lambda_n r} \tag{4.1.13}$$

and $Q_0(r)$ is only nonzero on the range $0 < r < \sigma$. Combining Eqs. (4.1.9) and (4.1.13)

$$\hat{Q}(k) = 1 - \sum_{n=-\infty}^{\infty}{}' \frac{\zeta_n}{i(\lambda_n - k)} - 2\pi\rho \int_0^\sigma dr\, Q_0(r) e^{ikr} \tag{4.1.14}$$

The last part of the solution is to determine the function $Q_0(r)$. It can be shown that

$$\begin{aligned} Q_0(r) + \frac{1}{2L}\int_{-L}^{L} dt\, Q_0(r-t) = 2J(r) - 4\pi\rho \int_0^\sigma dt\, Q_0(t) J(|r-t|) \\ + \frac{1}{4\pi\rho L}\theta(L-r)\left[1 - \sum_{n=-\infty}^{\infty}{}' \frac{\zeta_n}{i\lambda_n}(1 - e^{i\lambda_n(L-r)})\right] \\ - 2\sum_{n=-\infty}^{\infty}{}' \zeta_n \int_0^\infty dt\, J(|r-t|) e^{-i\lambda_n t} \end{aligned} \tag{4.1.15}$$

where $\theta(r)$ is the usual Heavyside step function. The appearance of a factor of density ρ in the denominator of the step function term is a little disconcerting. However, it can be shown that the density-independent part of the square-bracketed term is exactly zero. From a numerical point of view, the square-bracketed term is a Dirichlet series and hence difficult to evaluate accurately. Although there has been some numerical work published in which a small number of poles have been considered,[101] most progress has been made using the zero pole approximation (ZPA) in which ζ_n is set to zero for all values of n.[102]

It is instructive to consider the ZPA solution for diatomic symmetric molecules, as it is far simpler than continuing the development of the exact solution obtained above. Setting ζ_n to zero and differentiating with respect to r gives

$$\begin{aligned} Q_0'(r) + \frac{1}{2L}\{Q_0(r+L) - Q_0(r-L)\} + \frac{\delta(r-L)}{4\pi\rho L} \\ = -2rh(r) - 4\pi\rho \int_0^\sigma dt\, Q_0(t)(r-t)h(|r-t|) \end{aligned} \tag{4.1.16}$$

As we already know $Q_0(r)$ for much of its range, we consider the $0 < r < \sigma$. Here $h(r) = -1$, and as the delta function only contributes at $r = L$ we will exclude that point for the moment. This equation then reduces to

$$Q_0'(r) + \frac{1}{2L}\{Q_0(r+L) - Q_0(r-L)\} = ar + b \tag{4.1.17}$$

where

$$a = 2 - 4\pi\rho \int_0^\sigma dt\, Q_0(t) \tag{4.1.18}$$

$$b = 4\pi\rho \int_0^\sigma dt\, t Q_0(t) \tag{4.1.19}$$

There are two further conditions that the function $Q_0(r)$ must satisfy. First, $Q_0(r)$ must be continuous at $r = \sigma$, and second the delta function implies that $Q_0(r)$ has a jump discontinuity at $r = L$. The easiest case to solve is that where the bond length L is exactly half the hard core diameter, that is, $L/\sigma = \frac{1}{2}$. To do this, consider $0 < r < L$, then

$$Q_0'(r) + \frac{1}{2L} Q_0(r + L) = ar + b \tag{4.1.20}$$

as $r - L < 0$ the term $Q_0(r - L)$ is always zero. Also we have

$$Q_0'(r + L) - \frac{1}{2L} Q_0(r) = a(r + L) + b \tag{4.1.21}$$

Differentiating the first of these equations with respect to r and eliminating $Q_0'(r + L)$ we obtain

$$Q_0''(r) + Q_0(r) = -ar + a(\sigma - L) - b \tag{4.1.22}$$

This is a very simple second-order differential equation. The solution involves two constants of integration that can be determined using the continuity conditions at $r = L$ and $r = \sigma$. The function Q_0 thus obtained is valid only for $0 < r < L$, but the remainder of the range can easily be obtained from $Q_0(r + L)$ using Eq. (4.1.20). Once Q_0 is known, the site–site correlation function can be obtained by quadrature from Eq. (4.1.16).

The results obtained from the analytic solution are similar to those obtained previously using purely numerical methods. One of the advantages of the analytic solution is that it is possible to make changes to the closure relation which systematically improve the accuracy of the results. One method of doing this is by adding a Yukawa tail to the site–site direct correlation function. The two parameters in the Yukawa tail (that is, its amplitude and decay rate) can be chosen to give thermodynamic consistency and improve the accuracy of the site–site correlation function $g_{\alpha\gamma}$. This approach has been pursued Cummings and Morriss.[103–106]

B. Analytic Solution for Polar Hard Dumbbells

One of the most fruitful applications of the analytic solution technique has been to polar molecular fluids. Historically, the extension of atomic fluid techniques to more realistic and interesting systems has proceeded independently in two separate directions. That is, the added difficulties of shape effects and electrostatic interactions have most often been studied separately, rather than together, despite the fact that they usually appear together. It is common, however, to find that these two effects are complementary in the sense that, for example, a polar hard core diatomic molecule (polar hard dumbbell) has a dielectric constant which increases with increasing dipole moment, but decreases with increasing *shape* (here shape refers simply to bond length at fixed molecular volume).[107,60] The magnitude of these effects is the same so it is clear that in many cases the effects of shape and electrostatic interactions must be considered together. In this section, we review the solution of the SSOZ–MSA equation for polar hard dumbbells[50] and discuss some of the applications of the solution.

The molecule we consider is the same symmetric diatomic considered previously, except that each site has an equal and opposite charge. There are now two types of intermolecular correlation functions to consider, the correlation function between sites with the same charge h_{++}, and that between sites with opposite charge h_{+-}. For this system, it is convenient to define sum and difference correlation functions as follows:

$$\begin{aligned} h_s(r) &= \tfrac{1}{2}(h_{++}(r) + h_{+-}(r)) \\ h_d(r) &= \tfrac{1}{2}(h_{++}(r) - h_{+-}(r)) \end{aligned} \tag{4.2.1}$$

For the sum and difference correlation functions, the SSOZ equation decouples into two scalar equations. The equation involving the sum correlation functions h_s and c_s, and their closures, is identical to the scalar SSOZ equation for the nonpolar symmetric diatomic molecule, so its solution is known. The scalar equation for the difference correlation functions is given by

$$[1 - \hat{w}(k) + 2\rho\hat{h}_d(k)][(1 - \hat{w}(k))^{-1} - 2\rho\hat{c}_d(k)] = 1 \tag{4.2.2}$$

The closure is

$$\begin{aligned} h_d(r) &= 0 \quad \text{for} \quad r \leqslant \sigma \\ c_d(r) &= -\frac{\beta e^2}{r} \quad \text{for} \quad r > \sigma \end{aligned} \tag{4.2.3}$$

where the reduced charge e is given by $e = (\beta/\sigma)^{1/2} q$. This scalar SSOZ equation can also be solved using an extension of the method used previously. The details of the solution can be found in the literature for the full solution and also the ZPA solution. There is an extra ingredient that emerges in the polar molecule case which is of particular interest. To illustrate this, we consider the simplest ZPA case where $L = \sigma/2$. It is possible to show that $Q(r)$ for this system is

$$Q(r) = 2zJ_0 + p\cos r + q\sin r \qquad 0 < r < \tfrac{1}{2}$$

$$Q(r) = -2zJ_0 + \frac{z}{2\pi\rho} - p\sin(r - \tfrac{1}{2}) + q\cos(r - \tfrac{1}{2}) \qquad \tfrac{1}{2} < r < 1 \quad (4.2.4)$$

where $z = 6/L^2 + 8\pi\rho\beta^2 e^2 = (6/L^2)(1 + 3y)$, y is the familiar measure of dipole strength (in reduced units). The parameter J_0 is closely related to the electrostatic energy per particle $U^* = U/NkT = 24\eta e^2 J_0$ where

$$J_0 = \int_0^\infty dr\, rh_d(r) \quad (4.2.5)$$

$$576\eta^2 z(1 - \cos\tfrac{1}{2})J_0^2 + 12\eta[2z(\sin\tfrac{1}{2} + 3\cos\tfrac{1}{2} - 3) + (2 - \cos\tfrac{1}{2})]J_0$$
$$+ \frac{z}{2}(5 - 3\sin\tfrac{1}{2} - 4\cos\tfrac{1}{2}) + (\cos\tfrac{1}{2} - 1) = 0 \quad (4.2.6)$$

where $\eta = \pi\rho\sigma^3/3$. Some generalizations and extensions of the polar fluid solution can be found in reference 17.

The Helmholtz free energy A of the polar hard dumbbell system is given by $A = A_0 + \Delta A$ where A_0 is the hard core contribution and ΔA is the electrostatic contribution. The electrostatic contribution ΔA can be calculated using a *charging* technique

$$\beta\Delta A = \int_0^\beta d\beta'\, U(\eta, \beta') \quad (4.2.7)$$

This approach has been used[108] to map out the phase diagram (coexistence and spinodal curves) for the polar hard dumbbell fluid at $L = 0.5$. The major conclusion from this work is that the critical point, due to electrostatic interactions alone, is at a much lower temperature than the critical point of a corresponding real fluid. Thus, electrostatic interactions are a small perturbation to the thermodynamics of the Lennard-Jones interaction.

These same techniques can be applied to polar/nonpolar fluid mixtures[109] where the polar species is the polar hard dumbbell. In this case, the internal energy per particle is $U^* = 24\eta x^2 e^2 J_0$ where x is the number fraction of the polar species. The nonpolar species is reasonably arbitrary as the only information needed is its free energy as a function of density. The dumbbell/polar dumbbell mixture has been studied, its excess thermodynamic properties have been calculated and its solubility curves obtained. The main feature of this mixture is that it exhibits fluid–fluid immiscibility, ending in an upper critical solution temperature. No lower critical solution temperature would be expected for this mixture.

C. Numerical Solution

The pioneering numerical solutions of the SSOZ equation were restricted to hard core molecules with the PY closure and were performed using the variational procedure devised by Lowden and Chandler.[42] This requires an assumed functional form for the site–site direct correlation functions $c_{\alpha\gamma}$. It was some years before the first solutions for soft atom–atom potentials were obtained,[44] initially with the Picard iteration scheme which was then standard in integral equation solutions for atomic fluids. More recently, Gillan's improved technique for atomic fluids[99] has been adapted to solve the SSOZ equation by Monson,[46] and we favor this numerical method because of its stability and insensitivity to the initial guess. The method combines the standard Picard iterative scheme with a Newton–Raphson procedure in an optimal way. Using a small number of basis functions and a Newton–Raphson procedure on the coefficients reduces the order of the Jacobian matrix to a manageable size. The rapid convergence of the Newton–Raphson procedure ensures that the solution obtained from the basis functions alone is close to the exact solution so that the Picard scheme is only required to make fine adjustments to this. The other advantages of this method are several. As the method is completely numerical, it does not require an assumed functional form for the site–site direct correlation functions, as the original Lowden and Chandler functional minimization scheme does. It is also completely flexible with regard to the choice of potential and closure. A fully general description of the method and its application to the SSOZ equation has been given by Morriss and MacGowan.[112]

Our remarks so far concern the speeding of convergence in iterative solutions of the SSOZ equation with various closures, and are applicable to any site–site pair potentials. Additional considerations arise when some of the sites carry charges. Special techniques then have to be adopted to overcome difficulties associated with the long range of the Coulomb potential. The method conventionally used for this purpose is the renormalization introduced by Hirata and Rossky,[48] which is analogous to Allnatt's method for electrolytes.[113] This has been combined with the Gillan method by Morriss and Monson[102] and was later used[114] to obtain solutions for nonlinear triatomic molecules. Here we outline that method and contrast it with a simpler technique taken from Ng's work on the one-component plasma.[110] We find that the latter has significant advantages.

The preferred numerical method is then a combination of Newton–Raphson and Picard schemes first proposed by Gillan,[99] together with the Ng[110] method for handling long-range Coulomb potentials. It is a completely general technique and can be used with any closure or potential model to solve any integral equation for molecules of arbitrary symmetry.[112] We will demonstrate the application of the method to the SSOZ equation. For this purpose, it is convenient to write the SSOZ equation in the form

$$\hat{H} = \hat{W}\,\hat{C}(I - \hat{W}\hat{C})^{-1}\hat{W} \tag{4.3.1}$$

where the matrices $\hat{H}$, $\hat{C}$, and $\hat{W}$ are

$$(\hat{H})_{\alpha\eta} = \frac{4\pi(\rho_\alpha\rho_\eta)^{1/2}}{k}\int_0^\infty dr\, r\sin(kr)h_{\alpha\eta}(r) \tag{4.3.2}$$

$$(\hat{C})_{\alpha\eta} = \frac{4\pi(\rho_\alpha\rho_\eta)^{1/2}}{k}\int_0^\infty dr\, r\sin(kr)c_{\alpha\eta}(r) \tag{4.3.3}$$

$$(\hat{W})_{\alpha\eta} = \hat{\omega}_{\alpha\eta}(k) = \frac{\sin kL_{\alpha\eta}}{kL_{\alpha\eta}} \tag{4.3.4}$$

Here we have included a density factor of $(\rho_\alpha\rho_\eta)^{1/2}$ in the Fourier transforms of the site–site correlation functions $h_{\alpha\eta}(r)$ and $c_{\alpha\eta}(r)$, which is more convenient for mixture calculations. $L_{\alpha\eta}$ is the interamolecular distance between sites α and η. For mixtures, $\hat{\omega}_{\alpha\eta}(k) = 0$ when sites α and η are in molecules of different species.

The SSOZ equation simply relates the total correlation functions $\hat{h}_{\alpha\eta}(k)$ to the direct correlation functions $\hat{c}_{\alpha\eta}(k)$. A second relation is required to obtain a closed system of equations. Two of the commonly used closure relations are the Percus–Yevick (PY) and hypernetted chain (HNC) approximations. The PY closure is

$$c_{\alpha\eta}(r) = (1 + \gamma_{\alpha\eta}(r))(\exp\{\phi_{\alpha\eta}(r)\} - 1) \tag{4.3.5}$$

and the HNC closure is

$$c_{\alpha\eta}(r) = \exp(\phi_{\alpha\eta}(r) + \gamma_{\alpha\eta}(r)) - \gamma_{\alpha\eta}(r) - 1 \tag{4.3.6}$$

where $\gamma_{\alpha\eta}(r)$ is an element of the matrix Γ defined by $\Gamma = H - C$. The correlation function matrix, Γ, is usually used in numerical solutions of integral equations because its elements are continuous functions of r, even when the potential is discontinuous (such as for hard core potentials). $\phi_{\alpha\eta}(r) = -\beta u_{\alpha\eta}(r)$, where $\beta = (k_B T)^{-1}$ and $u_{\alpha\eta}(r)$ is a site–site pair potential.

It is now established that the combined Gillan–Ng procedure is a very general and flexible method of obtaining solutions to the molecular fluid integral equations for both polar molecular fluids and their mixtures. It may nevertheless prove useful to extend the variational solution method to a wider range of closures by the introduction of new functionals such as have been suggested for the SSOZ–HNC[115] and SSOZ–PY[18] approximations. Of course, such extensions would be intrinsically more complex than the Lowden–Chandler variational solution of the SSOZ–MSA approximation[42,43]. If any of the site–site pair potentials contains a Coulomb interaction, it may be necessary to reformulate the integral equation and closure so that the Fourier transforms of all correlation functions are well defined at $k = 0$. We consider two ways of doing this.

1. *Renormalization*

We define a set of renormalized potentials $q_{\alpha\eta}(r)$ through the matrix Fourier transform equation

$$\hat{Q} = \hat{W}\hat{\Phi}^e(I - \hat{W}\hat{\Phi}^e)^{-1}\hat{W} \tag{4.3.7}$$

where the elements of the matrix $\hat{\Phi}^e$ are the Fourier transforms of the Coulomb part of the potential $\phi^e_{\alpha\eta}(r)$, given by

$$\hat{\phi}^e_{\alpha\eta}(k) = -\frac{4\pi\beta}{k^2}(\rho_\alpha\rho_\eta)^{1/2}e_\alpha e_\eta \tag{4.3.8}$$

This identification of the renormalized potentials can be motivated more directly by r-space graphical arguments[116] or analogous k-space arguments.[102]

In general, the total site–site pair potential also includes a short-range part $\phi^0_{\alpha\eta}(r)$

$$\phi^0_{\alpha\eta}(r) = \phi_{\alpha\eta}(r) - \phi^e_{\alpha\eta}(r) \tag{4.3.9}$$

As the site–site direct correlation function at large r decays in the same way as the site–site pair potential (for the standard closure relations), we can write

$$c^0_{\alpha\eta}(r) = c_{\alpha\eta}(r) - \phi^e_{\alpha\eta}(r) \tag{4.3.10}$$

A convenient short-range function is

$$\begin{aligned}\gamma^0_{\alpha\eta}(r) &= h_{\alpha\eta}(r) - c^0_{\alpha\eta}(r) - q_{\alpha\eta}(r) \\ &= \gamma_{\alpha\eta}(r) + \phi^e_{\alpha\eta}(r) - q_{\alpha\eta}(r)\end{aligned} \tag{4.3.11}$$

and the solution is formulated in terms of γ^0 and c^0. In terms of the functions just defined, the (renormalized) SSOZ equation is

$$\hat{\Gamma} = (\hat{W} + \hat{Q})\hat{C}^0(I - (\hat{W} + \hat{Q})\hat{C}^0)^{-1}(\hat{W} + \hat{Q}) - \hat{C}^0 \tag{4.3.12}$$

and the renormalized HNC closure is

$$c^0_{\alpha\eta} = \exp[\phi^0_{\alpha\eta} + q_{\alpha\eta} + \gamma^0_{\alpha\eta}] - q_{\alpha\eta} - \gamma^0_{\alpha\eta} - 1 \tag{4.3.13}$$

This is precisely the same as Eq. (4.3.6), but the renormalized closure analogous to the PY closure

$$c^0_{\alpha\eta} = (1 + \gamma^0_{\alpha\eta})(\exp[\phi^0_{\alpha\eta} + q_{\alpha\eta}] - 1) - q_{\alpha\eta} \tag{4.3.14}$$

is different from Eq. (4.3.5). The normal PY closure is well-known to be inappropriate for charged systems, but when the non-Coulomb potential is purely hard core

$$\phi^0_{\alpha\eta}(r) = \begin{cases} -\infty & r < \sigma_{\alpha\eta} \\ 0 & r > \sigma_{\alpha\eta} \end{cases} \tag{4.3.15}$$

the mean spherical approximation (MSA), which reduces to PY for zero charges, is also of interest. The MSA closure is

$$c_{\alpha\eta} = (1 + \gamma_{\alpha\eta})(\exp \phi^0_{\alpha\eta} - 1) + \phi^e_{\alpha\eta} \tag{4.3.16}$$

or, in terms of renormalized functions,

$$c^0_{\alpha\eta} = (1 + \gamma^0_{\alpha\eta} + q_{\alpha\eta} - \phi^e_{\alpha\eta})(\exp \phi^0_{\alpha\eta} - 1) \tag{4.3.17}$$

In obaining the renormalized potentials $q_{\alpha\eta}(r)$ from $\hat{Q}(k)$, it is numerically expedient to subtract the $k \to \infty$ form of $\hat{Q}(k)$ first, as it can be transformed analytically to obtain

$$\hat{Q}^\infty = \hat{\Phi}^e(I - \hat{\Phi}^e)^{-1} = \frac{k^2\hat{\Phi}^e}{k^2 + \mu^2} \tag{4.3.18}$$

$$\mu^2 = 4\pi\beta \sum_{\alpha} \rho_\alpha e_\alpha^2 \tag{4.3.19}$$

Hence, for small r, it can be shown that the renormalized potential takes the form

$$q_{\alpha\eta}^{\infty}(r) = -\beta(\rho_\alpha \rho_\eta)^{1/2} e_\alpha e_\eta \frac{\exp(-\mu r)}{r} \tag{4.3.20}$$

The difference between $q_{\alpha\eta}^{\infty}$ and the full renormalized potential $q_{\alpha\eta}$, is a well-behaved function that is evaluated numerically. The interest in the renormalization procedure is now mainly a theoretical one as formal results regarding screening and other thermodynamic parameters can be obtained this way. Results applicable to both pure one-component fluids or mixtures can be obtained. The numerical solution of integral equations, such as the SSOZ and CSL equations, for sites with charge interactions should no longer use the renormalization method but rather the method we are about to describe.

2. *Ng Method*

The renormalization method is closely analogous to the techniques introduced by Allnatt[113] for electrolyte solutions and later applied by Cooper to the one-component plasma.[117] The renormalized method for the one-component plasma has since been superseded by improved numerical methods for the unrenormalized equation. The method of Springer et al.[118] has been used by Abernethy and Gillan[119] for electrolytes and also by Ohba and Arakawa[120] for polar interaction site molecules, but the best method available is that of Ng, which has more recently been applied to electrolytes by Rogers[121] and to the SSOZ equation by Morriss and MacGown.[112] We proceed to illustrate the application of the Ng method by considering the SSOZ equation with HNC and MSA closures. We do not give the PY closure since it known to be poor for charged systems.

We divide the electrostatic potentials into long- and short-range parts:

$$\phi_{\alpha\eta}^{1}(r) = -\beta e_\alpha e_\eta \frac{\operatorname{erf}(tr)}{r} \tag{4.3.21}$$

$$\phi_{\alpha\eta}^{s}(r) = \phi_{\alpha\eta}^{e}(r) - \phi_{\alpha\eta}^{1}(r) \tag{4.3.22}$$

It can be shown that the Fourier transform of the long-range part of the potential is given by

$$\hat{\phi}^1_{\alpha\eta}(k) = -\frac{4\pi\beta}{k^2}(\rho_\alpha\rho_\eta)^{1/2}e_\alpha e_\eta \exp(-k^2/4t^2) \tag{4.3.23}$$

Making the definitions

$$c^s_{\alpha\eta}(r) = c_{\alpha\eta}(r) - \phi^1_{\alpha\eta}(r) \tag{4.3.24}$$

and

$$\gamma^s_{\alpha\eta}(r) = \gamma_{\alpha\eta}(r) + \phi^1_{\alpha\eta}(r) \tag{4.3.25}$$

the HNC closure takes the form

$$c^s_{\alpha\eta}(r) = \exp\{\phi^0_{\alpha\eta}(r) + \phi^s_{\alpha\eta}(r) + \gamma^s_{\alpha\eta}(r)\} - \gamma^s_{\alpha\eta}(r) - 1 \tag{4.3.26}$$

and, when the non-Coulomb potential is purely hard core, the MSA closure can be written as

$$c^s_{\alpha\eta}(r) = (1 + \gamma^s_{\alpha\eta}(r) + \phi^s_{\alpha\eta}(r))(\exp\{\phi^0_{\alpha\eta}\} - 1) + \phi^s_{\alpha\eta}(r) \tag{4.3.27}$$

The parameter t is adjustable, but its value is irrelevant to the solution except for numerical convenience ($t = 1$ appears to be satisfactory for most computations). Numerical solution involves Fourier transformation of both $c^s_{\alpha\eta}(r)$ and $\hat{\gamma}^s_{\alpha\eta}(k)$. The advantage of Ng's scheme is that both of these functions are exponentially short-range and so can be accurately transformed numerically without the need to use a very large integration range. It is clear that Eq. (4.3.24) removes the long-range part from $c_{\alpha\eta}(r)$, but many different choices of $\phi^1_{\alpha\eta}(r)$ achieve that. The crucial point is that Eq. (4.3.25) does not lead to the introduction of a term in $\hat{\gamma}^s_{\alpha\eta}(k)$ which decays slowly as $k \to \infty$. $\hat{\gamma}_{\alpha\eta}(k)$ itself is already a rapidly decaying function as $k \to \infty$. This identifies clearly the superiority of Ng's method to that of Springer et al.[118] who replace Eq. (4.3.21) by

$$\phi^1_{\alpha\eta}(r) = -\beta e_\alpha e_\eta \frac{1 - \exp(-tr)}{r} \tag{4.3.28}$$

and so Eq. (4.3.23) by

$$\hat{\phi}^1_{\alpha\eta}(k) = -\beta e_\alpha e_\eta (\rho_\alpha\rho_\eta)^{1/2} \frac{t^2}{k^2(k^2 + r^2)} \tag{4.3.29}$$

This leads to a k^{-4} decay in $\hat{\gamma}^{s}_{\alpha\eta}(k)$ as $k \to \infty$ in contrast to the exponential decay in Ng's algorithm. The problem of inducing a slowly decaying term also occurs in the renormalized equation. In fact, from Eqs. (4.3.11), (4.3.18), and (4.3.20), it is easy to see that the large k decay of $\hat{\gamma}^{0}_{\alpha\eta}(k)$ is precisely that given by Eq. (4.3.29) except that the adjustable parameter t is replaced by the fixed value μ.

We give details of the numerical solution for uncharged molecules in the following subsections, after which the effect of adding charges is discussed briefly. The basic element of the iteration scheme is the Picard cycle: starting with given $\gamma_{\alpha\eta}$, we use the closure to get $c_{\alpha\eta}$, then Fourier transform and obtain $\hat{\gamma}_{\alpha\eta}$ from the SSOZ equation followed by inverse Fourier transformation. The additional Newton–Raphson step is just a good way of speeding up convergence.

3. *Basis Functions*

Given an arbitrary set of basis functions (such as the *roof functions* used by Gillan, and Morriss and MacGowan), which do not form a complete set, we divide each $\gamma_{\alpha\eta}(r)$ into two parts: a part that can be expressed in terms of the basis functions (a coarse part), and the remainder which is orthogonal to the basis set (a fine part). In the following, the argument i refers to the discretized distance variable $r_i = i\,\delta r$, $i = 0, 1, \ldots N$. This means that $\gamma_{\alpha\eta}(i)$ can be written as

$$\gamma_{\alpha\eta}(i) = \sum_{u} a_{u,\alpha\eta} P_u(i) + \Delta\gamma_{\alpha\eta}(i) \tag{4.3.30}$$

and, as the fine part is orthogonal to each of the basis functions. We have assumed for simplicity that the same basis functions are used for each of the $\gamma_{\alpha\eta}$. Given $\gamma_{\alpha\eta}(i)$ we wish to calculate the coefficients $a_{u,\alpha\eta}$. To do this, we look for conjugate functions $S_u(i)$ which project out the coefficients:

$$a_{u,\alpha\eta} = \sum_{i} S_u(i)\gamma_{\alpha\eta}(i) \tag{4.3.31}$$

The functions $S_u(i)$ are given by

$$S_u(i) = \sum_{v} B_{uv} P_v(i) \tag{4.3.32}$$

where

$$B_{uv} = \left[\sum_{i} P_u(i) P_v(i) \right]^{-1} \tag{4.3.33}$$

It is easily seen that the matrix of coefficients B is simply the inverse of the matrix P_uP_v. As the basis functions are positive for all values of i, the matrix P_uP_v is positive definite.

4. *Newton–Raphson Procedure*

Suppose an elementary Picard cycle acting on an initial $\gamma_{\lambda v}$ gives $\gamma'_{\alpha\eta}$, which then can be decomposed into a set of coefficients $a'_{u,\alpha\eta}$ and a fine part, $\Delta\gamma'_{\alpha\eta}(i)$.We can use a Newton–Raphson procedure to find a better estimate of the converged coefficients $a_{u,\alpha\eta}$ which we term $a''_{u,\alpha\eta}$. Then we have

$$a''_{u,\alpha\eta} = a_{u,\alpha\eta} - \sum_{v,\lambda v} (J^{-1})_{uv,\alpha\eta\lambda v}(a_{v,\lambda v} - a'_{v,\lambda v}) \tag{4.3.34}$$

where J is the Jacobian matrix

$$J_{uv,\alpha\eta\lambda v} = \frac{\partial}{\partial a_{v,\lambda v}}(a_{u,\alpha\eta} - a'_{u,\alpha\eta}) = \delta_{uv}\delta_{\alpha\lambda}\delta_{\eta v} - \frac{\partial a'_{u,\alpha\eta}}{\partial a_{v,\lambda v}} \tag{4.3.35}$$

Using Eqs. (4.3.30) and (4.3.31)

$$\frac{\partial a'_{u,\alpha\eta}}{\partial a_{v,\lambda v}} = \sum_{i,j} S_u(i)\frac{\partial \gamma'_{\alpha\eta}(i)}{\partial \gamma_{\lambda v}(j)}P_v(j) \tag{4.3.36}$$

and, using the chain rule

$$\frac{\partial \gamma'_{\alpha\eta}(i)}{\partial \gamma_{\lambda v}(j)} = \sum_m \frac{\partial \gamma'_{\alpha\eta}(i)}{\partial \hat{\gamma}_{\alpha\eta}(m)}\frac{\partial \hat{\gamma}_{\alpha\eta}(m)}{\partial \hat{c}_{\lambda v}(m)}\frac{\partial \hat{c}_{\lambda v}(m)}{\partial c_{\lambda v}(j)}\frac{\partial c_{\lambda v}(j)}{\partial \gamma_{\lambda v}(j)} \tag{4.3.37}$$

Here m refers to the discretized k-space variable $k_m = m\delta k$, where $m = 0, 1, \ldots N$, while i and j refer to the discretized distance already introduced. We notice that, in the general matrix case, $\gamma'_{\alpha\eta}$ couples to all other $\gamma_{\lambda v}$'s through the partial derivative $\partial\hat{\gamma}_{\alpha\eta}/\partial\hat{c}_{\lambda v}$ only, as the Fourier transform, its inverse, and the closure all relate correlation functions for the same pair of sites. Two of the required partial derivatives can be calculated from the discrete Fourier transform and its discrete inverse. These are

$$\frac{\partial \gamma'_{\alpha\eta}(i)}{\partial \hat{\gamma}_{\alpha\eta}(m)} = \frac{\delta k}{2\pi^2 r_i}(\rho_\alpha\rho_\eta)^{-1/2}k_m \sin k_m r_i \tag{4.3.38}$$

and

$$\frac{\partial \hat{c}_{\lambda v}(m)}{\partial c_{\lambda v}(j)} = \frac{4\pi\delta r}{k_m}(\rho_\lambda\rho_v)^{1/2}r_j \sin k_m r_j \tag{4.3.39}$$

5. *The Jacobian*

The general Jacobian element for the SSOZ equation can be derived in two different ways, depending upon whether or not the symmetry of the matrices H and C is explicitly included. In developing a computer program to solve the matrix equation for molecules with arbitrary symmetry and any number of sites, we can ignore any possible simplifications that could be made in specific cases because of molecular symmetry (that is, we have assumed that each element of the C matrix is independent, particularly that $c_{\alpha\eta}$ is independent of $c_{\eta\alpha}$). This symmetry, and the symmetry due to the equivalence of particular sites in the molecule could be used to reduce program storage and improve program efficiency,[122] but this would reduce the generality of its applicability. It would also change the structure of the Jacobian matrix. The most general form for the Jacobian has been obtained by Morriss and MacGowan.[112] The result is

$$\frac{\partial \hat{\gamma}_{\alpha\eta}}{\partial \hat{c}_{\lambda\nu}} = M_{\alpha\lambda} M_{\nu\eta} - \delta_{\alpha\lambda}\delta_{\nu\eta} \tag{4.3.40}$$

where

$$M = (I - \hat{W}\hat{C})^{-1}\hat{W} \tag{4.3.41}$$

Thus, the Jacobian is given as the product of two elements of a matrix that is already calculated to obtain $\hat{\gamma}$ from $\hat{c}$.

$$\frac{\partial \gamma'_{\alpha\eta}(i)}{\partial \gamma_{\lambda\nu}(j)} = \frac{\delta r}{\pi} \frac{r_j}{r_i} \left[\frac{\rho_\lambda \rho_\nu}{\rho_\alpha \rho_\eta}\right]^{1/2} (D_{\alpha\eta\lambda\nu}(i-j) - D_{\alpha\eta\lambda\nu}(i+j)) \frac{\partial c_{\lambda\nu}(j)}{\partial \gamma_{\lambda\nu}(j)} \tag{4.3.42}$$

where

$$D_{\alpha\eta\lambda\nu}(i) = \delta k \sum_n \cos k_n r_i (M_{\alpha\lambda}(n) M_{\nu\eta}(n) - \delta_{\alpha\lambda}\delta_{\nu\eta}) \tag{4.3.43}$$

This form of the general Jacobian element allows for the straightforward solution of the SSOZ equation for molecules of arbitrary symmetry. However, in the numerical solution using Gillan's methods, most of the computation time is involved in calculating the elements of the Jacobian matrix, rather than in the calculation of its inverse or in the calculation of transforms. Indeed, as the forward and backward Fourier transforms can be carried out using a fast Fourier transform routine, the time-limiting step is the double summation over i and j in Eq. (4.3.36). With this restriction in mind, it is clear that the exploitation of any symmetry in the Jacobian matrix calçulation will increase the efficiency of the numerical solution. First we observe that the symmetries of $D_{\alpha\eta\lambda\nu}$, with respect to $\alpha\eta\lambda\nu$, are determined by the symmetry of the matrix product $M_{\alpha\lambda}M_{\nu\eta}$. The matrix D is composed of square blocks whose size is determined by the number of basis functions. The elements of a given block are functions of the same M matrix product, and are determined by the individual basis functions. As would be expected, the symmetries within a block are determined by the character of the basis functions, and unless the S_u are the same as the P_v there will be no symmetry within the block. It is clear from the fact that the matrix M is symmetric, that D is symmetric. Although this symmetry is easily implemented using a pattern recognition matrix,[123] it does not lead to a large increase in effciency unless the number of sites is large. It is possible to show that many other equalities exist between various elements of D but little efficiency is to be gained by pursuing these. The generalization of the SSOZ jacobian matrix result to the CSL equation is straightforward. Essentially, each element in the SSOZ matrix equation is replaced by a 2×2 matrix of the form given in Eq. (3.4.8).

Almost the whole of the above method is valid for charged sites once $c_{\alpha\eta}$ and $\gamma_{\alpha\eta}$ are replaced by $c^o_{\alpha\eta}$ *and* $\gamma^o_{\alpha\eta}$ in the renormalized method and by $c^s_{\alpha\eta}$ and $\gamma^s_{\alpha\eta}$ in Ng's method. The principal change arises in the derivative $\partial c_{\lambda\nu}/\partial\gamma_{\lambda\nu}$. For the PY and HNC closures, respectively

$$\frac{\partial c_{\lambda\nu}}{\partial\gamma_{\lambda\nu}} = \exp(\phi_{\lambda\nu}) - 1 \tag{4.3.44}$$

$$\frac{\partial c_{\lambda\nu}}{\partial_{\lambda\nu}} = \exp(\phi_{\lambda\nu} + \gamma_{\lambda\nu}) - 1 \tag{4.3.45}$$

When charges are added, apart from the appropriate replacements of $c_{\lambda\nu}$ and $\gamma_{\lambda\nu}$, $\phi_{\lambda\nu}$ must be replaced by $\phi^0_{\lambda\nu} + q_{\lambda\nu}$ in the renormalized PY and HNC closures and by $\phi^0_{\lambda\nu} + \phi^s_{\lambda\nu}$ in the Ng form of the HNC closure. For the MSA closure in either renormalized or Ng form, $\phi_{\lambda\nu}$ should be replaced by (the purely hard core) $\phi^0_{\lambda\nu}$ in Eq. (4.3.44).

Although part of the attraction of Gillan's method is the fact that it can often be used with poor initial data, the required number of iterations is nonetheless significantly reduced by good starting values. For neutral hard core molecules, the initial $c_{\alpha\eta}(r)$ is taken to be the one-component hard sphere PY solution for the appropriate diameter $\sigma_{\alpha\eta}$. This is also used for Lennard–Jones soft cores with $\sigma_{\alpha\eta}$ equal to 9/10 of the distance of zero potential. For molecules with small charges on the sites, we add $\phi^e_{\alpha\eta}(r)$ to the neutral starting value. For large charges, the initial values are obtained by extrapolating linearly in e^2 from two solutions with proportionally scaled down site charges. One of these is usually the zero charge solution. This type of extrapolation was previously used by Ng; but Morriss and MacGowan only required it for a very highly quadrupolar dumbbell.

It is important, once a converged numerical solution is obtained, that some independent measure of the accuracy be made. As usual, a numerical solution is of high precision, but that has no bearing upon whether the solution is close to the correct solution or not. In general, comparison with a simulation value of the energy is a useful check, but this is clouded by the accuracy of the closure approximation. For polar systems, a good measure of the accuracy of a numerical solution is that it satisfy the charge neutrality condition

$$J_1 = 2\pi\rho^* \sum_{\alpha,\eta} e_\alpha^* e_\beta^* \int_0^\infty dr\, r^2 g_{\alpha\eta}(r) = 0 \tag{4.3.46}$$

The reduced density ρ^* and charge e^* are defined by $\rho^* = \rho\sigma_1^3$, $e^* = e/(kT\sigma_1)^{1/2}$ and r is in units of σ_1 (where σ_1 is taken to be the diameter of the largest site). A number of results for the polar hard bumbbell and quadrupolar dumbbell are given in Morriss and MacGowan, and these should be sufficient to check the accuracy of a computer program based on this method.

V. RESULTS FOR HOMOGENEOUS FLUIDS

In this section, we will review some of the results obtained for homogeneous fluids. The focus of the section strongly reflects the author's particular interest rather than a complete review of all work done in this area. To a large extent, we will concentrate on aspects that have not been reviewed previously,[3–6] or on areas that developed since those reviews. The first section deals with the influence of electrostatic interactions on the structure factor, and we stress the decoupling of dipole–dipole interactions from the structure factor, although there is a strong effect on particular $g_{\alpha\gamma}(r)$'s. In Section V.B we consider the dielectric constant obtained from the CSL equation with particular reference to the influence of shape forces in the dielectric properties. Section V.C considers the application of interaction site theories to calculate thermodynamic properties and fluid phase equilibria.

A. Structural Properties

The structure factor for an n-site molecular liquid is given by

$$S(k) = \frac{\sum_{\alpha,\gamma=1}^{n} a_\alpha(k) a_\gamma(k)[\hat{\omega}_{\alpha\gamma}(k) + \rho \hat{h}_{\alpha\gamma}(k)]}{\sum_{\alpha=1}^{n} a_\alpha^2(k)} \tag{5.1.1}$$

where $a_\alpha(k)$ describes the scattering characteristics of atom α. For neutron scattering $a_\alpha(k) = b_\alpha$, the neutron scattering length for the nucleus of atom α, while for x-ray scattering $a_\alpha(k)$ corresponds to the x-ray atomic form factor $f_\alpha(k)$. In all model system calculations, the factors $a_\alpha(k)$ are ignored that is, $a_\alpha(k) = 1$ for all sites); however, if comparison with experimental structure factors is needed, then such simplifications can not be made. Indeed, for many model systems, the structure has been displayed in terms of the site–site correlation functions alone.

The first calculations of site–site correlation functions were made by Lowden and Chandler.[42,43] The second of these papers was concerned with modeling such systems as CCl_4, CS_2, CSe_2, and C_6H_6. Further studies of N_2, O_2, and Br_2 were carried out by Hsu, Chandler, and Lowden,[124] and some difficulty in obtaining good results for Br_2 was observed. The first direct comparison between SSOZ–PY results and computer simulations appears to be the work of Chandler, Hsu, and Streett.[125] The difficulties with Br_2 were considered later by Pratt and Chandler.[126] In all of this work, the SSOZ–PY equation with hard core potentials gave a good representation of the structure of the system considered when the true interaction potential consisted of strong short-range repulsive forces. Whenever electrostatic interactions were an important part of the potential (particularly quadrupolar interactions), the results were not as good. The earliest calculations for more realistic potentials were done by Kojima and Arakawa.[44,127] These were followed by studies of polar diatomic fluids by Ohba and Arakawa,[128,120] ion–polar solvent mixtures,[129] and the solvophobic interaction[130] but this work is predated by the work of Pratt and Chandler.[111] Later, the same authors returned to consider the asymptotic behavior of SSOZ correlation functions and an early closure relation called RISM-2.[131]

The structure of site–site molecular fluids can now be obtained on a routine basis using either the SSOZ equation or CSL equation and an associated closure. The numerical method described in Section IV.C is generally applicable. It is fair to say that the results obtained are in general not as accurate as those obtained for atomic fluids using the analogous method. As a result we will not attempt to review the literature but refer the reader to the review by Rossky[5] which deals with structure exhaustively. We will, however, look at some general structural questions such as, why do strong dipolar interactions have little effect on the structure when quadrupolar interactions (such as those in Br_2) do effect the structure. These empirical observations have been made many times, but it was not until the work of Fraser et al.[114] that even a qualitative explanation of these effects was possible.

It had been observed previously that for polar hard dumbbells[102] the separation of the SSOZ–MSA equation into two independent scalar equations for the sum and difference correlation functions implied a separation of the structure and dielectric properties. The structure factor could be obtained from the sum correlation function alone, while the dielectric constant could be obtained from the difference correlation function alone. It was Fraser et al.[114] who showed that this result was not simply a consequence of the MSA closure, but was a special case of a much more general result. For the case of model systems where the site charges are placed on sites that have the same symmetry, they showed that SSOZ equation separates into an SSOZ equation with the correlation functions between charged sites, replaced by sum correlation functions. The closure relations for the sum correlation functions in the MSA are simply the usual hard core MSA closure relations; while, for the HNC closure, we find

$$c_s(r) = h_s(r) - \ln[(g_{++}(r)g_{+-}(r))^{1/2}] \tag{5.1.2}$$

This is very close to the HNC closure for hard core interactions, indeed the argument in the logarithm is the multiplicative mean of the two correlation functions g_{++} and g_{+}. If the multiplicative mean is replaced by the arithmetic mean, then this closure relation is the HNC approximation. Equation (5.1.2) can then be written as

$$c_s(r) = h_s(r) - \ln(g_s(r)) + \ln\left[\frac{\frac{1}{2}(g_{++}(r) + g_{+-}(r))}{(g_{++}(r)g_{+-}(r))^{1/2}}\right] \tag{5.1.3}$$

The question of how strong the decoupling of structural and dielectric properties is depends on the magnitude of the last term in Eq. (5.1.3). This effective decoupling has been explored numerically for acetonitrile.[123] In the next section, we will consider the second part of this problem, that is, the implication of structure on the dielectric properties of molecular fluids.

B. Dielectric Properties

The analytic solution of the SSOZ–MSA equation for polar hard dumbbells[50] came before any serious consideration was given to calculating the dielectric constants of such systems by computer simulation. At the time, there was considerable controversy about the simulation methods used to calculate the dielectric constant, and for the model systems then in vogue (dipolar hard spheres and the Stockmayer fluid) there was also debate about the correct value of the dielectric constant. Today, this problem is becoming better understood; in particular, the quality of the simulation work has improved greatly, and this has allowed meaningful conclusions to be drawn about the relative merits of simulation methods.[132]

The first calculations of the dielectric constant for fluids with shape forces were the polar hard dumbbell results of Morriss et al.[60,107] These simulations showed that the bond length of the molecule significantly affected the dielectric constant. Equivalent thermodynamic states were obtained by keeping the volume of the molecule constant regardless of the bond length. A two-center Lennard–Jones variation of the same system was considered by deLeeuw and Quirke,[133] which simply confirmed that the original results were unaffected by a change in short-range interactions. Unfortunately, it was already clear from the analytic solution for polar hard dumbbells that the dielectric constant obtained was independent of bond length and simply equal to the trivial ideal gas result (Eq. (3.3.14)). Clearly, the effects seen in the simulations could not be obtained from the SSOZ equation. Recently, it has become clear that the CSL equation does not lead to a trivial dielectric constant and there is a real possibility of observing a bond-length dependence. In this section, we will consider application of the CSL equation to the polar hard dumbbell fluid.

Lupkowski and Monson[96] have shown that for a dipolar fluid it is possible to decouple the CSL matrix equation into two separate matrix equations, one for the sum correlation functions H^{S} and C^{S}, and the other for the difference correlation functions H^{D} and C^{D}. In this section, we use the formalism developed by Fraser et al.[114] to make this separation. A convenient form of the CSL equation is

$$\hat{H} = (I + \hat{S}\bar{\rho})\hat{C}(I + \bar{\rho}\hat{S} + \bar{\rho}\hat{H}) \tag{5.2.1}$$

where I is the 4×4 identity matrix, and H and C are block matrices of the form

$$\hat{H} = \begin{pmatrix} \tilde{H}^{++} & \tilde{H}^{+-} \\ \tilde{H}^{+-} & \tilde{H}^{++} \end{pmatrix} \quad \text{and} \quad \hat{C} = \begin{pmatrix} \tilde{C}^{++} & \tilde{C}^{+-} \\ \tilde{C}^{+-} & \tilde{C}^{++} \end{pmatrix} \tag{5.2.2}$$

Each block of these matrices contains the four components of the particular site interaction, for example

$$\hat{H}^{++} = \begin{pmatrix} \hat{h}_{aa}^{++} & \hat{h}_{as}^{++} \\ \hat{h}_{sa}^{++} & \hat{h}_{ss}^{++} \end{pmatrix} \tag{5.2.3}$$

The connectivity matrix for this molecule is

$$\bar{\rho} = \begin{pmatrix} \tilde{\rho} & \tilde{0} \\ \tilde{0} & \tilde{\rho} \end{pmatrix} = \begin{pmatrix} \rho & \rho & 0 & 0 \\ \rho & 0 & 0 & 0 \\ 0 & 0 & \rho & \rho \\ 0 & 0 & \rho & 0 \end{pmatrix} \tag{5.2.4}$$

and the intramolecular correlation function is

$$\hat{S} = \begin{pmatrix} 0 & 0 & 0 & 0 \\ 0 & 0 & 0 & \omega/\rho \\ 0 & 0 & 0 & 0 \\ 0 & \omega/\rho & 0 & 0 \end{pmatrix} \tag{5.2.5}$$

The transformation into sum and difference equation is achieved using a generalization of the row/column interchange matrix F. Here, rather than interchanging rows (or columns), the matrix F interchanges blocks of rows (or columns). The particular F we use is

$$F = \begin{pmatrix} 0 & 1 \\ 1 & 0 \end{pmatrix} \tag{5.2.6}$$

where 1 is the 2×2 identity matrix and 0 is the 2×2 null matrix (all elements are zero). The first useful property of the F matrix is that

$$\tfrac{1}{2}(I + F)\hat{H} = \hat{H}^{\mathrm{S}} = \begin{pmatrix} \tilde{H}^{\mathrm{S}} & \tilde{H}^{\mathrm{S}} \\ \tilde{H}^{\mathrm{S}} & \tilde{H}^{\mathrm{S}} \end{pmatrix} \tag{5.2.7}$$

where each block of the H^{S} matrix is the same. The second property is that F is its own inverse. This means, in particular, that we have the identity

$$\tfrac{1}{2}(I + F) = \tfrac{1}{8}(I + F)^3 \tag{5.2.8}$$

The final property we need is that the matrix F commutes with each of the matrices (and matrix products) in the CSL equation. This can easily be verified for the two-site polar dumbbell considered here.

To obtain the sum correlation function form of the CSL equation, we multiply both sides of Eq. (5.2.1) by $\frac{1}{2}(I + F)$. This transforms the left-hand side into the H^{S} correlation function. We then use the identity in Eq. (5.2.8) to change $\frac{1}{2}(I + F)$ into the product of three identical terms $[\frac{1}{2}(I + F)]^3$. Using the fact that F commutes with each of the matrices, we apply one factor of $\frac{1}{2}(I + F)$ to each of the bracketed terms and one factor to the C matrix. The CSL equation becomes

$$\hat{H}^{\mathrm{S}} = (I^{\mathrm{S}} + (\hat{S}\bar{\rho})^{\mathrm{S}})\hat{C}^{\mathrm{S}}(I^{\mathrm{S}} + (\bar{\rho}\hat{S})^{\mathrm{S}} + \bar{\rho}\hat{H}^{\mathrm{S}}) \tag{5.2.9}$$

The 4×4 identity matrix I is transformed into a 2×2 block matrix I^{S} with each block equal to the 2×2 identity matrix times $\frac{1}{2}$. Similarly this transformation changes both of the products of the connectivity matrix and the intramolecular correlation matrix into 2×2 block matrices with each block being identical.

$$(\hat{S}\bar{\rho})^{\mathrm{S}} = \begin{pmatrix} 0 & 0 & 0 & 0 \\ \omega/2 & 0 & \omega/2 & 0 \\ 0 & 0 & 0 & 0 \\ \omega/2 & 0 & \omega/2 & 0 \end{pmatrix} \tag{5.2.10}$$

and

$$(\bar{\rho}\hat{S})^{\mathrm{S}} = \begin{pmatrix} 0 & \omega/2 & 0 & \omega/2 \\ 0 & 0 & 0 & 0 \\ 0 & \omega/2 & 0 & \omega/2 \\ 0 & 0 & 0 & 0 \end{pmatrix} \tag{5.2.11}$$

Each of the matrices in Eq. (5.2.9) is of the same form; each is a 2×2 block matrix with all blocks identical. Multiplying out the right-hand side, we can reduce Eq. (5.2.9) to a 2×2 matrix equation for the sum correlation function

$$\tilde{H}^{\mathrm{S}} = \tilde{W}^{\mathrm{T}}\tilde{C}^{\mathrm{S}}[\tilde{W} + 2\tilde{\rho}\tilde{H}^{\mathrm{S}}] \tag{5.2.12}$$

where $\tilde{W}$ is the matrix

$$\tilde{W} = \begin{pmatrix} 1 & \omega \\ 0 & 1 \end{pmatrix} \tag{5.2.13}$$

and $\tilde{W}^{\mathrm{T}}$ is its transpose.

The transformation to obtain the difference correlation function equation proceeds in the same way. The difference correlation function matrix is defined by

$$\tfrac{1}{2}(I - F)\hat{H} = \hat{H}^{\mathrm{D}} = \begin{pmatrix} \tilde{H}^{\mathrm{D}} & -\tilde{H}^{\mathrm{D}} \\ -\tilde{H}^{\mathrm{D}} & \tilde{H}^{\mathrm{D}} \end{pmatrix} \tag{5.2.14}$$

The matrix $\frac{1}{2}(I - F)$ satisfies an identity of the form of Eq. (5.2.8), that is

$$\tfrac{1}{2}(I - F) = \tfrac{1}{8}(I - F)^3 \tag{5.2.15}$$

Following the same set of steps which lead to Eq. (5.2.12), it can be shown that the difference correlation function is given by

$$\tilde{H}^{\mathrm{D}} = (\tilde{W}^{\mathrm{D}})^{\mathrm{T}}\tilde{C}^{\mathrm{D}}[\tilde{W}^{\mathrm{D}} + 2\tilde{\rho}\tilde{H}^{\mathrm{D}}] \tag{5.2.16}$$

where $\tilde{W}^{\mathrm{D}}$ is given by

$$\tilde{W}^{\mathrm{D}} = \begin{pmatrix} 1 & -\omega \\ 0 & 1 \end{pmatrix} \tag{5.2.17}$$

This completes the separation of the matrix integral equation into sum and difference correlation functions. Whether a complete separation is possible depends upon the chosen closure relation.

From Section 11.D, the dielectric constant of a polyatomic fluid can be calculated from the site–site correlation functions. The result is

$$\Delta\hat{h} = \sum_{\alpha\gamma} q_\alpha q_\gamma \hat{h}_{\alpha\gamma} = \frac{k^2\mu^2}{9\rho}\left(\frac{\varepsilon - 1}{y\varepsilon} - 3\right) \quad \text{as} \quad k \to 0 \tag{5.2.18}$$

For the polar dumbbell, the CSL equation gives

$$\Delta\hat{h} = 4q^2\hat{h}^D = 4q^2\{\hat{h}^D_{aa} + \hat{h}^D_{as} + \hat{h}^D_{sa} + \hat{h}^D_{ss}\} \tag{5.2.19}$$

Rearranging the difference CSL Eq. (5.2.16), it can be shown that

$$\hat{h}^D_{aa} = \frac{1}{D}\hat{c}^D_{aa} \tag{5.2.20}$$

$$\hat{h}^D_{as} = \frac{1}{D}[\hat{c}^D_{as} - \omega\hat{c}^D_{aa} + 2\rho(\hat{c}^D_{aa}\hat{c}^D_{ss} - \hat{c}^D_{as}\hat{c}^D_{sa})] \tag{5.2.21}$$

$$\hat{h}^D_{sa} = \frac{1}{D}[\hat{c}^D_{sa} - \omega\hat{c}^D_{aa} + 2\rho(\hat{c}^D_{aa}\hat{c}^D_{ss} - \hat{c}^D_{as}\hat{c}^D_{sa})] \tag{5.2.22}$$

$$\hat{h}^D_{ss} = \frac{1}{D}[\hat{c}^D_{ss} - \omega(\hat{c}^D_{as} + \hat{c}^D_{sa}) + \omega^2\hat{c}^D_{aa} - 2\rho(1+\omega)(\hat{c}^D_{aa}\hat{c}^D_{ss} - \hat{c}^D_{as}\hat{c}^D_{sa})] \tag{5.2.23}$$

where the determinant D is given by

$$D = 1 - 2\rho(\hat{c}^D_{aa}(1-\omega) + \hat{c}^D_{as} + \hat{c}^D_{sa}) - 4\rho^2(\hat{c}^D_{aa}\hat{c}^D_{ss} - \hat{c}^D_{as}\hat{c}^D_{sa}). \tag{5.2.24}$$

Combining Eqs. (5.2.20), (5.2.21), (5.2.22), (5.2.23), and (5.2.24) gives

$$\hat{h}^D = \frac{1}{D}\{\hat{c}^D_{aa}(1-\omega)^2 + (\hat{c}^D_{as} + \hat{c}^D_{sa})(1-\omega) + \hat{c}^D_{ss} + 2\rho(\hat{c}^D_{aa}\hat{c}^D_{ss} - \hat{c}^D_{as}\hat{c}^D_{sa})(1-\omega)\} \tag{5.2.25}$$

For a wide variety of closure relations the only long-range direct correlation function is c_{aa}; the remaining direct correlation functions are all short-range. This implies the following small k, forms for their Fourier transforms

$$\hat{c}^{D}_{aa} \sim -\frac{4\pi\beta q^2}{k^2} + O(k^0) \tag{5.2.26}$$

$$\hat{c}^{D}_{as} = \hat{c}^{D}_{sa} \sim \alpha_{as} + \beta_{as}k^2L^2 + O(k^3) \tag{5.2.27}$$

$$\hat{c}^{D}_{ss} \sim \alpha_{ss} + \beta_{ss}k^2L^2 + O(k^3) \tag{5.2.28}$$

$$1 - \omega = 1 - \frac{\sin kL}{kL} \sim \frac{k^2L^2}{6} + O(k^4) \tag{5.2.29}$$

It can be shown that if $\alpha_{ss} \neq 0$, then the dielectric constant has a trivial solution, $\varepsilon = \infty$. We eliminate this possibility by setting α_{ss} to zero. Substituting Eqs. (5.2.26), (5.2.27), (5.2.28), and (5.2.29) into Eq. (5.2.25), and substituting the result into Eq. (5.2.18), the dielectric constant ε can be shown to be given by

$$\varepsilon = \frac{1 + 3y - 4\rho\alpha_{as}(1 - \rho\alpha_{as}) + 36\rho y\beta_{ss}}{1 - 4\rho\alpha_{as}(1 - \rho\alpha_{as})} \tag{5.2.30}$$

This result is new and was derived in collaboration with Lupkowski. Preliminary numerical calculations based on this result have been disappointing. It appears that the coefficient α_{ss} evaluated numerically is not zero, although it is small. The problem is similar to that of obtaining exact charge neutrality from numerical simulations of either the SSOZ equation or the CSL equation. It may well be better to extract the dielectric constant from numerical solutions by estimating the slope of the graph of $\Delta\hat{h}$ versus k^2 and using Eq. (5.2.19), rather than using Eq. (5.2.30). In either case, this result is encouraging in that for the first time there is a real possibility of observing a bond-length dependence of the dielectric constant from integral equation theory.

C. Thermodynamic Properties and Fluid Phase Equilibria

There has been a trend towards rather more attention being given to theoretical calculations of the structure of interaction site fluids than to their thermodynamic properties, although in recent years this trend has been somewhat reversed. In this section, we describe the efforts that have been made to calculate the thermodynamic properties. From our point of view, the overriding motivation for such investigations is the identification and understanding of the nontrivial effects of molecular anisotropy upon the thermodynamics. This is perhaps most readily revealed in terms of the overall shape of the phase diagram. Thus, the present account will focus especially upon studies aimed at the computation of fluid phase equilibria. Most of this work has involved applications of the thermodynamic perturbation theories discussed in Sections III.E and III.F rather than the direct calculation of the properties from the site–site correlation functions predicted via the integral equation theories described in Sections III.A and III.B. Before moving on to a discussion of this work, it is nevertheless worthwhile to assess the utility of thermodynamic property calculations based upon the solution of integral equations.

Thermodynamic property calculations based upon the solution of the molecular OZ equation using spherical harmonic expansions have not been extensive. This is probably due to the computational effort required in the implementation of these methods. However, quite promising results have been obtained for the equation of state of hard sphere diatomics (Chen and Steele,[28] Freasier,[134] Lado[29]) and for 12–6 diatomics (Lado[39]). Again, as we noted earlier in Section III.A, more extensive tests of this approach need to be made.

Studies of the thermodynamic properties obtained from solutions of the SSOZ equation have been much more extensive but the success has been mixed. The first such calculations were those of Lowden and Chandler[42] who obtained the pressure of hard diatomic fluids from the RISM (SSOZ–PY) equation. They used two routes to the equation of state: a compressibility equation of state in which they integrated the bulk modulus calculated from the site–site correlation functions via

$$\frac{1}{kT}\left(\frac{\partial P}{\partial \rho}\right)_T = 1 - \rho \sum_{\alpha,\gamma} \int c_{\alpha\gamma}(r)\, d\mathbf{r} \tag{5.3.1}$$

and a virial route based on the coupling parameter calculation described in Section II.C. Both the virial and compressibility equations give results which differ markedly from Monte Carlo simulation results.[134] It has been shown subsequently[18] that the solutions to the SSOZ–PY equation satisfy a variational principle that leads to a closed-form expression for the pressure from the compressibility equation of state. We have

$$\begin{aligned}\frac{P}{kT} &= \rho + \frac{\rho^2}{2} \sum_{\alpha,\gamma} \int \left[\frac{c^2_{\alpha\gamma}(r)}{1 - \exp\left[u_{\alpha\gamma}(r)/kT\right]} - 2c_{\alpha\gamma}(r) \right] d\mathbf{r} \\ &\quad + \frac{1}{(2\pi)^3} \int \{ Tr\, \rho \hat{W} \hat{C} + \ln \det (I - \rho \hat{W} \hat{C}) \}\, d\mathbf{k}\end{aligned} \tag{5.3.2}$$

In a similar way the SSOZ–HNC equation satisfies a variational principle which leads to closed-form expressions for the Helmholtz free energy and the pressure.[115] These variational principles are straightforward extensions of those derived for the atomic OZ equation in the PY and HNC approximations by Baxter[135] and by Morita and Hiroike,[136] respectively.

An extensive study of the compressibility equation of state for 12–6 diatomics has been made on the basis of Equation (5.3.2).[18] The results for the pressure were found to be in poor agreement with the molecular dynamics results of Singer et al.[137] for dense fluid states. In addition, the critical densities and temperatures were obtained for various diatomic bond lengths. The critical temperatures were found to become very large as the bond length decreased, although the critical densities showed a quite smooth variation with bond length. However, the equation of state, when plotted in terms of reduced variables ($P/P_c, \rho/\rho_c, T/T_c$), showed reasonably systematic departures from the principle of corresponding states for larger anisotropies. The compressibility equation of state from the SSOZ–HNC (extended RISM) equation near the vapor–liquid critical point has been investigated by Lupkowski and Monson.[138] They have found that this equation exhibits anomalies beyond those encountered in the HNC results for atomic fluids (see, for example, Brey and Santos[139]). In particular, isotherms of the bulk modulus exhibit a large asymmetry with respect to the critical density and can have two minima at temperatures near critical. For shorter bond lengths, no numerical solutions could be found at low density for subcritical temperatures.

The most accurate route to the thermodynamic properties from the SSOZ equation seems to be the energy equation.[46] The integral from which the internal energy is obtained (see Eq. (2.3.1)) seems to be relatively insensitive to errors in the predicted site–site correlation functions. It might on this basis be reasonably assumed that calculations of the Helmholtz free energy via integration of the Gibbs–Helmholtz equation

$$-T^2\left(\frac{\partial \dfrac{A}{T}}{\partial T}\right)_v = U \tag{5.3.3}$$

would be a useful route to the thermodynamics, as was found by Henderson et al.[145] in their study of the PY theory for the 12–6 potential. In the case of the SSOZ–PY theory, this turns out also to be true for dense 12–6 diatomic fluids, but the approach has proved to be unsuitable for calculation of the vapor–liquid coexistence properties.[140] This is because the energy equation of state tends to predict a much lower critical temperature than the compressibility equation of state, especially for shorter bond lengths. In the

PY theory for atomic fluids, the energy equation of state predicts the higher critical temperature. Thus, to construct the vapor–liquid coexistence curve from the energy equation of state requires solutions of the SSOZ–PY equation for states where the compressibility equation of state may predict thermodynamic instability. Such solutions often cannot be found numerically and, where they can be found, are surely of dubious physical significance. It seems from these observations that accurate thermodynamic properties over a wide range of fluid states, suitable, for example, for the study of vapor–liquid equilibria, can not be obtained from solutions of the SSOZ–PY or SSOZ–HNC equations. The SSOZ–MSA theory seems to be an exception to this situation as we shall see below.

Partly because of the above considerations, but also because of experience derived from studies of atomic fluids,[9] first-order perturbation theories based upon the methods described in Section III.D have been more widely used in the context of thermodynamic property calculations. These methods seem to give very good results for the thermodynamic properties of dense 12–6 diatomic fluids.[83,69] They have been used to calculate vapor–liquid coexistence properties.[141] They have also been applied to the calculation of the structure and excess properties of diatomic mixtures.[142,143] As currently implemented, there are weaknesses in the approaches arising from two sources. First, the approximations used for the cavity distribution function of the hard diatomic reference fluid (either the SSOZ–PY theory if the site–site formalism is used, or the RAM theory if the molecular coordinate formalism is used) are expected to be the principal source of error in the implementation of the theories at first order. Errors in these approximations can affect the accuracy of the first-order term in the Helmholtz free energy (although this can be relatively insensitive to such errors), and also the estimation of the hard diatomic site diameter. These issues are discussed in some detail in a paper by Abascal et al.[87] The second weakness in these approaches is that they are limited to nonpolar molecules at relatively high fluid densities where the structure of the fluid is dominated by the repulsive part of the site–site interactions. For polar molecules and for lower densities, the longer-range forces influence the fluid structure, and first-order perturbation theory is no longer sufficient.

As we described in Section III.G, perturbation theories can be extended in a systematic way using cluster expansion techniques.[25] These techniques have recently been applied to the calculation of the thermodynamic properties and vapor–liquid equilibrium of 12–6 diatomics[95] and seem to offer a clear improvement over the first-order perturbation theories. To illustrate this point, Table I shows values of the critical density and critical temperature predicted by the ISF–ORPA theory[25,95] and the first-order perturbation theory[141] together with results recently obtained from molecular dynamics

TABLE 1
Critical Densities and Temperatures of 12–6 Diatomics[a]

L^*	ρ_c^*(MD)	T_c^*(MD)	ρ_c^*(CPT)	T_c^*(CPT)	ρ_c^*(PT1)	T_c^*(PT1)
0.329	0.230(20)	3.62(10)	0.237	3.74	0.241	3.94
0.670	0.176(20)	2.30(10)	0.171	2.33	0.174	2.54

[a]The reduced units are $L^* = L/\sigma$, $T^* = kT/\varepsilon$ and $\rho^* = \rho\sigma^3$. MD denotes the molecular dynamics results,[144] CPT denotes the results of McGuigan et al.[95] using cluster perturbation theory, and PT1 denotes the results from first-order perturbation theory.[141]

simulations[144] for two bond lengths representative of nitrogen and chlorine, respectively.

Both theoretical predictions of the critical density lie within the uncertainty of the simulation results. However, the first-order perturbation theory predictions of the critical temperature are much higher than the simulation results. McGuigan et al.[95] have shown that the critical densities and temperatures of the 12–6 diatomics predicted by their theory are to a good approximation linear in the molecular volume. They find that

$$T_c^* = 9.042 - 3.6208 v_m^* \tag{5.3.4}$$

$$\rho_c^* = 0.4937 - 0.1740 v_m^* \tag{5.3.5}$$

where v_m^* is the ratio of the volume of a diatomic to the volume of one of the atoms and is given by

$$v_m^* = 1.0 + 3/2L^* - 1/2L^{*3} \tag{5.3.6}$$

The critical properties of several real diatomic fluids show similar behavior. Both the work of Fischer et al.[141] and McGuigan et al.[95] show clearly how deviations from the principle of corresponding states arise from increases in the diatomic bond length. The most striking effects are an increase in the reduced saturated liquid density and a reduction in the slope of $\log(P_{vap}/P_c)$ vs. T_c/T.

Morriss and Isbister[108,109] have studied phase equilibrium of polar hard diatomic fluids, and mixtures of polar hard diatomic fluids with nonpolar hard diatomics using the analytic solution of the SSOZ–MSA[50] and the energy equation of state. They predict that vapor–liquid phase separation will occur in the polar hard diatomic fluid arising only from the dipolar interactions.[108] They also predict that a mixture of hard diatomic and dipolar hard diatomic fluids will phase separate into two dense fluid phases, one rich in the polar component and the other in the nonpolar component. This phase separation is analogous to the liquid–liquid phase separation in a mixture of polar and nonpolar components. Interestingly, it seems that the SSOZ–MSA is not subject to the difficulties of the SSOZ–HNC and SSOZ–PY theories near vapor–liquid coexistence. Cummings[106] reached similar conclusions in a study of hard diatomics with a site–site Yukawa potential in the SSOZ–MSA. However, Lupkowski and Monson[25] show that this approximation is theoretically unsound due to the resummation of infinite series of unallowed diagrams in the interaction site cluster expansion. This will presumably impinge upon the accuracy of the phase equilibrium predictions. Some computer simulation results for this type of system would be very useful at this point. Lupkowski and Monson[97] have applied their cluster-expansion-based perturbation theory to dipolar 12–6 diatomics. They have studied how molecular shape and polarity affect the vapor–liquid coexistence properties. They find that the influence of the dipole upon the coexistence properties is smaller as the molecular shape becomes more nonspherical.

VI. CHEMICAL REACTION EQUILIBRIUM

A challenging problem in the theory of molecular fluids is the treatment of chemical reactions. The theory of atomic fluids is well developed, and the treatment of molecular fluids is also advancing steadily. However, the introduction of chemical reactions into these theories is a new and promising area. The first work in this area is that of Andersen[146] who in two interesting papers examined incorporation of graph cancellation due to steric effects in the expansion of the singlet density ρ. This was followed by the work of Chandler and Pratt,[148–150] which is based on the idea of physical clusters. Cummings and Stell[155,156] have developed a theory of simple chemical reactions based on an extension of the SSOZ equation in which the delta function that gives rise to the rigid chemical bond in a molecular species is scaled to incorporate the effects of an incomplete chemical reaction. Similar ideas have been pursued by Høye and Olaussen.[147] Perhaps the most promising approach is that of Wertheim.[160–166] In a series of papers, Wertheim has developed a theory for fluids with highly directional attractive forces using the fugacity expansion of the grand partition function as a starting point. If the directional forces are attractive, they tend to promote association into dimers, trimers, and higher s-mers with conformations that depend on the geometry of the repulsion and the direction of attraction. The approach of Wertheim leads naturally to a formulation of statistical mechanics in terms of two density variables: ρ the product density, and ρ_0 the monomer density. There is a strong structural similarity to the graph theory based on ρ alone, and this gives analogues of the s-particle correlation function and the direct correlation functions. From a physical point of view, the two-density theory is superior as the correct low-density limit of a dimerizing gas is trivially obtained, whereas the Mayer ρ-expansion requires a formidable graph resummation.

In this section, we give a short summary of some of these theories and their applications, with an emphasis on contrasting the Pratt–Chandler, Cummings–Stell, and Wertheim approaches. The interested reader can find more technical details in the references.

A. Pratt–Chandler Theory

Chandler and Pratt[148–150] have developed a theory for describing conformational structures of nonrigid molecules dissolved in liquid solvents. One of the simplest systems to study is the *trans-gauche* conformational equilibria of a single n-butane molecule dissolved in a solvent (which may also be n-butane). The major simplifying feature of this reaction is that it is not diffusion-limited as only one reactant is involved, and the mechanism is a simple energy barrier crossing, so the reaction coordinate is easily identified. The ideal gas equilibria can easily be determined, and the emphasis of this approach is to estimate the influence of steric effects (due to the shape of both the n-butane molecule and the solvent) on the conformational equilibrium in the liquid state. There have been a number of molecular dynamics calculations of the conformational equilibria of n-butane dissolved in n-butane, but the most reliable of these have been done since the development of the Pratt, Hsu, and Chandler method.[151,152]

The Chandler–Pratt approach has been reviewed extensively before (Chandler[4]), so we restrict our attention to some interesting applications. Pratt, Hsu, and Chandler have considered n-butane dissolved in liquid carbon tetrachloride using two different approximations. One is the superposition approximation for the intramolecular structure, while the other is a scheme called the two-cavity model. They conclude that the two-cavity model is the more accurate approximation.

In the two-cavity model, the four hard spheres, representing the CH_3 and CH_2 groups, are replaced with a pair of spherical cavities. Each cavity is considered to be an ethyl group, and its center is determined as the midpoint of the ethyl carbon–carbon bond. As the conformation of the n-butane molecule changes, the separation between the centers of the two cavities changes. In the same paper, the conformational equilibria of n-butane dissolved in n-butane and n-hexane are also considered. In both cases, there is a self-consistency aspect. The structure of the solvent n-butane or n-hexane changes as the conformation of each dihedral angle changes. Indeed, for the conformation of n-butane in solvent n-butane, it is impossible to distinguish between solvent and solute molecules. Recent molecular dynamics calculations of the conformational equilibria in pure butane by Edberg et al.[153] suggest that the two-cavity theory accurately estimates the shift due to steric hindrance. A more recent study by Zichi and Rossky[154] using the HNC closure does not appear to be as accurate.

B. Cummings–Stell Theory

A somewhat different approach is taken by Cummings and Stell.[155,156] They consider a system of species A and B particles with interaction potentials given by

$$\phi_{AA}(r) = \phi_{BB}(r) = \begin{Bmatrix} \infty, & r < \sigma \\ 0, & r > \sigma \end{Bmatrix} \tag{6.2.1}$$

and

$$\frac{\phi_{AB}(r)}{k_B} = \begin{Bmatrix} \varepsilon_1, & 0 < r < L - \dfrac{w}{2} \\ -\varepsilon_2, & L - \dfrac{w}{2} < r < L + \dfrac{w}{2} \\ \varepsilon_1, & L + \dfrac{w}{2} < r < \sigma \\ 0, & r > \sigma \end{Bmatrix} \tag{6.2.2}$$

Here L is the resulting bond length and w is the width of the square well potential centered at L. A pair of A and B particles is said to be associated as a dimer if their separation lies in a prescribed range $R_L < r_{AB} < R_U$. The density of dimers can then be written as

$$\rho_{AB} = \rho_A \rho_B \int_{R_L < r < R_U} d\mathbf{r}\, g_{AB}(r) \tag{6.2.3}$$

If R_U is chosen to be $L + w/2$ and R_L to be in the range $[0, L + w/2)$, then the definition of the density of dimers prevents polymerization beyond the dimer level. There is some arbitrariness in the choice of these limits, particularly R_U, and the use of more realistic binding potentials would make this choice even more difficult.

To facilitate the analytic solution, the square well potential is taken to be infinitely narrow and infinitely strong, resulting in a delta function at $r = L$. This problem reduces to the solution of the particle–particle viewpoint of polyatomic liquids, for a hard sphere fluid with a delta function attraction in the unlike interaction potential. This approach has been applied to a number of systems by Cummings, Rasaiah, and Lee.[157,159]

C. Wertheim Theory

A model interaction potential which is capable of representing a wide variety of physical circumstances, in particular highly directional attractive forces, is

$$\phi(1,2) = \phi_R(1,2) + \sum_\alpha \sum_\beta \phi_{\alpha\beta}(|\mathbf{r}_2 + \mathbf{d}_\beta(\Omega_2) - \mathbf{r}_1 - \mathbf{d}_\alpha(\Omega_1)|) \tag{6.3.1}$$

where $\mathbf{r}_i$ is the position of the center of molecule i and $\mathbf{d}_\alpha$ is the vector from the center of the molecule to the αth interaction site on that molecule. The potential $\phi_{\alpha\beta}$ between sites on different molecules is assumed to be attractive, and ϕ_R is taken to be the interaction of two hard particles of given shape (that is, strongly repulsive). In his first paper, Wertheim[160] considers a one-component system of hard spheres of diameter σ, with a single interaction site located near the edge of the repulsive hard core. The simplest attractive potential ϕ_A is given by

$$\phi_A(x)\begin{bmatrix} <0 & \text{for} & x<a \\ =0 & \text{for} & x>a \end{bmatrix} \tag{6.3.2}$$

where $x = |\mathbf{r}_2 + \mathbf{d}(\Omega_2) - \mathbf{r}_1 - \mathbf{d}(\Omega_1)|$ and the magnitude of $\mathbf{d}$ must satisfy $(\sigma - a)/2 < d < \sigma/2$. The case of short-range attraction located near the edge of the hard core is realized by the additional restriction $a \ll \sigma$. In any allowed configuration of N molecules, each particle can take part in only one attractive interaction $\phi_A \neq 0$. If the pair (i, j) have an attractive interaction, then the repulsive cores of i and j prevent any particle k from coming close enough to feel the influence of the attraction sites of i or j.

In graph theories, the potential $\phi(1,2)$ appears in the Mayer f-function which is defined by

$$f(1,2) = e(1,2) - 1 = \exp[-\beta\phi(1,2)] - 1 \tag{6.3.3}$$

The full Mayer f-function can be decomposed into two parts: one is the hard core f-function, and the other contains the attractive part

$$f(1,2) = f_R(1,2) + F(1,2) = f_R(1,2) + e_R(1,2)f_A(1,2) \tag{6.3.4}$$

For uniform systems, the pressure is related to the grand partition function Ξ by $\beta pV = \ln \Xi$. The graph theoretic fugacity expansion for $\ln \Xi$ and $\rho(1)$ is given by

$\ln \Xi =$ sum of all connected graphs composed of z-points and f-bonds

$\rho(1) =$ graphs obtained from $\ln \Xi$ by taking all ways of turning one field point into a labeled point 1

where the fugacity $z = \exp(\beta\mu)$ and μ is the chemical potential. The first step is to generate a new set of z-graphs by replacing each $f(i,j)$ by $f_{\mathrm{R}}(i,j)$ or $F(i,j)$. Then consider the set of z-graphs on s-points with f_{R}-bonds or F-bonds. A subset of these is the set in which all s points are connected by paths of F-bonds. This subset can be constructed in two steps:

(1) construct all connected graphs on s points with all F-bonds;
(2) for any pair (i,j) not connected by a direct F-bond, insert an e_{R}-bond.

Since the s points are already connected after step (1), all pairs lacking the direct F-bonds receive the sum of no bond and a direct f_{R}-bond in step 2. The F-connected graphs filled with e_{R}-bonds on $2, 3, \ldots,$ few$, \ldots, s$ points as dimer, trimer,$\ldots$, oligomer,$\ldots$, s-mer graphs. Using this construction, the graphical expansion for $\ln \Xi$ can be written as

$\ln \Xi =$ sum of all connected graphs consisting of s-mer graphs $(s = 1, \ldots, \infty)$ and f_{R}-bonds between pairs of points in distinct s-mer graphs

The graphical expansion for $\rho(1)$ can be obtained from $\ln \Xi$ as before.

The process of filling connected F-graphs with e_{R}-bonds optimizes the cancellation due to steric incompatibility. For the short-ranged interaction considered, all s-mer graphs for $s > 2$ vanish because of steric incompatibility, because they contain either a triangle of two F-bonds and one e_{R}-bond, or a triangle of F-bonds. It is also possible to construct potentials that allow trimers and not tetramers, or potentials that have a cutoff at some higher s-mer.

If we consider the graphical expansion for $\rho(1)$, we can divide the graphs into two classes: if the labeled point 1 is a monomer point (that is, has no incident F-bond), then the graph is in $\rho_0(1)$; if 1 is an s-mer point with $s \geqslant 2$, then the graph is in $\rho_1(1)$. The physical interpretation of ρ_0 is as a monomer density.

The standard application of the usual statistical mechanical graph techniques of topological reduction lead to the definition of two correlation functions, $c_0(1)$ and $c_1(1)$, and eventually to an analogue of the Ornstein–Zernike equation,

$$h_{ij}(1,2) = c_{ij}(1,2) + \sum_{m=0}^{1} \sum_{n=0}^{1} \int d(3) c_{im}(1,3) \rho_{mn}(3) h_{nj}(3,2) \tag{6.3.5}$$

where the density matrix ρ_{ij} is given by

$$\rho = \begin{pmatrix} \rho & \rho_0 \\ \rho_0 & 0 \end{pmatrix} \tag{6.3.6}$$

Note that there is no simple relation between the ordinary direct correlation function $c(1,2)$ and the $c_{ij}(1,2)$ introduced here.

For the special case of a dimerizing gas, we can go further, and the restriction that at most, one F-bond is incident on each point makes it possible to achieve a substantial graph sum by combining the new OZ equation with relatively simple closure relations. Wertheim[161] has shown that an analogue of the PY equation can be obtained:

$$e_{\mathrm{R}}(1,2)c_{ij}(1,2) = f_{\mathrm{R}}(1,2)g_{ij}(1,2) + \delta_{i1}\delta_{j1}g_{00}(1,2)f_{\mathrm{A}}(1,2). \tag{6.3.7}$$

This closure reduces to the usual PY closure when $\phi_{\mathrm{A}} = 0$, and is similar in spirit to the PY closure for hard diatomics (Wertheim[162]).

The extension of these ideas to fluids with multiple attraction sites has been given by Wertheim[163] and used by him to develop a theory of equilibrium polymerization.[164,166] The difficulty with extension of the single-attraction site theory to multiple attraction sites is in incorporating the increasingly complex steric incompatibility (SI). In particular, the difficulty commonly encountered is SI3, which physically corresponds to the absence of self-hindrance, where the rigidity of an s-mer prevents two of its component molecules from encountering each other. In another paper, Wertheim[165] has considered fluids of dimerizing hard spheres and fluid mixtures of hard spheres and diatomics.

Wertheim's theory of fluids with highly directional forces has been compared with Monte Carlo calculations by Jackson, Chapman, and Gubbins.[167] They used the thermodynamic perturbation theory (rather than the OZ-like integral equation) to study hard spheres with one or two attraction sites. Excellent agreement with simulation was obtained for the compressibility factor, configurational energy, and monomer concentration. Thermodynamic perturbation-theory results for the phase equilibria were also obtained, and the coexistence curves were shown to be greatly affected by the degree of association. A more recent paper considers the phase equilibria of chain molecules with multiple bonding sites.[168]

VII. INHOMOGENEOUS SYSTEMS

The last decade has seen a number of interesting developments in the study of inhomogeneous molecular fluids. In particular, there have been several theoretical and computer-simulation studies of the vapor–liquid interface, fluids adsorbed at solid surfaces, and the freezing of liquids. A variety of intermolecular potential models has been used in this work, but in the spirit of the rest of the article, we will restrict the discussion to those based upon interaction site models. Again we shall focus principally upon the theoretical developments and discuss simulation results only as they impinge directly upon these developments. Each of the areas included in this survey is undoubtedly worthy of a review in its own right, and in some cases we can do little more than sketch the key accomplishments.

A. Theories of Freezing for Molecular Fluids

One of the most active areas of research in the statistical mechanics of interfacial systems in recent years has been the problem of freezing. The principal source of progress in this field has been the application of the classical density-functional theories (for a review of the fundamentals in these methods, see, for example, Evans[169]). For atomic fluids, such applications were pioneered by Ramakrishnan and Yussouff[170] and subsequently by Haymet and Oxtoby[171,172] and others (see, for example, Baret et al.[173]). Of course, such theories can also be applied to the vapor–liquid interface as well as to problems such as phase transitions in liquid crystals. Density-functional theories for these latter systems have not so far involved use of interaction site models for the intermolecular forces.

The general formulation of density-functional theory for molecular fluids has been described by Smithline et al.[174] They have applied a simplified version of the theory to the freezing of hard core diatomic fluids. A suitable starting point for such theories for rigid nonspherical molecules is the following expression for the grand potential

$$\Omega[n] = F[n] + \int n(\mathbf{r}, \omega)[\phi(\mathbf{r}, \omega) - \mu]d\mathbf{r}\, d\omega \tag{7.1.1}$$

In this expression, $n(\mathbf{r}, \omega)$ is the single-molecule density and is a function of both position, $\mathbf{r}$, and orientation, ω. $F[n]$ is the intrinsic Helmholtz free energy, μ is the chemical potential and $\phi(\mathbf{r}, \omega)$ is an external field. It can be shown (Evans[169]) that the equilibrium molecular density $\rho(\mathbf{r}, \omega)$ represents a global minimum in $\Omega[n]$ with respect to functional variations in the trial function $n(\mathbf{r}, \omega)$ at fixed $\phi(\mathbf{r}, \omega)$ and μ, a necessary condition for which is that

$$\left[\frac{\delta\Omega}{\delta n(\mathbf{r}, \omega)}\right]_{n=\rho} = 0 \tag{7.1.2}$$

The key to the practical application of the method is the development of suitable parametrization of the intrinsic Helmholtz free energy functional, $F[n]$, for the nonuniform system of interest. Now $F[n]$ may be written as a sum of an ideal gas contribution and one arising from the intermolecular forces in the system. We have

$$\begin{aligned} F[n] &= F_{\mathrm{id}}[n] + F_{\mathrm{ex}}[n] \\ &= kT \int n(\mathbf{r}, \omega)\{\ln[qn(\mathbf{r}, \omega)] - 1\} d\mathbf{r}\, d\omega + F_{\mathrm{ex}}[n] \end{aligned} \tag{7.1.3}$$

where q is the contribution to the molecular partition function from the internal molecular degrees of freedom. In treatments of the solid–fluid melt interface, the procedure used most often is to make a functional Taylor expansion of $F_{\mathrm{ex}}[n]$ about that of the bulk liquid in powers of $\Delta n(1) = n(\mathbf{r}_1, \omega_1) - \rho_{\mathrm{L}}$, where ρ_{L} is the density of the bulk liquid. To second order this expansion has the form

$$\begin{aligned} F_{\mathrm{ex}}[n] &= F_{\mathrm{ex}}(\rho_{\mathrm{L}}) + \frac{\mu_{\mathrm{ex}}(\rho_{\mathrm{L}})}{kT} \int \Delta n(1)\, d1 \\ &\quad - \frac{kT}{2} \int\int c(1,2;\rho_{\mathrm{L}}) \Delta n(1) \Delta n(2)\, d1\, d2 \end{aligned} \tag{7.1.4}$$

where $\mu_{\mathrm{ex}}(\rho_{\mathrm{L}})$ is the excess chemical potential for the uniform liquid phase, $c(1,2;\rho_{\mathrm{L}})$ is the direct correlation function of the uniform liquid, and we have for simplicity returned to the notation of Section III.A to describe molecular coordinates. The third-order correction to Eq. (7.1.4) involves the three-particle direct-correlation function for reference state, about which little or no reliable information is available. Thus, practical computations are limited to the second-order approximation. Smithline et al. have used the spherical harmonic expansion to treat the orientation dependence of $c(1,2;\rho_{\mathrm{L}})$, but instead of solving the PY equation using the techniques described in Section III.A, they adopt a much simpler empirical approximation due to Pynn.[175] Here $c(1,2)$ is replaced by the analytic PY result for hard spheres with diameter given by the collision diameter, $\sigma(1,2)$, for the given relative orientation of the molecules.

In addition to approximations for $c(1,2;\rho_{\mathrm{L}})$, a suitable parametrization of the trial function $n(\mathbf{r}, \omega)$ is required. The position dependence can be described in full detail by a Fourier expansion, the form of which is determined by the choice of crystal symmetry for the solid phase. More simply, a Gaussian distribution of molecular density about the sites of the crystal lattice may be assumed; the accuracy of this latter approximation has been verified for atomic systems. Superimposed upon this distribution is a suitable trial function for the molecular orientation in the crystal.

Chandler and co-workers[176,177] have developed an alternative approach to density-functional theories for molecular fluids which is based upon the interaction site formalism. The straightforward generalization of Eq. (7.1.1) to a single-component interaction site system is

$$\Omega[n] = F[n] - \sum_{\alpha} \int n_{\alpha}(\mathbf{r})[\mu_{\alpha} - \phi_{\alpha}(\mathbf{r})]\,d\mathbf{r} \tag{7.1.5}$$

where the summation is over the interaction sites α of the molecules, μ_{α} denotes the chemical potential of site α, and the site chemical potentials are governed by the constraint

$$\mu = \sum_{\alpha} \mu_{\alpha} \tag{7.1.6}$$

The key difficulty in developing this approach is the treatment of the influence of intramolecular correlations upon the site density, $n_{\alpha}(\mathbf{r})$. Recall from Section III.B that in the interaction site formalism, the explicit treatment of molecular orientation effects upon the correlation functions is replaced by a coupling of intramolecular and intermolecular correlations. For an inhomogeneous simple fluid, the molecular density may be written in the form[178,179]

$$\rho(\mathbf{r}) = \Lambda^{-3} \exp[\{\mu - \phi(\mathbf{r}) + c^{(1)}(\mathbf{r})\}/kT] \tag{7.1.7}$$

where Λ is a characteristic length, the value of which determines the reference-state chemical potential, and $c^{(1)}(\mathbf{r})$ is the single-particle direct-correlation function which describes the influence of the intermolecular forces upon $\rho(\mathbf{r})$. Neglect of $c^{(1)}(\mathbf{r})$ leads to the well-known *barometric law* describing the density variation of an ideal gas in an external field. The appropriate generalization of Eq. (7.1.7) to interaction site fluids is[176]

$$\rho_{\alpha}(\mathbf{r}) = \Lambda^{-3} \exp[\{\mu_{\alpha} - \phi_{\alpha}(\mathbf{r}) + D_{\alpha}(\mathbf{r}) + c_{\alpha}^{(1)}(\mathbf{r})\}/kT] \tag{7.1.8}$$

Here $D_{\alpha}(\mathbf{r})$ describes the influence of the intramolecular correlations, and $c_{\alpha}^{(1)}(\mathbf{r})$ gives the effect of intermolecular interactions upon the site density. Even in the absence of the intermolecular effects, the intramolecular correlations make a nontrivial contribution to the intrinsic Helmholtz free energy of the ideal gas. Chandler et al.[176,177] discuss two routes to the calculation of $D_{\alpha}(\mathbf{r})$. The first approach involves a second-order functional

Taylor expansion of the ideal gas free energy about that of the uniform fluid. This approach has the undesirable feature of leading to incorrect behavior in the limit where all the interaction sites are superimposed, the so-called *united atom* limit, but is expected to be useful in the limit of large intersite separations. A second approach, and one which gives essentially exact results, involves solving a set of m-coupled integral equations, where m is the number of sites in each molecule. In general, this is a quite complex procedure, although for the case of a diatomic fluid it has a particularly simple implementation. The expression used for the intermolecular contribution to the intrinsic Helmholtz energy corresponding to Eq. (7.1.4) is

$$F_{\mathrm{ex}}[n] = F_{\mathrm{ex}}(\rho_{\mathrm{L}}) + \mu_{\mathrm{ex}}(\rho_{\mathrm{L}}) \int \Delta n_{\alpha}(\mathbf{r}_{\alpha}) d\mathbf{r}_{\alpha} - \frac{kT}{2} \sum_{\alpha,\gamma} \int\int c_{\alpha\gamma}(|\mathbf{r}-\mathbf{r}'|; \rho_{\mathrm{L}}) \Delta n_{\alpha}(\mathbf{r}) \Delta n_{\gamma}(\mathbf{r}') d\mathbf{r}\, d\mathbf{r}' \tag{7.1.9}$$

where $c_{\alpha\gamma}(|\mathbf{r}-\mathbf{r}'|; \rho_{\mathrm{L}})$ is the site–site direct correlation function for the uniform liquid and the integral in the second term can be evaluated for any of the site densities.

The two approaches outlined above have been studied in greatest detail in relation to the freezing transitions of hard diatomic fluids[180,174] into orientationally disordered solid (plastic crystal) phases. Such investigations are of interest in understanding, for example, the transition between the α (orientationally ordered *fcc*) and β (orientationally disordered *hcp*) phases of solid nitrogen. Smithline et al. find that the plastic crystal is always stable with respect to orientationally ordered structures and that, within the accuracy of their method, there appears to be no difference in the stabilities of the *fcc* and *hcp* plastic crystals' structures. McCoy et al.[180] reached similar conclusions, although their study was restricted to orientationally disordered solid phases. Interestingly, both groups found that as the diatomic bond length was increased, a point was reached beyond which the fluid phase was stable with respect to all the solid phases considered. These studies should provide impetus for more extensive investigations via both refined versions of the density-functional theories and computer simulations. A preliminary study of the ice–water transition[181] has also been presented, based upon the interaction site density-functional theory, with correlation functions obtained from experimental data on liquid water. Qualitatively correct results for the melting temperature and solid density were obtained.

It appears from the above discussion that the density-functional approach offers considerable promise as a means of studying the freezing of molecular liquids. The increased variety of crystal structures produced by even small anisotropy in the intermolecular forces offers an especially interesting challenge to theory. It does seem, however, that the quantitative predictions of these theories may be quite sensitive to the approximations used; for example, the truncation at second order of the functional Taylor expansion of $F[n]$ or the treatment of the reference-fluid direct-correlation functions. A great deal of work remains to be done in assessing fully the impact of the approximations made.

B. The Vapor–Liquid Interface

Orientational structure at a liquid vapor–interface of diatomic interaction site fluids has been studied extensively by Gubbins and Thompson[182–185] using both thermodynamic perturbation theory and molecular dynamics simulation, and by Tarazona and Navascues[186,187] using perturbation theory. Chacon et al.[188,189] have applied density-functional theories to these systems. The theoretical methodology and results are reviewed in a comprehensive article by Gubbins,[190] to which the reader is directed for more complete details.

The theoretical techniques used most frequently in this context are perturbation theories of the type discussed in Section III.C based upon spherical reference systems. Thompson et al.[185] focus upon perturbation expansions of the molecular density, $\rho(\mathbf{r}, \omega)$, about that of the reference fluid, $\rho_0(\mathbf{r})$, using either the pair potential or the Mayer function to parametrize the expansion. If the pair potential is used, then to first order they obtain

$$\rho(\mathbf{r}_1, \omega_1) = \rho_0(\mathbf{r}_1) - \frac{\rho_0(\mathbf{r}_1)}{kT} \int\int \Delta u(1,2) \rho_0(\mathbf{r}_2) g_0(\mathbf{r}_1, \mathbf{r}_2)\, d\mathbf{r}_2 d\omega_2 \quad (7.2.1)$$

and if the Mayer function is used they obtain, again to first order,

$$\rho(\mathbf{r}_1, \omega_1) = \rho_0(\mathbf{r}_1) + \rho_0(\mathbf{r}_1) \int\int f_a(1,2) \rho_0(\mathbf{r}_2) g_0(\mathbf{r}_1, \mathbf{r}_2)\, d\mathbf{r}_2 d\omega_2 \quad (7.2.2)$$

where

$$f_a(1,2) = \exp\{-[u(1,2) - u_0(1,2)]/kT\} - 1 \quad (7.2.3)$$

and $g_0(\mathbf{r}_1, \mathbf{r}_2)$ is the pair distribution function for the inhomogeneous reference system. The density profile for the reference fluid is obtained by solution of the first equation in the Yvon–Born–Green hierarchy using an ansatz for $g_0(\mathbf{r}_1, \mathbf{r}_2)$ suggested by Toxvaerd.[191]

Tarazona and Navascues[186,187] have proposed a perturbation theory based upon the division of the pair potential given in Eq. (3.5.1). In addition, they make a further division of the reference potential into attractive and repulsive contributions in the manner of the WCA theory. The resulting perturbation theory for the interfacial properties of the reference system is constructed through adaptation of a method developed by Toxvaerd[192] in his extension of the BH perturbation theory to the vapor–liquid interface. The Tarazona–Navascues theory generates results for the Helmholtz free energy and surface tension in addition to the density profile. Chacon et al.[188,189] have shown how the perturbation theories based upon Eq. (3.5.1) may be developed by a series of approximations within the context of a general density-functional treatment.

As we stated at the beginning of this section, most of the work in this area has focused upon the study of molecular orientation at the liquid–vapor interface and how this is affected by details of the intermolecular forces. Thompson and Gubbins[182–185] have carried out molecular dynamics simulations of the vapor–liquid interface for homonuclear 12–6 diatomic molecules and for such molecules with point-charge quadrupoles. They find that in the case of the nonpolar molecules, there is a tendency for molecules in the liquid to align perpendicular to the surface and those in the vapor to align parallel to the surface. The addition of a quadrupole to the 12–6 diatomic[185] reverses this effect. A study of the vapor–liquid interface for an interaction site model of *n*-octance[193] leads to similar conclusions as for the nonpolar diatomic. These effects are reproduced qualitatively by all the theoretical approximations, with the exception of the influence of the quadrupole, which can only be predicted at first order within the context of the perturbation theory based upon division of the Mayer function Eq. (3.5.2).

C. Adsorption at a Fluid–Solid Interface

Adsorption at the fluid–solid interface for interaction site systems has been studied in two contexts: adsorption from dense fluids, and adsorption from the dilute gas phase. Work in the latter context has been focused particularly on submonolayer adsorption on graphite surfaces.

Sullivan et al.[194] and Thompson et al.[195] have studied the structure of hard diatomic fluids in contact with a hard wall and Lennard-Jones 12–6 diatomic fluids interacting with a wall via the Lennard-Jones 9–3 potential.[196] Computer simulations were carried out via the Monte Carlo method for the hard diatomic system and via molecular dynamics for the 12–6 diatomic system. In each case, the simulation results were compared with the results from solutions of the RISM or SSOZ–PY theory adapted to the fluid–wall problem. This adaptation can be achieved by noting that the site density profile for a diatomic fluid in contact with a plane surface can be related to the sphere–site distribution for a sphere solute at infinite dilution in a diatomic fluid after taking the limit where the sphere diameter becomes infinitely large. The SSOZ–PY theory gives qualitatively correct results for the site density profiles of hard diatomic fluids in contact with a hard wall. However, the site density profiles from the SSOZ–PY theory are seriously in disagreement with the simulation results for the 12–6 diatomics in contact with a 9–3 wall. Sokolowski[197] has generalized the CSL equation to the treatment of these systems. He shows the superior accuracy of the ISF–PY approximation over the SSOZ–PY approximation at low density.

Sokolowski and Steele[198,199] have adapted the spherical harmonic expansion technique described in Section III.A to the calculation of the density profiles of nonspherical molecules in contact with solid surfaces. They have used the results to investigate molecular orientation effects in high temperature adsorption from the gas phase. The RAM theory described in Section III.E has been extended to the adsorption of fluids on solid surfaces by Smith et al.[200] for hard-sphere interactions and by Sokolowski and Steele[198] to the case of more realistic fluid–solid interactions. The principal deficiency of the approach is the accuracy of the predicted correlation functions for the bulk fluid which are required as input to the theory.

Adsorption of small nonspherical molecules such as nitrogen,[201–206] oxygen[207,208] and ethylene[209] on graphite at low temperatures (below the bulk triple point) has been the subject of several recent investigations via computer simulation. The principal focus has been the rather rich variety of phases found in these systems. Even for simple spherical molecules, the coupling between the adsorbate–adsorbate interactions and the lateral variation of the adsorbate–surface interaction produced by the structure of the surface gives rise to adsorbed phases which may have the characteristics of two-dimensional vapors and liquids, or solids that are commensurate or incommensurate with the surface lattice.[210] The presence of anisotropy in the intermolecular forces greatly extends the range of possible surface phases. For example, in the case of nitrogen, there are commensurate solid phases in which the molecules are either orientationally ordered in a *herring-bone* structure or orientationally disordered.[201,202] An orientationally ordered incommensurate solid phase is also found.

VIII. OTHER APPLICATIONS

In this section, we will discuss some applications of interaction site models that do not fall so neatly within the scope of areas we have reviewed. In particular, we focus on applications to polymeric systems and to percolation phenomena. We conclude the section with some remarks suggesting where progress might be made in the future.

A. Polymeric Systems

Interaction site models are proving to have what might be regarded as an unexpected applicability in the context of polymer fluid theory. The notion of a polymer molecule as a large collection of interaction sites is an obvious one; the complexity of the rigorous treatment of such a system within the context of theories discussed in Section III is equally obvious. Nevertheless, recent work has shown that a certain amount of useful progress can be made by introducing some simplifying approximations into the SSOZ equation for such a system. These approximations have proved useful in studying both the structure and thermodynamics of polymer melts and blends,[211–213] as well as the properties of polyelectrolyte solutions.[214,215] However, the earliest application of these ideas was to the treatment of the classical hard sphere polymer in a hard sphere solvent which appears in the path integral representation of an electron dissolved in a hard sphere solvent.[216,217] A useful perspective on the different applications has been given in a recent paper by Chandler.[218]

In thinking about the extension of the SSOZ equation or any other approach in the theory of interaction site fluids to polymeric systems, two questions arise. First, it is possible to treat systems where the dimension of the matrices in Eq. (3.2.1) may be of order 10^4–10^6 or more? Second, is it possible to describe the influence of the environment of the polymer chain, in either solution or the melt, upon the intramolecular correlations? The answer to the first question appears to be: yes, within certain limits. However, the price paid for the approximations required is that the answer to the second question is apparently no. The dimensionality of the problem for a polymer melt can be reduced by noting that for molecules consisting of a ring of N equivalent interaction sites, the SSOZ equation becomes the scalar integral equation

$$h(r) = w * c * w(r) + \rho w * c * h(r) \tag{8.1.1}$$

where $h(r)$ and $c(r)$ are now the segment–segment total and direct-correlation functions, and

$$w(r) = \sum_{\alpha=1}^{N} W_{\alpha\gamma}(r) \tag{8.1.2}$$

For large N, this equation is expected to behave similarly to that for an infinite linear chain, although the convergence depends upon the form chosen for the intramolecular correlation functions, $W_{\alpha\gamma}(r)$.[218] Equation (8.1.1) can be generalized to the case of a single ring in a solvent of spherical molecules or to more complex cases where the solvent is also an interaction site fluid or a mixture. In the first case, we have

$$h_{\alpha s}(r) = w * c_{\alpha s}(r) + \rho w * c_{\alpha s} * h_{ss}(r) \tag{8.1.3}$$

where the subscripts αs and ss refer to segment–solvent and solvent–solvent correlations respectively, and w is again given by Eq. (8.1.2). This is the equation used by Chandler et al.[216,217] in their work on excess electrons in hard sphere solvents and is a special case of that used by Hirata and Levy[214,215] in their work on polyelectrolytes. A key feature of both Eqs (8.1.1) and (8.1.3) is that they are only correct when the sites are truly equivalent. Thus, if end effects are important in a linear chain, these equations will not be appropriate. A problem of perhaps greater importance is that the assumption of equivalent sites precludes the treatment of systems where the environment has a strong influence upon the chain conformation.[218] Hirata and Levy treat polyelectrolyte chains using the rigid molecule form of the intramolecular correlations (Eq. (3.2.3)). Schweizer and Curro[211–213] treat the intermolecular correlations using the results for a Gaussian chain and assume that these results are unchanged by the environment. It should also be remembered that Eqs. (8.1.1) and (8.1.3) will carry with them all theoretical difficulties associated with the SSOZ equation as described in Section III.B. Nevertheless, efforts of this type represent the first serious attempts at a realistic treatment of the influence of segment–segment and segment–solvent interactions upon the structure and thermodynamics of polymer blends and solutions in the context of off-lattice models. They should serve as a strong stimulus for further advances in this area.

B. Percolation Phenomena

Over the last decade, there has been considerable interest in the development of off-lattice models and theories describing percolation phenomena. Interest in percolation concepts has been spurred by the rather wide variety of applications for which such ideas are thought to be useful. These applications include the electrical conductivity and dielectric properties[219,220] or permeability[221] of composite materials, gelation,[222] analysis of hydrogen bond networks,[223] and reactions in porous catalysts.[224] Recent progress in the development of off-lattice or, as they are most often called, continuum models of percolation began with the work of Coniglio et al.[225,226] and Haan and Zwanzig,[227] who showed that it was possible to adapt a number of concepts and techniques from the modern theory of fluids, especially cluster expansions and integral equations, to the problem.

The key idea is the concept of the physical cluster originated by Hill.[228]

Hill suggested a division of the Mayer function into two contributions

$$f(r) = f^+(r) + f^*(r) \tag{8.2.1}$$

where $f^+(r)$ corresponds to the contribution from bound particles and $f^*(r)$ corresponds to the contribution from unbound particles. Hill described a procedure for making this division which was appropriate to the specification of physical clusters in, say, a Lennard-Jones fluid. More recently, Chandler and Pratt[148] have exploited the concept in their work on chemical reaction equilibrium. Within the context of percolation theory, Eq. (8.2.1) is viewed as a definition of interparticle connectivity and as such can be applied to a system where there are no physical interactions. For example, consider the problem of randomly distributed spheres of diameter, σ, in a matrix. An appropriate definition of $f^+(r)$ and $f^*(\mathrm{r})$ is

$$\begin{aligned} f^*(r) &= -1 \quad r < \sigma \\ &= 0 \quad r > \sigma \end{aligned} \tag{8.2.2}$$

$$\begin{aligned} f^+(r) &= 1 \quad r < \sigma \\ &= 0 \quad r > \sigma \end{aligned} \tag{8.2.3}$$

Hill also showed that the division of the Mayer function via Eq. (8.2.1) permits a decomposition of the grand partition function into contributions arising from bound particles in clusters and unbound particles. By introducing the division into the cluster expansion of the pair distribution function, Coniglio et al.[225,226] also showed that the set of nodeless diagrams in the cluster series continuous path of $f^+(r)$ bonds connecting the root points. This subset of diagrams is called the *pair connectedness function*, $h^+(r)$, and measures the probability that two particles are separated by a distance r and are members of the same cluster. An analogous quantity defined in the context of lattice models of percolation was introduced earlier by Essam.[229] Coniglio et al.[225,226] also showed that the set of nodeless diagrams in the cluster series for $h^+(r)$ could be related to $h^+(r)$ itself via the Ornstein–Zernike like equation

$$h^+(r) = c^+(r) + \rho c^+ * h^+(r) \tag{8.2.4}$$

The mean cluster size can be calculated from $h^+(r)$ or $c^+(r)$ using

$$S = 1 + \rho \int h^+(r)d\mathbf{r} = \left[1 - \rho \int c^+(r)d\mathbf{r}\right]^{-1} \tag{8.2.5}$$

which is, of course, reminiscent of the compressibility equation of state in fluid theory. The percolation threshold is the lowest density at which S diverges. A Percus–Yevick like approximation relating $c^+(r)$ and $h^+(r)$ has been solved for a variety of model systems,[230,231] and the results have been compared with computer simulation results.[232,233]

It is also of interest to develop theories for systems where the particles are nonspherical in shape. Recently, Lupkowski and Monson[234] have shown how the ideas described above may be extended to systems of nonspherical particles made up of interaction sites. They have formulated the interaction site cluster expansion of the site–site pair connectedness function, $h^+_{\alpha\gamma}(r)$, which measures the probability that two sites on different particles are separated by a distance r and are members of the same cluster. The extensions of the SSOZ and CSL integral equations to the site–site connectivity problem are readily obtained. For example, the connectivity SSOZ equation is

$$H^+ = WC^+W + \rho WC^+H^+ \tag{8.2.6}$$

where the notation is essentially the same as that described in connection with Eq. (3.2.1) Closure approximations based on the Percus–Yevick approximation are readily obtained to both this equation and the connectivity CSL equation obtained by adapting Eq. (3.4.3). Lupkowski and Monson describe the application of these equations to the system of dumbbells randomly distributed in a matrix. In particular, they show how these theories can predict the decrease in percolation threshold associated with an increase in particle anisotropy.

By changing the site–site Mayer function, this formalism can be applied to a number of different physical situations. For example, if the site–site Mayer functions are chosen as those of an interaction site model of water, then the formalism could be used to study hydrogen bonding as a percolation problem, as has been done in the lattice context by Stanley and Texeira.[223] Another possibility is to choose the Mayer functions to model polyfunctional chemical bonding. The formalism could then form the basis of treatment of gelation in polymer networks.

IX. CONCLUDING REMARKS

In this article, we have tried to provide a broad but compact survey of the progress made in the equilibrium theory of interaction site fluids. In conclusion, it is worthwhile to reflect upon the principal achievements and to attempt to identify the areas where progress needs to be made.

We think that the successes of theories of the equilibrium statistical mechanics of interaction site fluids have been mostly in their qualitative predictions, but even in this sense some remarkable progress has been achieved. Nevertheless, truly quantitative predictions are limited to just a few simple molecules which can be treated as nonpolar diatomics. Thermodynamic perturbation theories in the spirit of the Barker–Henderson and Weeks–Chandler–Andersen theories can be as successful for interaction site systems as they are for simple atomic fluids. However, this is only strictly true for nonpolar fluids, and only then in cases where accurate information about the reference fluid properties is available. Only in the case of homonuclear diatomics has it been shown that sufficiently reliable reference-fluid correlation functions are available to permit quantitative calculations of the free energy via perturbation theory. Substantial progress will have to be made in the calculation of hard core interaction site fluid correlation functions before perturbation theory can be applied on a routine basis with quantitative results. The importance of advances in this latter area is hard to overestimate.

The formal treatment describing the extension of perturbation theory to polar interaction site systems and to other situations where the perturbative forces are structure-determining is available[96] and has been applied with some success to some simple models of polar diatomics. More quantitative comparisons with computer simulations need to be made. Of course, qualitatively accurate information about the structure of polar interaction site fluids has been available for some time through solutions of the SSOZ–HNC equations.[5] However, this approach does not seem to be useful in the context of thermodynamics. Rather little attention has been paid to polarizable molecules, although these can be treated within the context of the interaction site formalism (see, for example, Chandler[52] and, more recently, Sprik and Klein[235]). Although the formal treatment of the dielectric constant within the interaction site formalism is now well established, no quantitative approximations seem to emerge from any of the theories available.

Some useful progress has been made in the theory of inhomogeneous interaction site systems. Again however, the results are only qualitatively accurate. Further improvements in the theory for the bulk phases will be crucial to future developments in this area. We believe that research into solid–fluid equilibria will be of great value in developing an understanding of the influence of molecular anisotropy on the global phase diagram. At the simplest level, some insight into the variation with molecular anisotropy of the ratio of the triple point temperature to the critical temperature might emerge from such efforts.

Some very good progress has been made in the understanding of reaction equilibria through the work of Chandler and Pratt, Cummings and Stell, and Wertheim. It does appear that the approach of Wertheim provides the simplest and most tractable route to the study of more complex systems. However, more widespread interest in all these approaches would be very beneficial to progress in this area. On a related note, we have seen in

Section VIII that some progress has been made in the extension of the interaction site approach to polymeric and other large molecular systems. The key difficulty here is the development of tractable theory describing how the environment influences intramolecular structure in macromolecules. This is clearly a very difficult problem but the solution would be of immense value.

Acknowledgments

The authors would like to thank the following colleagues and co-workers who have contributed to their understanding of interaction site systems: P. T. Cummings, K. Fraser, K. E. Gubbins, D. Isbister, M. Lupkowski, D. McGuigan, W. A. Steele, G. Stell and W. B. Streett. P. A. M. would like to thank J. S. Rowlinson and the staff of the Physical Chemistry Laboratory for their hospitality during the time when this article was completed. Research at University of Massachusetts on topics described here has been supported by the Gas Research Institute, National Science Foundation and the donors of the Petroleum Research Fund administered by the American Chemical Society.

References

1. C. G. Gray and K. E. Gubbins, *Theory of Molecular Fluids*, Vol. 1, Oxford Uniersity Press, 1984.
2. G. C. Maitland, M. Rigby, E. B. Smith, and W. A. Wakeham, *Intermolecular Forces*, Oxford University Press, 1981.
3. D. Chandler, *Ann. Rev. Phys. Chem.* **29**, 441 (1978).
4. D. Chandler, in *The liquid State of Matter: Fluids Simple and Complex*, E. Montroll and J. L. Lebowitz, Eds., North-Holland, 1982.
5. P. J. Rossky, *Ann. Rev. Phys. Chem.* **36**, 321 (1985).
6. E. Lippert, C. A. Chatzidimitriou-Dreismann, and K.-H. Naumann, *Adv. Chem. Phys.* **57**, 311 (1984).
7. G. Stell, in *The Equilibrium Theory of Classical Fluids*, H. L. Frisch and J. L. Lebowitz, Eds., Benjamin, New York, 1964.
8. H. C. Andersen, in *Modern Theoretical Chemistry, Vol. 5: Statistical Mechanics. Part A Equilibrium Techniques*, B. J. Berne, Ed., Plenum, New York, 1977.
9. J.-P. Hansen and I. R. McDonald, *Theory of Simple Liquids*, Academic Press, London, 1986.
10. J. K. Percus and G. J. Yevick, *Phys. Rev.* **110**, 1 (1958).
11. D. Chandler, *Molec. Phys.* **31**, 1213 (1976).
12. B. M. Ladanyi and D. Chandler, *J. Chem. Phys.* **62**, 4308 (1975).
13. P. T. Cummings, C. G. Gray, and D. E. Sullivan, *J. Phys.* **A14**, 1483 (1981).
14. D. Chandler and H. C. Andersen, *J. Chem. Phys.* **57**, 1930 (1972).
15. J. S. Høye and G. Stell, *SUNY CEAS Report No. 307* (1977); G. Stell, G. N. Patey, and J. S. Høye *Adv. Chem. Phys.* **48**, 183 (1981).
16. C. G. Gray, *Disc. Faraday Soc.* **66**, 178 (1978).
17. P. T. Cummings and G. Stell, *Molec. Phys.* **46**, 383 (1982).

18. P. A. Monson, *Molec. Phys.* **53**, 1209 (1984).
19. I. Aviram, D. J. Tildesley, and W. B. Streett, *Molec. Phys.* **34**, 881 (1977).
20. D. J. Tildesley, W. B. Streett, and D. S. Wilson, *Chem. Phys.* **36**, 63 (1979).
21. K. G. Honnell, C. K. Hall, and R. Dickman, *J. Chem. Phys.* **87**, 664 (1987).
22. J. S. Høye and G. Stell, *J. Chem. Phys.* **65**, 18 (1976); **66**, 795 (1977).
23. M. S. Wertheim, *J. Chem. Phys.* **55**, 4291 (1971).
24. J. S. Høye and G. Stell, *J. Chem. Phys.* **61**, 562 (1974).
25. M. Lupkowski and P. A. Monson, *J. Chem. Phys.* **87**, 3618 (1987).
26. W. A. Steele,*Faraday Discuss. Chem. Soc.* **66**, 138 (1978).
27. L. R. Pratt and D. Chandler, *J. Chem. Phys.* **66**, 147 (1977).
28. Y. D. Chen and W. A. Steele, *J. Chem. Phys.* **52**, 5284 (1970).
29. F. Lado, *Molec. Phys.* **47**, 283 (1982).
30. F. Lado, *Molec. Phys.* **47**, 299 (1982).
31. L. Blum, *J. Chem. Phys.* **57**, 1862 (1972) L. Blum and A. H. Narten, *Adv. Chem. Phys.* **34**, 203 (1976).
32. W. B. Streett and D. J. Tildesley, *Proc. Roy. Soc.* **A348**, 239 (1976).
33. W. B. Streett and D. J. Tildesley, *Proc. Roy. Soc.* **A355**, 239 (1977).
34. P. A. Monson and M. Rigby, *Molec. Phys.* **38**, 1699 (1979).
35. P. T. Cummings, I. Nezbeda, W. R. Smith, and G. P. Morriss, *Molec. Phys.* **43**, 1471 (1981).
36. P. A. Monson and W. A. Steele, *Molec. Phys.* **49**, 251 (1983).
37. J. R. Sweet and W. A. Steele, *J. Chem. Phys.* **47**, 3022 (1967).
38. T. W. Melnyk and W. R. Smith, *Molec. Phys.* **40**, 317 (1980).
39. F. Lado, *J. Chem. Phys.* **88**, 1950 (1988).
40. A. Perera, P. G. Kusalik, and G. N. Patey, *J. Chem. Phys.* **87**, 1295 (1987), errratum ibid., **89**, 5969 (1988); A. Perera and G. N. Patey, *J. Chem. Phys.* **89**, 5861 (1988).
41. D. Chandler, and H. C. Andersen, *J. Chem. Phys.* **57**, 1930 (1972); D. Chandler, *J. Chem. Phys.* **59**, 2742 (1973).
42. L. Lowden and D. Chandler, *J. Chem. Phys.* **59**, 6587 (1973); erratum and addenda, ibid., **62**, 4246 (1975).
43. L. Lowden and D. Chandler, *J. Chem. Phys.* **61**, 5228 (1974).
44. K. Kojima and K. Arakawa, *Bull. Chem. Soc. Japan* **51**, 1977 (1978).
45. E. Johnson and R. P. Hazoumé, *J. Chem. Phys.* **70**, 1599 (1979).
46. P. A. Monson, *Molec. Phys.* **47**, 435 (1982).
47. E. Enciso, *Molec. Phys.* **56**, 129 (1985).
48. F. Hirata and P. J. Rossky, *Chem. Phys. Lett.* **83**, 329 (1981).
49. F. Hirata, B. M. Pettitt, and P. J. Rossky, *J. Chem. Phys.* **77**, 509 (1982).
50. G. P. Morriss and J. W. Perram, *Molec. Phys.* **43**, 669 (1981).
51. D. Chandler, R. Silbey, and B. M. Ladanyi, *Molec. Phys.* **46**, 1335 (1982).
52. D. Chandler, *J. Chem. Phys.* **67**, 1113 (1977).
53. G. P. Morriss, Ph.D. Thesis, unpublished (1980).
54. D. E. Sullivan and C. G. Gray, *Molec. Phys.* **42**, 443 (1981).
55. P. T. Cummings, G. P. Morriss, and G. Stell, *J. Phys. Chem.* **86**, 1696 (1982).
56. P. T. Cummings and G. Stell, *Molec. Phys.* **44**, 529 (1981).

57. P. T. Cummings and D. E. Sullivan, *Molec. Phys.* **46**, 665 (1982).
58. J. C. Rasaiah and S. H. Lee, *J. Chem. Phys.* **83**, 6396 (1985).
59. P. J. Rossky, B. M. Pettitt, and G. Stell, *Molec. Phys.* **50**, 1263 (1983).
60. G. P. Morriss, *Molec. Phys.* **47**, 833 (1982).
61. D. Chandler, C. G. Joslin, and J. M. Deutch, *Molec. Phys.* **47**, 871 (1982).
62. P. J. Rossky and R. A. Chiles, *Molec. Phys.* **51**, 661 (1984).
63. J. W. Perram and L. R. White, *Molec. Phys.* **28**, 527 (1974).
64. W. R. Smith, *Can. J. Phys.* **52**, 2022 (1974).
65. D. Cook and J. S. Rowlinson, *Proc. Roy. Soc.* **A219**, 405 (1953).
66. S. Sung and D. Chandler, *J. Chem. Phys.* **56**, 4989 (1972).
67. W. A. Steele and S. I. Sandler, *J. Chem. Phys.* **61**, 1315 (1974).
68. I. Nezbeda and W. R. Smith, *Molec. Phys.* **45**, 681 (1982).
69. N. Quirke and D. J. Tildesley, *J. Phys. Chem.* **87**, 1972 (1983).
70. N. Quirke, J. W. Perram, and G. Jaccuci, *Molec. Phys.* **39**, 1311 (1980).
71. T.W. Melnyk, W. R. Smith, and I. Nezbeda, *Molec. Phys.* **46**, 629 (1982).
72. W. R. Smith and I. Nezbeda, *Molec. Phys.* **44**, 347 (1981).
73. F. Kohler, N. Quirke, and J. W. Perram, *J. Chem. Phys.* **71**, 4128 (1979).
74. J. Fischer, *J. Chem. Phys.* **72**, 5371 (1980).
75. J. L. Lebowitz and J. K. Percus, *J. Chem. Phys.* **79**, 443 (1983).
76. M. S. Shaw, J. D. Johnson, and B. L. Holian, *Phys. Rev. Lett.* **50**, 1141 (1983).
77. G. O. Williams, J. L. Lebowitz, and J. K. Percus, *J. Chem. Phys.* **81**, 2070 (1984).
78. I. Nezbeda, W. R. Smith, and S. Labik, *J. Chem. Phys.* **81**, 835 (1984).
79. D. MacGowan, J. D. Johnson, and M. S. Shaw, *J. Chem. Phys.* **82**, 3765 (1985).
80. D. MacGowan, D. B. Nicolaides, J. L. Lebowitz, and C.-K. Choi, *Molec. Phys.* **58**, 131 (1986).
81. J. A. Barker and D. Henderson, *J. Chem. Phys.* **47**, 4714 (1967).
82. J. D. Weeks, D. chandler, and H. C. Andersen, *J. Chem. Phys.* **54**, 5237 (1971).
83. D. J. Tildesley, *Molec. Phys.* **41**, 341 (1980).
84. M. Lombardero, J. L. Abascal, and S. Lago, *Molec. Phys.* **42**, 999 (1981).
85. T. Boublik and I. Nezbeda, *Coll. Czech. Chem. Comm.* **51**, 2301 (1986).
86. D. J. Tildesley and W. B. Streett, *Molec. Phys.* **37**, 985 (1980).
87. J. L. Abascal, C. Martin, M. Lombardero, J. Vasquez, J. Banon, and J. Santamaria, *J. Chem. Phys.* **82**, 2445 (1985).
88. J. O. Valderama, S. I. Sandler, and M. Fligner, *Molec. Phys.* **42**, 1041 (1981).
89. J. O. Valderama and S. I. Sandler, *Molec. Phys.* **49**, 925 (1983).
90. M. Wojcik and K. E. Gubbins, *Molec. Phys.* **51**, 951 (1984).
91. H. C. Andersen and D. Chandler, *J. Chem. Phys.* **57**, 1918 (1972).
92. H. C. Andersen, D. Chander, and J. D. Weeks, *Adv. Chem. Phys.* **34**, 105 (1976).
93. G. Stell, in *Statistical Mechanics* Part A, B. J. Berne, Ed., Plenum Press, 1977, p. 47.
94. W.G. Madden, *J. Chem. Phys.* **75**, 1984 (1981).
95. D. B. McGuigan, M. Lupkowski, D. M. Paquet, and P. A. Monson, *Molec. Phys.* **67**, 33 (1989).
96. M. Lupkowski and P. A. Monson, *Molec. Phys.* **63**, 875 (1988).

97. M. Lupkowski and P. A. Monson, *Molec. Phys.* **67**, 53 (1989).
98. G. P. Morriss, J. W. Perram, and E. R. Smith, *Molec. Phys.* **38**, 465 (1979); G. P. Morriss and J. W. Perram, *Molec. Phys.* **41**, 1463 (1980).
99. M. J. Gillan, *Molec. Phys.* **38**, 1781 (1979).
100. R. J. Baxter, *Aust. J. Phys.* **21**, 563 (1968).
101. G. P. Morriss and E. R. Smith, *J. Stat. Phys.* **24**, 611 (1981).
102. G. P. Morriss and P. A. Monson, *Molec. Phys.* **48**, 181 (1983).
103. P. T. Cummings, G. P. Morriss, and C. C. Wright, *Molec. Phys.* **43**, 1299 (1981).
104. G. P. Morriss and P. T. Cummings, *Molec. Phys.* **49**, 1103 (1983).
105. P. T. Cummings and G. P. Morriss, *Molec. Phys.* **43**, 1299 (1981).
106. P. T. Cummings, *Molec. Phys.* **53**, 849 (1984).
107. G. P. Morriss and P. T. Cummings, *Molec. Phys.* **45**, 1099 (1982).
108. G. P. Morriss and D. J. Isbister, *Molec. Phys.* **52**, 57 (1984).
109. G. P. Morriss and D. J. Isbister, *Molec. Phys.* **59**, 911 (1986).
110. K. C. Ng, *J. Chem. Phys.* **61**, 2680 (1974).
111. L. R. Pratt and D. Chandler, *J. Chem. Phys.*, **67**, 3683 (1977).
112. G. P. Morriss and D. MacGowan, *Molec. Phys.* **58**, 745 (1986).
113. A. R. Allnatt, *Molec. Phys.* **8**, 533 (1964).
114. K. J. Fraser, G. P. Morriss, and L. A. Dunn, *Molec. Phys.* **57**, 1233 (1986).
115. S. J. Singer and D. Chandler, *Molec. Phys.* **55**, 621 (1985).
116. P. J. Rossky and W. D. T. Dale, *J. Chem. Phys.* **73**, 2457 (1980).
117. M. S. Cooper, *Phys. Rev.* **A7**, 1 (1973).
118. J. F. Springer, M. A. Pokrant, and F. A. Stevens, *J. Chem. Phys.* **58**, 4863 (1972).
119. G. M. Abernethy and M. J. Gillan, *Molec. Phys.* **39**, 839 (1980).
120. M. Ohba and K. Arakawa, *Bull. Chem. Soc. Japan* **55**, 1387 (1982).
121. F. J. Rogers, *J. Chem. Phys.* **73**, 6272 (1980).
122. H. Bertagnolli, I. Hausleithner, and O. Steinhauser, *Chem. Phys. Lett.* **116**, 465 (1985).
123. K. J. Fraser, L. A. Dunn, and G. P. Morriss, *Molec. Phys.* **61**, 775 (1987).
124. C. S. Hsu, D. Chandler, and L. J. Lowden, *Chem. Phys.* **14**, 213 (1976).
125. D. Chandler, C. S. Hsu, and W. B. Streett, *J. Chem. Phys.* **66**, 5231 (1977).
126. L. R. Pratt and D. Chandler, *J. Chem. Phys.* **72**, 4045 (1980).
127. K. Kojima and K. Arakawa, *Bull. Chem. Soc. Japan* **53**, 1795 (1980).
128. M. Ohba and K. Arakawa, *J. Phys. Soc. Japan* **50**, 743 (1981).
129. M. Ohba and K. Arakawa, *Bull. Chem. Soc. Japan* **58**, 9 (1985).
130. M. Ohba and K. Arakawa, *Bull. Chem. Soc. Japan* **58**, 3068 (1985).
131. M. Ohba and K. Arakawa, *J. Phys. Soc. Japan* **55**, 2955 (1986).
132. P. G. Kusalik, private communication.
133. S. W. deLeeuw and N. Quirke, *J. Chem. Phys.* **81**, 880 (1984).
134. B. C. Freasier, *Chem. Phys. Lett.* **35**, 280 (1975).
135. R. J. Baxter, *J. Chem. Phys.* **47**, 4855 (1967).
136. T. Morita and K. Hiroike, *Prog. Theor. Phys.* **23**, 1003 (1960).
137. K. Singer, A. Taylor, and J. V. L. Singer, *Molec. Phys.* **33**, 1757 (1977).

138. M. Lupkowski and P. A. Monson, *Chem. Phys. Lett.* **136**, 258 (1987).
139. J. J. Brey and A. Santos, *Molec. Phys.* **57**, 149 (1986).
140. D. B. McGuigan, Ph.D. Thesis, University of Massachusetts (1989).
141. J. Fischer, R. Lustig, H. Breitenfelder-Manske, and W. Lemming, *Molec. Phys.* **52**, 485 (1985).
142. E. Enciso and M. Lombardero, *Molec. Phys.* **44**, 725 (1981).
143. D. J. Tildesley, E. Enciso, and P. Sevilla, *Chem. Phys. Lett.* **100**, 508 (1983).
144. S. Gupta, *J. Chem. Phys.* **92**, 7156 (1988).
145. D. Henderson, J. A. Barker, and R. O. Watts, *I.B.M. J. Res. Dev.* **14**, 688 (1970).
146. H. C. Andersen, *J. Chem. Phys.* **59**, 4714 (1973); **61**, 4985 (1974).
147. J. S. Høye and K. Olaussen, *Physica* **104A**, 435 (1980).
148. D. Chandler and L. R. Pratt, *J. Chem. Phys.* **65**, 2925 (1976).
149. L. R. Pratt and D. Chandler, *J. Chem. Phys.* **66**, 147 (1977).
150. L. R. Pratt and D. Chandler, *J. Chem. Phys.* **67**, 3683 (1977).
151. L. R. Pratt, C. S. Hsu, and D. Chandler, *J. Chem. Phys.* **68**, 4202 (1978).
152. C. S. Hsu, L. R. Pratt, and D. Chandler, *J. Chem. Phys.* **68**, 4213 (1978).
153. R. A. Edberg, D. J. Evans, and G. P. Morriss, *J. Chem. Phys.* **84**, 6933 (1986).
154. D. A. Zichi and P. J. Rossky, *J. Chem. Phys.* **84**, 1712 (1986).
155. P. T. Cummings and G. Stell, *Molec. Phys.* **51**, 253 (1984).
156. P. T. Cummings and G. Stell, *Molec. Phys.* **55**, 33 (1985).
157. S. H. Lee, J. C. Rasaiah, and P. T. Cummings, *J. Chem. Phys.* **83**, 317 (1985).
158. J. C. Rasaiah and S. H. Lee, *J. Chem. Phys.* **83**, 5870 (1985).
159. J. C. Rasaiah and S. H. Lee, *J. Chem. Phys.* **83**, 6396 (1985).
160. M. S. Wertheim, *J. Stat. Phys.* **35**, 19 (1984).
161. M. S. Wertheim, *J. Stat. Phys.* **35**, 35 (1984).
162. M. S. Wertheim, *J. Stat. Phys.* **78**, 4619 (1983).
163. M. S. Wertheim, *J. Stat. Phys.* **42**, 459 (1986).
164. M. S. Wertheim, *J. Stat. Phys.* **42**, 477 (1986).
165. M. S. Wertheim, *J. Chem. Phys.* **85**, 2929 (1986).
166. M. S. Wertheim, *J. Chem. Phys.* **87**, 7323 (1987).
167. G. Jackson, W. G. Chapman, and K. E. Gubbins, *Molec. Phys.* **65**, 1 (1988).
168. W. G. Chapman, G. Jackson, and K. E. Gubbins, *Molec. Phys.* **65**, 1057 (1988).
169. R. Evans, *Adv. Phys.* **28**, 143 (1979).
170. T. V. Ramakrishnan and M. Yussouff, *Phys. Rev. B* **19**, 2775 (1979).
171. A. D. J. Haymet and D. Oxtoby, *J. Chem. Phys.* **74**, 2559 (1981).
172. A. D. J. Haymet and D. Oxtoby, *J. Chem. Phys.* **84**, 1769 (1986).
173. J. L. Barat, J. P. Hansen, G. Pastore, and E. Waisman, *J. Chem. Phys.* **86**, 6360 (1987).
174. S. J. Smithline, S. W. Rick, and A. D. J. Haymet, *J. Chem. Phys.* **88**, 2004 (1988).
175. R. Pynn, *J. Chem. Phys.* **60**, 4579 (1974).
176. D. Chandler, J. D. McCoy, and S. J. Singer, *J. Chem. Phys.* **85**, 5971 (1986).
177. D. Chandler, J. D. McCoy, and S. J. Singer, *J. Chem. Phys.* **85**, 5977 (1986).
178. J. S. Rowlinson and B. Widom, *Molecular Theory of Capillarity*, Oxford University Press, 1982.

179. B. Widom, *J. Stat. Phys.* **19**, 563 (1978).
180. J. D. McCoy, S. J. Singer, and D. Chandler, *J. Chem. Phys.* **87**, 4853 (1987).
181. K. Ding, D. Chandler, S. J. Smithline, and A. D. J. Haymet, *Phys. Rev. Lett* **59**, 1698 (1987).
182. S. M. Thompson, *Faraday Discuss. Chem. Soc.* **66**, 107 (1978).
183. S. M. Thompson and K. E. Gubbins, *J. Chem. Phys.* **70**, 4947 (1979).
184. S. M. Thompson and K. E. Gubbins, *J. Chem. Phys.* **74**, 6467 (1981).
185. S. M. Thompson and K. E. Gubbins, and J. M. Haile, *J. Chem. Phys.* **75**, 1325 (1981).
186. P. Tarazona and G. Navasques, *Molec. Phys.* **47**, 145 (1982).
187. P. Tarazona and G. Navasques, *Molec. Phys.* **47**, 1021 (1982).
188. E. Chacon, P. Tarazona, and G. Navasques, *J. Chem. Phys.* **79**, 4426 (1983).
189. E. Chacon, P. Tarazona, and G. Navasques, *Molec. Phys.* **51**, 1475 (1984).
190. K. E. Gubbins, in *Fluid Interfacial Phenomena*, C. A. Croxton, Ed., Wiley, 1986, p. 469.
191. S. Toxvaerd, *Molec. Phys.* **26**, 91 (1973).
192. S. Toxvaerd, *J. Chem. Phys.* **55**, 3316 (1971).
193. T. A. Weber and E. Helfand, *J. Chem. Phys.* **65**, 2377 (1980).
194. D. E. Sullivan, R. Barker, C. G. Gray, W. B. Streett, and K. E. Gubbins, *Molec. Phys.* **44**, 597 (1981).
195. S. M. Thompson, K. E. Gubbins, D. E. Sullivan, and C. G. Gray, *Molec. Phys.* **51**, 21 (1984).
196. W. A. Steele, *The Interaction of Gases with Solid Surfaces*, Pergamon Press, Oxford, 1974.
197. S. Sokolowski, *Phys. Lett. A* **115**, 343 (1986).
198. S. Sokolowski and W. A. Steele, *Langmuir* **1**, 181 (1985).
199. S. Sokolowski and W. A. Steele, *Langmuir* **1**, 190 (1985).
200. W. R. Smith, I. Nezbeda, and M. R. Reddy, *Chem. Phys. Lett.* **106**, 575 (1984).
201. J. Talbot, D. J. Tildesley, and W. A. Steele, *Molec. Phys.* **51**, 1331 (1984).
202. J. Talbot, D. J. Tildesley, and W. A. Steele, *Surf. Sci.* **169**, 71 (1986).
203. A. V. Vernov and W. A. Steele, *Langmuir* **2**, 219 (1986).
204. A. V. Vernov and W. A. Steele, *Surf. Sci.* **171**, 83 (1986).
205. C. Peters and M. L. Klein, *Molec. Phys.* **54**, 985 (1985).
206. Y. Joshi and D. J. Tildesley, *Molec. Phys.* **55**, 999 (1985).
207. Y. Joshi and D. J. Tildesley, *Surf. Sci.* **166**, 169 (1986).
208. V. R. Bhethanabotla and W. A. Steele, *Langmuir* **3**, 581 (1987).
209. M. E. Moller and M. L. Klein, *Chem. Phys.* **129**, 235 (1989).
210. A. Thomy, X. Duval, and J. Regnier, *Surf. Sci. Reports* **1**, 1 (1980).
211. K. S. Schweizer and J. G. Curro, *Phys. Rev. Lett.* **58**, 246 (1987).
212. K. S. Schweizer and J. G. Curro, *Phys. Rev. Lett.* **60**, 809 (1988).
213. J. G. Curro and K. S. Schweizer, *J. Chem. Phys.* **87**, 1842 (1987).
214. F. Hirata and R. M. Levy, *Chem. Phys. Lett.* **136**, 267 (1987).
215. F. Hirata and R. M. Levy, *J. Phys. Chem.* **93**, 479 (1989).
216. D. Chandler, Y. Singh, and D. Richardson, *J. Chem. Phys.* **81**, 1975 (1984).
217. A. L. Nichols III, D. Chandler, Y. Singh, and D. M. Richardson, *J. Chem. Phys.* **81**, 510 (1984).
218. D. Chandler, *Chem. Phys. Lett.* **139**, 108 (1987).
219. P. Sheng, *Phys. Rev. Lett.* **45**, 60 (1980).

220. P. Sheng, *Phys. Rev.B.* **22**, 6364 (1980).
221. N. Shah, J. E. Sax, and J. M. Ottino, *Polymer* **26**, 1229 (1985).
222. D. Stauffer, A. Coniglio, and M. Adam, *Adv. Polym. Sci.* **44**, 103 (1982).
223. H. E. Stanley and J. Texeira, *J. Chem. Phys.* **73**, 3404 (1980).
224. N. Shah and J. Ottino, *Chem. Eng. Sci.* **42**, 63 (1987).
225. A. Coniglio, U. De Angelis, A. Forlani, and G. Lauro, *J. Phys. A* **10**, 219 (1977).
226. A. Coniglio, U. De Angelis, and A. Forlani, *J. Phys. A* **10**, 1123 (1977).
227. S. W. Haan and R. Zwanzig, *J. Phys. A* **10**, 1547 (1977).
228. T. L. Hill, *J. Chem. Phys.* **23**, 617 (1955).
229. J. W. Essam, in *Phase Transitions and Critical Phenomena*, Vol. 2, C. Domb and M. S. Green, Eds. Academic Press, 1972, p. 197.
230. Y. C. Chiew and E. D. Glandt, *J. Phys. A* **16**, 2599 (1983).
231. T. DeSimone, S. Demoulini, and R. M.Stratt, *J. Chem. Phys.* **85**, 391 (1986).
232. N. A. Seaton and E. D. Glandt, *J. Chem. Phys.* **86**, 4668 (1987).
233. E. M. Sevick, P. A. Monson, and J. M. Ottino, *J. Chem. Phys.* **88**, 1198 (1988).
234. M. Lupkowski and P. A. Monson, *J. Chem. Phys.* **89**, 3300 (1988).
235. M. Sprik and M. L. Klein, *J. Chem. Phys.* **89**, 7556 (1988).

ON THE THEORY OF THE ORIGIN OF SPATIALLY NONUNIFORM STATIONARY STATES (DISSIPATIVE STRUCTURES) IN HETEROGENEOUS CATALYTIC SYSTEMS

YU. E. VOLODIN, V. N. ZVYAGIN, A. N. IVANOVA, AND V. V. BARELKO

Institute of Chemistry
USSR Academy of Sciences
Moscow Region, USSR

CONTENTS

I. INTRODUCTION

The question of the existence of spatially inhomogeneous stationary states (in terms of[1] "dissipative structures," DSs) in heterogeneous catalysis was first raised in references 2–6. The type of DS studied in this series of works was named a standing wave, because such a stationary inhomogeneous structure is a particular case of the problem of autowave transition between two uniform steady thermal regimes of a catalytic process and is similar to the front of a traveling wave. The theoretical analysis in references 2–6 was performed within the scope of the simplest possible model that describes the process of exothermal conversion on a single catalyst wire in the uniform fields of external parameters, under the assumption of quasi-stationarity of a reactant mass balance equation for the gaseous phase, and without regard to diffusion transfer at the boundary layer along the wire axis. The model reduces to a single equation of thermal conductivity with a source nonlinear in catalyst temperature, which has enabled us, in the most physical obvious way, to (1) determine the existence conditions for DSs of the standing wave type and the means of their excitation, (2) reproduce experimentally the basic features of this phenomenon, (3) substantiate the fact of stabilization of the structure and its transition from the state of indifferent equilibrium to a stable steady state under controlled temperature conditions of the catalyst, and (4) explain the stabilization of nonuniform steady states and broadening of their region of existence in the case of reactions obeying a kinetic law with nonunique dependence of conversion velocity on catalyst temperature (within the framework of a model with a hysteretic source).

The one-phase model considered in references 2–6 describes adequately the generation of nonuniform steady states of thermal nature in systems characterized by a high thermal conductivity, which include the broad class of catalytic conversions on metals. However, widely occurring in catalysis are processes with nonmetal catalysts (e.g., oxides) or with metals on nonmetal supports. Belonging here are also processes in fixed-bed reactors, where contacts between separate particles are most imperfect, the thermal connection being, to a large extent, due to conduction through gaseous interlayers. In the description of such systems, the equation for diffusion reactant transfer in the gaseous phase along the catalyst surface cannot be disregarded. This work is an extension of the studies of DSs in heterogeneous catalysis based on models describing the processes in systems with a low thermal conductivity of the catalyst.

The present paper (just as previous ones, references 2–6) considers the thermal mechanism as being responsible for the formation of DSs. In other words, the factor of nonlinearity is here the exponential dependence of the reaction heat generation intensity on temperature, which is the commonest in chemistry, and the concentration–velocity relation corresponds to the linear case of a first-order reaction. Consideration of the chemically simplest case aims at forming a basis of the theory of DS in heterogeneous catalysis and its further development by consistently complicating the kinetic law of a reaction and introducing into the model nonlinearities (feedbacks) of both thermal and kinetic nature (such as those considered within the framework of a one-component approach in references 7–8).

Analogously to references 2–6, the present paper considers the process of a catalytic conversion on a single catalytic unit, which is a wire normal to the reactant stream. The choice of this physical model is due to its adequacy, simplicity, and the possibility it gives to verify experimentally the theoretical predictions by use of catalytic elements made of the materials of nonmetal catalysts or nonmetal supports of the active phase in the form of a fiber or rod. The methametical model for this reaction system, which is considered below, is of particular interest in view of its being a simplified one-dimensional approximation to the problem of finding the conditions for loss of stability (formation of DS) in uniform stationary regimes of processes in important industrial installations such as fixed-bed reactors, and also in the catalytic cells of flow-circulation reactors generally used in kinetic studies.

There are presently two approaches in chemical reactor stability studies: (1) considering a reactor as a lumped gradientless system (perfectly stirred reactor,[9] and (2) accounting for heat and mass transfer along a reactor in the reactant flow direction, while preserving the assumption of the reactant temperature and concentration uniformity on the transverse coordinate.[10]

The averaging technique characteristic of the second approach may apply to the case of a tubular reactor where the ratio of the characteristic catalyst particle size to the diameter of a single tube is close to unity, but it is invalid, as will be shown, in the general case of fixed-bed reactors. This approach keeps out of a researcher's field of vision the problem of the reactor stability to local perturbations. At the same time, the technologist is often faced with hot spots in the catalyst bed of a fixed-bed reactor, which make its operation imperfect and even lead to an emergency situation in a number of cases,[11,12] Until recently, nonuniformity of the fields of external parameters (e.g., nonuniform packing of the catalyst bed or nonuniformity of reactant stream velocity[12,13]) was considered the only cause of these phenomena. The question naturally arises whether the provision for uniformity of external conditions guarantees the uniformity of temperature and concentration profiles at the reactor cross-section. The present paper seeks to answer this question, which, as a matter of fact, has not yet been posed in such a form in the theory of chemical reactors.

At the first stage of the study, it would be logical to consider the simplest possible model that facilitates the analysis and unmasks the basic features of the phenomenon. In relation to the process in a catalytic reactor, this is a one-dimensional model of a fixed-bed reactor with a small thickness of the coating (i.e., with low degrees of conversion). The model accounts for temperature and concentration distributions only in the direction normal to the velocity vector of the entering reactant flow. Under certain conditions, a more general model describing the process in a plug flow reactor reduces to the above one. Analysis of these conditions is beyond the scope of this work, but recent numerical data enable us to conclude that the qualitative results obtained with the simplified model will be valid in the general case as well. Under the assumptions made, the model of a fixed-bed reactor is also fit for the case of a single catalyst wire and a flow-circulation reactor, which makes it possible to tie the consideration of these objects together.

II. DEFINITION OF THE PROBLEM. A MATHEMATICAL MODEL

Let us first write a model with reference to the process on a single catalytic element, and then discuss the properties of the system and the meaning of its parameters for the case of a reactor.

We consider the following physical situation. A first-order exothermic reaction runs on a smooth, nonporous catalytic element having a length L (thread, fiber, tube). Assume that there is no temperature distribution at the element cross-section, thus reducing the analysis to a one-dimensional case. Assume also that the reaction runs under conditions of transversal flow, and the heat and mass transfer between the catalyst surface and the bulk-flow are described by the effective coefficients α and β, respectively. Under these assumptions, the model can be written in the form of two equations, one describing the heat balance in the solid catalyst phase, and another the reactant balance in the gaseous phase of a certain characteristic layer adjacent to the catalyst surface:

$$m\frac{\partial T}{\partial t} = \lambda\frac{\partial^2 T}{\partial x^2} + \frac{a}{d}[qkC - \alpha(T - T_0)] \tag{1}$$

$$\frac{\partial C}{\partial t} = D\frac{\partial^2 C}{\partial x^2} + \frac{b}{\delta}[\beta(C_0 - C) - kC]$$

Here t is time, x the coordinate directed along the catalytic element axis, T, T_0 the temperature of the catalyst and the reaction medium, C, C_0 the reactant concentrations at the catalyst surface and in the bulk-flow, m, λ the heat capacity per unit volume and thermal conductivity of the catalyst, D the reactant diffusion coefficient for the gaseous phase, $k = k_0 \exp(-E/RT)$ the Arrhenius rate constant for a first-order reaction, q the thermal effect of the reaction, δ the characteristic size of the "effective" film,[14] a and b numerical coefficients of the order of unity related to the element geometry (for simplicity, let $a = b = 1$).

In a more accurate definition, system (1), when applied to a chemical reactor, must include equations of heat and mass balance on the gas flow. Under the assumption of a thin catalyst coating, the temperature T_0 and the concentration C_0 in the gas flow core are parameters. For such a model, it may be assumed to a first approximation that $d = \delta$ is the thickness of the coating, $\beta = V, d = Vm_g$ where V is the velocity of the reactant stream, m_g the gas heat capacity per unit volume, m and λ the effective thermal capacity and thermal conductivity of the catalyst bed, and L the reactor cross size. System (1) is considered simultaneously with the boundary conditions

$$\left.\frac{\partial T}{\partial x}\right|_{x=0} = \left.\frac{\partial T}{\partial x}\right|_{x=L} = \left.\frac{\partial C}{\partial x}\right|_{x=0} = \left.\frac{\partial C}{\partial x}\right|_{x=L} = 0 \tag{2}$$

Such conditions are inherent in the model of a real reactor (impermeability to gas and quasi-adiabaticity of heat-insulated reactor walls). In the case of a thread-shaped element, a special technique should be employed in practice for the conditions of Eq. (2) to be realized. However, a detailed study has shown that at large enough values of L/d the influence of boundary condition types on the characteristic properties of DS becomes insignificant (as a rule, in experimental runs $L/d > 10^3 \div 10^4$).

System (1, 2) reduces to a dimensionless form

$$\begin{cases} \dfrac{\partial \theta}{\partial \tau} = \varepsilon \dfrac{\partial^2 \theta}{\partial y^2} + \Phi(\theta, \eta) = \varepsilon \dfrac{\partial^2 \theta}{\partial y^2} + p \exp\left(\dfrac{\theta}{v + u\theta}\right)(1 - \eta) - \theta \\ \gamma \dfrac{\partial \eta}{\partial \tau} = \dfrac{\partial^2 \eta}{\partial y^2} + G(\theta, \eta) = \dfrac{\partial^2 \eta}{\partial y^2} + p \exp\left(\dfrac{\theta}{v + u\theta}\right)\cdot(1 - \eta) - \eta \end{cases} \tag{3}$$

$$\left.\frac{\partial \theta}{\partial y}\right|_{y=0} = \left.\frac{\partial \theta}{\partial y}\right|_{y=l} = \left.\frac{\partial \eta}{\partial y}\right|_{y=0} = \left.\frac{\partial \eta}{\partial y}\right|_{y=l} = 0 \tag{4}$$

Here $\tau = t/\tau_T$, $\tau_T = md/\alpha$ is the characteristic time of thermal relaxation in the absence of spatial temperature distribution, $y = x/L_c$, $L_c = (D \cdot \delta/\beta)^{1/2}$ is the characteristic diffusion length (stationary diffusion transfer front size), $\theta = (T - T_0)/(T_m - T_0)$, $T_m - T_0 = q\beta C_0/\alpha$ is the maximum possible catalyst temperature in the diffusion regime with no spatial temperature distribution (for a reactor, the maximum adiabatic temperature), $\eta = (C_0 - C)/C_0$ is the degree of conversion, $\varepsilon = L_T^2/L_c^2$, $L_T \sim (\lambda d/\alpha)^{1/2}$ is the characteristic thermal length (conductive heat transfer front size), $\gamma = \tau_c/\tau_T$, $\tau_c = \delta/\beta$ is the characteristic time for diffusion reconstruction of the process (for a reactor, residence time), $p = \tau_c/\tau_x = (k_0 \exp(-E/RT_0) \cdot C_0)/\beta C_0$ is the ratio of the characteristic reaction rates in the kinetic and diffusion regimes (for a reactor, p, an analogue of Damköhler's criterion, is the ratio of the characteristic residence time τ_c and the chemical reaction time $\tau_x = \delta \cdot [k_0 \exp(-E/RT_0)]^{-1}$, $v = RT_0^2/(E(T_m - T_0))$ is the ratio of the characteristic catalyst temperatures in the kinetic and diffusion regimes, $u = RT_0/E$ is Frank-Kamenetskii's criterion (that of the activatedness of a chemical reaction), and $l = L/L_c$ is the reactor cross size. With the selected scales, the dimensionless values of the temperature θ_s and the degree of conversion η_s for the uniform stationary solutions of Eqs. (3, 4) lie in the interval $0 \leqslant \theta_s, \eta_s \leqslant 1$, and $\eta_s = \theta_s$.

The study of the model was based on numerical experiments designed with allowance for the mathematical results obtained in references 15 and 16. The general trend of the present investigation has arisen from specific problems of the theory and practice of catalytic reactors. Among them are: (1) analysis of the effect of thermal and kinetic parameters on the conditions for loss of stability of the uniform stationary modes of chemical conversion on a single catalyst element and in the bed; (2) investigation of the effect of the system parameters on the characteristic linear dimensions of structures being formed and of their fragments, which is important in estimating the danger of loss of uniformity in the course of a reaction proceeding in a reactor with specific geometry; (3) analysis of the amplitude characteristics of DSs from the standpoint of both the break of a technologically advantageous, uniform regime and the abnormal spontaneous formation of hot spots; (4) analysis of the system sensitivity to local disturbances of large amplitude; and (5) elucidation of the peculiarities of some transient processes concomitant with the formation of DSs.

III. STABILITY OF UNIFORM STATES

Before considering the effect of thermal conductivity and diffusion on the stability of uniform solutions, we shall discuss some properties of the nullcurves $\Phi(\theta, \eta) = 0$, $G(\theta, \eta) = 0$ of system (3, 4). The identity of the nonlinear item in the functions $\Phi(\theta, \eta)$ and $G(\theta, \eta)$ enables us to use the pseudo-nullcurve $\Phi(\theta, \eta) - G(\theta, \eta) = \eta - \theta = 0$ for finding stationary values. Figure 1a–b presents a graphic illustration of possible uniform states in the system under study, which can be realized by varying the parameters p, u, and v. For example, the variation of p via the change in the mass transfer intensity, β, may give rise to any kind of stationary state, from the kinetic (curve 1, $\theta_s \to 0, \eta_s \to 0$) to the diffusion mode (curves 3, 4, $\theta_s \to 1, \eta_s \to 1$). Curve 2 corresponds to the passage of the system through the region of nonuniqueness. The nullcurve $\Phi(\theta, \eta) = 0$ (Fig. 1a, I) has two extrema located by the following expressions:

$$\theta_{\text{ext}}^1 = v\left(\frac{1 - 2u - \sqrt{1 - 4u}}{2u^2}\right); \quad \theta_{\text{ext}}^2 = v\left(\frac{1 - 2u + \sqrt{1 - 4u}}{2u^2}\right) \tag{5}$$

Depending on the parameter values, the steady states may belong to either the descending (AB, CD) or ascending (BC) branches of $\Phi(\theta, \eta) = 0$. As will

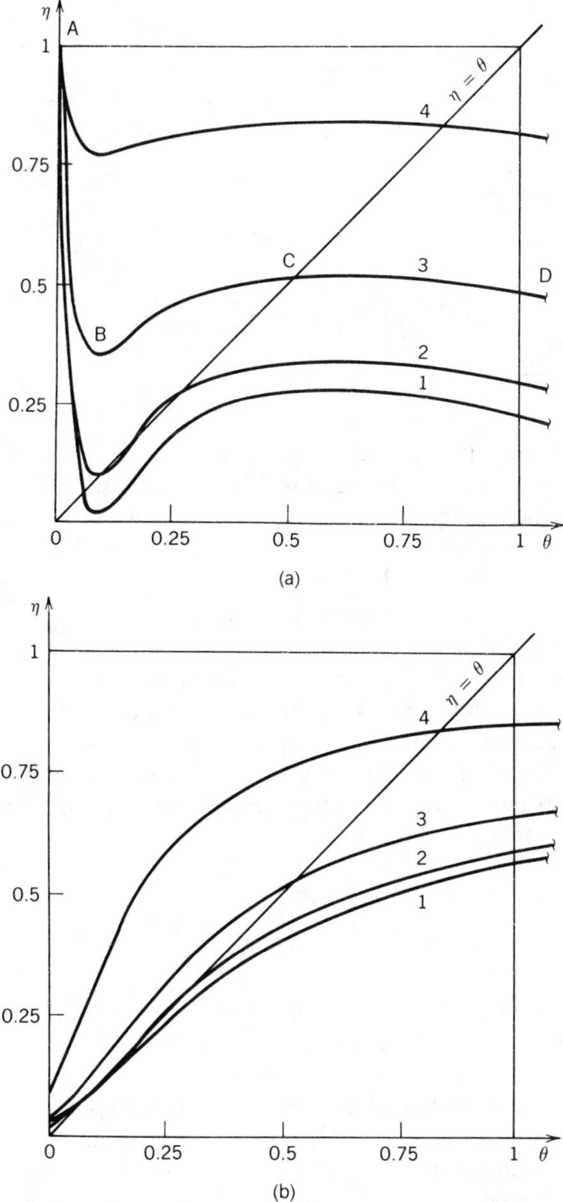

Figure 1. Nullcurves $\Phi(\theta, \eta) = 0$ (a) and $G(\theta, \eta) = 0$ (b) and pseudonullcurve $\eta = \theta$ in the phase plane. I (a) + (b) weakly activated reaction, $u = 0.2$, $v = 0.046$; curve 1, $p = 0.022$; 2, $p = 0.024$; 3, $p = 0.033$; 4, $p = 0.1$; II (a) and (b) strongly activated reaction, $u = 0.03$, $v = 0.2$; 1, $p = 0.083$; 2, $p = 0.1$; 3, $p = 0.125$.

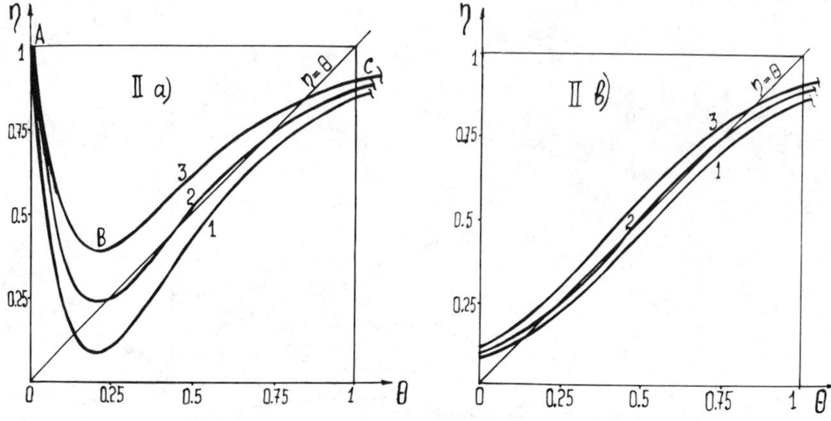

be shown later, the properties of the uniform stationary solutions corresponding to the descending branches and of those corresponding to the ascending branch differ principally in their stability to local perturbations. When the parameters u and v are varied (e.g., u is decreased, which corresponds to the increased activation energy E), the maximum of $\Phi(\theta, \eta) = 0$ may fall in the range $\theta > 1$ (provided that $v > 1 - 2u - \sqrt{1 - 4u}$), and the corresponding diffusion regime steady-state points will lie only on the ascending branch BC (Fig. 1a, II). In the limiting case of a strongly activated reaction (at $u = 0$), the maximum of the nullcurve goes to infinity and $\theta^1_{\mathrm{ext}} = v$.

The stability of the uniform stationary states in system (3, 4) to small spatially uniform perturbations (SUPs) was studied in detail by the theory of perfectly stirred reactors (see reference 9). It will be recalled that the system (3, 4) is stable to SUPs if

$$\sigma_1 = \left.\frac{\partial \Phi}{\partial \theta}\right|_{\theta_s, \eta_s} + \frac{1}{\gamma} \cdot \left.\frac{\partial G}{\partial \eta}\right|_{\theta_s, \eta_s} < 0 \tag{6}$$

$$\Delta = \left.\frac{\partial \Phi}{\partial \theta}\right|_{\theta_s, \eta_s} \cdot \left.\frac{\partial G}{\partial \eta}\right|_{\theta_s, \eta_s} - \left.\frac{\partial \Phi}{\partial \eta}\right|_{\theta_s, \eta_s} \cdot \left.\frac{\partial G}{\partial \theta}\right|_{\theta_s, \eta_s} > 0 \tag{7}$$

The violation of condition (6) results in an oscillatory (limit cycle) behavior, and the violation of condition (7) results in the realization of the critical ignition–extinction conditions (generation of an instability of the saddle type). The critical temperatures for ignition and extinction are found from the condition $\Delta = 0$:

$$\theta_{1,2} = 2v[1 - 2u \pm \sqrt{1 - 4(u + v)}]^{-1} \tag{8}$$

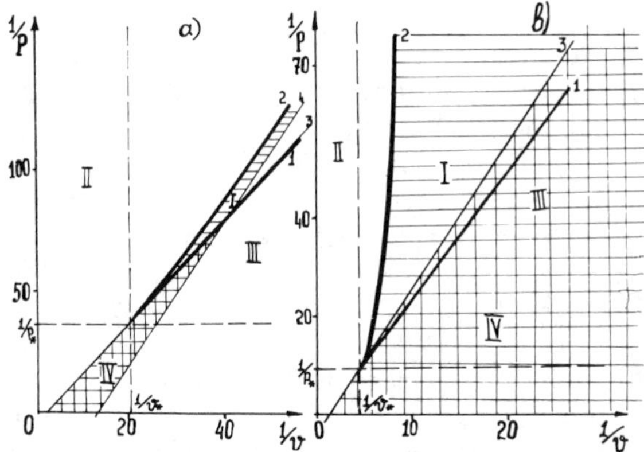

Figure 2. Location of regions with different stability characteristics in the $(1/p \div 1/v)$-plane; a, $u = 0.2$, b, $u = 0.03$; $1/p_* = e^2/(1-u^2)$, $1/v_* = 4/(1-4u)$, Regions: I, of multiple USS (bounded by curve 1 of uniform ignition and curve 2 of extinction); II, III, of unique USS (II, kinetic and III, diffusion regimes); IV, of instability to SNP (hatched area, curves 3, 4 meet the condition $\partial\Phi/\partial\theta = 0$; in b, line 4 corresponding to very large values of $1/v$ is not shown; vertical hatching corresponds to instability to SNP in the kinetic regime and horizontal hatching in the diffusion regime).

where the + sign corresponds to the ignition temperature θ_1, and the minus sign to the extinction temperature $\theta_2, \theta_1 < \theta_2$. On the steady-state temperature axis, θ_s, these temperatures separate the region of "saddles" $(\theta_1 < \theta_s < \theta_2)$ from those of "nodes" or focuses $(\theta_s < \theta_1, \theta_s > \theta_2)$. In the space of the parameters $1/p \div 1/v$, the condition $\Delta = 0$ defines the boundaries of the region of nonuniqueness of uniform stationary states (USSs), whose locations are shown in Fig. 2 for the case of a weakly activated (a) and strongly activated (b) reaction. These boundaries are given by

$$p_{1,2} = \frac{2v \exp\{2[1 \pm \sqrt{1-4(u+v)}]^{-1}\}}{1 - 2(u+v) \pm \sqrt{1-4(u+v)}}$$

(the plus sign corresponds to the critical conditions of ignition and the minus sign to the critical conditions of extinction).

A transition from the conditions approximating a perfectly mixed (lumped) system to the realistic conditions with both the diffusion and conductive factors being involved in the process of attaining a steady state (in a distributed system) may bring about a destabilization of the previously stable uniform state. A qualitative picture of the destruction of the spatially uniform mode of a catalytic process appears as follows. A small local perturbation resulting in a local overheating is accompanied by an increase in the reaction velocity and, hence, by a decrease in the reactant velocity and, hence, by a decrease in the reactant concentration at the surface of this locus. The concentration gradient thus induced causes a reactant flow from the undisturbed part of the system to the disturbed zone, which, in turn, results in further increase of the local reaction velocity and local temperature. If the conductive smoothing of temperature is less intense compared to the diffusive reactant influx, the described feedback mechanism may come into action, and the spontaneous growth of the perturbation amplitude will begin. The process terminates in the formation of a stable, spatially-inhomogeneous structure. This situation provides a physical explanation for the possible occurrence of DSs in systems with low thermal conductivity. The necessary condition for nonuniform solutions of system (3, 4) is the inequality $\varepsilon < \varepsilon_{cr}$. The value of ε_{cr}, which is a function of thermal and kinetic parameters, defines the boundary of the region in which the homogeneous regime becomes unstable to small nonuniform perturbations and disintegrates into DSs in objects of sufficient length.

IV. THE NECESSARY CONDITION FOR THE FORMATION OF DS (ANALYSIS OF THE CRITICAL VALUES OF THE PARAMETER ε)

By the classical bifurcation theory, the necessary condition for the birth of uniform stationary solutions is the passage through zero of the extreme right-hand side eigenvalue of the equation set linearized on the uniform solution θ_s, η_s (see references 15 and 16). For the system considered here (3, 4), the condition for the birth of a nonuniform stationary regime from a uniform one in the general form is written as

$$\varkappa_n^2 = \frac{\sigma_2 \pm \sqrt{\sigma_2^2 - 4\varepsilon\Delta}}{2\varepsilon} \tag{9}$$

Here

$$\sigma_2 = \left.\frac{\partial\Phi}{\partial\theta}\right|_{\theta_s,\eta_s} + \varepsilon\left.\frac{\partial G}{\partial\eta}\right|_{\theta_s,\eta_s}; \quad \varkappa_n = \pi\frac{n}{l}$$

$n = 1, 2, 3, \ldots$. The parameter n characterizes the number of half-periods of a periodic structure born on a catalytic element (or in a fixed-bed reactor) of a given length l. Since $\varkappa$ is, in its physical meaning, always positive and real, and $\Delta > 0$ (by (6)), the necessary conditions for the birth of DS in the catalytic processes considered are given by the following obvious relations:

$$\begin{cases} \sigma_2 > 0 & (10) \\ \sigma_2 - 4\varepsilon\Delta \geqslant 0 & (11) \end{cases}$$

Considering these relations simultaneously with Eq. (6) leads to a number of consequences of practical importance. From Eq. (10) it follows that nonuniform stationary states can arise only from the destruction of those uniform states which correspond to the ascending branch of $\Phi(\theta, \eta) = 0$, since the derivative $\partial G/\partial \eta < 0$ within the entire interval $0 < \eta_s < 1$, and $\partial \Phi/\partial \theta > 0$ only in the region of θ_s values lying just on this branch.

The critical value of the parameter ε can be roughly estimated from (10)

$$\varepsilon_{cr} = R(1 - \theta_s)(\theta_s - 1/R) \tag{12}$$

Here $R = v(v + u\theta_s)^{-2}$ is an integral parameter of nonlinearity. In what follows, we shall consider R independent of θ_s for greater obviousness and ease of presentation. This approximation becomes accurate in the limiting case $u = 0$ (then $R = 1/v$), but the basic qualitative conclusions will be also true at $u \neq 0$.

The function $\varepsilon_{cr}(\theta_s)$ has a maximum near $\theta_s = 1/2$, i.e., the central part of the interval $0 < \theta_s < 1$ is the most dangerous with respect to the loss of stability and the formation of DSs. The rigorous expression for ε_{cr} is found from the solution of the inequality set (Eqs. (6), (10), (11)):

$$\varepsilon_{cr} = R\theta_s^2(1 - \theta_s)(1 - \sqrt{1 - B})^2 \tag{13}$$

where $B = (R\theta_s - 1)/R\theta_s^2$. It is meaningful at $1 - B > 0$. When $B = 1$ (which corresponds to the condition $\Delta = 0$, or the occurrence of an instability of the saddle type), Eq. (13) takes the form

$$\varepsilon_{cr} = \theta_s \tag{14}$$

Hence it follows that $\varepsilon_{cr} < 1$ always.

Figure 3 shows a family of the curves $\varepsilon_{cr}(\theta_s)$ for different values of R. As seen, the greatest danger of the appearance of DS (or, in terms of the physics of semiconductors, "domain" instability[17] exists, in agreement with the above discussion, in the central part of the interval $0 < \theta_s < 1$. With the growth of R, the region of the domain instability rapidly broadens. However, at $R > 4$ in the vicinity of $\theta_s \simeq 0.5$ there exists a region of absolutely unstable stationary states of the saddle type, whose boundaries are defined by the points of intersection of the curves $\varepsilon_{cr}(\theta_s)$ (Eq. (13)) and the straight line $\varepsilon_{cr} = \theta_s$ (Eq. (14)).

It is seen from the Fig. 3 that destruction of a homogeneous state is more likely to occur in the diffusion than in the kinetic regime: the absolute values of ε_{cr} are greater on the right-hand than on the left-hand side of the interval $0 < \theta_s < 1$. A domain instability may also occur in the deep diffusion regime

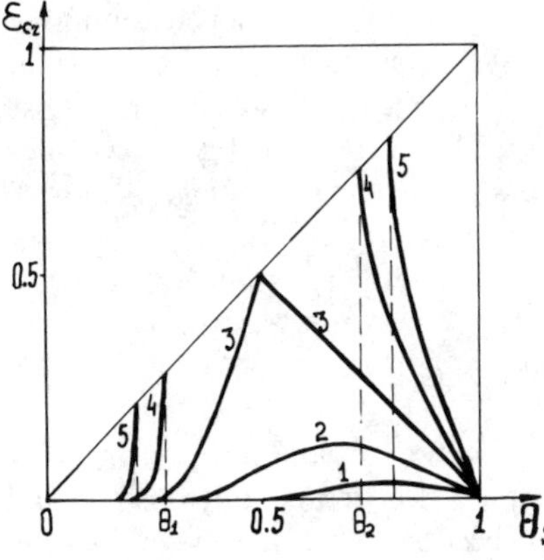

Figure 3. Critical value of the parameter ε_{cr} as a function of temperature θ_s of uniform stationary state (for curves 1–5, $R = 2, 3, 4, 5$, and 6, respectively).

(the maximum value of the stationary temperature θ_s at the boundary of the region of the existence of DS is equal to 1), whereas the minimum θ_s value, below which at no arbitrarily small value of ε_{cr} can a DS arise, is given by the ratio $\theta_s = 1/R$. This conclusion, however, is true for a strongly activated reaction, to be precise, for those parameter values with which the nullcurve $\Phi(\theta, \eta) = 0$ has a maximum at $\theta > 1$ (Eq. (13) turns to zero either at θ_s values corresponding to the extrema of $\Phi(\theta, \eta) = 0$ or at $\theta_s = 1$). If $\theta^2_{\text{ext}} < 1$, the diffusion regime has a temperature range $\theta^2_{\text{ext}} < \theta_s < 1$ within which no DS can be evoked by small nonuniform perturbations (SNPs).[a] The parameteric expression for the boundaries of the region of instability to SNP defined by the extrema of $\Phi(\theta, \eta) = 0$ has the form

$$p_{1,2} = \frac{2v \exp\{-2[1 \pm \sqrt{1-4u}]^{-1}\}}{(1 - 2u \pm \sqrt{1-4u}) - 2v}$$

where the + sign corresponds to the maximum of the nullcurve and the − sign to its minimum (see Fig. 2).

Using the results obtained, we shall analyze the effect of the dimensional parameters on the conditions for the generation of DS. Figure 4 presents two characteristic curves $\varepsilon_{cr} = \varepsilon_{cr}(p)$ plotted for the case of uniqueness (curve 1) and multiplicity (curve 2) of uniform solutions. As can be seen, in the region of uniqueness the function is continuous and in that of multiplicity it is discontinuous and not single-valued. In the diffusion region, the increase in p leads to a decrease in ε_{cr} (curve 2a) and in the kinetic region to its growth

[a]It will be recalled that Fig. 3 refers to the case $R = \text{const}$ which corresponds to a strongly activated reaction, as indicated above.

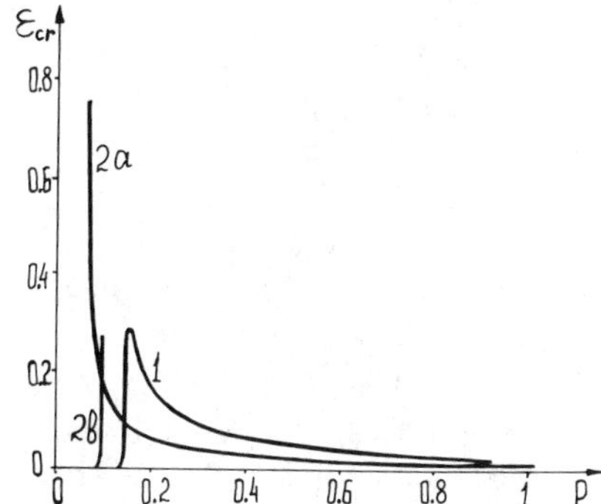

Figure 4. ε_{cr} as a function of the parameter p: 1, region of unique USS; 2, multiple USS; 2a, diffusion regime; 2b, kinetic regime.

(curve 2b). Analogous relationships hold also in the region of uniqueness. Since the parameter p is inversely proportional to the coefficient of the mass output β, it can be concluded that the increase in the intensity of heat and mass transfer facilitates the formation of DS in the diffusion regime and hinders it in the kinetic. Figure 5 shows the curve $\varepsilon_{cr} = \varepsilon_{cr}(1/v)$ analogous to that in Fig. 4. The reactant concentration in the stream C_0 as well as the thermal effect of the reaction q vary directly with $1/v$. Hence it follows that the increase in C_0 or q favors the formation of DS in the kinetic region and hinders it in the diffusion region. By similar lines of reasoning for $\varepsilon_{cr} = \varepsilon_{cr}(u)$, the increase in T_0 destabilizes the uniform mode in the kinetic region and, on the contrary, enhances the stability to SNP in the diffusion region. The activation energy has an opposite effect: the increase in E stabilizes the kinetic regime and reduces the stability of the diffusion regime.

The above data evidence that the necessary conditions for the origination of DS can be fulfilled in a broad range of parameter values, which makes their formation most probable. To illustrate this, we derive from Eq. (13) some estimates for the values of the dimensional parameters with which the appearance of DS should be anticipated. We consider the case $\theta_s = 0.9$; $R = 2$;

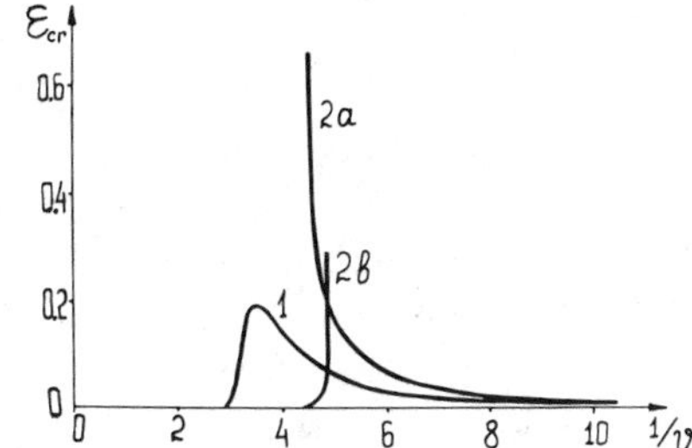

Figure 5. ε_{cr} as a function of the parameter $1/v$ (other notation as in Fig. 4).

$\varepsilon_{cr} = 1.35 \cdot 10^{-2}$; $u = 0.1$. With these values and at a temperature of $T_m - T_0 \simeq 300°C$ characterizing the maximum catalyst heating, the nonlinearity corresponding to as low an activation energy as $\simeq 2$ kcal/mole is dangerous. For a value of $E \simeq 20$ kcal/mole which is quite realistic, and for the same value of $T_m - T_0$, $R = 9$, ε_{cr} approaches unity and, as seen from Fig. 3, it reaches the value $\varepsilon_{cr} \simeq 0.9$ at $\theta_s = 0.9$. In other words, even an insignificant excess of the characteristic diffusion mass transfer length over the characteristic thermal length provides the necessary conditions for failure of the uniform mode of a catalytic process. From simple physical considerations, it is evident that in the case of a reactor, the processes of heat and mass transfer in the porous catalyst coating are nonidentical: in view of the medium being biphasic, the thermal resistance of the catalyst bed exceeds the diffusional.

V. SUFFICIENT CONDITIONS FOR THE FORMATION OF DS (ANALYSIS OF THE CRITICAL VALUES OF THE PARAMETER l)

The condition $\varepsilon < \varepsilon_{cr}$ is insufficient for a nonuniform mode to occur in system (3, 4) in response to SNP. A DS emerges if the length of a catalytic element (reactor cross-size) exceeds a certain critical value l_{cr}. At $l > l_{cr}$, the uniform solution loses stability and a nonuniform stationary profile of temperature and concentration results, which has the form of a fragment of a cosinusoid or "domain wall". The expression for the critical length is derived from Eqs. (9) and (13):

$$l_{cr\,n}^{1,2} = \pi n \left\{ \frac{(1-\theta_s)[M(\theta_s) - \tilde{\varepsilon}]}{2T(\theta_s)} \left[1 \pm \sqrt{1 - \frac{4\tilde{\varepsilon}T(\theta_s)}{[M(\theta_s) - \tilde{\varepsilon}]^2}} \right] \right\}^{1/2} \tag{15}$$

where

$$M(\theta_s) = \frac{B}{(1 - \sqrt{1-B})^2}; \quad T(\theta_s) = (1-B)(1 - \sqrt{1-B})^{-2}$$

and $\tilde{\varepsilon} = \varepsilon/\varepsilon_{cr}$ is a new parameter characterizing the degree of remoteness from the boundary of the existence region of DS. As seen from Eq. (15), corresponding to each value $n = 1, 2, 3, \ldots$ are two critical lengths for any $\tilde{\varepsilon} < 1$. This situation is illustrated by Fig. 6 which shows that when ε differs from ε_{cr}, there appear segments of instability on the straight line representing the catalytic element length. As $\tilde{\varepsilon}$ decreases, these segments grow and at a certain $\tilde{\varepsilon} = \tilde{\tilde{\varepsilon}}$ they join together. As a result, the length axis turns out to be broken into intervals of stable ($0 < l < l_{cr}^1$) and unstable ($l_{cr}^1 < l < \infty$) uniform steady state. The value of $\tilde{\varepsilon}$, at which this occurs, is close to unity for a broad

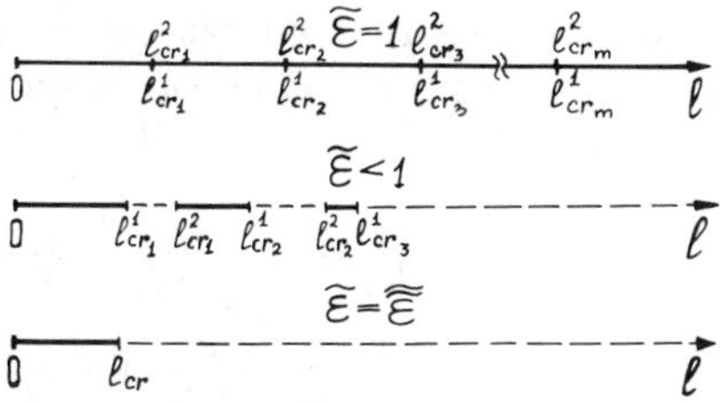

Figure 6. Transformation of the first $l^1_{cr_n}$ and second $l^2_{cr_n}$ sequences of critical lengths with decreasing $\tilde{\varepsilon}$: regions of uniform steady states stable to SNP (solid segments) and unstable to SNP (dashed segments).

temperature range. Below, we confine the consideration just to this case, where it is important to know only the very first critical value $l^1_{cr_1}$ determined by[b]

$$l_{cr} = l^1_{cr_1} = \pi\left\{\frac{1-\theta_s}{2}\frac{M(\theta_s)-\tilde{\varepsilon}}{T(\theta_s)}\left[1-\sqrt{1-\frac{4\tilde{\varepsilon}T(\theta_s)}{(M(\theta_s)-\tilde{\varepsilon})^2}}\right]\right\}^{1/2} \tag{16}$$

The basic qualitative conclusions of the behavior of l_{cr} and the relationships for evaluating this quantity can be derived from consideration of a number of limiting cases. Inside the region of the existence of DS (at $\tilde{\varepsilon} \ll 1$), the relationship (16) transforms to

$$l_{cr} \simeq \sqrt{\tilde{\varepsilon}} \tag{17}$$

From Eq. (17) it follows that the critical size of an object practically coincides in this case with the characteristic thermal length L_T. The simplicity of the physical meaning of this result enables us to use Eq. (17) in estimating the size of an element (reactor) dangerous with respect to loss of stability of a uniform mode within the region of the existence of DS.

Equation (16) is considerably simplified for arbitrary $\tilde{\varepsilon}$ in the temperature range θ_s corresponding to the extrema of $\Phi(\theta, \eta) = 0$ ($B \ll 1, \theta_s \approx \theta^{1,2}_{\text{ext}}$):

$$l_{cr} \simeq \pi[B(1-\theta_s)(1-\sqrt{1-\tilde{\varepsilon}})]^{1/2} \tag{18}$$

It is seen that the dimensional critical length in this case may have the same order of magnitudes as the thermal length L_T (or be smaller) even near the boundary of the existence region of DS ($l_{cr} \simeq \sqrt{B}$ at $\tilde{\varepsilon} \simeq 1$).

For temperatures θ_s corresponding to the critical conditions for ignition and extinction ($B \simeq 1, \theta_s \approx \theta_{1,2}$) we have

$$l_{cr} \simeq \pi[(1-\theta_s)(\tilde{\varepsilon}/(1-\tilde{\varepsilon}))]^{1/2} \tag{19}$$

[b] Mathematical aspects of the behavior of system (3, 4) at $\tilde{\varepsilon} > \tilde{\tilde{\varepsilon}}$ were considered in reference 16.

Also in this case, the dimensional critical length is determined at $\tilde{\varepsilon} \ll 1$ by the thermal length L_T. However, it grows significantly as $\tilde{\varepsilon}$ increases and at $\tilde{\varepsilon} \to 1$, $l_{cr} \to \infty$.

For an arbitrary value of θ_s and $\tilde{\varepsilon} = 1$, Eq. (16) becomes

$$l_{cr}(\varepsilon_{cr}) = \left[\frac{\pi^2(1-\theta_s)(1-\sqrt{1-B})}{\sqrt{1-B}} \right]^{1/2} \tag{20}$$

This relationship is, in essence, a sufficient condition for the occurrence of DS. Figure 7 shows its form for different values of R. As a matter of fact, Eq. (20) defines the second boundary separating the region of nonexistence of DS at any ε from the region of existence of DS at $\varepsilon < \varepsilon_{cr}$. It is analogous to Eq. (13) in being continuous at $R < 4$ and discontinuous at $R > 4$. In the region of uniqueness of USS, the maximum critical length is of the order of unity and it goes to infinity only when $R = 4$. This implies that in the given case the dimensional critical length is primarily defined by the diffusion length L_c. In the region of multiplicity of USS, l_{cr} ranges from zero at $\theta_s = \theta_{\text{ext}}^{1,2}, \theta_s = 1$ to infinity at $\theta_s = \theta_{1,2}$, i.e., at temperature values corresponding to the boundaries of the region of a saddle-shaped instability. Hence it follows that, although the necessary conditions for loss of stability of USS are easiest to satisfy in the temperature interval $\theta_{1,2}$, the probability of the occurrence of DS is the smaller, the closer the saddle-shaped instability is approached. With respect to the appearance of DS, the most dangerous are intermediate temperature intervals, in which the dimensionless critical length again takes the value of unity, i.e., the dimensional length coincides in the order of

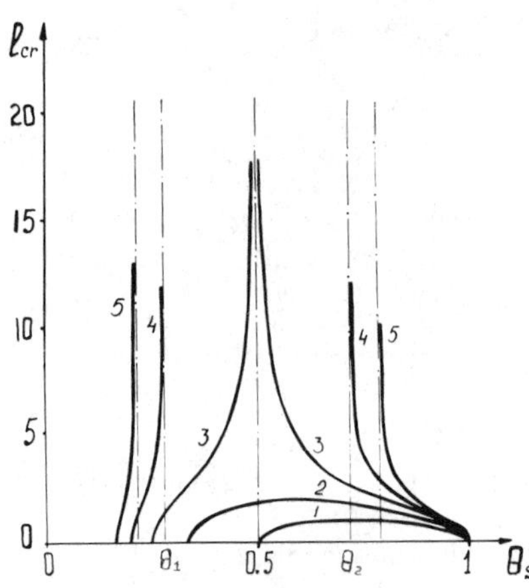

Figure 7. Critical length l_{cr} corresponding to $\varepsilon = \varepsilon_{cr}$ as a function of the temperature θ_s of USS: 1, $R = 2$; 2, $R = 3$; 3, $R = 4$; 4, $R = 5$; 5, $R = 6$.

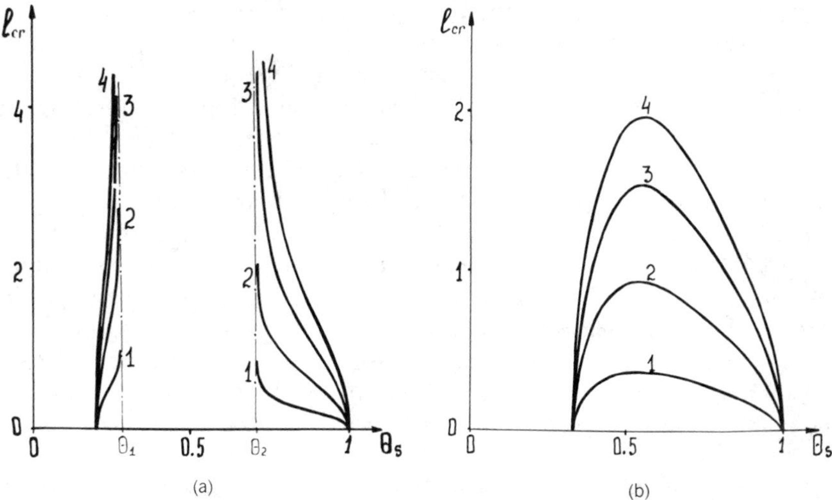

Figure 8. Critical length l_{cr} as a function of θ_s for different values of $\tilde{\varepsilon} = \varepsilon/\varepsilon_{cr}$: (a) $R = 5$; 1, $\tilde{\varepsilon} = 0.1$; 2, $\tilde{\varepsilon} = 0.5$; 3, $\tilde{\varepsilon} = 0.9$; 4, $\tilde{\varepsilon} = 1$; (b) $R = 3$; 1, $\tilde{\varepsilon} = 0.1$; 2, $\tilde{\varepsilon} = 0.5$; 3, $\tilde{\varepsilon} = 0.9$; 4, $\tilde{\varepsilon} = 1$.

magnitude with the diffusion length L_c. Thus, near the boundary of the region of existence of DS the diffusion length should be used in roughly estimating the critical size of an element (reactor).

With arbitrary values of the stationary temperature θ_s and the parameter $\tilde{\varepsilon}$, the estimated values of the critical size range between the diffusion and thermal lengths and the exact values are found from Eq. (16). To illustrate this, Figs. 8 and 9 plot the critical lengths against the parameters θ_s, R, and $\tilde{\varepsilon}$.

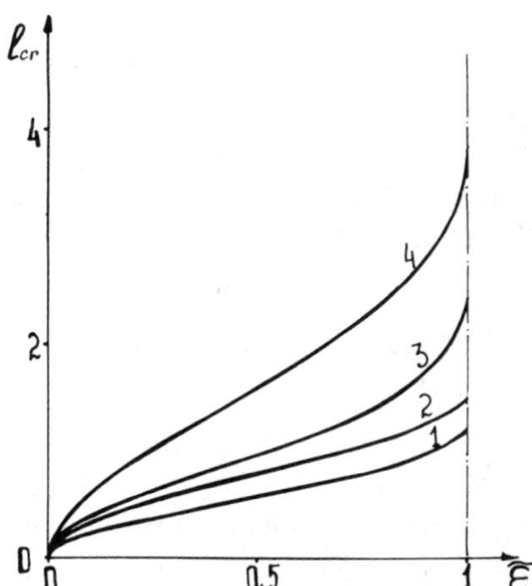

Figure 9. Critical length l_{cr} as a function of the parameter $\tilde{\varepsilon}$ for different values θ_s: 1, $\theta_s = 0.26$; 2, $\theta_s = 0.8$; 3, $\theta_s = 0.22$; 4, $\theta_s = 0.9$.

Of some interest is the limiting case $\tilde{\varepsilon} = 0$. Two types of physical conditions can be thought of as corresponding to this case: first, letting L_T become zero, and second, letting L_c tend to infinity. The first type reflects a hypothetical situation with complete absence of thermal contacts between separate catalyst particles in the bed (or between segments of a catalytic element), the gas phase included. At $L_T = 0$, the uniform mode of the process is unstable with any reactor size (element length), i.e., the critical size for loss of stability equals zero. In this case, some portion of the catalyst particles will act in the low-temperature (kinetic) regime and the other in the high-temperature (diffusion) regime, i.e., "contrasting" structures similar to those described in Eq. (18) will occur. The second type is characteristic of catalytic processes in which care is taken to provide vigorous stirring of the gaseous phase in the direction perpendicular to the reactant stream. In this case, model (3, 4) becomes unsuitable because the system was put into a dimensionless form on the scale $L_c \to \infty$. Transition to the scale L_T results in the new model reducing to a single integral-differential equation. The case is analogous in many ways to the process of "barreting" described in reference 19, particularly, to a reaction proceeding on the current-heated catalyst wire under voltage-controlled thermal conditions.[5] The expression for the critical length assumes a very simple form

$$l_{cr} = \frac{L_{cr}}{L_T} = \left[\frac{\pi^2}{(R\theta_s - 1)}\right]^{1/2} = \left\{\frac{\pi^2}{[v\theta_s/(v + u\theta_s)^2 - 1]}\right\}^{1/2} \tag{21}$$

The simplicity of the expression allows an examination of the dependence of l_{cr} not only on the generalized parameter of nonlinearity R, but also separately on each of the parameters u and v. We note, first of all, that the dimensional critical length for strongly activated reactions is comparable to or even less than the thermal length L_T. The squared critical length as a function of θ_s is a curve having a minimum and tending to infinity on the ends of the interval of existence of DS (at $\theta_s = \theta_{\text{ext}}^{1,2}$ or $\theta_s = 1$). It turns out that the minimum value of the function that occurs at θ_s corresponding to the inflection point of the nullcurve $\Phi(\theta, \eta) = 0$ is independent of the parameter v and is equal to

$$l_{cr\,\min} = 4\pi u \cdot (1 - 4u)^{-1}$$

This makes it possible to plot a universal curve relating the critical length to the normalized temperature θ_n for each value of u ($\theta_n = \theta_s/\theta_{\text{infl}}$), where θ_{infl} is the temperature at the inflection point of $\Phi(\theta, \eta) = 0$) on which curves differing in the values of the parameter v fall. A family of universal curves

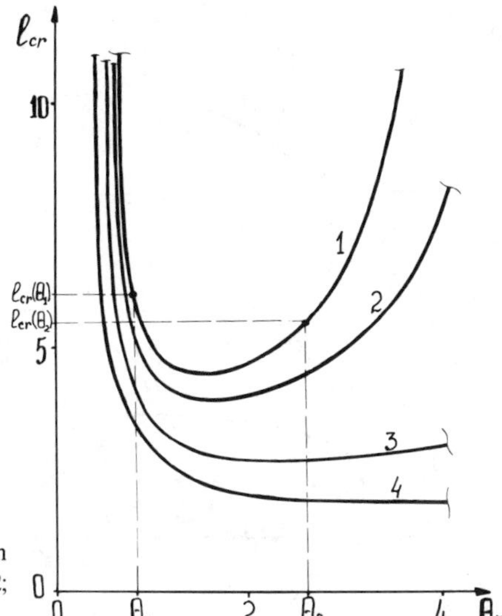

Figure 10. Critical length l_{cr} as a function of θ_n for the limiting case $L_c \to \infty$: 1, $u = 0.2$; 2, $u = 0.19$; 3, $u = 0.15$; 4, $u = 0.1$.

for different values of u is presented in Fig. 10. It is seen that in the region of uniqueness of USS, a DS most easily occurs in the temperature interval corresponding to the minimum of a curve, and in the region of multiplicity it occurs at temperatures corresponding to the critical conditions for ignition and extinction.

VI. ON THE FORMATION OF DS IN RESPONSE TO NONUNIFORM PERTURBATIONS OF FINITE AMPLITUDE

The investigation performed is insufficient for the boundaries of the region of existence of DS to be exactly defined with respect to the parameter l. To define them more exactly, it is necessary that the mode of birth of DS be considered. To this end, we consider different types of the bifurcation diagrams constructed, for example, on the $\Delta\theta = (\theta_{max} - \theta_{min}) \div l$ coordinate, where $\theta_{max}, \theta_{min}$ are a maximum and minimum in the nonuniform temperature profile (see Fig. 11). If, at the bifurcation point l_{cr}, the derivative $d(\Delta\theta)/dl > 0$, the birth is soft (i.e., the uniform mode is replaced by the nonuniform without a jump in amplitude, Fig. 11a). But if $d(\Delta\theta)/dl < 0$, the birth is hard (i.e., at the critical point, the uniform mode is replaced by the nonuniform when the amplitude reaches some finite value, see Fig. 11b). In the second case, as seen from the figure, along with the stable USS there is also a nonuniform state in the length interval $\bar{l}_{cr} < l < l_{cr}$. In other words, the condition $l < l_{cr}$ is no longer sufficient for nonexistence of nonuniform solutions and the formation

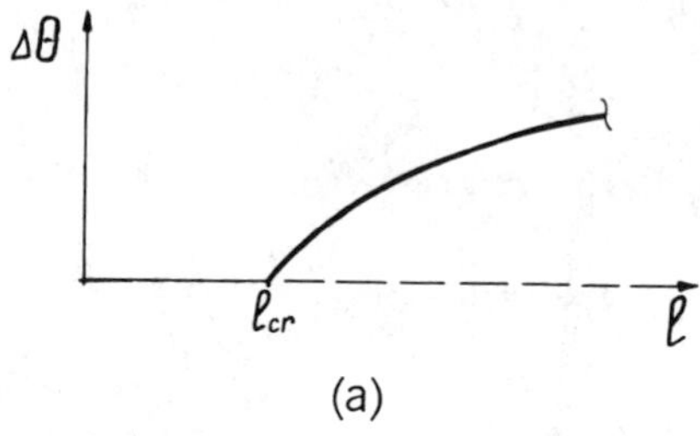

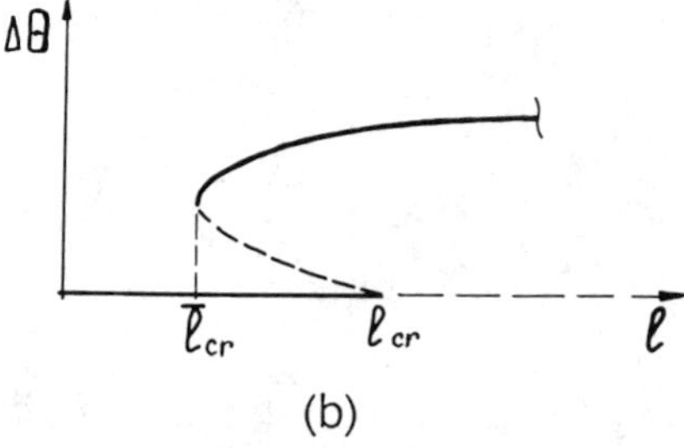

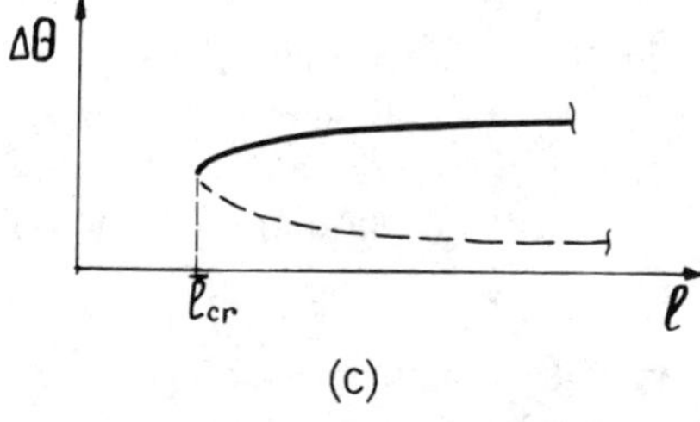

Figure 11. Character of bifurcations when a nonuniform solution is born: (a) hard birth; (b) soft birth; (c) hard birth, $l_{cr} \to \infty$.

of DSs becomes possible within the $\bar{l}_{cr} < l < l_{cr}$ interval, but their realization requires introducing of local perturbations of finite amplitude into the system and the less l the greater the amplitude. At the critical points l_{cr} and $\bar{l}_{cr}$, the system exhibits jump-like transitions between the branches of uniform and nonuniform stationary states. The bifurcation analysis, along with determination of necessary and sufficient conditions for loss of stability, allows also ascertaining the mode of DS birth. In the general case, the branching equation is clumsy and is presented in its complete form in reference 16. In the limiting case $L_c \to \infty$, $u = 0$, the branching equation is greatly simplified and the mode of birth depends on the sign of the expression

$$\Psi(\theta_s) = l_{cr}^2(\theta_s) - \varphi(\theta_s) \tag{22}$$

where

$$\varphi(\theta_s) = 2\pi^2\left[1 - \frac{3(1-\theta_s)}{R\theta_s^2 - R\theta_s + 1}\right]$$

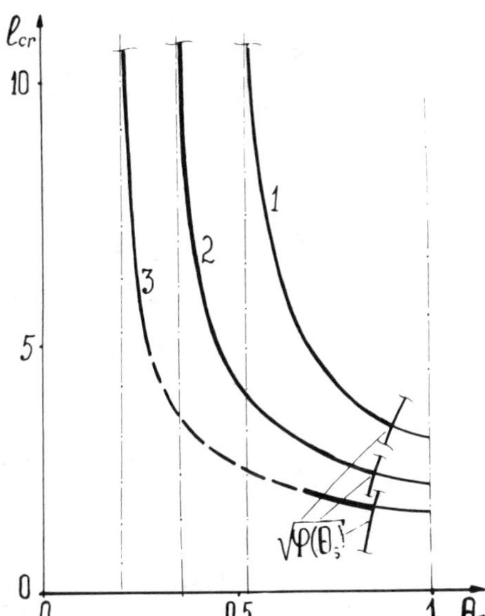

Figure 12. Determination of the regions of soft and hard birth by solving graphically Eq. (22) $l_{cr}(\theta_s) = \sqrt{\varphi(\theta_s)}$: $R = 2(1)$, $3(2)$, $5(3)$ (dashed lines, region of saddle-shaped instability).

and $l_{cr}^2(\theta_s)$ is found from Eq. (21). If $\Psi(\theta_s) > 0$ (the critical length l_{cr} as a function of the stationary temperature is greater than $\sqrt{\varphi(\theta_s)}$), the birth is hard; if $\Psi(\theta_s) < 0$ ($l_{cr}(\theta_s) < \sqrt{\varphi(\theta_s)}$), it is soft. Figure 12 presents a graphical solution to the equation $\Psi(\theta_s) = 0$. The thick, solid line corresponds to the hard birth, and the thin line to the soft. It is seen from the figure that as θ_s increases, the hard mode is replaced by the soft for any R, the hard birth being characteristic of the kinetic regime and the soft birth of the diffusion regime only, no matter whether the system is in the region of multiplicity ($R > 4$) or uniqueness ($R < 4$) of USS. The diffusion mode near the extinction temperatures of USS is characterized by the hard birth of DS as well. It also follows from the figure that l_{cr} corresponding to the transition from the hard to the soft mode is of the order of unity. In a more general case of $u \neq 0$, the above indicated properties are retained, but if the temperature θ_{ext}^2 corresponding to the maximum of $\Phi(\theta, \eta) = 0$ is less than unity, there appears a region of hard birth in the vicinity of θ_{ext}^2. The qualitative character of the considered relationship is illustrated by Fig. 13. The described behavior is observed within a wide range of kinetic and thermal-diffusion parameters. As an example, in Fig. 14 are drawn regions of soft and hard birth in the space of the parameter $1/p$ and $1/v$ for a weakly and strongly activated reaction.

It follows from the above that system (3, 4) has rather a broad region of hard birth of DS. This raises an important question of the value of the second

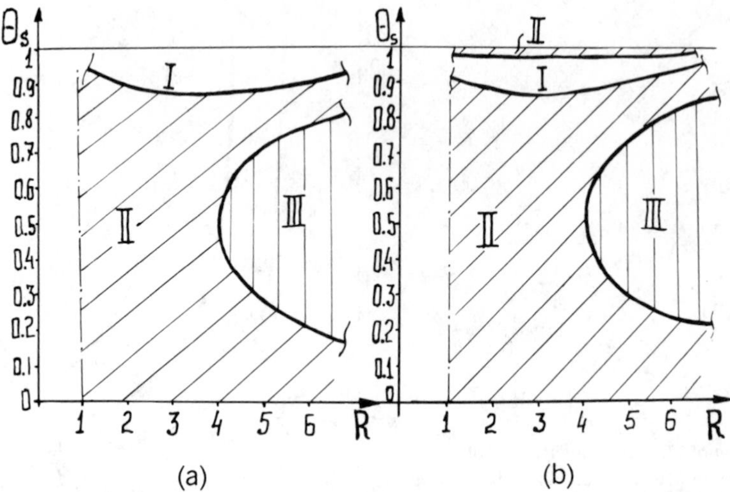

Figure 13. Locaion of the regions of soft (I) and hard (II) birth; III—region of saddle-shaped instability: (a) $L_c \to \infty$, $u = 0$; (b) $L_c \to \infty$, $u \neq 0$.

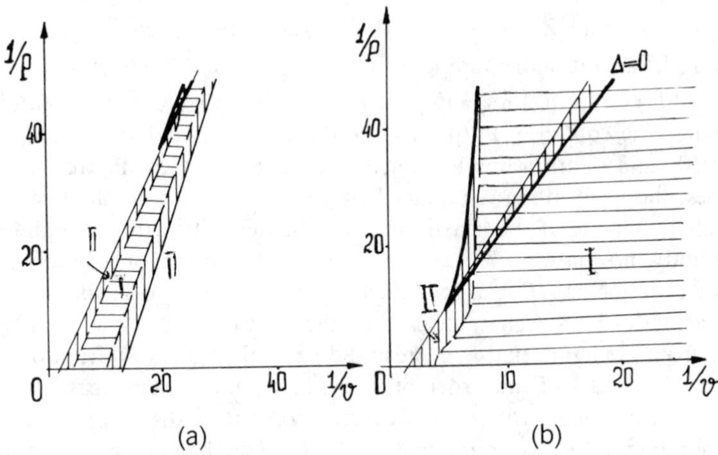

Figure 14. Location of the regions of soft (I, horizontal shading) and hard (II, vertical shading) birth of DS: (a) $u = 0.2$; (b) $u = 0.03$.

critical length $\bar{l}_{cr}$ with which the formation of DS becomes possible when a perturbation of finite amplitude (FAP) is introduced (in a strict sense, the question of the birth "from nothing" of a pair of nonuniform solutions, a stable and unstable one[15]. Analytical methods to determine $\bar{l}_{cr}$ are absent at present. Numerical experiments have shown that the value of the second critical length $\bar{l}_{cr}$ is primarily determined by the thermal length $(2 \div 5 l_T)$ and is weakly dependent on the degree of reaction activation, the stationary temperature θ_s and on the critical value of the parameter ε_{cr}.

Here again, the thermal length acts as an estimate for the critical reactor size with respect to the danger of the occurrence of nonuniform regimes, this time in response to FAP. Such an estimation has an advantage of being valid in a broad interval of stationary temperatures, including values near the extrema of $\Phi(\theta, \eta) = 0$, i.e., in the deep kinetic and diffusion modes.

As already noted, the first critical length l_{cr} is essentially dependent on θ_s and ε_{cr}, especially in the vicinity of the extrema of $\Phi(\theta, \eta) = 0$, but at the very extremal points both the necessary ($\varepsilon_{cr} = 0$) and sufficient conditions for the origination of DS in response to FAP[c] are no longer fulfilled. At the same time, the possibility for DS to occur is weakly dependent on θ_s and ε_{cr}. When considering the stability problem with respect to FAP, it would be appropriate, in this connection, to raise the question of the existence of DS in the parametric region that corresponds to USS stable to SNP, i.e., falling on the descending branches of the nullcurve $\Phi(\theta, \eta) = 0$ (where $d\eta/d\theta < 0$). In other words, is it possible to extend the solution from the region of instability (ascending branch of $\Phi(\theta, \eta) = 0$) to that of stability of SNP (descending branches of $\Phi(\theta, \eta) = 0$) with respect to the parameter θ_s (for $l > l_{cr}$)? Strict mathematical formalism to analyze this sort of problem has not yet been developed.[d]

The numerical experiment performed by us has made it possible to show the existence of DSs in the indicated regions of USS stability to SNP and to examine a number of their properties. First, it is noteworthy that the necessary conditions for the existence of DS are fulfilled within a broader range as compared with those calculated from Eq. (13) (see Fig. 15). The DSs realizable in the stability region can be divided qualitatively into two groups. Belonging to the first group are structures completely analogous to those arising in the regions of instability on the ascending branch of the nullcurve $\Phi(\theta, \eta) = 0$ in the vicinity of its extrema. The basic feature such structures is the possibility for them to grow in number by doubling their elementary fragments with increasing size of a catalytic element. (In this case, the original DS loses stability, and a reconstruction of the temperature and concentration

[c]It should be noted here that the dimensional critical length $L^1_{cr\,1}$ determined by expression (9) behaves differently, depending on whether ε_{cr} is decreased (i.e., the fulfillment of the necessary condition for the birth of DS is hindered) via decrease in L_T or via an increase in L_c. In the first case $L^1_{cr\,1}$ tends to zero and in the second it tends to infinity.

[d]The problem of constructing nonuniform solutions for the regions of stability to SNP was qualitatively considered in[20] in relation to models describing the processes in semiconductor, gas plasma, and some others. The publications[21] reporting numerical studies of models of the Belousov–Zhabotinskii reaction should also be mentioned here.

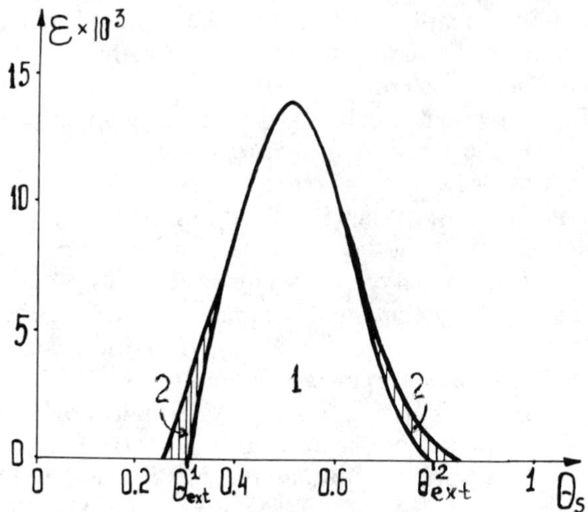

Figure 15. Enlargement of the region of necessary conditions for the existence of DS when perturbations of finite amplitude are taken into account: 1—instability to SNP, 2—instability to FAP; $u = 0.2$, $p = 0.11$.

profiles begins. As a result, the system produces a stable structure with a greater number of elements.) The second group includes structures that remain stable (incapable of increasing in number) as the catalytic element size is increased. The two groups of structures are exemplified in Figs. 16 and 17. In both cases, the structures being formed in the kinetic region are nonuniform temperature distributions, in which the smaller fraction of the sample is in the high-temperature condition close to the diffusion mode and the greater one is in the low-temperature kinetic regime near the uniform distribution that has lost stability. The group I structures arise in the temperature interval of θ_s adjaicent to θ^1_{ext}. The region of the existence of group II structures is in the lower temperature interval (see Fig. 18). Note that the latter are characteristic only of weakly activated reactions. As the parameter u decreases, the region of the existence of group II structures contracts, and at $u = 0$ only group I structures remain.

In the diffusion regime, structures arising in the region of stability to SNP are nonuniform temperature distributions in which the smaller fragment of the profile is in the low-temperture state and the greater part is in the high-temperatue state. In the entire temperature range $\theta_s > \theta^2_{\text{ext}}$, only group II

Figure 16. Evolution of a DS, formed in the region of stability to SNP, with the change in the element length (disintegrating structures): (a) $u = 0.03$, $\varepsilon = 2.5 \cdot 10^{-3}$, $v = 0.23$, $p = 0.11$, $\theta_s = 0.21$ (initial condition is given by step 1); (b) $u = 0.03$, $\varepsilon = 2 \cdot 10^{-4}$, $v = 0.23$, $p = 0.11$, $\theta_s = 0.21$.

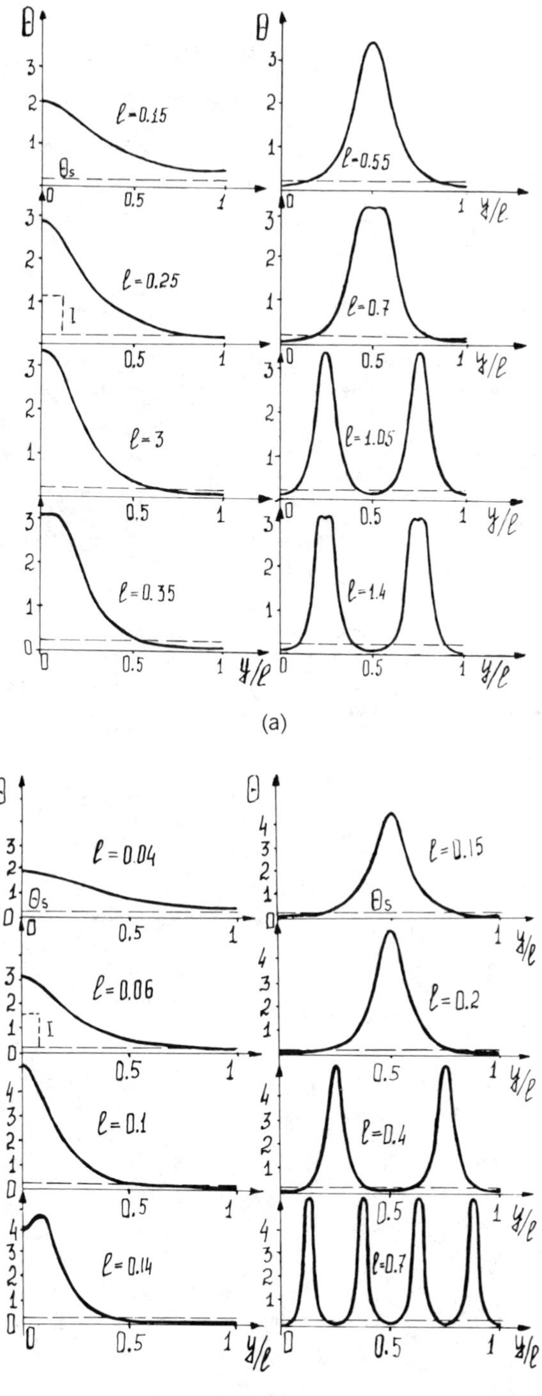
θ
ℓ=0.15
θs
0.5
1
ℓ=0.55
y/ℓ
ℓ=0.25
I
ℓ=0.7
ℓ=3
ℓ=1.05
ℓ=0.35
ℓ=1.4
(a)
ℓ=0.04
ℓ=0.15
ℓ=0.06
ℓ=0.2
ℓ=0.1
ℓ=0.4
ℓ=0.14
ℓ=0.7
(b)

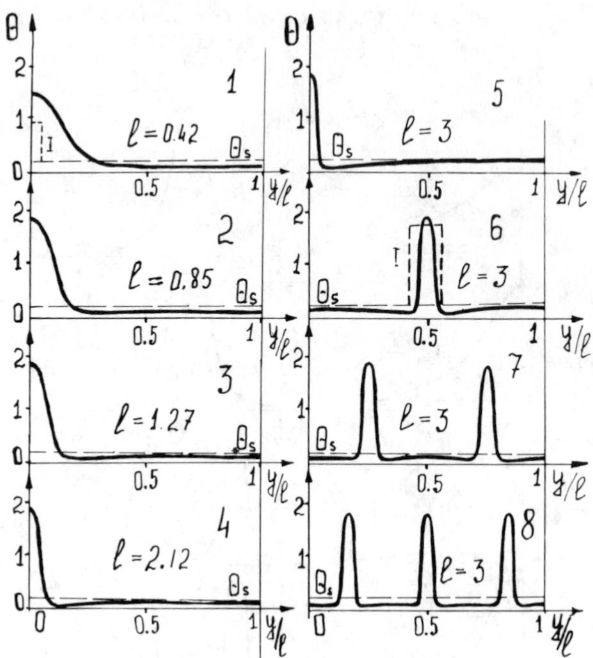

Figure 17. Evolution with the element length of a DS formed in the region of stability of SNP (nondisintegrating structures): (a) $u = 0.2$, $\varepsilon = 2 \cdot 10^{-4}$, $v = 0.165$, $p = 0.11$, $\theta_s = 0.3$ (initial condition is given by step 1; the increased number of DS fragments in Section 8 and 9 is due to additional perturbations in the profile of the DS formed).

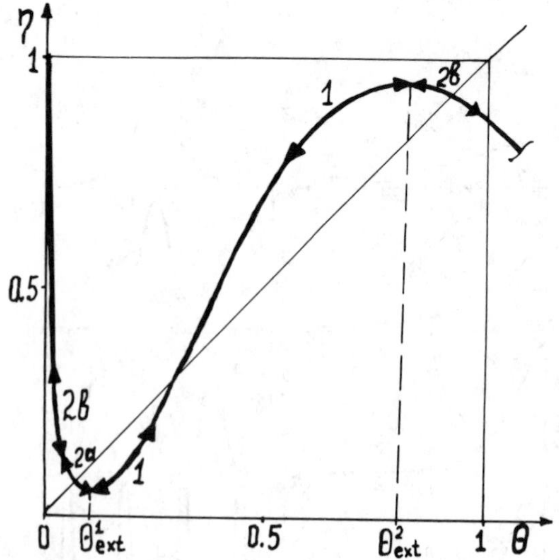

Figure 18. Location of the regions of instability on the nullcurve $\Phi(\theta, \eta) = 0$: 1—instability to SNP, 2—stability to SNP, 2a—disintegrating DS, and 2b—nondisintegrating DS.

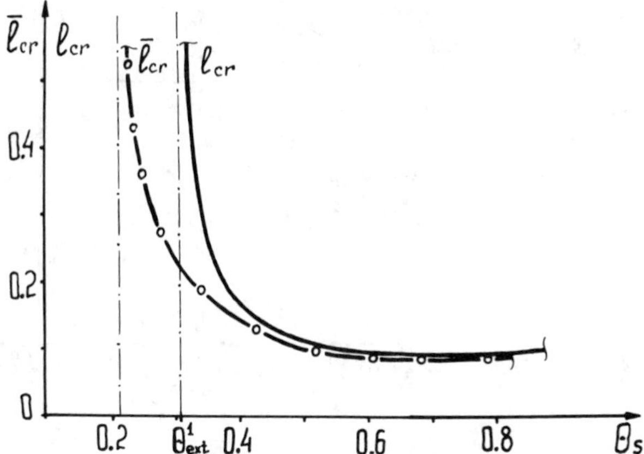

Figure 19. Enlargement of the region of sufficient conditions for the existence of DS when finite amplitude perturbations are taken into account: $u = 0.2$, $\varepsilon = 2 \cdot 10^{-4}$, $p = 0.011$, $v = \text{var}$.

structures are realized. To conclude this section, we present Fig. 19 which illustrates the broadening of the region of sufficient conditions for the existence of DSs.

VII. SOME PECULIARITIES IN THE FORMATION OF DS IN THE REGION OF NONUNIQUENESS OF USS

To conclude our discussion of the conditions for the origination of DS in catalytic systems, we consider the peculiarities of the process in the region of multiplicity of USS. It is known that in a model with a single equation of thermal conductivity (i.e., for $\varepsilon \gg 1$), a perturbation of certain amplitude will switch the system over from one USS to another (for details of the conditions and the dynamics of such a transition, see[3]). What will then happen in the model considered ($\varepsilon < 1$) where nonuniform steady states exist along with uniform ones? To which of them will the system transit? On which parameters do the dynamics of such transitions depend? This complicated, versatile problem will be treated here only at the definition level.

Our numerical experiments have led us to two important qualitative conclusions. The first is that introducing a local perturbation cannot switch a system over from one USS to another-there always takes place a transition to DS. The second conclusion is that the DS formed (a fragment or a set of fragments) is unique, no matter which of the USSs is perturbed.[e] This situation

[e]Naturally, the question of the generality of these conclusions calls for further study.

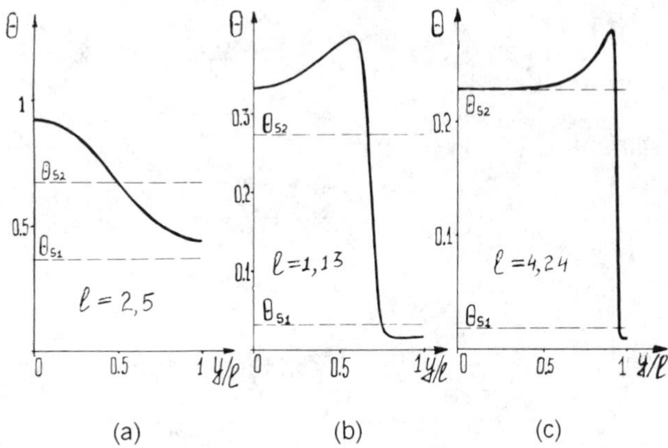

Figure 20. Examples of DS in the region of multiple USS (a) $u = 0.03$, $\varepsilon = 0.25$, $v = 0.21$, $p = 0.11$, $\theta_{s1} = 0.36$, $\theta_{s2} = 0.67$; (b) $u = 0.2$, $\varepsilon = 2 \cdot 10^{-4}$, $v = 0.0221$, $p = 0.011$, $\theta_{s1} = 0.03$, $\theta_{s2} = 0.27$; (c) $u = 0.2$, $\varepsilon = 2 \cdot 10^{-4}$, $v = 0.0217$, $p = 0.011$, $\theta_{s1} = 0.036$, $\theta_{s2} = 0.29$.

is particularly evident in the case where both USSs are unstable to SNP (as follows from the above, there are cases where both USSs are stable to SNP and where one is stable and the other is not). By the general bifurcation theory, each USS losing stability (with different critical lengths l_{cr}) must give way to a nonuniform solution. Since the USSs are different, the DSs born must also differ. In practice, however, a transition to one and the same DS always occurred. Figure 20 presents DSs corresponding to all three cases: (a) both USSs are unstable to SNP; (b) the low-temperature (θ_{s_1}) USS is stable and the high-temperature (θ_{s_2}) USS is unstable; (c) both USSs are stable.

VIII. ANALYSIS OF THE DIMENSIONS OF THE CHARACTERISTIC FRAGMENTS OF DS

Having considered the conditions for the origin of DSs and their existence regions, we pass on to an analysis of their geometric characteristics. The results are illustrated by Figs. 16, 17, 21, 22, and 23, which show examples of DSs occurring at different parameter values and their transformation due to the change in the catalytic element length l (reactor cross-size).

First, it is necessary to consider the effect on this characteristic of the prameter ε. When ε is close to unity, both the temperature and concentration profiles of DS are smooth (not infrequently quasi-harmonic) functions (see Fig. 24). As ε decreases, the concentration profile continues to be harmonic, but the temperature profile reveals intervals of sharp change of the type of

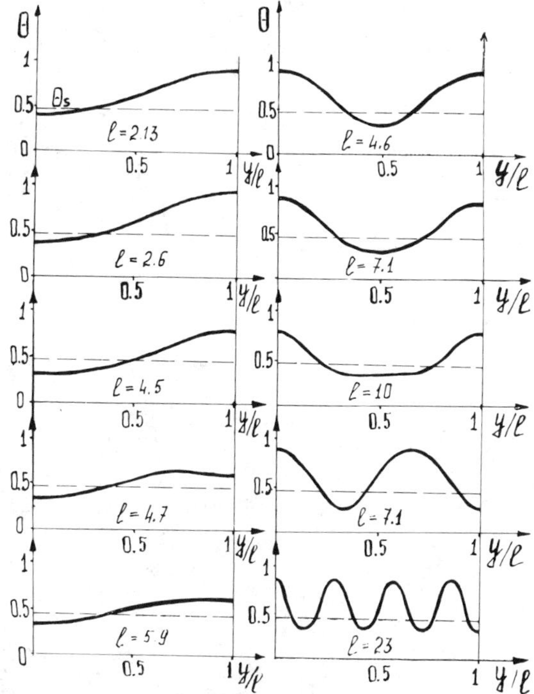

Figure 21. Evolution with the element length of a DS formed in the region of instability to SNP (hard birth, kinetic regime): $u = 0.03$, $\varepsilon = 0.22$, $v = 0.23$, $p = 0.13$, $\theta_s = 0.46$, $l_{cr} = 2.58$.

a wave front or "domain" wall, demarcating quasi-isothermal regions of the maximum, θ_{max}, and minimum, θ_{min}, temperatures. At the limit, when $\varepsilon = 0$ and $L_T = 0$, the temperature profile becomes discontinuous.

As seen from the figure, several characteristic regions (or dimensions) can be differentiated in a DS profile. The first of them is the characteristic size of the thermal front l_{fr}. It is reasonable to define this quantity as

$$l_{fr} = \frac{\Delta\theta}{(d\theta/dy)_{max}}$$

where $(d\theta/dy)_{max}$ is the value of the derivative at the inflection point of the DS profile. From physical considerations, it follows that the thermal size L_T must serve as a characteristic scale for l_{fr}, i.e. $l_{fr} \sim \sqrt{\varepsilon}$. Numerical analysis has shown that $l_{fr} \simeq (1 \div 5)\sqrt{\varepsilon}$ in a broad interval of ε values.

The second dimension, l_*, is the half-period length of a structure (or of an elementary fragment, $l_* = l/n$). Along this length the main change in the

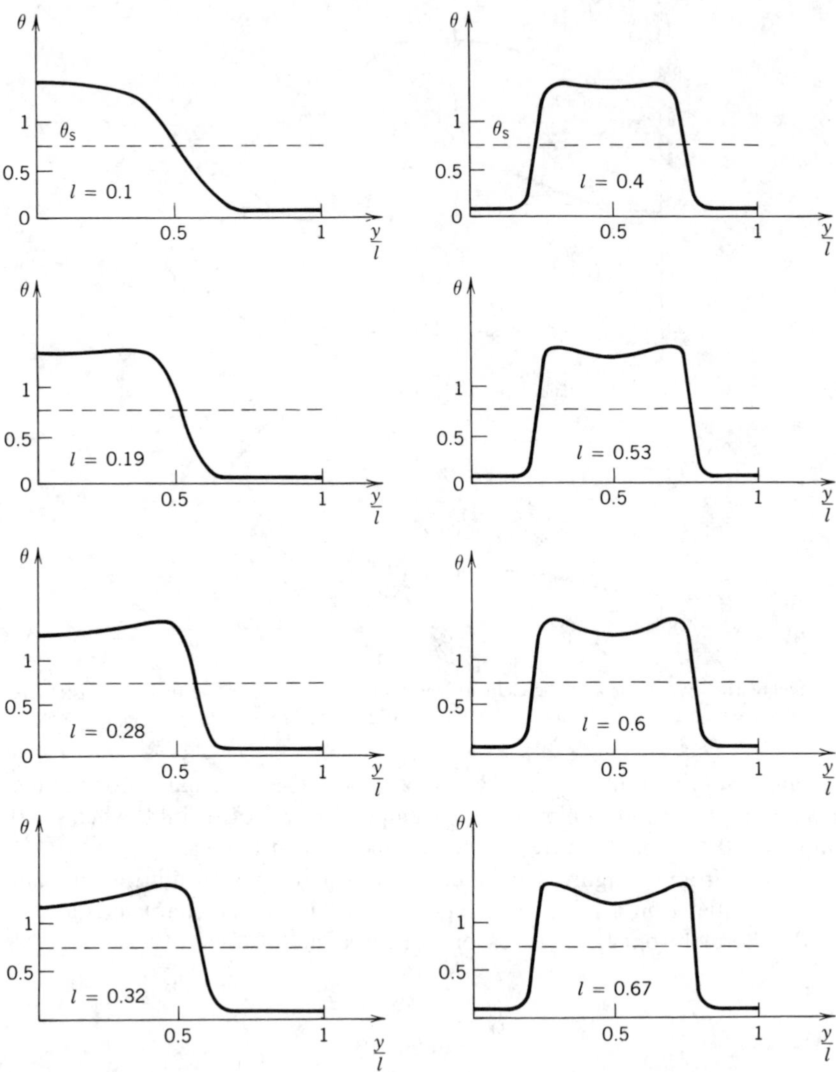

Figure 22. Evolution with the element length of a DS formed in the region of instability to SNP (hard birth, diffusion regime): $u = 0.2$, $\varepsilon = 10^{-5}$, $v = 0.075$, $p = 0.11$, $\theta_s = 0.75$, $l_{cr} = 0.03$.

Figure 23. Evolution with the element length of a DS formed in the region of instability to SNP (soft birth, diffusion regime): $u = 0.2$, $\varepsilon = 10^{-5}$, $v = 0.11$, $p = 0.11$, $\theta_s = 0.6$, $l_{cr} = 0.019$.

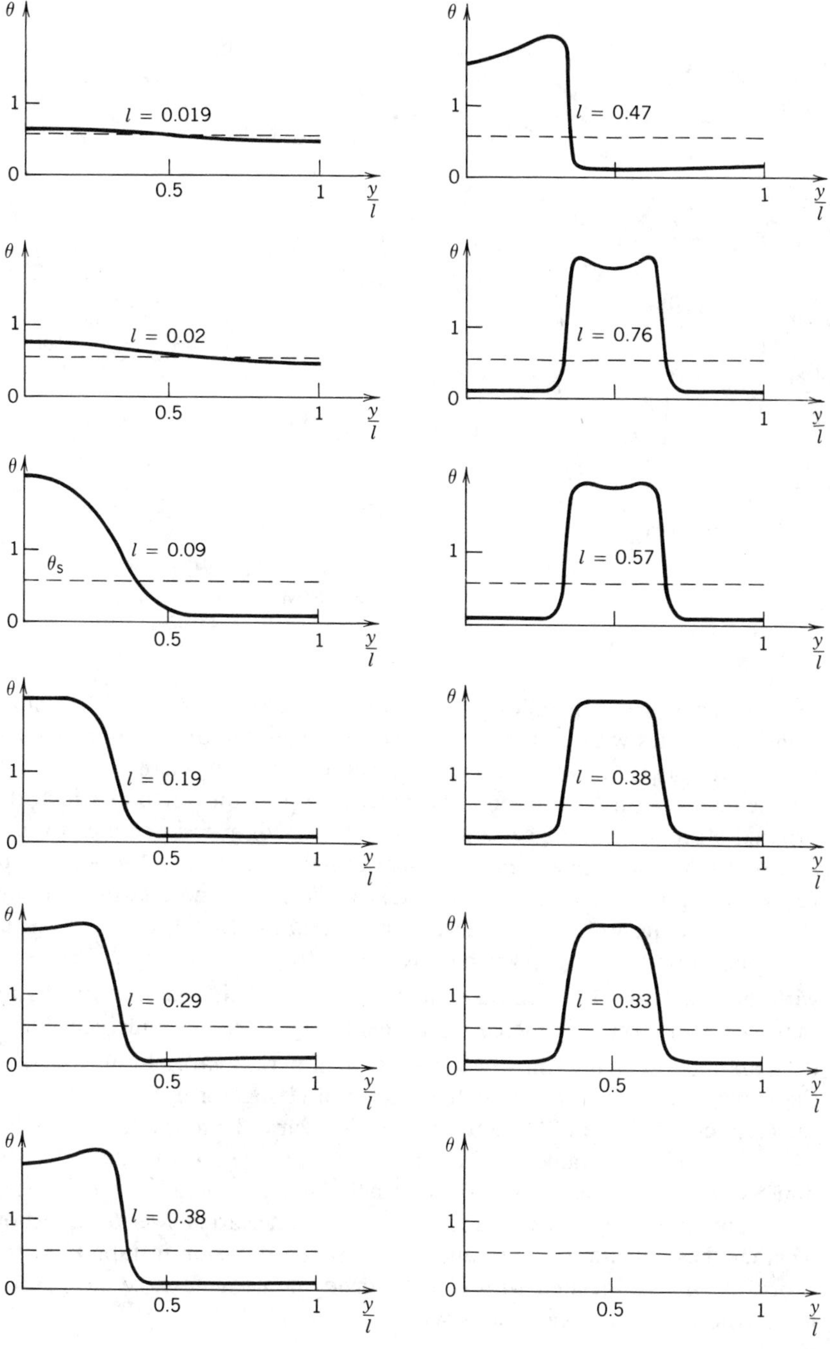
θ
1
0
0.5
1
y/l
l = 0.019
l = 0.02
l = 0.09
θs
l = 0.19
l = 0.29
l = 0.38
l = 0.47
l = 0.76
l = 0.57
l = 0.38
l = 0.33

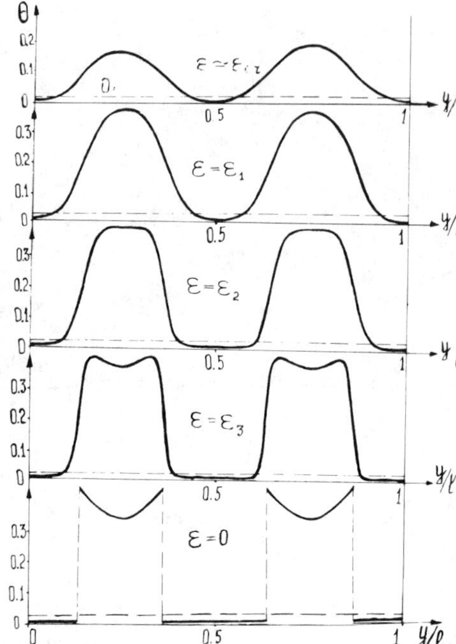

Figure 24. Scheme of DS transformation with decreasing ε, $0 < \varepsilon_3 < \varepsilon_2 < \varepsilon_1 < \varepsilon_{cr}$, $u = 0.2$, $v = 0.024$, $p = 0.011$, $l = 6.53$, $\theta_s = 0.025$.

reactant concentration takes place. Roughly estimated, $l_* \simeq 1$. The actual value of l_* varies within broad limits. Therefore, an important characteristic of DS is the degree of its deformability (characterizing the change in l_*) when the element length l monotonously increases up to the value at which the original structure with a certain set of half-periods n disintegrates to give way to a new one with a greater number of half-periods ($n + 1$, $n + 2$ or $2n$), i.e., a multiplication of a structure takes place. The peculiarities of this characteristic are most pronounced at $\varepsilon \ll 1$ and $\tilde{\varepsilon} \ll 1$. We shall analyze an example shown in Fig. 25, which corresponds to the soft birth of a structure with the clearly pronounced thermal front $l_{fr} \simeq \sqrt{\varepsilon}$. It is seen from the figure that with monotonously increasing element length, the DS undergoes linear deformation (it extends), in which case the half-period varies from $l_* \simeq \sqrt{\varepsilon}$ at the instant of birth to $l_* \approx 1$ at the instant of change of n. In other words, the characteristic size of DS during extension varies from the thermal value L_T to the diffusion value L_c. The obtained estimates of the characteristic dimensions of chamical domains may help the technologist, who is faced with nonuniform regimes in the reactor performance, to answer the question whether these phenomena are due to the spatial diffusion instability of the process or are associated with some hydrodynamic factors, i.e., to control the phenomena on a scientific basis.

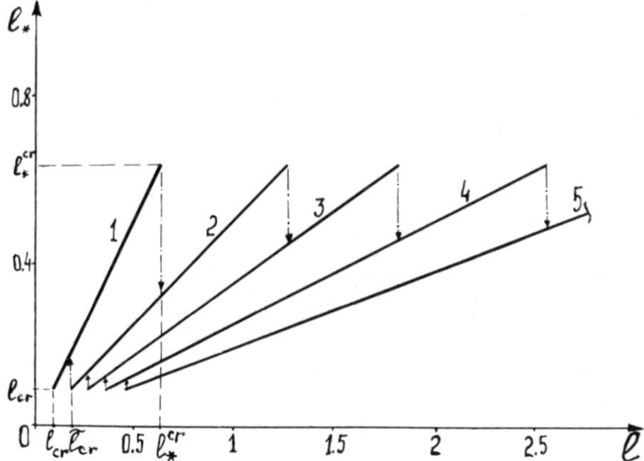

Figure 25. Dependence of the characteristic size of a fragment of DS l_* on element length l: $u = 0.2$, $\varepsilon = 2 \cdot 10^{-4}$, $v = 0.11$, $p = 0.11$, $\theta_s = 0.6$ (for curves 1 through 5, $n = 1 \div 5$, respectively).

The plot of Fig. 25 relating the characteristic dimension of a structure fragment, l_*, to the catalytic element length l has a pronounced hysteretic character: critical lengths for transitions between structures with a different number of elementary fragments do not coincide. The possibility for hysteresis to occur is independent of the way a DS arises from USS (Fig. 25 shows an example of a softly born structure), and its magnitude depends on the parameter ε: the less ε the wider is the hysteretic loop. The critical length $\tilde{l}_{cr}$, at which a transition to a structure with a smaller number of elementary fragments takes place, depends on the character of DS birth: when the birth is soft, it is close to l_{cr} and when it is hard, close to $\bar{l}_{cr}$.

The described hysteretic behavior being realized in a broad parameter range is nevertheless not a general rule. Nonhysteretic transitions between DSs with different numbers of elementary fragments are also possible (see reference 16).

The characteristic dimensions to be considered next are those of the high-temperature, l_h, and low-temperature, $l_c = l_* - l_h$, zones to which a USS splits when a DS is born. The ratio of these lengths is of importance in practice because it characterizes the void portion of a catalyst (or part of the catalyst bed).[f] We shall analyze this ratio and its dependence on the

[f] Analysis of the shape of DS formed in the kinetic region shows that in the low-temperature range the deviation from a uniform distribution is insignificant, whereas in the high-temperature range it may be very large. The reverse is true for DS in the diffusion region: while in the high-temperature range the deviation is small, there is a well-defined trough in the low-temperature range.

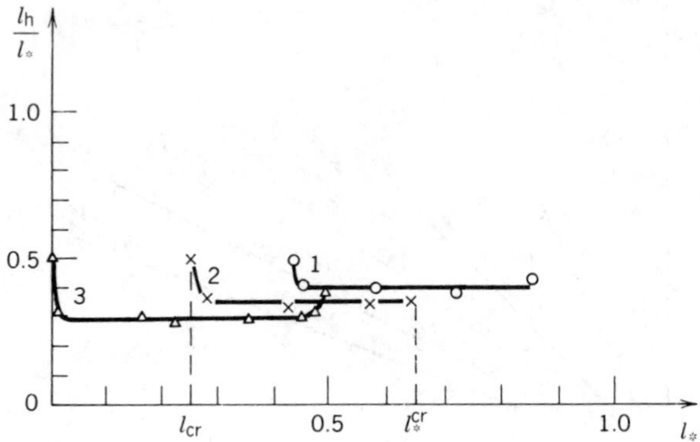

Figure 26. Dependence of the characteristic size of the hot zone of DS on reactor length l_*: 1, $\varepsilon = 2 \cdot 10^{-3}$; 2, $\varepsilon = 2 \cdot 10^{-4}$; 3, $\varepsilon = 10^{-5}$; $u = 0.2$, $v = 0.11$, $p = 0.11$, $\theta_s = 0.6$.

parameters for the kinetic regime, taking as a characteristic the relationship between the length of the hot zone l_h and that of an elementary fragment, l_*.

If ε approaches unity, $l_h/l_* \simeq 0.5$, and if ε decreases, the ratio l_h/l_* decreases too. An interesting property of DS is that this ratio is independent of the catalytic element size, as illustrated by Fig. 26. This is not true only in the vicinity of the critical lengths for the formation of DS and multiplication of elementary fragments. The ratio is also weakly dependent on ε when $\tilde{\varepsilon} \ll 1$.

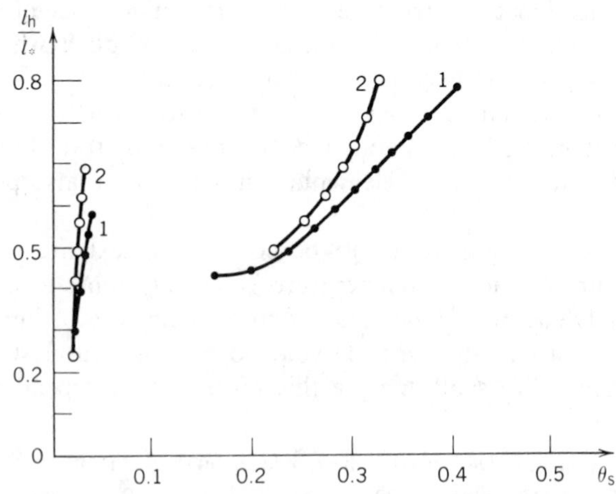

Figure 27. Dependence of the characteristic size of the hot zone of DS on the stationary temperature θ_s of USS: 1, $v = 2.5 \cdot 10^{-2}$, $p = 0.0108 \div 0.0149$; 2, $p = 2.5 \cdot 10^{-2}$, $v = 0.02 \div 0.025$; $\varepsilon = 2 \cdot 10^{-4}$; $u = 0.2$.

However, the relation between the hot and cold zones varies significantly with USS temperature, θ_s. Figure 27 provides examples for two cases: in the first, θ_s is changed via variation of the parameter v at fixed values of the two other ones, and in the second, via variation of p. A break in the curves corresponds to the region of instability of USS of the saddle type. In both cases, the length of the hot zone varies within wide limits, from 0.2 to 0.9. The corresponding DSs are shown in Fig. 28. In both cases, their transformation reduces to a change with θ_s of the relation between the hot and cold zones. The size of the hot zone l_h/l_* is a linear function of the parameters $1/p$ and $1/v$; the function is a decreasing one in the first case and an increasing one in the second (see Fig. 29).

IX. ANALYSIS OF THE AMPLITUDE CHARACTERISTICS OF DS

From a technological standpoint, the previous section considered the question of the proportion of a catalytic element (catalyst bed) which is dropped out of operating condition due to the formation of DS. The next

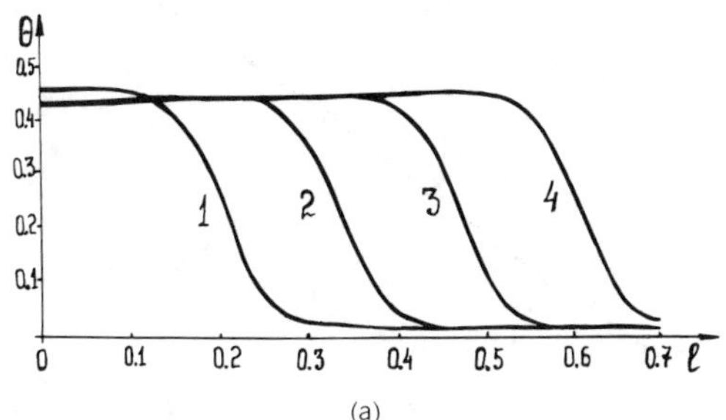

(a)

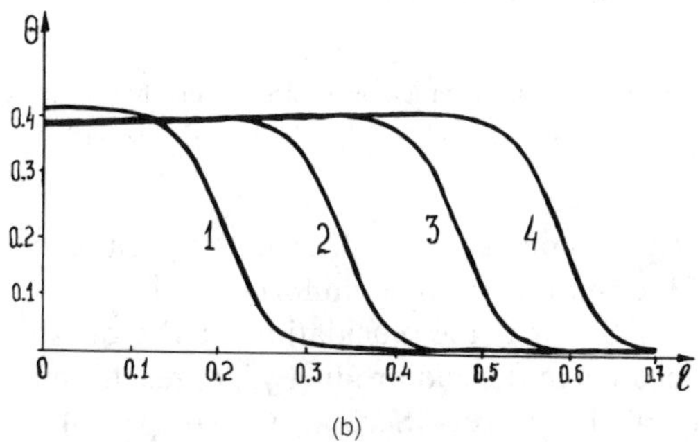

(b)

Figure 28. Evolution of DS profile with parameter change: $u = 0.2$; (a) $v = 2.5 \cdot 10^{-2}$; $p = 0.0108$ (1), 0.0118 (2), 0.0133 (3), 0.0149 (4). (b) $p = 0.01093$; $v = 0.025$ (1), 0.0233 (2), 0.0217 (3), 0.0204 (4).

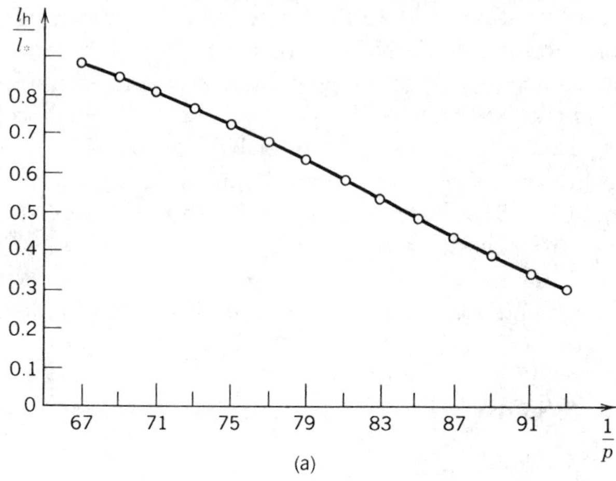

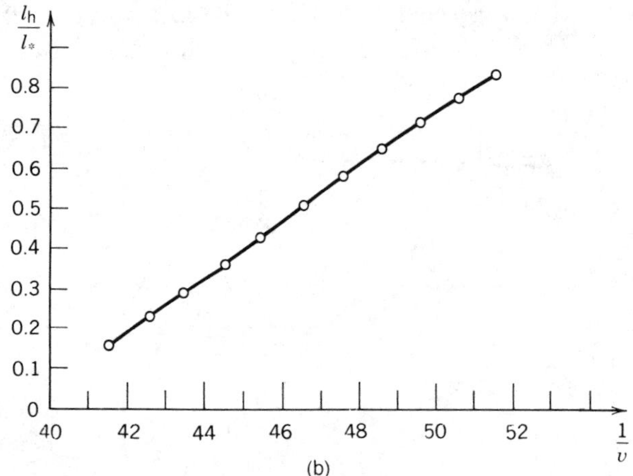

Figure 29. Dependence of the characteristic size of the hot zone l_h/l_* of DS on the parameters $1/p$ (a) and $1/v$ (b); $u = 0.2$; $\varepsilon = 2 \cdot 10^{-4}$; (a) $v = 2.5 \cdot 10^{-2}$; (b) $p = 2.5 \cdot 10^{-2}$.

question of principal importance for practical applications is naturally that of the degree of deviation from a uniform mode, i.e., of the amplitude characteristics of DS. First, the elucidation of the question may provide estimates of the change in the end product yield, reactor efficiency, and even in the selectivity of the process. Second, it may permit evaluation of the danger of an abnormal situation, since the hot regions of DS may develop high temperatures, which results in sintering and irreversible degradation of a catalyst. Third, it offers the possibility of taking advantage from disintegration of a uniform mode: in some cases, a nonuniform mode may turn out to be more optimal, leading, for example, to greater yields.

We consider first some qualitative estimates. It is physically evident that the reactant concentration at the catalyst surface can in no case exceed C_0 (i.e., $\eta(y) < 1$ always). No such constraint is laid on temperature. Using the method of thermal flow,[14] it is possible to estimate the maximum temperature in a DS profile. In the diffusion regime, it has the form

$$\theta_{\max} \simeq \frac{2}{1+\sqrt{\varepsilon}} \tag{23}$$

In the kinetic regime,

$$\theta_{\max} \simeq \frac{p\exp(\bar{\theta}/(v+u\bar{\theta}))}{1+\sqrt{\varepsilon}} \tag{24}$$

where $\bar{\theta}$ is the temperature at the inflection point of the temperature profile. It follows from Eq. (23) that the maximum temperature in the hot zone of DS may be twice the maximum catalyst temperature in the diffusion region for a uniform mode, whose value is widely used in practice to determine the maximum possible heating of the catalyst bed. Such overheating is connected to the fact that besides the diffusion reactant fluxes directed to the hot zone from the gas phase flux core, which are normal to the catalyst surface, there are also tangential fluxes from the neighboring low-temperature layers enriched with reactant.

By Eq. (24), a relatively weak heating will take place at $p \ll 1$. For the case of a reactor, this can be illustrated as follows. Since p has the meaning of the ratio between the residence time τ_r and chemical time τ_x (analogous to Damköhler's criterion), the effect of lateral reactant transfer is small at small residence times and therefore can not afford considerable rise of temperature. An increase in the residence time will lead to a rise in the maximum temperature. Analogously, it may be anticipated that the maximum heating will be relatively small for $v \simeq 1$ and large for $v \ll 1$. It may also be anticipated that the greatest effect on the maximum heating in the DS profile will be that of the parameter u. The effect of ε will be weak for $\varepsilon \ll 1$ and strong for $\varepsilon \simeq 1$.

The results of numerical experiments support the above observations. Figure 30 plots the maximum heating $\theta_{\max}$ against the temperature of the initially uniform steady state, θ_s, that could be changed either by varying p at a fixed v or by varying v at a fixed p. There are two maximum temperature intervals corresponding to large and small values of the parameters p and

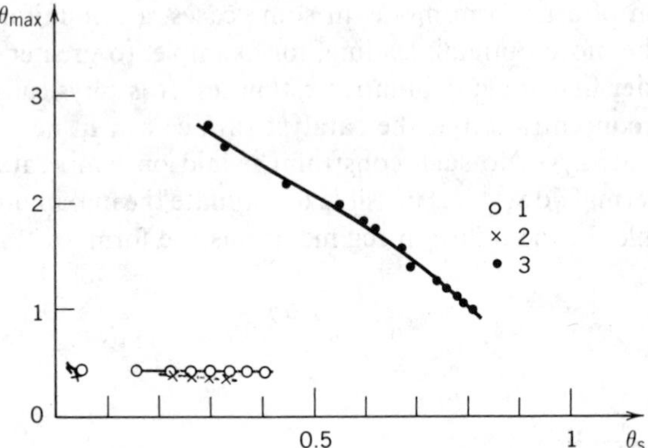

Figure 30. Dependence of the maximum temperature in DS profile, θ_{max}, on the steady-state temperature θ_s of USS: 1, $u = 0.2$, $\varepsilon = 2 \cdot 10^{-4}$, $v = 0.025$, $p = 0.011 \div 0.015$, $l = 0.707$; 2, $u = 0.2$, $\varepsilon = 2 \cdot 10^{-4}$, $p = 0.011$, $v = 0.02 \div 0.025$, $l = 0.707$; 3, $u = 0.2$, $\varepsilon = 2 \cdot 10^{-4}$, $p = 0.11$, $v = 0.061 \div 0.154$, $l = 778$.

v, respectively. The maximum heating may be expected to exceed the value of 2 estimated to a first approximation in view of system (3, 4) being severely nonlinear.

For small values of ε, the dependence of θ_{max} on ε is weak (see Fig. 31). The greatest influence on θ_{max} is exerted by the parameter u: at $u = 0$, $\theta_{max} \simeq 3$. (Fig. 32). For $u \neq 0$, the maximum values of θ_{max} are attained in the region of the stability to SNP (nondisintegrating structures). Such overheating should naturally be treated as an emergency situation (a two-fold increase in the

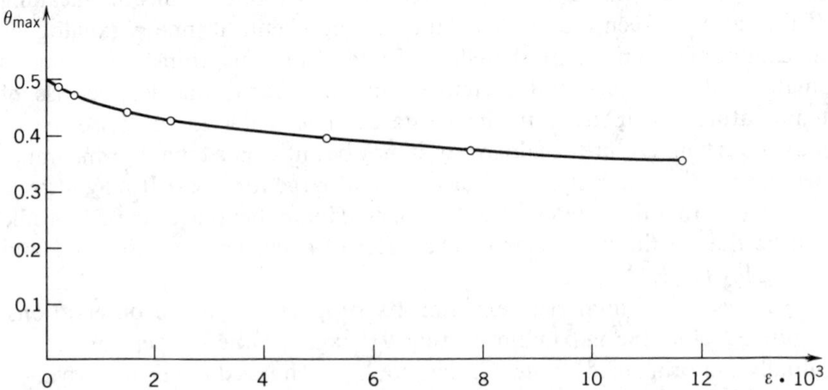

Figure 31. Dependence of the maximum temperature in DS profile on the parameter ε: $u = 0.2$, $v = 2.5 \cdot 10^{-2}$, $p = 0.011$, $\theta_s = 2.5 \cdot 10^{-2}$, $l = 0.7$.

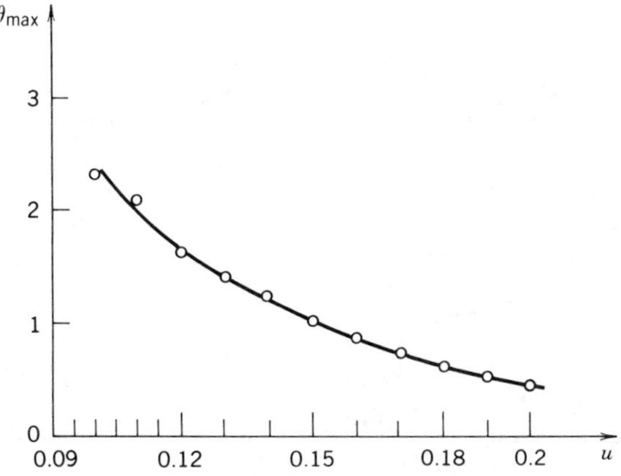

Figure 32. Dependence of the maximum temperature in DS profile on the parameter u: $\varepsilon = 2 \cdot 10^{-4}$, $v = 2.5 \cdot 10^{-2}$, $p = 0.011$, $l = 0.7$.

average dimensional maximum temperature in the diffusion region $T_m - T_0$ may result in temperatures as high as 900–1000°C, destructive for most catalytic system). It should be noted that although $\theta_{\max}$ of the order of 2 is characteristic of small ε, the temperatures exceeding considerably those of the uniform mode were obtained also for ε close to unity. For instance, for the values of $\theta_s = 0.46$, $u = 0.03$, $p = 0.13$, $v = 0.23$, $\tilde{\varepsilon} = 0.82$, and $\varepsilon_{cr} = 0.26$, the maximum heating is 0.94, i.e., it is twice that in the uniform mode (see Fig. 21).

Qualitatively, the amplitude characteristics can be divided into three groups (see Fig. 33). The first group corresponds to disintegration of the uniform state, when the temperature in one part of the nonuniform profile is above and in the other part is below θ_s of the USS that has lost stability (see Fig. 33a). The second group corresponds to the case where the whole nonuniform profile is above the uniform one (the situation is possible for the kinetic regime, Fig. 33b). The apparent paradoxicalness of this situation is due to the fact that there is an additional heat flow from the hot zone of the catalyst bed to the cold one and, as a result, the temperature of the latter exceeds that of the initial USS. Characteristic of the third group is that the greater part of a reactor exhibits a quasi-uniform mode, while the smaller one shows an essentially nonuniform temperature distribution (Fig. 33c). The amplitude characteristics of the first and second groups are typical of DSs which correspond to USSs unstable to SNP and those of the third group correspond to USSs stable to SNP. With respect to practical applications, the most interesting is the situation where the characteristics of the second group are realized. On one hand, this may be an obvious case of emergency if the maximum temperature in the DS profile attains a value of 2 or more.

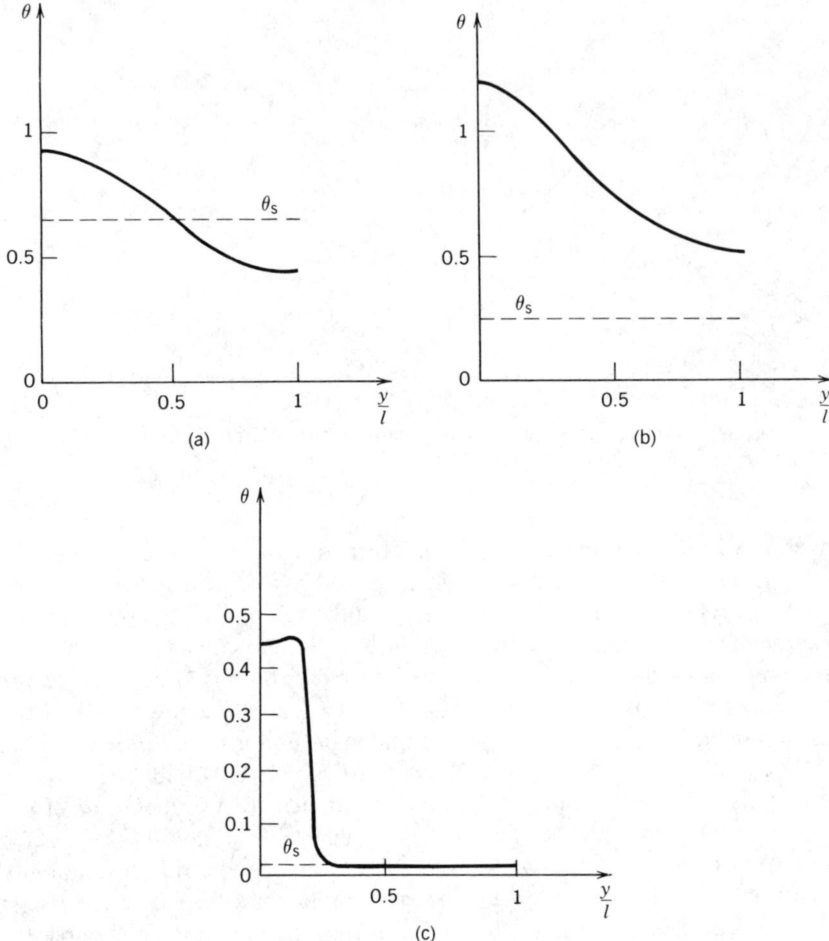

Figure 33. Different types of destruction of uniform states: (a) $u = 0.03$, $v = 0.21$, $p = 0.11$, $\varepsilon = 0.25$, $\theta_s = 0.65$, $l = 2.5$; (b) $u = 0.03$, $v = 0.216$, $p = 0.11$, $\varepsilon = 0.01$, $\theta_s = 0.258$, $l = 0.255$. (c) $u = 0.2$, $v = 0.163$, $p = 0.11$, $\varepsilon = 2 \cdot 10^{-4}$, $\theta_s = 0.293$, $l = 2.12$.

On the other hand, for reactions with simple kinetics, such a nonuniform mode may prove to be technologically more advantageous than a spatially uniform mode. This is illustrated by Fig. 34 which plots the integral reaction velocity

$$W = \frac{p}{l} \int_0^l [\exp(\theta/(v + u\theta))(1 - \eta)] \, dy$$

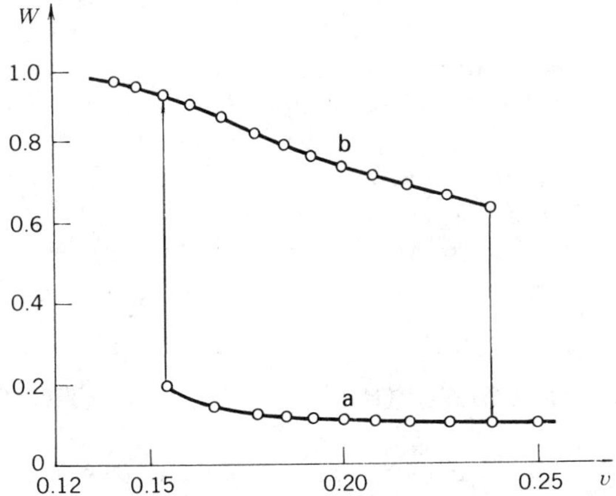

Figure 34. Dependence of the integral reaction velocity W on the parameter v for uniform (a) and nonuniform (b) temperature distributions: $u = 0.03$, $\varepsilon = 0.1$, $p = 0.071$, $l = 3.2$.

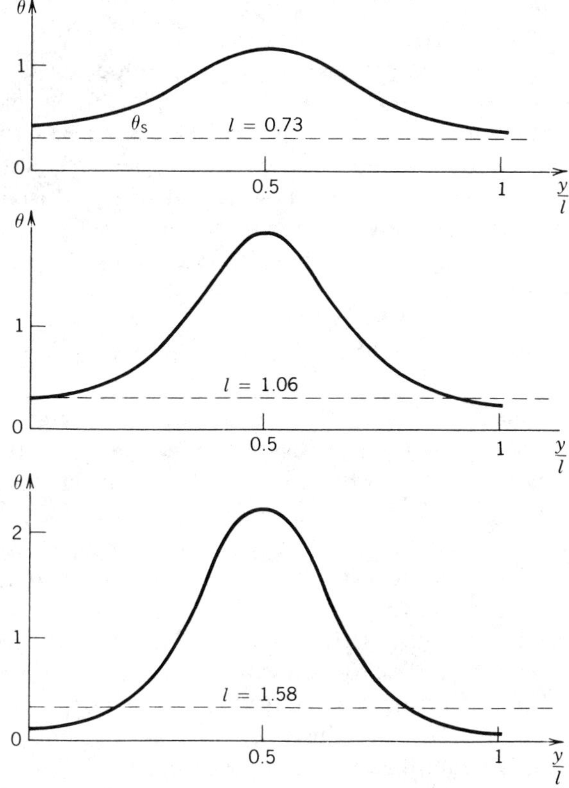

Figure 35. Evolution of DS with the characteristic length l_*: $u = 0.03$, $\varepsilon = 0.025$, $v = 0.24$, $p = 0.13$, $\theta_s = 0.312$.

as a function of the parameter v in the region of existence of DS for uniform and nonuniform temperature distributions. In the entire range of v values, the reaction velocity for the nonuniform distribution is 7 times that for the uniform. The data obtained indicate that this situation is not unique. In the course of evolution of a structure under the change in l_*, the characteristics of the first and second groups are realized in a broad range of kinetic and thermal-diffusional parameters replacing each other as shown in Fig. 35.

X. SOME ASPECTS OF THE DYNAMIC BEHAVIOR OF THE DISTRIBUTED HETEROGENEOUS CATALYTIC SYSTEMS

The problem of the dynamical behavior of systems such as (3, 4) is a complicated one. We shall consider here only some of its most important aspects at the level of definition.

A. Existence of Spatially Nonuniform Self-Oscillatory Modes

In this work, the numerical experiment was carried out for $\gamma = 1$, i.e., in the region where uniform oscillations in system (3, 4) are impossible. Can the distributedness of the system act, under such conditions, as a factor capable of inducing a spatially nonuniform self-oscillatory process? There is no strict criterion at the present time for the existence of this new type of self-oscillatory instability. Numerical experiments, however, provide a positive answer to this question. Figure 36 shows a time display of a self-oscillatory process discovered in the course of DS transformation as the catalytic element length l was varied. Other parameters were:

$$\theta_s = 0.26, \quad u = 0.03, \quad p = 0.11, \quad v = 0.21, \quad \varepsilon_{cr} = 1.14 \cdot 10^{-2}, \quad \varepsilon = 1.11 \cdot 10^{-2}.$$

The oscillations were stable, remaining undamped for several tens of periods (actual computation time) and consisted of alternate symmetric transitions from a stationary "one-humped" ($n = 2$) to "two-humped" ($n = 4$) structure and vice-versa. Figure 37 presents time dependences of the integral (in the element length) temperature and heat generation velocity. The oscillation process has a rather complicated form, including rapid and slow stages. During the rapid stages having times of the order of τ_T, one of the one-humped structures disappears to give way to a two-humped structure. The slow stages with times of the order $10 \div 10^2 \tau_T$ involve the processes of transformation of the latter. The experimentally observed self-oscillations in heterogeneous catalytic systems are, as a rule, registered by the average integral signal. The data presented illustrate the idea expressed previously[4] that oscillations in real distributed systems may be due to conductive-diffusional rather than kinetic factors. The example considered does not, naturally, exhaust the

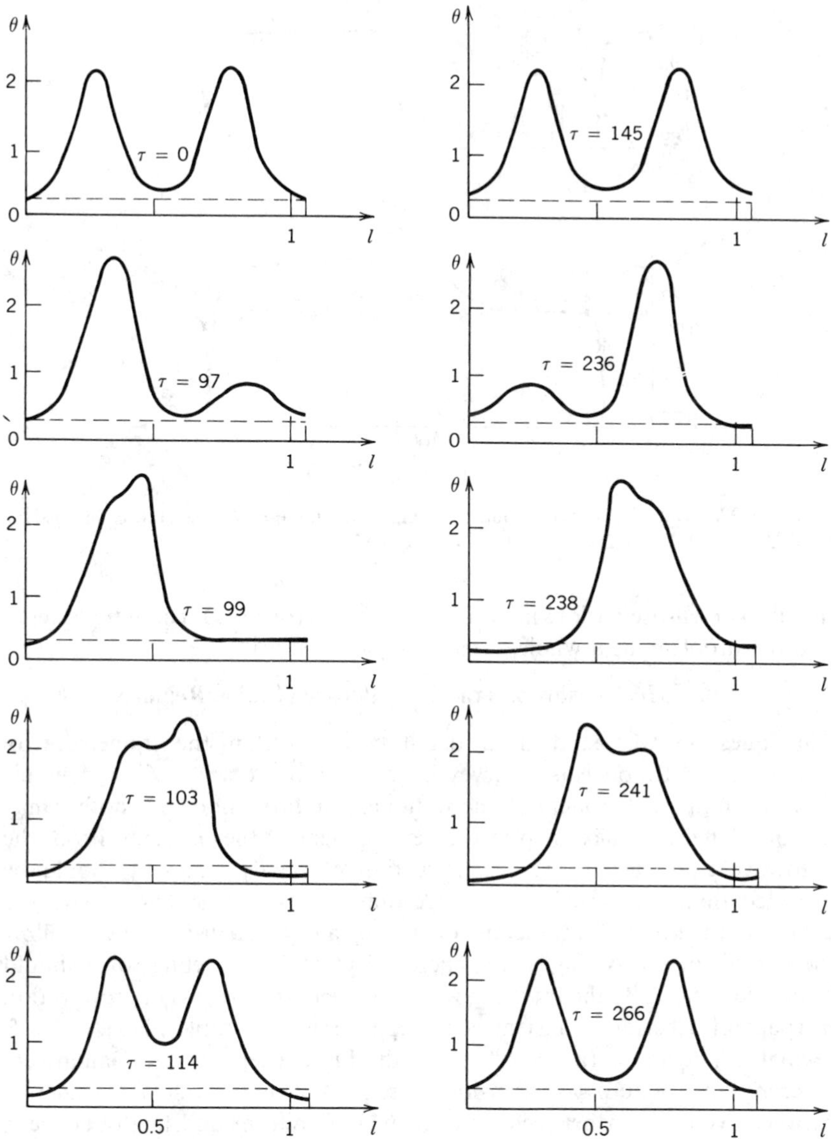

Figure 36. Time display of a nonuniform self-oscillatory process: $u = 0.03$, $\varepsilon = 0.011$, $v = 0.216$, $p = 0.11$, $l = 1.33$, $\theta_s = 0.26$.

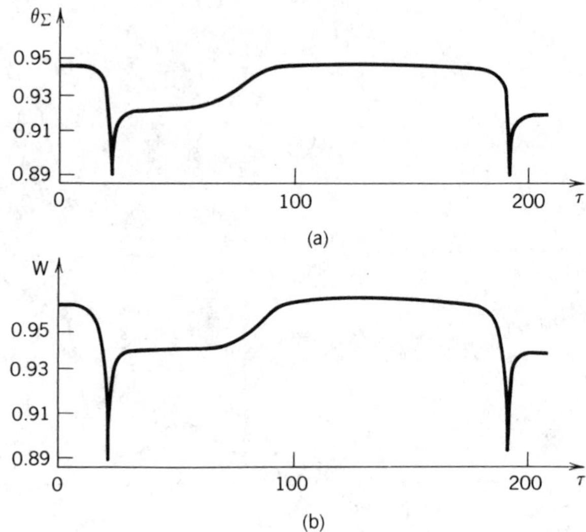

Figure 37. Time dependence of integral temperature (a) and heat generation velocity (b): $u = 0.03$, $\varepsilon = 0.011$, $v = 0.216$, $p = 0.11$, $l = 1.33$, $\theta_s = 0.26$.

question of self-oscillations in the distributed heterogeneous catalytic systems, and the problem, as a whole, calls for separate study.

B. Mechanisms of Transition Between Stable Regimes

This question was studied in detail earlier within the framework of one-component models (see survey in reference 3). A study of the dynamic patterns in the replacement of one stationary uniform mode by another may be considered as a next step in the development of the problem. From the autowave character of the reconstruction of steady states, it should be expected that, given sufficiently large dimensions of a system ($l \gg l_*$), the change of the nonuniform regimes evoked by a local perturbation will follow the pattern of a traveling wave. The concept of the traveling wave should be extended here. By the traveling wave, we mean the process of propagation in space of a certain transient zone separating two stable (including self-oscillatory) regimes. In accordance with this concept, three qualitatively different types of autowave transitions are possible in the system considered: between two uniform, between uniform and nonuniform, and, finally, between two nonuniform stationary states.

It was noted above that in the region of nonuniquenee of USS, the DS occurring at a fixed parameter set is unique. Therefore, a question arises whether the system can be switched over from one uniform state to another, or a transition to a DS will take place.

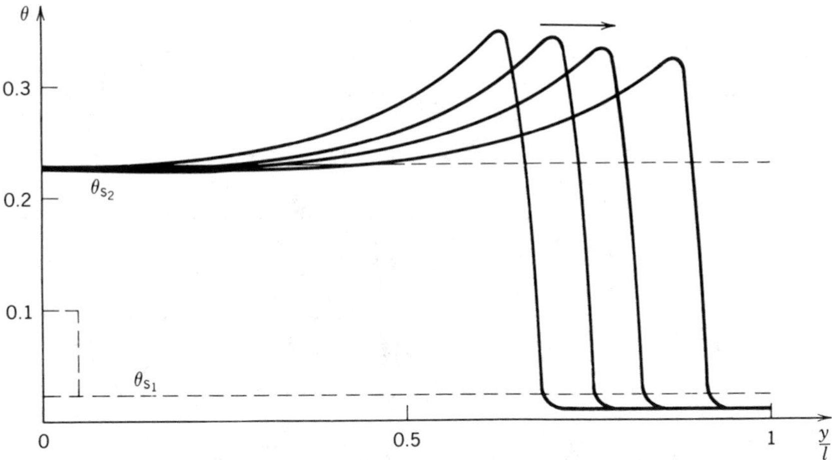

Figure 38. Autowave transition from uniform state to DS when both USSs are stable to SNP: $u=0.2$, $\varepsilon=2\cdot10^{-4}$, $v=1.5\cdot10^{-2}$, $p=0.7\cdot10^{-2}$, $\theta_{s1}=0.02$, $\theta_{s2}=0.23$, $v_{fr}=0.54\ v_T$.

Our numerical experiments have demonstrated a possibility of autowave transitions between the low-temperature uniform state and DS in the region of the stability of SNP for a broad interval of the initial conditions. A spatial–temporal pattern of temperature distribution during the autowave reconstruction is presented in Fig. 38. Over the greater part of a catalytic element, the wave profile propagates at the constant velocity $v_{fr}=0.54v_T$, $v_T=l_T/\tau_T$. A deviation from constant velocity occurs only at the initiation stage and in the transition to a stationary DS. The principal difference between this process and the waves in one-component systems lies in the fact that in the latter case, the wave is a spatially nonmonotonous temperature profile which is arrested when approaching the boundary of a catalytic element. The resulting DS is a nonmonotonous transition between the low-temperature and high-temperature USS, a kind of standing wave. Another feature of such an autowave reconstruction is that only the transition from the low-temperature USS to a DS is realized. The dynamics of the transition from the high-temperature USS to a DS is not of the autowave type. This seems to be connected to the fact that the existence of the reactant concentration gradient along the catalytic element facilitates, due to an additional feed to the reaction zone, the propagation of a wave of transition from the low-temperature to the high-temperature state (ignition wave) and inhibits the reverse transition (extinction wave).

Of peculiar character is the autowave transition in the parameter range where the USS is unique, stable and gives way to a disintegrating structure

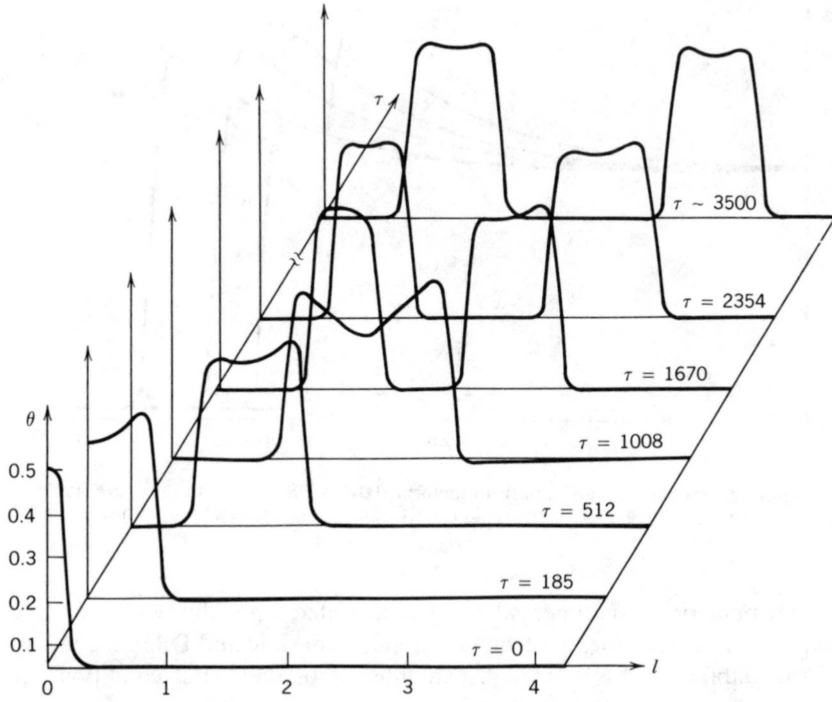

Figure 39. Autowave transition from USS to DS (disintegrating) for the case of unique USS stable to SNP: $u = 0.2$, $\varepsilon = 2 \cdot 10^{-4}$, $v = 2.4 \cdot 10^{-2}$, $p = 0.011$, $\theta_s = 0.024$, $l = 4.24$, $v_{fr} = (0.1 \div 0.2)\, v_T$.

(see Fig. 39 and also Fig. 16). The initial step-shaped perturbation applied at the left boundary of the element then undergoes a modification and converts to a propagating pulse taking the form of DS. While propagating, the latter undergoes successive doubling, until the catalytic element is complately filled up with stationary structures. The propagation velocity of such a structure does not remain constant, varying from $\sim 0.075 v_T$ at instants of disintegration of $\sim 0.2 v_T$. Figure 40 shows the dynamics of the transition to DS in the case of a unique USS unstable to SNP, the loss of stability being soft. Here, given sufficiently large element length l, a transition to a multihumped DS takes place. The specificity of such a transition consists in the fact that the applied perturbation evokes first a "leading" wave which runs along the sample, dividing the whole region into fragments from which a stationary DS gradually develops. The characteristic time of the leader's run is $\sim 10^2 \tau_T$ and the time for the development of a stationary DS is $10^4 \tau_T$.[g]

[g]A similar dynamics for the development of a DS in the "brusselator" model was reported in reference 22.

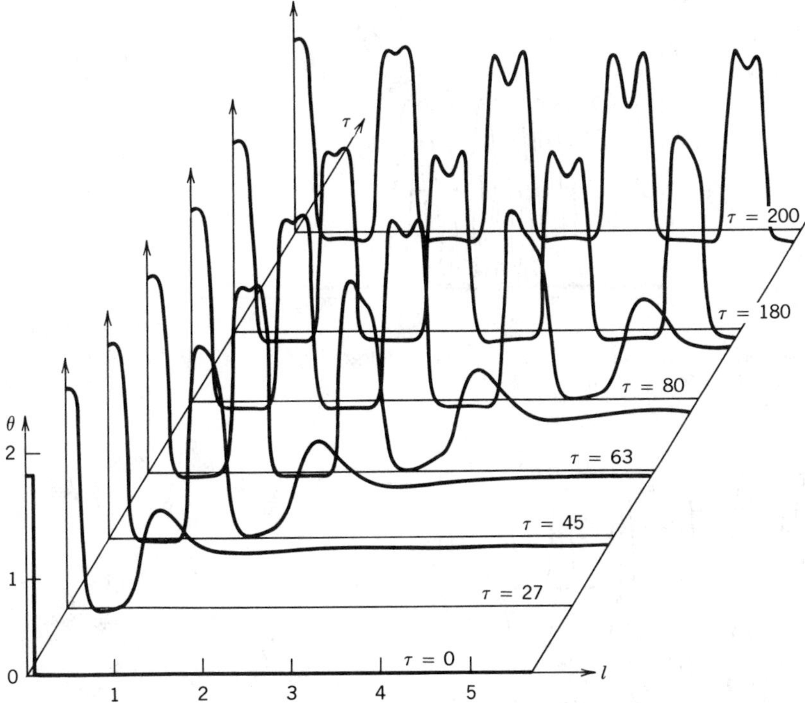

Figure 40. Autowave transition to DS for the case of unique USS unstable to SNP: $u = 0.2$, $v = 0.11$, $p = 0.11$, $\varepsilon = 10^{-5}$, $\theta_s = 0.6$, $l = 1.27$, $v_{fr} = 10\,v_T$.

We also observed the propagation of a nondisintegrating DS as a whole. If the initial perturbation is set up as a single pulse of large amplitude applied nonsymmetrically at some distance from one of the boundaries of an element, the pulse in the course of its conversion to a DS begins to shift to the center of the element at a velocity of $\simeq 10^{-3} v_T$ until the distribution becomes symmetric (Fig. 41). Since such structures do not undergo spontaneous doubling while propagating, their number can be increased by introducing additional perturbations of large amplitude. These can be applied in any part of the catalytic element, but the new-formed DSs always undergo a displacement, resulting in symmetric temperature and concentration distributions (Fig. 41).[h]

[h]Similar nondisintegrating DSs were earlier observed in models of a Fitz-Hugh–Nagumo type[21] widely used for simulating biological, chemical, and even physical processes. However, these structures were always immovable and resided at the place where a perturbation was introduced. This was interpreted as a "memory" effect in an active medium. Studying such effects as well as comparing the properties of different models is, undoubtedly, of interest but is beyond the scope of the present paper.

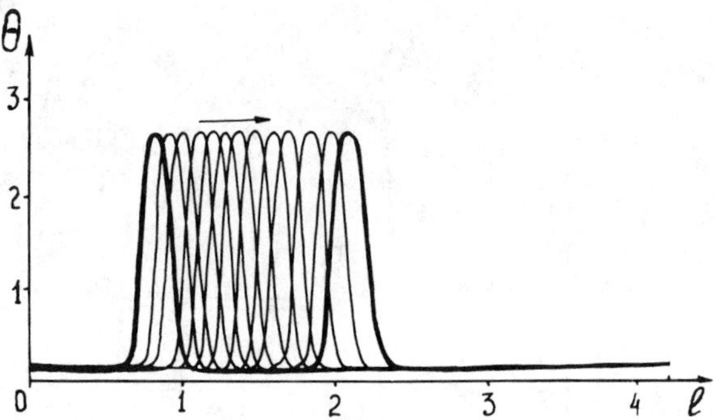

Figure 41. Autowave propagation of nondisintegrating DS: $\varepsilon = 2\cdot 10^{-4}$, $v = 0.165$, $p = 0.11$, $\theta_s = 0.3$, $l = 4.24$, $v_{fr} = 8\cdot 10^{-3}\ v_T$, $u = 0.2$.

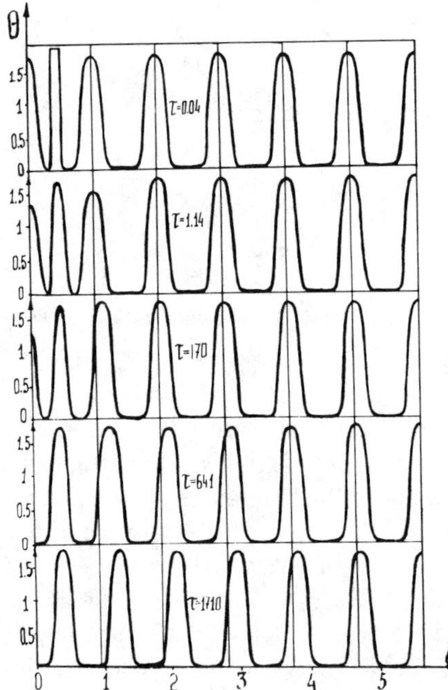

Figure 42. Dynamics of the transition between nonuniform stationary states: $u = 0.2$, $\varepsilon = 2\cdot 10^{-4}$, $v = 0.11$, $p = 0.11$, $\theta_s = 0.6$, $l = 5.66$.

A dynamic transition of the autowave type between two nonuniform stationary states in the stability region to SNP is shown in Fig. 42.

C. Features of the Wave Formation Stage Preceding Quasi-Stationary Propagation and Development of a Stable Stationary DS

Analysis of this stage involves a question of the values of the parameters of an excitatory perturbation capable of inducing a transient process leading to the formation of a new stationary state. Phenomenologically, this question is close to the problems studied by the theory of fuel combustion[23] as well as to similar problems considered earlier in relation to one-component heterogeneous catalytic systems.[24] The approach developed in the above works will serve as a basis for studying the systems considered here. There is no doubt, however, that due to their principal novelty, qualitatively new features of the process of initiation of autowave reconstruction will be brought to light.

XI. CONCLUSION

The data obtained in this work show convincingly that the spontaneous loss of uniformity leading to the "domain" regime of a heterogeneous catalytic conversion should not be considered as an occasional and exotic event. On the contrary, it is a very common occurrence in catalysis. Nonuniformities of loading and hydrodynamic conditions in actual reactors are factors that only reinforce the stratification of the regime in most cases.

The ease of the occurrence of nonuniform stationary states in the extremely simplified model considered is far from being the greatest as compared to other kinetic models typical of catalysis. The region of the existence of DS in this model is limited by the value of the heat-diffusion parameter $\varepsilon = 1$. Although this limitation is not very rigid and holds for a broad class of processes (packed nonmetal catalyst beds, tubular catalytic elements with separated heat and mass transfer), it can also be removed by complicating the model (introducing kinetic nonlinearities), which broadens the regions of the existence of DS to such an extent that the uniform modes of a catalytic reaction may become even more uncommon than the nonuniform.

To be more specific, we refer to some known systems obeying the nonlinear kinetic laws mentioned above: one of them involves a wide variety of the reactions of catalytic combustion of carbon oxide and hydrocarbons described by models with substrate inhibition (see, for example, reference 25); another one is connected with ammonium oxidation on platinum characterized by a chain-branched multiplication of active centers.[26]

The first type of reaction takes place in the reburning of exhaust gases. Therefore, the stratification of catalytic processes may decrease dramatically the efficiency of reburning reactors or lead to incomplete ammonium conversion (adverse admixtures at a reactor outlet). We may refer to a well-known example from the nitrogen production industry relating to the existence of nonuniform temperature fields in reactors of high-temperature methane purification of the released nitrogen oxides on palladium-plated catalysts.[11] Attempts to avoid these phenomena by increasing hydrodynamic uniformity of the gas flow and uniformity of the bed packing were, as a rule, unsuccessful because stratification of the process is most probably due to the spatial instability of uniform operating conditions of the catalyst. Among reactors of catalytic purification, in which the feasibility of uniform regimes is particularly doubtful, devices for reburning car engine exhausts should be mentioned separately. In these systems, a factor that hinders the formation of a uniform conversion velocity field in the active zone is the essentially nonstationary operation mode of reburners. Spatial stabilization of the reaction in the catalyst bed would allow both the increased purification quality and reduced consumption of platinum, which is commonly the major component of the active phase in the processes of car exhaust purification.

Another example illustrating possible applications of the results obtained is particularly instructive due to its great industrial importance. It reinforces arguments for the necessity of developing the concept of DS in solving practical problems of heterogeneous catalysis. To begin with, the features of the kinetic nonlinearities in the reaction of ammonium oxidation on platinum enable us right now to predict from references 5 and 6 that the domain regime of this contact reaction is feasible not only in he range $\varepsilon < 1$, but also at very large values of this parameter, $\varepsilon \gg 1$. In other words, the stratification phenomena cannot be eliminated, perhaps, in processes occurring on catalytic elements with high thermal conductivity (widely employed platinoid grids included). The most risky in this respect are platnium–palladium alloys: there are grounds to suppose[27] that in ammonium oxidation reactors among first-in-the-stream grid packets, there are some that operate in the essentially nonuniform regime, which is likely to be one of the causes of the reduced conversion as compared to others alloys (for example, alloys with a small admixture of palladium).

We shall consider next the problem of nonplatinum catalysts at the contact stage of nitric acid production.[28] The problem at large (complete replacement of platinum) has had no solution for several decades. The reason for the lack of success in one of the most important problems of modern chemical technology seamingly should be sought in the stratification phenomena in the packed nonplatinum catalyst bed. Let us clarify this point. The literature cites a great many examples with granulated catalysts, mainly with ferric oxides as the base, which are distinguished for their high activity (up to 98% ammonium conversion) as well as satisficatory stability and service life. These properties are stably reproduced under laboratory conditions. However, conducting the process in an industrial fixed-bed reactor would immediately show inapplicability of such systems for practical purposes. The main fact is that the catalyst bed could not be made to operate uniformly throughout the reactor cross-section, and the act of kindling always terminated in the excitation of activity only in a local zone of the bed. In other words, a domain regime of the process developed, which led to a reduced conversion and incomplete ammonium oxidation, which is absolutely prohibitive for the given technological stage. Because of this, the granulated nonplatinum catalyst can find a practical application only as a second contacting step behind the platinoid grid packet: the greater portion of reactant undergoes conversion prior to this stage, i.e., the effect of the thermal and kinetic nonlinearities is significantly weakened and the danger of stratification of the process in the bed reduces to a minimum. This explains physically the failure in designing nonplatinum catalytic systems for the contact stage in nitric acid production within the framework of the traditional approach to the design as a fixed-bed reactor. New technical approaches should be sought with due regard to the basic points of the theory of catalytic reactor stability to local perturbations.

The theory mentioned is still in an embryonic state. From the papers we are familiar with, we should indicate references 29–31 as bearing a relation to the problem. The data presented in the present work should be considered as an introduction to the problem, and the specific examples from technological practice as an illustration of its importance for technical catalysis.

To sum up, it would be useful to formulate some problems outlined at the first stage of the study. Their understanding is important for further development of this new trend in the theory of reactors, although they do not, naturally, exhaust its complexity. They are:

(a) studying the effect of nonunique dimensionality of a catalytic element on the boundary location conditions for the region of the existence of DS;
(b) involving a richer variety of the kinetic laws obeyed by the catalytic systems considered, using three-component models to account for active center balance on the contact surface;
(c) developing theoretical foundations for the initiation stage of DS formation due to a local perturbation of finite amplitude (problems of the critical characteristics of a perturbation dangerous to a reactor);
(d) investigating the peculiarities of unstationary and transient processes, in particular, spatially nonuniform, self-oscillating regimes in the chemical reactor performance.

To conclude the discussion of some technological aspects of the theory of DS, we shall touch upon the question of its role in the catalytic reaction kinetics. Since Langmuir's time, the kinetic laws of a heterogeneous catalytic process have been described exclusively by models involving ordinary differential equation sets. Our results indicate also that under experimental conditions, the researcher is most likely to run into the stratification phenomena, the domain structure formation in a kinetic reactor (stationary, propagating, pulsating structures). In other words, in the overwhelming majority of cases, zero-dimensional models, which take no account of energy, matter, and activity transfer in three dimensions, cannot be used to interpret kinetic results without providing evidence for uniformity and synchrony of the development of a chemical process in the entire catalytic cell volume.

The kinetic methods almost universally used at the present time are those of integral registration. The measured signal is a superposition of signals going from different parts of the catalytic element of a reaction cell, and an attempt to describe an experimental curve by traditional lumped kinetic models would be incorrect in the general case. Naturally, the experimentalist is not infrequently faced with nonuniform regimes in the processes studied. It is with the aim of eliminating these phenomena that the flow-circulation technique with so-called gradientless reactors[32,33] was proposed about 40 years ago, and since then it has had wide use. However, such a reactor, as shown above, is not safe against the occurrence at its cross-section of domain structures, and the intense stirring by a circulating flow is a factor that facilitates rather than hinders the formation of DSs (at the expense of an enhanced reactant transfer in the gas phase, or decreased parameter ε).

To confirm this, we refer to some experiments specially designed to unmask the stratification phenomena in heterogeneous catalytic systems. Reference 4 describes the spontaneous appearance of a hot domain in carbon oxide oxidation on a single platinum wire in an electro-thermographic cell, i.e., under conditions where the high degree of uniformity of the external parameter fields is ensured. The structure arises as a result of disintegration of a uniform mode in the deep diffusion region; in this case, in accordance with the conclusions of Section IX of the present work, the domain is overheated to temperatures exceeding significantly those in the diffusion region, which were treated in the classical kinetics as the maximum possible.

The same paper describes various self-oscillatory, spatially nonuniform regimes in ethylene oxidation in the catalytic cell of an electrothermograph. The oscillatory process manifests as a pulsatory change in the location of the boundaries demarcating the zones of the hot and cold domains. The researcher, who is unaware of the possibility of stratification of the process and observes oscillations by the integral signal, would traditionally describe these phenomena in terms of lumped models, which would obviously have no relation to the actual kinetics. The kinetic studies on heterogeneous catalysis 34, 35 cite a great many examples of an oscillatory instability in the conversion regimes being interpreted in terms of homogeneous models, but none of them contains an analysis of the spatial patterns of these phenomena.

Reference 5 presents data on stratification in conversion processes due to factors of electrochemical nature. These phenomena are most typical of the kinetic measurements using as sensitive elements incandescent filaments and should be taken into consideration in their interpretation. Such sensitive elements are widely used in the cells of the kinetic devices intended for studying various catalytic and adsorption–desorption processes.

Unfortunately, at this stage, the kinetics of heterogeneous catalytic reactions as a whole remains within the limits of the classical approach operating with homogeneous schemes and models. Studies aimed at working out a new methods of the kinetic theory and new experimental approaches in investigating kinetic features of chemical conversions with due regard to transfer factors are still few. It may be stated that in this respect, the chemical kinetics is far behind the physical kinetics that is significantly involved in studying self-organizing phenomena in various systems, such as plasma, semiconductors, biologically active media, and other fields.

The authors acknowledge Professor Krinski for his help in discussion and support.

References

1. Nicolis, G. and Prigogine, I., *Self-Organization in Non-Equilibrium Systems*, Wiley, New York, 1977.
2. Barelko, V. V., Kurochka, I. I. and Merzhanov, A. G., *Doklady AN SSSR* 1976 **229**(4) 298.
3. Barelko, V. V., *Problems in Kinetics and Catalysis* (in Russian), Nauka, Moscow, 1981 **18** 61.
4. Zhukov, S. A. and Barelko, V. V., *Khimicheskaya fizika* 1982 **1**(4) 516.
5. Volodin, Yu. E. and Barelko, V. V., *Khimicheskaya fizika* 1982 **1**(5) 670.
6. Pechatnikov, E. L. and Barelko, V. V., *Khimicheskaya fizika* 1985 **4**(9) 1272.
7. Barelko, V. V. and Volodin, Yu. E., *Doklady AN SSSR* 1975 **223**(1) 112.
8. Zhukov, S. A. and Barelko, V. V., *Doklady AN SSSR* 1978 **238**(1) 135.
9. Volter, B. V. and Salnikov, I. E., *Stability of the Performance of Chemical Reactors* (in Russian), Khimiya, Moscow, 1981.
10. Perlmutter, D., *Stability of Chemical Reactors* (in Russian), Khimiya, Leningrad, 1976.
11. Olevskii, V. M., Ed., *Production of Nitric Acid in Aggregates of High Unit Power* (in Russian), Khimiya, Moscow 1985.
12. Matros, Yu. S., Ed., *Aerodynamics of a Fixed-Bed Chemical Reactor* (in Russian), Nauka, Novosibirsk, 1985.
13. Boreskov, G. K., Matros, Yu. S., Klenov, O. P. et al., *Doklady AN SSSR* 1981 **258**(6) 1418.
14. Frank-Kamenetskii, D. A., *Diffusion and Heat Transfer in Chemical Kinetics* (in Russian), Nauka, Moscow, 1967.
15. Volpert, A. I. and Ivanova, A. N., *On the Spatially Nonuniform Solutions of Nonlinear Diffusion Equations*, Preprint (in Russian), Chernogolovka Branch of Inst. Chem. Phys., USSR Acad. Sci., Chernogolovka, 1981.
16. Ivanova, A. N. and Maganova, N. E., *Investigation of the Behavior of Nonuniform Solutions with Increasing Spatial Domain of Diffusion-Kinetic Systems*, Preprint (in Russian), Chernogolovka Branch of Inst. Chem. Phys., USSR, Acad. Sci., Chernogolovka, 1983.
17. Bonch-Bruevich, V. L., Zvyagin, I. P. and Mironov, A. G., *Electrical Domain Instability in semiconductors* (in Russian), Nauka, Moscow, 1972.

18. Vasilyev, V. A., Romanovskii, Yu. M. and Yakhno, V. G., *Uspekhi Fizicheskikh Nauk*, 1979 **128**(4) 625.
19. Barelko, V. V., Beibutyan, V. M., Volodin, Yu. E. and Zeldovich, Ya. B., *Doklady AN SSSR* 1981 **257**(2) 339.
20. Kerner, B. S. and Osipov, V. V., *ZhETF* 1978 **74** 1965; *ZhETF* 1980 **79** 2218; *ZhETF* 1982 **83** 2201.
21. Balkarei, Yu. I., Evtikhov, M. G. and Elinson, M. I., *Mikroelektronika* 1979 **8**(6) 428; 1982 **11**(1) 25; 1983 **12**(2) 171.
22. Elenin, G. G., Krylov, V. V., Polezhaev, A. A. and Chernavskii, D. S., *Doklady AN SSSR* 1983 **271**(1) 84.
23. Zeldovich, Ya. B., Barenblatt, G. I., Librovich, V. B. and Makhviladze, G. M., *A Mathematical Theory of Combustion and Explosion* (in Russian), Nauka, Moscow, 1980.
24. Pechatnikov, E. L., Barelko, V. V., Shkadinskii, K. G. and Brikenstein, H.-M. A., *Khimicheskaya Fizika* 1985 **4**(7) 981.
25. Barelko, V. V. and Zhukov, S. A., *On the Stability of Carbon Oxide Oxidation on Platinum in a Lumped and in a Distributed System*, Preprint (in Russian), Chernogolovka Branch of Inst. Chem. Phys., USSR Acad. Sci., Chernogolovka 1979.
26. Barelko, V. V. and Volodin, Yu. E., *Kinetika i kataliz* 1976 **17** 683.
27. Barelko, V. V., Khalzov, P. I. and Chernyshov, V. I., *Khimicheskaya promyshlennost* 1987 **8** 506.
28. Karavaev, M. M., Zasorin, A. N. and Klestchev, N. F., *Catalytic Oxidation of Ammonium*, Karavaev, M. M., Ed., Khimiya, Moscow, 1983.
29. Shyldrot, H. and Ross, J., *J. Chem. Phys.* 1985 **82**(1) 113.
30. Sheintuch, M., *Chem. Engng. Sci.* 1981 **36** 893.
31. Sheintuch, M. and Pismen, L. M., *Chem. Engng. Sci.* 1981 **36** 489.
32. Temkin, M. J., Kiperman, S. L. and Lukyanova, L. I., *Doklady AN SSSR* 1950 **74** 763.
33. Boreskov, G. K., Slinko, M. G. and Filippova, A. L., *Doklady AN SSSR* 1953 **92** 353.
34. Yablonskii, G. S., Bykov, V. I. and Elokhin, V. I., *Kinetics of Model Reactions in Heterogeneous Catalysis* (in Russian), Nauka, Novosibirsk, 1984.
35. Slinko, M. M. and Slinko, M. G., *Kinetika i kataliz* 1982 **23**(6) 1421.

NOTATION

Dimensional values

L	length of catalytic element
α, β	effective coefficients of heat and mass transfer
t	time
x	coordinate
T, T_0	temperatures of catalyst and reaction flow
C, C_0	reactant concentrations at the catalyst surface and in the bulk flow

λ, D	effective coefficients of thermal conductivity and diffusion
$k = k_0 \exp(-E/RT)$	reaction rate Arrhenius number
m, m_g	thermal capacities: averaged over layer and in gas phase
δ	thickness of effective film
d	diameter of catalyst wire or height of catalyst bed
a, b	coefficients related to the geometry of a catalytic element
$v_T = L_T/\tau_T$	characteristic thermal velocity
V	velocity of feed stream
v_{fr}	thermal front propagation velocity
$L_T = (\lambda d/\alpha)^{1/2}$	characteristic thermal length
$L_c = (D\delta/\beta)^{1/2}$	characteristic diffusion length
$\tau_T = md/\alpha, \tau_c = \delta/\beta$	characteristic times of thermal and diffusion relaxation
$T_m - T_0 = qkC/\alpha$	maximum possible heating in the diffusion regime for a lumped system
$\tau_x = \delta[k_0 \exp(-E/RT)]^{-1}$	characteristic time of chemical reaction
q	thermal effect of chemical reaction
E	activation energy
R	universal gas constant

Dimensionless values

$\theta = (T - T_0)/(T_m - T_0)$	temperature
$\eta = (C_0 - C)/C_0$	degree of conversion
$\tau = t/\tau_T$	time
$y = x/L_c$	coordinate
$l = L/L_c$	length of catalytic element
$p = \tau_c/\tau_x = k_0 \exp(-E/RT) \cdot C_0/\beta$	Damköhler number for catalyst bed
$v = RT_0^2/E(T_m - T_0)$	ratio of the characteristic catalyst temperatures in the kinetic and diffusion regimes
$u = RT_0/E$	Frank-Kamenetskii number for the degree of reaction activation
$\gamma = \tau_c/\tau_T,\ \varepsilon = L_T^2/L_c^2$	parameters evaluating the ratios of the characteristic thermal and diffusion times and lengths

$\varepsilon_{cr}, \tilde{\varepsilon} = \varepsilon/\varepsilon_{cr}$	critical and normalized parameter values
$R = v/(v + u\theta)^2$	generalized parameter of nonlinearity
θ_s	temperature of uniform steady state
$\theta^1_{\text{ext}}, \theta^2_{\text{ext}}$	temperatures at the nullcurve minimum and maximum
θ_1, θ_2	temperatures of uniform ignition and extinction
θ_{infl}	temperature at the nullcurve inflection point
$\theta_n = \theta_s/\theta_{\text{infl}}$	normalized temperature
$\theta_{\text{max}}, \theta_{\text{min}}, \bar{\theta}$	temperatures at the maximum, minimum, and inflection point of nonuniform temperature distribution
$l_* = l/n$	size of an elementary fragment (halfperiod) of DS
l_{fr}	thermal front width
l_h, l_c	dimensions of the hot and cold zone of DS
$l_{cr}, \bar{l}_{cr}$	critical lengths for loss of the stability of a uniform distribution with respect to SNP and FAP
$\theta_\Sigma = 1/l \int_0^l \theta(y)\,dy$	average temperature
$W = p/l \int_0^l \times \{\exp[\theta/(v + u\theta)](1 - \eta)\}\,dy$	average velocity of heat generation

AUTHOR INDEX

Numbers in parentheses are reference numbers and indicate that the author's work is referred to although his name is not mentioned in the text. Numbers in *italic* show the pages on which the complete references are listed.

SUBJECT INDEX